WILLIAM F. MAAG LIBRARY
YOUNGSTOWN STATE UNIVERSITY

ANNUAL REVIEW OF
PHYSICAL CHEMISTRY

EDITORIAL COMMITTEE (1984)

HANS C. ANDERSEN
DONALD M. CROTHERS
GEORGE W. FLYNN
DONALD H. LEVY
B. SEYMOUR RABINOVITCH
J. MICHAEL SCHURR
HERBERT L. STRAUSS
JOHN T. YATES

Responsible for the organization of Volume 35
(Editorial Committee, 1982)

ANDREAS C. ALBRECHT
DONALD M. CROTHERS
GEORGE W. FLYNN
ROBERT GOMER
B. SEYMOUR RABINOVITCH
J. MICHAEL SCHURR
WILLIAM STEELE
HERBERT L. STRAUSS
ERNEST R. DAVIDSON (Guest)
BRUCE E. EICHINGER (Guest)

Production Editor   SUZANNE PRIES COPENHAGEN
Indexing Coordinator   MARY A. GLASS
Subject Indexer   STEVEN SORENSEN

ANNUAL REVIEW OF PHYSICAL CHEMISTRY

VOLUME 35, 1984

B. SEYMOUR RABINOVITCH, *Editor*
University of Washington

J. MICHAEL SCHURR, *Associate Editor*
University of Washington

HERBERT L. STRAUSS, *Associate Editor*
University of California, Berkeley

ANNUAL REVIEWS INC   4139 EL CAMINO WAY   PALO ALTO, CALIFORNIA 94306 USA

ANNUAL REVIEWS INC.
Palo Alto, California, USA

COPYRIGHT © 1984 BY ANNUAL REVIEWS INC., PALO ALTO, CALIFORNIA, USA. ALL RIGHTS RESERVED. The appearance of the code at the bottom of the first page of an article in this serial indicates the copyright owner's consent that copies of the article may be made for personal or internal use, or for the personal or internal use of specific clients. This consent is given on the condition, however, that the copier pay the stated per-copy fee of $2.00 per article through the Copyright Clearance Center, Inc. (21 Congress Street, Salem, MA 01970) for copying beyond that permitted by Sections 107 or 108 of the US Copyright Law. The per-copy fee of $2.00 per article also applies to the copying, under the stated conditions, of articles published in any Annual Review serial before January 1, 1978. Individual readers, and nonprofit libraries acting for them, are permitted to make a single copy of an article without charge for use in research or teaching. The consent does not extend to other kinds of copying, such as copying for general distribution, for advertising or promotional purposes, for creating new collective works, or for resale. For such uses, written permission is required. Write to Permissions Dept., Annual Reviews Inc., 4139 El Camino Way, Palo Alto, CA 94306 USA.

International Standard Serial Number: 0066–426X
International Standard Book Number: 0-8243-1035-7
Library of Congress Catalog Card Number: A-51-1658

Annual Review and publication titles are registered trademarks of Annual Reviews Inc.

Annual Reviews Inc. and the Editors of its publications assume no responsibility for the statements expressed by the contributors to this Review.

TYPESET BY AUP TYPESETTERS (GLASGOW) LTD., SCOTLAND
PRINTED AND BOUND IN THE UNITED STATES OF AMERICA

PREFACE

The collected testimony in each issue of this series never fails to impress us anew with the range and beauty of physical chemical phenomena and with the devotion, ingenuity, and insight demonstrated by the practitioners in this field. The pace and diversity of the advances reviewed here attest again to the robust state of physical chemistry. From the fundamental physical laws themselves to the remarkable properties of various synthetic materials, virtually no topic seems impervious to the scrutiny of physical chemists.

The Editorial Committee gratefully acknowledges the scholarly and critical contributions of the authors and thanks them for their willing cooperation. We acknowledge with gratitude, also, the special contribution of our Production Editor, Suzanne Pries Copenhagen, as well as the efforts of Steven Sorensen, who compiled the index, and of all those who contributed to the production of this volume.

Annual Review of Physical Chemistry
Volume 35, 1984

CONTENTS

WHEN POLYMER SCIENCE LOOKED EASY, *Walter H. Stockmayer and Bruno H. Zimm*	1
STIFF-CHAIN MACROMOLECULES, *Hiromi Yamakawa*	23
ELECTRON ENERGY LOSS SPECTROSCOPY IN THE STUDY OF SURFACES, *Phaedon Avouris and Joseph Demuth*	49
DIELECTRIC PROPERTIES OF POLYELECTROLYTE SOLUTIONS, *M. Mandel and T. Odijk*	75
STATE-RESOLVED MOLECULAR REACTION DYNAMICS, *Stephen R. Leone*	109
INTERACTIONS AND KINETICS IN MEMBRANE MIMETIC SYSTEMS, *Janos H. Fendler*	137
VARIATIONAL TRANSITION STATE THEORY, *Donald G. Truhlar and Bruce C. Garrett*	159
AB INITIO VIBRATIONAL FORCE FIELDS, *Géza Fogarasi and Péter Pulay*	191
CHARACTERIZATION OF SURFACES THROUGH ELECTRON AND PHOTON STIMULATED DESORPTION, *Theodore E. Madey, David E. Ramaker, and Roger Stockbauer*	215
METALLIC GLASSES, *Frans Spaepen and David Turnbull*	241
EFFECTS OF SATURATION ON LASER-INDUCED FLUORESCENCE MEASUREMENTS OF POPULATION AND POLARIZATION, *Robert Altkorn and Richard N. Zare*	265
TRANSITION METAL MOLECULES, *William Weltner, Jr. and Richard J. Van Zee*	291
DIFFERENTIAL ABSORPTION AND DIFFERENTIAL SCATTERING OF CIRCULARLY POLARIZED LIGHT: APPLICATIONS TO BIOLOGICAL MACROMOLECULES, *Ignacio Tinoco, Jr. and Arthur L. Williams, Jr.*	329
EFFECTIVE POTENTIALS IN MOLECULAR QUANTUM CHEMISTRY, *Morris Krauss and Walter J. Stevens*	357
VELOCITY MODULATION INFRARED LASER SPECTROSCOPY OF MOLECULAR IONS, *Christopher S. Gudeman and Richard J. Saykally*	387
SIMULATION OF POLYMER MOTION, *A. Baumgärtner*	419
ELECTRON TRANSFER REACTIONS IN CONDENSED PHASES, *Marshall D. Newton and Norman Sutin*	437
HUMAN EFFECTS ON THE GLOBAL ATMOSPHERE, *Harold S. Johnston*	481
THE ELECTROMAGNETIC THEORY OF SURFACE ENHANCED SPECTROSCOPY, *Horia Metiu and Purna Das*	507

CONTENTS (*continued*)

ELECTRONIC PROPERTIES OF SURFACES, *Marvin L. Cohen and Steven G. Louie*	537
QUANTUM ERGODICITY AND SPECTRAL CHAOS, *E. B. Stechel and E. J. Heller*	563
RELAXATION AND VIBRATIONAL ENERGY REDISTRIBUTION PROCESSES IN POLYATOMIC MOLECULES, *V. E. Bondybey*	591
ELECTRONIC PROCESSES IN ORGANIC SOLIDS, *Martin Pope and Charles E. Swenberg*	613
SELECTIVE EXCITATION STUDIES OF UNIMOLECULAR REACTION DYNAMICS, *F. F. Crim*	657
INDEXES	
Author Index	693
Subject Index	715
Cumulative Index of Contributing Authors, Volumes 31–35	728
Cumulative Index of Chapter Titles, Volumes 31–35	730

SOME RELATED ARTICLES IN OTHER *ANNUAL REVIEWS*

From the *Annual Review of Astronomy and Astrophysics*, Volume 22 (1984)
  Origin and History of the Outer Planets: Theoretical Models and Observational Constraints, J. B. Pollack

From the *Annual Review of Biochemistry*, Volume 53 (1984)
  The Molecular Structure of Centromeres and Telomeres, E. H. Blackburn and J. W. Szostak
  Pyruvoyl Enzymes, P. A. Recsei and E. E. Snell
  Principles That Determine the Structure of Proteins, C. Chothia
  Three-Dimensional Structure of Membrane and Surface Proteins, D. Eisenberg
  Polyamines, C. W. Tabor and H. Tabor
  The Chemistry and Biology of Left-Handed Z-DNA, A. Rich, A. Nordheim, and A. H.-J. Wang

From the *Annual Review of Biophysics and Bioengineering*, Volume 13 (1984)
  Characterization of Transient Enzyme-Substrate Bonds by Resonance Raman Spectroscopy, P. R. Carey and A. C. Storer
  Multifrequency Phase and Modulation Fluorometry, E. Gratton, D. M. Jameson, and R. D. Hall
  Solid State NMR Studies of Protein Internal Dynamics, D. A. Torchia
  NMR Studies of Intracellular Metal Ions in Intact Cells and Tissues, R. K. Gupta, P. Gupta, and R. D. Moore
  Fluctuations in Protein Structure from X-Ray Diffraction, G. A. Petsko and D. Ringe
  Biophysical Applications of Quasi-Elastic and Inelastic Neutron Scattering, H. D. Middendorf
  Detailed Analysis of Protein Structure and Function by NMR Spectroscopy: Survey of Resonance Assignments, J. L. Markley and E. L. Ulrich

From the *Annual Review of Earth and Planetary Sciences*, Volume 12 (1984)
  Applications of Accelerator Mass Spectrometry, L. Brown
  Physics and Chemistry of Biomineralization, K. E. Chave

From the *Annual Review of Materials Science*, Volume 14 (1984)
  Inorganic Ion Exchangers with Layered Structures, A. Clearfield
  Application of the Pseudopotential Model to Solids, M. L. Cohen
  Photoeffects at the Semiconductor/Liquid Interface, N. S. Lewis
  Synchrotron Radiation Photoemission Spectroscopy of Semiconductor Surfaces and Interfaces, G. Magaritondo and A. Franciosi
  Electronic Properties of Small Particles, R. Kubo, A. Kawabata, and S.-I. Kobayashi
  Polymer Modification of Electrodes, R. W. Murray
  Ion Implantation in Metals, S. T. Picraux
  Polymer Fatigue, M. T. Takemori

ANNUAL REVIEWS INC. is a nonprofit scientific publisher established to promote the advancement of the sciences. Beginning in 1932 with the *Annual Review of Biochemistry*, the Company has pursued as its principal function the publication of high quality, reasonably priced *Annual Review* volumes. The volumes are organized by Editors and Editorial Committees who invite qualified authors to contribute critical articles reviewing significant developments within each major discipline. The Editor-in-Chief invites those interested in serving as future Editorial Committee members to communicate directly with him. Annual Reviews Inc. is administered by a Board of Directors, whose members serve without compensation.

1984 Board of Directors, Annual Reviews Inc.

Dr. J. Murray Luck, Founder and Director Emeritus of Annual Reviews Inc.
  Professor Emeritus of Chemistry, Stanford University
Dr. Joshua Lederberg, President of Annual Reviews Inc.
  President, The Rockefeller University
Dr. James E. Howell, Vice President of Annual Reviews Inc.
  Professor of Economics, Stanford University
Dr. William O. Baker, *Retired Chairman of the Board, Bell Laboratories*
Dr. Winslow R. Briggs, *Director, Carnegie Institute of Washington, Stanford*
Dr. Sidney D. Drell, *Deputy Director, Stanford Linear Accelerator Center*
Dr. Eugene Garfield, *President, Institute for Scientific Information*
Dr. Conyers Herring, *Professor of Applied Physics, Stanford University*
Mr. William Kaufmann, *President, William Kaufmann, Inc.*
Dr. D. E. Koshland, Jr., *Professor of Biochemistry, University of California, Berkeley*
Dr. Gardner Lindzey, *Director, Center for Advanced Study in the Behavioral Sciences, Stanford*
Dr. William D. McElroy, *Professor of Biology, University of California, San Diego*
Dr. William F. Miller, *President, SRI International*
Dr. Esmond E. Snell, *Professor of Microbiology and Chemistry, University of Texas, Austin*
Dr. Harriet A. Zuckerman, *Professor of Sociology, Columbia University*

Management of Annual Reviews Inc.

John S. McNeil, Publisher and Secretary-Treasurer
William Kaufmann, Editor-in-Chief
Mickey G. Hamilton, Promotion Manager
Donald S. Svedeman, Business Manager
Richard L. Burke, Production Manager

ANNUAL REVIEWS OF
Anthropology
Astronomy and Astrophysics
Biochemistry
Biophysics and Bioengineering
Earth and Planetary Sciences
Ecology and Systematics
Energy
Entomology
Fluid Mechanics
Genetics
Immunology
Materials Science
Medicine
Microbiology
Neuroscience
Nuclear and Particle Science
Nutrition
Pharmacology and Toxicology
Physical Chemistry
Physiology
Phytopathology
Plant Physiology
Psychology
Public Health
Sociology

SPECIAL PUBLICATIONS
Annual Reviews Reprints:
  Cell Membranes, 1975–1977
  Cell Membranes, 1978–1980
  Immunology, 1977–1979

Excitement and Fascination
of Science, Vols. 1 and 2

History of Entomology

Intelligence and Affectivity,
  by Jean Piaget

Telescopes for the 1980s

For the convenience of readers, a detachable order form/envelope is bound into the back of this volume.

WHEN POLYMER SCIENCE LOOKED EASY

Walter H. Stockmayer

Department of Chemistry, Dartmouth College, Hanover, New Hampshire 03755

Bruno H. Zimm

Department of Chemistry, University of California at San Diego, La Jolla, California 92093

INTRODUCTION

By the eve of America's entry into World War II, both industrial and academic activity in polymer science had grown to a stage permitting the identification of a number of basic problems in macromolecular physical chemistry and their solution without radical advances in existing experimental or theoretical techniques. Both of us were at Columbia University at that time, loyal to quite other branches of physical chemistry, when to each the "call" came in a different way. In this memoir we hope to convey some notion of the prevailing atmosphere, and to recall and transmit as best we can the way certain subfields looked and how they developed. This is a highly subjective business that lays us open to criticism by historians of science, one of whom in a recent review (1) remarked with some justice, "For scientists, history is not the field upon which they wrestle for truth, but principally their field of celebration and self-congratulation." And again, "Though recollection may add vividness and color, it cannot reliably be used except as embellishment of a picture delineated by written sources from the period." So, even with our bibliography, *caveat lector*.

To set the historical stage slightly, we remind readers that the struggle of Hermann Staudinger and Wallace H. Carothers to establish the macromolecular hypothesis was mainly waged and won in the 1920s and early 1930s (2). The relevant areas of physical chemistry, however, actually go back further than this, since the ultimate constitution and nature of the binding forces within the particles, be they association colloids or true single macromolecules, are not primary considerations. Thus, for example, Einstein's calculation of the viscosity of a suspension of spheres in a hydrodynamic continuum appeared in 1906 (with a correction five years later), and Svedberg developed the ultracentrifuge in the early 1920s. But in the 1930s the rate of progress increased markedly. Among the most provocative developments were Staudinger's announcement of his (now long abandoned) viscosity-molecular weight rule (3) and the first paper on the molecular theory of rubber elasticity by Eugene Guth and Herman Mark (4). It was also in this decade that Werner Kuhn, Paul J. Flory, G. V. Schulz, and Maurice L. Huggins first appeared on the polymeric scene. Kuhn, a versatile theoretician, was already well known for his work on optical activity and a classical spectral sum rule. His is the classic paper on the conformational statistics of flexible chains (5), in which he foresaw the excluded volume effect; he also considered such distinctly chemical problems as the statistics of random hydrolytic degradation of cellulose (6). Flory at the start of his still highly active career was also much concerned with polymerization reactions, producing seminal works on the statistics and kinetics of polycondensation (7) and free-radical chain polymerization (8). In the latter area we also find the pioneer kinetic investigations of Schulz (9); while Huggins (10) established definitively that the viscosities of polymer solutions showed them to be not the rigid rods envisaged by Staudinger but the random flexible coils of Kuhn, Guth, and Mark.

The Columbia chemistry department in the early 1940s was an exciting place for physical chemists. The chairman was Harold C. Urey, and the staff numbered such luminaries as Louis P. Hammett, George E. Kimball, Victor K. LaMer, and Joseph E. Mayer. The last-named was then at the height of his fame and prowess as a creator of the modern statistical mechanics of fluids (11, 12) and coauthor with his wife of a pioneering text (13). Each of us had gone to Joe Mayer for primary inspiration: W. H. S., one year after taking his degree at the Massachusetts Institute of Technology, had accepted an instructorship in the Columbia Extension Division (requiring evening lecturing in general, analytical and physical chemistry) mainly to be near him; and B. H. Z., after graduating from Columbia College in 1941, chose him as mentor of his doctoral research. Maria Goeppert Mayer was then professor of physics at Sarah Lawrence College, but she managed to spend several days a week at Columbia. She was working on a variety of problems, and had recently published a Thomas-Fermi calculation predicting (14) the start of the actinide series at element 93, but this was some years before her elucidation of the "magic numbers" for nuclear stability which earned her a share of the 1963 Nobel prize in physics. Also, the war already raging in Europe had diverted to the

Columbia neighborhood a number of refugees, of whom Francis Perrin and Ronald W. Gurney frequently dropped by the chemistry department.

Immediately after Pearl Harbor in December, 1941, Urey called the entire Department together to stress that the war would be fought with applied science and that quixotic enlistment in the armed forces would be neither helpful nor wise. In the months that followed, some of the staff and many of the students went off to other locations to take part in war-related activities. By the summer of 1942, Joe Mayer was spending most of each week at Aberdeen Proving Ground, but would return for one or two days of lecturing and counseling. The famed Monday Lunch Club, which he and George Kimball ran with their research groups, now generally met on Tuesdays, and gave us a chance to meet informally many of the now frequent great visitors, including Enrico Fermi, Eugene P. Wigner, Edward Teller, and John G. Kirkwood. Those of us who stayed on at Columbia were of course also drawn into war-related work: B. H. Z. into a project that is described in the later section on Molecular Weights and Light Scattering, and W. H. S. into calculations with George Kimball on deuterium exchange equilibria, commandeered by Urey. The remaining graduate students were permitted to keep their doctoral programs barely alive, and time was somehow stolen to think now and then about other types of research, play a little chamber music or softball, or even to hike in New Jersey or the Catskills.

Although it was Joe Mayer who drew us to the eighth floor of Chandler laboratory, where he and Kimball had their offices and labs, it was another member of that colony, Charles O. Beckmann, who must take most of the blame for our conversion into (not quite full-time) polymer physical chemists. The ways in which this occurred is recounted below. Charlie Beckmann deserved more recognition than he received during his lifetime (1904-1968) for his researches on optical activity and on polymers, particularly starch derivatives. But, like many of us, he didn't enjoy writing papers and preferred to think about new problems. He and his students worked with a home-built ultracentrifuge of which he was justly proud. His informal manner laid him open to tricks from his juniors. One evening when Beckmann addressed the departmental colloquium, we managed to doctor the first slide, a photo of Beckmann at the helm of the centrifuge, by adding a scantily clad bathing beauty—an "MCP" maneuver that neither of us would dare to attempt nowadays!

We left Columbia at different times. W. H. S. returned to MIT in late 1943, shifting his war-related efforts to the Laboratory for Insulation Research under Arthur von Hippel. As LaMer confided to Beckmann, "It's a shame Stockmayer is leaving. His wife is the best dancer in the department!" B. H. Z., on the other hand, stayed on nearly till the end of the war, when he migrated to the Brooklyn Polytechnic Institute to join Paul Doty, Turner Alfrey, and Arthur Tobolsky as a member of Herman Mark's team in polymer science.

The selection of topics for this recollection was perhaps more difficult than the actual writing. The limitations of space, plus the hope of interesting a wider circle of readers than just polymer specialists, were probably the determining factors in our arbitrary choice.

CHEMICAL STATISTICS: TREES, RINGS AND TRANSITIONS (W. H. S.)

In the November, 1941 issue of JACS there appeared a now famous set of three papers by Paul Flory (15) discussing the molecular statistics and gelation of polymerizing systems containing ingredients with more than two functional groups per monomer molecule. It had been known for some time that many such systems (for example, "Glyptal" resins made from equivalent quantities of glycerol and phthalic anhydride) change suddenly, during the course of the polymerization reaction, from liquids of moderate fluidity to infusible, insoluble gels with infinite Newtonian viscosity. Carothers had suggested (16) that the phenomenon was due to the formation of essentially macroscopic molecules—giant trees—and discussed the process in a simple stoichiometric way. For example, a mixture of two moles of glycerol, $CH_2OHCHOHCH_2OH$, with three moles of adipic acid, $HOCO(CH_2)_4COOH$, will form a single gigantic tree (provided cyclic structures are forbidden) at the point at which 5/6 of the carboxyl or hydroxyl groups have been esterified. If some ring structures are permitted, the event will be postponed to higher degrees of reaction. This simple picture is, however, quantitatively inadequate, as the observed gel point always occurs *earlier*; in the above example, it comes at about 75% esterification, corresponding to a number-average molecular weight of only about 10^3 g mol^{-1}. The notion was therefore current that a physical logjam among the relatively small but bristly branched molecules could drive up the viscosity. What Flory had now done was to rescue the Carothers picture by showing that statistically the very polyfunctionality of the reagents produces extremely broad distributions of molecular size, a few structures becoming really huge even though the number-average size remains modest.

Quantitatively, Flory's well-known criterion for the critical degree of reaction at which giant trees first form is

$$\alpha_c = 1/(f-1), \qquad \qquad 1.$$

where f is the number of functional groups on the multifunctional monomer branch units and α is the probability that a chosen functional group on one branch unit is connected, through the necessary sequence of one or more reactions, to another branch unit. The crucial quantity α can be simply evaluated from the reaction stoichiometry; in the above example, again taking equivalent quantities of hydroxyl and carboxyl functions, we have simply $\alpha = p^2$, where p is the fractional extent of esterification, since two ester links are needed to connect a pair of glycerol units. Since $f = 3$ in our example, Eq. 1 predicts $\alpha_c = 1/\sqrt{2} = 0.707$, as compared to the observed 0.75. The difference is attributable (15) partly to a reactivity difference between the primary and secondary hydroxyl groups, and partly to ring formation. In any case, the prediction is startlingly good. Flory has given two excellent expositions (17, 18) of the story we have just sketched.

Early in 1942 Flory's papers were reviewed in Charlie Beckmann's group seminar by a graduate student, Thomas G. Fox, Jr., who went on to a notable career (19) in polymer science, including extensive collaborations with Flory. The effect of Tom Fox's presentation on me was electrical: Here was a kind of phase transition, predictable almost quantitatively from stoichiometry and relatively simple statistics, and not requiring estimates of high-order cluster integrals or intractable lattice combinatory factors! The techniques used by Flory had been atypical, but I got the notion that the sol-gel transition could be treated by methods more imitative of standard statistical thermodynamics. Through the attempt to do this, I fell into polymer science.

To appreciate the impact of Flory's work at that time, it should be realized that knowledge of phase transitions was not then highly developed or organized. Although there was experimental evidence suggesting that the critical region was somehow anomalous (20, 21), most physical chemists did not go beyond the classical treatment of vapor-liquid equilibrium used by Maxwell, Gibbs, and van der Waals: The pressure and the Helmholtz free energy were supposed to be analytic functions of temperature and density everywhere on the primitive thermodynamic surface, and two-phase equilibrium was only to be determined by knocking out the metastable or essentially unstable regions with the requirement of equal chemical potentials. In the later 1930s Joe Mayer had rocked the boat (12) by championing the belief that the canonical partition function itself must contain all the secrets of equilibrium, including the singularities on the binodal locus. It may be conjectured that this view was inspired by the ideal Bose-Einstein condensation and that it was shared by Max Born, earlier a patron of both the Mayers in Göttingen (22). Indeed Born (23), perhaps influenced by the much earlier and cruder notions of Lindemann (24), had proposed that the melting point of a crystalline solid coincided with the vanishing of its shear modulus, a theory that commanded some respect till it was scotched by Joseph Slepian, obviously a man of the world, who remarked (25) that the ice cubes in his highball glass were perceptibly rigid.

Mayer's ideas about vapor-liquid condensation were not so easily disposed of. Developing the thermodynamic functions in cluster series, he and his students observed that singularities might in principle arise from either of two apparently distinct mathematical conditions, which need not be coincident. Thus there emerged the "derby hat" picture of the critical region, which though now long abandoned was an important stepping stone to modern theories of the critical region. For details of this picture the 1940 textbook by the Mayers (13) may be consulted.

The Mayer theory eludes quantitative test because of the necessity of evaluating big cluster integrals, and a similar remark could probably be made about any off-lattice treatment of systems with additive pairwise interactions. As to lattice models, the exact critical temperature for the unmagnetized two-dimensional Ising model had indeed been located by Kramers & Wannier (26), but Onsager's evaluation of the partition function did not appear until 1944. The Flory gelation model, with its inherently tree-like structure ("Bethe lattice"), was much simpler, and its success invited generalization, with the hope of relating it theoretically to other more formidable transitions, as well as of producing a method for finding the molecular species distribution for an arbitrary stoichiometric system.

For ease, I first considered the so-called RA_f model, in which a single kind of f-functional monomer is capable of reacting with itself, and in which case α is just the fractional extent of the bonding reaction. Flory had treated other special systems and developed the statistics mainly through appropriate recursion relationships. With the RA_f model, with exclusion of rings, one can go straight for the goal with the microcanonical ensemble, maximizing the entropy at a fixed degree of bond formation (i.e. constant energy); the details can be read in the original paper (27). The combinatory factor caused a few headaches, but it was finally obtained through conscious mimicry of a similar problem occurring in the Mayer cluster theory, by invoking imaginary Erector-set constructions with frames, bolts, and washers (13, pp. 455–59). Since Joe had just taken on a new graduate student named Harris Mayer, there was a strong temptation to prepare a joint manuscript by Mayer, Mayer, Mayer & Stockmayer, but Joe and Maria never yielded.

The microcanonical treatment can be replaced by a canonical ensemble, if desired (29). In either case, the size distribution for the RA_f model without rings can be derived without any mathematical approximations. It yields the following formula for the weight fraction w_x of x-mers:

$$w_x = K_x(1-\alpha)^2\alpha^{-1}\beta^x, \qquad 2.$$

with

$$K_x = f(fx-x)!/(fx-2x+2)!(x-1)!$$

and

$$\beta = \alpha(1-\alpha)^{f-2}.$$

The number-average size is

$$\langle x \rangle_n = 1/\Sigma x^{-1} w_x = 1/(1-\tfrac{1}{2}\alpha f) \qquad 3.$$

while the weight-average is

$$\langle x \rangle_w = \Sigma x w_x = (1+\alpha)/(1+\alpha-f\alpha). \qquad 4.$$

The breadth of the distribution is evident from the fact that $\langle x \rangle_w$ diverges at Flory's gel point, Eq. 1, while $\langle x \rangle_n$ remains small there.

Before the final polishing, I wrote to Flory (then at Esso Laboratories in New Jersey) and received a cordial invitation to come over and compare ideas. We didn't completely see eye to eye about the theory beyond the gel point, but that didn't prevent our becoming friends.

The Journal of Chemical Physics had gotten thin because of the war. (The minimum was reached in 1944, when Volume 12 had just 531 pages.) Publication was therefore rapid: My manuscript was submitted in late October 1942 and appeared in February. A second paper followed a year later (30). In an Appendix to the earlier work, a simple set of kinetic equations was formulated for the RA_f model, excluding reverse reactions as well as rings, and it was observed that these were satisfied by Eq. 2 if α were properly expressed as a function of time, though the uniqueness of the solution was not established. In recent years such "generalized Smoluchowski equations" have become objects of interest (31, 32).

The old methods of treating tree models have been superseded by the elegant applications of cascade theory due to Manfred Gordon (33), Walther Burchard (34), and their associates. Kinetic schemes of great complexity can be treated, including chain mechanisms (35); radii of gyration of Gaussian molecules can be calculated; and unequal reactivities of functional groups can be readily accommodated (36). Allowance can even be made for ring structures within the cascade approach by the so-called spanning tree approximation (37, 38).

In the last few years theoretical interest in the sol-gel transition has risen to a high level, which unfortunately we do not have space to discuss or document to any significant extent. Three major addenda to the classical tree theories must be addressed: (*a*) inclusion of excluded volume effects, perhaps not only on the chain dimensions but also on the chemical rates or equilibria themselves; (*b*) realistic description of ring formation at all stages of the reaction; and (*c*) full description of the system past the gel point. These problems are by no means mutually independent. A typically modern question is that of the critical exponents in the immediate vicinity of the transition point. For example, according to Eq. 4 the weight-average degree of polymerization $\langle x \rangle_w$ scales as $(\alpha_c - \alpha)^{-1}$ as the gel point is approached. An alternative approach near the transition is that of the *percolation* model, which allows in a prescribed way for excluded volume and ring formation, and which makes $\langle x \rangle_w$ vary as $(\alpha_c - \alpha)^{-\gamma}$ with $\gamma \simeq 1.7$. Existing experimental data have not provided a clear decision between these alternatives. For two contrasting points of view, the reviews of Stauffer (39) and Burchard (34) may be consulted.

It still wasn't clear in the summer of 1942 that polymers would become my main research interest. I was working on a simple quantum-mechanical calculation of the non-pairwise-additive overlap energy of three He atoms,[1] when chance again intervened. I had struck up a friendship with Lester Weil, a graduate student doing organic synthesis in the lab directly across the corridor from my office. His advisor rather suddenly went off to a war job, and Weil opted to begin a new research problem, his old one having proved unfruitful. We soon concocted a program directly inspired by Flory's papers: to measure the effects of dilution (and thus presumably of increasing intramolecular reaction) on gel points, and then to assay the effects of branching on solution and melt viscosities. The green light was soon given, thanks to the liberal policies of Urey and the Columbia department, which imposed no formal segregation of the various branches of chemistry and which allowed mere instructors to be research advisers. Thus suddenly there was a graduate student collaborator, and he was to work on polymers.

The gel point work went very well indeed. Taking pentaerythritol, $C(CH_2OH)_4$ with $f = 4$, and adipic acid, $HOCO(CH_2)_4COOH$, in equivalent amounts and adding varying quantities of the inert polyether diluent, $CH_3(OCH_2CH_2)_4OCH_3$, Weil found (41) the critical degree of esterification p_c [determined by titration, as in Flory's experiments (15)] to increase monotonically with dilution, as would qualitatively be expected if there were increasing amounts of intramolecular ring formation. For the undiluted system Weil measured $p_c = 0.630$, a few percent larger than the theoretical value $1/\sqrt{3} = 0.577$ for a system of pure trees. Quantitatively, we assumed that p_c would increase linearly with the relative $1/c$, where c is

---

[1] This work was abandoned in an almost finished state. A similar calculation was later published by Rosen (40).

the volume concentration of the active ingredients, and argued that no rings would exist at vanishing $1/c$. Extrapolating p_c linearly to zero against $1/c$, Weil obtained the least-squares intercept $p_c(0) = 0.578 \pm 0.005$, in remarkable support of the tree theory. The elation we then felt still seems justified: For the first time (as far as we then knew and now know) a "phase transition" (though admittedly not a physical separation into two distinct phases) had been located in the laboratory, when the conditions of the model were satisfied, at exactly the point predicted *a priori*. The results deserved more complete and rapid publication than they got (42).

In retrospect, it has to be admitted that there is no proof that the extrapolation of p_c against $1/c$ should be linear. Today the same system could of course also be studied by additional techniques, such as equilibrium and quasi-elastic light scattering, to determine weight-average molecular weights, mean square radii of gyration, and hydrodynamic radii, and there are many other chemical systems that would be more suitable for precise studies of this type.

Weil made many measurements on viscosities of branched polyesters in solution and in the melt, but publication was deferred and eventually postponed indefinitely when it became clear that end-group titrations alone (the only method then available to us) could not yield reliable molecular weights, even when corrected for ring formation in an approximate way to force agreement with observed gel points. There was moreover no sound viscosity theory yet at hand with which to discuss the results. Today the influence of branching on solution and melt properties, including viscosity, is again an active research area, in which the ability to synthesize almost monodisperse star polymers (and some other structures) by anionic polymerization has played a key role (43, 44).

Our own interest in the dimensions and properties of branched polymers did not altogether die out with the termination of Weil's efforts. A few years later, when B. H. Z. was at Berkeley and I at MIT, we collaborated at long range in some calculations of mean square radii of gyration for various branched Gaussian structures (45). The extensions of such calculations by the cascade technique came subsequently at the hands of Gordon (33) and Burchard (34). When Marshall Fixman was an MIT graduate student, he and I made some calculations of excluded volume effects and hydrodynamic radii in star polymers (46). These problems are also under active scrutiny today.

In the fall or winter of 1942 at Columbia, Homer Jacobson joined the group as my second graduate student, bringing a variety of talents, including musical aptitude and imagination: When studying music theory for recreation, he satisfied a course requirement by writing classical four-part harmony to a ground bass of "Pistol Packin' Mama," a juke-box pest of 1943 which the instructor failed to recognize. After some preliminary work on gel-point theory for systems polymerizing by chain mechanisms (47) (e.g. vinyl acetate + divinyl adipate, etc), it was decided that Jacobson should pursue both theoretical and experimental studies of ring-chain equilibrium in linear condensation polymers. This seemed like a logical first step before confronting the immeasurably harder problem of ring formation in nonlinear systems—a problem on which we never managed to produce any dents.

Let R_x denote the number of closed-ring x-mers and C_y the number of open-chain y-mers in the system. Then contemplate the reaction

$$C_{x+y} \rightleftarrows C_y + R_x, \qquad 5.$$

which involves no change in the number of bonds and hence no appreciable isothermal change in energy. The essential component of the equilibrium constant of the above reaction is therefore the conformational entropy change accompanying ring closure. This must depend on the length of the bond to be formed. However, a similar bond is broken when the (x + y)-mer chain is cleaved, and so the bond length disappears from the calculation. Assuming Gaussian chain statistics, Jacobson was able to formulate the equilibrium constant (here written only for the simplest case of a self-condensing single monomer species) by following a method taught by Mayer & Mayer (13, pp. 213–17), with the result (49)

$$K \equiv C_y R_x / V C_{x+y} = (3/2\pi \langle r_x^2 \rangle)^{3/2}/2x \qquad 6.$$

where $\langle r_x^2 \rangle$ represents the equilibrium mean square end-to-end distance of the x-meric open chain, which for a Gaussian chain is proportioned to x, and V is the volume of the system. The factor $1/2x$ comes from the symmetry number $2x$ of the ring, or can equally well be rationalized kinetically as due to the x different bonds that could be cleaved in the reverse of Reaction 5, followed by its being joined to the y-mer chain at either end.

It should surprise nobody that the physical argument leading to Eq. 6 had been anticipated by Kuhn (50), but this fact was not known to us for many years.

When the above result is applied to the over-all statistics of an equilibrium Gaussian ring-chain system, the number of ring x-mers takes the form

$$R_x = VBx^{-5/2} p^x \qquad 7.$$

where p is the fractional degree of condensation in the open-chain molecules, and the symbol B subsumes the details of Eq. 6. The total number of ring molecules and the total number of monomer units contained in them are then given by sums ΣR_x and $\Sigma x R_x$, which are identical to those for the pressure and number density of a perfect Bose-

Einstein gas in terms of fugacity. The latter sums have singularities at unit fugacity, as the polymer sums do at $p = 1$. Thus the Bose-Einstein condensation is exactly reproduced in the polymer problem with the following physical meaning: Below a certain critical density, $\rho_c = B\zeta(3/2)$, where ζ is the zeta function, the system can be driven to 100% condensation to produce a system of rings only.

Readers conversant with double-helix theory (51) will be aware that the "loop factor" in the two-strand partition function is in essentials the same as that of Eq. 6. Indeed, the analogy to the Bose-Einstein condensation was rediscovered (52) in that context. Useful application to formation of large DNA rings has also been made (53).

Experimentally, Jacobson was able to give support to the theory by observing changes in solution viscosity of the predicted magnitude upon dilution and reequilibration (54). In this part of his work he was advised by Charlie Beckmann, for I had by then left Columbia. The proportionality of K to $x^{-5/2}$ for Gaussian chains (theta solvent conditions) has been directly confirmed by Semlyen (55), notably for poly(dimethylsiloxane) but also for a number of other polymers. His results also clearly document the increase of the exponent of $1/x$ above 5/2 in good solvents where excluded-volume effects come into play.

The theory has been refined for "unperturbed" rings too small to be Gaussian, with regard for bond-angle and internal-rotational angle restrictions in closed rings (56), and improved agreement with experiment is seen. A more challenging theoretical problem concerns the effects of intermolecular excluded volume on polymerization equilibria, including ring-chain equilibria, producing departures from Flory-Huggins thermodynamics (which in their simple form require no "activity coefficient" corrections to Eq. 6). Recent progress is due to Wheeler & Kennedy (57), who have exploited the now well-known correspondence (58) between self-avoiding lattice walks and the $n = 0$ magnetic lattice model. Aside from shifts of detail and exponents, there is at least one striking *qualitative* result: For the ring-chain transition in molten sulfur at 160°C a small hump is predicted on the density-temperature curve (as contrasted to a mere discontinuity of slope in the classical theory), and this is in fact found experimentally (59).

MOLECULAR WEIGHTS AND LIGHT SCATTERING (B. H. Z.)

In the early 1940s the absolute molecular weights of most polymers were either unknown or known only within rough limits, and such data as existed were not considered highly reliable. This situation persisted because of difficulties of measurement. Polymers do not have a gas phase, so that they must be measured in dilute solution, but the ideal, Raoult's-Law, term in the free energy, on which the determination of molecular weight depends, decreases at a given mass fraction as the reciprocal of the molecular weight, while the nonideal terms are roughly independent of the molecular weight. Thus, to suppress the nonideal terms, which are proportional to the second and higher powers of the concentration, it is necessary to measure at lower and lower concentrations as the molecular weight increases, and at some point one runs out of measurement accuracy. With the traditional methods of organic chemistry, such as the depression of the freezing point of a camphor solution, this point comes when the molecular weight is about ten thousand, much less than the molecular weight of almost any significant polymer.

The one thermodynamic method with sufficient sensitivity is osmometry, but here the chemical and mechanical stability, as well as the permeability and selectivity, of the membrane is critical. [There is a good discussion in an article by Wagner & Moore (60).] At that time the membrane usually used with organic solvents was partially denitrated nitrocellulose, which the experimenter prepared himself by treating a film of partially dried collodion, cast on a mercury surface, with ammonium sulfide. The permeability and selectivity of such a membrane depended on the extent to which it had been dried and the solvent to which it had been transferred, while the amount of soluble impurity depended on the thoroughness of the sulfide treatment and the subsequent washing. The spurious osmotic pressure that appeared when a new membrane of this kind was first put into an osmometer with pure solvent on both sides, a pressure presumably arising from soluble components of the membrane, could be spectacular, and frequently decayed only slowly with time. Also, the porosity of such a membrane could easily be great enough to let some macromolecular components pass through. Thus osmotic molecular-weight measurements could, and did, give results that varied considerably from one experiment to another.

The only other known absolute methods also had difficulties. Determination of chain ends by chemical analysis worked with only a few polymers where the end groups were definitely known, such as polyesters, and then only at rather low molecular weights, and in any case the determination was sensitive to small amounts of impurities. Svedberg had invented the ultracentrifuge, but only a few of the instruments were available in the whole world until the Spinco corporation went into commercial production of electrically driven machines at the end of the decade. Moreover, synthetic polymers were polydisperse in molecular weight, and their solutions were highly nonideal, so that molecular-weight measurements on them with the ultracentrifuge were complicated, and data reduction was time-consuming with the laborious methods then in use. [See, for example, papers by Wales and co-workers (61, 62).]

The most widely used measure of molecular size, then as now, was the intrinsic viscosity, but this is a relative, not an absolute, measure. The nature of this relation had been the subject of a vigorous dispute between Staudinger and Mark in the previous decade, and the fundamental theory of it was not satisfactorily developed until the work of Debye & Bueche (63) and of Kirkwood & Riseman in 1948 (64) and of Flory & Fox in 1950 (65). Before that, the intrinsic viscosity was mainly a quantity of empirical significance only.

Thus in the 1940s determination of the values of the absolute molecular weights of synthetic polymers was a subject of fundamental interest, and when P. J. W. Debye (66) introduced a new and very different method, based on measurement of the light scattered from solutions, it was a major event.

Debye's method was actually a new extension of an old theory, one with which the names of Rayleigh and Einstein were primarily associated. After Rayleigh's development in the nineteenth century of the theory of the scattering of light from individual small particles, and following a preliminary discussion by von Smoluchowski in 1908 (67), Einstein in 1910 (68) derived a formula for the scattering from a pure liquid on the basis of a Fourier analysis of density fluctuations. In his 1944 publication, Debye (66) extended this formula to include composition fluctuations in a mixture, and related these to thermodynamics, to the dilute-solution laws, and hence, finally, to the molecular weight of a solute. This remarkable formula is

$$\tau = \frac{32\pi^3 k T n^2}{3\lambda^4} \left[\frac{\rho(\partial n/\partial \rho)^2}{(\partial p/\partial \rho)_{T,c}} + \frac{c(\partial n/\partial c)^2}{(\partial P/\partial c)_{T,p}} \right]. \qquad 8.$$

Here τ is the turbidity, i.e. the fraction of light scattered per unit length of path, kT as usual, n is the refractive index, λ the wavelength in vacuo, ρ the density, p the pressure, c the concentration of solute, and P the osmotic pressure. The first term is Einstein's original term, and represents the scattering from density fluctuations, while the second term represents the scattering from composition fluctuations. The first term is practically independent of concentration and for dilute solutions can be replaced by τ_0, the scattering from pure solvent. If we then introduce van't Hoff's law for P, we get

$$\tau - \tau_0 = \frac{32\pi^3 n^2 (\partial n/\partial c)^2 cM}{3\lambda^4 N_a}, \qquad 9.$$

where c is now the concentration of the solute in mass per unit volume, N_a is Avogadro's number, and M is the molecular weight of the solute. Thus

Debye could say (66, p. 340), referring to a slightly altered form of the above: "Equation 5' can therefore be interpreted as showing how by the combination of two measurements, the first of the turbidity, the second of the difference in refraction of solution and solvent, the molecular weight of the substance in solution can be evaluated, without introducing any kind of empirical constants."

The effective introduction of light scattering as a method for measurement of molecular weight and size of polymers occurred during World War II, and much of the original work was never published in the usual journals, or was published only much later. For that reason it seems best to insert some personal reminiscences.

Light scattering and I (B. H. Z.) had become acquainted during the summer of 1942, at the end of my first year of graduate school, when I worked on a project investigating the optical properties of smokes for possible military use as smoke screens. This project was located in the chemistry department's laboratories and was under the direction of Victor K. LaMer. Paul Doty, also a first-year graduate student, and I were hired for the summer to help with this. The smokes in question consisted of fine spherical particles of dyes with strong absorption bands in the visible, and correspondingly complex dependences of scattering on wavelength. The object was to see whether a smoke could be found that would scatter white light strongly but that would become transparent at a specific wavelength. We were not very successful at finding such a smoke—the features of the scattering-versus-wavelength curve were not pronounced enough—but the project was a good introduction to the optical theory of light scattering. This I remember studying in the best book then available, the original German edition of Born's *Optik* (70), which was lent to me by David Sinclair, a physicist working on the project (and son of the novelist Upton Sinclair). (In addition to learning some excellent physics from this book, I profited from the practice of reading Born's elegant German; nearly everything that I have since had to read in that language has seemed easy.)

Later Doty and I both did research for our Ph.D. theses with Joe Mayer at Columbia. A third research student at the same time with Mayer was William G. McMillan, Jr., who was beginning a study of the vapor pressure of mixtures of triethylamine and water near their lower critical mixing point. His aim was to see whether experimental evidence could be found for the "derby-hat picture" of the critical point that Mayer had proposed on theoretical grounds (71). The derby hat was supposed to be a region adjacent to the critical point within which the vapor-pressure isotherms had zero slope. McMillan was facing the unrewarding task of measuring the vapor pressures so precisely that the almost vanishing slopes of curves through these data could be confidently said to be "zero" over a finite range.

Then one of us, I do not remember who, noticed a section in Fowler's book on statistical mechanics (72) that discussed the Einstein-Smoluchowski theory of light scattering and its relation to fluctuations. This theory showed that the intensity of scattering is inversely proportional to the slope of the isotherm of vapor pressure against concentration, and so was obviously useful for McMillan's problem.

Probably nothing would have come of this, if our conversations had not been overheard by Charlie Beckmann, who occupied the laboratories adjoining Mayer's on the eighth floor of Chandler Hall. Beckmann, as we have mentioned, was interested in the physical chemistry of starch; he had one of the first ultracentrifuges to be built in the United States, and among his other instruments was a turbidimeter from Carl Zeiss of Jena. When Beckmann, who was a friendly man, heard us talking about light scattering, he immediately offered the use of his turbidimeter. This was a simple machine: a visual differential (Pulfrich) photometer, a tungsten-filament bulb with a colored filter, and a cell holder with a water jacket as thermostat. There was also a most important accessory, a turbidity standard in the form of a piece of beautiful smoky glass with the value of its turbidity engraved by the Zeiss firm on the brass holder. [I found the published description of the calibration of this standard (73) for the first time while preparing this account.] To make a measurement, one had to balance the illumination intensities in two halves of the visual field of a telescope eyepiece by adjusting calibrated drums that controlled the aperture stops of two objectives, one aimed at the scattering solution or the standard and the other at a piece of opal glass in the same light beam. Late in 1943, McMillan, Doty, and I made a number of measurements of triethylamine and water in this way, but we did not find any sign of the derby hat. Not knowing what to make of this, we never did anything with the data; I still have them in a file folder. If the modern theory of critical points had been available, we would probably have tried to see what exponents the data exhibited; in fact, we actually plotted them on log-log paper, found straight lines, and noted the slopes, but we had no idea of the significance of the latter. Instead of pursuing the triethylamine work, McMillan developed a thesis on statistical mechanics, which became the well-known McMillan-Mayer theory of multicomponent systems (74).

At the beginning of 1944, Paul Doty finished the work for his thesis on the electron affinity of bromine and took a position with a research project directed by Professor Herman Mark at the Polytechnic Institute of Brooklyn. The project was concerned with various aspects of polymer chemistry and physics, especially their application to processing of plastics for military applications such as covers for guns and packages for supplies. Shortly after joining the project, Doty came back one day with the report that Debye at Cornell had developed an as yet unpublished method for measuring molecular weights of polymers by light scattering. Being familiar with the Einstein-Smoluchowski theory, we understood immediately that combining the theory with Raoult's Law would give a molecular-weight method; since we already had the necessary apparatus, we became interested in actually trying the method out. Doty obtained three samples of polystyrene from Professor Mark, samples that two of Mark's former students, Turner Alfrey and Al Bartovics, had prepared and had measured by osmometry (75). In a few weeks of work we made solutions and measured the scattering and the refractive index increment of these samples in both toluene and methyl ethyl ketone. After some struggles with the units of the constants in the Einstein-Smoluchowski formula, we were pleased to find that the molecular weights of the same sample in the two different solvents not only agreed with each other, but that they also agreed with the osmotic values. Doty, Mark, and I published a short communication in the April, 1944, issue of the *Journal of Chemical Physics* describing these results (76), and we published a longer paper the following year (77). Debye's first paper had appeared shortly before in the *Journal of Applied Physics* (66).

It is obvious that we were extremely lucky in the availability of a calibrated instrument as well as of samples of polystyrene, probably the most suitable of all polymers for light scattering because of its complete solubility and its high refractive index, and in the fact that Alfrey & Bartovics had already measured the samples by osmometry. Later we realized that we had been lucky in some more subtle ways too. The light-scattering molecular weights agreed too well with the osmometric results. Since light scattering gives a weight average and osmometry a number average, and since the samples were only roughly fractionated, the former molecular weights should have been higher than the latter by about 50%. What had probably happened was that Alfrey & Bartovics' osmotic membranes were too permeable, on the one hand, so that their molecular weights were too high, and on the other hand, we had overlooked the necessity of applying corrections to measured luminosities when the light has passed from a medium of one refractive index (organic liquid) to another (air) (78); these corrections would have raised the scattering values. Also we had not taken the angular dependence of the scattering into account; the Zeiss turbidimeter measured at only one angle (135 degrees from the incident beam), and this correction would have raised the results somewhat further.

It took a number of years to sort all these problems out. At first we continued to work with the Zeiss Pulfrich photometer but with two new cell holders, one that allowed measurements at 90 degrees from the incident beam, and one that allowed measurements of the ratio of the scattering intensities at 135 and 45 degrees (the "dissymmetry"). This work was done by Doty and several students at Brooklyn, where I also went late in 1944 after finishing my thesis (far from polymer solutions; it was on the vapor pressures of alkali halides) and working for most of the year making smokes again for LaMer's project. Gradually it became evident that there were difficulties with the calibration of the new light-scattering instrument, for which we no longer had Carl Zeiss to rely on. The Zeiss firm was inaccessible, of course, in an enemy country. The calibrations were different, depending on whether we measured the turbidity of a strongly scattering solution directly by the attenuation of transmission in a spectrophotometer, or whether we attempted to calibrate with the scattering from a highly reflecting magnesium-carbonate surface. The task was not made any easier by the size of the ratio of intensities of the incident beam and of the scattering from liquids at the photometer aperture; this ratio was of the order of one million.

To further confuse the situation, some writers in the older literature proposed that Einstein's basic scattering formula was defective and should have a factor of $[(n^2+2)/3]^2$ included, where n is the refractive index, allegedly to take account of the modification of the electric field of the light by the cavity containing the scattering molecule. There was considerable discussion of this in a French book by Cabannes (79). Depending on which calibration method one favored for the instrument, one could easily convince oneself that the experiments verified one form of the theory or the other. That there was no obvious place to include such a factor in Einstein's elegant 1910 derivation tended to be overlooked. Einstein used a phenomenological (optical) dielectric constant in his derivation, and assumed it and its derivatives to be equal to the corresponding macroscopic property; this assumption led to the debate. Of course, differences would be expected at the scale of the molecular dimensions of the liquid. However, the Fourier components of the density fluctuations of interest in light scattering have wavelengths of hundreds of nanometers at the least, much larger than the molecules of ordinary liquids; thus one would expect the bulk dielectric constant to apply. In fact, later careful derivations based on molecular theory lead to the same result; see papers by Fixman (80) and by Zwanzig (81).

In 1946 I went to the University of California at Berkeley where I built a scattering photometer using the newly available multiplier phototube instead of a visual device (82). My student, Clide I. Carr, Jr., elected to do a thesis on the absolute scattering power of pure liquids and solutions and the relation of the scattering to Einstein's theory. In the course of this he rediscovered the effect of refraction at a surface on the apparent brightness of an object, a relation that was well known in optics (78), but which we had overlooked. With this taken into account, everything fell into place; three independent methods of calibrating the photometer agreed, and the measured scattering from pure liquids and from solutions of simple substances was in accordance with theory (83). Using this methodology, Paul Outer, a post-doctoral fellow from Belgium, was able to make an extensive series of measurements on polystyrene solutions and to get all the numbers well pinned down (84).

Confusion about the "high values" and "low values" of the magnitude of the scattering from liquids persisted in the literature for several years until the accumulating experimental evidence came out heavily in favor of the high values. Even in 1953 and 1954 the debate was still going on (85, 86); see some discussion in a review by Stockmayer, Billmeyer & Beasley (87). Later, I found that Debye's group had been aware of the refraction effect at the time of their first work, but mention of it was buried inconspicuously in their writings, and had been missed by everyone else.

In fact, the circumstances surrounding the first publications of the determination of the molecular weight of polymers are curious from a bibliographic point of view. The first work of Debye and his associates at Cornell was published in reports to the Office of Rubber Reserve, Reconstruction Finance Corporation, and were given only limited circulation because of wartime security restrictions. Debye's first paper in the open literature in 1944 (66) omits mention of any previous derivation of his formula for getting the molecular weight. I have often wondered whether he derived it independently, which certainly would not have been difficult for him; much of his previous work had been on other aspects of scattering. Actually the formula, an extension of Einstein's 1910 formula for a pure liquid, had been published long before in extensive discussions by Gans (88) and Raman & Ramanathan (89) in 1923, and these were the references cited by Doty, Mark, and me in our first work (76). Einstein himself had given a short treatment of composition fluctuations, but had unnecessarily limited himself to the case where the vapors of the constituents were ideal gases. Einstein's comment (68, p. 1297) had been: "This formula, which contains only experimentally accessible quantities, completely determines the opalescence properties of binary liquid mixtures, insofar as one may treat their saturated vapors as ideal gases, up to a small region in the immediate neighborhood of the critical point." So the essence of the theory had been available in 1910, if any one had wished to use it. It was not until 1927 that Raman first pointed out the applicability of the formula to colloidal solutions in a paper (91) that we all overlooked; it had appeared in the *Indian Journal of Physics*. In about 1946 several of us unexpectedly received reprints of this paper; they arrived in the mail from India without explanation or comment.

There had also been a few studies in which light scattering was used to compare molecular weights of macromolecules, but in which the full power of the method for determining absolute values was not utilized. Such were papers published by Putzeys & Brosteaux in 1935 (92), by Staudinger & Haenel-Immendörfer in 1943 (93), and by Schulz in 1944 (94); referring to the first two of these, Debye (95) said: "The authors do not yet realize that the [constants in the molecular-weight formula] can be determined experimentally without making assumptions about the particles or their so-called optical constants."

In most of the preceding we have not mentioned the angular dependence of the scattering, which is essentially an interference phenomenon, and which gives information about the linear dimensions of the scattering particles if the particles are sufficiently large. Nor have we said anything about the depolarization of the scattering, which depends on, and gives information about, the anisotropy of the scattering particles. These phenomena complicate the determination of molecular weights, and they are of interest in their own right, but discussion of them would go too far in prolonging what is already a long story. Also we have omitted discussion of the fluctuations of the intensity of the scattering, which have played such a prominent role in "dynamic light scattering" in the last two decades. In the 1940s and 1950s we viewed these fluctuations simply as a nuisance to be averaged over. Even if someone had thought otherwise, it would have been hard to exploit them with the poorly coherent high-intensity light sources that were available before the invention of the laser.

Literature Cited

1. Forman, P. 1983. *Science* 220:824–27
2. Flory, P. J. 1953. *Principles of Polymer Chemistry*, Chapt. 1, pp. 3–28. Ithaca, NY: Cornell Univ. Press. 672 pp.
3. Staudinger, H., Heuer, W. 1980. *Berichte* 63:222–34
4. Guth, E., Mark, H. 1934. *Monatsh.* 63:93–121
5. Kuhn, W. 1934. *Kolloid-Z.* 68:2–15
6. Kuhn, W. 1932. *Z. Phys. Chem. A* 159:363–73
7. Flory, P. J. 1936. *J. Am. Chem. Soc.* 58:1877–85
8. Flory, P. J. 1937. *J. Am. Chem. Soc.* 59:241–53
9. Schulz, G. V., Dinglinger, A., Husemann, E. 1939. *Z. Phys. Chem. B* 43:385–408
10. Huggins, M. L. 1938. *J. Phys. Chem.* 42:911–20
11. Mayer, J. E. 1937. *J. Chem. Phys.* 5:67–73
12. Mayer, J. E., Harrison, S. F. 1938. *J. Chem. Phys.* 6:87–104
13. Mayer, J. E., Mayer, M. G. 1940. *Statistical Mechanics*. New York: Wiley. 495 pp.
14. Mayer, M. G. 1941. *Phys. Rev.* 60:184–87
15. Flory, P. J. 1941. *J. Am. Chem. Soc.* 63:3083–3100
16. Carothers, W. H. 1936. *Faraday Soc. Trans.* 32:39–53
17. Flory, P. J. 1946. *Chem. Rev.* 39:137–97
18. Flory, P. J. 1953. See Ref. 2, Chapter 9, pp. 347–98
19. Casassa, E. F., Mark, H., Markovitz, H., Overberger, C. G., Pearce, E. M., Flory, P. J. 1979. *J. Polym. Sci. Polym. Phys. Ed.* 17:1815–24
20. Lowry, H. H., Erickson, W. R. 1927. *J. Am. Chem. Soc.* 49:2729–34
21. Maass, O. 1938. *Chem. Rev.* 23:17–28
22. Mayer, J. E. 1982. *Ann. Rev. Phys. Chem.* 33:1–23
23. Born, M. 1939. *J. Chem. Phys.* 7:591–603

24. Lindemann, F. A. 1910. *Phys. Z.* 11:609–12
25. Slepian, J., quoted by Siegel, S., Cummerow, R. 1940. *J. Chem. Phys.* 8:847
26. Kramers, H. A., Wannier, G. H. 1941. *Phys. Rev.* 60:252–62
27. Stockmayer, W. H. 1943. *J. Chem. Phys.* 11:43–55
28. Deleted in proof
29. Cohen, R. J., Benedek, G. B. 1982. *J. Phys. Chem.* 86:3696–3714
30. Stockmayer, W. H. 1944. *J. Chem. Phys.* 12:125–31
31. Ziff, R. M., Stell, G. 1980. *J. Chem. Phys.* 73:3492–99
32. Ziff, R. M., Hendriks, E. M., Ernst, M. H. 1982. *Phys. Rev. Lett.* 49:593–95
33. Gordon, M. 1962. *Proc. R. Soc. London Ser. A.* 268:240–59
34. Burchard, W. 1983. *Adv. Polym. Sci.* 48:1–124
35. Whitney, R. S., Burchard, W. 1980. *Makromol. Chem.* 181:869–90
36. Müller, M., Burchard, W. 1978. *Makromol. Chem.* 179:1821–35
37. Gordon, M., Scantlebury, G. R. 1965. *J. Polym. Sci. Pt. C* 16:3933–42
38. Dušek, K. 1979. *Makromol. Chem. Suppl.* 2:35–49
39. Stauffer, D., Coniglio, A., Adam, M. 1982. *Adv. Polym. Sci.* 44:103–58
40. Rosen, P. 1953. *J. Chem. Phys.* 21:1007–12
41. Weil, L. L. 1945. Ph.D. dissertation, Columbia Univ., NY
42. Stockmayer, W. H. 1945. Molecular size distribution in high polymers. In *Advancing Fronts in Chemistry*, ed. S. B. Twiss, 1:61–73. New York: Reinhold. 196 pp.
43. Rempp, P., Decker-Freyss, D. 1965. *J. Polym. Sci. Pt. C* 16:4027–34
44. Bywater, S. 1979. *Adv. Polym. Sci.* 30:89–116
45. Zimm, B. H., Stockmayer, W. H. 1949. *J. Chem. Phys.* 17:1301–14
46. Stockmayer, W. H., Fixman, M. 1953. *Ann. NY Acad. Sci.* 57:335–52
47. Stockmayer, W. H., Jacobson, H. 1943. *J. Chem. Phys.* 11:393
48. Deleted in proof
49. Jacobson, H., Stockmayer, W. H. 1950. *J. Chem. Phys.* 18:1600–6
50. Kuhn, W. 1949. *Helv. Chim. Acta* 32:735–43
51. Poland, D., Scheraga, H. A. 1970. *Theory of Helix-Coil Transitions in Biopolymers.* New York: Academic. 797 pp.
52. Poland, D., Scheraga, H. A. 1966. *J. Chem. Phys.* 45:1464–69
53. Wang, J. C., Davidson, N. 1966. *J. Mol. Biol.* 15:111–23
54. Jacobson, H., Beckmann, C. O., Stockmayer, W. H. 1950. *J. Chem. Phys.* 18:1607–12
55. Semlyen, J. A. 1976. *Adv. Polym. Sci.* 21:41–75
56. Flory, P. J., Semlyen, J. A. 1966. *J. Am. Chem. Soc.* 88:3209–12
57. Kennedy, S. J., Wheeler, J. C. 1983. *J. Chem. Phys.* 78:953–62
58. DeGennes, P. G. 1979. *Scaling Concepts in Polymer Physics*, pp. 265–281. Ithaca, NY: Cornell Univ. Press. 324 pp.
59. Kennedy, S. J., Wheeler, J. C. 1983. *J. Chem. Phys.* 78:1523–27
60. Wagner, R. H., Moore, L. D. Jr. 1959. In *Physical Methods of Organic Chemistry*, ed. A. Weissberger, Vol. 1, pt. 1, pp. 815–94. New York/London: Interscience. 894 pp. 3rd ed.
61. Wales, M. 1948. *J. Phys. Colloid Chem.* 52:235–48
62. Wales, M., Williams, J. W., Thompson, J. O., Ewart, R. H. 1948. *J. Phys. Colloid Chem.* 52:984–98
63. Debye, P., Bueche, A. M. 1948. *J. Chem. Phys.* 16:573–79
64. Kirkwood, J. G., Riseman, J. 1948. *J. Chem. Phys.* 16:565–73
65. Flory, P. J., Fox, T. G. Jr. 1950. *J. Polym. Sci.* 5:745–47
66. Debye, P. J. W. 1944. *J. Appl. Phys.* 15:338–42
67. von Smoluchowski, M. 1908. *Ann. Phys. Leipzig* 25:205–26
68. Einstein, A. 1910. *Ann. Phys. Leipzig* 33:1275–98
69. Deleted in proof
70. Born, M. 1933. *Optik*. Berlin: Springer, 591 pp.
71. Mayer, J. E., Mayer, M. G. 1940. *Statistical Mechanics*, p. 312. New York: Wiley. 495 pp.
72. Fowler, R. H. 1936. *Statistical Mechanics*. Cambridge: Univ. Press. 864 pp.
73. Sauer, H. 1931. *Z. Tech. Phys.* 12:148–62
74. McMillan, W. G. Jr., Mayer, J. E. 1945. *J. Chem. Phys.* 13:276–305
75. Alfrey, T., Bartovics, A., Mark, H. 1943. *J. Am. Chem. Soc.* 65:2319–23
76. Doty, P. M., Zimm, B. H., Mark, H. 1944. *J. Chem. Phys.* 12:144–45
77. Doty, P. M., Zimm, B. H., Mark, H. 1945. *J. Chem. Phys.* 13:159–66
78. Born, M., Wolf, E. 1964. *Principles of Optics*, p. 189. New York: MacMillan. 808 pp.
79. Cabannes, J. 1929. *La Diffusion Moléculaire de la Lumière*. Paris: Presses Universitaires de France
80. Fixman, M. 1955. *J. Chem. Phys.* 23:2074–79
81. Zwanzig, R. 1964. *J. Am. Chem. Soc.* 80:3489–93

82. Zimm, B. H. 1948. *J. Chem. Phys.* 16: 1099–1116
83. Carr, C. I. Jr., Zimm, B. H. 1950. *J. Chem. Phys.* 18:1616–26
84. Outer, P., Carr, C. I. Jr., Zimm, B. H. 1950. *J. Chem. Phys.* 18:830–39
85. Rousset, A., Lochet, R. 1953. *J. Polym. Sci.* 10:319–32
86. Zimm, B. H. 1953. *J. Polym. Sci.* 10:351–52
87. Stockmayer, W. H., Billmeyer, F. W., Beasley, J. K. 1955. *Ann. Rev. Phys. Chem.* 6:359–80; p. 370
88. Gans, R. 1923. *Z. Phys.* 17:353–97
89. Raman, C. V., Ramanathan, R. 1923. *Philos. Mag. Ser. 6* 45:213–24
90. Deleted in proof
91. Raman, C. V. 1927. *Indian J. Phys.* 2:1–6
92. Putzeys, P., Brosteaux, J. 1935. *Trans. Faraday Soc.* 31:1314–25
93. Staudinger, H., Haenel-Immendörfer, I. 1943. *J. Makromol. Chem.* 1:185–96. (English abstr. in 1945; *Chem. Abstr.* 40:1719)
94. Schulz, G. V. 1944. *Z. Phys. Chem.* 194: 1–27
95. Debye, P. 1947. *J. Phys. Colloid Chem.* 51:18–32

STIFF-CHAIN MACROMOLECULES

Hiromi Yamakawa

Department of Polymer Chemistry, Kyoto University, Kyoto 606, Japan

INTRODUCTION

Ordinary flexible-chain polymers such as polymethylene and polystyrene may be characterized by the proportionality of the mean-square end-to-end distance to the molecular weight (or the number of skeletal bonds) over a wide range in the unperturbed state with vanishing excluded volume effects (1). This property characteristic of the random-flight chain is violated by (static) chain stiffness arising from structural constraints and hindrances to internal rotation. This leads to the definition of semiflexible- or stiff-chain macromolecules in a broad sense (1, 2); they include not only typical stiff chains such as DNA, α-helical polypeptides, and cellulose derivatives but also short chains of ordinary flexible polymers. The present review is intended to cover various aspects of equilibrium and nonequilibrium properties of such stiff chains without excluded volume in dilute solution. Emphasis is focused on molecular models, theoretical methods, and adaptation to real chains (or determination of model parameters).

Among a number of models presented for chain molecules, the rotational isomeric state model (3) can best mimic the equilibrium conformational behavior of real chains of arbitrary length since it takes account of the details of the chain structure. However, for many equilibrium and steady-state transport problems on stiff chains, such details are not amenable to mathematical treatments, and moreover are often unnecessary to consider. Some coarse-graining may then be introduced to replace this discrete chain by continuous models, although the discreteness must be, to some extent, retained in a study of dynamic properties, especially local chain motions. The foremost of these models is the worm-like chain proposed by Kratky & Porod (4) in 1949 and its numerous, subsequent modifications. Those theoretical developments made before the early 1970s have been reviewed previously (1, 2, 5). Since then, many investigations have been performed on the worm-like chain and also on its generalization to a continuous model, now called the helical worm-like chain (6, 7). The latter includes the former as a special case, and bridges a gap between it and real chains, both flexible and stiff. In the present review, major attention is therefore given to these advances made during the past decade.

In the sections that follow, stiff-chain models, equilibrium properties, transport properties, and dynamic properties are discussed in order. In particular, the next section presents some fundamentals of both the worm-like chain and the helical worm-like chain with a rather detailed discussion of their adaptation to real chains; it is also introductory to nonspecialists.

STIFF-CHAIN MODELS

Worm-like Chains

Consider a freely rotating chain composed of n bonds of length l joined with supplementary bond angle θ. The Kratky-Porod (KP) worm-like chain (4) is defined as the continuous limit $n \to \infty$, $l \to 0$, and $\theta \to 0$ of this discrete chain under the restriction that $nl \equiv L$, the total chain contour length, and $l/(1 - \cos \theta) \equiv (2\lambda)^{-1}$ remain constant. Let $\langle R^2 \rangle$ be the mean-square end-to-end distance. For this model, the dimensionless ratio $\lambda \langle R^2 \rangle / L$ increases monotonically from 0 to 1 as λL is increased from 0 to ∞, so that it mimics those real chains for which the characteristic ratio $C_n = \langle R^2 \rangle / nl^2$ increases with increasing n. In other words, it is an interpolation from the two extremes, random coil limit $\lambda L \to \infty$ ($\langle R^2 \rangle = L/\lambda$) and rigid rod limit $\lambda L \to 0$ ($\langle R^2 \rangle = L^2$). For it, therefore, λ^{-1} is equal to the Kuhn segment length A_K, as defined by $\langle R^2 \rangle / L$ in the limit $L \to \infty$, $(2\lambda)^{-1}$ being its persistence length (4).

The KP chain may be regarded as a differentiable space curve. Let $\mathbf{r}(s)$ be the radius vector of an arbitrary point of the curve as a function of the contour distance s ($0 \leq s \leq L$) from one end to that point. The unit vector $\mathbf{u}(s) = \partial \mathbf{r}/\partial s \equiv \dot{\mathbf{r}}(s)$ tangential to the curve at the point s is subject to the constraint $\mathbf{u}^2 = 1$. Both $\mathbf{r}(s)$ and $\mathbf{u}(s)$ may then be considered Markov processes, with s being regarded as "time." It is therefore possible to define the conditional distribution function, i.e. the Green's function $G(\mathbf{R}, \mathbf{u} | \mathbf{u}_0; s)$ of $\mathbf{r}(s) = \mathbf{R}$ and $\mathbf{u}(s) = \mathbf{u}$ when $\mathbf{r}(0) = \mathbf{0}$ and $\mathbf{u}(0) = \mathbf{u}_0$, $\mathbf{R}$ being the end-to-end vector of the chain of contour length s. The original definition of the KP chain above yields the transition moment $\langle (\Delta \mathbf{u})^2 \rangle = 4\lambda \Delta s$ ($\Delta \mathbf{r} = \mathbf{u} \Delta s$). A Fokker-Planck diffusion equation for G is then found to be (8, 9)

$$(\partial/\partial s - \lambda \nabla_u^2 + \mathbf{u} \cdot \nabla_R) G(\mathbf{R}, \mathbf{u} | \mathbf{u}_0; s) = \delta(s)\delta(\mathbf{R})\delta(\mathbf{u} - \mathbf{u}_0), \qquad 1.$$

so that the Fourier transform of G, i.e. the characteristic function

$I(\mathbf{k}, \mathbf{u} | \mathbf{u}_0; s)$, satisfies

$$(\partial/\partial s - \lambda \nabla_u^2 - i\mathbf{k} \cdot \mathbf{u}) I(\mathbf{k}, \mathbf{u} | \mathbf{u}_0; s) = \delta(s)\delta(\mathbf{u} - \mathbf{u}_0) \qquad 2.$$

with i the imaginary unit.

Equation 2 is seen to be equivalent to the Schrödinger equation (in units of $\hbar$) for a rigid dipole $\mathbf{u}$ in an electric field $\mathbf{k}$. Therefore, I may be represented in terms of the path integral over the path $\mathbf{u}(s')$ (10),

$$I(\mathbf{k}, \mathbf{u} | \mathbf{u}_0; s) = \int_{\mathbf{u}(0)=\mathbf{u}_0}^{\mathbf{u}(s)=\mathbf{u}} \exp\left[i \int_0^s (iU/k_B T + \mathbf{k} \cdot \mathbf{u}) \, ds'\right] \mathscr{D}[\mathbf{u}(s')], \qquad 3.$$

where k_B is the Boltzmann constant, T is the absolute temperature, and U is given by

$$U = \tfrac{1}{2}\alpha \dot{\mathbf{u}}^2 \qquad 4.$$

with $\alpha = k_B T/2\lambda$. Note that the integrand of the integral over s' in Eq. 3 is the "Lagrangian" (in units of $\hbar$). It is now recognized that since $\dot{\mathbf{u}}$ is the curvature vector, U is just an elastic energy of bending per unit contour length with α the bending force constant if the KP chain is regarded as an elastic wire. In fact, from this point of view, Saito et al (11) first derived Eq. 2 in the case of $\mathbf{k} = 0$ conversely from Eq. 3, noting that the exponential in Eq. 3 is then the Boltzmann factor. Subsequently this approach has been elaborated by Freed (12, 13).

Because of the constraint $\mathbf{u}^2 = 1$, however, it is impossible to obtain the exact solution of Eq. 1 or 2 in a closed form. Therefore, various attempts (12, 14–18) have been made to relax this constraint. These modifications have already been critically reviewed (5), and no further mention is made of them here. Indeed, in recent years, there have been no significant advances in equilibrium theories along these lines.

Helical Worm-like Chains

For some real chains, the characteristic ratio C_n increases to its coil limiting value C_∞ more rapidly than expected for the KP chain [e.g. polydimethylsiloxane (19)], while for others, it decreases to C_∞ with increasing n [e.g. poly-DL-alanine (20)] or even exhibits a maximum [e.g. syndiotactic poly(methyl methacrylate) (21)]. Such a breakdown of the KP model is probably due to the fact that the chain possesses randomly complete or incomplete helical conformations or spiral configurations. These local helices may be traces of those present in the crystalline state or they may be spirals of relatively large radius arising from certain preferred conformations in the chain with different bond angles. Further, even when the behavior of C_n or of the chain contour can be interpreted by the KP chain, it is impossible to assign, for instance, local dipole moments and polarizabilities to it unless they are cylindrically symmetric about the chain contour. These circumstances make one recognize a need to extend it to a more general elastic wire model such as a hybrid of the three extreme forms of rigid rod, random coil, and regular helix (6, 7).

This may be achieved by considering an elastic wire having both bending and torsional energies. Indeed, Yamakawa & Fujii (22) adopted the Bugl-Fujita potential (23) as such, but relaxed a certain constraint inherent in it in the course of the derivation of a Fokker-Planck equation. The model that resulted was called the helical worm-like (HW) chain; and strictly, it should be distinguished from the Bugl-Fujita chain. Subsequently, Yamakawa & Shimada (24) have given completely the foundation of the HW model with its exact corresponding potential. Now affix a localized Cartesian coordinate system (e_ξ, e_η, e_ζ) to the wire at the contour point s with $e_\zeta = u(s)$ and with e_ξ and e_η being in the directions of the principal axes of inertia of its cross section at s. The localized system at $s + \Delta s$ is obtained by an infinitesimal rotation $\Delta\bar{\Omega}$ of the system at s, and the deformed state of the wire may be determined by the "angular velocity" vector $\omega(s) = (\omega_\xi, \omega_\eta, \omega_\zeta)$ $= \Delta\bar{\Omega}/\Delta s \, (\Delta s \to 0)$. The potential energy of the HW chain per unit contour length at s is then given by (24–26)

$$U = \tfrac{1}{2}\alpha[\omega_\xi^2 + (\omega_\eta - \kappa_0)^2] + \tfrac{1}{2}\beta(\omega_\zeta - \tau_0)^2, \qquad 5.$$

where α and β are the bending and torsional force constants, respectively, and κ_0 and τ_0 are constants independent of s. This U is seen to become a minimum zero at $\omega = (0, \kappa_0, \tau_0)$. The differential-geometrical curvature $\kappa(s)$ and torsion $\tau(s)$ of the chain contour as a space curve (27) are given by (24)

$$\kappa \equiv |\dot{u}| = (\omega_\xi^2 + \omega_\eta^2)^{1/2}, \qquad 6.$$

$$\tau \equiv (u \times n) \cdot \dot{n} = \omega_\zeta - (d/ds)\tan^{-1}(\omega_\xi/\omega_\eta) \qquad 7.$$

with $n = \kappa^{-1}\dot{u}$ being the unit curvature vector. At the minimum of U, therefore, the contour becomes a regular helix defined by $\kappa = \kappa_0$ and $\tau = \tau_0$ (constant curvature and torsion); and it is now called the characteristic helix of the HW chain. Its radius ρ and pitch h are given by $\rho = \kappa_0/\nu^2$ and $h = 2\pi\tau_0/\nu^2$ with $\nu = (\kappa_0^2 + \tau_0^2)^{1/2}$, respectively, it being right-handed for $h > 0$ and left-handed for $h < 0$. The Bugl-Fujita chain is subject to the unphysical constraint $\omega_\xi = 0$, so that then $\omega_\eta = \kappa$ and $\omega_\zeta = \tau$. It should also be noted that although the above U may be formally generalized to its most general form (26), as done by Miyake & Hoshino (28, 29), it is then almost impossible to adapt the model to real chains because of redundant model parameters. Thus the discussion that follows is confined to the HW chain.

While $e_\zeta = u$ is the unit tangent vector as before, $e_\xi \equiv a$ is equal to the unit mean curvature vector $\kappa_0^{-1}\langle \dot{\mathbf{u}}\rangle$. It is then possible to define the Green's function $G(\mathbf{R},\mathbf{u},\mathbf{a}\,|\,\mathbf{u}_0,\mathbf{a}_0;s)$ of $\mathbf{r}(s) = \mathbf{R}$, $\mathbf{u}(s) = \mathbf{u}$, and $\mathbf{a}(s) = \mathbf{a}$ when $\mathbf{r}(0) = \mathbf{0}$, $\mathbf{u}(0) = \mathbf{u}_0$, and $\mathbf{a}(0) = \mathbf{a}_0$, or simply $G(\mathbf{R},\Omega\,|\,\Omega_0;s)$, where $\Omega = (\theta,\phi,\psi)$ ($0 \leq \theta \leq \pi$, $0 \leq \phi \leq 2\pi$, $0 \leq \psi \leq 2\pi$) denotes the Euler angles defining the orientation of the localized coordinate system with respect to an external coordinate system with $\mathbf{u} = (1,\theta,\phi)$ in spherical polar coordinates. Its characteristic function $I(\mathbf{k},\Omega\,|\,\Omega_0;s)$ may be represented in terms of the path integral over the paths $\mathbf{u}(s')$ and $\mathbf{a}(s')$, or $\Omega(s')$, as in Eq. 3 with Eq. 5 for U. The HW chain is then seen to be equivalent to a symmetric top in a gravitational field $\mathbf{k}$ and an angular-velocity-dependent potential. Therefore, the Schrödinger equation for I may readily be written down, and Fourier inversion leads to the Fokker-Planck equation for G,

$$(\partial/\partial s + \mathscr{A} + \mathbf{u}\cdot\nabla_R)G(\mathbf{R},\Omega\,|\,\Omega_0;s) = \delta(s)\delta(\mathbf{R})\delta(\Omega-\Omega_0) \qquad 8.$$

with

$$\mathscr{A} = \kappa_0 L_\eta + \tau_0 L_\zeta - \lambda L_\xi^2 - \lambda L_\eta^2 - \lambda(1+\sigma)L_\zeta^2, \qquad 9.$$

where $\mathbf{L} = (L_\xi, L_\eta, L_\zeta)$ is the angular momentum operator (in units of $-i\hbar$) (30), λ is related to α as before, and $\sigma = \alpha/\beta - 1$ is Poisson's ratio.

Integration of the above G over $\mathbf{R}$ yields the fundamental (free-particle) Green's function $G(\Omega\,|\,\Omega_0;s) = I(\mathbf{0},\Omega\,|\,\Omega_0;s)$. It may be expanded in terms of the simultaneous eigenfunctions of $\mathbf{L}^2$ and L_ζ, i.e. the normalized Wigner functions (rotation matrices) $\mathscr{D}_l^{mj}(\Omega)$, as follows (30),

$$G(\Omega\,|\,\Omega_0;s) = \sum_{l,m,j,j'} g_l^{jj'}(s)\mathscr{D}_l^{mj}(\Omega)\mathscr{D}_l^{mj'*}(\Omega_0), \qquad 10.$$

where the asterisk indicates the complex conjugate, and the sums are taken over $l \geq 0$, $|m| \leq l$, $|j| \leq l$, and $|j'| \leq l$ with l being nonnegative integers. The above $\mathscr{D}$ function, when unnormalized, i.e. $\bar{\mathscr{D}}_l^{mj} = [8\pi^2/(2l+1)]^{1/2}\mathscr{D}_l^{mj}$, is identical with Davydov's D_{mj}^l (31) and Edmonds' $\mathscr{D}_{jm}^{(l)}$ (32). The expansion coefficients $g_l^{jj'}$ as functions of s, for which analytical expressions have been derived (30, 33), are equal to $8\pi^2\langle\mathscr{D}_l^{mj*}(\Omega)\mathscr{D}_l^{mj'}(\Omega_0)\rangle$, where $\langle\ \rangle$ denotes an average over Ω and Ω_0, so that they have the meaning of time-independent angular correlation functions (34). All kinds of equilibrium moments or properties may in principle be expressed in terms of them, and they are the fundamental quantities in the statistical mechanics of the HW chain.

It is important to mention the particular case of $\kappa_0 = 0$ in some detail. In this case, the first term on the right of Eq. 5 becomes equal to the bending energy given by Eq. 4 for the KP chain, and the characteristic helix becomes a straight line. Further, all terms with $j \neq j'$ vanish in Eq. 10. Then, if integration over ψ is carried out, Eq. 8 reduces to Eq. 1, and Eq. 10 becomes the expansion of $G(\mathbf{u}\,|\,\mathbf{u}_0;s)$ in terms of the normalized spherical harmonics

$Y_l^m(\theta, \phi)$ with the expansion coefficients $g_l^{00} = \exp[-\lambda l(l+1)s]$ (11, 35); i.e. the HW chain reduces to the KP chain. This may be restated as follows: The HW chain with $\kappa_0 = 0$ is nothing other than the KP chain as far as only the behavior of the chain contour, e.g. $\langle R^2 \rangle$, is concerned, and then such properties are independent of τ_0. In general, however, the parameter τ_0 is required when those properties, e.g. dipole moments and polarizabilities, whose definitions require the localized coordinate systems, are considered. Then it proves convenient to classify the HW chains with $\kappa_0 = 0$ into two types: one with $\tau_0 \neq 0$ (KP1) and the other with $\tau_0 = 0$ (KP2) (36). The localized coordinate systems affixed to them at the minimum of energy are depicted in Figure 1. All these chains with $\kappa_0 = 0$, both original and generalized (KP1 and KP2), are referred to simply as the KP chain unless necessary to specify.

For the HW chain as well as the KP chain (4, 37), analytical expressions can readily be derived for $\langle R^2 \rangle$ and also the mean-square radius of gyration $\langle S^2 \rangle$ (22, 38). In order to show its typical behavior, values of the ratio of $\langle R^2 \rangle$ to its coil limiting value $\langle R^2 \rangle_C = A_K L$ are plotted against the logarithm of the reduced contour length λL in Figure 2 for two cases: (a) $\kappa_0' = 5$, $\tau_0' = 1$, and $\sigma = 0$ and (b) $\kappa_0' = 30$, $\tau_0' = 8$, and $\sigma = 0$, where the prime indicates that the quantity is reduced by λ ($\kappa_0' = \lambda^{-1}\kappa_0$ and $\tau_0' = \lambda^{-1}\tau_0$). The corresponding values for the KP chain and the random coil (C) are also included. The results have been shown to be insensitive to variation of σ ($0 \leq \sigma \leq 0.5$) (22), and therefore it is usually set equal to zero, for simplicity.

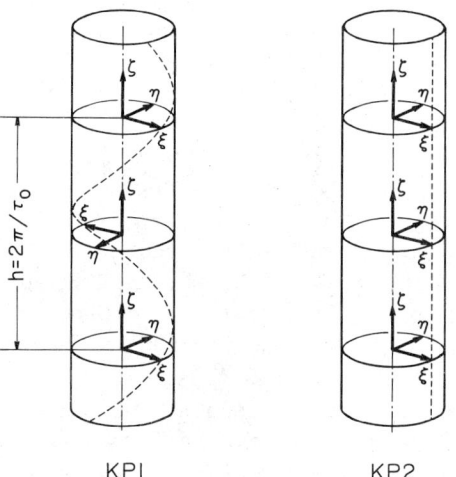

Figure 1 The localized coordinate systems affixed to the KP1 ($\kappa_0 = 0, \tau_0 \neq 0$) and KP2 ($\kappa_0 = \tau_0 = 0$) chains at the minimum of energy.

In order to apply the model to a real chain, the total contour length L of the former must be converted to the number of repeat units in the latter or the molecular weight M. This is done conveniently by introducing the shift factor M_L, as defined by M/L. Note that it is somewhat different from the one originally introduced by Maeda et al (39). Thus, κ'_0, τ'_0, λ^{-1}, and M_L may be chosen as the basic model parameters (with $\sigma = 0$). Then the HW chain can mimic rather well the equilibrium conformational behavior of all kinds of real chains if proper length scales are assumed and if the localized coordinate systems affixed to the former are made to correspond properly to those affixed to the latter. [The agreement is not very good for short branched chains (40, 41).] Indeed, these parameters may be determined either from a comparison with the rotational isomeric state model with respect to $g_l^{ij'}$, $\langle R^2 \rangle$, $\langle S^2 \rangle$, and the persistence vector $\mathbf{A}$ (42, 43), as defined as the mean $\mathbf{R}$ with the initial localized coordinate system fixed, or from an analysis of experimental data for $\langle S^2 \rangle$, the temperature coefficient of $\langle R^2 \rangle_C$ (44), optical and electrical properties, and steady-state transport coefficients (see the following sections).

The results thus obtained, although not all, are summarized in Table 1 (38). It is pertinent to note here that κ'_0, τ'_0, and λ^{-1} are somewhat dependent on temperature (and also on solvent for typical stiff chains), and that most of the important properties are independent of the sign of τ_0 or h. In general, λ^{-1} is larger for syndiotactic chains than for isotactic chains. For amylose, the parameters for Code 26 in the Table have been determined from

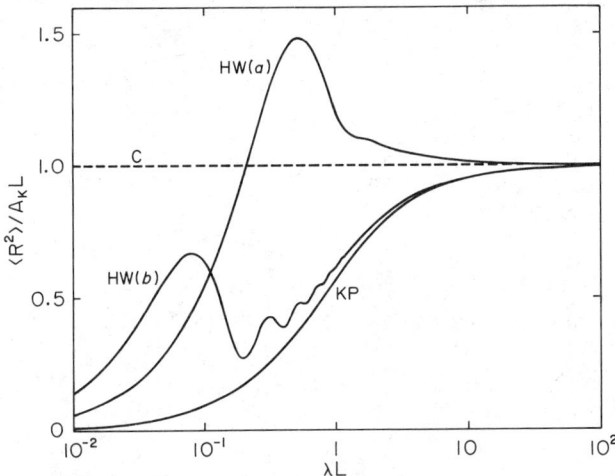

Figure 2 The ratio of $\langle R^2 \rangle$ to its coil limiting value plotted against the logarithm of the reduced contour length λL for two cases (*a*) and (*b*) of the HW chain, the KP chain, and the random coil (C) (see text).

Table 1 Values of the model parameters for various polymer chains

Code	Polymer	Temp. (°C)	κ_0'	$|\tau_0'|$	λ^{-1}(Å)	M_L(Å^{-1})
1	Polymethylene	140	0.3	0	14.5	10.1
2	Polydimethylsiloxane	25	2.6	0	20.7	19.6
3	Polyoxymethylene	30	17	25	22.3	13.1
4	Polyoxyethylene	25	2.4	0.5	12.0	8.8
5	Polypropylene, isotactic	140	3.7	5.0	11.9	16.1
6	syndiotactic	140	0.6	2.4	23.9	16.8
7	Polystyrene, isotactic	27	11	15	26.4	41.2
8	syndiotactic	27	0.8	2.3	37.5	41.9
9	Poly(methyl acrylate), isotactic	27	6.5	11	20.0	33.4
10	syndiotactic	27	0.35	2.0	35.8	33.9
11	Poly(methyl vinyl ketone), isotactic	27	22	33	53.2	29.5
12	syndiotactic	27	0.1	2.0	65.1	27.6
13	Poly(vinyl acetate), isotactic	27	4.0	8.6	15.1	34.4
14	syndiotactic	27	0.4	2.5	42.0	34.7
15	Poly(vinyl chloride), isotactic	25	6.0	16.5	20.4	26.1
16	syndiotactic	25	0.14	2.0	78.0	24.8
17	Poly(vinyl bromide), isotactic	30	17	25	37.6	41.1
18	syndiotactic	30	0.7	1.9	22.9	43.5
19	Poly(methyl methacrylate), isotactic	27	1.7	1.4	32.7	33.5
20	syndiotactic	27	4.4	0.8	65.6	35.7
21	Poly(α-methyl styrene), isotactic	27	2.3	1.0	17.8	42.7
22	syndiotactic	27	4.4	1.0	76.5	43.5
23	Bisphenol A polycarbonate	25	0	0	20.0	18.3
24	Poly(p-phenylene oxide)	27	0	0	14.8	20.4
25	Poly-DL-alanine	27	6.5	0	28.3	14.7
26	Amylose[a]	25	4.9	1.0	630	79
27	Amylose[b]	25	0.2	2.0	56.5	82.7
28	Amylose[c]	25	28	12	141	33.4
29	Poly(n-butyl isocyanate)[d]	23	0	1890	1440	55.1
30	Poly(γ-benzyl L-glutamate)[e]	25	0	3640	3130	146
31	DNA[f]	20	0	200	1100	195
32	Schizophyllan[g]	25	0	1390	4000	215

[a] In dimethyl sulfoxide.
[b] In aqueous solution (on large length scales).
[c] In aqueous solution (on small length scales).
[d] In carbon tetrachloride.
[e] α-Helix in N,N-dimethyl formamide.
[f] Double helix in 0.2 M NaCl.
[g] Triple helix in aqueous solution.

experimental data for transport coefficients in dimethyl sulfoxide (45), and those for Codes 27 and 28 from the persistence vectors in aqueous solution (38), where the chain contours have been taken along the actual helix axis and the actual helical sequence, respectively. For the last four polymers, Codes 29–32, the KP1 chain has been adopted, the actual helical sequence corresponding to the broken curve in Figure 1. For poly(n-butyl isocyanate), the 8_3 helix (46, 47) has been assumed. A (κ'_0, τ'_0) diagram has also been constructed that may be used to guess the values of κ'_0 and τ'_0 for a given polymer or the behavior of $\langle R^2 \rangle / A_K L$ as displayed in Figure 2 when they have been determined from other quantities (38, 48).

Generally, for the HW chain, the component of **A** in the direction of $\mathbf{u}_0$ in the limit $L \to \infty$ is equal to $A_K/2$ and still has the meaning of the persistence length, but it is not equal to $(2\lambda)^{-1}$ unless $\kappa_0 = 0$. A_K is nearly proportional to C_∞ but not necessarily to λ^{-1} (6). The parameter λ^{-1} may be regarded as a measure of chain stiffness for all types of real chains and is referred to as the stiffness parameter; for typical stiff chains, $\lambda^{-1} \gtrsim 100$ Å. It is reasonable to assume that κ_0, τ_0, and α are independent of T, so that λ^{-1} is proportional to T^{-1}. This assumption leads to rather good agreement between observed and calculated values of the temperature coefficient of $\langle R^2 \rangle_C$, cellulose derivatives being an exception (44). The adaptation of the model to a given polymer (in a given solvent) depends on the length scales to be adopted, which depend on what property is studied; global or slow processes are associated with the large scales, and local or fast processes with the small scales. The model parameters in Table 1 have been determined rather on the large length scales for Codes 23, 24, 26, and 27, and on the small scales for the remaining codes (38). Those thus determined from some properties for a given polymer may be used for an analysis of others, especially dynamic properties.

EQUILIBRIUM PROPERTIES

Some equilibrium properties (observables) may be evaluated simply with the fundamental Green's function $G(\Omega | \Omega_0; s)$ for the chain of contour length s, while evaluation of others requires in general the distribution function $G(\mathbf{R}, \Omega | \Omega_0; s)$ or its moments $\langle R^{l''+2n} \mathscr{D}_l^{mj*}(\Omega) \mathscr{D}_{l'}^{m'j'}(\Omega_0) Y_{l''}^{m-m'}(\Theta, \Phi) \rangle$ $\equiv \langle R \mathscr{D}^* \mathscr{D} Y \rangle$, where $\mathbf{R} = (R, \Theta, \Phi)$ in spherical polar coordinates (49). The former G is given by Eq. 10, but no exact solution of Eq. 8 for the latter G can be obtained in a closed form. However, these moments themselves may be evaluated efficiently (30, 49) by an extension of the operational method developed for the KP chain (35). The results may then be expressed in terms of convolution integrals of $g_l^{jj'}(s)$. Note that $g_l^{jj'}$ and the even moments $\langle R^{2m} \rangle$ of $G(\mathbf{R}; s)$ are special cases of $\langle R \mathscr{D}^* \mathscr{D} Y \rangle$ except for the numerical coefficients, and that similar distribution functions and moments can be defined also for the rotational isomeric state model. Thus some devices are required to obtain approximations to those properties, e.g. $G(\mathbf{R};s)$, $I(\mathbf{k};s)$, and $\langle R^{-1}\rangle$, which cannot be expressed simply or in closed forms in terms of $g_l^{jj'}$.

These approximations may be classified into three categories: (a) asymptotic expansions from the coil limit, (b) asymptotic expansions from the rod limit, and (c) least-squares polynomial approximations. The first category includes the Daniels approximation and the polynomial expansion. The former, which was originally developed for the KP chain (8, 50, 51), consists of expanding the deviation of $G(\mathbf{R}, \Omega | \Omega_0; s)$ from its asymptotic Gaussian form in inverse powers of λs (49), and the latter is its expansion in terms of Laguerre polynomials with its moments (49). A special case of the latter is the Hermite polynomial expansion of $G(\mathbf{R};s)$, originally developed by Nagai (52) and Jernigan & Flory (53) for the rotational isomeric state model. The corresponding I and $\langle R^{-1}\rangle$ may also be evaluated. However, it is clear that both expansions are invalid near the rod limit $\lambda s \to 0$, although the convergence of the Daniels approximations is better near the coil limit $\lambda s \to \infty$ (49).

The second category of approximations, called the ε method, consists of expanding, for instance, quantities related to $G(\mathbf{R};s)$ in terms of the moments $\langle \varepsilon^m \rangle$ of the relative deviation ε of R^2 either from $\langle R^2 \rangle$ or from s^2 near the rod limit; $\langle \varepsilon^m \rangle$ may be expressed in terms of $\langle R^{2n} \rangle$ ($n \leq m$) and is at most of order $(\lambda s)^m$ for $m \geq 2$ (48, 54). It is particularly effective for the evaluation of $I(\mathbf{k};s)$ and $\langle R^{-1}\rangle$. In any case, at the final stage, if desired, it can yield expansions of $s^{-2m}\langle R^{2m}\rangle$, and therefore also of $s\langle R^{-1}\rangle$, in powers of λs to terms of desired order. The coefficients of their linear terms are constants independent of κ'_0 and τ'_0, those of square and cubic terms are functions of κ'_0, and those of higher terms are functions of κ'_0 and τ'_0. Thus they include as special cases the WKB approximations (first-order terms) (55) and also the expansions derived by Norisuye et al (56) for the KP chain by an application of the elementary recurrence formula of Hermans & Ullman (9).

The third category, called the weighting function method, is a modification of the polynomial expansion, including the original procedure of Fixman & Skolnick (57) and its generalizations (48, 58). This method provides a least-squares approximation of G by a polynomial with a suitable weighting function, in which the coefficients are determined to give exactly the first several (more than three) moments. The weighting function itself is chosen in such a way that it alone gives exactly the first two or three moments. The Hermite or Laguerre polynomial expansion of G, when truncated, may be regarded as an approximation by this method with a Gaussian weighting function. In general, the convergence of this method is much better than that of the expansions in the first category, although it still breaks down at small λs, especially for chains close to the KP chain, i.e. with small κ'_0. Thus, for example, an interpolation formula for $s\langle R^{-1}\rangle$ as a function of λs, κ'_0, and τ'_0, which includes as a special case the modification (59) of the Hearst-Stockmayer equation (60) for the KP chain, has been constructed on the basis of its values from the Daniels approximation, the ε method, and the weighting function method (61). This interpolation formula is useful for the evaluation of transport and dynamic properties and takes account of hydrodynamic interactions.

As for other moments, e.g. the higher moments $\langle S^{2m}\rangle$ ($m > 1$) of the radius of gyration and the moments of inertia tensor (62, 63), it is noted only that evaluation of them requires in general the multivariate distribution function $P(\mathbf{R}_1, \mathbf{R}_2, \ldots, \mathbf{R}_p, \Omega, \Omega_0; s)$ of $\mathbf{R}_1, \ldots, \mathbf{R}_p$, $\Omega(s) = \Omega$, and $\Omega(0) = \Omega_0$ for the chain of contour length s, where $\mathbf{R}_j$ is the distance between the contour points s_j and s'_j ($0 \leq s_j < s'_j \leq s; j = 1, 2, \ldots, p$) (41, 64, 65). In the paragraphs that follow, attention is given to experimentally observable quantities.

The isotropic scattering function, or form factor, $P(\theta)$ for the chain of total contour length L is given, in the point-scatterer approximation, by the integral of $2L^{-2}(L-s)I(\mathbf{k}; s)$ over s from 0 to L, where $\mathbf{k}$ has the meaning of scattering vector. It has been evaluated numerically by the Hermite polynomial expansion in the light scattering range of small $k' = \lambda^{-1}k$ (66, 67), and by the ε and weighting function methods in the small-angle X-ray and neutron scattering range of large k' (68). On the basis of these values, an interpolation formula for $P(\theta)$ as a function of k' and λL for various values of κ'_0 and τ'_0 has also been constructed to cover almost all possible ranges (68). In particular, in the light scattering range, it can be evaluated analytically in the first Daniels approximation near the coil limit (36) and by the ε method near the rod limit (54), including as special cases the equation of Sharp & Bloomfield (69) and that of Norisuye et al (56), respectively, for the KP chain. The results show that the square-root plot of $P^{-1}(\theta)$ is better than the conventional plot at small k' and any λL as in the case of random-flight chains (70). Indeed, in most cases, light scattering data have been treated by that plot to determine $\langle S^2\rangle$. Its analysis is one of standard experimental methods for the determination of the model parameters, especially λ^{-1} and M_L for the KP chain, as done for various stiff chains (71–76). For amylose in dimethyl sulfoxide, the observed decrease in $\langle S^2\rangle/M$ with increasing M (77, 78) may well be reproduced by the model parameters for Code 26 in Table 1 (45).

The model can also explain the first maximum and minimum that $k^2P(\theta)$ as a function of k (or scattering angle θ) exhibits (67, 68), as observed in the small-angle X-ray and neutron scattering by syndiotactic poly(methyl methacrylate) (79, 80). However, it fails to explain the second maximum and minimum at larger k. This is rather natural in the point-scatterer approximation for the continuous model. Such oscillations at very large k may be predicted on the basis of discrete models, e.g. the rotational isomeric state model (21, 81). (Of course, random-flight chains cannot explain any of them.) The plateau of $k^2 P(\theta)$ in the transition range, as observed in the small-angle neutron scattering by polystyrene (82, 83) and polymethylene (84), may be interpreted by the use of the model parameters in Table 1 (68). This is not the case with the KP chain, whose $k^2 P(\theta)$ is a monotonically increasing function of θ (68). In this connection, it is important to note that earlier theories (85, 86) predict the existence of the plateau region for the KP chain, but this happens only to arise from the approximations used.

The effects of optical anisotropy are often remarkable for stiff or short chains. The intensity of scattered light may then be evaluated by arraying (continuously) local optical polarizabilities on the chain contour. According to Nagai's theorem (87), the scattered intensity for arbitrary polarizations of the incident and scattered waves may in general be expressed as a linear combination of four independent scattered components, for instance, R_{Vv}, R_{Hv}, R_{Hh}, and R_{Qq}, where the subscripts $v(V)$ and $h(H)$ indicate that the incident (scattered) polarizations are vertical and horizontal, respectively, with respect to the scattering plane, and Qq indicates that both polarizations make the angle $\pi/4$ with that plane. Each of these four components may in general be expressed as a linear combination of five fundamental quantities Z_{000}, Z_{202}, Z_{220}, Z_{222}, and Z_{224}, where $Z_{ll'l''}$ may be evaluated by the use of $I(\mathbf{k},\Omega|\Omega_0;s)$ in the Laguerre series approximations with the corresponding moments $\langle R\mathscr{D}^*\mathscr{D}Y \rangle$, the isotropic part Z_{000} being related to $P(\theta)$ (36). The results include as special cases the evaluation of Nagai (88) for the KP chain with cylindrically symmetric local polarizabilities, that of Horn et al (89) for its rod limit ($\lambda \to 0$), and that of Tagami (90) for the rod limit of the KP2 chain with diagonal polarizabilities. [For any type of rods, analytical evaluation is possible (36).] The results derived by Horn (91) and Utiyama (92) for the Gaussian chain without correlation between orientations of the scatterers are written in terms of only Z_{000} and Z_{220}, but the coil limit with the short-range correlation must also include Z_{202} (36). [Tagami's results (90) for the same Gaussian chain are incorrect (36).]

Clearly it is impossible to express Z_{000} inversely in terms of the four independent scattered components by solving four equations for them with respect to the five quantities $Z_{ll'l''}$, and therefore to extract the isotropic Rayleigh ratio R_θ or $P(\theta)$ from observed independent components, as pointed out first by Nagai (87). [For the Gaussian chain above, R_θ may be expressed in terms of R_{Vv} and R_{Hv} (92).] Nagai (87) has claimed that as far as terms of $0(k^2)$ are concerned, R_θ may be expressed as a linear combination of the four independent components. However, it does not give the correct coefficient of $\sin^2(\theta/2)$ (36). Moreover, some coefficients in Nagai's linear combination are singular at $\theta = 0$, and therefore extrapolation to $\theta = 0$ from his R_θ determined at finite θ may involve appreciable errors. A more satisfactory procedure is established by constructing a linear combination of R_{Vv}, R_{Hv}, and R_{Qq}, although it is not very accurate near the rod limit (36). As for very stiff chains, Berry and co-workers (93, 94) have treated their data for R_{Uv} (with U indicating the unpolarized wave) by the use of the equation for rods (89). However, a detailed theoretical analysis (54) indicates that data for R_{Vv} rather than R_{Uv} should then be treated.

In particular, the polarized (Vv) and depolarized (Hv) components at $\theta = 0$ are related to the mean-square optical anisotropy of the entire chain. It may be evaluated analytically with $G(\Omega | \Omega_0; s)$ and expressed in terms of the integrals of $(L-s)g_2^{jj'}(s)$ over s from 0 to L (36). The result includes as a special case the equation derived by Nagai (88) and Arpin et al (95) for the KP chain with cylindrically symmetric polarizabilities. Similarly, if permanent dipole moments and electrical polarizabilities are also attached, it is possible to express the mean-square dipole moment $\langle \mu^2 \rangle$ in terms of $g_1^{jj'}$, and the electric birefringence (Kerr constant) and the electric linear dichroism in terms of $g_l^{jj'}$ ($l = 1, 2$) (96). These are also useful for the determination of the model parameters (36, 38, 96).

The ring closure probabilities $G(0; L)$ and $G(0, \Omega_0 | \Omega_0; L)$ (97, 98) obtained from $G(\mathbf{R}; L)$ and $G(\mathbf{R}, \Omega | \Omega_0; L)$ are related to the so-called J factor (3, 99) as defined as the ratio of equilibrium constants for cyclization and bimolecular association. For large λL, evaluation of them may be carried out by the use of the Daniels approximation and the weighting function method. For small λL, it is in general very difficult but is possible for the KP1 chain. Indeed, this can be done by replacing the continuous chain by an equivalent discrete chain to treat directly the configuration integral, especially for $G(0, \Omega_0 | \Omega_0; L)$ through linking-number-dependent ring closure probabilities (98). The topological linking number Lk (100, 101) may be defined as follows. Suppose that a given closed KP1 chain with $\Omega = \Omega_0$ is deformed smoothly so that its contour may form a circle in a plane. Then Lk is the number of rotations the localized coordinate system at s completes about the contour as s is changed from 0 to L. (It is independent of chain configuration.) In particular, $G(0, \Omega_0 | \Omega_0; L)$ thus evaluated as a function of λL stays at zero for very small λL, then increases oscillating, and finally decreases monotonically. Analysis of the experimental data of Shore et al (102, 103) for DNA leads to $\sigma \simeq 0$ (or <0) and the estimate of λ^{-1} consistent with the value in Table 1 (98). Although it is known that $\sigma \simeq 0.5$ for most bulk polymeric materials, the assumption of $\sigma \simeq 0$ for DNA, and generally for the HW chain, is not necessarily surprising since local elasticity is being considered on the atomic or molecular level.

The specification of Lk is closely related to the imposition of nonperiodic boundary conditions on the distribution functions. Evaluation of them with such boundary conditions is possible near the rigid rod limit (104) and also near the rigid ring limit (98), but it is difficult when large fluctuations are allowed. Indeed, the path integral representations, and therefore also the differential equations, have been formulated for the Green's functions with periodic boundary conditions, so that they may be expanded in terms of $\mathscr{D}_l^{mj}$ with l being integers, as in Eq. 10. Of course, nonperiodic boundary conditions can be imposed in mechanical (not statistical) problems such as a determination of the most stable chain configuration under constraints (105–107).

TRANSPORT PROPERTIES

There are two approaches to the problem of evaluating steady-state transport coefficients such as the sedimentation coefficient s and the intrinsic viscosity $[\eta]$: (a) the well-known Kirkwood procedure (108, 109) on the basis of bead models, (b) the Oseen-Burgers (OB) procedure (2, 110) on the basis of cylinder models. Earlier theories (60, 111–113) for the KP chain belong to the first category. An application of the second procedure to KP worm-like cylinder models was initiated in 1973 by Yamakawa and co-workers (59, 114–117), and it has recently been extended to HW cylinders (45, 118, 119). Both procedures use the Kirkwood-Riseman approximations (108) in the formulation of the frictional force: (a) preaveraging of the Oseen hydrodynamic interaction tensor, (b) neglect of the coupling between translational and rotational motions (118, 120) if any, and (c) assumption of the uniform rotation of the molecule in a steady shear flow, although the approximations (b) and (c) are valid in the rod limit. The first approach is never unphysical except for the occurrence of the Zwanzig singularities (121), in spite of Ullman's criticism (122, 123). If touched bead models are adopted, the first approach is valid as well as the second, although somewhat different numerical results are obtained. In the remainder of this section, major attention is given to the latter.

In the long coil limit, the theory in this category becomes equivalent to the Kirkwood-Riseman theory (108) for nondraining Gaussian chains. In the rod limit, however, the preaveraging can be avoided. For a long straight cylinder of length L and diameter d, the OB procedure can then give expressions for the reduced sedimentation coefficient $\bar{s}$ (or the mean translational diffusion coefficient D_t), the rotatory diffusion coefficient $D_{r,1}$

about the transverse axis, and $[\eta]$ (115, 124), e.g.

$$\bar{s} = 3\pi\eta_0 LD_t/k_B T = \ln p + \gamma_s(p), \qquad 11.$$

$$[\eta] = (2\pi N_A L^3/45M)[\ln p + \gamma_\eta(p)]^{-1} \qquad 12.$$

with $\gamma_s(\infty) = 2 \ln 2 - 1$ and $\gamma_\eta(\infty) = 2 \ln 2 - 25/12$, where $p = L/d$, η_0 is the solvent viscosity, and N_A the Avogadro number. Now recall that while the exact analytical solution of the linearized Navier-Stokes equation with the stick boundary condition has not been obtained for any cylinder, the exact expressions for the transport coefficients above are well known for spheroids (125–128). For a prolate spheroid of major axis L and minor axis d, their asymptotic forms for $p \gg 1$ are coincident with those for the cylinder above, and the values of $\gamma_s(\infty)$ and $\gamma_\eta(\infty)$ obtained from the exact solutions and from the OB procedure with the non-preaveraged Oseen tensor agree with each other (124). Therefore, the corresponding values for the cylinder above, which differ from those for the spheroid, may be regarded as correct. Thus the OB procedure is certainly valid for large p, but does not give accurate results for $p \lesssim 50$ (129).

For small p, the exact numerical solutions have been obtained for a spheroid cylinder, i.e. a straight cylinder with oblate or prolate hemispheroid caps at the ends, by solving an integral equation having the Green's function of the linearized Navier-Stokes equation, i.e. the Oseen tensor, as the kernel (129). It is equivalent to solving the latter with the stick boundary condition. An expression for $\bar{s}$ derived for the spheroid cylinder by the OB procedure with the preaveraged Oseen tensor and with the Kirkwood-Riseman approximation to a solution of the OB integral equation happens to be consistent with Eq. 11 for $p \gg 1$ and to give values that are in good agreement with the exact ones for small p (124, 129). Thus it may be regarded as a good interpolation formula. Note that the equation of Norisuye et al (130) for the spherocylinder is a special case. As for $D_{r,1}$ and $[\eta]$, empirical interpolation formulas have been constructed on the basis of the exact numerical solutions for small p and the asymptotic solutions for large p (129). The expressions of Broersma (131–133) for the cylinder underestimate $\bar{s}$ and $D_{r,1}$ appreciably except for large p. It is important to note that the expressions for s and $[\eta]$ for KP or HW cylinders extrapolate well to the spheroid cylinder in spite of the use of the preaveraged Oseen tensor (45, 129; see also below).

For rigid rod-like molecules, evaluation is also possible by an application of the Kirkwood procedure in discrete space to bead models provided that the number of beads is not very large. Then the preaveraging can also be avoided, and the modified Oseen tensor (134, 135) may be used. Its use is not necessarily at the same level of approximation for translation and rotation (136), but must be much better than the use of the original Oseen tensor for discrete models. Indeed, there have been a number of investigations (137–146) in this category, including those of complex, rigid, biological macromolecules. These are reviewed by García de la Torre & Bloomfield (147) in some detail. The numerical results for such rigid rods are somewhat different from those for the spheroid cylinder. This may be regarded as arising mainly from the difference between the discrete and continuous models in the shape of the surface. It is pertinent to quote here the computations of Hagerman & Zimm (148) along a similar line but by a Monte Carlo method for the KP chain near the rod limit in the rigid-body ensemble approximation.

As mentioned above, the length scales associated with the steady-state transport coefficients are relatively large. Under such circumstances, typical examples for which transport data may be certainly better explained by the HW chain (45) are amylose in dimethyl sulfoxide (78), i.e. Code 26 in Table 1, and cellulose acetate in trifluoroethanol (149). Indeed, the expressions for s and $[\eta]$ for KP worm-like cylinders have been applied to analyses of transport data to determine the model parameters λ^{-1}, d, and M_L for a wide variety of stiff chains, i.e. cellulose derivatives (114, 150, 151), polycarbonate (152), polyamides (73, 94, 153–156), poly(n-hexyl isocyanate) (72, 74), α-helical polypeptides (75, 157), DNA (59, 71, 114, 158, 159), schizophyllan (76, 160), and some others (93, 94, 161–163).

Figure 3 shows examples of a comparison between theory and experiment with plots of $\bar{s}$ and the logarithm of $12M_L[\eta]/5\pi N_A d^2$ against the logarithm of p ($= M/dM_L$). The triangles and the open and filled circles represent observed values for poly(n-hexyl isocyanate) (PHIC) in n-hexane at 25°C (74), poly(γ-benzyl L-glutamate) (PBLG) in N,N-dimethyl formamide at 25°C (75, 157, 164), and schizophyllan in aqueous solution at 25°C (160), respectively. The model parameters determined from a best fit of the theoretical values to the observed ones of $[\eta]$ are $\lambda^{-1} = 840$ Å, $d = 16$ Å, and $M_L = 71.5$ Å^{-1} for PHIC, $\lambda^{-1} = 3130$ Å, $d = 18.8$ Å, and $M_L = 146$ Å^{-1} for PBLG, and $\lambda^{-1} = 4000$ Å, $d = 26$ Å, and $M_L = 215$ Å^{-1} for schizophyllan. These values of d and M_L have been used to plot the data points. (The values of λ^{-1} and M_L for PBLG and schizophyllan are listed in Table 1.) The full curves in Figure 3 represent the theoretical values calculated for the KP chain from Eqs. 23 and 25 of Ref. (45) with $\varepsilon = 1$ for $[\eta]$ and from Eqs. 49 and 51 of Ref. (59), Eq. 28 of Ref. (119) with $\kappa_0 = 0$, and Eq. 116 of Ref. (129) with $\varepsilon = 1$ for $p \lesssim 5$ for $\bar{s}$, both with the indicated values of λd, where the length of the intermediate cylinder part of the spheroid cylinder is equal to $L - \varepsilon d$. The dotted curves R represent the theoretical values calculated for the spheroid cylinder from Eq. 26 of Ref.

(45) with $\varepsilon = 0.6$, 1.0, and 1.3 for $[\eta]$ and from Eq. 116 of Ref. (129) with $\varepsilon = 1$ for $\bar{s}$. Note that the end effects are negligibly small for $\bar{s}$, and appreciable for $[\eta]$ only for $p \lesssim 5$. It is seen that with the model parameters determined from $[\eta]$, there is good agreement between theory and experiment also for $\bar{s}$ for schizophyllan, but not for PHIC and PBLG, especially near the rod limit. A best fit for $\bar{s}$ leads to $d = 25$ and 30 Å for PHIC and PBLG, respectively, values that are definitely larger than d from $[\eta]$, the values of λ^{-1} and M_L remaining unchanged. A similar inconsistency is known for DNA (71). For most cases, however, the cylinder model can explain transport data consistently.

Nevertheless, it is necessary to inquire into possible sources of the disagreement between the diameters above, which is beyond experimental uncertainty. At present, the most probable one seems to be the replacement of a rough surface of a real macromolecular chain by a smooth cylinder surface (124). Then the breakdown of the stick boundary condition that may occur on the rough surface (165, 166) has no significant effect on the

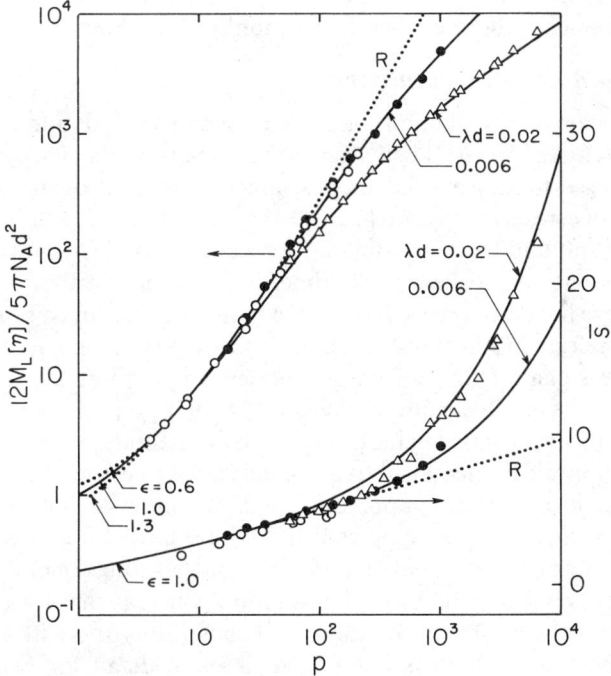

Figure 3 Comparison between theory and experiment for $\bar{s}$ and $[\eta]$ for PHIC (△), PBLG (○), and schizophyllan (●). The *full curves* represent the theoretical values for the KP chain and the *dotted curves* (R) those for the spheroid cylinder (see text).

results (124). Another possibility may be the preaveraging of the hydrodynamic interaction. Since the recent work of Zimm (167), there has been some renewal of activity in the study of its effects (168–173), especially for Gaussian chains, giving attention to Zimm's rigid-body ensemble approximation. However, the effects seem rather small near the rod limit (124, 173), although the problems have not completely been solved as yet.

DYNAMIC PROPERTIES

The development of polymer dynamics is usually confined to the classical diffusion limit, i.e. the Smoluchowski level (2, 174). The classical treatment of the vibrational degrees of freedom is only an approximation (175, 176), though not a bad one for the equilibrium problem (177). The vibrational degrees of freedom may then be constrained, and such structural constraints on bond lengths and bond angles are the main source of chain stiffness, as mentioned in the Introduction. Therefore, attention is generally given to the dynamics of constrained systems in that limit, although there have been some attempts (11, 14, 178–182) to approach the problem of stiff-chain dynamics without explicit imposition of constraints.

Conventional Bond Chains

Consider a conventional bond chain composed of N beads and $N-1$ bonds, and let $\mathbf{q} = (q^1, q^2, \ldots, q^{3N})$ be its generalized coordinates. The subscripts s and h are used to indicate the unconstrained (soft) and constrained (hard) subspaces of $\mathbf{q}$, respectively, so that $\mathbf{q}_s = (q^1, \ldots, q^m)$ and $\mathbf{q}_h = (q^{m+1}, \ldots, q^{3N})$ denote the soft and hard coordinates, respectively. In the derivation of the diffusion equation satisfied by the time t-dependent distribution function $\Psi(\mathbf{q}_s; t)$ there have been considered three types of constrained bond chains, which are referred to as types 1, 2, and 1'. For the type-1 chain, called also the Kramers chain (183), the constraints are imposed at the Lagrangian level so that the hard velocities $\dot{\mathbf{q}}_h$ vanish (184, 185). [It is in general different from a chain with vanishing hard conjugate momenta $\mathbf{p}_h = (p_{m+1}, \ldots, p_{3N})$, which is unphysical since $\mathbf{p}$ is the covariant velocity vector.] For the type-2 chain, the constraints are imposed at the Smoluchowski level so that the hard drift velocities $\mathbf{u}_h = \langle \dot{\mathbf{q}}_h \rangle_q$ vanish (186–189), where $\langle \ \rangle_q$ denotes an average over $\mathbf{p}$ and the solvent phase variables. A starting equation for type-1 and type-2 chains is the Liouville equation, whereas that for the type-1' chain (190) is the Langevin equation with constraints and without inertia.

First, the type-2 chain is mentioned in some detail for convenience. In the diffusion limit, the Liouville equation is reduced to the continuity equation for the distribution function $\Psi(\mathbf{q}; t)$ in the full $\mathbf{q}$ space (191–193), i.e. $\partial \Psi / \partial t = -\nabla \cdot \mathbf{J}$, where $\nabla \cdot$ is the divergence operator defined by $\nabla \cdot = g^{-1/2}(\partial/\partial \mathbf{q})g^{1/2} \cdot$ with g the metric determinant in this space, and $\mathbf{J} = (J^1, \ldots, J^{3N}) = \Psi \mathbf{u}$ is the (contravariant) flux vector. Note that this Ψ is normalized so that the integral of $g^{1/2}\Psi$ over $\mathbf{q}$ is unity. In the field-free case, $\mathbf{J}$ or $\mathbf{u}$ may be determined from the force balance equation,

$$\zeta \cdot \mathbf{u} = -\nabla(k_B T \ln \Psi + U) + \mathbf{P}, \qquad 13.$$

or

$$\mathbf{u} = (k_B T)^{-1} \mathbf{D} \cdot [-\nabla(k_B T \ln \Psi + U) + \mathbf{P}], \qquad 14.$$

where $\nabla = \partial/\partial \mathbf{q}$ is the gradient operator, U the soft potential, $\mathbf{P}$ the constraining force vector, and ζ and $\mathbf{D} = k_B T \zeta^{-1}$ the friction and diffusion tensors, respectively. Note that in the $3N$-dimensional Cartesian space $\mathbf{D} = k_B T(\zeta^{-1}\mathbf{I} + \mathbf{T})$, where ζ is the translational friction coefficient of the bead, $\mathbf{I}$ the unit tensor, and $\mathbf{T}$ the Oseen hydrodynamic interaction tensor. Following Ikeda (186) and Erpenbeck & Kirkwood (187, 188), the soft components of $\mathbf{J}$ are then obtained from Eq. 13 by projection of $\zeta \cdot \mathbf{u}$ onto the s subspace with $\mathbf{u}_h = \mathbf{0}$ and $\mathbf{P}_s = \mathbf{0}$,

$$\mathbf{J}_s = -(\mathbf{D}_{ss} - \mathbf{D}_{sh} \cdot \mathbf{D}_{hh}^{-1} \cdot \mathbf{D}_{hs}) \cdot [\nabla_s \Psi + (k_B T)^{-1} \Psi \nabla_s U]. \qquad 15.$$

Alternatively, $\mathbf{J}_s$ is obtained from Eq. 14 by projection of $\mathbf{u}$ onto the s and h subspaces and elimination of $\mathbf{P}_h$ following Fixman & Kovac (189). ($\mathbf{P}$ may be suppressed from the outset in the former route but not in the latter.) The continuity equation with Eq. 15 and $\mathbf{J}_h = \mathbf{0}$ gives the diffusion equation for $\Psi(\mathbf{q}_s; t)$ for the type-2 chain with the constraints $\mathbf{q}_h = \mathbf{q}_h^0$ if the δ-function $\delta(\mathbf{q}_h - \mathbf{q}_h^0)$ is separated from $\Psi(\mathbf{q}; t)$. These constraints may be considered the so-called "flexible" constraints (194), although with infinitely large force constants.

The diffusion equation for the type-1 chain has been derived by Bird and co-workers (184, 185) only in the free-draining case. The result is equivalent to that for the type-2 chain (with $\mathbf{T} = \mathbf{0}$) except the metric determinant. In general, the metric determinant for the type-1 chain depends on the bead masses since the constraints are imposed at the Lagrangian level. However, in the case of identical beads, it becomes the metric determinant g_s in the s subspace. On the other hand, the diffusion equation for the type-1' chain, which has been derived by Fixman (190) in its Brownian dynamics simulation study, does not involve the bead masses because of the suppression of the inertia term, and is equivalent to that for the type-2 chain with g_s in place of g. The chain treated by Kirkwood (109, 195) is also of type 1', although the term $\mathbf{D}_{sh} \cdot \mathbf{D}_{hh}^{-1} \cdot \mathbf{D}_{hs}$ has been erroneously dropped in his original expression for $\mathbf{J}_s$ (1, 186). In the free-draining case with identical beads, these two types of chains are identical. The diffusion equations for them may be converted to that for the type-2 chain by addition of the metric potential $k_B T \ln (g'_s)^{1/2}$ to U, where g'_s is that part of g_s which depends on the internal soft coordinates (190). The implication is that the simulation of the type-1' chain with this potential is equivalent to that of the type 2. The constraints on the type-1 and 1' chains are the so-called "rigid" constraints (194).

Now there arises the question: Which type of chain is best? The answer is that the type 2 seems better than the type 1 or 1' since there is a statistical-mechanical ground for it (176). Indeed, the rotational isomeric state model in the equilibrium conformational study belongs to the type 2, and the diffusion equations of this type have been standard in polymer dynamics. Also in the Brownian dynamics simulation (based on the Langevin equation), Helfand and co-workers (196, 197) have adopted chains with flexible constraints, and Weiner and co-workers (198, 199) have used type-1' chains with the metric potential (and with inertia). Further, the evaluation of g_s required for the type 1 or 1' is a difficult problem for large N (190, 200). Although for the type 2 there is of course a difficulty in inversion of some matrices, it is greatly diminished by choosing soft coordinates expressed in an external frame (33, 201). However, it is pertinent to note that the dynamics of the type-1' rod (202) are the same as those of type 2, and that these two types seem almost equivalent for long enough, ordinary flexible chains (190).

Mention should be made of some remarkable results for ordinary flexible chains with constraints.

1. Relaxation times for local motions are predicted to be much greater than the smallest Rouse-Zimm relaxation time, or one or two orders of magnitude greater than that for the rotation of a small molecule (203).
2. It is now possible to understand the high-frequency behavior of the dynamic viscosity (203, 204).
3. The simulation studies show that the activation energy for local conformational transitions is only about 10 kJ mol^{-1}, nearly corresponding to the single trans-gauche barrier height, whether the constraints are flexible (196) or rigid (199), although the transition rate itself is larger for the former with finite force constants. This indicates the nonexistence of the crankshaft motion, and is also consistent with experimental results (205–207).

These results should be considered in the development of the dynamics of any other model.

Helical Worm-like Chains

The diffusion equation for the conventional bond chain with constraints is quite difficult to treat, and not much has actually been done regarding the evaluation of various dynamic properties with its solution. Thus Yamakawa & Yoshizaki (33) have presented a discrete HW chain, which is more tractable and is also suitable for a study of local and global motions of both flexible and stiff chains. It is a chain of $N-1$ bonds of fixed length and N beads, each having three rotational degrees of freedom. Apart from its location, its configuration may be specified by N sets of Euler angles $\Omega_1, \ldots, \Omega_N$ representing the bead orientations, i.e. by $3N$ soft coordinates, so that its full coordinate space is $6N$-dimensional. The equilibrium distribution of these angles is made to obey HW statistics, as defined by Eq. 10. It is therefore equivalent to a system of coupled symmetric tops with constraints such that the rotation axis of each points to the center of mass of its successor with the fixed distance between them.

The diffusion equation derived for this chain (33) is formally the same as that for the type-2 bond chain with the flux $\mathbf{J}_s$ given by Eq. 15 except that in the $6N$-dimensional Cartesian space the diffusion tensor is given by $\mathbf{D} = \mathrm{diag}(\mathbf{D}_t, \mathbf{D}_r)$, where $\mathbf{D}_t$ is the previous $\mathbf{D}$ for the bond chain, and $\mathbf{D}_r$ is defined by $\mathbf{D}_r = (k_B T/\zeta_r)\mathbf{I}$ with ζ_r the rotatory friction coefficient of the bead and with $\mathbf{I}$ the unit tensor in the $3N$-dimensional, infinitesimal-rotation space. The advantage of this equation is that the problem can be reduced to an eigenvalue problem that may be decoupled conveniently by an application of the angular momentum theory in quantum mechanics (201). Some numerical results for certain time-correlation functions and correlation times have been obtained (61). In the particular case of the (discrete) KP1 chain near the rod limit, its torsion dynamics (61) have been shown to be equivalent to those (104, 208) developed more directly for DNA.

There have been numerous experimental investigations of dynamic properties such as dielectric relaxation, NMR relaxation, fluorescence depolarization, viscoelasticity, flow and electric birefringence, and dynamic light scattering. (There is no space to cite the extensive literature on them.) In the near future the dynamic theory should be developed to provide a consistent interpretation of these properties on the basis of the HW chain above.

Literature Cited

1. Yamakawa, H. 1971. *Modern Theory of Polymer Solutions*. New York: Harper & Row. 419 pp.
2. Yamakawa, H. 1974. *Ann. Rev. Phys. Chem.* 25:179–200
3. Flory, P. J. 1969. *Statistical Mechanics of Chain Molecules*. New York: Interscience. 432 pp.
4. Kratky, O., Porod, G. 1949. *Rec. Trav. Chim.* 68:1106–22
5. Yamakawa, H. 1976. *Pure Appl. Chem.* 46:135–41
6. Yamakawa, H. 1977. *Macromolecules* 10:692–96
7. Yamakawa, H. 1978. *Math. Sci.* No. 178, pp. 38–44. Tokyo: Science-sha (In Japanese)
8. Daniels, H. E. 1952. *Proc. R. Soc. Edinburgh A* 63:290–311
9. Hermans, J. J., Ullman, R. 1952. *Physica* 18:951–71
10. Feynman, R. P., Hibbs, A. R. 1965.

Quantum Mechanics and Path Integrals. New York: McGraw. 365 pp.
11. Saito, N., Takahashi, K., Yunoki, Y. 1967. *J. Phys. Soc. Jpn.* 22:219–26
12. Freed, K. F. 1971. *J. Chem. Phys.* 54:1453–63
13. Freed, K. F. 1972. *Adv. Chem. Phys.* 22:1–128
14. Harris, R. A., Hearst, J. E. 1966. *J. Chem. Phys.* 44:2595–2602
15. Tagami, Y. 1969. *Macromolecules* 2:8–13
16. Noda, I., Hearst, J. E. 1971. *J. Chem. Phys.* 54:2342–54
17. Fixman, M., Kovac, J. 1973. *J. Chem. Phys.* 58:1564–68
18. Hoshikawa, H., Saito, N., Nagayama, K. 1975. *Polym. J.* 7:79–85
19. Flory, P. J., Semlyen, J. A. 1966. *J. Am. Chem. Soc.* 88:3209–12
20. Miller, W. G., Brant, D. A., Flory, P. J. 1967. *J. Mol. Biol.* 23:67–80
21. Yoon, D. Y., Flory, P. J. 1975. *Polymer* 16:645–48
22. Yamakawa, H., Fujii, M. 1976. *J. Chem. Phys.* 64:5222–28
23. Bugl, P., Fujita, S. 1969. *J. Chem. Phys.* 50:3137–42
24. Yamakawa, H., Shimada, J. 1978. *J. Chem. Phys.* 68:4722–29
25. Landau, L. D., Lifshitz, E. M. 1970. *Theory of Elasticity.* Oxford: Pergamon. 165 pp. 2nd ed.
26. Love, A. E. H. 1927. *A Treatise on the Mathematical Theory of Elasticity.* New York: Dover. 643 pp. 4th ed.
27. Struik, D. J. 1961. *Differential Geometry.* Reading, Mass: Addison-Wesley. 232 pp. 2nd ed.
28. Miyake, A., Hoshino, Y. 1979. *J. Phys. Soc. Jpn.* 46:1324–32
29. Miyake, A., Hoshino, Y. 1979. *J. Phys. Soc. Jpn.* 47:942–46
30. Yamakawa, H., Fujii, M., Shimada, J. 1976. *J. Chem. Phys.* 65:2371–76
31. Davydov, A. S. 1976. *Quantum Mechanics.* Oxford: Pergamon. 636 pp. 2nd ed.
32. Edmonds, A. R. 1974. *Angular Momentum in Quantum Mechanics.* Princeton, NJ: Princeton Univ. 146 pp. 2nd ed.
33. Yamakawa, H., Yoshizaki, T. 1981. *J. Chem. Phys.* 75:1016–30
34. Yamakawa, H., Shimada, J. 1979. *J. Chem. Phys.* 70:609–21
35. Yamakawa, H. 1973. *J. Chem. Phys.* 59:3811–15
36. Yamakawa, H., Fujii, M., Shimada, J. 1979. *J. Chem. Phys.* 71:1611–29
37. Benoit, H., Doty, P. 1953. *J. Phys. Chem.* 57:958–63
38. Fujii, M., Nagasaka, K., Shimada, J., Yamakawa, H. 1983. *Macromolecules* 16:1613–23
39. Maeda, H., Saito, N., Stockmayer, W. H. 1971. *Polym. J.* 2:94–100
40. Mansfield, M. L., Stockmayer, W. H. 1980. *Macromolecules* 13:1713–15
41. Fujii, M., Nagasaka, K., Shimada, J., Yamakawa, H. 1982. *J. Chem. Phys.* 77:986–1004
42. Flory, P. J. 1973. *Proc. Natl. Acad. Sci. USA* 70:1819–23
43. Yamakawa, H., Fujii, M. 1977. *J. Chem. Phys.* 66:2584–88
44. Yamakawa, H., Yoshizaki, T. 1982. *Macromolecules* 15:1444–45
45. Yamakawa, H., Yoshizaki, T. 1980. *Macromolecules* 13:633–43
46. Troxell, T. C., Scheraga, H. A. 1971. *Macromolecules* 4:528–39
47. Shmueli, U., Traub, W., Rosenheck, K. 1969. *J. Polym. Sci. A-2* 7:515–24
48. Yamakawa, H., Shimada, J., Fujii, M. 1978. *J. Chem. Phys.* 68:2140–50
49. Shimada, J., Yamakawa, H. 1977. *J. Chem. Phys.* 67:344–52
50. Gobush, W., Yamakawa, H., Stockmayer, W. H., Magee, W. S. 1972. *J. Chem. Phys.* 57:2839–43
51. Nagai, K. 1973. *Polym. J.* 4:35–48
52. Nagai, K. 1963. *J. Chem. Phys.* 38:924–33
53. Jernigan, R. L., Flory, P. J. 1969. *J. Chem. Phys.* 50:4185–4200
54. Fujii, M., Yamakawa, H. 1980. *J. Chem. Phys.* 72:6005–14
55. Yamakawa, H., Fujii, M. 1973. *J. Chem. Phys.* 59:6641–44
56. Norisuye, T., Murakami, H., Fujita, H. 1978. *Macromolecules* 11:966–70
57. Fixman, M., Skolnick, J. 1976. *J. Chem. Phys.* 65:1700–7
58. Freire, J. J., Rodrigo, M. M. 1980. *J. Chem. Phys.* 72:6376–81
59. Yamakawa, H., Fujii, M. 1973. *Macromolecules* 6:407–15
60. Hearst, J. E., Stockmayer, W. H. 1962. *J. Chem. Phys.* 37:1425–33
61. Yamakawa, H., Yoshizaki, T. 1983. *J. Chem. Phys.* 78:572–87
62. Šolc, K., Stockmayer, W. H. 1971. *J. Chem. Phys.* 54:2756–57
63. Šolc, K. 1971. *J. Chem. Phys.* 55:335–44
64. Shimada, J., Yamakawa, H. 1980. *J. Chem. Phys.* 73:4037–44
65. Shimada, J., Nagasaka, K., Yamakawa, H. 1981. *J. Chem. Phys.* 75:469–76
66. Yamakawa, H., Fujii, M. 1974. *Macromolecules* 7:649–54
67. Fujii, M., Yamakawa, H. 1977. *J. Chem. Phys.* 66:2578–83
68. Yoshizaki, T., Yamakawa, H. 1980. *Macromolecules* 13:1518–25
69. Sharp, P., Bloomfield, V. A. 1968. *Biopolymers* 6:1201–11

70. Berry, G. C. 1966. *J. Chem. Phys.* 44: 4550–64
71. Godfrey, J. E., Eisenberg, H. 1976. *Biophys. Chem.* 5:301–18
72. Rubingh, D. N., Yu, H. 1976. *Macromolecules* 9:681–85
73. Motowoka, M., Fujita, H., Norisuye, T. 1978. *Polym. J.* 10:331–39
74. Murakami, H., Norisuye, T., Fujita, H. 1980. *Macromolecules* 13:345–52
75. Itou, S., Nishioka, N., Norisuye, T., Teramoto, A. 1981. *Macromolecules* 14: 904–9
76. Kashiwagi, Y., Norisuye, T., Fujita, H. 1981. *Macromolecules* 14:1220–25
77. Everett, W. W., Foster, J. F. 1959. *J. Am. Chem. Soc.* 81:3459–64
78. Fujii, M., Honda, K., Fujita, H. 1973. *Biopolymers* 12:1177–95
79. Kirste, R. G. 1967. *Small-Angle X-Ray Scattering*, ed. H. Brumberger, pp. 33–61. New York: Gordon & Breach. 509 pp.
80. Kirste, R. G., Kruse, W. A., Ibel, K. 1975. *Polymer* 16:120–24
81. Yoon, D. Y., Flory, P. J. 1976. *Macromolecules* 9:299–303
82. Wignall, G. D., Ballard, D. G. H., Schelten, J. 1974. *Eur. Polym. J.* 10: 861–65
83. Cotton, J. P., Decker, D., Benoit, H., Farnoux, B., Higgins, J., et al 1974. *Macromolecules* 7:863–72
84. Fischer, E. W., Wendorff, J. H., Dettenmaier, M., Lieser, G., Voigt-Martin, I. 1976. *J. Macromol. Sci. B* 12:41–59
85. Heine, S., Kratky, O., Porod, G., Schmitz, P. J. 1961. *Makromol. Chem.* 44–46:682–726
86. Koyama, R. 1973. *J. Phys. Soc. Jpn.* 34:1029–38
87. Nagai, K. 1972. *Polym. J.* 3:563–72
88. Nagai, K. 1972. *Polym. J.* 3:67–83
89. Horn, P., Benoit, H., Oster, G. 1951. *J. Chim. Phys.* 48:530–35
90. Tagami, Y. 1971. *J. Chem. Phys.* 54: 4990–5010
91. Horn, P. 1955. *Ann. Phys. Paris* 10: 386–434
92. Utiyama, H. 1965. *J. Phys. Chem.* 69: 4138–51
93. Berry, G. C. 1978. *J. Polym. Sci. Polym. Symp.* 65:143–72
94. Wong, C.-P., Ohnuma, H., Berry, G. C. 1978. *J. Polym. Sci. Polym. Symp.* 65:173–92
95. Arpin, M., Strazielle, C., Weill, G., Benoit, H. 1977. *Polymer* 18:262–64
96. Yamakawa, H., Shimada, J., Nagasaka, K. 1979. *J. Chem. Phys.* 71:3573–85
97. Yamakawa, H., Stockmayer, W. H. 1972. *J. Chem. Phys.* 57:2843–54
98. Shimada, J., Yamakawa, H. 1984. *Macromolecules* 17: In press
99. Jacobson, H., Stockmayer, W. H. 1950. *J. Chem. Phys.* 18:1600–6
100. Fuller, F. B. 1971. *Proc. Natl. Acad. Sci. USA* 68:815–19
101. Crick, F. H. C. 1976. *Proc. Natl. Acad. Sci. USA* 73:2639–43
102. Shore, D., Langowski, J., Baldwin, R. L. 1981. *Proc. Natl. Acad. Sci. USA* 78: 4833–37
103. Shore, D., Baldwin, R. L. 1983. *J. Mol. Biol.* 170:957–81
104. Barkley, M. D., Zimm, B. H. 1979. *J. Chem. Phys.* 70:2991–3007
105. Benham, C. J. 1977. *Proc. Natl. Acad. Sci. USA* 74:2397–2401
106. Benham, C. J. 1979. *Biopolymers* 18: 609–23
107. Le Bret, M. 1979. *Biopolymers* 18: 1709–25
108. Kirkwood, J. G., Riseman, J. 1948. *J. Chem. Phys.* 16:565–73
109. Kirkwood, J. G. 1954. *J. Polym. Sci.* 12:1–14
110. Burgers, J. M. 1938. *Second Report on Viscosity and Plasticity of the Amsterdam Academy of Sciences*, Chap. 3, pp. 113–84. Amsterdam: North-Holland. 287 pp.
111. Peterlin, A. 1952. *J. Polym. Sci.* 8:173–85
112. Ptitsyn, O. B., Eizner, Yu. E. 1961. *Vysokomol. Soedin.* 3:1863–69 (In Russian)
113. Hearst, J. E., Beals, E., Harris, R. A. 1968. *J. Chem. Phys.* 48:5371–77
114. Yamakawa, H., Fujii, M. 1974. *Macromolecules* 7:128–35
115. Yamakawa, H. 1975. *Macromolecules* 8:339–42
116. Fujii, M., Yamakawa, H. 1975. *Macromolecules* 8:792–99
117. Shimada, J., Yamakawa, H. 1976. *Macromolecules* 9:583–86
118. Yamakawa, H., Yoshizaki, T., Fujii, M. 1977. *Macromolecules* 10:934–43
119. Yamakawa, H., Yoshizaki, T. 1979. *Macromolecules* 12:32–38
120. Happel, J., Brenner, H. 1973. *Low Reynolds Number Hydrodynamics*. Leyden: Noordhoff. 553 pp. 2nd ed.
121. Zwanzig, R., Kiefer, J., Weiss, G. H. 1968. *Proc. Natl. Acad. Sci. USA* 60: 381–86
122. Ullman, R. 1968. *J. Chem. Phys.* 49:5486–97
123. Ullman, R. 1970. *J. Chem. Phys.* 53:1734–40
124. Yamakawa, H. 1983. *Macromolecules* 16:1928–31
125. Jeffery, G. B. 1922. *Proc. R. Soc. London A* 102:161–79
126. Perrin, F. 1936. *J. Phys. Rad.* 7:1–11

127. Simha, R. 1940. *J. Phys. Chem.* 44:25–34
128. Saito, N. 1951. *J. Phys. Soc. Jpn.* 6:297–301
129. Yoshizaki, T., Yamakawa, H. 1980. *J. Chem. Phys.* 72:57–69
130. Norisuye, T., Motowoka, M., Fujita, H. 1979. *Macromolecules* 12:320–23
131. Broersma, S. 1960. *J. Chem. Phys.* 32:1626–31
132. Broersma, S. 1960. *J. Chem. Phys.* 32:1632–35
133. Broersma, S. 1981. *J. Chem. Phys.* 74:6989–90
134. Rotne, J., Prager, S. 1969. *J. Chem. Phys.* 50:4831–37
135. Yamakawa, H. 1970. *J. Chem. Phys.* 53:436–43
136. Yoshizaki, T., Yamakawa, H. 1980. *J. Chem. Phys.* 73:578–82
137. Yamakawa, H., Tanaka, G. 1972. *J. Chem. Phys.* 57:1537–42
138. McCammon, J. A., Deutch, J. M. 1976. *Biopolymers* 15:1397–1408
139. Nakajima, H., Wada, Y. 1977. *Biopolymers* 16:875–93
140. Nakajima, H., Wada, Y. 1978. *Biopolymers* 17:2291–2307
141. García de la Torre, J., Bloomfield, V. A. 1977. *Biopolymers* 16:1747–63
142. García de la Torre, J., Bloomfield, V. A. 1977. *Biopolymers* 16:1765–78
143. García de la Torre, J., Bloomfield, V. A. 1978. *Biopolymers* 17:1605–27
144. Swanson, E., Teller, D. C., de Haën, C. 1978. *J. Chem. Phys.* 68:5097–5102
145. Tirado, M. M., García de la Torre, J. 1979. *J. Chem. Phys.* 71:2581–87
146. Tirado, M. M., García de la Torre, J. 1980. *J. Chem. Phys.* 73:1986–93
147. García de la Torre, J., Bloomfield, V. A. 1981. *Q. Rev. Biophys.* 14:81–139
148. Hagerman, P. J., Zimm, B. H. 1981. *Biopolymers* 20:1481–1502
149. Tanner, D. W., Berry, G. C. 1974. *J. Polym. Sci. Polym. Phys. Ed.* 12:941–75
150. Meyerhoff, G. 1958. *J. Polym. Sci.* 29:399–410
151. Fried, F., Searby, G. M., Seurin-Vellutini, M. J., Dayan, S., Sixou, P. 1982. *Polymer* 23:1755–58
152. Tsuji, T., Norisuye, T., Fujita, H. 1975. *Polym. J.* 7:558–69
153. Motowoka, M., Norisuye, T., Fujita, H. 1977. *Polym. J.* 9:613–24
154. Sadanobu, J., Norisuye, T., Fujita, H. 1981. *Polym. J.* 13:75–84
155. Arpin, M., Strazielle, C. 1977. *Polymer* 18:591–98
156. Bianchi, E., Ciferri, A., Preston, J., Krigbaum, W. R. 1981. *J. Polym. Sci. Polym. Phys. Ed.* 19:863–75
157. Fujita, H., Teramoto, A., Yamashita, T., Okita, K., Ikeda, S. 1966. *Biopolymers* 4:781–91
158. Record, M. T. Jr., Woodbury, C. P., Inman, R. B. 1975. *Biopolymers* 14:393–408
159. Kovacic, R. T., Van Holde, K. E. 1977. *Biochemistry* 16:1490–98
160. Yanaki, T., Norisuye, T., Fujita, H. 1980. *Macromolecules* 13:1462–66
161. Helminiak, T. E., Berry, G. C. 1978. *J. Polym. Sci. Polym. Symp.* 65:107–23
162. Millaud, B., Strazielle, C. 1979. *Polymer* 20:563–70
163. Motowoka, M., Norisuye, T., Teramoto, A., Fujita, H. 1979. *Polym. J.* 11:665–70
164. Doty, P., Bradbury, J. H., Holtzer, A. M. 1956. *J. Am. Chem. Soc.* 78:947–54
165. Hynes, J. T. 1977. *Ann. Rev. Phys. Chem.* 28:301–21
166. Zwanzig, R. 1978. *J. Chem. Phys.* 68:4325–26
167. Zimm, B. H. 1980. *Macromolecules* 13:592–602
168. Wilemski, G., Tanaka, G. 1981. *Macromolecules* 14:1531–38
169. Fixman, M. 1981. *Macromolecules* 14:1706–9
170. Fixman, M. 1981. *Macromolecules* 14:1710–17
171. Fixman, M. 1983. *J. Chem. Phys.* 78:1588–93
172. Fixman, M. 1983. *J. Chem. Phys.* 78:1594–99
173. Tanaka, G., Yoshizaki, T., Yamakawa, H. 1984. *Macromolecules* 17: In press
174. Stockmayer, W. H. 1976. *Molecular Fluids-Fluides Moleculaires*, ed. R. Balian, G. Weill, pp. 107–49. New York: Gordon & Breach. 459 pp.
175. Gō, N., Scheraga, H. A. 1969. *J. Chem. Phys.* 51:4751–67
176. Gō, N., Scheraga, H. A. 1976. *Macromolecules* 9:535–42
177. Flory, P. J. 1974. *Macromolecules* 7:381–92
178. Soda, K. 1973. *J. Phys. Soc. Jpn.* 35:866–70
179. Bixon, M., Zwanzig, R. 1978. *J. Chem. Phys.* 68:1896–1902
180. Gotlib, Yu. Ya., Svetlov, Yu. Ye. 1979. *Polym. Sci. USSR* 21:1682–90
181. Iwata, K. 1979. *J. Chem. Phys.* 71:931–43
182. Iwata, K. 1980. *Biopolymers* 19:125–45
183. Kramers, H. A. 1946. *J. Chem. Phys.* 14:415–24
184. Bird, R. B., Hassager, O., Armstrong, R. C., Curtiss, C. F. 1977. *Dynamics of Polymeric Liquids*, Vol. 2. New York: Wiley. 290 pp.
185. Curtiss, C. F., Bird, R. B., Hassager, O. 1976. *Adv. Chem. Phys.* 35:31–117

186. Ikeda, Y. 1956. *Bull. Kobayashi Inst. Phys. Res.* 6:44–48 (In Japanese)
187. Erpenbeck, J. J., Kirkwood, J. G. 1958. *J. Chem. Phys.* 29:909–13
188. Erpenbeck, J. J., Kirkwood, J. G. 1963. *J. Chem. Phys.* 38:1023–24
189. Fixman, M., Kovac, J. 1974. *J. Chem. Phys.* 61:4939–49
190. Fixman, M. 1978. *J. Chem. Phys.* 69:1527–37
191. Doi, M., Okano, K. 1973. *Polym. J.* 5:216–26
192. Yamakawa, H., Tanaka, G. 1975. *J. Chem. Phys.* 63:4967–69
193. Wilemski, G. 1976. *J. Statist. Phys.* 14:153–69
194. Helfand, E. 1979. *J. Chem. Phys.* 71:5000–7
195. Kirkwood, J. G. 1949. *Rec. Trav. Chim.* 68:649–60
196. Helfand, E., Wasserman, Z. R., Weber, T. A. 1980. *Macromolecules* 13:526–33
197. Helfand, E., Wasserman, Z. R., Weber, T. A., Skolnick, J., Runnels, J. H. 1981. *J. Chem. Phys.* 75:4441–45
198. Pear, M. R., Weiner, J. H. 1979. *J. Chem. Phys.* 71:212–24
199. Perchak, D., Weiner, J. H. 1981. *Macromolecules* 14:785–92
200. Fixman, M. 1974. *Proc. Natl. Acad. Sci. USA* 71:3050–53
201. Yamakawa, H., Yoshizaki, T., Shimada, J. 1983. *J. Chem. Phys.* 78:560–71
202. Kirkwood, J. G., Auer, P. L. 1951. *J. Chem. Phys.* 19:281–83
203. Fixman, M., Evans, G. T. 1978. *J. Chem. Phys.* 68:195–208
204. Fixman, M. 1978. *J. Chem. Phys.* 69:1538–45
205. Stockmayer, W. H. 1973. *Pure Appl. Chem. Suppl. Macromol. Chem.* 8:379–91
206. Matsuo, K., Kuhlmann, K. F., Yang, H. W.-H., Gény, F., Stockmayer, W. H., et al. 1977. *J. Polym. Sci. Polym. Phys. Ed.* 15:1347–61
207. Liao, T.-P., Morawetz, H. 1980. *Macromolecules* 13:1228–33
208. Allison, S. A., Schurr, J. M. 1979. *Chem. Phys.* 41:35–59

ELECTRON ENERGY LOSS SPECTROSCOPY IN THE STUDY OF SURFACES

Phaedon Avouris and Joseph Demuth

IBM, Thomas J. Watson Research Center, Yorktown Heights, New York 10598

INTRODUCTION

Inelastic electron scattering provides a powerful technique for the study of the excitations of free atoms, molecules (1, 2), and solids (3). However, its most recent and very significant application has been in surface science, where electron energy loss spectroscopy (EELS) is used extensively to study the excitations of clean surfaces and adsorbates. The first such study was reported by Propst & Piper in 1967 (4), who studied the vibrational spectra of W(100) surfaces exposed to various gases with a resolution of 50 meV. The resolution of EELS was significantly improved by Ibach, who in 1970 reported the study of the surface phonons of ZnO with a resolution <20 meV (5). Due to the efforts of Ibach and others the resolution has been improving ever since (currently a 2–3 meV resolution can be achieved), and the scope and applications of EELS have been constantly widening (6).

Electron excitation has several features that make it particularly appropriate for surface spectroscopy. First, it covers a very wide spectral range. The Fourier decomposition of the electric field at the surface produced by an electron at a distance d from the surface and moving with velocity v has frequency components up to a "cut off" frequency $\omega_c \sim v/d$. Thus the electron behaves like a source of continuum radiation—a "poor man's synchrotron." Therefore, low frequency phonons, high frequency intra-adsorbate vibrational modes, and electronic excitations can be monitored under the same scattering conditions, thus allowing a more complete understanding of the system under study. Second, the strong electron-matter interaction provides both a high surface selectivity and high absolute sensitivity (e.g. $\sim 10^{-3}$ of a monolayer). The dominant electron scattering mechanism at surfaces involves long range dipole scattering and provides information about optically allowed excitations. However, other scattering mechanisms involving the direct short range interaction of the electron with the atomic potentials (impact scattering) or the actual capture of the electron to form a temporary negative ion (resonance scattering) are currently used to obtain information about all excitations of the system of interest, not only the dipole-allowed ones. In addition to the spectroscopic information, the application of the selection rules pertinent to the particular excitation mechanism provides important structural (orientational) information regarding the surface or adsorbates.

In this review, we try to present a representative picture of the applications of EELS in the study of clean and adsorbate covered surfaces of solids, particularly metals. Due to the limited space, an encyclopedic coverage of the field is impossible and we apologize to our colleagues whose work we were unable to include. A more detailed discussion of the subject can be found in the monograph by Ibach & Mills (6). Here we discuss briefly the basic physics of the scattering mechanisms in surface EELS, followed by discussions of the vibrational excitations of clean surfaces, of adsorbates, and of the corresponding electronic excitations.

EXCITATION MECHANISMS AND THEORETICAL MODELING OF SURFACE EELS

In response to the pioneering experiment of Ibach on the electron scattering by the surface phonons of ZnO (5), several theoretical treatments appeared. Lucas & Sunjic (7) presented a semiclassical theory in which the incident electron was treated as a classical particle acting as a time-dependent perturbation, linearly coupled to the quantized field of elementary excitations such as optical photons and plasmons. The model explained successfully the phonon spectra of ZnO as well as plasmon loss spectra reported by Powell (8). The semiclassical treatment of Lucas & Sunjik did not consider the surface reflectivity and the conservation of the parallel component of electron momentum. However, its conclusions were confirmed by the fully quantum mechanical model of Evans & Mills (9). The quantum mechanical model was later extended by Evans & Mills (10) to include multiphonon processes and by Mills (11) to obtain a general expression for scattering from electric field fluctuations without resorting to a detailed picture of the elementary excitation responsible for the scattering. After the first applications of EELS to the study of clean surfaces, most experimental studies concentrated on adsorbates, and the theoretical modeling of such systems became necessary. Newns (12) developed a semiclassical theory to describe long range electron scattering by the localized vibrations of a single adsorbate on a metal surface. The same problem was considered soon after in a quantum mechanical way by Persson (13). Delanaye et al (14) considered the scattering from an ordered distribution of point molecular vibrating dipoles on a metal. The model was further refined by Lenac et al (15). Persson (16) described a scattering model in which he included the Coulomb interaction between the incident electrons and the induced charge density that they give rise to in the surface region of the metal.

Because of the importance of the adsorbate studies in the fields of catalysis, semiconductor surface processing, and physical chemistry in general, we describe in more detail below some of the conclusions of the theory regarding electron scattering by adsorbates.

Dipole Scattering

In Figure 1 we summarize four distinct trajectories of electrons scattering from a point scatterer M located above a semi-infinite metal surface. There are two categories of inelastic interactions represented in Figure 1. In processes (a) and (b) the scattering is due to the long range dipole potential of the vibrating adsorbate. In process (a) the incident electron is nearly forward scattered by the adsorbate and then specularly scattered by the substrate. In (b) the incident electron is first specularly reflected and is then scattered by M. Both processes lead to small angle, near specular scattering. In contrast, processes (c) and (d) involve scattering via short range

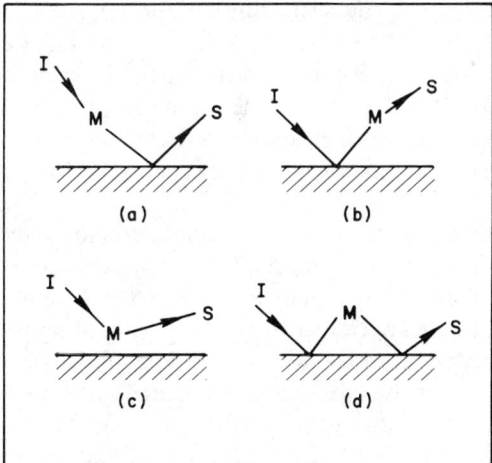

Figure 1 Schematic representation of four electron scattering paths from an adsorbate M located above a metal surface.

interactions. Such short range interactions result in large angle, more isotropic, scattering distributions. We first discuss the long range scattering. In this case the electron-target (molecule) interaction is treated within the Born-Oppenheimer approximation and the transition rate for the excitation process from vibrational (or electronic) state $|i\rangle$ to state $|f\rangle$ is given by the Golden Rule using a perturbation of the form $H' = -e\boldsymbol{\mu}(R) \cdot (\mathbf{r}/r^3)$ to describe the coupling of the electron to the dipole field of the oscillating nuclei (or electrons). The dipole moment $\mu(R)$ can be expanded as $\mu(R) = \mu_s + s(d\mu/ds)\boldsymbol{\mu}_0$, where s is the vibrational normal coordinate, $\boldsymbol{\mu}_0$ is a unit vector in the direction of the molecular dipole and μ_s is the static dipole moment. Only the dynamic dipole is important in the scattering and $d\mu/ds$ is usually referred to as the effective dynamic charge, e^*. Since the interaction potential is long range the approximation is usually made that the electronic wavefunctions are independent of the internal coordinate and the electron wavefunctions above the metal surface are thus written as "reflected plane waves." Such wavefunctions are appropriate if the incident electrons have a sufficiently high kinetic energy and the angle of incidence is not too large. If these conditions are not met, Persson has shown (16) that it becomes important to consider the force on the external electrons from the induced charge density that they give rise to in the surface region of the metal. The differential cross section (defined as the ratio of the one-quantum loss intensity to the intensity of the elastic beam) is given by (15):

$$\frac{d^2\sigma}{d\Omega\, dE'} = \frac{1}{R^2 S_{\text{eff}}} \left(\frac{E'}{E_0}\right)^{1/2} \frac{m}{M} \frac{e^2}{E_0 \hbar \omega_0} \left(\frac{d\mu}{ds}\right)^2 |F(k_s, k_I)|^2 \delta(E_0 - E' - \hbar\omega_0). \quad 1.$$

In Eq. 1 $F(k_s, k_I)$ is the scattering amplitude given by $F(\mathbf{k}_s, \mathbf{k}_I) = (k_I/4\pi) \langle \psi_{k_s}^-|(\boldsymbol{\mu}_0 \cdot \mathbf{r}/r^3)|\psi_{k_I}^+\rangle$, $E' = E_0 - \hbar\omega_0$, $S_{\text{eff}} = S \cos\alpha$ is the effective normalization area, where S is the metal surface per adsorbed molecule and α is the electron impact polar angle. The factor R^2 (reflectivity) normalizes the distribution per electron elastically reflected in the specular direction.

The main characteristics of surface dipole scattering can be deduced from Eq. 1:

1. Only modes with non-zero dynamic dipole $d\mu/ds$ (i.e. infrared active modes) can scatter via this mechanism. In addition, only modes with a non-zero component of the dynamic dipole perpendicular to the metal surface are excited. There are two ways of viewing this absence of scattering by the parallel component of the dynamic dipole. If we consider the electron as being scattered by the dipole field of the oscillating nuclei, then the image of a dipole parallel to the surface will cancel out this dipole field (15). If we consider the interaction as due to the field of the incident electron acting on the dynamic dipole, then the image of the electron in the metal cancels out the parallel component of the field experienced by the adsorbate (16). This

result is usually referred to as the *surface dipole selection rule*. By analogous arguments, we can deduce that the scattering by a dipole normal to a metal surface is four times stronger (net dipole twice as large) than that produced by the same dipole in free space.

2. If the metal reflectivity R^2 for the incident electrons is high, processes (*a*) and (*b*) of Figure 1 dominate the scattering, and therefore the inelastically scattered electrons form a lobe near the specular direction. The scattering process can be pictured as a nearly forward scattering (small momentum transfer) process by an isolated molecule preceded or proceeded by elastic backscattering from the metal surface. The displacement of the lobe from specular is a function of $\hbar\omega_0/2E_0$. Another characteristic of the specular scattering process is that for a slow varying metal reflectivity, in particular if $R^2(E_0) \simeq R^2(E')$, the ratio of the inelastic signal to the elastic signal is independent of the reflectivity.

3. When Eq. 1 is integrated over the solid angle Ω the loss intensity increases with decreasing incident electron energy as $E_0^{-1/2}$ if $E_0 \gg \hbar\omega_0$.

The discussion up to this point has involved dipole scattering from an isolated adsorbate. As the surface coverage increases, we may expect the interaction between the adsorbates to manifest themselves in the spectra. This is the case, for example, for chemisorbed CO, where both frequency shifts (of a few meV) and a nonlinear loss intensity as a function of surface coverage were observed. Mahan & Lucas (17) explored the effect of dipole-dipole interactions among the adsorbates to the observed frequency shifts. In their treatment, they took the total field acting on a molecule to be the sum of the external field and the dipole field from all the other dipoles and their images, and derived the following expression for the excitation frequency $\Omega(\mathbf{k}_\parallel)$

$$(\Omega(\mathbf{q}_\parallel)/\omega_0)^2 = 1 + c[\alpha_v U(\mathbf{q}_\parallel)]/[1 + \alpha_e U(\mathbf{q}_\parallel)] \qquad 2.$$

where c is the surface coverage, $U(\mathbf{q}_\parallel)$ is the spatial Fourier transform of the dipole field, α_e is the electronic polarizability, and α_v the vibrational polarizability. Andersson & Persson (18), using angle-dependent EELS, studied the dispersion of the C-O stretching vibration in the Cu(100)+c(2 × 2)CO system and found it in accord with the predictions of the dipole-dipole coupling model. However, the experimentally obtained α_v was four times higher than that of free CO. Similarly, the differential scattering cross section was found to be reduced over that for isolated adsorbates due to the screening effect of the polarizability of the monolayer.

Higher multipole, particularly quadrupole, contributions to the scattering potential were investigated by Thomas & Weinberg (19). It was shown that the quadrupole term allows the excitation of parallel modes and leads to more isotropic scattering. However, for typical quadrupole derivatives of

1–3 au, the expected scattering is two to three orders weaker than specular dipole scattering.

Impact Scattering

When the incident electrons are in close proximity (~ 1 Å) to the solid surface they can also be scattered directly by the localized atomic potentials of the adsorbates and surface. The electrons may also undergo multiple scattering while in the surface region of the solid. The resulting angular distribution of the scattered electrons, unlike in the case of dipole scattering, is broad and the momentum transfers are large. Since the electrons are scattered by the atomic potentials, not the dipole field of the vibrating adsorbates, all vibrational modes, including parallel modes, are now active. The description of the scattering in this "impact" regime is somewhat analogous to the theoretical description of low energy electron diffraction (LEED). While LEED from a rigid periodic structure leads to Bragg diffraction, the thermal motion of the nuclei can inelastically scatter the incident electrons away from the Bragg directions, and this forms the thermal diffuse background. Li, Tong & Mills (20) have developed a theoretical description of impact scattering along these lines. The differential cross section for the excitation of a phonon (single loss) with wavevector $q_\parallel$ and polarization p was written as:

$$\frac{d\sigma(k_\mathrm{I}, k_\mathrm{s})}{d\Omega} = \frac{mE_0 \cos^2 \theta_\mathrm{s}}{2\pi^2 \hbar^2 \cos \theta_\mathrm{I}} A |M(k_\mathrm{I}, k_\mathrm{s}; q_\parallel, p)|^2. \qquad 3.$$

In Eq. 3, θ_I and θ_s are the incidence and scattering angles with respect to the surface normal and A is the surface area of the crystal. To calculate M, Li, Tong & Mills (20) utilized "muffin tin" potentials, which are rigidly displaced during the thermal motion of the nuclei.

Although the integrated cross section of impact scattering can be comparable to or larger than that of dipole scattering, under angle-resolved specular scattering conditions the impact scattering intensity is typically ~ 100 times lower. Impact scattering is favorable for high frequency modes (e.g. C-H stretch modes) and, as can be seen from Eq. 3, it is also most favorable for higher energy (E_0) electrons incident normal to the surface and for near grazing collection. The energy dependence of the scattering cross section, however, is not monotonic but shows significant structure, which is a result of interference effects between the scattered beams. In principle, the energy variation of the impact scattering cross section can provide important structural information although its calculation is very difficult.

Although impact scattering in general can excite vibrational modes of all symmetries, the excitation probability may vanish for certain symmetries and scattering geometries. The selection rules for impact scattering were derived by Tong, Li & Mills (21). These selection rules were confirmed by Blanchet, DiNardo & Plummer (22) and Davies & Erskine (23) for H/W(110) and, more recently, by DiNardo, Demuth & Avouris (24), who used them to determine the molecular symmetry and to assign the vibrational modes of acetylene on Ni(001).

Resonance Scattering

In addition to impact scattering, a qualitatively different short range scattering mechanism involving the temporary capture of the incident electron by the target to form a *compound state* or *resonance* has long been recognized in the gas phase (1). When the incident electron is trapped in the potential associated with the ground electronic state, we speak of *shape resonances*. In this case, the centrifugal, polarization, and exchange forces combine to create a potential with a penetrable barrier surrounding the target. Incident electrons of the appropriate (resonance) energy penetrate the barrier and are temporarily trapped. The attachment time of an electron with a few eV energy is $\sim 10^{-16}$ s, whereas the resonance lifetime varies widely from $\sim 10^{-10}$–10^{-15} s. Excitation results from the following sequence of events: (a) electron capture by the target in state ϕ_i^0 to form the ϕ_j^- ion state, (b) nuclear relaxation (motion) in ϕ_j^-, and (c) electron emission and formation of an excited (vibrationally, rotationally, or electronically) neutral target state, ϕ_f^0. If we consider vibrational (overtone) excitation, the cross section for the $\phi_i^0 \to \phi_f^0$ process will be determined by matrix elements of the form: $|\sum_j \langle \phi_f^0 | \phi_j^- \rangle \langle \phi_j^- | \phi_i^0 \rangle \exp(iE_j\tau_j)|^2$, i.e. it depends on the displacement of the neutral and ion potential energy curves and the resonance lifetime.

Although resonance scattering is particularly strong for electron scattering by gaseous species (1), no firm evidence for its occurrence in the condensed or adsorbed phases was reported until recently. The absence of such resonances was associated with a strong perturbation of the negative ion state upon adsorption. The surface may change the lifetime for electron emission to the vacuum (by screening the polarization interaction and distorting the centrifugal barrier) and can also introduce a new decay channel, namely electron emission to the empty band states of the substrate. Thus, the resonance may be severely broadened or completely eliminated.

The first experimental results suggesting involvement of a resonance in surface scattering was that of Andersson & Davenport for OH on NiO (25). More recently, unambiguous evidence for the occurrence of resonance scattering by a variety of diatomics (N_2, O_2, H_2, CO) adsorbed on silver at low temperatures was provided by Demuth, Schmeisser & Avouris (26). The evidence included the characteristic energy dependence of the inelastic scattering cross section and the observation of high overtone and other optically forbidden excitations (27, 28). The presence of the surface was found to have two general effects on the adsorbate resonances: (a) a shift to lower energy of the resonance maximum, which was explained in terms of image effects on the temporary negative ion, and (b) a broadening of the resonance and a reduction of the relative intensity of successive overtones. Both of these observations were ascribed to a reduction of the resonance lifetime on the surface (29). Figure 2 shows (a) the vibrational overtone spectrum of N_2 on Ag, (b) the resonance profiles for the $0 \to 1$ excitation of gaseous, solid, and adsorbed N_2, and (c) the relative intensities of successive overtones for gaseous and adsorbed CO. Gadzuk (30), using a simple model, estimated a resonance lifetime for the surface N_2 of $\sim 40\%$ of that of free N_2. For chemisorbed adsorbates, evidence for possible resonance scattering contributions was provided recently in the case of CO on Pt(111)

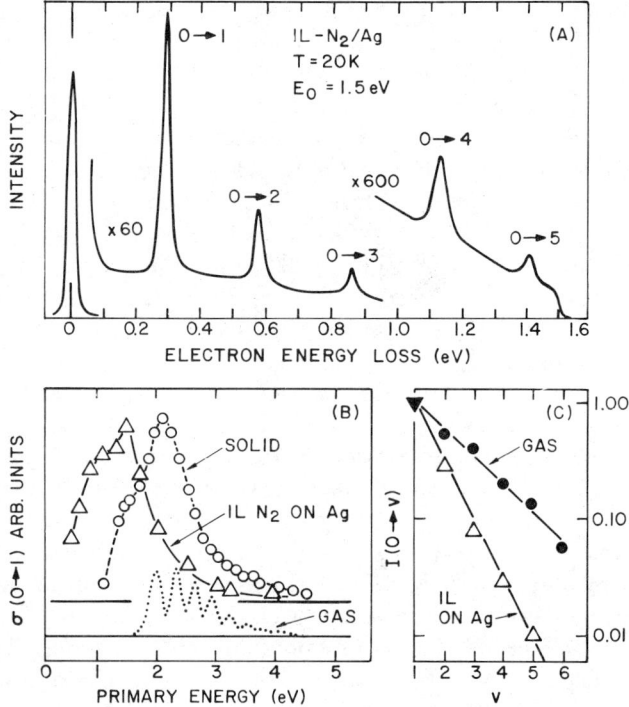

Figure 2 (A) The vibrational spectrum (fundamental and several overtones) of N_2 on a silver surface at 20K obtained via resonance electron scattering. (B) Resonance profiles for the fundamental vibrational excitation of N_2 in the gaseous, solid, and adsorbed (on Ag) phases. (C) Normalized intensity of the vibrational overtones of a monolayer of N_2 on Ag compared to the corresponding intensities for gaseous N_2. Adapted from Refs. (26, 28).

by Ibach (31). It is obvious that there is need for more systematic study of resonance scattering by adsorbates. It is important to look for higher lying resonances, since low energy resonances by a combination of image and bonding shifts may be actually shifted below the vacuum level and thus be unobservable by electron scattering studies. Resonance scattering has several unique and useful features, such as very high sensitivity, excitation of dipole forbidden transitions, as well as the ability to provide important symmetry information regarding unoccupied levels and bonding configurations. The latter possibility was illustrated by model X-α calculations by Davenport, Ho, and Schrieffer (32), who considered the scattering by oriented adsorbates but with electronic structure unaffected by the substrate.

We conclude by considering a different type of resonance. In EELS studies of H/W(100) it was observed that for certain incident electron energies and scattering geometries enhancements appear in the inelastic cross section. The explanation given by Ho, Willis & Plummer (33) involves the creation of diffraction beams with grazing emergence that allow the electrons to be trapped at the surface in a state characteristic of a one-dimensional image potential well. Tuning into such a resonance produces two dimensional electron waves that propagate along the surface and thus enhance the excitation cross section of the adsorbates.

THE VIBRATIONS OF CLEAN SURFACES

Surface Phonons of Ionic Compounds: Fuchs-Kliewer Phonons

Ionic compounds have both infrared (dipole)-active bulk transverse optical (TO) and surface optical (SO) phonons. The frequency, ω_{SO}, of the surface optical mode is defined by the maximum of the surface loss function $-Im[1+\varepsilon(\omega)]^{-1}$ that occurs when $\varepsilon(\omega) = -1$ (here $\varepsilon(\omega)$ is the bulk dielectric function). One thus obtains: $\omega_{SO} = \omega_{TO}[(\varepsilon(0)+1)/(\varepsilon(\infty)+1)]^{1/2}$; ω_{SO} is higher than ω_{TO} but lower than the corresponding longitudinal optical (LO) mode frequency. The surface phonons of ionic compounds have been studied in detail by Fuchs & Kliewer (34) and are commonly referred to as Fuchs-Kliewer phonons. The Fuchs-Kliewer phonons are of mixed transverse-longitudinal character and are accompanied by a long range electric field that decays exponentially in the bulk of the crystal and in the vacuum above it. The penetration depth $1/q_{\parallel}$ becomes particularly large as $q_{\parallel} \to 0$; therefore, the Fuchs-Kliewer phonons are not strictly localized in the surface region.

Due to their long range electric field, Fuchs-Kliewer phonons are expected to be strongly active in dipole electron scattering. In fact, the first high resolution EELS study was that of the Fuchs-Kliewer phonons of ZnO by Ibach (5). Ibach's observations can be summarized as follows:

1. Both one-phonon loss $(E_0 - \hbar\omega)$ and gain $(E_0 + \hbar\omega)$ peaks were observed with $\hbar\omega = 59$ meV in accord with the frequency estimate obtained using bulk dielectric data. The intensity ratio of the one-phonon gain to the one-phonon loss peaks is given by the Boltzmann factor $\exp(-\hbar\omega/kT)$.
2. The one-phonon loss intensity divided by the intensity of the elastic beam falls off as $E_0^{-1/2}$ (E_0 incident electron energy) and varies as $(\cos\theta_i)^{-1}$, with varying incidence angle θ_i in accord with the predictions of the dipole scattering theory of Evans & Mills (9).
3. Besides the one-phonon loss, multiphonon $(n\hbar\omega)$ losses were observed up to $n = 5$. The peak intensities were found to obey a Poisson distribution law.

Since Ibach's original paper on ZnO, several other studies of Fuchs-Kliewer phonons have appeared, such as those of Matz & Lüth (35) on GaAs (110), of Kesmodel, Gates & Chung (36) on $TiO_2(100)$, and of Baden et al (37) on $SrTiO_3(100)$.

When compound semiconductors such as GaAs are doped to increase the free carrier concentration, a new range of phenomena are observed. The doping results in low frequency surface plasmon modes that near $q_{\parallel} = 0$ have fields that, like the Fuchs-Kliewer modes, extend deeply into the bulk and the vacuum and, therefore, can be observed by dipole scattering EELS. Moreover, when the surface phonons and plasmons have comparable frequencies, they mix and cannot be considered anymore as separate excitations.

This coupling of surface phonons and surface plasmons was demonstrated by Matz & Lüth (35), who studied cleaved GaAs(110) surfaces doped with Te or Si and with or without adsorbates. Losses due to these coupled modes were observed and studied as a function of doping n. They found that the carrier density needed to explain the data is not necessarily the bulk density n but some effective surface density n'. For Te-doping it was found that $n' \simeq n$, while for Si-doping or hydrogen adsorption $n' < n$. These observations were explained in terms of band-bending and the formation of depletion layers. While Te does not lead to band-bending, both Si-doping and hydrogen chemisorption result in an upward band-bending, thus forming a depletion layer. EELS experiments of this type can provide, therefore, quantitative information about charge carrier densities, space charge layers, and transport properties, which is particularly valuable to electronic device technology.

Surface Phonons of Elemental Semiconductors

In the case of elemental semiconductors such as Si, Ge, etc, due to the cubic symmetry of the bulk, bulk optical phonons are not infrared active, and

Fuchs-Kliewer type phonons are not present. At the surface, however, the situation is different, the symmetry is reduced, and a nonzero effective dynamic charge may arise. In the Si(111) 2 × 1 superstructure, for example, the two atoms of the surface unit mesh are at different local environments and the vibration of the two sub-lattices has a nonzero effective dynamic charge. Ibach studied the Si(111) 2 × 1 surface and found a dipolar phonon loss at ~ 56 meV (38). The surface character of the loss was inferred from the fact that it is very sensitive to surface contamination. This loss is also specific to the 2 × 1 superstructure. Heating the 2 × 1 surface to form the 7 × 7 reconstruction also eliminates the 56 meV loss. From the theory of Evans & Mills (9), an effective dynamic charge associated with the vibrational motion perpendicular to the surface $\varepsilon_\perp^* \sim 0.7e$ was obtained.

Surface Phonons of Metals

The phonons of clean flat metal surfaces have not been observed by dipole scattering of low energy electrons. However, such phonons can be observed in non-flat, e.g. stepped, or adsorbate covered flat surfaces. So, for example, Ibach & Bruchmann (39) observed a phonon loss at 205 cm^{-1} from a clean, stepped Pt[6(111) × ($\bar{1}$11)] surface. This loss is observed only under specular scattering conditions; i.e. it has dipolar character and disappears upon small exposures to CO, which adsorbs preferentially at the steps. The 205 cm^{-1} loss was attributed, therefore, to a surface phonon localized near the step edge.

Surface phonon losses from flat surfaces can be observed by their coupling to adsorbate vibrations. For example, Ibach & Bruchmann (40) studied Ni(111) with $p(2 \times 2)$ and $\sqrt{3} \times \sqrt{3} R30°$ overlayers of oxygen and found dipole losses at 135, 265, 580 cm^{-1} and at 240 and 580 cm^{-1}. The 580 cm^{-1} loss is due to the Ni-O stretch for O atoms in a three-fold hollow site, but all the other losses are due to Ni surface phonons. Similar results were obtained by Andersson for CO/Cu(100) (41). The fact that metal surface phonons can be observed by EELS has to be considered when interpreting adsorbate spectra. In addition to the one-phonon losses, multiple losses and losses corresponding to combinations of adsorbate vibrations and surface phonons can occur, thus complicating the interpretation of the adsorbate spectra. The appearance of the surface phonon modes as a result of the coupling to the adsorbates can, however, give important information about the adsorption site symmetry, as has been demonstrated by Andersson & Persson (42).

As we discussed above in the section on excitation mechanisms and theoretical modeling, the cross section for impact scattering increases with increasing E_0. Lehwald et al (43) took advantage of this fact to study the phonons of flat, clean Ni(100) surfaces with high energy (180 and 320 eV) electrons. Single phonon losses were thus observed and the surface phonon dispersion was measured along the $\bar{\Gamma}-\bar{X}$ direction. The dispersion curve could be fit very well by a simple nearest-neighbor central-force model in which the force constant coupling atoms in the first layer to atoms in the second layer is increased by 20% to simulate a small inward relaxation of the surface layer. Experiments of this type will provide very valuable information necessary for the understanding of surface lattice dynamics.

Finally, it should be mentioned that EELS studies of the single particle (*e-h* pair) excitations of clean Cu(100) were reported recently by Andersson & Persson (44). As discussed by Persson (45), such studies provide a direct experimental measurement of the response function of the surface to an external charge density, a quantity very important in the description of surface dynamical processes.

VIBRATIONAL SPECTROSCOPY OF ADSORBATES

The study of the vibrational spectra of adsorbates has been the most active area of EEL spectroscopy. Here we discuss only some general aspects of this field in order to indicate to the reader the types of information that can be obtained by vibrational EELS. For a detailed discussion on specific adsorbate systems the reader is referred to the original literature, which has been tabulated recently by Thiry (46).

The simplest case involves monatomic adsorbates. Such an adsorbate can have three vibrational modes: one stretch mode perpendicular to the surface (z) and two modes (x, y) parallel to the surface. As dictated by the surface dipole selection rule, only the perpendicular mode will be dipole active and therefore observable by optical techniques. By non-specular EELS measurements (impact scattering), however, the parallel modes are also observable. Important information regarding the nature of the binding site can be obtained by simply noting the number of distinct modes. For example, Willis (47) observed three modes for H/W(100); this implies that H cannot sit on a site of four-fold symmetry (i.e. $x \neq y$). Similarly for H/Pt(111), Baro et al (48) observed two modes; this implies that H is not in a bridge site. The vibrational frequencies can be used to obtain force constants from which (through empirical relations) quantities such as bond dissociation energy or bond order can be obtained (49). The same quantities can be obtained in a more direct manner by using isotopic substitution or observing vibrational overtones (50). Vibrational frequencies can also be used to determine binding sites. For example, H/GaAs(110) has been observed by Lüth & Matz (51) to give bands at 1890 cm^{-1} and 2150 cm^{-1}, which can be assigned as due to Ga-H and As-H sites, respectively. By combining vibrational data on H/GaAs(110) and D/GaAs(110), the bond dissociation energies are determined to be 1.3 eV (Ga-H) and 0.73 eV (As-H).

Changes in adsorbate vibrational frequencies can also be used to probe changes in substrate reconstruction. Thus, W(100) at saturated hydrogen coverage has a 1×1 superstructure (β_1 phase); at low coverages, however, the substrate reconstructs and a new $c(2 \times 2)$ H-phase (β_2 phase) is formed. In both cases H is bridge bonded, but the W-H-W angle changes from $\sim 100°$ (β_1 phase) to $\sim 80°$ (β_2 phase) as the W atoms come closer together by ~ 0.2 Å, and thus affect in a characteristic manner the H stretching frequencies (52). Hydrogen spectra have also been studied on other substrates such as Ni(100) (53, 54), Pd(100) (55), Ni(111) (56), Si(111) (57–59), GaP(100) (60), InP(100) (60), and ZnO(0001) (61). Studies on other monatomic adsorbates include: O/Ru(001) (62, 63), O/Ni(111) (40), O/W(100) (64), O/Fe(110) (65), O/Ni(100) (60, 43, 66), O/Pt(111) (67, 68), O/Cu(100) (69), O/Ag(110) (70), O/Al(111) (71), Al(100) (72), Mo(100) (73), and several semiconductor surfaces. As a general trend, the stretching frequency is lower than that of the corresponding diatomic molecule, and when the same adatom can occupy more than one site on the same surface, the symmetric stretch frequency is lowest at the highest ligancy site. As an example, O/W(100) (64) has a stretch frequency of 610 cm^{-1} in the four-fold site and 740 cm^{-1} in the two-fold site. The frequency in free WO is 1047 cm^{-1}. Subsurface oxygen vibrations in Al(111) have also been reported (71).

EELS studies of atoms and molecules adsorbed at low temperatures ($T \sim 15K$) have provided information on previously inaccessible weakly bound adsorbates and precursor states. Thus at $\sim 15K$, molecular hydrogen was observed by Avouris, Schmeisser & Demuth on Ag(111) (74) and by Andersson & Harris on Cu(100) (75). H_2 does not have dipole-allowed excitations but, by tuning the primary electron energy to the $^2\Sigma_u^+$ negative ion resonance (26), the above workers were able to observe not only the H-H stretching vibration but also, for the first time, pure rotational and ro-vibrational excitations. In fact, it was concluded that H_2 on these surfaces behaves as a nearly unhindered three-dimensional rotor, thus providing a striking demonstration of the molecular mobility in the physisorbed state.

In chemisorbed diatomic adsorbates, one observes the intramolecular vibrational stretching mode, vibrations of the molecule against the surface, and bending modes. Since bonding to the surface involves charge transfer or rehybridization, the intramolecular mode frequency is altered and in itself can be used to obtain information about the nature of the chemisorption bond. This is for example the case of molecularly adsorbed O_2 on metals. The O-O stretching frequency is ~ 630 cm^{-1} on Ag(110) (76, 77), 870 cm^{-1} on Pt(111) (78, 79), and 1548 cm^{-1} on Ag(111) at 20K (28); the frequency in free O_2 is 1580 cm^{-1}. The binding geometry, by analogy to chemical compounds like $Pt(PPh_3)_2O_2$, is taken to be parallel to the surface. The lowering of the O-O frequency on the surface is ascribed to charge transfer to the $1\pi_g^*$ orbital of O_2. Although the motion of the O-atoms is parallel to surface, this motion modulates the charge transfer and thus acquires strong dipole character. From the O-O frequency the amount of charge transferred to O_2 can be estimated to be $\sim 1.7e$ on Ag(110), $\sim 1.3e$ on Pt(111), and $\sim 0.1e$ on Ag(111).

Chemisorbed CO has been the best studied adsorbate in surface science. These investigations include numerous vibrational studies using EELS and infrared spectroscopy. The EELS studies include CO on Ni(100) (80), Ni(111) (81), Ru(111) (82), Rh(111) (83), Pt(111) (84), Fe(110) (85), W(100) (86), Cu(100) (87), Pt(110) (88), Ni[5(111)×($\bar{1}$10)] (89), and Pt[6(111)×($\bar{1}$11)] (39, 90) and Pt(321) (91). The generally accepted model of CO chemisorption was originally proposed by Blyholder (92), and involves charge donation to the metal via the highest occupied 5σ orbital of CO (as evidenced in photoemission) (93) and charge backdonation to the unoccupied $2\pi^*$ orbital. The strongest arguments in favor of charge backdonation have in fact come from the observed lowering of the C-O stretch upon adsorption. Since the 5σ orbital is essentially nonbonding (lone pair like on C), the (partial) occupation of the antibonding $2\pi^*$ orbital is required [see, however, Ref. (94)]. This backdonation is higher at high ligancy sites and therefore the C-O stretch frequency shows a minimum in this case. The metal-molecule stretch shows the opposite trend, having the highest frequency for an on-top binding site. As an example, Andersson (80) found for on-top (terminal) CO on Ni(100) that $v(Ni-C) = 480$ cm^{-1} and $v(C-O) = 2105$ cm^{-1}, while for bridge bonded CO $v(Ni-C) = 380$ cm^{-1} and $v(C-O) = 1870$ cm^{-1} [for free CO $v(C-O) = 2144$ cm^{-1}]. At this point we should remind the reader that the arguments regarding the frequency shifts apply to low surface coverages ("singleton" frequencies). At high coverages, dipole-dipole coupling can lead to large shifts, as discussed above in the first section. Finally, a small (< 50 cm^{-1}) upward shift in C-O frequency is expected as a result of mechanical renormalization. NO is another interesting adsorbate; it has a single electron in the $2\pi^*$ and as a result it has been observed to bond in both linear and bent configurations [see for example Refs. (95, 96)].

EELS has also found wide application in the study of polyatomic adsorbates on single crystal surfaces, since it can provide direct structural and chemical information not easily accessible by other surface science methods. Hydrocarbon molecules have attracted considerable attention due to their role as prototypes for the understanding of C-H and C-C bond breaking and reformation at surfaces. Studies have been performed on alkanes, alkenes (primarily ethylene), alkynes (primarily acetylene), and aromatic hydrocarbons (primarily benzene) on such surfaces as Ni, Pt, Rh, Ru, W, Cu, and Ag. On the basis of frequency shifts of C-C and C-H vibrations upon adsorption, information has been obtained on the nature of the chemisorption bond rehybridization of the molecule as well as charge transfer. The binding geometry was inferred using the surface dipole selection rule, and the surface reactivity has been followed as a function of crystal face, temperature, presence of steps, or preadsorbed substances (e.g. oxygen). Here we are able to comment only on some of the general conclusions.

Saturated hydrocarbons appear to interact weakly with metals and in a UHV environment, adsorb on clean surfaces at relatively low temperatures ($T \lesssim 80-180K$). An interesting exception involves cyclohexane on Ni (97), Pt (97), and Ru (98) surfaces, where, in addition to the normal C-H stretching modes, new low frequency (2700–2600 cm^{-1}) broad and intense C-H bands are observed. These "soft modes" have been attributed to hydrogen bonded simultaneously to the molecule and the metal (multicenter bonding). The magnitude of the shift to lower frequencies (mode softening) appears to correlate with the tendency of cyclohexane to dehydrogenate on the particular surface. Soft modes have also been observed for cyclopentane on Ru(001) (99) but not for cyclopropane (100). Unsaturated hydrocarbons show a varied behavior. Whereas they interact only weakly and maintain their molecular identity on noble metals, they decompose in a thermally activated manner on most transition metals to give a variety of chemical intermediates. For example, acetylene adsorbs on Ni(111) (101), Pt(111) (102), Pd(111) (103), and Rh(111) (104) at low temperatures to give a rehybridized molecular form of C_2H_2. From the frequencies of the $v(CC)$ and $v(CH)$ modes determined by EELS, it was inferred that the hybridization changed from sp in the gas phase to between sp^2 and sp^3 for the adsorbed species. An even more strongly distorted ($\sim sp^3$) species was observed by Kesmodel (105) for C_2H_2 on Pd(100). Ethylene is also adsorbed molecularly at low temperatures on the same surfaces and has an $\sim sp^3$ hybridization. Upon heating of ethylene and acetylene on Pt(111) (102), Pd(111) (103), and Rh(111) (104), a new chemical species first observed by Ibach & Lehwald (102) and now identified as ethylidyne ($\equiv$C-CH$_3$) is formed. Further heating results in fragmentation; EELS studies of the acetylene fragmentation have shown the production of species such as CCH$_2$ [Pt(111), Pd(111)], CH [Pt(111), Pd(111), Ni(111)], CCH [Pd(111), Pd(100)] (106), and CH$_2$ [Ru(001)] (107).

Stepped surfaces are found to have a different reactivity than the flat surfaces. An example is provided by the EELS study of Lehwald & Ibach (108) of C_2H_2 on a stepped Ni surface: Ni[5(111) × ($\bar{1}$10)]. This surface contains flat (111) regions with a monatomic step along a [$\bar{1}\bar{1}2$] direction every ~10 Å. As seen in Figure 3, at 150K the spectrum shows the presence of chemisorbed C_2H_2 (890, 1200, and 2920 cm^{-1}) but also decomposition products H (at 690 cm^{-1}) and C_2 (350 and 2210 cm^{-1}). In contrast, flat Ni(111) does not decompose C_2H_2 up to ~350K and no C_2 has been observed on the flat surface. The C_2 fragment dissociates at ~180K and a new loss due to isolated C atoms appears at 520 cm^{-1}. At higher temperatures (e.g. 300K) only atomic H and C are present. No CH is

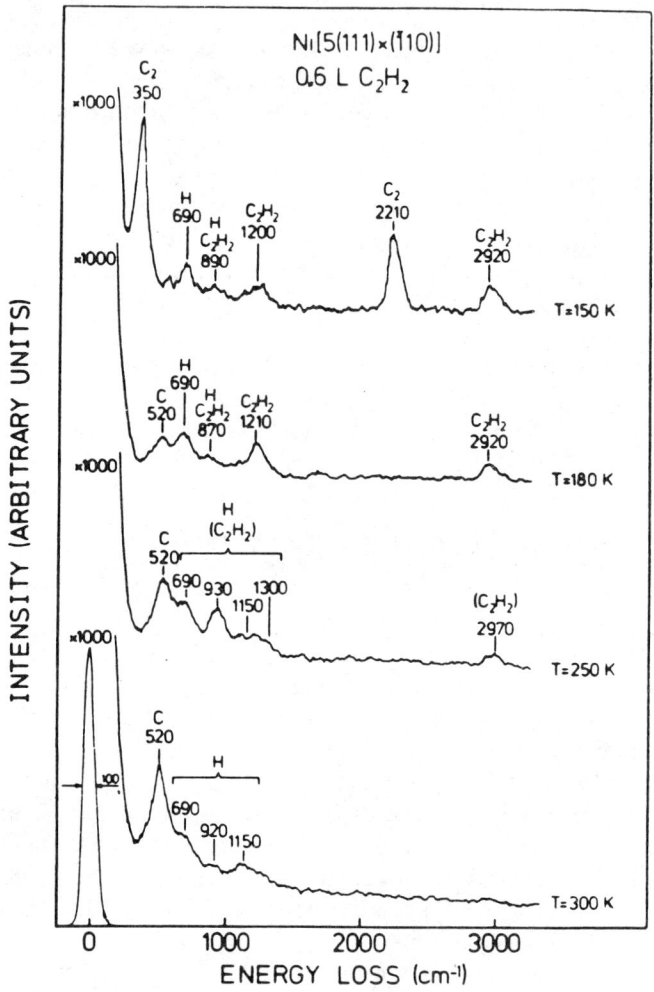

Figure 3 Vibrational spectra of acetylene (0.6 L) and its decomposition products on a stepped nickel surface (Ni[5(111) × ($\bar{1}$10)]) as a function of surface temperature. From Ref. (108).

observed on the stepped surface, while this species is observed at higher coverages on the flat surface (101).

Aromatic hydrocarbons are found to adsorb associatively on most metals. For example, benzene is found to be π-bonded with the ring parallel to the surface on Ni(111) (109), Ni(100) (110), Pt(111) (109), Rh(111) (111), and Ag(111) (112). In all cases, a $C_{3v}(\sigma_d)$ symmetry was inferred. On the basis of a splitting of the out-of-plane CH bonding mode, two phases were inferred for C_6H_6 on Pt and Ni. Other types of EELS studies of hydrocarbons include studies of the interaction with co-adsorbed oxygen relevant to oxidation reactions—in particular, ethylene + O_2 on Pt(110) (113) and on Ag(110) (114).

EELS studies have also considered simple oxygen-containing organics again because of their role as intermediates or products in a wide range of synthetic or oxidation reactions. Particular attention has been paid to the adsorbed surface products of methanol, formic acid, acetic acid, formaldehyde, and acetone on a variety of transition metals Ni, Pd, Pt, Ru, Mo, and noble metals Cu and Ag. For example, methanol at low temperatures adsorbs molecularly on all surfaces. On Ni(111) (115) and Mo(100) (116) conversion to a stable methoxide species occurs between 180–200K. Further heating decomposes the methoxide to $CO + H_2$. On Pd (117), Pt (118), and Cu (119) methanol largely adsorbs reversibly; in the presence of coadsorbed oxygen, stable methoxide intermediates are also formed. On Mo(100) (116), however, saturation coverages of oxygen completely prohibit methoxide formation. On the stepped Ni[5(111) × ($\bar{1}$10)], methanol decomposes to methoxide species even at 150K (6). On all surfaces studies the methoxide species is believed to sit upright or very nearly so on the surface.

Formic acid decomposition occurs at temperatures > 400K on Cu(100) (120), and > 200K on Mo(100) (121) and Ru(100) (122) to form a stable formate species. The formate species decomposes to produce CO_2 and H_2 at higher temperatures. On Ag(110) (123) co-adsorption of oxygen is required to produce the formate species, while on Mo(100) (121) co-adsorption prohibits its formation. In several cases a symmetric bidentate orientation is observed, whereas on a carbon-covered Ni(110) (124) surface a monodentate configuration is found. Acetic acid has also been studied on Cu(100), where it forms an analogous acetate species bound in a bidentate configuration (125). Acetone was found (126) to adsorb on Pt(111) in an end-on fashion through the oxygen atom, but in a side-on configuration on Ru(001). When the Lewis acidity of Ru(001) is increased by oxygen co-adsorption, both configurations coexist. The side-on configuration is a precursor to dissociation on the surface.

STUDIES OF ELECTRONIC EXCITATIONS

Clean Surfaces

Most of the published studies of the electronic excitations of clean surfaces have been performed on semiconductors [for a review see Froitzheim (127)]. The interest in most cases has been the identification of interband transitions between surface states characteristic of the particular surface superstructure. A typical case is provided by the cleaved Si(111) 2×1 surface. Froitzheim, Ibach & Mills (128) observed a low energy, ~ 0.5 eV, electronic excitation of the 2×1 surface which on the basis of its sensitivity to surface contamination was attributed to excitation of electrons from one sub-band of surface states to another. The band gap between these surface states was estimated to be ~ 0.25 eV and to be intimately connected with the 2×1 reconstruction. More information regarding the detailed nature of this reconstruction has been provided by the wavevector-resolved low energy EELS studies of Matz, Lüth & Ritz (129). The surface state transition observed at ~ 0.5 eV with 50 eV electrons (128) was found to vary between 0.5 and 0.8 eV at lower incident electron energies. The studies as a function of E_0, and also off-specular studies, allowed the measurement of the dispersion of the surface transition along the Γ-J symmetry line of the 2×1 surface Brillouin zone and found it to be very substantial. This observation has significance regarding the nature of the reconstruction. The buckling model of Haneman (130) provides weak interaction between the dangling bonds and thus predicts a small dispersion along Γ-J. Contrary to this, in a model proposed by Pandey (131) the dangling bonds located at nearest neighbor atoms form π-bonded zig-zag chains along [110]. The occupied and empty surface states are simply π and π^* states of the chain, respectively. Because of the high degree of delocalization within the π-chains, Pandey's model requires high dispersion along Γ-J and is supported by the EELS measurements of Matz et al (129).

Electronic Excitations of Adsorbates

Although traditionally surface science has dealt with the interaction of ground state atoms and molecules with solid surfaces, there are a great number of surface processes that involve electronically excited states. [For a review see Ref. (132).] As a result, increasing attention has been focused on understanding the nature and decay mechanisms of excited adsorbates. In addition, when species in their ground states chemisorb on surfaces, the mixing of the adsorbate and substrate states creates new levels, some of them partially occupied or totally unoccupied. To completely understand the nature of the chemisorption bond we need to probe not only the filled levels but also the empty ones, and this can be done by studying the electronic excitations of the adsorbate-substrate complex.

Adsorbates which in their ground states interact weakly with the substrate (i.e. are physisorbed) may also do so in their excited states. This is the case for N_2 physisorbed on Al(111). Avouris, Schmeisser & Demuth (133) concluded that the interaction of the $^3\Pi_u$ excited state of N_2 with the metal is determined primarily by electromagnetic field coupling with the Al electron-hole pair excitations. The situation is different in the case of noble gases physisorbed on metals. Their excitations are Rydberg states, the lowest one having the alkali-like $np^5(n+1)s^1$ configuration. Alkalis adsorb ionically on metals and therefore the excited states of adsorbed noble gases may also be ionic. For this to happen to $(n+1)s$ level must lie above the Fermi level of the metal. Demuth, Avouris & Schmeisser (134) found that the free noble gas atom excitations are observed in the adsorbed phase (even when $E_{(n+1)s} > E_F$), but are significantly broadened and shifted to higher energies. For example, the $5p^6 \rightarrow 5p^5_{3/2}6s$ excitation of Xe on Cu is broadened to a FWHM of ~ 0.6 eV and blue shifted by ~ 0.25 eV (135). These observations can be accounted for by the theoretical model of Persson & Avouris (136). The Franck-Condon excited states of adsorbed noble gases are weakly ionic but decay fast ($\sim 10^{14}$ s^{-1}) via resonant ionization to the empty conduction band states of the substrate.

A large number of electron energy loss studies of adsorbate electronic excitations have considered the excitations of chemisorbed CO. In free CO, the lowest energy electronic excitations are centered at ~ 6 eV and at ~ 8.5 eV and are due to triplet and singlet coupled $5\sigma \rightarrow 2\pi^*$ transitions, respectively. As we have discussed above, the Blyholder model of CO chemisorption involves both 5σ and $2\pi^*$ levels and therefore the monitoring of the $5\tilde{\sigma} \rightarrow 2\tilde{\pi}^*$ (the $\sim$ implies chemisorption) excitation can provide important information regarding the detailed nature of the chemisorption bond on different substrates. A list of studies of the excitations of chemisorbed CO includes CO on the following surfaces: Ni(100) (137–139), Ni(110) (140), Ni(111) (141), Pt(100) (142a,b), Pd(100) (143, 144), W(100) (145), Ru(001) (146), [Cu(100), Cu(110), Cu(111), Cu] (147), Cu(311) (148), Ag(111) (149), and Au(110) (150). Although there are some variations from system to system and some disagreements between different studies of the same system, in most cases excitations are seen at ~ 6 eV, ~ 9 eV, and at ~ 14 eV.

There are two main assignment schemes: In one, the ~ 6 and ~ 9 eV excitations are assigned as $d_{\sigma,\pi} \rightarrow 2\tilde{\pi}^*$ charge-transfer excitations respectively, while the ~ 14 eV [or an ~ 12 eV (137)] excitation is assigned as the $^1(5\tilde{\sigma}, 1\tilde{\pi} \rightarrow 2\tilde{\pi}^*)$ intra-adsorbate excitation (142, 143, 147, 139). In the other scheme the ~9 eV excitation is assigned as $^1(5\tilde{\sigma} \to 2\tilde{\pi}^*)$ (138, 141), while the ~6 eV loss observed in low energy EELS is assigned as the corresponding $^3(5\tilde{\sigma} \to 2\tilde{\pi}^*)$ excitation (138). The $^3(5\tilde{\sigma} \to 2\tilde{\pi}^*)$ excitation is most likely overlapped by a charge transfer excitation.

The assignment of the ~9 eV excitation as a $^1(5\tilde{\sigma} \to 2\tilde{\pi}^*)$ transition is based on angle resolved photoemission results regarding the energy of $5\tilde{\sigma}^*$, and on inverse photoemission results regarding the energy $2\tilde{\pi}^*$ [for CO on Ni(111)(151) and Cu(152)]. The analysis should take proper account of the differences in screening between the ionization spectroscopies and the neutral excitation, and the effect of the substrate on the intraadsorbate Coulomb and exchange interactions as discussed by Avouris et al (138).

If the ~14 eV excitation is assigned as $^1(5\tilde{\sigma} \to 2\tilde{\pi}^*)$, however, the same analysis places the $2\tilde{\pi}^*$ at its free CO position (i.e. ~2 eV above vacuum). There is, however, evidence from a variety of sources that the $2\tilde{\pi}^*$ level is below the vacuum level (138). Further insight into the nature of the "9 eV" and "14 eV" excitations can be obtained by noting that the exact energy of the "14 eV" excitation correlates well with the $5\tilde{\sigma}$ ionization energy and also with the threshold for electron stimulated desorption (ESD) of CO^+. The "9 eV" transition, on the other hand, correlates with the peak for ESD of neutral CO. For example, in the case of α-CO/W(100) the energy loss spectrum shows peaks at 9.1 eV and 14 eV(145), the $5\tilde{\sigma}$ ionization energy is 14.1 eV (153), the CO ESD shows a resonance at 9 eV (154), and finally the threshold for CO^+ ESD is at 14.6 eV (155). On the basis of the Blyholder model (92) of the CO chemisorption bond it is easy to see why the $(5\tilde{\sigma} \to 2\tilde{\pi}^*)$ and $(5\tilde{\sigma}^{-1})$ states could lead to unbound CO and CO^+, respectively.

The discovery of and subsequent interest in the surface-enhanced Raman (SER) effect (156, 157) also has focused attention on the electronic states of aromatic adsorbates. A detailed study of the excitations of benzene, pyridine, and pyrazine on Ag(111) was performed by Avouris & Demuth (158). Using low energy EELS they observed the $\pi \to \pi^*$ excitations of the aromatic rings (in both a parallel configuration and the tilted one with respect to the surface in the case of pyridine) and found them to be broadened and shifted by no more than ~0.1 eV from their free molecule positions. In addition to these intraadsorbate excitations, new low energy excitation bands were observed for pyridine and pyrazine on Ag(111), and were assigned as charge transfer (CT) transitions from the Ag (sp band) to unoccupied (π^*) levels of the adsorbates (158). These CT bands can be resonantly excited by the lasers (Ar^+, Kr^+) used in SER spectroscopy and thus have been implicated in short ranged surface Raman enhancement mechanisms (157).

The behavior of chemisorbed aromatics on transition metals can be significantly different. Netzer and co-workers did not observe the characteristic $\pi \to \pi^*$ excitations of the aromatic ring for benzene on Pt(111) (159) or benzene on Ir(111) (160) but did observe them for pyridine on Ir(111) (160). It was proposed that benzene assumes a flat adsorption geometry and thus the in-plane polarized $\pi \to \pi^*$ excitations are heavily screened by the substrate, while pyridine assumes a tilted configuration and the excitations are not screened. When, however, benzene, pyridine, and their derivatives were adsorbed on Pd(111), Netzer & Mack (161) failed to observe any intraadsorbate excitations, although the screening in this case is rather weak. From the Pd (161) and Ag (158) results it appears that screening may not be the dominant factor in determining the ability to observe the electronic excitations. On the basis of parallel vibrational and electronic EELS studies of pyridine on Ni(100), DiNardo, Avouris & Demuth (162, 138) concluded that pyridine, depending on the surface temperature and coverage, binds in a flat or tilted configuration to the surface. In the flat (π-bonded) configuration no π-π^* excitations could be observed. In the tilted (n-bonded) configuration, however, the $\pi \to \pi^*$ excitations were clearly observed along with a $d \to \pi^*(3b_1)$ charge transfer band. This behavior is shown in Figure 4. It was suggested that in the flat phase, π or π^* bonding interactions with the substrate both broaden and reduce the integrated intensity of the $\pi \to \pi^*$ excitations so that in combination with screening effects they render them unobservable. It thus appears that in some cases molecular chemisorption can eliminate the free molecule excitations [see also Ref. (163)].

Finally, we note that dispersion of adsorbate electronic excitations has been observed for Br on Pd(111) (164) and ordered Xe layers on Ag(111) and Al(111) (165).

CONCLUSIONS—OUTLOOK

In this review, we have tried to present a general picture of the varied applications of electron energy loss spectroscopy in the study of clean and adsorbate-covered surfaces of solids. As we have seen, EELS has provided us with valuable physical information about surface phonons, surface reconstruction, binding site, etc, as well as chemical information regarding surface reactivity and identification of reaction intermediates.

We may expect that EELS will continue to be one of the most valuable surface analysis techniques. Several improvements and new applications may be expected in the near future. We may certainly expect the instrumental resolution to be further improved. Application of multichannel detection schemes to EELS would allow not only the detection of weaker signals and therefore allow better resolution but would also permit the performance of time-resolved EELS experiments. Such efforts are

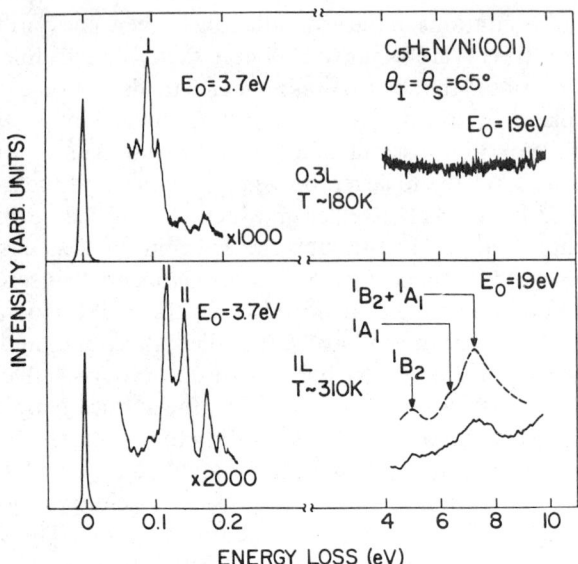

Figure 4 The vibrational and electronic loss spectra of pyridine on Ni(100). The *top spectra* are obtained under conditions where the "flat" (π-bonded) adsorption geometry is stable, whereas the *bottom spectrum* results from the "inclined" (*n*-bonded) pyridine phase. The vibrational modes are designated as parallel or perpendicular depending on their polarization with respect to the molecular plane. The vibrational spectra through the surface dipole selection rule establish the molecular orientation. The *dashed spectrum* gives the $\pi \to \pi^*$ excitations of the "flat" pyridine phase on Ag(111) from Ref. (158). Adapted from Ref. (138).

currently under way in Jülich (H. Ibach et al) and at Cornell (W. Ho et al). While most of the published EELS studies utilize dipole scattering under specular conditions, we expect increasing attention to be paid to short range large-angle scattering. By using higher primary beam energies, the cross-section for impact scattering will be increased and will allow more detailed studies via this scattering mechanism. We may also expect more careful and systematic investigation of resonance scattering by chemisorption systems. Although studies on single crystals have provided us with important fundamental information, we also expect more attention to be paid to more "real life" surfaces, e.g. polycrystalline films, wafers, supported catalysts, etc. Further improvements in the theoretical description of EELS are also needed. In particular, a unified description of both the short and long range scattering is still lacking.

Literature Cited

1. Schulz, G. J. 1973. *Rev. Mod. Phys.* 45: 378, 423
2. Trajmar, S. 1980. *Science* 208: 247
3. Schnatterly, S. E. 1979. *Solid State Phys.* 34: 275
4. Propst, F. M., Piper, T. C. 1967. *J. Vac. Sci. Technol.* 4: 53
5. Ibach, H. 1970. *Phys. Rev. Lett.* 24: 1416
6. Ibach, H., Mills, D. L. 1982. *Electron*

Energy Loss Spectroscopy and Surface Vibrations. New York: Academic
7. Lucas, A. A., Sunjic, M. 1971. *Phys. Rev. Lett.* 26:229
8. Powell, C. J. 1968. *Phys. Rev.* 175:952
9. Evans, E., Mills, D. L. 1972. *Phys. Rev. B* 5:4126
10. Evans, E., Mills, D. L. 1973. *Phys. Rev. B* 7:853
11. Mills, D. L. 1975. *Surf. Sci.* 48:59
12. Newns, D. M. 1977. *Phys. Lett.* A60:461
13. Persson, B. N. J. 1977. *Solid State Commun.* 24:573
14. Delanaye, F., Lucas, A., Mahan, G. D. 1978. *Surf. Sci.* 70:629
15. Lenac, Z., Sunjic, M., Sokcevic, D., Brako, R. 1979. *Surf. Sci.* 80:602
16. Persson, B. N. J. 1980. *Surf. Sci.* 92:265
17. Mahan, G. D., Lucas, A. A. 1978. *J. Chem. Phys.* 68:1344
18. Andersson, S., Persson, B. N. J. 1980. *Phys. Rev. Lett.* 45:1421
19. Thomas, G. E., Weinberg, W. H. 1979. *J. Chem. Phys.* 70:1000
20. Li, C. H., Tong, S. Y., Mills, D. L. 1980. *Phys. Rev. B* 21:3057
21. Tong, S. Y., Li, C. H., Mills, D. L. 1980. *Phys. Rev. Lett.* 44:407
22. Blanchet, G. B., DiNardo, N. J., Plummer, E. W. 1982. *Surf. Sci.* 118:496
23. Davies, B. M., Erskine, J. L. 1983. *J. Electron Spectrosc.* 29:323
24. DiNardo, N. J., Demuth, J. E., Avouris, Ph. 1983. *Phys. Rev. B* 27:5832
25. Andersson, S., Davenport, J. W. 1978. *Solid State Commun.* 28:677
26. Demuth, J. E., Schmeisser, D., Avouris, Ph. 1981. *Phys. Rev. Lett.* 47:1166
27. Avouris, Ph., Schmeisser, D., Demuth, J. E. 1982. *Phys. Rev. Lett.* 48:199
28. Schmeisser, D., Demuth, J. E., Avouris, Ph. 1982. *Phys. Rev. B* 26:4857
29. Demuth, J. E., Avouris, Ph., Schmeisser, D. 1983. *J. Electron Spectrosc.* 29:163
30. Gadzuk, J. W. 1983. *J. Chem. Phys.* 79:3982
31. Ibach, H. 1982. *J. Mol. Struct.* 79:129
32. Davenport, J. W., Ho, W., Schrieffer, J. R. 1978. *Phys. Rev. B* 17:273
33. Ho, W., Willis, R. F., Plummer, E. W. 1980. *Phys. Rev. B* 21:4202
34. Fuchs, R., Kliewer, K. L. 1965. *Phys. Rev.* 140:2076; 1966. 150:573
35. Matz, R., Lüth, H. 1981. *Phys. Rev. Lett.* 46:500
36. Kesmodel, L. L., Gates, J. A., Chung, Y. W. 1981. *Phys. Rev. B* 23:489
37. Baden, A. D., Cox, P. A., Egdell, R. G., Orchard, A. F., Willmer, R. J. D. 1981. *J. Phys. C* 14:L1081
38. Ibach, H. 1971. *Phys. Rev. Lett.* 27:253
39. Ibach, H., Bruchmann, D. 1978. *Phys. Rev. Lett.* 41:958
40. Ibach, H., Bruchmann, D. 1980. *Phys. Rev. Lett.* 44:36
41. Andersson, S. 1982. *Vibrations at Surfaces*, ed. R. Caudano, J. M. Gilles, A. A. Lucas. New York: Plenum
42. Andersson, S., Persson, M. 1981. *Phys. Rev. Lett.* 45:1421
43. Lehwald, S., Szeftel, J. M., Ibach, H., Rahman, T. S., Mills, D. L. 1983. *Phys. Rev. Lett.* 50:518
44. Andersson, S., Persson, B. N. J. 1983. *Phys. Rev. Lett.* 50:2028
45. Persson, B. N. J. 1983. *Phys. Rev. Lett.* 50:1089
46. Thiry, P. A. 1983. *J. Electron Spectrosc. Relat. Phenom.* 30:261
47. Willis, R. F. 1979. *Surf. Sci.* 89:457
48. Baro, A. M., Ibach, H., Bruchmann, H. D. 1979. *Surf. Sci.* 88:384
49. Ibach, H., Hopster, H., Sexton, B. 1977. *Appl. Surf. Sci.* 1:1
50. Lehwald, S., Ibach, H., Steininger, H. 1982. *Surf. Sci.* 117:342
51. Lüth, H., Matz, R. 1981. *Phys. Rev. Lett.* 46:1652
52. Barnes, M. R., Willis, R. F. 1978. *Phys. Rev. Lett.* 41:1729
53. Andersson, S. 1978. *Chem. Phys. Lett.* 55:185
54. DiNardo, N. J., Demuth, J. E., Avouris, Ph. 1984. *J. Chem. Phys.* In press
55. Nyberg, C., Tengstal, C. G. 1983. *Surf. Sci.* 126:163
56. Ho, W., DiNardo, N. J., Plummer, E. W. 1980. *J. Vac. Sci. Technol.* 17:134
57. Froitzheim, H., Ibach, H., Lewald, S. 1975. *Phys. Lett.* A55:247
58. Wagner, H., Butz, R., Backes, U., Bruchmann, D. 1981. *Solid State Commun.* 38:1155
59. Backes, U., Ibach, H. 1981. *Solid State Commun.* 40:575
60. Dubois, L. H., Schwartz, G. P. 1982. *Phys. Rev. B* 26:794
61. Many, A., Wagner, I., Rosenthal, A., Gersten, J. I., Goldstein, Y. 1981. *Phys. Rev. Lett.* 46:1648
62. Thomas, G. E., Weinberg, W. H. 1978. *J. Chem. Phys.* 69:3611
63. Rahman, T. S., Anton, A. B., Avery, N. R., Weinberg, W. H. 1983. *Phys. Rev. Lett.* 51:1979
64. Froitzheim, H., Ibach, H., Lehwald, S. 1976. *Phys. Rev. B* 14:1362
65. Erley, W., Ibach, H. 1981. *Solid State Commun.* 37:937
66. Andersson, S. 1979. *Surf. Sci.* 79:385
67. Gland, J. L., Sexton, B. A., Fisher, G. B. 1980. *Surf. Sci.* 95:587
68. McClellan, M. R., McFeeley, F. R., Gland, J. L. 1983. *Surf. Sci.* 124:188
69. Sexton, B. A. 1979. *Surf. Sci.* 88:299
70. Sexton, B. A., Madix, R. J. 1980. *Surf. Sci.* 95:587

71. Erskine, J. L., Strong, R. L. 1982. *Phys. Rev. B* 26:3483
72. Wendelken, J. F., Propst, F. M. 1976. *Rev. Sci. Instrum.* 47:1069
73. Miles, S. L., Bernasek, S. L., Gland, J. L. 1983. *J. Electron Spectrosc. Related Phenom.* 29:239
74. Avouris, Ph., Schmeisser, D., Demuth, J. E. 1982. *Phys. Rev. Lett.* 48:199
75. Andersson, S., Harris, J. 1982. *Phys. Rev. Lett.* 48:545
76. Backx, C., deGroot, C. P. M., Biloen, P. 1981. *Surf. Sci.* 104:300
77. Sexton, B. A., Madix, R. J. 1980. *Chem. Phys. Lett.* 76:294
78. Gland, J. L., Sexton, B. A., Fisher, G. B. 1980. *Surf. Sci.* 95:587
79. Steininger, H., Lehwald, S., Ibach, H. 1982. *Surf. Sci.* 123:1
80. Andersson, S. 1979. *Solid State Commun.* 21:75
81. Erley, W., Wagner, H., Ibach, H. 1979. *Surf. Sci.* 80:612
82. Thomas, G. E., Weinberg, W. H. 1979. *J. Chem. Phys.* 70:954
83. Dubois, L. H., Somorjai, G. A. 1980. *Surf. Sci.* 91:514
84. Baro, A. M., Ibach, H. 1979. *J. Chem. Phys.* 71:4812
85. Erley, W. 1981. *J. Vac. Sci. Technol.* 18:472
86. Froitzheim, H., Ibach, H., Lehwald, S. 1977. *Surf. Sci.* 63:56
87. Andersson, S. 1979. *Surf. Sci.* 89:477
88. Hoffmann, P., Bare, S. R., King, D. A. 1982. *Surf. Sci.* 117:245
89. Erley, W., Ibach, H., Lehwald, S., Wagner, H. 1979. *Surf. Sci.* 83:585
90. Hopster, H., Ibach, H. 1978. *Surf. Sci.* 77:109
91. McClellan, M. R., Gland, J. L., McFeeley, F. R. 1981. *Surf. Sci.* 112:63
92. Blyholder, G. 1964. *J. Phys. Chem.* 68:2772
93. Plummer, E. W., Eberhardt, W. 1982. *Adv. Chem. Phys.* 49:533
94. Heskett, D., Plummer, E. W., Messmer, R. P. 1984. *Surf. Sci.* In press
95. Pirug, G., Bonzel, H. P., Hopster, H., Ibach, H. 1979. *J. Chem. Phys.* 71:593
96. Avouris, Ph., DiNardo, N. J., Demuth, J. E. 1984. *J. Chem. Phys.* 80:491
97. Demuth, J. E., Ibach, H., Lehwald, S. 1978. *Phys. Rev. Lett.* 40:1044
98. Hoffmann, F. M., Felter, T. E., Thiel, P. A., Weinberg, W. H. 1983. *Surf. Sci.* 130:173
99. Hoffmann, F. M., O'Brien, E. V., Hrbek, J., DePaola, R. A. 1983. *J. Electron Spectrosc.* 29:301
100. Felter, T. E., Hoffmann, F. M., Thiel, P. A., Weinberg, W. H. 1983. *Surf. Sci.* 130:163
101. Demuth, J. E., Ibach, H. 1978. *Surf. Sci.* 78:L238
102. Ibach, H., Lehwald, S. 1978. *J. Vac. Sci. Technol.* 15:407
103. Gates, J. A., Kesmodel, L. L. 1983. *Surf. Sci.* 124:68
104. Dubois, L. H., Castner, D. G., Somorjai, G. A. 1980. *J. Chem. Phys.* 72:5234
105. Kesmodel, L. L. 1983. *J. Chem. Phys.* 79:4646
106. Kesmodel, L. L., Waddill, G. D., Gates, J. A. 1984. To be published
107. George, P. M., Avery, N. R., Weinberg, W. H., Tebbe, F. N. 1983. *J. Am. Chem. Soc.* 105:1393
108. Lehwald, S., Ibach, H. 1979. *Surf. Sci.* 89:425
109. Lehwald, S., Ibach, H., Demuth, J. E. 1978. *Surf. Sci.* 78:577
110. DiNardo, N. J., Avouris, Ph., Demuth, J. E. 1984. *J. Chem. Phys.* In press
111. Koel, B. E., Somorjai, G. A. 1983. *J. Electron Spectrosc.* 29:287
112. Avouris, Ph., Demuth, J. E. 1981. *J. Chem. Phys.* 75:4783; 79:488
113. Steininger, H., Ibach, H., Lehwald, S. 1982. *Surf. Sci.* 117:685
114. Backx, C., deGroot, C. P. M., Biloen, P. 1980. *Appl. Surf. Sci.* 6:256
115. Demuth, J. E., Ibach, H. 1979. *Chem. Phys. Lett.* 60:395
116. Miles, S. L., Bernasek, S. L., Gland, J. L. 1983. *Surf. Sci.* 127:271
117. Gates, J. A., Kesmodel, L. L. 1983. *J. Catal.* 83:437
118. Sexton, B. A. 1981. *Surf. Sci.* 102:271
119. Sexton, B. A. 1979. *Surf. Sci.* 88:299
120. Sexton, B. A. 1979. *Surf. Sci.* 88:319
121. Miles, S. L., Bernasek, S. L., Gland, J. L. 1983. *Surf. Sci.* 127:271
122. Avery, N. R., Toby, B. H. Anton, A. B., Weinberg, W. H. 1982. *Surf. Sci.* 122:L574
123. Sexton, B. A., Madix, R. J. 1981. *Surf. Sci.* 105:177
124. Madix, R. J., Gland, J. L., Mitchell, G. E., Sexton, B. A. 1983. *Surf. Sci.* 125:481
125. Sexton, B. A. 1979. *Chem. Phys. Lett.* 65:469
126. Avery, N. R., Weinberg, W. H., Anton, A. B., Toby, B. H. 1983. *Phys. Rev. Lett.* 51:682
127. Froitzheim, H. 1977. *Topics in Current Physics*, ed. H. Ibach, 4:205. New York: Springer-Verlag
128. Froitzheim, H., Ibach, H., Mills, D. L. 1975. *Phys. Rev. B* 11:4980
129. Matz, R., Lüth, H., Ritz, A. 1983. *Solid State Commun.* 46:343
130. Haneman, D. 1961. *Phys. Rev. A* 21:1093

131. Pandey, K. C. 1981. *Phys. Rev. Lett.* 47:1913
132. Avouris, Ph., Persson, B. N. J. 1984. *J. Phys. Chem.* 88:837
133. Avouris, Ph., Schmeisser, D., Demuth, J. E. 1983. *J. Chem. Phys.* 79:488
134. Demuth, J. E., Avouris, Ph., Schmeisser, D. 1983. *Phys. Rev. Lett.* 50:600
135. Avouris, Ph., Demuth, J. E., DiNardo, N. J. 1984. *J. Phys.* 44:C10–451
136. Persson, B. N. J., Avouris, Ph. 1983. *J. Chem. Phys.* 79:5156
137. Akimoto, K., Sakisaka, Y., Nishijima, M., Onchi, M. 1979. *Surf. Sci.* 88:109
138. Avouris, Ph., DiNardo, N. J., Demuth, J. E. 1984. *J. Chem. Phys.* 80:491
139. Koel, B. E., Peebles, D. E., White, J. M. 1983. *Surf. Sci.* 125:739
140. Küppers, J. 1973. *Surf. Sci.* 36:53
141. Rubloff, G. W., Freeouf, J. L. 1978. *Phys. Rev. B* 17:4680
142a. Netzer, F. P., Willie, R. A., Matthew, J. A. D. 1977. *Solid State Commun.* 21:97
142b. Netzer, F. P., Matthew, J. A. D. 1979. *Surf. Sci.* 81:L651
143. Bader, S. D., Blakely, J. M., Brodsky, M. B., Friddle, R. J., Panosh, R. L. 1978. *Surf. Sci.* 74:405
144. Netzer, F. P., El Gomati, M. M. 1983. *Surf. Sci.* 124:26
145. Chesters, M. A., Hopkins, B. J., Winton, R. I. 1976. *Surf. Sci.* 59:46
146. Hesse, R., Staib, P., Menzel, D. 1979. *Appl. Phys.* 18:227
147. Spitzer, A., Lüth, H., 1981. *Surf. Sci.* 102:29
148. Papp, H. 1977. *Surf. Sci.* 63:182
149. McElhiney, G., Papp, H., Pritchard, J. 1976. *Surf. Sci.* 54:617
150. McElhiney, G., Pritchard, J. 1976. *Surf. Sci.* 60:397
151. Fauster, Th., Himpsel, F. J. 1983. *Phys. Rev.* 27:1390
152. Rogozik, J., Scheidt, H., Dose, V., Prince, K., Bradshaw, A. 1984. *Surf. Sci.* Submitted
153. Baker, J. M., Eastman, D. E. 1973. *J. Vac. Sci. Technol.* 10:223
154. Rubio, J., López-Sancho, J. M., López-Sancho, M. P. 1982. *J. Vac. Sci. Technol.* 20:217
155. Menzel, D. 1968. *Ber. Bunsenges. Phys. Chem.* 72:591
156. Otto, A. 1983. *Light Scattering in Solids*, ed. M. Cardona, G. Guntherodt. New York: Springer-Verlag
157. Avouris, Ph., Demuth, J. E. 1983. *Surface Studies with Lasers*, ed. F. R. Aussenegg, A. Leitner, M. E. Lippitsch. New York: Springer-Verlag
158. Avouris, Ph., Demuth, J. E. 1981. *J. Chem. Phys.* 75:4783
159. Netzer, F. P., Matthew, J. A. D. 1979. *Solid State Commun.* 29:209
160. Netzer, F. P., Bertel, E., Matthew, J. A. D. 1980. *Surf. Sci.* 92:43
161. Netzer, F. P., Mack, J. V. 1983. *J. Chem. Phys.* 79:1017
162. DiNardo, N. J., Avouris, Ph., Demuth, J. E. 1984. *J. Chem. Phys.* In press
163. Robota, H. J., Whitmore, P. M., Harris, C. B. 1982. *J. Chem. Phys.* 76:1692
164. Netzer, F. P., El Gomati, M. M. 1983. *Surf. Sci.* 124:26
165. Schmeisser, D., Weinert, C. M., Avouris, Ph., Demuth, J. E. 1984. *Chem. Phys. Lett.* 104:263

DIELECTRIC PROPERTIES OF POLYELECTROLYTE SOLUTIONS

M. Mandel and T. Odijk

Division of Physical and Macromolecular Chemistry, Gorlaeus Laboratories, University of Leiden, 2300 RA Leiden, The Netherlands

INTRODUCTION

Dielectric properties of polyelectrolyte solutions have been studied experimentally for more than thirty years, but progress in understanding these properties has been rather slow. A particularly unfavorable combination of experimental difficulties and unsettled theoretical problems has noticeably influenced both the extent of the work devoted to this subject and the pace in which explanation and prediction proceeds in this field. The latter is not surprising in view of the difficulties encountered in dielectric theory in general (see e.g. Refs. 1, 2) and taking into account the additional problems arising through the presence of free charges in the solution. Moreover, in the complementary field of the dielectric properties of ordinary electrolyte solutions, the situation is hardly better, and even there many problems remain unsolved in spite of recent progress (3).

In recent years new theoretical work directly or indirectly related to the dielectric properties of aqueous polyelectrolyte solutions has appeared (see the section on Theory), focusing attention again on the rather unsatisfactory state of affairs in this field. Hence, the appearance of this review seems justified and we hope it will stimulate further investigations, theoretical as well as experimental, in which new possibilities available through developments in other areas might be exploited. A further justification is that in recent times, only a few review articles of limited extent in this field have appeared (4–7).

We shall limit this survey to aqueous solutions of true polyelectrolytes, excluding proteins, but with an occasional reference to related systems, without attempting to be exhaustive. We discuss only linear dielectric effects. First, some general remarks regarding polyelectrolyte solutions and their dielectric properties will be made. In the second part experimental results as described in the literature are summarized, whereas the section after that deals with the theoretical aspects. Finally, the review ends with some general conclusions and suggestions for further investigations.

EXPERIMENTS

General Remarks

Before reviewing the state of the art of the field of dielectric properties of polyelectrolyte solutions, it is useful to point out certain facts that may be important in understanding some of the problems that are discussed later on.

We shall here define polyelectrolytes as macromolecules which in solution carry a relatively large number of charged groups. Sometimes it is possible to control the amount of charge, e.g. in polyelectrolytes with weakly acidic or basic groups, but in many cases all chargeable groups are fully ionized in water at ordinary temperatures. It is not uncommon to have more than 10^3 elementary charges, either positive or negative, distributed more or less uniformly along the macromolecular chain, together with an equivalent number of small counterions in the solution irrespective of the presence of small ions provided by added low molar mass electrolytes. The probably strong cooperative interactions between the charges on the macromolecule and the counterions has led to speculations concerning the association between the polyelectrolyte molecule and a certain fraction of the latter at equilibrium. Some experimental facts support the hypothesis of association but, contrary to what some people believe or even tend to assert, association or other related phenomena such as condensation are far from unambiguously established or theoretically well justified (8).

Charged macromolecular chains cannot be represented as point charges but rather are characterized by a spatial charge distribution that depends on the flexibility of the chain and on all interactions within the solution: intra- and intermacromolecular interactions, interactions of the chains with the small ions, and interactions with the solvent. The problem of the average macromolecular conformation has not yet been solved in a completely satisfactory way but recently new insight has been gained by considering the polyelectrolyte as a worm-like chain (9–11). This applies particularly to systems in which the macromolecular charges are considerably screened by an excess of small ions arising from added low molar mass electrolyte. Under these circumstances, the electrostatic interactions may be considered as perturbations only so that the average conformation problem may, with appropriate modifications, be related to that of a neutral macromolecule in a good solvent, the latter problem having been extensively studied (12, 13). Conformational aspects are much more difficult to understand in polyelectrolyte solutions when either the simple electrolyte has a concentration much smaller than the equivalent concentration of the polyelectrolyte or is completely absent. In that case the screening of the macromolecular charges is still unsatisfactorily understood. Therefore, the theoretical treatment of polyelectrolyte solutions under these conditions and the subsequent explanation of their physical properties lag behind those for polyelectrolyte solutions with an excess of salt. This is particularly so for the macromolecular concentration effects. Whereas in the excess salt case the concentration regions can be distinguished in analogy with solutions of neutral polymers (14), this cannot be easily done in the salt-free case (14–16), especially because the polyelectrolyte concentration may affect the influence of electrostatic interactions on the flexibility of the chain. Experimental evidence for the occurrence of various concentration regimes has been obtained by quasi-elastic light scattering measurements (17–20). These results also confirmed the difference in concentration behavior of polyelectrolyte solutions with or without added salt. For the latter the question has even been raised as to whether there is an experimentally accessible concentration where intermacromolecular interactions are negligible (14, 15, 21). It should be clear that the interpretation of physical quantities of polyelectrolyte solutions without low molar mass electrolyte needs proper consideration if no concentration effects have been studied.

For dielectric measurements on polyelectrolyte solutions, an additional and considerable difficulty of purely experimental nature has to be taken into account, as is also the case for ordinary electrolyte solutions. This difficulty arises from two sources, which both find their origin in the electric conductivity of the solutions considered. The first one is due to the small value of the intrinsic phase angle θ of a conducting solution, which decreases with increasing conductivity and decreasing frequency according to the following formula.

$$tg\ \theta = 5.65 \times 10^{-13}\ (\varepsilon' f/\sigma) \qquad \qquad 1.$$

Here ε' is the relative permittivity (dielectric constant) and σ the specific conductivity (in $\Omega^{-1}\ cm^{-1}$) of the solution, f the frequency (in Hz). For $f = 10^3$ Hz and $\varepsilon' = 10^3$ (a high but not rare value for a polyelectrolyte solution) and assuming $\sigma = 10^{-5}\ \Omega^{-1}\ cm^{-1}$, the phase angle θ is only 6×10^{-2} rad (less than 2 degrees). Such a low value of the phase angle makes an accurate determination of the relative permittivity by conventional capacitance measurements quite difficult if not impossible. For frequencies below 1 MHz, commercial general purpose instruments cannot be used for a reliable study of dielectric properties of conducting solutions without a careful and critical examination of possible complications arising from the conductivity of these systems. At frequencies above 1 MHz the phase angle is generally no longer of major importance, but it is replaced by more conventional problems, which are by no means diminished in view of the persistence of an important conductance contribution (22–24). Special measuring devices have been proposed to determine the capacitance of a two or three electrode cell with small phase angles in the more difficult frequency region (25–28).

The second difficulty encountered in these dielectric measurements arises in capacitance determinations performed with blocking electrodes between which electric current circulates. The blocking electrodes give rise to a frequency-dependent polarization impedance also called *electrode effect* (22, 29, 30). This impedance affects the measured capacitances up to relatively high frequencies (of the order 10–100 kHz) and generally depends on the electrode material and electrode surface, on the conductivity of the solution and on some other factors which are not so well established. Different correction procedures for electrode effects in convention impedance/admittance measurements using a two or three electrode sample cell have been proposed. They are based either on an empirically established simple frequency dependence of the electrode impedance contribution (31–34) or on the assumption of the electrode effect being independent of the electrode spacing. For the latter some direct experimental evidence has been found (29). The former procedures can only be applied in a frequency range where the dielectric properties of the solution itself are not frequency dependent, or necessitate fitting of experimental permittivity data at various frequencies to an empirical equation, involving a large number of adjustable parameters (35). In the other procedures special sample cells are needed with variable but accurately known electrode spacings. Here the electrode effect corrections are applied at every single frequency, and no interference with dielectric dispersion of the solution has to be feared (22, 36–38). Other complications may arise, however, from the nonideality of the plane parallel capacitor used as the sample cell (39, 40).

Both correction methods tend to break down below 10 kHz when the contribution of the electrode admittance, which increases rapidly with decreasing frequency, dominates the measured capacitance, although measurements at much lower frequencies have been published, in which a conventional two electrode cell at two different spacings was used in conjunction with a fast digital signal processing technique (41). Therefore special low frequencies techniques have been developed involving a four electrode cell in which the conductivity and the permittivity of a given solution are measured under such experimental conditions that (most of) the electrode effect cannot interfere. This method, apparently proposed but not applied by Schwan (22, 38), was first used for measurement of the complex conductivity from which the low frequency dependence of the relative permittivity was calculated (42, 43). Techniques for the direct determination of the relative permittivity together with the conductivity were developed later on (44–47). With these techniques, determinations of ε' have been claimed down to 1 Hz or even lower, depending on the conductivity of the solution; the upper bound lies in the kHz range. These four terminal methods seem to give quite reliable results in the frequency range covered, provided the conductivity is rather low. It remains to be proven that at the low frequency end, electrode polarization effects really do not contribute, particularly for those systems for which no plateau value of ε' is reached at the lowest frequencies attainable.

Another rather recent technique for determining the relative permittivity and dielectric loss of conducting systems at very low frequencies (of the order 0.1 to 10 Hz) is based on the combination of Fourier synthesized pseudorandom noise (used as a superposition of electric perturbations) and a four electrode cell (48, 49). This method also seems to avoid the influence of electrode effects, but the frequency range is limited and the results are noisy.

It should be realized that all the techniques mentioned have a rather limited sensitivity in as far as detecting an increase of the relative permittivity of the solution with respect to the pure solvent is concerned. This sets a lower limit to the polyelectrolyte concentration accessible for ε' determinations. On the other hand, except at the higher frequency end, there is an upper bound to the concentration imposed by the conductivity (in general $\sigma < 10^{-3} \Omega^{-1} \text{cm}^{-1}$). The latter limitation makes it also difficult to perform dielectric measurements on polyelectrolyte solutions with a large excess of low molar mass electrolyte. Most of these techniques have a rather limited frequency range that is generally much smaller than the total dispersion region of polyelectrolyte solutions (see below). Used alone they will as a rule contain only partial information concerning the dielectric dispersion considered, and some of the dielectric parameters will have to be obtained by extrapolation procedures, which themselves need careful consideration.

It is important to stress that the accuracy of the ε' measurements may be different for the various techniques discussed and it is sometimes difficult to establish how accurate the determinations are in view of the many difficulties involved. Nevertheless, as the dielectric increments of polyelectrolyte solutions are large ($\Delta\varepsilon' > 20$ at most frequencies and for concentrations not too low), most of the measurements may yield results that in a semiquantitative way are reasonably reliable. In some cases, however, different authors have obtained quite different results for the same system under comparable conditions. It is much more difficult to determine the imaginary part of the dielectric increment $\Delta\varepsilon''$, which corresponds to the dielectric loss of the polyelectrolyte dispersion. The measured conductivity of the solution $\sigma = \sigma(\omega)$ will contain two contributions, one from the static conductivity σ_0 and another from the dielectric loss, according to the general relation

$$\sigma(\omega) = \sigma_0 + \frac{\omega \Delta\varepsilon''(\omega)}{4\pi}. \qquad 2.$$

At low frequencies the first term may dominate the second. Note that the dielectric loss contribution cannot be distinguished or measured separately from the frequency dependent part of the conductivity if the conductivity itself also presents a dispersion in the frequency range considered. This stems from the well-known general correspondence between the complex relative permittivity ε^* and complex conductivity σ^*

$$\sigma^*(\omega) = \sigma'(\omega) + i\sigma''(\omega) = \frac{i\omega\varepsilon^*(\omega)}{4\pi} = \frac{i\omega}{4\pi}(\varepsilon' - i\varepsilon'') = \frac{\omega}{4\pi}\varepsilon''(\omega) + \frac{i\omega}{4\pi}\varepsilon'(\omega).$$
3.

(If for the frequency dependence of the field one prefers $\exp(-i\omega t)$ to $\exp(i\omega t)$ as used here, then $i\omega$ in the third and fourth member of Eq. 3 should be replaced by $-i\omega$ and ε^* should be written $\varepsilon' + i\varepsilon''$.)

Review of Results

We do not present here a historical survey of papers giving experimental data on the dielectric properties of aqueous polyelectrolyte solutions. Rather, we try to review these properties in a general way with references to the relevant publications. Only a few papers are mentioned in the beginning to establish how progress in this field has taken place. We make in principle no distinction between solutions containing polyelectrolytes of synthetic origin or natural products (DNA, polyelectrolytic polysaccharides, etc). In as far as it is possible to make a comparison of all results, no specific qualitative differences appear between these two classes of polyelectrolytes.

The first investigations of polyelectrolytes appeared around 1950 with high frequency measurements on DNA solutions by Allgen and co-workers (51, 52), followed by measurements on a synthetic polyelectrolyte at somewhat lower frequencies in 1954 by Dintzis et al (53). The latter conjectured the existence of two separate dispersion regions. In the same year Allgen and co-workers (54) established that neutral polysaccharide analogues of charged carboxy-methylcellulose did not show any comparable dielectric increment. The first systematic investigation of a synthetic polyelectrolyte in which concentration, molar mass, linear charge density, and the influence of salt were studied in the low frequency region was published in 1963 (55, 56), followed in 1969 by an analogous study at higher frequencies (57). In the former it was also shown, at least qualitatively, that the dielectric behavior of positively and negatively charged polyelectrolytes is identical. The first paper in which both the high and the low dispersion regions were studied simultaneously appeared in 1972 (58). The frequency range was limited to 300 Hz–6 MHz. The upper bound was extended in 1974 to 100 MHz (59), whereas the first direct determinations of ε' for polyelectrolyte solutions below 100 Hz (down to a few Hz or less) were published in 1976 (50).

From all the published evidence so far, a certain number of general features seem to emerge for the dielectric properties of polyelectrolytes in water or in solutions with low concentrations of mono-monovalent salts ($c_0 \leq 10^{-3}$ M). We summarize these below.

At room temperature, for degrees of polymerization not too low ($DP > 300$) and for sufficiently high charge densities (for weak polyacids $0.3 < \alpha_D < 1$, with α_D the degree of dissociation), the measured relative permittivity exceeds the value of the pure solvent for frequencies below 100 MHz. Near 100 MHz, in practically all cases investigated, the dielectric increment with respect to the solvent becomes negligible, indicating that at this frequency no significant contribution of the charged macromolecule to the polarization of the system persists, nor is the polarization of the solvent affected by the polyions (59–63). Below 100 MHz the dielectric increment increases with decreasing frequency, generally exhibiting two dispersion regions that are more or less well separated (53, 58–60, 62–65). The dispersion region in the higher frequency range has a critical frequency $f_{c,2}$ between 1 and 10 MHz that is independent of the molar mass of the polyelectrolyte at all concentrations investigated (57–60, 63). The same is true for the total amplitude of this dispersion region. The lower frequency dispersion region, on the other hand, is characterized by a critical frequency that decreases with increasing molar mass, whereas the total amplitude increases on the contrary (55, 59, 60, 63, 66–69).

Both dispersion regions do not obey the simple Debye relaxation equation but can be fitted to the empirical Cole-Cole curve, which for ε' is given by (70)

$$\Delta\varepsilon'(\omega) = \Delta\varepsilon'(\infty) + \tfrac{1}{2}(\Delta\varepsilon'(0) - \Delta\varepsilon'(\infty))S1 - \left[1 - \frac{\sinh(\beta_c x)}{\cosh(\beta_c x) + \cos(\beta_c \pi/2)}\right].$$

4.

Here $\Delta\varepsilon'(\omega)$, $\Delta\varepsilon'(\infty)$, $\Delta\varepsilon'(0)$ represent the frequency dependent dielectric increment at circular frequency $\omega \equiv 2\pi f$, the high frequency limit and the low frequency limit of the dielectric increment, respectively, $x \equiv \ln(\omega\tau)$ where $\tau = (2\pi f_c)^{-1}$ stands for the mean relaxation time of the dispersion and β_c ($0 < \beta_c < 1$) is a parameter characteristic of the deviations from the Debye relaxation curve for which $\beta_c = 1$. The total amplitude of the dispersion region is given by $\Delta\varepsilon'(0) - \Delta\varepsilon'(\infty)$. The value of $\varepsilon''(\omega)$, given by an analogous equation, can be expressed as function of ε' at each frequency, which yields the well known semicircular Cole-Cole plot (70). Evidence of dielectric dispersions of polyelectrolyte solutions fitting either representations can be found in the literature (45, 50, 55–57, 59, 60, 62, 63, 69).

The amplitudes of both dispersion regions are strongly concentration dependent and in many cases it was found that the specific values of the dielectric increments $\Delta\varepsilon_s'/C_p$ decreased with increasing concentration at constant M and charge density (45, 53–55, 58–60, 62, 63, 67, 71). Here C_p is the concentration of the polyelectrolyte expressed in g dm^{-3} and $\Delta\varepsilon_s'$ the (extrapolated) static dielectric increment, $\Delta\varepsilon_s' = \Delta\varepsilon_1'(0) + \Delta\varepsilon_2'(0)$; the indices 1 and 2 refer to the lower and the higher dispersion region, respectively. We return to this point below.

There is also evidence that the specific increments increase with increasing charge density at constant C_p and M in both dispersion regions (55, 58, 59, 62, 72). It should be noted, however, that Sachs et al (57) found for poly(acrylic acid) partially neutralized by NaOH that $\Delta\varepsilon_2'(0)/C_p$ did not change with α_D in the interval $0.25 < \alpha_D < 0.75$. This is in contrast to the results of van der Touw et al (59) with the same system. The origin of this discrepancy is not clear and may be due to the fitting procedure used in the former paper. In this respect it should be realized that the absolute amplitude of the higher frequency dispersion is generally small, often even smaller than 10 (contrary to what is observed for the lower dispersion where, depending on the molar mass and concentration, very high amplitudes—sometimes even larger than several hundreds—are observed), and that there is some overlap between the high frequency part of the lower frequency dispersion and the low frequency part of the higher frequency dispersion.

There is ample evidence that $f_{c,2}$ is an increasing function of the concentration (54, 57–60, 62, 63). Not much evidence is available for a possible charge density dependence of $f_{c,2}$. It was reported that for poly(acrylic acid), $f_{c,2}$ did not change in the range $0.25 < \alpha_D < 0.75$ (57) but, on the other hand, $f_{c,2}$ was found to be higher at $\alpha_D = 0.65$ than at 0.10 for the maleic acid-ethylene-copolymer (62). Finally, for poly(glutamic acid) $f_{c,2}$ only very slightly increased with α_D (60). In the first work cited here the constancy of $f_{c,2}$ within 2% for the four degrees of dissociation investigated is too remarkable not to raise again questions concerning the fitting procedure.

For the lower frequency dispersion the α_D-dependence of the critical frequency has not been investigated systematically either. Only for poly(glutamic acid) was such a study undertaken, but this is rather a particular case as this polyelectrolyte undergoes a conformational transition upon increase of the pH. Takashima (73) found approximately the same $f_{c,1}$ at low and at high pH but a minimum in the pH region where the macromolecule undergoes its helix-coil transition. Later on, analyzing results over a broader frequency range, Muller et al (60) found that $f_{c,1}$ yields a more or less sigmoidal curve with increasing pH. It should be noted that Takashima studied also the heat-induced denaturation of DNA (67) and found a different behavior in the transition region between double helix and coil as compared to later, more extensive, work by Tung et al (45).

The concentration dependence of $f_{c,1}$ has been the object of many studies but, unfortunately, just as is the case for $\Delta\varepsilon'_1(0)$, the results obtained for different systems by different investigators are contradictory. No definite concentration effect for $f_{c,1}$ was found for K-poly(methacrylate) at $\alpha_D = 1$ in the same concentration range where $\Delta\varepsilon'_1(0)$ decreased with concentration (55). This agrees with the findings for $(C_4H_9)_4N$-poly(acrylate) (65), with results obtained with maleic acid-ethylene-copolymer at low charge density ($\alpha_D = 0.1$) (62), and with low molar mass DNA (63). It disagrees, however, with results obtained with Na-poly(styrenesulfonate) (59) and poly(glutamate) in the coil form ($\alpha_D = 0.9$) (60) where $f_{c,1}$ was found to increase with concentration, and with results on higher molar mass DNA where $f_{c,1}$ decreases with increasing concentration (74).

A word of warning against simple comparison of concentration effects in solutions of different polyelectrolytes is perhaps appropriate here. As pointed out above, polyelectrolyte solutions without salt or with a very low salt concentration may exhibit a considerable number of concentration regimes or transition regions, the location of which on the concentration scale may depend on the nature of the macromolecule (e.g. the flexibility of its chain), its contour length, which is proportional to M (14), and the interaction with the counterions (21). A given concentration range expressed in g dm^{-3} may therefore cover different situations for two different polyelectrolytes, such as a low molar mass poly(acrylate) and a high molar mass DNA, and thus yield different experimental concentration dependencies. In as far as concentration effects depend strongly on molar mass, the use of heterodisperse macromolecular samples also can yield confusing results. In most of the investigations published so far it is unfortunately difficult to reach any conclusions concerning the molar mass distribution of the samples used. We have already indicated above that for

$\Delta\varepsilon'_1(0)/C_p$ generally a decrease with increasing C_p is found. For probably heterodisperse salmon testis DNA ($M \simeq 2 \times 10^6$ g mol^{-1}) Sakamoto et al (74) found that the specific increment remained constant up to 0.1 g dm^{-3} but increased thereafter, whereas Molinari et al (69), using a fractionated calf thymus DNA sample (of unknown degree of polydispersity, however), found a decrease of the specific static increment for a few concentrations up to 0.1 g dm^{-3} with an indication for a subsequent increase. With heterodisperse sonicated calf thymus DNA ($M_w \simeq 3 \times 10^5$ g mol^{-1}) Vreugdenhil et al (63) found that the specific static increment steadily decreased from 0.03 to 1 g dm^{-3}. It is difficult at present to draw any definite conclusions regarding DNA, the more so because in practically all cases, $\Delta\varepsilon'_1(0)$ had to be obtained by extrapolation, as even for the lowest frequencies at which the DNA solutions had been investigated, the increment was not yet frequency independent. In view of the fact that for practically all other polyelectrolytes investigated only specific increments decreasing with increasing concentrations have been found, we wonder whether the opposite effect observed in some cases with high molar mass DNA should not be attributed to some other origin. Note that Sakamoto et al also claimed not to have found evidence for a high frequency dispersion for heterodisperse calf thymus DNA (50); some of their Cole-Cole plots show definite deviations at the high frequency end, probably caused by the second dispersion region.

It is clear anyway that there is an urgent need for a systematically performed investigation of the concentration dependence of the dielectric properties of polyelectrolytes over as broad a concentration range as possible. This may or may not confirm the empirical relations proposed by van der Touw & Mandel (59) for the concentration dependence of both the specific increments and relaxation times, which have been (partially) confirmed in some cases (60, 63),

$$\Delta\varepsilon'_s/C_p = \frac{A_s}{1+B_s C_p} \qquad \text{5a.}$$

$$\Delta\varepsilon'_2(0)/C_p = \frac{A_2}{1+B_2 C_p} \qquad \text{5b.}$$

$$\tau_i = (2\pi f_{c,i})^{-1} = \frac{A'_i}{1+B'_i C_p} \quad i=1,2 \qquad \text{6.}$$

but not in others where a decrease of $\Delta\varepsilon'/C_p$ with C_p had been observed indeed (45). In these equations A_s, B_s, A_2, B_2, A'_1, A'_2, B'_1 and B'_2 are assumed to be constants. Equations 5 and 6, if they can be applied, make it possible to perform extrapolations to zero concentration where intermolecular interactions should be absent, and to compare the extrapolated values

$$A_s = (\Delta\varepsilon'_s/C_p)_{C_p \to 0}, \quad A_2 = (\Delta\varepsilon'_2/C_p)_{C_p \to 0},$$
$$A'_1 = (\tau_1)_{C_p \to 0} \quad \text{and} \quad A'_2 = (\tau_2)_{C_p \to 0}$$

to theoretical expressions (which have been derived without taking into account such interactions; see next section). It has been noted, however, that the extrapolated values may well differ from the values of the corresponding parameters at the real infinite dilution situation, because of the occurrence of different concentration regimes (6), and therefore such extrapolations should be considered with proper care.

It is necessary to point out that agreement on certain general features of the dielectric dispersion of polyelectrolyte solutions (as illustrated in Table 1) is only of a qualitative nature. Two analogous systems studied in different laboratories often yield different values both for the dielectric increments and the critical frequencies. As an example we may quote the differences in $\Delta\varepsilon'_s$ and τ_1 observed by Tung et al (45) between their results and Takashima's (67b). Here again it is necessary to take into account eventual differences in the systems investigated although both were calf thymus DNA. But of course there also remains to be considered the differences in the experimental techniques used as well as in their accuracy and sensitivity,

Table 1 Comparison of the value of some dielectric parameters for different polyelectrolyte systems

Polyion and Ref.	DP	α_D	C_p (g dm^{-3})	C_0^a (M)	$\Delta\varepsilon'_s/C_p$ (g^{-1} dm^3)	$\Delta\varepsilon'_2/C_p$ (g^{-1} dm^3)	τ_1 (s)	τ_2 (s)
K-PSS[b] (59)	500	1	0.3	0	118	41	6.3×10^{-6}	4.8×10^{-8}
Na-PAA[c] (59)	7500	0.4	0.5	0	73	26	8×10^{-6}	2.8×10^{-8}
Na-PGA[d] (60)	620	1	0.13	0	246	77	8.9×10^{-6}	5×10^{-8}
Na-DNA[e] (50)	12000	1	0.1	10^{-4}	6800	—[f]	6.8×10^{-2}	—
Na-DNA[e] (50)	12000	1	0.1	10^{-3}	620	—	1.6×10^{-2}	—
Na-DNA[e] (63)	900	1	0.1	3×10^{-4}	110	37	3.3×10^{-6}	1.2×10^{-7}
Na-DNA[e]	900	1	0.5	3×10^{-4}	54	22	4.3×10^{-6}	5.8×10^{-8}

[a] Mono-monovalent salt, concentration in molarity (M).
[b] Potassium poly(styrenesulfonate).   [c] Sodium poly(acrylate).
[d] Sodium poly(glutamate).   [e] Sodium DNA.
[f] (—) stands for "not measured."

combined with the application of different fitting and extrapolation procedures. A good appraisal of such differences is sometimes difficult, as the details about the way the experiments have been performed and the experimental data have been analyzed are sometimes only fragmentary.

Such quantitative discrepancies combined with the problems related to the concentration effects make comparison with theory difficult and may explain why different theories have been quoted to be consistent with experimental results. Another illustrative example of such discrepancies is shown by the molar mass dependence of $\Delta\varepsilon'_s$ and $f_{c,1}$. Although only a few investigations have been performed, quantitative agreement is lacking. In some papers it is claimed that $\Delta\varepsilon'_s$ and $f_{c,1}^{-1}$ are proportional to M (55, 66, 69) but in others that $\Delta\varepsilon'_s$ and $f_{c,1}^{-1}$ increase as M^2 (67, 68). Here also the results refer to different systems under different experimental conditions and the remarks formulated above for the concentration dependence also apply to these cases.

There is nearly complete agreement on the qualitative effect of added salt on the dielectric increments of polyelectrolyte solutions. In all cases investigated, low molar mass electrolyte was found to have a depressing effect on $\Delta\varepsilon'$ (45, 50, 55, 58, 71, 75, 76). Salts with divalent counterions gave a larger decrease than salts with monovalent counterions. Some specificity on the nature of the counterions has been observed. This particularly applies to polyelectrolyte solutions in the absence of salt but with a mixture of monovalent and divalent counterions in various ratios (56, 64, 65; unpublished results from this laboratory with different polyelectrolyte solutions).

Some isolated effects have been briefly discussed in the literature and are emphasized here because they need further confirmation and more extensive investigations. The only study on the influence of the temperature on the dielectric properties of polyelectrolyte solutions of which we are aware was performed on high molar mass DNA (45). In a range where no denaturation was expected (between 5 and 25°C), $\Delta\varepsilon'_s$ was found to increase with temperature; for temperatures above 64°C, denaturation caused $\Delta\varepsilon_s$ to decrease with increasing temperature. Above 90°C the increment became very small indeed. The other observation concerns the dielectric properties of polyelectrolytes of small dimensions. It was reported (61) that for poly(phosphates) in aqueous solution, a sample of $DP = 338$ still exhibited the two dispersion regions observed normally. For $DP = 112$, however, the experimental results could best be fitted to a single dispersion curve. The intermediate sample of $DP = 198$ was found to be a borderline case as the frequency dependence of $\Delta\varepsilon'$ could be fitted to either one or two dispersion curves. As these measurements have been performed at a single concentration, it is not yet clear whether these differences depend only on M or also on the difference in concentration regime of the three solutions studied.

There remains to be discussed what interpretation was given to the dielectric effects observed. In the earlier period there had been much confusion about the nature and the origin of the increments and the dispersions observed. Specific permanent dipole moments, Maxwell-Wagner effects, dipoles arising from the charged groups and their counterions or from impurities attached to the macromolecular chain, have been invoked to explain the relatively large dielectric increments, even at high frequencies, as compared to uncharged analogues of the polyelectrolytes studied or to ordinary electrolyte solutions. It has been shown in several publications that these explanations cannot be upheld. This became particularly evident when DNA in its double helix conformation was found to yield qualitatively the same dielectric results as other polyelectrolyte solutions. It is indeed to be excluded that the two antiparallel strands of DNA forming the double helix can give rise to a net permanent dipole moment, and careful purification of DNA samples did not result in qualitatively different dielectric behavior (74). As already pointed out by Dintzis et al in 1954 (53), a Debye-Falkenhagen type of effect due to the distribution of the small ions around the charged macromolecules should rather be thought to be responsible for the dielectric response of polyelectrolyte solutions. The absence of spherical symmetry for the macromolecular chain at low salt concentrations down to $c_0 = 0$ has considerably retarded the more rigorous theoretical treatment of this problem, but different simple models based on a one-dimensional approach have been proposed particularly to explain $\Delta\varepsilon'_s$ and the lower frequency dispersion. These associated counterion polarization models are discussed briefly in the next section.

It was even more difficult to account for the higher frequency dispersion. Van der Touw & Mandel (76) introduced a model in which the polyelectrolyte was pictured as a nonlinear sequence of identical rod-like subunits and the mobility of associated counterions along a subunit was assumed to be different from the mobility of the counterion when going from subunit to subunit. Induced dipole moments due to distribution of counterions along the subunits would be responsible for the amplitude of the higher frequency dispersion region, whereas $\Delta\varepsilon_s$ would be determined by the induced dipole moment of the total macromolecule and its associated counterions. The higher frequency dispersion relaxation time, $\tau_2 = (2\pi f_{c,2})^{-1}$, would essentially be determined by the relaxation mechanism of the counterion distribution along a subunit and would therefore, just as also $\Delta\varepsilon'_2(0)$, be independent of M. The lower frequency dispersion would be characterized by an average relaxation time containing possibly contributions from the relaxation mechanism of the overall counterion distribution along the polyelectrolyte and from the rotational diffusion of the polyion (both contributions depending on the average size, and therefore on the molar mass, of the macromolecule). This "*ad hoc*" theory leads to the simple relations

$$\Delta\varepsilon'_2(0)/C_p \sim \tau_2 \sim b^2 \qquad \qquad 7a.$$

$$\Delta\varepsilon'_s/\Delta\varepsilon'_2(0) = \frac{\langle R_G^2 \rangle}{b^2} \qquad \qquad 7b.$$

with $\langle R_G^2 \rangle$ the mean square radius of gyration and b the length of a subunit. Some evidence for these relations has been found for quantities extrapolated to zero concentration according to Eqs. 5 and 6 (5, 6, 59, 60, 62, 63). It has also been shown by Odijk (14) that the length of the subunit could be related to the correlation distance ξ between the macromolecular chains in solutions of not too high concentrations where interactions have to be taken into account. Consequently the higher frequency dispersion region is expected to be altogether absent for a completely rigid, cylindrical polyelectrolyte or for a more flexible chain at very low concentrations both of macromolecule and salt, where the macromolecule is practically completely stretched and no overlap between different chains occurs. Evidence for this is too scarce to permit any definite conclusions.

Another possible explanation for the higher frequency dispersion proposed by Minakata (65) is related to a radial component of the induced dipole moment arising from the perturbed ion distribution around a rigid polyelectrolyte. If this should be the origin of $\Delta\varepsilon'_2(0)$ and its dispersion, it is to be expected that they should disappear neither for low values of the molar mass nor at very small concentrations. As pointed out above, not enough experimental evidence is available to decide about the validity of this mechanism. It is perhaps also appropriate to point out that under the conditions mentioned, the effects sought may become too small to be observed with enough accuracy.

Neither approach finally can explain in a satisfactory way why the relaxation mechanism deviates from the simple Debye-type and, in particular, why the Cole-Cole parameter β_c is so insensitive to the exact nature of the polyelectrolyte or its concentration and why both dispersion regions do not differ with respect to that parameter.

We now present a more detailed discussion of the theoretical background for the dielectric properties of solutions containing charged macromolecules.

THEORY

Introductory Remarks

The rather complicated picture that emerges from the dielectric experiments performed with polyelectrolyte solutions has not yet found a satisfactory explanation. Although, as briefly outlined at the end of the preceding section, different semiempirical attempts have been made to understand at least in a semiquantitative way these experimental results, they are far from serious approaches and cannot be justified by sound theoretical arguments.

A series of papers were published that tried to give a theoretically sounder treatment of the dielectric response of solutions of charged particles. Unfortunately, these treatments were based on models that did not satisfy the experimental conditions of the systems discussed in the preceding section. Most of these treatments considered compact spherical particles of a radius much larger than the Debye length, which cannot be assumed to be a reasonable approximation of a macromolecule with a high linear charge density in a solution of low ionic strength. As interparticle interactions were not taken into account, these theories would probably apply to concentrations outside the range usually studied. In the few cases in which cylindrical particles were discussed the situation with respect to the experimental conditions would be even worse. In particular, no treatment of either case is available for systems in the absence of low molar mass electrolyte. Therefore one has to conclude that there exists a regrettable gap between the systems accessible to experimental investigation and the theoretical treatments that can presently be handled. The semiempirical treatments mentioned above can thus not be disregarded; they still have a certain utility because they predict rather satisfactory order of magnitudes for the experimental quantities. This may be due to certain compensating effects not yet entirely understood, and hence these semiempirical treatments may obscure certain essential aspects of the polarization effects arising from the perturbation of the ion distribution in polyelectrolyte systems.

In order to give a feeling for the more fundamental approach to these polarization aspects and the effects that physically accompany the application of oscillating external electric fields, we shall outline in detail some of the more recent theoretical approaches that have been used for the case of compact, spherical particles in particular within the "thin double layer approximation." After some further elaborations of these treatments a short discussion of the cruder approaches is presented under the heading "Associated Counterion Polarization."

Prior to the work of Dukhin & Shilov, summarized in their monograph (77), workers attempting to calculate a polyion's dipole moment induced by an electric field routinely neglected fluid convection, ion currents, and the compensating dipole moment due to the ion atmosphere. Moreover, in the case of an oscillating field, the out-of-phase component of the dipole moment makes an appreciable if not dominant contribution to the real part of the dielectric response, but this out-of-phase component used to be disregarded altogether. In recent years a number of theorists (78–84) have corroborated and extended the analysis of Dukhin & Shilov.

First of all, we summarize several of the assumptions and starting equations pertaining to a single spherical polyelectrolyte in an infinite mono-monovalent electrolyte solution. Related discussions of these have appeared (77, 78, 80, 85–90). In the case of simple electrolytes, the equations were written down by Onsager (91) and by Debye & Falkenhagen (92) nearly sixty years ago. A lucid account of Onsager's conductance theory has recently appeared in the Landau & Lifshitz course (93). The reader should be aware of the fact that the work on dielectric response (77–84) is intimately related to the theories of electrophoresis (85–88, 94–97), conductivity (98–101), sedimentation (102–104), and electroviscous effects (105–110). In general, the same or analogous equations need to be solved so approximation schemes used in one area can be fruitfully carried over to another. For instance, O'Brien's conductivity theory of charged suspensions (99, 100) has been applied by Hinch & Sherwood to dielectric response (84) and the primary electroviscous effect (110).

Double-Layer Polarization

BASIC EQUATIONS Let us consider an impenetrable, nonducting sphere of radius a immersed in a mono-monovalent electrolyte solution of bulk concentration c_0. The sphere has a uniform permittivity ε_I as well as a uniform surface charge density σ_q, assumed positive for definiteness. Because of the externally applied, uniform, oscillating electric field $\mathbf{E}_{\text{ex}}\, e^{i\omega t}$ (ω is the angular frequency and t is time) an electric body force within the solution will cause solvent flow via the quasistatic version of the Navier-Stokes equation, also known as the Stokes equation (111)

$$\eta \nabla^2 \mathbf{v} - \nabla P + \rho_q \mathbf{E} = 0 \qquad 8.$$

valid everywhere within the fluid. Here the charge density ρ_q, the local electric field $\mathbf{E}$, the solvent velocity $\mathbf{v}$ and pressure P are functions of position $\mathbf{r}$ and time t. The solvent viscosity η is assumed to be uniform. In Eq. 8 it is supposed that the inertial terms can be neglected, i.e. $\rho \partial \mathbf{v}/\partial t$ is small compared to the viscous term (ρ is the fluid density), and the nonlinear terms do not come into play—the Reynolds number is small: $Re = (\rho a u/\eta) \ll 1$ where u is a typical velocity of the sphere. It is customary to regard the fluid as incompressible (111).

$$\nabla \cdot \mathbf{v} = 0 \qquad 9.$$

The local electric potential ϕ that determines the electric field $\mathbf{E} = -\nabla\phi$, is related to the charge density via Poisson's equation (112)

$$\varepsilon_0 \nabla^2 \phi = -4\pi \rho_q, \qquad 10.$$

valid within the solution (c.g.s.e.s. units are used for convenience). Here we have assumed the solvent permittivity ε_0 to be uniform. At any instant there are nonuniform concentrations of the ions c_i ($i = 1$, counterions; $i = 2$, coions). They determine the charge density $\rho_q = -qc_1 + qc_2$, where q is the (positive) elementary charge, and satisfy conservation equations (111)

$$\partial c_i/\partial t = -\mathbf{\nabla}\cdot\mathbf{J}_i = -\mathbf{\nabla}\cdot c_i\mathbf{v} \quad i = 1,2 \qquad 11.$$

where the ion fluxes $\mathbf{J}_i = c_i\mathbf{v}_i$ with $\mathbf{v}_i = \mathbf{v}_i(\mathbf{r}, t)$ denoting the ion velocities. Because $\mathbf{v}_i(\mathbf{r}, t)$ generally differ from the fluid velocity $\mathbf{v}(\mathbf{r}, t)$, frictional forces will be exerted on the ions by the fluid. These must be balanced by diffusional forces arising from the nonuniformity of the ion distributions as well as by the electrical forces (86)

$$\beta_i(\mathbf{v}_i - \mathbf{v}) = -k_B T \mathbf{\nabla} \ln c_i + q_i \mathbf{E} \qquad 12.$$

where $k_B T$ signifies Boltzmann's constant times temperature and β_i and q_i represent the frictional constant and the charge of the ions, respectively ($q_1 = -q; q_2 = q$). Equation 12 is certainly not exact. Besides inertial effects, hydrodynamic interactions between ions are neglected, the diffusional force is approximated by an ideal term, and polarization of the surrounding fluid by the movement of the ions is not taken into account (3). If $\mathbf{E}_{ex} = 0$, $\mathbf{J}_i = 0$ ($i = 1, 2$) and we end up with the equilibrium Poisson-Boltzmann equation (8)

$$\nabla^2 \psi_e = \kappa^2 \sinh \psi_e \qquad 13.$$

where the dimensionless equilibrium potential $\psi_e \equiv q\phi_e/k_B T$ has been introduced, $\kappa^2 = 8\pi Q c_0$ defines the Debye screening length κ^{-1} and $Q = q^2/\varepsilon_0 k_B T$ represents the Bjerrum length. Although of limited validity, Eq. 13 is known to be quite accurate under certain specific conditions for highly charged polyelectrolytes (8). From Eqs. 11 and 12 we can eliminate the ion velocities to obtain the Smoluchowski-type equation.

$$\partial c_i/\partial t = -\mathbf{\nabla}\cdot c_i\mathbf{v} + \frac{k_B T}{\beta_i}\nabla^2 c_i - \frac{q_i}{\beta_i}\mathbf{\nabla}\cdot c_i\mathbf{E} \qquad 14.$$

The terms on the right hand side represent ion convection, ion diffusion, and conduction, respectively.

Before discussing boundary conditions we want to make some general remarks. It is evident that a substantial number of explicit as well as implicit assumptions are needed in order to establish Eqs. 8–14. In the literature one sometimes sees an *ad hoc* adjustment or "improvement" of one or several of these assumptions without concomitant modification of the others. For example, one might let ε_0 be nonuniform. However, in that case Eq. 12 becomes even more approximate—the static version of Eq. 11 would have to be reassessed. Furthermore, effects of discreteness would now have to be taken into account. It is also equally inconsistent to let η be uniform. Besides these considerations it is not at all clear that nonuniformity is adequately represented by letting ε_0 be only a function of $\mathbf{r}$, i.e. a nonlocal response would be equally plausible.

BOUNDARY CONDITIONS   Far away from the sphere, the fields determined by Eqs. 8–14 assume the following limiting forms.

$$\mathbf{v} \to 0 \quad \text{as} \quad \mathbf{r} \to \infty \qquad \text{15a.}$$

$$P \to \text{constant} \qquad \text{15b.}$$

$$c_i \to c_0 \qquad \text{15c.}$$

$$-\nabla \phi \to \mathbf{E}_0 \, e^{i\omega t} \qquad \text{15d.}$$

In order to specify boundary requirements near the sphere, we are forced to make additional assumptions. One viewpoint, common to polyelectrolyte physical chemists, is to regard the surface charge density as a fixed quantity (measurable in principle), especially if the macromolecule is fully charged. Furthermore, it is supposed that the solvent surrounding the sphere has uniform properties right up to the polyion surface. In this case we have

$$\mathbf{v} = \mathbf{u}(t) \quad \mathbf{r} \in S \qquad \text{16a.}$$

$$\mathbf{n}_S \cdot [-k_B T \nabla c_i + q_i c_i \mathbf{E}_+]_S = 0 \qquad \text{16b.}$$

$$\mathbf{n}_S \cdot [\varepsilon_0 \mathbf{E}_+ - \varepsilon_I \mathbf{E}_-]_S = 4\pi \sigma_q. \qquad \text{16c.}$$

Equation 16a expresses the assumption that the fluid sticks to the sphere moving with an electrophoretic velocity $\mathbf{u}(t)$ under influence of the electric field; in the context of this survey we do not specify $\mathbf{u}(t)$ (see e.g. refs. 83, 84). Equation 16b states that the ions cannot penetrate the polyion sphere, $\mathbf{n}_S$ denoting a unit vector at point S normal to the surface. Equation 16c is the usual condition for the continuity of the normal component of the electric displacement vector on the boundary between two media and follows from electrostatics (112), $\mathbf{E}_{+,S}$ signifying the value of the electric field at S viewed from the fluid side and $\mathbf{E}_{-,S}$ the field as seen inside the particle. Inside the sphere Laplace's equation is satisfied by ϕ; we assume $\varepsilon_I \ll \varepsilon_0$.

The other extreme disseminated by colloid scientists (113) is to consider the surface potential ϕ_S fixed (but very difficult to measure). It is supposed that a number of ions are adsorbed onto the surface, the ill-defined layer just outside the polyion and including these ions being termed the Stern layer. The potential ϕ_d at the outer surface of the Stern layer is not necessarily constant. Finally the so-called surface of shear S' at which the fluid is postulated to stick may be located beyond the Stern layer. Hence, it is asserted that electrokinetic measurements yield not ϕ_d but ζ, the zeta potential at the shearing layer. The colloid model has several adjustable parameters. Since dielectric theories are at present in only semiquantitative agreement with experiments, it seems apt to adopt the simplification of setting $\phi_d = \zeta$, i.e. the outer Stern layer corresponds to the surface of shear. The equivalent of Eqs. 16 is now

$$\mathbf{v} = \mathbf{u}(t) \quad \mathbf{r} \in S' \qquad 17a.$$

$$\mathbf{n}_{S'} \cdot [-k_B T \nabla c_i + q_i c_i \mathbf{E}_+]_{S'} = 0 \qquad 17b.$$

$$\phi(S') = \phi_d = \zeta. \qquad 17c.$$

The choice of boundary conditions (Eqs. 16 or Eqs. 17 or perhaps a hybrid set) is not yet clear-cut in practice. An important example (although not for a spherical polyelectrolyte) is which set of boundary conditions one should pick for double-stranded DNA. In the literature the hydrodynamic diameter of DNA has often been set equal to its approximate geometrical diameter (see e.g. 114) whereas the surface charge density has been estimated from the actual number of phosphate charges per turn of the helix. Nevertheless, Schellman & Stigter (115) argue in favor of a colloid-type model, the hydrodynamic diameter, the kinetic charge and ζ potential being adjustable parameters within limits.

DIELECTRIC RESPONSE Once we have solved Eqs. 8–14 together with Eqs. 15 and either Eqs. 16 or 17, we still have to relate these fields to the dielectric response function $\varepsilon^*(\omega)$. This is certainly not trivial. One measures macroscopic parameters, whereas the underlying expressions are microscopic. Accordingly, some kind of averaging procedure must be employed. In the uncharged case, Batchelor (116) developed a formalism for so-called statistically homogeneous suspensions (see also 117 for a review). Stigter (98) was apparently the first to realize that the older formalism (77) did not take adequate account of the nonuniformity of the microscopic equations. Inspired by Batchelor's work (116), O'Brien (99) developed an elegant formalism for calculating the electrical conductivity of a dilute suspension of spheres under the influence of a constant, uniform electric field. De Lacey & White (80) extended this to the case of oscillating fields, but it seems imperative to survey O'Brien's work first of all.

Batchelor (116) and others (111) have argued that for statistically uniform suspensions, macroscopic averages are just spatial averages of the relevant microscopic quantities. For instance, one may use volume averages (99)

$$\langle \mathbf{i} \rangle = \frac{1}{V} \int_V \mathbf{i} \, d\mathbf{r} \qquad 18a.$$

$$\langle \mathbf{E} \rangle = \frac{1}{V} \int_V \mathbf{E} \, d\mathbf{r} \qquad 18b.$$

where **i** is the locally defined microscopic current and V is the volume of the solution. Ohm's law pertains to these averaged quantities

$$\langle \mathbf{i} \rangle = \sigma \langle \mathbf{E} \rangle \qquad 19.$$

with σ the steady-state conductivity of the system. Equation 19 holds only for low enough fields. The microscopic current is derived from Eq. 12

$$\mathbf{i} = \sum_{i=1,2} q_i \mathbf{J}_i = \rho_q \mathbf{v} - \sum_{i=1,2} \beta_i^{-1} q_i [k_B T \nabla c_i + q_i c_i \nabla \phi]. \qquad 20.$$

O'Brien (99) simply neglected the convection term. After rearrangement of Eq. 20, he eventually arrived at the following expression for the average current.

$$\langle \mathbf{i} \rangle = \sum_{i=1,2} \beta_i^{-1} q_i^2 c_0 \langle \mathbf{E} \rangle$$

$$+ \frac{1}{V} \sum_k \int_{V_k} \left[\mathbf{i} + \sum_{i=1,2} \beta_i^{-1} q_i (k_B T \nabla \delta c_i + c_0 q_i \nabla \delta \phi) \right] d\mathbf{r} \qquad 21.$$

The first term is just the average current of the pure salt solution and is equal to $\sigma_0 \langle \mathbf{E} \rangle$. The k summation is over all charged spheres, V_k is a volume enclosing the kth sphere and its double layer, δc_i and $\delta \phi$ are the (small) deviations from equilibrium that are induced by the electric field. The suspension is dilute, i.e. $NV_k \ll V$ where N is the number of spheres in the solution. Note that ion diffusion contributes to the macroscopic current, a fact that is neglected by Dukhin (77) as well as Fixman (78), although their formulas agree with Eq. 21 whenever $\beta_1 = \beta_2$. If the ion mobilities β_1^{-1} and β_2^{-1} differ appreciably, the ion diffusion terms play a substantial role. It should be stressed that theories assuming equilibrium of the double layer are hopeless since O'Brien (99) showed that the term $k_B T \nabla \delta c_i + c_0 q_i \nabla \delta \phi$ is not zero in general. The integrals in Eq. 21 are related to the asymptotic properties of the concentration and electric potential "far fields."[1]

O'Brien's results were corroborated by Saville (104) in his treatment of the sedimentation potential in dilute, charged suspensions. Saville showed that one is required to incorporate diffusion terms like those above, in order to be consistent with earlier work on linear, irreversible thermodynamics (90).

Finally, the same procedure was used by de Lacey & White (80) to derive the frequency dependent, complex conductivity $\sigma^*(\omega)$ and hence the dielectric response $\varepsilon^*(\omega)$ if the field is oscillatory. Accordingly, Eq. 19 is generalized to

$$\mathbf{I}_0 e^{i\omega t} \equiv \langle \mathbf{i} \rangle = \sigma^*(\omega) \langle \mathbf{E} \rangle = \sigma^*(\omega) \langle \mathbf{E} \rangle_0 e^{i\omega t}. \qquad 22.$$

---

[1] Note that this discussion pertains to the $\omega = 0$ case. The works of Dukhin (77) and Fixman (78) are correct if $\omega \neq 0$.

Furthermore, one has to add a displacement current to Eq. 20. After several manipulations, de Lacey & White (80) ended up with the following equation

$$\langle \mathbf{i} \rangle = \left(\sigma_0 + \frac{i\omega\varepsilon_0}{4\pi} \right) (\langle \mathbf{E} \rangle + 3ga^{-3}\varepsilon_0^{-1}\boldsymbol{\mu}^*) \qquad 23.$$

where σ_0 is assumed constant, the volume fraction ($g \equiv 4\pi a^3 N/3V$) of the charged spheres is small, and the induced dipole moment is complex: $\boldsymbol{\mu}^* = \boldsymbol{\mu}' + i\boldsymbol{\mu}''$. The latter arises because the integrals of the frequency dependent analogue of Eq. 21 are related to the asymptotic form of the far fields (see e.g. 78, 80) pertaining to the one sphere problem, e.g.

$$\phi(\mathbf{r}) \sim \delta\phi(\mathbf{r}) \sim -\langle \mathbf{E} \rangle \cdot \mathbf{r} - \varepsilon_0^{-1}\boldsymbol{\mu}^* \cdot \nabla r^{-1} \quad (r \to \infty). \qquad 24.$$

The δc_i far fields do not contribute if $\omega \neq 0$, however. The physical reason for this is that the left-hand side of Eq. 11 is unequal to zero when ω is likewise, so the concentration fields decay essentially exponentially. They simply do not have time enough to build up macroscopic currents. From a mathematical point of view, the $\omega \to 0$ limit is nonuniform or singular as pointed out by de Lacey & White (80).

If we introduce a complex polarizability defined as $\alpha^*(\omega)\langle \mathbf{E} \rangle = (\alpha' + i\alpha'')\langle \mathbf{E} \rangle = \boldsymbol{\mu}^*$, we can deduce the real part of the dielectric increment from Eqs. 22 and 23 (see also Eqs. 2 and 3)

$$\Delta\varepsilon'(\omega) = \frac{4\pi}{\omega} \Delta\sigma''(\omega) = \frac{3g}{a^3} \left(\alpha'(\omega) + \frac{4\pi\sigma_0 \alpha''(\omega)}{\omega\varepsilon_0} \right) \qquad 25.$$

where $\Delta\sigma^*(\omega) = \sigma^*(\omega) - \sigma_0$. The first term corresponds to the usual in-phase polarization. However, there is also an out-of-phase polarization related to α'' of the sphere because the polarization lags behind the applied field (77). Hence, an out-of-phase field results that acts on the ions in the surrounding fluid. Out-of-phase conduction currents appear, but these in turn can give rise to in-phase polarization. Note that the α'' term can be large if the solution is highly conductive. As a matter of fact, it usually overwhelms the first term in the theory of double layer polarization (77, 80).

In order to calculate $\Delta\varepsilon'(\omega)$, one has to solve Eqs. 8–17, and then derive the far field terms, in particular, for $\delta\phi(\mathbf{r})$ (see Eq. 24). As a rule, this is a difficult scheme analytically, but even numerically, since the various equations are highly nonlinear. Nevertheless, relevant values of $a\kappa$ are generally much larger than unity, because measurements are carried out on quite large colloidal spheres usually. In this limit at least, the theory is tractable even if the dimensionless surface potential is high. In conclusion, it seems worthwhile to point out that the α'' term in Eq. 25 corresponds to the Debye-Falkenhagen dielectric increment in the theory of simple electrolyte solutions (92).

Spherical Particles: Neglect of Solvent Flow

THIN DOUBLE LAYER APPROXIMATION Even though Fixman's 1980 paper (78) is not always easy to follow, it presents a physically motivated and efficient way of attacking the aforementioned set of differential equations in the limit of large $a\kappa$. At first Fixman (78) implicitly disregarded solvent flow (see his discussion following his Eq. 3.22), but he incorporated it in a later paper (83). The fact that neglect of fluid flow is not always allowed can be discerned from the following rough argument. If $a\kappa \gg 1$, i.e. the double layer is very thin so the sphere is seen as a plane on the scale of a Debye length, we have from Eqs. 8 and 10

$$v = \mathcal{O}\left(\frac{\varepsilon_0 \zeta E_0}{\eta}\right)$$

where the absolute value of the Maxwell field $\langle \mathbf{E} \rangle = \mathbf{E}_0 \, e^{i\omega t}$ has been chosen as a typical value of the electric field strength. Next, in Eq. 14 we compare the ratio $R_{c,i}$ of the convection to the conduction terms

$$R_{c,i} = \mathcal{O}\left(\frac{\beta_i v}{q_i \langle E \rangle}\right) = \mathcal{O}\left(\frac{\varepsilon_0 \beta_i \zeta}{q_i \eta}\right) = \mathcal{O}\left(\frac{Z d_i}{Q}\right) \qquad 26.$$

where $\beta_i = 6\pi\eta d_i$ defines an effective ion radius and $Z \equiv q\zeta/k_B T$ is a dimensionless zeta potential. Equation 26 is derived by writing $c_i = c_{i,e} + \delta c_i$ with c_e the equilibrium concentration and δc_i the perturbation, splitting up the vectors into normal and tangential components with respect to the sphere surface, $\mathbf{v} = \mathbf{v}_n + \mathbf{v}_t$ and $\mathbf{E} = \mathbf{E}_n + \mathbf{E}_t$, and comparing tangential terms. For H_2O at 298 K, the Bjerrum length $Q = 7.14$ Å, so $R_{c,i}$ can easily be of order unity and solvent flow must be accounted for.

It is, nevertheless, of methodological interest to study the $v = 0$ case even if $Z = \mathcal{O}(1)$. In that case Fixman (78) proceeded as follows.

1. If fluid flow is neglected, the charged sphere is simply considered to be motionless ($\mathbf{u}(t) = 0$), which is certainly all right in a calculation of the dielectric response. The origin of our coordinate system is conveniently placed at the center of the sphere. It is also expedient to split up the potentials $\ln c_i$ and ϕ into equilibrium (indicated by subscript e) and perturbation parts,[2] viz $\ln c_i = \ln c_{i,e} + h_i$ and $\phi = \phi_e + \delta\phi$. The perturbation terms that arise because of the applied field add up to a velocity potential determining the velocity of the ions

$$\mathbf{v}_i = -k_B T \beta_i^{-1} \nabla p_i$$
$$p_i \equiv h_i + (q_i/k_B T)\, \delta\phi. \qquad 27.$$

The complex dipole moment of the charge distribution will be obtained

---

[2] If h_i is small, it is equal to $\delta c_i / c_{i,e}$.

from the asymptotic behavior of the $\delta\phi$ far field. (This is identical to the asymptotic form of the ϕ far field. We specify the "far fields" below.)

2. The total solution is divided into a double layer region surrounding the charged sphere ($= V_k - 4\pi a^3/3$) and the remaining outer region. The double layer thickness δ is equal to, say several times κ^{-1}. If the applied field is sufficiently small, the region beyond the double layer is electroneutral because of screening. This implies $\rho_q = 0$, $h_1 = h_2$ and $\nabla^2 \delta\phi = 0$ in the outer region. Clearly, the quantities h_i, p_i, and $\delta\phi$ vary rather slowly outside the double layer. Accordingly, it makes sense to call them far fields in that case. They will have to obey certain boundary conditions at the outer surface of the double layer.

The rate of change of Γ_i the number of ions i per unit area within the double layer is obtained by integrating Eq. 11 across the double layer.

$$\Gamma_i = \int_0^\delta c_i \, dy \qquad 28a.$$

$$\frac{\partial \Gamma_i}{\partial t} = -[\mathbf{n}_s \cdot c_i \mathbf{v}_i]_\delta - \int_0^\delta dy \, \nabla_t \cdot (c_i \mathbf{v}_i)_t \qquad 28b.$$

Here, the divergence of $c_i \mathbf{v}_i$ has been split up into a component in the radial direction and a tangential divergence $\mathbf{n}_s$ is a unit vector in the outward radial direction y, and all integrations are performed from the surface of the sphere ($y = 0$) to the boundary of the double layer. The first term on the right hand side of Eq. 28b is defined on this same boundary. The reader should note that Eqs. 28 and 29 differ somewhat from Fixman's original paper (78). He also defined far fields within the double layer, and considered quantities like $c_i - c_{i,f}$. To within all approximations considered, his analysis is just the same as ours to leading order (e.g. his $\varepsilon_i \simeq$ our Γ_i).

3. The restriction $a\kappa \gg 1$ is now introduced (the so-called thin double layer approximation). The aforementioned boundary conditions for the far fields on the outer surface of the double layer can now be regarded as effective requirements on the sphere's surface itself (i.e. we are replacing $a + \delta$ by a). Inside the double layer there is an excess of counterions compared to the bulk, whereas there is a deficiency in coions. In what follows we assume that either the surface charge density or the surface potential is so high that c_1 within the double layer is much larger than c_0. To the same approximation, c_2 is disregarded inside the double layer. These simplifications as well as Fixman's approximation (78) of assuming the tangential component of $\mathbf{v}_i$ to be constant across the double layer serve to reduce Eqs. 28b to the following expressions

$$\frac{k_B T}{\beta_1} \left[c_1 \frac{\partial p_1}{\partial y} + \nabla_t \cdot \Gamma_1 \nabla_t p_1 \right]_\delta = \frac{\partial \Gamma_1}{\partial t} \qquad 29a.$$

$$\left(\frac{k_B T}{\beta_2}\right) c_2 \frac{\partial p_2}{\partial y} = 0. \qquad 29b.$$

Assuming Γ_1 varies as $\exp(i\omega t)$, we are able to set $\partial \Gamma_1/\partial t$ equal to zero because the Maxwell relaxation time $\tau_M \simeq \varepsilon_0/\sum_i q_i^2 c_0 \beta_i^{-1}$ is much shorter than ω^{-1} for $\omega < 10^7$ Hz (78). The first term on the left-hand side of Eq. 29a signifies the radial flux (into or out of the double layer), whereas the second term is the tangential flux along the sphere surface. In view of electroneutrality outside the double layer, Eq. 14 reduces to the following diffusion equations for the far fields h_1 and h_2 (78):

$$\frac{\partial h_i}{\partial t} = \left(\frac{k_B T}{\beta_i}\right) \nabla^2 h_i \quad r > a + \delta \simeq a. \qquad 30.$$

These fields tend to zero as r goes to infinity.

The physical meaning of these equations is that the polarization of the double layer by the external field manifests itself in a counterion flow within the double layer tangential to the sphere and in the opposite direction to that of the applied field, but also in a counterion flow into and out of the double layer. Because in the outer region electroneutrality has to be maintained ($p_q = 0$; $h_1 = h_2$), coions will diffuse in order to compensate for any counterions entering or leaving this region.

4. Since the applied field varies periodically with circular frequency ω, one can safely assume the perturbed variables to have the same frequency dependence $\exp(i\omega t)$. (The equations become linear if the perturbed quantities are small.) The dipole solution h^* of Eq. 30 will be proportional to $e^{i\omega t} \partial(r^{-1} e^{-\lambda r})/\partial r$ where $\lambda^{-1} = (i\omega \beta/k_B T)^{-1/2}$ is a characteristic length scale of the ion diffusion (78). Thus, a typical time scale for this diffusion is $\tau_s \equiv \beta a^2/2k_B T$. Here, we have set β_1 equal to β_2 ($\equiv \beta$), for convenience. In view of the electroneutrality condition for the outer region and the linear relations between h_1, h_2, and $\delta\phi$, the dipole part of the far field $\delta\phi$ will be characterized by the same relaxation time τ_s. As $\text{Im}(h^* e^{-i\omega t})$ is proportional to λ^2 at very low frequencies, the term in Eq. 25 containing σ_0 reaches a well-defined limit as $\omega \to 0$.

5. A convenient small parameter for large spheres turns out to be Γ_1/ac_0, where we have set $c_1 \simeq c_0$ at the outer boundary of the double layer (see Eq. 29a). From Fixman's paper (78) one can derive the following expression for the dielectric increment, using Eq. 25.

$$\Delta\varepsilon'(\omega) = \frac{9}{16} g(a\kappa)^2 \varepsilon_0 (\Gamma_1/ac_0)^2 \{1/[(1+\omega^{1/2}\tau_s^{1/2})(1+\omega\tau_s)]\} \qquad 31.$$

It can be shown that the α' term is much smaller than the α'' term (80). The decay of $\Delta\varepsilon'(\omega)$ is not of the Debye-type but is quite gradual. Note also that $\Delta\varepsilon'(0)$ can easily be very large provided that Γ_1 is large enough.

SINGULAR PERTURBATION ANALYSIS A more systematic though quite complicated route to solving the set of differential equations is by singular perturbation theory (for reviews see 118 and 119). Neglecting ion convection, Chew & Sen (82) attempted to find uniformly valid series expansions to Eqs. 8–15, 17 in terms of powers of $(a\kappa)^{-1}$. Obviously one needs an analytical approach to the Poisson-Boltzmann equation (Eq. 13) to start with, but Chew & Sen solved this (120) also by the method of matched asymptotic expansions, one of the best known singular perturbation techniques (118, 119). This method entails identifying a region (or regions) of nonuniformity, often enough a region in which the fields vary markedly—the thin double layer in this case—expanding the variables systematically within the chosen inner and outer regions, respectively, and matching the inner to the outer expansions, again systematically (i.e. matching order by order in terms of the chosen expansion parameter). The outer limits of the inner expansions must coincide asymptotically with the respective inner limits of the outer expansions. In this way the resulting composite solution is uniformly valid within the entire region.

In the theory of the dielectric response of spheres (neglecting convection), it turns out that the relevant small parameter is not $(a\kappa)^{-1}$ but rather $(a\kappa)^{-1} \exp Z/2$ (82). If the latter is small and $Z \gtrsim 2$, Chew & Sen (82) derived the following expression using basically the same approximations as Fixman (78).

$$\Delta\varepsilon'(\omega) = \frac{9}{4} g\varepsilon_0 \, e^Z \{1/[(1+\omega^{1/2}\tau_s^{1/2})(1+\omega\tau_s)]\}. \qquad 32.$$

From the viewpoint of the colloid physical chemist, the surface charge density σ_q can be related to $\phi_d \simeq \zeta$ when $a\kappa \gg 1$, for in that case we just need the flat plate solution (113): $\sigma_q = 2q\kappa^{-1}c_0 \exp Z/2$ valid for $Z \gtrsim 2$. Since we have $\sigma_q \simeq q\Gamma_1$, Eq. 32 is virtually identical to Eq. 31.

The parameter $(a\kappa)^{-1} \exp Z/2$ first turns up in Dukhin & Shilov's work (77). If it is assumed to be small and the fluid convection is neglected also [the latter is the case when their parameter $m = \mathcal{O}(dQ^{-1})$ is set equal to zero; m^{-1} is analogous to a Hartmann number in electro- and magnetohydrodynamics], their Eq. III.103 corresponds to Eq. 32 if we take into consideration a number of misprints. We emphasize that Dukhin & Shilov were the first authors to attack Eqs. 8–17 in a serious way. We have outlined Fixman's approach mainly because it yields results more directly.

Spherical Particles: Further Developments

Not only did Dukhin & Shilov (77) initiate the theory of dielectric dispersion based on double layer polarization, but they also attempted a more general analysis than the one surveyed above. Unfortunately, their book (77) abounds with typographical errors, so that it is frustrating trying to figure out the correct formulas. Dukhin & Shilov gave a much more elaborate derivation of a surface boundary differential equation analogous to Fixman's (see Eq. 29), but their exceedingly lengthy account is not particularly clear. In our view, other papers (78–84) are much more lucid in this respect. Still, Dukhin & Shilov (77) discuss clearly the physical picture of the processes occurring.

THIN DOUBLE LAYER APPROXIMATION   Fixman (83) incorporated convection by solving the Stokes equation (Eq. 8) in a thin-double layer approximation in a manner analogous to O'Brien's analysis of electrophoresis (100). Typically, $\Delta\varepsilon'(0)$ increases by about 25% because of convective polarization (for $Z \simeq 4$). From Eqs. 8 and 12 one can see that the counterions within the double layer are swept along by the fluid in the same direction as the electric field forces them to move. Fixman (83) also showed numerically that his approximation of constant tangential ion velocities across the double layer is highly accurate. The perturbation potentials h_i and $\delta\phi$ change far less slowly than p_i (83). Accordingly, Fixman remarked that different perturbation expansions for p_i versus h_i and $\delta\phi$ will have to be devised in order to set up a complete theory incorporating convection. Fixman's analytical theory (83) agrees rather well with his own numerical work for $a\kappa = 50$ and $Z = 4$. The latter, in turn, is in reasonable agreement with the numerical results of de Lacey & White (80), allowing for the possibly different values of parameters chosen.

PERTURBATION THEORY   Hinch et al (84) have extended the methods of previous workers (77–83) to include effects due to differences in the mobilities of co- and counterions (called "asymmetry"), constant σ_q versus constant ϕ_d boundary conditions, and varying σ_q and ϕ_d because of chemical reactions taking place very close to the surface of the sphere. In general, although these refinements have a small but significant influence on $\Delta\varepsilon'(\omega)$ (about the same as convective polarization), we are not certain whether they will show up in measurements. Hinch et al (84) assert that their analytical results diverge from the numerical work (80), but, again, this might be due partly to their choice of parameter values.

LOW POTENTIAL LIMIT   A theory for arbitrarily thick double layers has been developed by O'Brien (81) for small Z ($\lesssim 2$). It incorporates solvent flow as well as asymmetry of the electrolyte, and it agrees well with numerical work (80). This interesting theory will be difficult to test because the predicted $\Delta\varepsilon'(\omega)$ values are not very large.

FINAL REMARKS   The above-mentioned theories have assumed the applied field to be low but beg the question of how low it should be. In a sophisticated numerical analysis, Fixman & Jagannathan (121) showed that nonlinear effects already start occurring at field strengths of the order of 500 V cm^{-1}, a figure certainly lower than one might first expect.

Finally, we stress that even though quite a few elaborations to Eq. 32 have been made, experiments are carried out at rather high volume fractions (usually $g > 0.1$). Significant concentration effects should show up in that event, since ion diffusion (Eq. 30) occurs on a scale of a particle radius. The recently developed theories of reaction diffusion in a suspension of spheres will be useful in elucidating g dependent effects, e.g. (122a,b). Really reliable experiments on charged sphere suspensions at low volume fractions (at $g \lesssim 0.05$) have not, to our knowledge, been performed. Various authors' (77–84) comparisons of their theoretical results to Schwan's data (123) are at best tentative since $g \simeq 0.3$ in his case.

Rod-like Polyions

Dukhin & Shilov (77) summarized some of the Russian work on the polarization of long rod-like polyions. Instead of τ_s, we now have $\tau_l = L^2\beta/8k_BT$ where L is the rod length (generally one has $\tau_D \simeq D^2\beta/k_BT$ for a polyelectrolyte with D the largest, average, dimension). They did not give explicit expressions for $\Delta\varepsilon'(\omega)$, which they apparently derived using unrealistic approximations, however.

THIN-DOUBLE LAYER APPROXIMATION The steady-state dipole moment (neglecting solvent flow) can also be calculated for charged rods from Laplace's equation for h_1 and h_2, and $\delta\phi$, together with the boundary conditions Eqs. 29 in the steady state, Eq. 16c with $\varepsilon_I \ll \varepsilon_0$, Eq. 15d with $\omega = 0$, and $h_{1,2} \to 0$ as $r \to \infty$. Dukhin & Shilov (77) gave a complicated analysis of an analogous boundary value problem but without deriving analytical formulas. Fixman (79) employed a convenient trick based on Landau & Lifshitz's calculation of the induced dipole moment of a conducting rod in an insulating medium (112). Within the thin-double layer approximation ($a_c\kappa \gg 1$, a_c = cylinder radius), Fixman's formula (79) for the dipole moment of a long rod ($L \gg a_c$) in a mono-monovalent electrolyte solution reads

$$\mu' = \frac{\varepsilon_0 L \mathbf{E}_0}{8\gamma^2 G}\left[1 - \frac{\tanh \gamma L}{\gamma L}\right] \qquad 33.$$

where the constant $G = 2\ln(2L/a_c) - \frac{14}{3}$ and the characteristic decay length $\gamma^{-1} = [2a_c\Gamma_1 G/c_0]^{1/2}$. Note that γ^{-1} scales as the Debye radius κ^{-1}. Its parameter dependence can be inferred from Eq. 29a. As soon as the rod length is about γ^{-1} or longer, the counterion flux out of the double layer starts having an appreciable influence. For $\gamma L = \mathcal{O}(1)$ or larger, μ' must have a lesser dependence on L than the limiting form of Eq. 33 for $\gamma L \ll 1$.

$$\mu' = \frac{\varepsilon_0 L^3 \mathbf{E}_0}{24G}. \qquad 34.$$

This has the well-known L^3 dependence (see Eq. 37), which is not really surprising since the underlying equations become similar. In fact, $\gamma \to 0$ implies that no ionic currents can occur out of the double layer. It appears feasible and worthwhile to extend Fixman's calculations (79) for worm-like polyelectrolytes. His integral equation might then be profitably resolved by Kirkwood-Riseman or Yamakawa-Fujii type arguments (12, 114). Fixman (124) corrected for solvent flow approximately in a later calculation, especially focusing attention on short, rod-like polyelectrolytes. (Polyions have to be smaller than about a persistence length in order that one can treat them as rod-like.) He tabulated his numerical results for values of the parameters pertaining to DNA.

A convenient paper has been published by Teubner (125), who showed that one can calculate the force and torque on a rigid, charged macromolecule of, in principle, arbitrary shape, in an electric field, without explicitly solving Eqs. 8–15. He generalized the so-called reciprocal theorem well-known in the field of low Reynolds number hydrodynamics (126). If fluid flow is neglected in Eq. 15, the hydrodynamic part of the problem amounts to solving the flow around the hypothetically uncharged particle in the absence of the external field. As an application, Teubner showed how Eqs. 33 and 34 could be rederived.

Associated Counterion Polarization

Many workers have simply neglected out-of-phase polarization altogether, i.e. they neglected the $\alpha''(\omega)$ term in Eq. 25. In a formal sense this corresponds to setting σ_0 equal to zero. Earlier ideas started from a two-phase model; the inner phase of "associated" counterions surrounding the charged macromolecule was regarded as very easily polarizable, the outer phase was assumed to not take part in the polarization. It should now be clear that neglect of $\alpha''(\omega)$ is just not allowed, as has been stressed by Dukhin & Shilov (77) and Fixman (78). Moreover, the neglect of double layer polarization, various ionic currents, and solvent flow makes most of the theories of $\alpha'(\omega)$ extremely crude. Still, some of the models outlined below do incorporate physical features that have been disregarded entirely in the work mentioned above. Thus, besides being useful in a rough, semi-empirical way, the older theories should at least point to areas where more fundamental research might take place in the future. Note that the counterions that we have called "associated" are also termed "condensed," "accumulated," "adsorbed," "domain bound," etc.

SPHERES   Overbeek (86) calculated $\alpha'(0)$ for charged spheres long ago and showed the resulting $\Delta\varepsilon'(0)$ to be small. Subsequently, Schwarz (127) tried to explain the experimentally large values of $\Delta\varepsilon'(0)$ by assuming polarization of adsorbed mobile counterions. An order of magnitude estimate of the canonical-ensemble averaged induced dipole moment is given by

$$\langle \mu' \rangle_c \simeq \frac{Nq^2 D^2 \mathbf{E}_0}{k_B T} \qquad 35.$$

where D is the largest dimension of the particle and N is the number of associated counterions. This result is easily obtained via the canonical ensemble if $qE_0 D/k_B T$ is small enough. Equations 25 and 35 with $D = a$ and $\alpha'' = 0$ yield

$$\Delta\varepsilon'(0) \simeq \frac{gNq^2}{ak_B T} \simeq gaQ\varepsilon_0 \Gamma_1. \qquad 36.$$

This result is approximately a factor $ac_0 \Gamma_1^{-1}$ larger than $\Delta\varepsilon'(0)$ from Eq. 31. However, leakage through the double layer (first term on the left-hand side Eq. 29a) will drastically decrease Schwarz's result (127). Schurr (128) attempted a modification in this direction, but his calculation was heavily criticized by Dukhin & Shilov (77). Their own version of Schwarz's theory leads to values much lower than those from Eq. 31. It seems that a reassessment of these calculations may be called for. As we mentioned above, no experiments at sufficiently low volume fractions have been published with which we could put these theories to a stringent test.

The relaxation behavior in Schwarz's theory stems from an equation analogous to Eq. 29a but without the normal component flux and retaining the time derivative. Hence, the relaxation of the associated counterions is diffusional and the relaxation time scales as $D^2\beta/k_B T = a^2\beta/k_B T$, i.e. the same as in double layer polarization theories (77–84).

ROD-LIKE POLYIONS   Mandel's result (129) for rod-like polyelectrolytes is obtained by setting $D = L$ in Eq. 35. If we assume N to be the number of "condensed" counterions from naive condensation theory (130), Eq. 35 becomes

$$\langle \mu' \rangle_c \simeq (1 - 1/Q)\varepsilon_0 L^3 Q \mathbf{E}_0/l \qquad 37.$$

with l the average spacing between the elementary charges along the polyelectrolyte contour and $l \leq Q$. Equation 37 is almost identical to Eq. 34, which again is not startling. Oosawa (131) and others (132, 65) tried tampering with the assumption that the associated counterions have no interactions. Oosawa assumed Debye-Hückel screening due to counterions within the adsorbed sheath. This dubious procedure has been severely criticized by Fixman (78, 79).

Equations similar to Eq. 29a, but not within the Poisson-Boltzmann approach, have been postulated by three groups of workers. Van Dijk et al (133, 134) derived an equation almost identical to Eq. 33 by postulating a flux of counterions into the bulk proportional to the deviation from equilibrium of the number of associated counterions per unit length. Their length parameter γ_D^{-1} analogous to γ^{-1} had to be estimated numerically. It is in fact also approximately proportional to κ^{-1} (see their figure 3) and of almost the same magnitude as Fixman's γ^{-1}. A more detailed but nevertheless approximate treatment by Rau & Charney (135) leads to identical results. In effect, their very complicated function F corresponds all but exactly to the function in brackets in Eq. 33. Their conclusion that the polarizability is proportional to $L^{1.85}\kappa^{-1.15}$ ($\simeq L^2\kappa^{-1}$) in regions where $0.2 \lesssim \kappa L \lesssim 2$ is also applicable to Eq. 33 and van Dijk's result (133). Meyer et al (136–138) generalized Eq. 29 by including association-dissociation kinetics in the manner of Berg & Blomberg (139) as well as unscreened repulsive interactions. The Poisson-Boltzmann treatment does not yield these effects but, of course, it is deficient on a very local scale in the vicinity of the polyion surface. Unfortunately, the numerically calculated dielectric relaxation (136–138) depends on as yet vague, adjustable parameters. The numerical work of Wesenberg & Vaughan (138) shows the enormous influence of postulated unscreened interactions between counterions on the dielectric behavior of polyelectrolytes. Their calculations represent an upper bound on this effect (which may conceivably be greatly overestimated).

Whereas Fixman's formula for μ' (see Eq. 33) is very similar to van Dijk's expression (133), the relaxation behavior of μ' for the respective models is quite disparate. In the theories of double layer polarization this relaxation is determined by electroneutral concentration fluctuations (see Eq. 23), the characteristic time scale being $L^2\beta/k_B T$ for rods. In van Dijk's analysis (133) the time dependent behavior is solely derived from an equation analogous to Eq. 29a, including the time derivative term. In that case the ion flux out of the associated counterion layer dominates the ion relaxation if $\gamma_D L \gg 1$. This fixes the time scale as $\beta/k_B T \gamma_D^2$ as they calculate explicitly. However, in the model used, somewhere outside the associated counterion layer, regions of nonzero charge density should apparently be built up. Fixman (78) reasoned that at not too high frequencies this just cannot occur. The ion flow out of the double layer can only take place in such a way that electroneutrality exists outside the double layer. Hence, the rate determining step is the relaxation of electroneutral concentration fluctuations (Eq. 30). A similar discussion applies to Oosawa's relaxation mechanism (131) or its variants (65, 132).

FLEXIBLE POLYELECTROLYTES   In reality, flexible linear polyelectrolytes are not usually rod-like and solutions of them are invariably semidilute (14, 15). Hence, the above-mentioned theories are very crude approximations in general. Moreover, at least two dispersion regions are found experimentally (see previous section). An *ad hoc*, semiempirical approach to explaining the experiments is to assign some characteristic dimension to D and see how far one gets. Thus, the root-mean square extension length R of a polyelectrolyte has been associated with the first dispersion region by van der Touw & Mandel (76). In other words, the relaxation time is approximated by $\tau_1 \simeq R^2\beta/k_BT$. Because the concentration behavior of τ_1 and τ_2 (the relaxation time of the second region) seems to be similar (59), van der Touw & Mandel (76) also postulated that the length defined by $\tau_2 \simeq b^2\beta/k_BT$ is a basic scale pertaining to the macromolecules. Specifically, they speculated that it could be the average distance between potential barriers along the chains. Subsequently Odijk (14) took their approach one step further, conjecturing that b might be identified with the correlation length ξ, i.e. the average distance between entanglements in the semidilute solution. From scaling theory (14), definite predictions can be given for R and ξ. What is perhaps rather startling is that this certainly crude treatment of polyelectrolyte dielectrics yields *a priori* estimates of the parameters A and B (see previous section) that are in reasonable agreement with certain experimental results.

CONCLUSIONS

As has been emphasized several times in this review, the situation regarding dielectric properties of aqueous polyelectrolyte solutions is far from satisfactory. Experimental results seem to be rather fragmentary and sometimes contradictory, and progress is seriously hampered by experimental difficulties that could be overcome if recent technical developments were fully exploited. What is needed is a series of systematic investigations over a broad frequency range, extending from at least 1 Hz to 10^7–10^8 Hz, in which, one at a time, several parameters of a given system are varied.

Ideally these parameters should include molar mass (of well specified, homodisperse samples), concentration, nature and charge of the counterions, concentration of additional salt, and temperature; but a systematic study of the influence of at least the first two is presently imperative. In choosing the range of composition, recent ideas about concentration effects in polyelectrolyte solutions should be kept in mind. Experiments with more or less flexible polyelectrolytes should, if possible, be supplemented by parallel investigations in the same sense of rigid charged particles of spherical and other well defined shapes. This should help test existing theories at low concentrations and also help to find indications for the extension of the latter to higher volume fractions and other geometries.

The present gap between theory and experiment should be reduced from the theoretical side. Extension of the actual approaches to systems that are more readily accessible to experimental investigations would be a very valuable development. This would apply to treatments in which some polyion-polyion interactions are taken into account, particularly at lower ionic strengths with the subsequent abandonment of the thin-double layer approximation. In parallel, it is important to include chain flexibility, e.g. by using theoretical models based on a wormlike coil approach. We hope a (partial) solution to the problems posed above will be discussed in a future review.

Literature Cited

1. Stell, G., Patey, G. N., Høye, J. S. 1981. *Adv. Chem. Phys.* 48:183–328
2. Deutch, J. M. 1977. *Faraday Symp. Chem. Soc.* 11:26–32
3. Wolynes, P. G. 1980. *Ann. Rev. Phys. Chem.* 31:345–76
4. Mandel, M., van der Touw, F. 1974. In *Polyelectrolytes*, ed. E. Sélégny, M. Mandel, U. P. Strauss, pp. 285–300. Dordrecht/Boston: Reidel
5. Mandel, M. 1977. *Ann. NY Acad. Sci.* 303:74–87
6. Mandel, M. 1982. In *Dynamic Aspects of Biopolyelectrolytes and Biomembranes*, ed. F. Oosawa, pp. 235–44. Tokyo: Kodansha
7. Sorriso, S., Surowiec, A. 1982. *Adv. Mol. Relax. Interact. Processes* 22:259–79
8. Fixman, M. 1979. *J. Chem. Phys.* 70:4995–5003
9. Odijk, T. 1977. *J. Polym. Sci. Polym. Phys. Ed.* 15:477–83
10. Skolnick, J., Fixman, M. 1977. *Macromolecules* 10:944–48
11. Odijk, T. 1982. In *Ionic Liquids, Molten Salts and Polyelectrolytes*, ed. K. H. Bennemann, F. Brouers, D. Quitmann, pp. 184–98. Berlin: Springer
12. Yamakawa, H. 1971. *Modern Theory of Polymer Solutions*, New York/Evanston/San Francisco/London: Harper & Row
13. de Gennes, P. G. 1979. *Scaling Concepts in Polymer Physics*. Ithaca/London: Cornell Univ. Press
14. Odijk, T. 1979. *Macromolecules* 12:688–93
15. de Gennes, P. G., Pincus, P., Velasco, R. M., Brochard, F. 1976. *J. Phys. Paris* 37:1461–73
16. Hayter, J., Jannick, G., Brochard-Wyart, F., de Gennes, P. G. 1980. *J. Phys. Lett. Paris* 41:L451–54
17. Koene, R. S., Mandel, M. 1983. *Macromolecules* 16:220–27
18. Koene, R. S., Nicolai, T., Mandel, M. 1983. *Macromolecules* 16:227–31
19. Koene, R. S., Nicolai, T., Mandel, M. 1983. *Macromolecules* 16:231–36
20. Koene, R. S., Mandel, M. 1983. *Macromolecules* 16:973–78
21. Mandel, M. 1983. *Eur. Polym. J.* 19:911–18
22. Schwan, H. P. 1963. In *Physical Techniques in Biological Research*, ed. W. L. Nastuk, Chap. 6, p. 6B. New York: Academic
23. van Beek, W. M., van der Touw, F., Mandel, M. 1976. *J. Phys. E* 9:385–91
24. Hall, D. G., Cole, R. H. 1981. *J. Phys. Chem.* 85:1065–69
25. Cole, R. H., Gross, P. M. 1949. *Rev. Sci. Instr.* 20:252–58
26. Mandel, M., Jung, P. 1952. *Bull. Soc. Chim. Belg.* 61:553–68
27. Schwan, H. P., Sittel, K. 1953. *Trans. Am. Inst. Elect. Eng.* 72:114–20
28. van der Touw, F., de Goede, J., van Beek, W. M., Mandel, M. 1975. *J. Phys. E* 8:840–44
29. Mandel, M. 1956. *Bull Soc. Chim. Belg.* 65:308–42
30. Scheider, W. 1975. *J. Phys. Chem.* 79:127–36
31. Oncley, J. L. 1938. *J. Am. Chem. Soc.* 60:1115–23
32. Oncley, J. L. 1940. *J. Phys. Chem.* 44:1103–13

33. Ferry, J. D., Oncley, J. L. 1941. *J. Am. Chem. Soc.* 63:272–78
34. Johnson, J. F., Cole, R. H. 1951. *J. Am. Chem. Soc.* 73:4536–40
35. Sasaki, S., Ishikawa, A., Hanai, T. 1981. *Biophys. Chem.* 14:45–53
36. Shaw, T. H. 1942. *J. Chem. Phys.* 10:609–17
37. Mandel, M., Jenard, A. 1958. *Bull. Soc. Chim. Belg.* 67:575–98
38. Schwan, H. P. 1968. *Ann. NY Acad. Sci.* 148:191–98
39. van der Touw, F., Mandel, M. 1971. *Trans. Faraday Soc.* 67:1336–42
40. van der Touw, F., Mandel, M., Honijk, D. D., Verhoog, H. G. F. 1971. *Trans. Faraday Soc.* 67:1343–54
41. Umemura, S., Hayakawa, R., Wada, Y. 1980. *Biophys. Chem.* 11:317–20
42. Hanss, M., Guermonprez, R. 1966. *J. Chim. Phys.* 63:663–68
43. Hanss, M., Bernengo, J. C. 1973. *Biopolymers* 12:2151–59
44. Berberian, J. G., Cole, R. H. 1969. *Rev. Sci. Instrum.* 40:811–19
45. Tung, M. S., Molinari, R. J., Cole, R. H., Gibbs, J. H. 1977. *Biopolymers* 16:2653–69
46. Hayakawa, R., Kanda, H., Sakamoto, M., Wada, Y. 1975. *Jpn. J. Appl. Phys.* 14:2039–52
47. Umemura, S., Sakamoto, M., Hayakawa, R., Wada, Y. 1979. *Biopolymers* 18:25–34
48. Nakamura, H., Husimi, Y., Wada, A. 1977. *Jpn. J. Appl. Phys.* 16:2301–10
49. Nakamura, H., Husimi, Y., Wada, A. 1981. *J. Appl. Phys.* 52:3053–58
50. Sakamoto, M., Kanda, H., Hayakawa, R., Wada, Y. 1976. *Biopolymers* 15:879–92
51. Jugner, G., Jugner, I., Allgen, L. G. 1949. *Nature* 163:849–50
52. Allgen, L. G. 1950. *Acta Physiol. Scand. Suppl.* 22:76
53. Dintzis, H. M., Oncley, J. L., Fuoss, R. M. 1954. *Proc. Natl. Acad. Sci. USA* 40:62–67
54. Allgen, L. G., Roswall, S. 1954. *J. Polym. Sci.* 12:229–36
55. Mandel, M., Jenard, A. 1963. *Trans. Faraday Soc.* 59:2158–69
56. Mandel, M., Jenard, A. 1963. *Trans. Faraday Soc.* 59:2170–77
57. Sachs, S. B., Raziel, A., Eisenberg, H., Katchalsky, A. 1969. *Trans. Faraday Soc.* 65:77–90
58. Minakata, A., Imai, N. 1972. *Biopolymers* 11:329–46
59. van der Touw, F., Mandel, M. 1974. *Biophys. Chem.* 2:231–41
60. Muller, G., van der Touw, F., Zwolle, S., Mandel, M. 1974. *Biophys. Chem.* 2:242–54
61. van Beek, W. M., Odijk, T., van der Touw, F., Mandel, M. 1976. *J. Polym. Sci. Polym. Phys. Ed.* 14:773–81
62. Paoletti, S., van der Touw, F., Mandel, M. 1978. *J. Polym. Sci. Polym. Phys. Ed.* 16:641–51
63. Vreugdenhil, T., van der Touw, F., Mandel, M. 1979. *Biophys. Chem.* 10:67–80
64. Minakata, A. 1972. *Biopolymers* 11:1567–82
65. Minakata, A. 1977. *Ann. NY Acad. Sci.* 303:107–20
66. Takashima, S. 1963. *J. Mol. Biol.* 7:455–67
67a. Takashima, S. 1966. *J. Phys. Chem.* 70:1372–80
67b. Takashima, S. 1966. *Biopolymers* 4:663–67
68. Sakamoto, M., Hayakawa, R., Wada, Y. 1978. *Biopolymers* 17:1507–12
69. Molinari, R. J., Cole, R. H., Gibbs, J. H. 1981. *Biopolymers* 20:977–90
70. Cole, K. S., Cole, R. H. 1941. *J. Chem. Phys.* 9:341–51
71. Allgen, L. G., Roswall, S. 1957. *J. Polymer Sci.* 23:635–50
72. Allgen, L. G. 1954. *J. Polym. Sci.* 14:281–86
73. Takashima, S. 1963. *Biopolymers* 1:171–82
74. Sakamoto, S., Hayakawa, R., Wada, Y. 1979. *Biopolymers* 18:2769–82
75. Takashima, S. 1967. *Biopolymers* 5:899–913
76. van der Touw, F., Mandel, M. 1974. *Biophys. Chem.* 2:218–30
77. Dukhin, S. S., Shilov, V. N. 1974. *Dielectric Phenomena and the Double Layer in Disperse Systems and Polyelectrolytes*. Jerusalem: Keter
78. Fixman, M. 1980. *J. Chem. Phys.* 72:5177–86
79. Fixman, M. 1980. *Macromolecules* 13:711–16
80. de Lacey, E. H. B., White, L. R. 1981. *J. Chem. Soc. Faraday Trans. 2* 77:2007–39
81. O'Brien, R. W. 1982. *Adv. Colloid Interface Sci.* 16:281–320
82. Chew, W. C., Sen, P. N. 1982. *J. Chem. Phys.* 77:4683–93
83. Fixman, M. 1983. *J. Chem. Phys.* 78:1483–91
84. Hinch, E. J., Sherwood, J. D., Chew, W. C., Sen, P. N. 1984. *J. Chem. Soc. Faraday Trans. 2*. In press
85. Saville, D. A. 1977. *Ann. Rev. Fluid Mech.* 9:321–36
86. Overbeek, J. T. G. 1943. *Kolloidchem. Beihefte* 54:287–364
87. Wiersema, P. H., Loeb, A. L., Overbeek,

87. J. T. G. 1966. *J. Colloid Interface Sci.* 22: 78–99
88. O'Brien, R. W., White, L. R. 1978. *J. Chem. Soc. Faraday Trans. 2* 74: 1607–26
89. Derjaguin, B. V., Dukhin, S. S. 1974. In *Surface and Colloid Science*, ed. E. Matijevic, 7: 273–335. New York: Wiley
90. de Groot, S. R., Mazur, P. 1969. *Non-Equilibrium Thermodynamics*. Amsterdam: North Holland
91. Onsager, L. 1927. *Phys. Z.* 28: 277–98
92. Debye, P., Falkenhagen, H. 1928. *Phys. Z.* 29: 401–26
93. Lifshitz, E. M., Pitaevskii, L. P. 1981. *Physical Kinetics*. Oxford: Pergamon
94. Stigter, D. 1978. *J. Phys. Chem.* 82: 1417–23
95. Stigter, D. 1978. *J. Phys. Chem.* 82: 1424–29
96. Sherwood, J. D. 1982. *J. Chem. Soc. Faraday Trans. 2* 78: 1091–1100
97. van der Drift, W. P. J. T., de Keizer, A., Overbeek, J. T. G. 1979. *J. Colloid Interface Sci.* 71: 67–78
98. Stigter, D. 1979. *J. Phys. Chem.* 83: 1663–70
99. O'Brien, R. W. 1981. *J. Colloid Interface Sci.* 81: 234–48
100. O'Brien, R. W. 1983. *J. Colloid Interface Sci.* 92: 204–16
101. Saville, D. A. 1983. *J. Colloid Interface Sci.* 91: 34–50
102. Stigter, D. 1980. *J. Phys. Chem.* 84: 2758–62
103. Stigter, D. 1982. *J. Phys. Chem.* 86: 3553–58
104. Saville, D. A. 1982. *Adv. Colloid Interface Sci.* 16: 267–79
105. Russel, W. B. 1978. *J. Fluid Mech.* 85: 673–83
106. Russel, W. B. 1978. *J. Fluid Mech.* 85: 209–32
107. Sherwood, J. D. 1980. *J. Fluid Mech.* 101: 609–29
108. Sherwood, J. D. 1981. *J. Fluid Mech.* 111: 347–66
109. Hinch, E. J. 1981. In *Continuum Models of Discrete Systems 4*, ed. O. Brulin, R. K. T. Hsieh, pp. 411–22. Amsterdam: North Holland
110. Hinch, E. J., Sherwood, J. D. 1983. *J. Fluid Mech.* 132: 337–47
111. Landau, L. D., Lifshitz, E. M. 1959. *Fluid Mechanics*. Reading, Mass: Addison Wesley
112. Landau, L. D., Lifshitz, E. M. 1960. *Electrodynamics of Continuous Media*. Reading, Mass: Addison–Wesley
113. Dukhin, S. S. 1974. In *Surface and Colloid Science*, ed. E. Matijevic, 7: 1–47. New York: Wiley
114. Yamakawa, H., Fujii, M. 1973. *Macromolecules* 6: 407–15
115. Schellman, J. A., Stigter, D. 1977. *Biopolymers* 16: 1415–34
116. Batchelor, G. K. 1970. *J. Fluid Mech.* 41: 545–70
117. Batchelor, G. K. 1974. *Ann. Rev. Fluid Mech.* 6: 227–55
118. Nayfeh, A. 1973. *Perturbation Methods*. New York: Wiley
119. van Dyke, M. 1975. *Perturbation Methods in Fluid Mechanics*. Stanford: Parabolic Press
120. Chew, W. C., Sen, P. N. 1982. *J. Chem. Phys.* 77: 2042–44
121. Fixman, M., Jagannathan, S. 1983. *Macromolecules* 16: 685–99
122a. Bedeaux, D., Kapral, R. 1983. *J. Chem. Phys.* 79: 1783–88
122b. Kayser, R. F., Hubbard, J. B. 1983. *Phys. Rev. Lett.* 51: 79–82
123. Schwan, H. P., Schwarz, G., Maczuk, J., Pauly, H. 1962. *J. Phys. Chem.* 66: 2626–35
124. Fixman, M., Jagannathan, S. 1981. *J. Chem. Phys.* 75: 4048–59
125. Teubner, M. 1982. *J. Chem. Phys.* 76: 5564–73
126. Happel, J., Brenner, H. 1965. *Low Reynolds Number Hydrodynamics*. Englewood Cliffs, NJ: Prentice Hall
127. Schwarz, G. 1962. *J. Phys. Chem.* 66: 2636–42
128. Schurr, J. M. 1964. *J. Phys. Chem.* 68: 2407–13
129. Mandel, M. 1961. *Mol. Phys.* 4: 489–96
130. Manning, G. S. 1969. *J. Chem. Phys.* 51: 924–33
131. Oosawa, F. 1970. *Biopolymers* 9: 677–88
132. McTague, J. P., Gibbs, J. H. 1966. *J. Chem. Phys.* 44: 4295–301
133. van Dijk, W., van der Touw, F., Mandel, M. 1981. *Macromolecules* 14: 792–4
134. van Dijk, W., van der Touw, F., Mandel, M. 1981. *Macromolecules* 14: 1554–58
135. Rau, R. C., Charney, E. 1981. *Biophys. Chem.* 14: 1–9
136. Meyer, P. I., Vaughan, W. E. 1980. *Biophys. Chem.* 12: 329–39
137. Meyer, P. I., Wesenberg, G. E., Vaughan, W. E. 1981. *Biophys. Chem.* 13: 265–73
138. Wesenberg, G. E., Vaughan, W. E. 1983. *Biophys. Chem.* 18: 381–90
139. Berg, O. G., Blomberg, C. 1976. *Biophys. Chem.* 4: 367–81

STATE-RESOLVED MOLECULAR REACTION DYNAMICS[1]

Stephen R. Leone[2]

Joint Institute for Laboratory Astrophysics, National Bureau of Standards and University of Colorado, and Department of Chemistry, University of Colorado, Boulder, Colorado 80309

INTRODUCTION

Gas phase molecular reaction dynamics is a mature field, but one that continues to offer exciting new perspectives on the fundamental nature of chemical transformations. Chemists have in their vocabulary such familiar concepts as "early" and "late" reaction barriers, the "harpoon mechanism," and "direct" and "complex" reaction dynamics. We also have a modicum of understanding about which forms of energy, i.e. vibrational, translational, electronic, or rotational, will successfully carry a reaction to completion. Much of this understanding comes from ingenious state-resolved experimental measurements on elementary chemical reactions, coupled with the excellent insight provided by detailed theoretical calculations.

To give the reader an appreciation for the magnitude of the field, a comprehensive review in 1979 contained 1144 references (1). Numerous textbooks are also available on the subject (2–4). A number of other recent reviews are of special relevance. These include excellent chapters on the reactivity of selectively excited states (5, 6), energy disposal in simple reactions (7), and thorough introductions to the experimental methods (1, 8). For theoretical discussions, the reader is referred to the recent articles in the book series edited by Henderson (9) and to earlier excellent reviews (10, 11).

Because so much has already been discovered and said about reaction dynamics, it is important to try to identify significant new developments

---

[1] The US Government has the right to retain a nonexclusive, royalty-free licence in and to any copyright covering this paper.
[2] Staff Member, Quantum Physics Division, National Bureau of Standards.

and results. This review focuses on recent experimental investigations in state-resolved molecular reaction dynamics, involving primarily work carried out from 1980 to 1983. A common trend is toward investigations that involve a higher degree of sophistication to learn about geometrical and orientation effects, bimodal state distributions that result from multiple reaction pathways, resonance effects, and the competition between reactive and inelastic channels. In addition, well-known techniques are being applied to many new systems, including investigation of reaction dynamics on catalytic metal surfaces, the study of atom-radical reactions and ion-molecule reactions, and determination of product branching in reactions that have a manifold of pathways. As one might suspect, some old controversies have been settled, but some new ones are under way.

In the following sections, recent results are grouped and discussed according to the types of excitation principally involved, including vibrational, rotational, translational, and electronic excitation. Each section considers the results of excitation in either the reagents or products. The two last parts consider orientation effects in reagents or products and new methods to interrogate the transition states of reactions.

VIBRATIONAL EXCITATION

Infrared Chemiluminescence

Infrared chemiluminescence is used extensively to determine the vibration-rotation states of products in simple chemical reactions. Atom-polyatomic reactions have received much attention recently. In particular there have been numerous studies of F atom reactions with polyatomics such as CH_4, NH_3, H_2CO, CH_3OH, CH_3OD, $DCOOH$, and B_2H_6 to form vibrationally excited HF or DF products. If the reaction is a direct H atom abstraction, the fraction of energy partitioned into vibration, f_v, is often uniformly a high value, approximately 50%, and the states are highly inverted. If the reaction involves a long-lived attack on some other functional group, then the vibrational distribution can show evidence for statistical partitioning of the energy and the f_v is much lower. This latter type of reaction might be considered an addition/elimination mechanism.

The classic example that shows both a direct and indirect behavior is the reaction of F with DCOOH (12). The DF product is observed to be highly excited, indicative of a direct abstraction, whereas the HF product is formed with a statistical distribution in vibration at a temperature of 4300 K, attributed to a long-lived attack at the carboxylic acid site. In an analogous type of study on the reaction $F + CD_3COOH$ (13), abstraction at the CD_3 group produces a highly excited diatomic product, whereas abstraction at the COOH group is again nearly statistical. The results for $F + CH_3OD$

and other substituted alcohols show much less of a difference in the abstraction at the two different sites (14–16). Indeed there is even some question in the literature as to which product, the DF or HF, is more highly excited. Nevertheless, it does not appear that abstraction at the OH group necessarily involves a strongly attractive, long-lived intermediate.

The values for the f_v's in some reactions of substituted hydrocarbons are very similar to that for an unsubstituted species such as CH_4. An excellent example is the f_v for the series (17): CH_4 (0.56), CH_3Cl (0.50), CH_3Br (0.58), CH_3I (0.60). The uniformly high amount of vibrational excitation in each case is attributed to the fact that nearly identical forces must be involved and to the effect of the light atom transfer: a high f_v typically results from the transfer of a light atom between two heavier masses. This latter effect seems to be an overriding consideration in nearly all H atom abstraction reactions. It has been suggested that the greatly diminishing values of f_v first observed for the series CH_3Cl (0.50), CH_2Cl_2 (0.23), $CHCl_3$ (0.13) may be due to long-lived encounters that occur with the increasingly bulky radicals and remove some of the HF vibrational excitation before the products can separate (17). However, these results are now in question because of the discovery of obscuring secondary atom-radical reactions (see below). When the equilibrium geometry of the polyatomic radical fragment is very different from the geometry in the parent molecule, a radical "stabilization" energy is required that can effectively reduce the amount of excitation energy available to the diatomic product. This has been clearly observed in F atom reactions with CH_3CHO and C_6H_5CHO (18), where the highest levels excited in the HF product are well below the available exothermicity.

There are a number of other reactions in which strong attractive forces or the possibility of attack on a lone electron pair on the polyatomic molecule are definitely thought to proceed by an addition/elimination mechanism. The result of this type of mechanism is a vibrational state distribution that is not highly inverted. Most notable among these is the reaction of F with NH_3 (19–22). There is not yet complete agreement on the exact initial vibrational distribution, because of complications such as rapid vibrational relaxation of HF by NH_3 and secondary reaction of F with NH_2. However, in general the distribution in HF for $v = 1:2:3$ is thought to be decreasing, 0.6:0.4:0.0. Other examples of similar behavior include the reactions of $Cl + PH_3$ (23) and $F + HNCO$ (22).

Sometimes the conclusion concerning a direct or indirect type of reaction appears to be somewhat arbitrary. For example, the reaction $Cl + GeH_4$ (23) exhibits a cold $v = 1:2:3$ distribution of 0.79:0.19:0.0, and yet the authors favor a direct abstraction mechanism based on the extremely rapid reaction rate constant. Perhaps MacDonald et al (14) summed up the dilemma very well when they remarked that observation of substantial vibrational excitation does not necessarily guarantee direct reaction dynamics. There are reasons that the converse will also be true, for example, if in forming the radical fragment there are geometry changes that necessitate substantial vibrational excitation of the radical fragment.

Since HF vibrational emission is relatively easy to detect, there previously has not been much work to measure the internal excitation of the other fragment in the reaction. Recently there has been an attempt to do this (24). The HS(DS) radical in the $F + H_2S(D_2S)$ reactions and the CH_3O radical in the $F + CH_3OH$ reaction have been probed by laser-induced fluorescence. In both reactions it is concluded that very little energy is partitioned into the radical fragment, thus confirming the direct nature of the H atom abstractions in these cases.

What is most interesting about the work on polyatomic atom abstraction reactions is the discovery of efficient secondary reactions of F atoms with the polyatomic radical products (18, 21, 22, 25–29). Although these secondary reactions present a difficulty when one wishes to obtain the nascent vibrational state distributions in the main parent reaction, they offer exciting new possibilities to study the product distributions involved in atom-radical interactions. Because the atom-radical systems involve two open shell species, the resulting product state distributions may reflect the strong attractive character and the chemically different sites of attack involved in the reaction intermediates. Secondary radical reactions have been investigated by incorporating a prereaction zone into the infrared chemiluminescence apparatus in order to generate the radicals of interest. In this way it has been shown that the reaction of $F + OH$ produces a statistical distribution of HF vibrational states (25), suggesting complex formation rather than direct abstraction. Similarly, the highly exothermic reactions of $F + C_2H_3$ and $F + HCO$ produce vibrational distributions that agree with statistical phase space calculations (27, 29). It is clear that these results are just the beginning of what can be learned about long-lived complex formation in radical-radical reactive collisions.

Over the past years there has been a growing controversy concerning a number of vibrational state distributions obtained in flow tube devices versus those determined in low pressure "arrested relaxation" apparatuses. The problem is discussed in detail in a number of publications (16, 30, 31). Vibrational populations obtained in the low pressure reactors for relatively simple reactions like $F + HBr$ are found to change dramatically with flow rates, and they exhibit very little excitation at the lowest flows achievable. In the flow tube environment or at higher flows of reagents in the low pressure reactors the distributions are much more highly excited. In many of the reactions in which discrepancies have been noted, explanations for the differences can be found among the possibilities that there is some rapid vibrational relaxation of HF(v) or a substantial occurrence of secondary reactions of excess F atoms with radical products. The focal point for the controversy, then, has come down to a reaction like F+HBr, which is simple enough that these other problems should not interfere.

A summary of all of the results for F+HBr has been published (32), and the overwhelming majority favors the more highly excited distribution with an f_v of 0.6. There have been numerous suggested explanations for why the low pressure reactor gives an anomalously cold distribution. These explanations range from various forms of wall deactivation, through the formation of complexes in the nozzle mixing zone, to the possibility that the reaction really does produce a very cold distribution and that the flow tube measurements are wrong because of some unusual collisional process that interferes with the normal reaction. Reactor wall problems have been ruled out, and it is unlikely that there is really some microscopic oddity that causes the basic reaction to exhibit different distributions in different environments. There is evidence that the problem might lie with the specific types of inlet arrangements in the low-pressure reactors and the possibility that atoms may back-diffuse into the molecular reagent inlet, react there, and produce relaxed products because of wall collisions in the inlet tube (D. J. Donaldson, private communication). Until this problem is resolved, the evidence seems to suggest that the flow tube results are somewhat more reliable at this time.

Infrared chemiluminescence methods have recently been applied to the study of thermal energy ion-molecule reactions in a flow tube environment. An example is a reaction of the type (34–36):

$$F^- + HX \rightarrow HF(v) + X^- \quad (X = Cl, Br, I).$$

The similarity of these ion reactions to their neutral counterparts, F+HCl, HBr, HI, allows some interesting conclusions to be drawn concerning the influence of long-range attractive forces on simple atom abstractions. The ion reactions have strong, long-range ion-dipole and ion-induced dipole forces in the entrance and exit channels, which are absent in the neutrals. These forces could significantly alter the vibrational state distributions compared to their neutral counterparts. In the F, F^- +HCl, HBr, HI series, the exothermicities are nearly identical for each corresponding reaction, and the mass combinations are essentially identical. The results (35) show that in each case the vibrational distribution of the ion reaction is slightly colder (Figure 1), but that the partitioning of energy into vibration is typically very high (f_v^{ion} = 0.4–0.5, $f_v^{neutral}$ = 0.5–0.6). The high vibrational excitation, in spite of the long-range attractive forces in the ion case, is attributed to the fact that a light H atom is being transferred and that the rather large exothermicity and small number of degrees of freedom in the three-atom system preclude a very long-lived complex. However, since the data for another three-atom system, F+OH, do show a statistical distribution of states (25), there are clearly circumstances in which the attractive potential can lead to complex formation and indirect reaction dynamics. A key consideration may be the availability of an initial strong binding site somewhere on the molecule that takes precedence over the direct abstraction pathway.

The infrared chemiluminescence method has been applied to the study of several other new kinds of ion reactions. Reactions of CN^- with HX species yields substantial excitation in the CH stretch mode of the HCN product, and in the case of $CN^- + HI$, excited HNC is detected (36). Product state vibrational excitation has been observed for the first time for associative detachment reactions, for example $F^- + H \to HF(v) + e^-$ (37–39). The results show a propensity to populate the highest accessible vibrational states ($f_v = 0.7$); this indicates that the nonadiabatic transfer to the HF plus free electron continuum occurs on the incoming, outer part of the HF^- potential. The reaction of $SF_6^- + H$ forms SF_5^- and vibrationally excited HF with an f_v of 0.4 (40). This suggests a direct reaction mechanism, with very

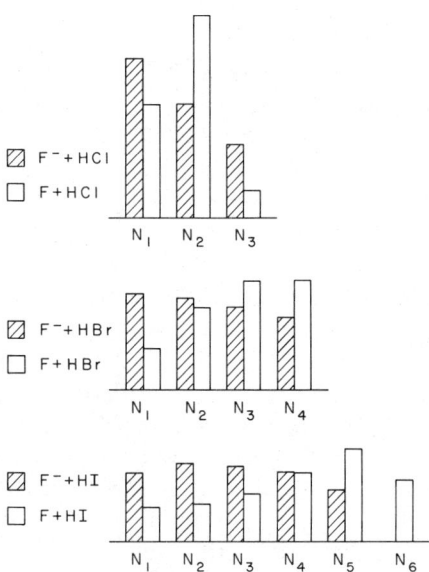

Figure 1 Comparison of the vibrational distributions for the ion reactions, $F^- + HCl$, HBr, HI, with their corresponding neutral analogues, F + HCl, HBr, HI. All reactions populate HF(v) up to their limit of exothermicity, yet the products of the ion reactions are noticeably less excited. Reproduced with permission from (35).

little influence by the long-range ion-dipole forces in the exit channel. A bimodal distribution in the vibrationally excited NO^+ product of the $N^+ + O_2$ reaction suggests that this reaction forms a substantial fraction of electronically excited atomic products at thermal energies (41).

Whereas discrepancies do exist between the low pressure arrested relaxation and flow tube methods for vibrational distributions of F atom reactions, there do not appear to be similar discrepancies for reactions of H atoms (42, 43). Unfortunately, a disconcerting result has been obtained with spectral and time-resolved mapping of the DF product for the $D + F_2$ reaction (44), which suggests that conventional measurements (45) may have seriously underestimated the extent of the high vibrational levels ($v = 10$–13) populated in the $DF(v)$. The authors (44) have suggested that estimates of residence times in low pressure reactors may be too short and that radiative cascade or other relaxation mechanisms in conventional measurements may be a more serious problem than thought. The quality of the signal levels in this first time-resolved mapping experiment leave something to be desired; it is also unlikely that the estimates of residence times are that far wrong. The authors have recently also found other problems with their data that make this result less reliable (D. C. Tardy, private communication). Their more recent data are in better agreement with the other results. Nevertheless, the problems brought to light by this study will have to be reckoned with in future investigations of very high vibrational levels, where unusually rapid relaxation processes could take place.

Several other results of a general nature should be noted. Experiments have been performed using infrared chemiluminescence detection for reactions with elevated translational kinetic energies, for example, $H + Cl_2$, SCl_2, and PCl_3 at 0.45 eV (46). It is found that the excess translational energy is predominantly transformed into product translation and rotation. This effect has long been known to be generally true. The thermal reaction of H with SF_5 gives a vibrational distribution that declines with increasing v, which is taken as indicative of formation of an HSF_5 complex followed by elimination (47). Reactions of O atoms with HI and GeH_4 bear close resemblance to most F atom abstraction reactions (48). The reactions of $H + BrF$ and $H + IF$ show bimodal vibrational distributions, indicative of direct and migratory pathways, to form the HF product (49). There are a few recent studies to determine the energy disposal in the polyatomic products of reactions such as $CH_3 + F_2$ (50) and $CH_3 + Cl_2$ and Br_2 (51). Approximately 50–60% of the available energy goes into vibrational excitation of the $CH_3 X$ products, but the vibrational energy is rapidly randomized among the vibrational modes before detection. In the addition/elimination reactions of $F + C_2D_3Br$ and C_2D_3Cl, the energy in the C_2D_3F product is also statistically distributed, indicating a long-lived complex intermediate (52).

In the past few years, infrared chemiluminescence has been applied for the first time to study surface catalytic reaction dynamics (53–55). The reaction of surface adsorbed CO and O atoms on platinum was found to produce gaseous CO_2 product molecules with a vibrational temperature of about 2000 K (53), far above the surface temperature. Similarly, the reaction of C and O on platinum produces $CO(v)$ up to $v = 7$ (Figure 2) (55). These studies, together with earlier investigations using electron beam excitation/fluorescence to probe the vibrational and rotational states of the N_2 product in the N atom recombination on an iron surface (56, 57), are the very first attempts at product state analysis of reactive events on surfaces. Although this work is in a very primitive state, it is clear that it offers a tremendous future for the study of reactive dynamics involving species in novel, constrained geometries.

Laser-Induced Fluorescence

Laser-induced fluorescence has been used extensively for analysis of vibrational and rotational states, especially for the investigation of reactions that produce OH, metal halides, the halogens, and more recently

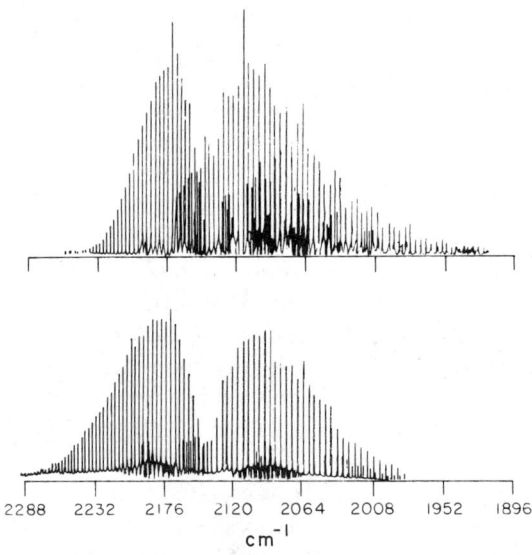

Figure 2 Emission from vibrationally excited CO produced in the reaction of C+O on a platinum surface (*upper*). Emission from thermally accommodated $CO(v)$ on collision with the same surface for comparison (*lower*). Reproduced with permission from (55).

for ion-molecule reactions that yield products such as N_2^+ and CO^+. One popular type of study involves the reactions of $O(^1D)$ atoms, which are generated by laser photolysis of O_3. Laser-induced fluorescence is used to detect the vibrational, rotational, spin, and lambda doublet components in the OH product upon reactions of $O(^1D)$ with H_2, NH_3, HCl, H_2O, hydrocarbons, NO_2, alcohols, and thiols. Reactions of $O(^1D)$ can proceed via insertion or abstraction. In principle, the vibrational and rotational state distributions should shed some light on the nature of the mechanisms; however, an analysis of the available data tends to dispel that notion.

Combining infrared chemiluminescence results with laser-induced fluorescence measurements, a relatively complete vibrational distribution is obtained for the reaction of $O(^1D)$ with H_2 (58, 59). For OH $v = 0$–4 the populations are somewhat unusual, 0.22:0.20:0.23:0.22:0.13, which corresponds to an f_v of 0.5. A crossed molecular beam, angular and velocity measurement (60) has shown that there is a forward/backward symmetry in the products. Because of the symmetry involved in the C_{2v} approach of $O(^1D)$ with a homonuclear diatomic, the authors suggest that this is probably not indicative of a long-lived complex, but rather of a short-lived insertion mechanism. There is substantial speculation and theoretical work concerning the fact that there will be contributions of both direct abstraction and insertion pathways in this reaction, and since the vibrational and rotational state distributions (59) do not fit either a simple statistical or a simple abstraction mechanism, it is not possible to conclude that one mechanism dominates.

There is evidence that some other $O(^1D)$ reactions probably do proceed predominantly by insertion followed by fragmentation in a time short compared to energy randomization. For example, in the $O(^1D)+HCl$ reaction, the vibrational and rotational states are highly nonstatistical (61), and in particular the unusually large population in high rotational states is suggestive of a fragmentation of a short-lived HOCl complex, rather than a direct abstraction. The reaction of $O(^1D)$ with isotopically labeled $D_2^{18}O$ provides one of the most complete analyses of the product states of an $O(^1D)$ reaction (62–66). The most recent experiments summarize the selectivity of the energy disposal quite well (66). The ^{16}OD product, which comes from the newly formed bond, is found to have 41% of the available energy in its vibration. In contrast, the ^{18}OD fragment has only 2% of the available energy in its vibration. Both radicals, however, have nearly identical energy in rotation (17%). It has been speculated, therefore, that this reaction is a direct abstraction; however, it would appear that a mechanism of insertion to form HOOH* followed by a decomposition before energy randomization, much like a photofragmentation process, could also explain these data.

A similar lack of conclusive evidence for insertion versus abstraction exists for $O(^1D) + NH_3$. There are definite bimodal distributions in rotation and highly nonstatistical distributions in vibration, as well as the formation of electronically excited NH_2 and NH. These suggest that the reaction may have components of both insertion and abstraction (67–69). Similar bimodal distributions in vibration and rotation for $O(^1D)$ reactions with hydrocarbon compounds also prompt the conclusion that both mechanisms are occurring (70). However, a recent study on alcohols and thiols concluded that insertion into the COH group must predominate and that the lifetime of the intermediate is short compared to the rate of energy redistribution (71). Thus, after a tremendous amount of work on $O(^1D)$ reactions, it does not appear possible to give a consistent summary concerning the insertion-versus-abstraction behavior of this electronically excited atom.

A few other highlights concerning vibrational state information obtained by laser-induced fluorescence deserve to be mentioned. Reactions of F atoms with ICl, IBr, and I_2 to produce IF show dramatic bimodal vibrational distributions that are attributed to direct and indirect microscopic branching pathways (72). It has been suggested (72) that formation of the $^2P_{1/2}$, $^2P_{3/2}$ halogen atomic states may play a role in the two different pathways; however, for $F + I_2$, it has recently been demonstrated that the reaction produces only the ground state I atom (72a). An earlier report that obtained a very different vibrational distribution for $F + I_2$ appears to be less reliable (73). Reactions of $F + CH_3I$ and CF_3I at low kinetic energies produce statistical vibrational and rotational distributions in the IF product (74), which is in qualitative agreement with the observations of long-lived complexes in these reactions by molecular beam scattering experiments. The reaction of $Ba + CH_3Br$ shows a decrease in BaBr vibrational excitation with increasing kinetic energy of the reagents (75). The authors speculate that this may be due to a change in the reactive pathway, since in several other Ba reactions with HF, HCl, and HBr there is typically very little change in the vibrational excitation with kinetic energy. In a series of papers on $O(^3P)$ reactions with hydrocarbons, amines, and aldehydes (76–79), the OH product has been detected by laser-induced fluorescence. It is concluded that direct H atom abstraction is occurring for the hydrocarbons, but that for the unsaturated double bonds and amine groups, an addition mechanism followed by fragmentation or displacement is important.

Laser-induced fluorescence has been applied for the first time to study simple ion-molecule reaction dynamics at thermal energies. Vibrational distributions in the OH product of the $O^- + HF$ (80) reaction and the CO^+ product of the $N^+ + CO$ (81) reaction have been obtained in a flow tube and a single collision beam device, respectively. In the thermal energy charge transfer reaction of $N^+ + CO$, the vibrational distribution in $v = 0-2$ is essentially Franck-Condon, favoring the $v = 0$ state. Optical state detection provides a complementary technique to other methods, such as the determination of translational energies of the products in ion cyclotron resonance traps (82). There is tremendous interest in the mechanism of simple charge transfer reactions and whether the disposal of energy proceeds according to a Franck-Condon principle or an energy resonance consideration; the latter means that the excess energy is channeled into vibrational or electronic internal degrees of freedom. At least in this first example of $N^+ + CO$, there is moderately good agreement between the optical probe and ion cyclotron resonance results (81, 83). In some cases, for example in the $Ar^+ + CO$ reaction, the degree of vibrational selectivity reported for the CO^+ product by the cyclotron resonance method is remarkable (all in $v = 4$) (82). However, attempts to detect the $v = 4$ state by laser-induced fluorescence in the ion trap have been unsuccessful, possibly due to rapid relaxation of CO^+ by CO (84). Further work is needed to determine definitively the product state distributions in these reactions.

Vibrational Enhancement of Reactivity

A high degree of vibrational selectivity and energy resonance effects are observed in a number of ion-molecule reactions, especially charge transfer reactions. Many recent results are obtained by threshold photoelectron/product ion coincidence techniques. Thus it is possible to measure the cross sections of an ion reaction for a great many vibrational or electronic states of the reagents. The reactions that have been studied can be organized into a few important groups: $Ar^+ + N_2$ and its reverse (85–87), both of which have also been studied by laser-induced fluorescence (88, 89); $H_2^+ + Ar$ and its reverse (90–100); $H_2^+ + H_2$ (101); $O_2^+ + H_2$ and its reverse (102–104). In a number of cases there are dramatic increases in reactive cross sections when a particular vibrational state is excited (Figure 3). The energy of that particular state is in resonance with the energy of a specific product state. The cross sections show very little dependence on kinetic energy up to values typically of 20 eV, but have a remarkable dependence on vibrational state. One of the most dramatic examples involves a factor of ten enhancement of the cross section for reaction of N_2^+ ($v = 1$) with Ar as compared to $v = 0$, and with much smaller enhancements for successively higher vibrational levels (86, 87). This result is confirmed in the reverse direction by laser-induced fluorescence probing of the N_2^+ $v = 0$ and 1 states (89).

Similarly, a clear resonance enhancement of the charge transfer channel in the $H_2^+ + Ar$ reaction occurs specifically for H_2^+ ($v = 2$) (93, 94, 98) (Figure 3).

Interestingly, there is no similar enhancement of the proton transfer channel of the $H_2^+ + Ar$ reaction with vibrational excitation (92). It has been observed that the charge transfer and proton transfer channels are strongly coupled at higher energies and resemble each other, giving rise to oscillations in the cross section for the $v = 0$ state as a function of kinetic energy at about 3 eV (93). At extremely high kinetic energies, up to 1000 eV, the effect of vibrational selectivity in the reaction cross section is eventually diminished (97). A remarkable enhancement of the total cross section also takes place with spin-orbit excitation of Ar^+ in the $Ar^+ + H_2$ case, but no similar enhancement occurs for $Ar^+ + D_2$ (100), presumably because of a greater mismatch in the energies of the reactant and product vibrational states.

The reactions of Sr, Ca, and Ba with HF offer some of the best recent examples of the tremendous enhancements in reactivity that can be obtained with vibrational excitation (105–108). The reactions can be enhanced by as much as a factor of 10^4 with just one quantum of HF vibrational excitation. Vibrational excitation is far more effective than a corresponding amount of reagent translational excitation. Analysis of the

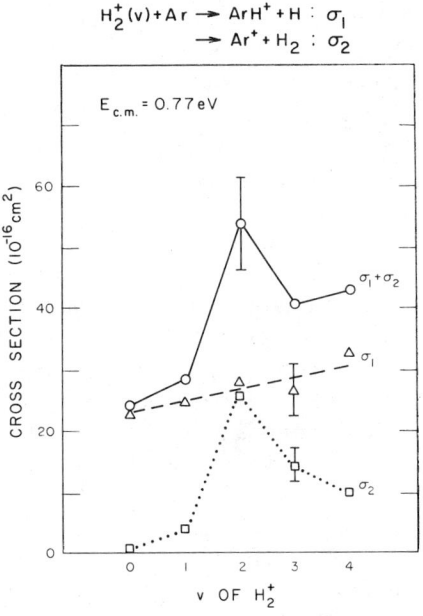

Figure 3 State-selected cross sections for the reactions of $H_2^+(v) + Ar$ at 0.77 eV, showing the dramatic resonant enhancement for $H_2^+(v = 2)$ in the charge transfer channel. Reproduced with permission from (98).

metal halide product shows that excitation in the product extends out to higher v values when the HF reactant is excited to higher v's. Reaction of H_2 with OH is enhanced by a factor of 120 with H_2 ($v = 1$) (109). In the reactions of Br with CH_3I and CF_3I, reagent vibration appears to be less effective at promoting the reaction than reagent translation (110). In contrast, the threshold kinetic energy behavior of the reaction of CH_3Br with $Xe(^3P_{2,0})$ shows a pronounced enhancement by vibrational excitation of CH_3Br (111).

For many years there has been a continuing controversy over the existence of vibrationally driven four-center reactions, for example $H_2 + D_2$ to yield 2 HD. In an experiment to excite HD directly to $v = 5$ by direct overtone pumping of HD with a laser, no product D_2 was detected (112). In a recent shock tube experiment specifically designed to eliminate impurities, no products of four-center reactions were observed (113). Thus it appears that four-center reactions between H_2 and D_2 are indeed prohibited by a large barrier, as has been suggested theoretically all along. In contrast, a careful search for IF chemiluminescence in a crossed beam reaction of $I_2 + F_2$ reveals a small extent of directly formed, electronically excited IF (114). It is postulated that the reaction forms predominantly I_2F + F, but that approximately one reaction in a thousand undergoes an involved encounter in which the departing F atom abstracts an I atom from I_2F to yield the electronically excited IF product. It would appear that this is a true four-center reaction, but perhaps not quite in the typical way that four-center transition states have been envisioned.

ROTATIONAL EFFECTS

Rotational state distributions are an important means of elucidating mechanisms of chemical reactions. The observation of bimodal rotational distributions are often indicative of a duality of direct and migratory pathways, sometimes referred to as microscopic branching. Generally, it appears that atom abstraction reactions with polyatomic molecules produce less rotational excitation than atom abstraction with diatomics (22). For example, the reactions of F with CH_4 and a series of halogenated methanes give rotational excitations very similar to a thermal distribution (17). Similarly, reactions of $O(^3P)$ with saturated and unsaturated hydrocarbons, amines, and aldehydes uniformly give low rotational excitation, which is attributed to a near-collinear attack in a simple atom abstraction process (76–79). In contrast, reactions of (O^1D) with H_2, HCl, and CH_4 give highly inverted rotational states (58–61, 70), and with other molecules, such as H_2O and NH_3, the rotational distributions are noticeably bimodal (62–66, 67–69). In the case of NH_3, the bimodality can possibly be attributed to the formation of different electronically excited states of NH_2 (macroscopic branching), rather than to microscopic branching effects. Bimodality has also been reported for several vibrational states in the reactions of translationally fast H atoms with Cl_2 and SCl_2 (46).

The reaction of $Mg(^1P_1)$ with H_2 is an excellent example of a reaction in which a bimodal rotational distribution (Figure 4) is indicative of microscopic branching effects (115). There are two clear peaks in the rotational distribution of the MgH product, occurring at $N \geq 30$ and $N = 6$. The authors suggest that reaction out of a linear transition state would not favor rotational excitation, whereas a perpendicular geometry, such as in an insertion reaction, would give access to very high rotational states. Similar rationalizations have been discussed for the reaction of (O^1D) with H_2. In the $Mg(^1P_1) + H_2$ case, it appears that these two pathways both have high probability, and hence a remarkably clear bimodal distribution is observed.

The effect of reagent rotation on reactivity has been studied in particular for reactions of rotationally excited HCl and HF with K, Ba, Sr, and Ca (107, 116–119). In general, for an endothermic reaction of this type, which probably involves a long-range electron transfer, rotational excitation seems to inhibit the reaction. However, it also turns out that reagent rotational excitation is more effective in promoting these reactions than is translational energy (119). This latter effect is also true for vibrational

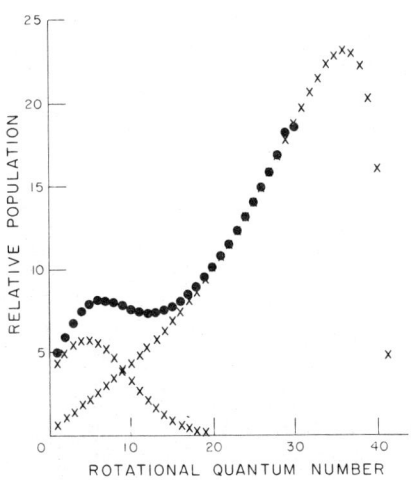

Figure 4 Bimodal rotational state distribution of $MgH(v = 0)$ in the reaction of $Mg(^1P_1)$ with H_2. The *solid circles* are the experimental result and the *crosses* are distributions obtained by fitting a linear surprisal to the higher rotational states and by subtraction. Reproduced with permission from (115).

excitation of the hydrogen halide reagent. The increased rotational energy also produces higher vibrational levels in the metal halide product (118).

An interesting kinematic constraint occurs for the mass combination $H + H'L \rightarrow HH' + L$, such as in the reaction of $Ba + HCl$. This effect, first described by Herschbach (120), requires that the orbital angular momentum in the collision be channeled specifically into rotational excitation of the diatomic product. This has led to the suggestion that the rather direct mapping of impact parameter into rotational state might allow information to be obtained about the cross section as a function of impact parameter (121, 122). Extreme examples of the $H + H'L$ system are the associative detachment reactions of $H + F^-$ and Cl^-, which release the electron as the light product (37–39). This effect results in a severe reduction in the population of the higher vibrational states, due to the total energy and angular momentum constraints (37–39, 123). In addition, theoretical treatment of the specific associative detachment problem (123) shows that each vibrational state in the product gives access to a certain group of rotational states. As each vibrational state turns on, a different group of rotational states becomes populated.

TRANSLATIONAL EXCITATION AND ANGULAR DISTRIBUTIONS

Classic molecular beam scattering methods continue to be used to make investigations of reaction dynamics. Experiments have been carried out with greater ranges of collision energies, with excitation of vibrational states of reagents, and with new supersonic sources of atoms and radicals. For example, the reaction of $Hg + I_2$ to form $HgI + I$ has been reinvestigated over a greater range of collision energies (124). Previously this reaction was observed to occur predominantly via a long-lived complex. At energies above 2–2.3 eV, a back-scattered component is observed that can be attributed to another channel opening up, that of a direct reaction via a collinear approach. The back-scattered product accounts for most of the scattering at higher energies. Classic measurements of the excitation function, that is the cross section versus energy, have been accurately obtained for the $K + C_2H_5I$ reaction (125–127). Simple models, assuming that the reaction occurs by electron transfer to an ionic surface, fit the data extremely well. A study of the $Br + HAt$ reaction reveals a startlingly low translational energy in the products (128). It has been suggested that this is indicative of a potential energy surface with a pronounced widening in the neighborhood of the transition state. Investigation of the effect of translational excitation upon the reactions $Ba + HF$ (129) and $K + HCl$ and HF (130) shows that these reactions are moderately enhanced by translation. However, vibrational excitation in the K+HCl and HF cases is orders of magnitude more effective in promoting the reaction. At very high collision energies, the effectiveness of vibration is lessened substantially. The excess translational energy in the Ba+HF reaction is partitioned approximately one half into product translation and the remainder equally between vibration and rotation. Reactions of D atoms with Br_2 and alkyl iodide molecules shows predominantly back-scattered products with substantial energy disposed into product translation, indicative of reaction at small impact parameters and with significant repulsion (131). Only a small fraction of the repulsive release goes into exciting the radical fragment.

A large number of elementary reactions of supersonic O atoms with halogens, alkyl iodides, and H_2S have been studied (132–141). For $O+Cl_2$ the reaction exhibits a stripping type of behavior, but for many of the other halogens and interhalogens, approximately equal amounts of forward/backward products are observed, indicative of a longer-lived, complex mechanism. Similar isotropic distributions are observed for the alkyl iodides, and the substantial internal excitation suggests a complex reaction model (132, 134). In the reaction of $O+H_2S$, the HSO radical has been detected unambiguously (135, 140), indicating the importance of the atom displacement reaction via initial attack on the lone pair electrons of the sulfur. In many cases, there is not an unambiguous mass to detect in order to ascertain the primary products prior to fragmentation and ionization in the mass spectrometer detector. Lee and co-workers have taken advantage of parent/daughter-ion correlations in the time-of-flight analysis to show which ion fragments come from initial parent species (142–144). In this way they have determined the nascent products of a variety of reactions for which there have been tremendous controversies, including $O+C_6H_6$, C_2H_4, and C_2H_3Br. In the case of benzene (142) they observe two channels only, C_6H_5O and C_6H_5OH, but not formation of CO or OH, which has been previously reported. In $O+C_2H_4$, the predominant channel is addition/elimination to produce C_2H_3O. This latter result has now been confirmed in many independent experiments using laser-induced fluorescence.

One of the truly exciting topics of discussion in molecular beam reactive scattering concerns the possibility of observing quantum mechanical "resonances." These resonances effectively sample the vibrational structure of the transition state. They would show up as a sharp feature in the translational energy dependence of the reaction probability with respect to formation of a particular vibrational state. The energies at which the resonance would be observed correspond to the bound states in the vibrationally adiabatic potential of the transition state. There are numerous theoretical (145) and experimental (146, 147) references to this type of effect for the reaction $F+H_2$. In the early experimental reports on this effect, it was thought that a resonance was observed in the sideways peaking of the $v = 2$ product density (146, 147). More recently, however, a re-evaluation of these data and new data suggest that a resonance may be observed in the $v = 3$ state, which is forward scattered over a range of energies (148). Future work will help to ascertain the exact way in which these quantum mechanical resonances will show up.

The use of lasers to produce translationally hot atoms by photofragmentation of small molecules is one of the exciting new directions in reactive dynamics. The technique has been used with time-resolved infrared chemiluminescence and laser-induced fluorescence to study a host of reactive events at higher collision energies. Two studies that have received much attention involve the vibrational and rotational state distributions of the basic $H+D_2$ reaction (149, 150). Translationally energetic H atoms with 1.3 eV in the center of mass are produced by excimer laser photolysis of HI. In this case the HD product is detected by either multiphoton ionization or CARS spectroscopy (Figure 5). Significant population of HD ($v = 1$) and rotational states up to $J = 12$ in $v = 0$ are observed. However, both experiments are preliminary. Preliminary results have also been obtained for the reaction $H+HBr$ to produce vibrationally and rotationally excited H_2 (151). Detailed measurements are reported for the reactions $H+O_2$ (152, 153), H_2O (154), and CO_2 (155, 156). In the $H+O_2$ and CO_2 reactions the rotational excitation in the OH product is very high, whereas in $H+H_2O$ the rotational excitation is low. A recent study of the collisions of translationally hot H and D on HCl and DCl reveals that the inelastic excitation channel predominates (157). Of those molecules that do undergo reactive atom exchange, the vibrational excitation in the product is hotter than the inelastically excited molecules (no exchange). Such studies now

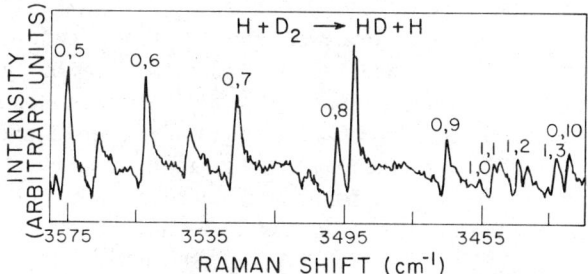

Figure 5 Vibrational Q-branch CARS spectrum of HD produced in the reaction of $H+D_2 \rightarrow HD+D$ with 1.3 eV center-of-mass collision energy. The states are labeled in a (v, J) format. Unlabeled peaks are due to D_2 or HI. Reproduced with permission from (150).

allow the competition between inelastic and reactive pathways to be probed with a high degree of state-selected detail and will lead to many exciting new results and insights.

A novel method to produce translationally fast metal atoms takes advantage of the vaporization of a metal film by a short pulsed laser (158). From chemiluminescent signals as a function of time-of-flight, reactive cross sections as a function of kinetic energy have been readily obtained for the reaction of Zn with N_2O. Finally, a conventional supersonic beam source of atoms coupled with tunable vacuum ultraviolet laser detection was used to probe the efficacy of reaction in the H(D)+HBr(DBr) system (159). The combination of the slower velocity D atoms with the HBr molecule was the most favorable pair for the reaction. This is attributed to the longer interaction time obtained with the slower velocity D atom and to the more rapid rotation of the HBr, compared to DBr, which facilitates achieving a favorable geometry for reaction within the interaction time.

ELECTRONIC STATES

Much of the interest in electronically excited species has centered on reactions of metal atoms with various oxidizers or halogen molecules to produce electronically excited products, the reactions of electronically excited metastable rare gas atoms, and reactions of electronically excited diatomics. Typical metal-oxidant systems that have been studied are $Ca(^1D)+O_2$ (160); Ti, $V+O_2$ (161); Sr, Ca, Ba and their electronically excited states $+N_2O$ (162–164); $B+O_2$, N_2O and other oxidizers (165–167); and $Mg(^3P$ and $^1P_1)+N_2O$ and CO_2 (168–170). A common feature of many of these reactions is that they involve an electron jump mechanism. Adiabatic correlation rules do not seem to be very helpful in predicting the outcome of the electronically excited product states. The large number of electronic potential surfaces and surface crossings makes these reactions exceedingly complex and difficult to predict. In many cases electronically excited product states are observed and the cross sections for their production have been obtained. In several cases (167) it is thought that the slowness of a reaction may be due to the fact that the adiabatic electron affinities of oxidizers such as CO_2 and N_2O are negative, and that it is necessary to bend the molecule to enhance its ability to accept the electron.

There are a similarly large number of reactions of ground state and excited state metal atoms with halogens. These include Sn, Ge, Si, Al, Ga, In, and Cu with F_2 (171–174); $Hg(^3P_2)$, $Mg(^3P)$, and $Cu(^2D)$ with halogens (175–177); and Y with halomethanes (178). Typically the total cross sections are a few hundred $Å^2$, whereas the chemiluminescent channels that are detected are only a few percent of the total. In most cases these electronically excited channels come from nonadiabatic transitions. The fraction of the remaining energy that is partitioned into vibration after the formation of an electronically excited product is typically quite high, 60–90%, and the vibrational levels are highly inverted. The high degree of vibrational excitation indicates that the reactions are most likely direct. In a few cases chemi-ions are detected. These also correspond to only a few percent of the total reactive cross section.

The reactions of electronically excited metastable rare gas atoms bear a remarkable resemblance to metal atom reactions, in that an electron jump mechanism typically explains the reactive behavior. However, since the electronic energy is exceedingly great, many other pathways can be observed, such as electronic quenching followed by fragmentation of the quenching partner or chemi-ionization. Quenching or reaction is nearly always gas-kinetic. There are a variety of recent studies of electronically excited rare gases with the halogens, alkyl halides, H_2O, and other hydrogen-containing compounds (179–186). The alignment properties of the products of such reactions are discussed in the next section. One of the most interesting results involves the reaction of Ar* with H_2O; part of the quenching results in complete fragmentation of both bonds of the water to produce total atomization (184). In quenching by other hydrogen-containing molecules, elimination of H atoms is often favored over elimination of molecular H_2 (185). Just as in the metal atom reactions with halogens, the f_v's for reactions that produce the electronically excited rare gas halides are typically very large, 70–80% (186).

Reactions of electronically excited diatomic halogens also bear a resemblance to metal-halogen reactions, in that an electron jump mechanism seems to be important. The involvement of ion-pair states is also significant. Reactions of high-lying electronically excited states of IBr* and Br_2* with Xe are observed to produce electronically excited XeBr with mechanisms analogous to the reactions of the rare gas metastables plus halogens (187–188). In contrast, however, searches for reactions of the lower-lying electronically excited B state of I_2* with In and Hg did not detect any enhancements in the formation of the InI or HgI products (189, 190). Collisions of $I_2(D)$ states with polyatomic molecules typically lead to quenching, but there is some evidence for product formation (191, 192).

An interesting new area is the investigation of the reactivity of different spin-orbit states, especially when the energies of the states are small compared to the overall energetics of the reaction of interest. The reaction of $F(^2P_{1/2})$ and $F(^2P_{3/2})$ with HBr shows no preference to populate the corresponding Br spin-orbit states in the product, thus ruling out the notion of adiabatic correlations (193). Instead, the extent of Br* formed in the product is thought to be governed by a vibration-to-electronic energy transfer pathway from vibrationally excited HF, which occurs in the exit channel of the reaction. Similarly, it has been suggested that the reaction of Br* with H_2 ($v = 1$) may be enhanced by a near resonant electronic-to-vibrational energy transfer pathway in the entrance channel of the reaction, which generates the reagents on the necessary reactive potential surface (194). The reaction of different $Ca(^3P_{2,1,0})$ spin-orbit states with Cl_2 shows remarkable differences in reactivity, considering the small splittings of the energy in this case (195). A similar result has been observed for the Penning ionization reaction of $Ne(^3P_{2,0})+CO$ which produces $CO^+(A)$ (196). The reactivity ratio for $^3P_2/^3P_0$ is 0.4. Clearly these effects are able to give subtle information about the electronic potential surfaces and the total angular momentum correlations in the molecular transition states.

ORIENTATION EFFECTS

There are two excellent recent reviews on the preparation and reaction of oriented and aligned molecular reagents (197, 198). A simple way to distinguish these two effects is to think of a linear molecule with two different ends. Alignment allows the experimenter to arrange the linear axis either parallel to perpendicular to a collision partner. Orientation allows the system to be arranged such that either of the two different ends of the linear molecule can point toward its collision partner. It has been amply pointed out that the use of hexapole fields can provide both alignment and orientation, and hence a complete geometrical arrangement of reagents, whereas optical preparation of states can only provide alignment conditions. The classic example of oriented reaction dynamics is the reaction of $Rb+CH_3I$. This reaction has recently been studied in more detail (199, 200). A large reactive asymmetry (0.44) is observed in the back-scattered product, with the decrease of reactivity occurring when the Rb approaches the CH_3 end of the molecule. The steric factor for forward scattering is much less, only 0.94, indicating that large impact parameter collisions show a much lower preference for reagent orientation.

Perhaps one of the most complete studies of reagent orientation is for the reaction of NO with O_3 (201, 202). There are rather puzzling facts concerning this reaction: for example, the rate is enhanced by vibrational excitation in either the NO or the O_3. These results can now be explained by the insights of elegant orientation experiments. It appears that the reaction is direct but essentially exhibits two different modes, characterized by two different transition states. One transition state involves NO attack with the nitrogen end approaching the central O atom of the O_3; the other involves attack of NO to strip off one of the outer O atoms on the O_3.

In spite of the fact that optical state preparation lacks the ability to select different ends of the molecules, these methods have also yielded some exciting results recently. An HF reagent has been aligned in the reaction of Sr + HF; a small enhancement of the $v = 2$ SrF product state is detected for the broadside attack (203). Reaction of the Ca($^1P_1$) electronically excited state with HCl yields a branching ratio in the CaCl A and B electronically excited product states that is sensitive to the alignment of the Ca P orbital (204, 204a). There is a two-fold increase in the B state versus the A state for the σ symmetry; this is so even though all factors, such as beam angle spread, are not fully controlled yet. For molecular reagents, the degree of alignment for a simple photoabsorption is not perfect. However, under conditions of a strong dissociating pulse, the remaining undissociated molecules can have a high degree of alignment. This principle has been used to prepare rotationally aligned reagents of IBr (205, 206). In collision with Xe* a preferential reaction is observed when the Xe* approaches parallel to the plane of rotation of the IBr.

Alignment in product molecule rotation has also been extensively studied for reactions of metastable Xe and Kr atoms with molecules such as BrCN and CH_3I to produce electronically excited rare gas halide products (111, 179–182). The degree of rotational alignment can typically be quite high and increases with increasing kinetic energy. An appealingly simple model to explain this effect is a long-range electron jump followed by an immediate and large repulsion, thus preserving the rotational alignment, very much like in a photofragmentation process. It has been observed that the competition between reactive and inelastic channels in the case of Kr* + Br_2 serves to lower the degree of alignment in the reactive channel (181).

Alignment information in the orbital configurations of transition states is also obtained by an analysis of the lambda doublet sublevels of the OH products of numerous reactions (61, 76–79, 152–155, 207). The OH spin doublets rarely show any preferential excitation, but the lambda doublet components are often populated nonstatistically. For high K rotational quantum numbers, the lambda doublet states can be uniquely associated with directed orbitals. For example, in the reaction of translationally fast H atoms with O_2, the π^+ state is preferred by 6:1 over π^- (152). The π^+ lambda doublet is the energetically lower component and represents the case in which the singly occupied orbital lies in the plane of the OH rotation. Preferential population of this particular orbital configuration has been observed for many other reactions that produce OH. This preferential population is taken as indicative of a planar transition state in which the bond that is broken leaves the lone electron in an orbital that lies in the plane of rotation of the OH product. There is an incorrect interpretation of this orbital alignment in one publication (207), which might be misleading. The configuration has been properly assigned in other studies. The spectroscopic details of this effect are discussed in detail in a recent paper on H_2O photofragmentation, in which the only case of the opposite orbital configuration has been observed, namely, that of the lone electron appearing in an orbital perpendicular to the plane of rotation (208), and in a recent theoretical paper (208a). The photofragmentation of water may be responsible for the interstellar OH maser that operates from the preferentially excited π^- component.

PROBING THE TRANSITION STATE

Tremendous excitement has been generated recently concerning the prospects to drive chemical reactions in new and selective ways by optical excitation of the transition state in the course of the reaction. An early demonstration of this possibility is the laser-induced Penning and associative ionization of two Na atoms during a collision (209). About the same time, there appeared a report of a laser-assisted reaction of $K + HgBr_2$ to form KBr and electronically excited HgBr (210). Subsequent attempts to reproduce this reaction have not been successful (211). However, a number of other examples present convincing evidence that the transition state of a reaction can be manipulated by optical light fields. There are three reports of laser-assisted reactions between Xe and Cl_2 (212–214). While the first of these used a photon of sufficient energy that the reaction can plausibly be driven by the light (212), the second experiment used a photon of too long a wavelength to involve a single photon absorption in the transition state without deriving energy from subsequent collisions. In the third experiment (214), the authors provide convincing evidence for a two-photon assist. A

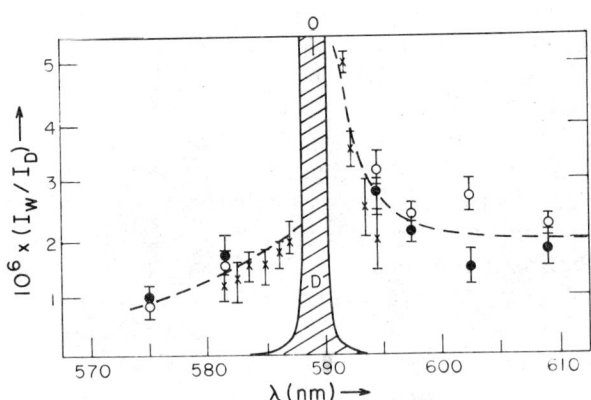

Figure 6 Wing emission from the reaction intermediate FNaNa* in the reaction of $F + Na_2$ normalized to the Na D line intensity. Reproduced with permission from (217).

more interesting experiment involves first the formation of a van der Waals cluster of Hg and Cl_2, followed by a photoabsorption that gives rise to prompt formation of electronically excited HgCl (215). The idea of manipulating a transition state with a light field has been turned around to detect emission from the transition state. In an elegant series of experiments (216, 217), optical emission in the far wings around the Na D line has been detected from the NaNaF* electronically excited transition state in the reaction of $F+Na_2$ (Figure 6). While these experiments are only the mere beginnings, it is clear that many imaginative and exciting results concerning the understanding and manipulation of chemical reaction dynamics are yet to come.

ACKNOWLEDGMENTS

The author is grateful to his students and colleagues for all their work in advancing the state of molecular reaction dynamics. He would like to acknowledge the continued support of the National Bureau of Standards and grants from the National Science Foundation, Department of Energy, Air Force Office of Scientific Research, and Army Research Office. Acknowledgment is made to the donors of the Petroleum Research Fund, administered by the American Chemical Society, for partial support of this research. The author would especially like to thank P. Krog, L. Volsky, G. Romey, and L. Haas for expert technical assistance with the literature search and manuscript preparation. This review was prepared with the assistance of the JILA Atomic Collisions Data Center.

Literature Cited

1. Levy, M. R. 1979. *Prog. Reaction Kinetics* 10:1–252
2. Levine, R. D., Bernstein, R. B. 1974. *Molecular Reaction Dynamics.* New York: Oxford Univ. Press
3. Smith, I. W. M. 1980. *Kinetics and Dynamics of Elementary Gas Reactions.* London/Boston: Butterworths
4. Bernstein, R. B. 1982. *Chemical Dynamics Via Molecular Beam and Laser Techniques.* New York: Oxford Univ. Press
5. Smith, I. W. M. 1980. In *Physical Chemistry of Fast Reactions*, ed. I. W. M. Smith, 2:1–82. New York: Plenum
6. Kneba, M., Wolfrum, J. 1980. *Ann. Rev. Phys. Chem.* 31:47–79
7. Holmes, B. E., Setser, D. W. 1980. In *Physical Chemistry of Fast Reactions*, ed. I. W. M. Smith, 2:83–214. New York: Plenum
8. Geddes, J. 1982. *Contemp. Phys.* 23: 233–55
9. Henderson, D., ed. 1981. *Theoretical Chemistry. Theory of Scattering: Papers in Honor of Henry Eyring*, Vols. 6A, B. New York: Academic
10. Polanyi, J. C., Schreiber, J. L. 1974. In *Physical Chemistry, An Advanced Treatise*, ed. W. Jost, 6:383–487. New York: Academic
11. Kuntz, P. J. 1975. In *Modern Theoretical Chemistry*, ed. W. H. Miller, Pt. B, Chapt. 2. New York: Plenum
12. MacDonald, R. G., Sloan, J. J. 1978. *Chem. Phys.* 31:165–76
13. Parsons, J. M., Watson, D. G., Sloan, J. J. 1981. *Chem. Phys.* 58:175–83
14. MacDonald, R. G., Sloan, J. J., Wassel, P. J. 1979. *Chem. Phys.* 41:201–8
15. Dill, B., Heydtmann, H. 1980. *Chem. Phys.* 54:9–20

16. Dill, B., Hildebrandt, B., Vanni, H., Heydtmann, H. 1981. *Chem. Phys.* 58:163–74
17. Nazar, M. A., Polanyi, J. C. 1981. *Chem. Phys.* 55:299–311
18. Bogan, D. J., Setser, D. W., Sung, J. P. 1977. *J. Phys. Chem.* 81:888–98
19. Douglas, D. J., Sloan, J. J. 1980. *Chem. Phys.* 46:307–12
20. Sloan, J. J., Watson, D. G., Williamson, J. 1980. *Chem. Phys. Lett.* 74:481–85
21. Donaldson, D. J., Parsons, J., Sloan, J. J., Stolow, A. 1984. *Chem. Phys.* 85:47–62
22. Manocha, A. S., Setser, D. W., Wickramaaratchi, M. A. 1983. *Chem. Phys.* 76:129–46
23. Wickramaaratchi, M. A., Setser, D. W. 1983. *J. Phys. Chem.* 87:64–72
24. Agrawalla, B. S., Setser, D. W. 1984. *J. Phys. Chem.* 88:657–60
25. Sloan, J. J., Watson, D. G. Williamson, J. M., Wright, J. S. 1981. *J. Chem. Phys.* 75:1190–1200
26. Sloan, J. J., Watson, D. G., Wright, J. S. 1981. *Chem. Phys.* 63:283–92
27. Sloan, J. J., Watson, D. G. 1981. *J. Chem. Phys.* 74:744–45
28. Donaldson, D. J., Watson, D. G., Sloan, J. J. 1982. *Chem. Phys.* 68:95–107
29. Donaldson, D. J., Sloan, J. J. 1983. *Can. J. Chem.* 61:906–11
30. Tamagake, K., Setser, D. W., Sung, J. P. 1980. *J. Chem. Phys.* 73:2203–17
31. Jonathan, N. B. H., Sellers, P. V., Stace, A. J. 1981. *Mol. Phys.* 43:215–28
32. Dzelzkalns, L. S., Kaufman, F. 1983. *J. Chem. Phys.* 79:3836–44
33. Deleted in proof
34. Zwier, T. S., Bierbaum, V. M., Ellison, G. B., Leone, S. R. 1980. *J. Chem. Phys.* 72:5426–36
35. Weisshaar, J. C., Zwier, T. S., Leone, S. R. 1981. *J. Chem. Phys.* 75:4873–84
36. Maricq, M. M., Smith, M. A., Simpson, C. J. S. M., Ellison, G. B. 1981. *J. Chem. Phys.* 74:6154–70
37. Zwier, T. S., Maricq, M. M., Simpson, C. J. S. M., Bierbaum, V. M., Ellison, G. B., Leone, S. R. 1980. *Phys. Rev. Lett.* 44:1050–53
38. Zwier, T. S., Weisshaar, J. C., Leone, S. R. 1981. *J. Chem. Phys.* 75:4885–92
39. Smith, M. A., Leone, S. R. 1983. *J. Chem. Phys.* 78:1325–34
40. Hamilton, C. E., Bierbaum, V. M., Leone, S. R. 1984. *J. Chem. Phys.* 80:1831–38
41. Smith, M. A., Bierbaum, V. M., Leone, S. R. 1983. *Chem. Phys. Lett.* 94:398–403
42. Wickramaaratchi, M. A., Setser, D. W., Hildebrandt, B., Korbitzer, B., Heydtmann, H. 1983. *Chem. Phys.* In press
43. Berquist, B. M., Bozzelli, J. W., Dzelzkalns, L. S., Piper, L. G., Kaufman, F. 1982. *J. Chem. Phys.* 76:2972–83
44. Bittenson, S., Tardy, D. C., Wanna, J. 1981. *Chem. Phys.* 58:313–23
45. Jonathan, N. B. H., Liddy, J. P., Sellers, P. V., Stace, A. J. 1980. *Mol. Phys.* 39:615–27
46. Hildebrandt, B., Vanni, H., Heydtmann, H. 1984. *Chem. Phys.* 84:125–37
47. Malins, R. J., Setser, D. W. 1980. *J. Chem. Phys.* 73:5666–80
48. Agrawalla, B. S., Manocha, A. S., Setser, D. W. 1981. *J. Phys. Chem.* 85:2873–77
49. Brandt, D., Polanyi, J. C. 1980. *Chem. Phys.* 45:65–84
50. Moss, M. G., Hudgens, J. W., McDonald, J. D. 1980. *J. Chem. Phys.* 72:3486–89
51. Kovalenko, L. J., Leone, S. R. 1984. *J. Chem. Phys.* In press
52. Moss, M. G., Ensminger, M. D., Stewart, G. M., Mordaunt, D., McDonald, J. D. 1980. *J. Chem. Phys.* 73:1256–64
53. Mantell, D., Ryali, S., Halpern, B. L., Haller, G. L., Fenn, J. B. 1981. *Chem. Phys. Lett.* 81:185–87
54. Bernasek, S. L., Leone, S. R. 1980. *Chem. Phys. Lett.* 84:401–4
55. Kori, M., Halpern, B. L. 1983. *Chem. Phys. Lett.* 98:32–36
56. Thornman, R. P., Anderson, D., Bernasek, S. L. 1980. *Phys. Rev. Lett.* 44:743–46
57. Thornman, R. P., Bernasek, S. L. 1981. *J. Chem. Phys.* 74:6498–6504
58. Smith, G. K., Butler, J. E. 1980. *J. Chem. Phys.* 73:2243–53
59. Butler, J. E., MacDonald, R. G., Donaldson, D. J., Sloan, J. J. 1983. *Chem. Phys. Lett.* 95:183–87
60. Buss, R. J., Casavecchia, P., Hirooka, T., Sibener, S. J., Lee, Y. T. 1981. *Chem. Phys. Lett.* 82:386–91
61. Luntz, A. C. 1980. *J. Chem. Phys.* 73:5393–95
62. Gericke, K.-H., Comes, F. J. 1980. *Chem. Phys. Lett.* 74:63–66
63. Butler, J. E., Talley, L. D., Smith, G. K., Lin, M. C. 1981. *J. Chem. Phys.* 74:4501–8
64. Comes, F. J., Gericke, K.-H., Manz, J. 1981. *J. Chem. Phys.* 75:2853–63
65. Gericke, K.-H., Comes, F. J., Levine, R. D. 1981. *J. Chem. Phys.* 74:6106–12
66. Guillory, W. A., Gericke, K.-H., Comes, F. J. 1983. *J. Chem. Phys.* 78:5993–6001
67. Sanders, N. D., Butler, J. E., McDonald, J. R. 1980. *J. Chem. Phys.* 73:5381–83
68. Rettner, C. T., Cordova, J. F., Kinsey, J. L. 1980. *J. Chem. Phys.* 72:5280–82

69. Cordova, J. F., Rettner, C. T., Kinsey, J. L. 1981. *J. Chem. Phys.* 75:2742–48
70. Luntz, A. C. 1980. *J. Chem. Phys.* 73:1143–52
71. Goldstein, N., Wiesenfeld, J. R. 1983. *J. Chem. Phys.* 78:6725–31
72. Trickl, T., Wanner, J. 1983. *J. Chem. Phys.* 78:6091–6101
72a. Agrawalla, B. S., Singh, J. P., Setser, D. W. 1983. *J. Chem. Phys.* 79:6416–18
73. Donovan, R. J., Fernie, D. P., Fluendy, M. A. D., Glen, R. M., Rae, A. G. A., Wheeler, J. R. 1980. *Chem. Phys. Lett.* 69:472–75
74. Stein, L., Wanner, J., Walther, H. 1980. *J. Chem. Phys.* 72:1128–37
75. Munakata, T., Matsumi, Y., Kasuya, T. 1983. *J. Chem. Phys.* 79:1698–1707
76. Andresen, P., Luntz, A. C. 1980. *J. Chem. Phys.* 72:5842–50
77. Kleinermanns, K., Luntz, A. C. 1982. *J. Chem. Phys.* 77:3537–39
78. Kleinermanns, K., Luntz, A. C. 1982. *J. Chem. Phys.* 77:3533–36
79. Kleinermanns, K., Luntz, A. C. 1982. *J. Chem. Phys.* 77:3774–75
80. Hamilton, C. E., Duncan, M. A., Zwier, T. S., Weisshaar, J. C., Ellison, G. B., et al. 1983. *Chem. Phys. Lett.* 94:4–9
81. Guyer, D. R., Hüwel, L., Leone, S. R. 1983. *J. Chem. Phys.* 79:1259–71
82. Marx, R. 1984. In *Ionic Processes in the Gas Phase*, ed. M. A. Almoster Ferreira, pp. 67–86. Boston: Reidel
83. O'Keefe, A., Parent, D., Mauclaire, G., Bowers, M. T. 1984. *J. Chem. Phys.* In press
84. Danon, J., Marx, R. 1982. *Chem. Phys.* 68:255–60
85. Kato, T., Tanaka, K., Koyano, I. 1982. *J. Chem. Phys.* 77:337–41
86. Kato, T., Tanaka, K., Koyano, I. 1982. *J. Chem. Phys.* 77:834–38
87. Govers, T. R., Guyon, P. M., Baer, T., Cole, K., Fröhlich, H., Lavollee, M. 1984. Submitted for publication
88. Mahan, B. H., Martner, C., O'Keefe, A. 1982. *J. Chem. Phys.* 76:4433–38
89. Hüwel, L., Guyer, D. R., Lin, G. H., Leone, S. R. 1984. *J. Chem. Phys.* Submitted for publication
90. Bilotta, R. M., Preuninger, F. N., Farrar, J. M. 1980. *Chem. Phys. Lett.* 74:95–100
91. Bilotta, R. M., Preuninger, F. N., Farrar, J. M. 1980. *J. Chem. Phys.* 73:1637–48
92. Bilotta, R. M., Farrar, J. M. 1981. *J. Chem. Phys.* 74:1699–1706
93. Houle, F. A., Anderson, S. L., Gerlich, D., Turner, T., Lee, Y. T. 1982. *J. Chem. Phys.* 77:748–55
94. Houle, F. A., Anderson, S. L., Gerlich, D., Turner, T., Lee, Y. T. 1981. *Chem. Phys. Lett.* 82:392–95
95. Campbell, F. M., Browning, R., Latimer, C. J. 1980. *J. Phys. B* 13:4257–62
96. Campbell, F. M., Browning, R., Latimer, C. J. 1981. *J. Phys. B* 14:1183–95
97. Latimer, C. J., Campbell, F. M. 1982. *J. Phys. B* 15:1765–71
98. Tanaka, K., Kato, T., Koyano, I. 1981. *J. Chem. Phys.* 75:4941–45
99. Tanaka, K., Durup, J., Kato, T., Koyano, I. 1980. *J. Chem. Phys.* 73:586–88
100. Tanaka, K., Durup, J., Kato, T., Koyano, I. 1981. *J. Chem. Phys.* 74:5561–71
101. Anderson, S. L., Houle, F. A., Gerlich, D., Lee, Y. T. 1981. *J. Chem. Phys.* 75:2153–62
102. Tanaka, K., Kato, T., Guyon, P.-M., Koyano, I. 1982. *J. Chem. Phys.* 77:4441–46
103. Tanaka, K., Kato, T., Guyon, P.-M., Koyano, I. 1983. *J. Chem. Phys.* 79:4302–5
104. Anderson, S. L., Turner, T., Mahan, B. H., Lee, Y. T. 1982. *J. Chem. Phys.* 77:1842–54
105. Karny, Z., Zare, R. N. 1978. *J. Chem. Phys.* 68:3360–65
106. Gupta, A., Perry, D. S., Zare, R. N. 1980. *J. Chem. Phys.* 72:6250–57
107. Torres-Filho, A., Pruett, J. G. 1980. *J. Chem. Phys.* 72:6336–42
108. Torres-Filho, A., Pruett, J. G. 1982. *J. Chem. Phys.* 77:740–47
109. Zellner, R., Steinart, W. 1981. *Chem. Phys. Lett.* 81:568–72
110. Krajnovich, D., Zhang, Z., Huisken, F., Shen, Y. R., Lee, Y. T. 1982. In *Physics of Electronic and Atomic Collisions*, ed. S. Datz, pp. 733–44. Amsterdam: North-Holland
111. Hennessy, R. J., Simons, J. P. 1981. *Mol. Phys.* 44:1027–34
112. Herman, I. P. 1980. *J. Chem. Phys.* 72:5777–78
113. Lifshitz, A., Bidani, M., Carroll, H. F. 1983. *J. Chem. Phys.* 79:2742–47
114. Kahler, C. C., Lee, Y. T. 1980. *J. Chem. Phys.* 73:5122–30
115. Breckenridge, W. H., Umemoto, H. 1981. *J. Chem. Phys.* 75:4153–55
116. Dispert, H. H., Geis, M. W., Brooks, P. R. 1979. *J. Chem. Phys.* 70:5317–19
117. Man, C.-K., Estler, R. C. 1981. *J. Chem. Phys.* 75:2779–85
118. Altkorn, R., Bartoszek, F. E., Dehaven, J., Hancock, G., Perry, D. S., Zare, R. N. 1983. *Chem. Phys. Lett.* 98:212–16
119. Hoffmeister, M., Potthast, L., Loesch, H. J. 1983. *Chem. Phys.* 78:369–80

120. Herschbach, D. R. 1966. *Adv. Chem. Phys.* 10:319-93
121. Siegel, A., Schultz, A. 1980. *J. Chem. Phys.* 72:6227-36
122. Siegel, A., Schultz, A. 1982. *J. Chem. Phys.* 76:4513-27
123. Gauyacq, J. P. 1982. *J. Phys. B* 15:2721-39
124. Oprysko, M. M., Aoiz, F. J., McMahan, M. A., Bernstein, R. B. 1983. *J. Chem. Phys.* 78:3816-31
125. Aoiz, F. J. L., Herrero, V. J., Gonzalez Urena, A. 1980. *Chem. Phys. Lett.* 74:398-99
126. Aoiz Moleres, F. J., Herrero, V. J., Gonzalez Urena, A. 1981. *Chem. Phys.* 59:61-73
127. Herrero, V. J., Saez Rabanos, V., Gonzalez Urena, A. 1982. *Mol. Phys.* 47:725-40
128. Grover, J. R., Malloy, D. E., Mitchell, J. B. A. 1982. *J. Chem. Phys.* 76:362-77
129. Gupta, A., Perry, D. S., Zare, R. N. 1980. *J. Chem. Phys.* 72:6237-49
130. Heismann, F., Loesch, H. J. 1982. *Chem. Phys.* 64:43-67
131. Davidson, F. E., Duncan, G. L., Grice, R. 1981. *Mol. Phys.* 44:1119-29
132. Browett, R. J., Hobson, J. H., Gorry, P. A., Nowikow, C. V., Grice, R. 1980. *Mol. Phys.* 40:1315-24
133. Browett, R. J., Hobson, J. H., Davidson, F. E. 1981. *Mol. Phys.* 43:113-21
134. Browett, R. J., Hobson, J. H., Grice, R. 1981. *Mol. Phys.* 42:425-34
135. Clemo, A. R. Davidson, F. E., Duncan, G. L., Grice, R. 1981. *Chem. Phys. Lett.* 84:509-11
136. Durkin, A., Smith, D. J., Grice, R. 1982. *Mol. Phys.* 46:1251-60
137. Durkin, A., Smith, D. J., Grice, R. 1982. *Mol. Phys.* 46:55-66
138. Durkin, A., Smith, D. J., Hoffmann, S. M. A., Grice, R. 1982. *Mol. Phys.* 46:1261-70
139. Fernie, D. P., Smith, D. J., Durkin, A., Grice, R. 1982. *Mol. Phys.* 46:41-54
140. Davidson, F. E., Clemo, A. R., Duncan, G. L., Browett, R. J., Hobson, J. H., Grice, R. 1982. *Mol. Phys.* 46:33-40
141. Durkin, A., Smith, D. J., Grice, R. 1983. *Mol. Phys.* 48:1137-38
142. Sibener, S. J., Buss, R. J., Casavecchia, P., Hirooka, T., Lee, Y. T. 1980. *J. Chem. Phys.* 72:4341-49
143. Buss, R. J., Baseman, R. J., He, G., Lee, Y. T. 1981. *J. Photochem.* 17:389-96
144. He, G., Buss, R. J., Baseman, R. J., Tse, R., Lee, Y. T. 1982. *J. Phys. Chem.* 86:3547-53
145. Wyatt, R. E., McNutt, J. F., Redmon, M. J. 1982. *Ber. Bunsenges. Phys. Chem.* 86:437-48
146. Sparks, R. K., Hayden, C. C., Shobatake, K., Neumark, D. M., Lee, Y. T. 1980. *Horizons of Quantum Chemistry*, ed. K. Fukui, B. Pullman, pp. 91-105. Boston: Reidel
147. Lee, Y. T. 1982. *Ber. Bunsenges. Phys. Chem.* 86:378-86
148. Neumark, D. M., Wodtke, A. M., Robinson, G. N., Hayden, C. C., Lee, Y. T. 1984. *Phys. Rev. Lett.* Submitted
149. Rettner, C. T., Marinero, E. E., Zare, R. N. 1983. In *13th Int. Conf. on Phys. of Electronic At. Collisions, Berlin, FRG, Jul. 27-Aug., 1983*. Amsterdam: North Holland. In press
150. Gerrity, D. P., Valentini, J. J. 1983. *J. Chem. Phys.* 79:5202-3
151. Quick, C. R. Jr., Moore, D. S. 1983. *J. Chem. Phys.* 79:759-64
152. Kleinermanns, K., Wolfrum, J. 1984. *J. Chem. Phys.* 80:1446-50
153. Kleinermanns, K., Schinke, R. 1984. *J. Chem. Phys.* 80:1440-45
154. Kleinermanns, K., Wolfrum, J. 1984. *Appl. Phys. B.* In press
155. Kleinermanns, K., Wolfrum, J. 1984. *Chem. Phys. Lett.* 104:157-59
156. Quick, C. R. Jr., Tiee, J. J. 1983. *Chem. Phys. Lett.* 100:223-26
157. Wight, C. A., Leone, S. R. 1984. *J. Chem. Phys.* Submitted for publication
158. Wicke, B. G. 1983. *J. Chem. Phys.* 78:6036-44
159. Hepburn, J. W., Klimek, D., Liu, K., MacDonald, R. G., Northrup, F. J., Polanyi, J. C. 1981. *J. Chem. Phys.* 74:6226-41
160. Irvin, J. A., Dagdigian, P. J. 1980. *J. Chem. Phys.* 73:176-82
161. Parson, J. M., Geiger, L. C., Conway, T. J. 1981. *J. Chem. Phys.* 74:5595-5605
162. Cox, J. W., Dagdigian, P. J. 1982. *J. Phys. Chem.* 86:3738-45
163. Irvin, J. A., Dagdigian, P. J. 1981. *J. Chem. Phys.* 74:6178-87
164. Hsu, Y. C., Pruett, J. C. 1982. *J. Chem. Phys.* 76:5849-55
165. Green, G. J., Gole, J. L. 1980. *Chem. Phys. Lett.* 69:45-49
166. Dehaven, J., O'Connor, M. T., Davidovits, P. 1981. *J. Chem. Phys.* 75:1746-51
167. DiGuiseppe, T. G., Davidovits, P. 1981. *J. Chem. Phys.* 74:3287-91
168. Dagdigian, P. J. 1982. *J. Chem. Phys.* 76:5375-84
169. Bourguignon, B., Rostas, J., Taieb, G. 1982. *J. Chem. Phys.* 77:2979-87
170. Breckenridge, W. H., Umemoto, H. 1983. *J. Phys. Chem.* 87:476-79
171. Rosano, W. J., Parson, J. M. 1983. *J. Chem. Phys.* 79:2696-2709

172. Schwenz, R. W., Geiger, L. C., Parson, J. M. 1981. *J. Chem. Phys.* 74:1736–44
173. Schwenz, R. W., Parson, J. M. 1982. *J. Chem. Phys.* 76:4439–44
174. Schwenz, R. W., Parson, J. M. 1980. *J. Chem. Phys.* 73:259–67
175. Dreiling, T. D., Setser, D. W. 1983. *J. Chem. Phys.* 79:5439–44
176. Kowalski, A., Menzinger, M. 1983. *J. Chem. Phys.* 78:5612–20
177. Schwenz, R. W., Parson, J. M. 1980. *Chem. Phys. Lett.* 71:524–28
178. Dirscherl, R., Lee, H. U. 1980. *J. Chem. Phys.* 73:3831–37
179. Hennessy, R. J., Ono, Y., Simons, J. P. 1980. *Chem. Phys. Lett.* 75:47–51
180. Hennessy, R. J., Ono, Y., Simons, J. P. 1981. *Mol. Phys.* 43:181–92
181. Hennessy, R. J., Simons, J. P. 1980. *Chem. Phys. Lett.* 75:43–46
182. Simons, J. P. 1982. *Chem. Phys. Lett.* 91:484–86
183. Golde, M. F., Ho, Y.-S., Ogura, H. 1982. *J. Chem. Phys.* 76:3535–42
184. Balamuta, J., Golde, M. F. 1982. *J. Chem. Phys.* 76:2430–40
185. Balamuta, J., Golde, M. F., Ho, Y.-S. 1983. *J. Chem. Phys.* 79:2822–30
186. Tamagake, K., Setser, D. W., Kolts, J. H. 1981. *J. Chem. Phys.* 74:4286–4305
187. Ehrlich, D. J., Osgood, R. M. Jr. 1980. *J. Chem. Phys.* 73:3038–45
188. Wilkinson, J. P. T., MacDonald, M., Donovan, R. J. 1983. *Chem. Phys. Lett.* 101:284–86
189. Rettner, C. T., Woste, L., Zare, R. N. 1981. *Chem. Phys.* 58:371–83
190. Oprysko, M. M., Aoiz, F. J., Bernstein, R. B., McMahan, M. A. 1983. *Chem. Phys.* 79:341–50
191. O'Grady, B. V., Lain, L., Gower, M. C., Donovan, R. J. 1982. *Chem. Phys. Lett.* 91:491–93
192. Donovan, R. J., O'Grady, B. V., Lain, L., Fotakis, C. 1983. *J. Chem. Phys.* 78:3727–31
193. Hepburn, J. W., Liu, K., MacDonald, R. G., Northrup, F. J., Polanyi, J. C. 1981. *J. Chem. Phys.* 75:3353–64
194. Nesbitt, D. J., Leone, S. R. 1980. *J. Chem. Phys.* 73:6182–90
195. Yuh, H.-J., Dagdigian, P. J. 1983. *J. Chem. Phys.* 79:2086–88
196. Bruno, J. B., Krenos, J. 1983. *J. Chem. Phys.* 78:2800–1
197. Stolte, S. 1982. *Ber. Bunsenges. Phys. Chem.* 86:413–21
198. Zare, R. N. 1982. *Ber. Bunsenges. Phys. Chem.* 86:422–25
199. Parker, D. H., Chakravorty, K. K., Bernstein, R. B. 1981. *J. Phys. Chem.* 85:466–68
200. Parker, D. H., Chakravorty, K. K., Bernstein, R. B. 1982. *Chem. Phys. Lett.* 86:113–17
201. van den Ende, D., Stolte, S. 1980. *Chem. Phys. Lett.* 76:13–15
202. van den Ende, D., Stolte, S., Cross, J. B., Kwei, G. H., Valentini, J. J. 1982. *J. Chem. Phys.* 77:2206–8
203. Karny, Z., Estler, R. C., Zare, R. N. 1978. *J. Chem. Phys.* 69:5199–5202
204. Rettner, C. T., Zare, R. N. 1981. *J. Chem. Phys.* 75:3636–37
204a. Rettner, C. T., Zare, R. N. 1982. *J. Chem. Phys.* 77:2416–29
205. deVries, M. S., Srdanov, V. I., Hanrahan, C. P., Martin, R. M. 1982. *J. Chem. Phys.* 77:2688–89
206. deVries, M. S., Srdanov, V. I., Hanrahan, C. P., Martin, R. M. 1983. *J. Chem. Phys.* 78:5582–89
207. Murphy, E. J., Brophy, J. H., Arnold, G. S., Dimpfl, W. L., Kinsey, J. L. 1981. *J. Chem. Phys.* 74:324–30
208. Andresen, P., Ondrey, G. S., Titze, B. 1984. *J. Chem. Phys.* 80:2548–69
208a. Alexander, M. H., Dagdigian, P. J. 1984. *J. Chem. Phys.* In press
209. Polak-Dingels, P., Delpech, J. F., Weiner, J. 1980. *Phys. Rev. Lett.* 44:1663–66
210. Hering, P., Brooks, P. R., Curl, R. F., Judson, R. S., Lowe, R. S. 1980. *Phys. Rev. Lett.* 44:687–90
211. Brooks, P. R., Curl, R. F., Maguire, T. C. 1982. *Ber. Bunsenges. Phys. Chem.* 86:401–7
212. Grieneisen, H. P., Xue-Jing, H., Kompa, K. L. 1981. *Chem. Phys. Lett.* 82:421–26
213. Wilcomb, B. E., Burnham, R. E. 1981. *J. Chem. Phys.* 74:6784–86
214. Ku, J. K., Inoue, G., Setser, D. W. 1983. *J. Phys. Chem.* 87:2989–93
215. Jouvet, C., Soep, B. 1983. *Chem. Phys. Lett.* 96:426–28
216. Arrowsmith, P., Bartoszek, F. E., Bly, S. H. P., Carrington, T., Charters, P. E., Polanyi, J. C. 1980. *J. Chem. Phys.* 73:5895–97
217. Arrowsmith, P., Bly, S. H. P., Charters, P. E., Polanyi, J. C. 1983. *J. Chem. Phys.* 79:283–301

INTERACTIONS AND KINETICS IN MEMBRANE MIMETIC SYSTEMS

Janos H. Fendler

Department of Chemistry and Institute of Colloid and Surface Science, Clarkson College of Technology, Potsdam, New York 13676

INTRODUCTION

Membrane mimetic chemistry has become a vitally important area of research (1–5). It is directed to the development of chemistries based on membrane-mediated processes in organized surfactant assemblies and molecular hosts. Aqueous and reversed micelles, microemulsions, monolayers, organized multilayers, bilayer or black lipid membranes (BLMs), and vesicles are considered to constitute the organized surfactant assemblies (1, 6). Molecular hosts include naturally occurring cyclodextrins (7) and synthetic crown ethers, cryptands, and spherands, collectively referred to as cavitands (8, 9). Organization and compartmentalization in membrane mimetic systems are exploited for reactivity control (2), transport (1, 10), recognition (1, 9, 11), drug delivery (12–14), and artificial photosynthesis (15–18).

Reactivities in the microheterogeneous environments of membrane mimetic systems cannot be described in terms of homogeneous kinetics. Attention has to be given to the partitioning of the reactants between the organized assembly and the bulk phase and to the structural and dynamic features of the system. Different kinetic treatments have been proposed for "fast" and "slow" reactions occurring in the different membrane mimetic systems. The important factor is, of course, the rate of reaction relative to the rates of reactant(s) and surfactant assembly reorganizations. A large variety of diverse approaches have been taken. Physical organic chemists have focused their attention on reactions whose rates were appreciably slower than the time scale needed for the reorganization of the membrane mimetic system used. They treated reactivities in terms of simple partitioning of the reactants (19–22) and, subsequently, in terms of a pseudophase ion exchange model (23–27). Physical chemists, particularly those concerned with artificial photosynthesis, have considered free radical and excited state processes, occurring at timescales similar to those needed for the aggregate reorganization, in terms of Poisson's distributions and residence times (28–32). Theoreticians have approached reactivities in membrane mimetic systems in terms of stochastic and lattice statistical theories (33–43).

Different treatments of reactivities in membrane mimetic systems is the major subject of the present review. The review begins with a brief summary of the properties of the different surfactant assemblies. This will orient the neophyte. Emphasis is placed on recent experimental results that demand mechanistic interpretations and theoretical analyses. The highly interdisciplinary nature of membrane mimetic chemistry should be recognized at the onset. Synthetic, organic, inorganic, colloid, physical, polymer, biological, electro-, and photochemists have contributed to this area. Naturally, their methodologies, interpretations, and even terminologies have been quite different. Reduction of this interdisciplinary communication barrier, an important goal of the present review, will, I hope, challenge and stimulate the reader to contribute creatively to membrane mimetic chemistry.

ORGANIZED SURFACTANT ASSEMBLIES

Surfactants, amphiphiles, or detergents are molecules having distinct hydrophobic and hydrophilic regions. Depending on their chemical structures they can be neutral [e.g. polyoxyethylene-(9.5)-octylphenol, triton X-100] cationic (e.g. hexadecyl trimethylammonium bromide, CTAB) anionic (e.g. sodium dodecyl sulfate, SDS), and zwitterionic [e.g. 3-(dimethyldodecylammonio)-propane-1-sulfonate]. The hydrophobic part of the surfactant can be of various lengths (typically 8–20 carbon atoms), contain multiple bonds, or consist of two (e.g. dioctadecyldimethylammonium chloride, DODAC, or dihexadecyl phosphate, DHP) or more hydrocarbon chains. Surfactants have also been functionalized to contain desired reactive and/or reporter groups. Aggregation behavior depends on the nature and the concentration of the surfactants, the nature of the solvent, and the method of preparation (1, 44–47).

In water, 50–100 surfactant molecules, above a characteristic concentration (the critical micelle concentration, or CMC), dynamically and spontaneously associate to aggregates known as micelles. Formation of aqueous micelles is a cooperative process. Opposing forces of repulsion between the polar headgroups and association between the hydrophobic chains of the surfactants are responsible for micellization (48). In general, the longer the hydrocarbon tails, the lower the CMC. Aqueous micelles are the most disorganized organized surfactant assemblies. They rapidly break up and reform by two known processes. The first occurs on the microsecond time scale. It releases a single surfactant from the micelle and subsequently reincorporates it. The second process occurs on the millisecond time scale and is due to (*a*) the dissolution of the micelle and (*b*) the subsequent reassociation of the monomers. Hydrodynamic diameters of micelles vary between 40–100 Å.

Micelle solubilization of water insoluble compounds has been known, of course, for a long time (49). Solubilizates are typically distributed in the relatively aqueous environment of the micelle. Referring to "deep substrate penetration" or "rigid substrate location" in the micellar interior is erroneous.

Increasing the concentration of the surfactant leads to the formation of rod-like micelles and subsequently to liquid crystals.

Surfactants having an appropriate hydrophobic lipophilic balance [e.g. bis-(2-ethylhexyl) sodium sulfosuccinate, AOT] undergo self-association in apolar solvents (50–54). In the absence of water, the association is sequential rather than cooperative. There is no equivalent of a critical micelle concentration (CMC). Small (containing 4–10 monomers) concentration-dependent aggregates are formed. A useful property of hydrocarbon-soluble surfactants is that they are able to solubilize a large number of water molecules (1 AOT molecule is able to solubilize up to 60 molecules of water in a hydrocarbon solvent). Surfactant-solubilized water pools in the hydrocarbon solvent are referred to as reversed or inverted micelles. Reversed micelles have diameters of 40–100 Å and their sizes are independent of the temperature. Increasing the concentration of the surfactant-entrapped water, i.e. the size of the water pools, at a given surfactant concentration results in the formation of water-in-oil (w/o) microemulsions. The hydrodynamic diameter of w/o microemulsions are strongly dependent on the temperature. Reversed micelle-entrapped water pools are unique. At relatively small water-to-surfactant ratios (w-values, $w = 8$–10), all water molecules are strongly bound to the surfactant head groups and cannot be frozen. These immobilized water pools provide the medium for very large and highly specific rate enhancements and resemble the hydrophilic pockets of enzymes (50, 51).

Spreading an organic solution of a surfactant or phospholipid on water leads to monolayer formation at the water-air interface (55–57). Spreading is accomplished in a Langmuir trough which controls the surface area and surface pressure. Scrupulous cleanliness is a must in monolayer studies.

Dipping a dry plate or glass through the monolayer results in its transfer to a solid support. Successive dipping in and out of a monolayer-covered liquid results in the build-up of multilayers. Depending on the mode of dipping, plate-tail-head-tail (X-type multilayers), plate-tail-head-head-tail (Y-type multilayers), or plate-head-tail-head-tail (Z-type multilayers) can be formed. More recently monolayers have been formed by adsorbing surfactants on the polar solids, typically via silination (58). Appropriate chemical treatments allow the build up of successive multilayers. Monolayers are ideal media for investigating two-dimensional processes.

Bilayer or black lipid membrane (BLMs) are formed upon brushing an organic solution of lipids across a pinhole separating aqueous compartments (10). The initially formed film is rather thick and reflects the light with interference colors. Thinning of the film to a single bilayer occurs within minutes and can be monitored by capacitance measurements and by the graying of the reflected light. Although BLMs are good models for the biological membrane, their relatively short lifetime is a distinct disadvantage.

Closed bilayer aggregates, formed from phospholipids (liposomes) (59) or from surfactants (vesicles) (60), represent one of the most sophisticated models of the biological membrane. Swelling of thin lipid films in water results in the formation of onion-like multilamellar vesicles (MLVs) of 1000–8000 Å diameter. Sonication of MLVs above the temperature at which they are transformed from a gel to a liquid (phase transition temperature) leads to the formation of fairly uniform single bilayer vesicles of 300–600 Å diameter (small unilammelar vesicles, SUVs). SUVs can also be prepared by injecting an alcohol solution of the surfactant through a small bore syringe into water, or by detergent dialysis, or gel filtration. Uniform SUVs can be obtained by ultracentifugation, and by membrane or gel filtration. Both SUVs and MLVs undergo temperature-induced phase transitions. At low temperature, phospholipids are arranged in tilted one-dimensional lattices. At temperatures just below the transition, two-dimensional arrangements form. Above the main phase transition temperature, phospholipids revert to one-dimensional arrangements, separated somewhat from each other, and assume liquid-like configurations. The kinetic stability of vesicles is considerably greater than that of micelles. There is no equivalent of a critical micelle concentration. Once formed, vesicles cannot be destroyed by dilution. Typically, they are stable for weeks. Their stability can be enhanced further by polymerization (61–63). Vesicles have a large number of solubilization sites. Hydrophobic molecules can be distributed among the hydrocarbon bilayers of the vesicles. Polar molecules may move about relatively freely in vesicle-entrapped water pools, particularly if they are electrostatically repelled from the inner surface. Small charged ions can be electrostatically attached to the oppositely charged outer or inner surfaces of vesicles. Species having charges identical to those of the vesicles can be anchored onto the vesicle surface by a long hydrocarbon tail.

Table 1 summarizes the most important properties of surfactant assemblies utilized in membrane mimetic chemistry.

PSEUDOPHASE PARTITIONING AND ION EXCHANGE MODELS FOR HETEROLYTIC REACTIONS

Kinetic treatments were developed initially for heterolytic reactions occurring in the environments of aqueous micelles in terms of a pseudophase model (19–22). This model assumes a separate phase for micelles, the micellar pseudophase, even though aqueous micellar solutions are macroscopically homogeneous and optically transparent (1, 64, 65). The micellar pseudophase appears at the CMC, above which all additional surfactants are assumed to be in the form of micelles. Further assumptions are that the reactants do not disturb the monomer-micelle equilibrium and that the distribution of the reactants between the micellar pseudo- and bulk phases are unaltered during the reaction. Reactivities are then considered to be the sums of reactions occurring in the bulk aqueous (R_w) and in the micellar pseudophases (R_M):

$$R_{total} = R_w + R_M. \qquad 1.$$

For unimolecular reactions, partitioning of only one substrate needs to be considered. Reactivities are given by

$$S + M \underset{}{\overset{K_S}{\rightleftharpoons}} SM$$
$$\downarrow k_w \qquad \downarrow k_M \qquad\qquad 2.$$
products products

where S is the substrate, M is the micelle, SM is the micelle-substrate complex, and k_w and k_M are the first order rate constants in the aqueous and micellar pseudophases, respectively. The observed first order rate constant for the overall reaction, k_{obs}, is described by:

$$k_{obs} = \frac{k_w + k_M K_s[M]}{1 + K_s[M]}. \qquad 3.$$

Recognizing the analogy between Eq. 3 and the Michaelis-Menten equation for enzyme-catalyzed reactions (66) allowed the treatment of data

Table 1 Properties of organized surfactant assemblies

	Aqueous micelles	Reversed micelles	Microemulsions	Monolayers	BLMs	Vesicles
Method of preparation	Dissolving appropriate concentration of the surfactant in water (above the CMC)	Dissolving appropriate concentration of the surfactant in an apolar solvent and adding small amounts of H_2O	Dissolving appropriate concentrations of the surfactant and cosurfactant in water or in oil	Spreading the surfactant (or a dilute solution of it in an organic solvent) on water surface	Painting a dilute surfactant on a teflon pinhole	Shaking thin films of lipids (or surfactants) in water, or ultrasonication, or alcohol injection, or detergent dilution
Weight averaged molecular weight	2000–6000	2000–6000	10^4–10^7	Depends on area covered and density of coverage	Depends on area covered and density of coverage	$>10^7$
Hydrodynamic diameter Å	40–100	40–100	50–5000	Depends on area covered and density of coverage	Depends on area covered and density of coverage	300–10,000
Timescale of monomer aggregate formation, breakdown	10^{-4}–10^{-6} sec	10^{-4}–10^{-6} sec	10^{-4}–10^{-6} sec	Monomer to subphase minutes-hours	Monomer to plateau border minutes	Monomer to bulk minutes-hours
Stability	Months	Months	Months	Days, weeks	Hours	Weeks-months
Dilution by H_2O	Destroyed	Water pools, enlarges: w/o microemulsions are formed	Depends on the phase diagram	—	—	Unaltered
Number of reactants solubilized per aggregate	Few	Few	Large	Large	Large	Large
Solubilization sites	Distributed around and within the Stern layer. No "deep penetration"	Aqueous inner pool, inner surface, surfactant tail	Inner pool, inner surface, surfactant tail	Intercalation and surface	Either or both sides of the bilayer or within the bilayer	Inner pool, inner, outer surface, bilayer

by

$$\frac{1}{k_w - k_{obs}} = \frac{1}{k_w - k_M} + \frac{1}{(k_w - k_M)K_s[M]}, \qquad 4.$$

which is similar to the Lineweaver-Burke equation of enzyme kinetics. Equation 4 describes well the kinetics of both inhibited and catalyzed unimolecular processes (6). Micellar effects on unimolecular reactions can be rationalized in a manner analogous to that marshalled for solvent effects on reactivities (6). An overall decrease in the free energy can be accomplished either by ground state destabilization or by transition state stabilization, or by a combination of both. Equation 4 describes well (4) the kinetics of both micelle-inhibited and catalyzed reactions. Typical sigmoidal plots were obtained as function of surfactant concentration (6).

Equation 3 does not adequately describe the kinetics of bimolecular or higher order reactions in micelles. Rate enhancements with increasing concentrations often pass through maxima. Partitioning of both reactants (A and B) between bulk and micellar pseudophases has to be considered.

$$(A+B)_w \xrightarrow{k_w} \text{Products}$$

$$(A+B)_M \xrightarrow{k_M} \text{Products} \qquad 5.$$

where the subscripts w and M refer to the bulk water and micellar pseudophases. The overall rate, described by Eq. 1, is modified to

$$R_{total} = k_M[B]_M[A]_M[M]\bar{V} + k_w[A]_w[B]_w(1-[M]\bar{V}) \qquad 6.$$

where $\bar{V}$ is the molar volume of the micelle and the concentrations of reagents A and B are given by material balances:

$$[A]_{Total} = [A]_M[M]\bar{V} + [A]_w(1-[M]\bar{V}) \qquad 7.$$

$$[B]_{Total} = [B]_M[M]\bar{V} + [B]_w(1-[M]\bar{V}). \qquad 8.$$

If the chemical reaction 5 does not affect the partition equilibria

$$[A]_w \xrightleftharpoons{P_A} [A]_M \qquad 9.$$

$$[B]_w \xrightleftharpoons{P_B} [B]_M \qquad 10.$$

the observed second order rate constant for reactions in the presence of micelles is given by

$$k_2 = \frac{k_M P_A P_B[M]\bar{V} + k_w(1-[M]\bar{V})}{(1+(P_A-1)[M]\bar{V})(1+(P_B-1)[M]\bar{V})}. \qquad 11.$$

If both A and B bind strongly to the host ($P_A \gg 1$ and $P_B \gg 1$) and if $[M]\bar{V} \ll 1$ then Eq. 11 simplifies to

$$k_2 = \frac{(k_M/\bar{V})K_A K_B [M] + k_w}{(1 + K_A[M]\bar{V})(1 + K_B[M])} \qquad 12.$$

where the binding constants are expressed by

$$K_A = (P_A - 1)\bar{V}$$
$$K_B = (P_B - 1)\bar{V}. \qquad 13.$$

Equations 11 and 12 have been found to describe well the kinetics of bimolecular reactions in micelles (1, 22, 67). Dependence of the observed second order rate constants, k_2, on the surfactant concentration have been used to calculate the "true" rate constants for the reaction in the micellar pseudophase, k_M, and to assess substrate binding constants K_A and K_B. Agreements between kinetically determined K_A and K_B values and those obtained independently (from solubility experiments, for example) have lent credence to the treatment and justified the assumptions made in deriving Eqs. 11 and 12.

Equation 11 has been shown to be inadequate for treating ionic reactions in the presence of charged micelles and for accounting for electrolyte effects on the rates of micelle-catalyzed reactions (67). Two modifications have been proposed to overcome these difficulties. The first, developed by Romsted (21, 22), incorporated counterion distributions in the micellar Stern layer. The second, the ion exchange model (24–27), viewed the micellar surface as an ion exchange resin.

The Romsted model considers the micellar Stern layer to be saturated by counterions (21, 22). The degree of ionization, (i), reflecting the counterion distribution between the aqueous and micellar pseudophases is assumed to be independent of the surfactant concentration and the ionic strength at constant temperature. The hydrophilic ionic reagent, X, and the nonreactive micellar counterion, Y, exchange rapidly between the two phases

$$X_M + Y_w \rightleftharpoons X_w + Y_M \qquad 14.$$

where the subscripts M and w refer to micelles and water. The selectivity coefficient for the ionic reagent at the Stern layer

$$K_{X/Y} = \frac{X_w Y_M}{X_M Y_w} \qquad 15.$$

determines the concentration of the hydrophilic reagent in the micellar phase. It is quite feasible, therefore, to obtain high X_M concentrations even when the stoichiometric concentration of Y far exceeds that of X. The observed second order rate constant between a neutral reagent A and a hydrophilic reagent in the presence of an ionic micelle is described by the modified Eq. 12 (21, 22)

$$k_2 = \frac{K_M i' D K_A[M]}{(1-K_A[M])(X_t + Y_t K_{X/Y})} + \frac{k_w}{K_A[M]+1} \qquad 16.$$

where i' is the degree of counterion binding to the Stern layer ($i' = 1-i$), d is the molar density of the micellar phase expressed in moles of surfactant per liter of micellar phase, and $X_t = [M] + [BY]$ is the concentration of added salts. Equation 16 has successfully predicted the kinetic behavior of a large number of second order reactions in ionic micelles as well as the salt effects therein (68). This model cannot, however, be applied to buffered systems.

The ion exchange model has provided quantitative rationalizations for the binding of a reactive ion to the micelle both in the absence and in the presence of buffers and for both first order (ionic substrate) and second order (neutral substrate with an ionic reagent) reactions. Dissociation of weak acids and bases and the binding of OH^- to cationic micelles have also been treated by this model (23–26). The ion exchange model assumes that (a) the distribution of aggregate sizes can be presented in terms of most probable aggregation N; (b) ion-ion and ion-headgroup-headgroup interactions are noncooperative; (c) degrees of ionization (i) of the individual micellar species are the same; (d) ion-ion exchange rates are rapid compared to the lifetime of the micelle; and (e) activities of micellar and ionic species can be treated in terms of their concentrations. With these assumptions, the selectivity coefficient for a reactive counterion X^- in B^+X^-, in micelle forming detergent D^+Y^-, in the absence or in the presence of an added common salt, B^+Y^-, is given by

$$K_{X/Y} = \frac{X_M^-}{(X_T^- - X_M^-)} \frac{i[M] + CMC + X_M^- + [B^+Y^-]_T}{(1-i)[M] - X_M^-} \qquad 17.$$

where the subscripts T and M refer to total concentrations of the appropriate species and to those present in the micellar pseudophase. At high detergent concentration

$$K_{X/Y} = \frac{i}{1-i} \frac{X_M^-}{X_w^-} \qquad 18.$$

equation 18 predicts that X_M^-/X_w^- tends to a limiting value and it allows, therefore, the assessment of K_{X^-/Y^-}. Addition of a buffer maintains X_w^- rather than X_M^-. Rate constants for the reaction of a univalent ionic substrate, S^-, an an oppositely charged micelle, D^+Y^-, is given by

$$k_{obs} = \frac{k_M K_{S^-/Y^-}(Y_M^-/Y_w^-) + k_w}{1 + K_{S^-/Y^-}(Y_M^-/Y_w^-)}. \qquad 19.$$

Equation 19 should be compared to Eq. 3. The selectivity of micelle bound ions rather than substrate partitioning is expressed in Eq. 19. Similarly, incorporation of the concepts of ion exchange theory in treating bimolecular reactions in micelles (see Eq. 12) leads to Eqs. 20 and 21 for reactions between S and an oppositely charged X in the absence and in the presence of buffers, respectively (23–26):

$$k_2 = \frac{X_T[(k_M/\bar{V})(K_S K_{X/Y})(Y_M^-/Y_w^-) + k_w]}{(1 + K_S[M])[1 + K_{X/Y}(Y_M^- Y_w^-)]} \qquad 20.$$

$$k_2 = \frac{X_w[(k_M/\bar{V})(K_S K_{X/Y})(Y_M^-/Y_w^-) + K_w]}{(1 + K_S[M])}. \qquad 21.$$

It should be noted that the assumptions are the same in the Romsted-Bunton (67) and the ion exchange (23–26) treatments. They only differ in the mathematical treatments and experimental tests. The main difference between the two approaches is that equations in the former are derived in terms of stoichiometric quantities of materials including hydrophilic ions, whereas the ion exchange equations are expressed in terms of the concentration of the hydrophilic ions in the aqueous phase, which have to be determined independently. Not unexpectedly, the pseudophase model breaks down for reactions carried out in surfactants near their CMCs (27).

Distribution of reactive counterions, discussed in terms of micellar surface potentials (36, 69–71), has led to equations similar to those based on the ion exchange model, although the surface potential model cannot account for salt effects.

Kinetic treatments, derived for heterolytic reactions in aqueous micelles, have been extended to reversed micelles and vesicles (72–74) and are likely to be applicable to monolayers, microemulsions, and polyelectrolytes. Counterion condensation theory (75–79) can be used to account for reactivities in polyelectrolytes.

Regardless of the model used, rate enhancements for bimolecular reactions appear to be the mere consequence of concentrating the reagents in the membrane mimetic agents. Media effects can only be operational for unimolecular reactions.

STATISTICAL TREATMENTS OF EXCITED STATE AND RADICAL REACTIONS

Excited state and radical reactions occurring in small micellar compartments have stimulated many theoretical treatments (1). The lifetimes (typically, 10^{-9}–10^{-4} sec) and volumes of reactants relative to those of the micelles are the important parameters. Both intra- and intermicellar reactions have been realized experimentally and treated theoretically.

Three types of intramicellar kinetic processes can be visualized. In the first type, reactant A is localized on the micellar surface subsequent to the diffusion of reactant B (typically photoinitiated) to the site of reactant A. In the second type, irreversible or reversible reactions take place between reactants A and B, both distributed within the micelle. Finally, in the third type of reaction, diffusion of both A and B is restricted to the micellar surface. Careful control of conditions allowed the experimental realization of all of these intramicellar reactions (38).

Photosensitized energy transfer from a micelle-solubilized donor to an acceptor, localized on the micellar surface (38), illustrates the first type of intramicellar reaction. Diffusion of the excited donor to the micellar surface is described by a random walk process. The average concentration of donor at time t and at a distance r from the center of the micelle, $C(r, t)$, is given by

$$\frac{\partial C(r, t)}{\partial t} = D\nabla^2 C(r, t) \qquad 22.$$

where D is the diffusion coefficient and ∇^2 is the Laplacian operator with spherical symmetry. Making the assumptions that the surface of the micelle is a perfectly absorbing boundary, that the donor is free to move about in all positions at $t = 0$, and that its concentration remains finite for all times and at all positions allowed, the exact solution of Eq. 22, which expressed in terms of the spacially averaged donor concentration, $\langle C(r,t) \rangle_r = \langle C(t) \rangle$, becomes (38)

$$\langle C(t) \rangle = \frac{6C_0}{\pi^2} \sum_{n=1}^{\infty} \frac{1}{n^2} \exp\left(\frac{-n^2 \pi^2 D t}{r_0^2}\right). \qquad 23.$$

An important consequence of Eq. 23 is that on a time scale for which $t > r_0^2/\pi^2 D$, the $n = 1$ term in the summation dominates the kinetic behavior. Since for micelles $r_0^2/\pi^2 D$ is in the order of 10 nsec, any processes occurring on time scales greater than 10 nsec will be governed by pseudo-first-order kinetics.

Energy transfers between hydrophobic donors and acceptors provide an example of the second type of intramicellar reaction. Both irreversible and reversible energy transfers have been treated (33, 34, 38). Precise theoretical treatments are hampered by the comparable lengths of reactants and micelles. The problem has been approached by considering the distribution of the reactants according to Poisson's statistics and by placing an absorbing barrier into the path of the excited donor undergoing random walk. Equations have been derived both for two and three dimensional random walks. Pseudo-first-order kinetics are observed when there are not more than one donor and acceptor molecule micelle. Rates of electron transfer from a single donor to several acceptors occupying the same micelle were found to increase linearly with increasing acceptor concentration (78, 79).

Electron, energy and proton transfers occur on micellar surfaces at an extremely fast rate (80–82). In most cases the diffusion of the reacting partners is minimal. Special implications of reduced dimensionality are discussed in the following section.

A more general stochastic master equation has been recently derived for the kinetic analysis of intramicellar processes that incorporates both physical diffusion and chemical reactions (33–35). This has the potential of being applied to microemulsions and vesicles. Lattice statistical theories have also been utilized for treating intramicellar reactions (42, 43).

Quenching the fluorescence of a completely micellized probe, P, by a quencher, Q, provides the means for determining intermicellar reactivities. The system is described by Eq. 24 (31):

$$\begin{array}{ccccc}
(P^*) + Q_{aq} \underset{k_-}{\overset{k_+}{\rightleftharpoons}} (P^* \cdot Q) & +Q \underset{2k_-}{\overset{k_+}{\rightleftharpoons}} & (P^*2Q) & & \\
h\nu \uparrow\downarrow \tau_0^{-1} & h\nu \uparrow\downarrow \tau_0^{-1} + k_r & h\nu \uparrow\downarrow \tau_0^{-1} + 2k_r & & 24. \\
(P) + Q_{aq} \underset{k_-}{\overset{k_+}{\rightleftharpoons}} (P^*Q) & +Q \underset{2k_-}{\overset{k_+}{\rightleftharpoons}} & (P \cdot 2Q) & &
\end{array}$$

where Q may be in the aqueous, Q_{aq}, or in the micellar phase, $P \cdot Q$. The circles symbolize the micelles, M, without implying any preferred sites for P, P^*, $P^* \cdot Q_{aq}$ and so on; τ_0 is the fluorescence lifetime in the absence of quenchers; k_+ and k_- are rate constants for entry into and exit from the micelles; and k_r is the intramicellar quenching rate constant. Four limiting cases have been recognized:

1. Case 1 is static quenching for systems in which both the probe and the quencher are exclusively localized in the micelles and, therefore, emission occurs only from that fraction of P^* containing micelles that are free of quencher molecules. Case 1 is characterized by a single exponential decay of the fluorescence intensity, which is independent of $[Q]$. The steady state luminescence intensity, under these conditions, is given by Eq. 25

$$I/I_0 = \exp(-[Q]/[M]) \qquad 25.$$

where I and I_0 are fluorescence intensities in the presence and in the absence of Q. The micelle concentration is expressed by

$$[M] = \frac{[\text{Total Surfactant}] - [\text{Monomers}]}{\bar{n}}. \qquad 26.$$

Combination of Eqs 25 and 26 allows the determination of the average aggregation number, $\bar{n}$, by means of Eq. 27 from fluorescence quenching (83):

$$\ln (I_0/I) = [Q]\bar{n}/(\text{Total Surfactant}] - [\text{Monomers}]). \qquad 27.$$

2. Case 2 is that while P resides exclusively in the micellar phase, Q is distributed in the micellar and bulk aqueous phases, and the quenching is static. Steady state emission still occurs only from the fraction of micellized P^* that is free of Q and for steady state illumination

$$\frac{I_0}{I} = \frac{(1+k_+\tau_0[Q_{aq}]) \exp K[Q]}{(1+K[M])} \qquad 28.$$

where the first and second terms are contributions to intensity quenching from dynamic and static quenching, respectively. Equation 28 reduces to Eq. 27 when Q is completely in the micellar phase ($K[M] \gg 1$). The lifetime of emission, τ, is however, reduced by dynamic diffusional quenching by the water solubilized quenchers, Q_{aq}

$$\tau^{-1} = \frac{\tau_0^{-1} + k_+[Q_{aq}]}{(1+K[M])}. \qquad 29.$$

Since increasing concentrations of surfactants Q_{aq}, Case 2 is characterized by increased luminescence lifetimes with increasing concentrations of micelles. Steady state and time resolved data for Case 2 afford values for K, k_+, k_-, and $\bar{n}$.

3. Dynamic quenching with complete micellization of Q represents the third mode of quenching (Case 3). For a given micelle the quenching probability is proportional to the number of quenchers in the micelle with the resultant nonexponential decay of the fluorescence intensity. Fluorescence intensity under steady state irradiation is given by

$$\frac{I}{I_0} = e^{-\langle q \rangle} \sum_{q=0} \frac{\langle q \rangle^q}{[1+q(k_r\tau_0)]q!} \qquad 30.$$

where $q = [Q]/[M]$. The luminescence decay function is expressed as

$$\frac{I(t)}{I_0} = \exp\left\{-\left[\frac{I}{\tau_0} + \langle q \rangle (1-e^{-k_rt})\right]\right\}. \qquad 31.$$

Equation 31 reduces to Eq. 25 when $k_r\tau_0 \to 0$ (i.e. when quenching becomes static). Treatment of data under conditions corresponding to Case 3 provides information on the intramicellar quenching constant, k_r (84).

4. Dynamic quenching with partially micellized quencher represents the typical situation (Case 4). Fluorescence decay for Case 4 is also nonexponential (85).

Kinetics of substrate solubilization in reversed micelles have been investigated by relaxation techniques (86–88).

Dimensionality Reductions

Reduction of a three dimensional reaction volume to a two dimensional surface in which molecules react by diffusion has been recognized to contribute to the efficiency of biochemical processes (89, 90). Membrane mimetic systems, particularly micelles (80–82), monolayers (91), and bilayers (92), have provided ideal media for examining the effects of changes in dimensionality and geometry on the rates of diffusion controlled reactions (92a).

Pulse radiolytic examination of Br_2^- disproportionation (Eq. 32)

$$Br_2^- + Br_2^- \rightarrow Br_3^- + Br^- \qquad 32.$$

in the presence of micellar CTAB led to the identification of two kinetic processes—a two dimensional surface diffusion and a three dimensional intermicellar reaction (80). Rate constants of $k_{2d} = 2.1 \times 10^6$ sec^{-1} and $k_{3d} = 1.46 \times 10^9$ M^{-1} sec^{-1} have been determined for these processes. The ratio of these rate constants ($k_{2d}/2k_3$) has been substantiated by calculating the ratios of steps a random walker needs to take in a three dimensional, $\langle n \rangle_3$, vs in a two dimensional, $\langle n \rangle_2$, space. Using Montroll's lattice geometry (93) and estimating the concentration of traps (2.0×10^{-2} traps/total number of lattice sites on the CTAB surface and 4.44×10^{-5} traps/total number of sites in the micelle) led to $\langle n \rangle_3 \sim 3.4 \times 10^4$ and $\langle n \rangle_2 \sim 82$. Taking the time for the reactions of two Br_2^- radicals in two or three dimensional space to be $\langle \tau \rangle_i = \langle n \rangle_i t_i$ (where t_i is the time required for a diffusional jump from one lattice point to the next and the subscript i is the dimensionality) and assuming the time and distance between jumps to be the same in two and three dimensional space, $\langle \tau \rangle_3 / \langle \tau \rangle_2 = \langle n \rangle_3 / \langle n \rangle_3$ is obtained and hence

$$k_{2d}/2k_{3d} = \ln 2[Br_2^-]_0 \langle n \rangle_3 / \langle n \rangle_2. \qquad 33.$$

Substitution of $[Br_2^-] = 2.69 \times 10^{-7}$ M and $\langle n \rangle_3 = 3.4 \times 10^4$ and $\langle n \rangle_2 = 82$ into Eq. 33 leads to $k_{2d}/2k_{3d} = 7.7 \times 10^{-3}$, which is in good agreement with the experimentally obtained ratio of two to three dimensional rate constants, 1.5×10^{-3} (80). As seen from this example, dimensionality reduction manifested in significant kinetic effects.

Reduction of dimensionality has been treated more exactly in terms of a lattice theory of reaction efficiency (42, 43). Group theoretical arguments have been used for determining the average number of steps required ($\langle n \rangle$) for a diffusing coreactant A to react with stationary target molecule B. Changes in $\langle n \rangle$ as a function of dimensionality reductions have been calculated for systems having symmetrical and tubular geometries (43). These calculations provided evidence for the complexity of the interplay among dimensionality, geometry, and reactivity (38).

REACTIVITY CONTROL IN SURFACTANT VESICLES

Advantage has been taken of the different solubilization sites in surfactant vesicles to organize reactants and hence to control reactivities (60). Both inter- and intravesicular reactions have been demonstrated. Intravesicular reactions may occur at the interior (endovesicular), exterior (exovesicular), or across the bilayer (transvesicular) of vesicles (94–103). Alteration of vesicle fluidities (104–108) and polymerization (109) have also been utilized for reactivity control.

Distinct exovesicular reactions provide an entry (98, 100, 110), in addition to direct synthesis (101), for generating chemically dissymmetrical surfactant vesicles. Dissymmetrical vesicles are not only inherently interesting but have important potential applications in membrane mimetic chemistry (99).

Reaction rates, under appropriate conditions, can be controlled by the transit of one of the reagents across the bilayer. An example is the oxidation of $2C_{16}N^+2C_1Br^-$ vesicle-entrapped 2-nitro-5-thiolatobenzoate, Ellman's anion, Ell$^-$, by exovesicular o-iodosobenzoate ion, ArlO$^-$ (96). In water, oxidation of Ell$^-$ by ArlO$^-$ is rapid; $k = 675$ M^{-1} sec^{-1}. When Ell$^-$ was added to empty $2C_{16}N^+2C_1Br^-$ vesicles, followed by the addition of ArlO$^-$, the reaction rate was found to be similar to that in water. It was attributed, therefore, to exovesicular reactivity. Cosonication of $2C_{16}N^+2C_1Br^-$ and Ell$^-$, followed by ion exchange chromatography, led to vesicles that contained Ell$^-$ only in their inside. Reaction in this vesicle, occurring 440-fold slower than that observed for the corresponding exovesicular process, was suggested to be limited by the rate of ArlO$^-$ permeation across the vesicle bilayer (96).

Polymerized surfactant vesicles (1, 5, 61–63) provide additional possibilities for reactivity control. Aminolysis of p-nitrophenyl laurate (PNPL) added to vesicles prepared from a styrene containing surfactant, $[C_{15}H_{31}CO_2(CH_2)_2]_2N^+(CH_3)CH_2C_6H_4CH=CH_2$, $Cl^-:2C_{18}N^+C_1Sty$, by ethylene diamine (EDA) resulted in biphasic kinetics that were resolved into k_{fast} and k_{slow} components (108). Increasing EDA concentrations increased both k_{fast} and k_{slow}. This rate enhancement was two orders of magnitude more effective for nonpolymerized than for polymerized vesicles. In nonpolymerized vesicles 90% of PNPL reacted with EDA. Although only 50% of PNPL could be converted to products in polymerized $2C_{18}N^+C_1Sty$ vesicles, the kinetics were still governed by two consecutive (k_{fast} and k_{slow}) first-order steps. These results could only be accommodated by assuming reaction sites additional to endo- and exovesicular locations and that PNPL is distributed differently in nonpolymerized and polymerized vesicles (108).

Kinetic examination of $2C_{18}N^+C_1Sty$ photopolymerization has provided an interesting example for two dimensional surface reactions (111). Photolysis of argon-bubbled vesicular solutions of $2C_{18}N^+C_1Sty$ by 15 nsec bursts of 266 nm laser pulses led to the first order disappearance of styrene absorbances. Polymerization rates for vesicular $2C_{18}N^+C_1Sty$ were found to be independent of the vesicle concentration but depended linearly on the applied laser energy. Conversely, rates in ethanol depended on the concentration of monomeric $2C_{18}N^+C_1Sty$. This observation and the fact that sizes of vesicles remained unaltered upon polymerization supported the proposed intravesicular surface polymerization. Photopolymerization has been treated according to Eq. 34 (111)

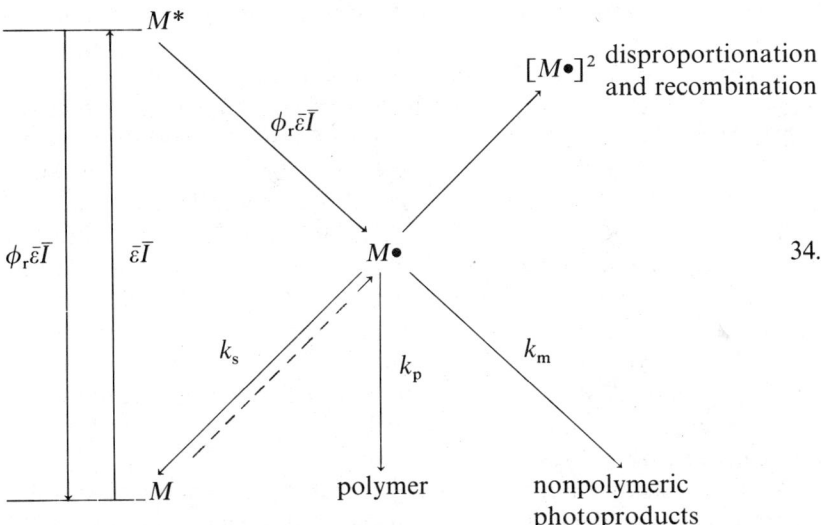

34.

where the styrene free radical, $M\bullet$, is assumed to form via the excited state (M^*) of the monomer, M; Φ_f and Φ_r are quantum efficiencies for nonproductive excited state depletion and $M\bullet$ formation, respectively; $\bar{\varepsilon}$ is the average molecular extinction coefficient; and $\bar{I}$ is the average intensity of exciting photons. Several possible fates await the free radical. Most interestingly, it can link successively to other monomers and thus propagate a polymer chain (a process governed by k_p). $M\bullet$ may also be deactivated and return to the ground state ("suicide," by k_s); or react with oxygen, impurity, and the wall of the vessel (by k_m); or undergo radical disproportionations and recombinations; or restore a styrene double bond and partially maintain chain propagation by conserving a free radical. This negligible term is represented by $[M\bullet]^2$. Considering photopolymerization on a per vesicle rather than on a concentration or unit volume basis leads to differential Eqs. 35 and 36

$$\frac{dM}{dt} = (\Phi_f - 1)\bar{\varepsilon}\bar{I}M + k_s M\bullet - k_p w(t)M\bullet + f[M\bullet]^2 \qquad 35.$$

$$\frac{dM\bullet}{dt} = \bar{\varepsilon}\bar{I}\Phi_r M - (k_m + k_s)M\bullet - (1-f)[M\bullet]^2 \qquad 36.$$

where $M(t)$ and $M\bullet(t)$ are the number of double-bond-containing monomers remaining and free radicals present at time t, f is the fraction of free radicals involved in combinations and disproportionations that restore double bonded monomers; and $w(t)$ indicates the average number of nearest monomeric neighbors of M at time t. Notice that propagation, $k_p w(t)M\bullet$, present only in Eq. 35, does not affect the $M\bullet$ population, represented in Eq. 36, as the free radical is conserved upon each successful polymerization link. Notice also that the term k_m does not appear in the equation for M (see Eq. 35), as photoproduct formation does not restore any M.

Experimental data showed the completion of $2C_{18}N^+C_1Sty$ vesicle laser photopolymerization within the 1 msec to 2 sec time domain (111). Vesicle solutions are restored, therefore, to a new equilibrium of monomers, polymers, and photoproducts prior to the arrival of the next laser pulse at 2 Hz repetition rate. Since the energy is applied in successive, essentially equal, increments, the most useful way of expressing the data is in the form of a graph of monomer concentration vs total energy deposited [i.e. number of equi-energy laser pulses, $M(n)$]. Simplifying and solving Eqs. 35 and 36 led to

$$M(n) = Mo(1-\eta)^n \qquad 37.$$

and

$$\frac{dM(n)}{dn} = \ln(1-\eta)M(n) \qquad 38.$$

where Mo is the number of nonpolymerized surfactants in $2C_{18}N^+C_1Sty$ vesicles ($Mo = 1.4 \times 10^5$) and η is defined as the fraction of the double bonded monomers consumed after the photochemical events induced by a single laser pulse have subsided:

$$\eta = \frac{\Phi_r \bar{\varepsilon} E k_p}{k_m + k_s} \qquad 39.$$

where E is the average energy/cm^2 of the laser pulse. The good linearity of absorbance vs laser intensity plots (111) justified the assumptions involved in deriving Eq. 38. The average chain length, $k_p/(k_m + k_s)$, is related to η by

$$\frac{k_p}{k_m + k_s} = \frac{\eta}{\Phi_r \varepsilon E}. \qquad 40.$$

Substituting appropriate experimental values into Eq. 40 resulted in a value of 23 for the average chain length obtained in the laser photopolymerization of $2C_{18}N^+C_1Sty$ vesicles. Formation of relatively low molecular weight polymers is in accord with the observed fluidities of polymerized vesicles (61). Extending the kinetic investigations to vesicles prepared from mixtures of polymerizable and nonpolymerizable surfactants will provide information on lateral surfactant's mobilities and domain formation.

The importance of phase changes and reagent-catalyst cluster formation in surfactant vesicles has been demonstrated (104–108). Advantage has been taken of an azobenzene-bearing surfactant, $CH_3(CH_2)_{11}OC_6H_4 N=NO(CH_2)_{10}N^+(CH_3)_3$, Br^- : $C_{12}AzOC_{10}N^+$, to report phase separation (112). The absorption maximum of monomeric $C_{12}AzOC_{10}N^+$ shifts from 355 nm to 316 nm upon cluster formation. A similar behavior has been found for $C_{12}AzOC_{11}His$, a surfactant bearing histidine moiety as catalyst in addition to the azobenzene reporter group. Cluster formation in vesicles prepared from mixtures of dialkylammonium surfactants and $C_{12}AzOC_{11}His$ were promoted by neutralizing the anionic histidine moiety either by pH changes or complex formation with Cu^{2+}, by increasing the $C_{12}ZzOC_{11}His$ concentration and by decreasing the temperature below the phase transition of the vesicles. Arrhenius plots for reactivities in vesicles were found to be linear or to have breaks around the phase transition temperatures. Linear Arrhenius plots imply that phase transition changes do not influence the reactivity of vesicle-incorporated substrates. Arrhenius plots with steeper slopes in the phase transition range (larger activation energy) indicate the formation of catalyst or substrate clusters in the solid phase with a resultant decrease in reactivity. Plots with shallower slopes in the phase transition region (smaller activation energy) correspond to the segregation of substrates and catalysts in the rigid vesicle matrix, with a resultant increase in reactivity (112).

CONCLUDING REMARKS

Within the last five years, synthetic surfactant vesicles have emerged as the most versatile membrane mimetic systems. Organization of reactants in the different compartments provided by vesicles have led experimentalists to postulate intervesicular, exovesicular, endovesicular, and transversicular reactions. Changes in fluidities and polymerization have also been shown to affect intravesicular reactivities. Systematic investigations of all parameters that control reactivities and the development of all encompassing theories are needed for the realization of the full potentials of surfactant vesicles as media for reactivity control. Initial approaches could well be based on treatments developed for reactions in micelles. The field is ripe for physical chemists and theoreticians to make their much needed contributions.

ACKNOWLEDGMENTS

Research in our own laboratories could not have succeeded without the enthusiastic, dedicated, and skillful work of the co-workers whose names appear in the references listed. I am most grateful to them and to the agencies (National Science Foundation, Department of Energy, Army Research Office, and Petroleum Research Foundation) who have provided financial support for different aspects of our research.

Literature Cited

1. Fendler, J. H. 1982. *Membrane Mimetic Chemistry.* New York: Wiley-Interscience. 522 pp.
2. Fendler, J. H. 1982. *Pure Appl. Chem.* 54:1809–17
3. Fendler, J. H. 1983. *J. Chem. Ed.* 60:872–76
4. Fendler, J. H. 1984. *Chem. Eng. News* 62:25–38 (Jan. 2)
5. Fendler, J. H. 1984. *Science* 223:888–94
6. Fendler, J. H., Fendler, E. J. 1975. *Catalysis in Micellar and Macromolecular Systems.* New York: Academic. 545 pp.
7. Bender, M. L., Komiya, M. 1978. *Cyclodextrin Chemistry.* Berlin: Springer-Verlag. 96 pp.
8. Cram, D. J. 1983. *Science* 219:1177–83
9. Lehn, J. M. 1978. *Acc. Chem. Res.* 11:49–57
10. Tien, T. H. 1974. *Bilayer Lipid Membranes (BLM). Theory and Practice.* New York: Dekker. 655 pp.
11. Lehn, J. M. 1973. *Struct. Bonding Berlin* 16:1–69
12. Gregoriadis, G., Allison, A. C. 1980. *Liposomes in Biological Systems.* New York: Wiley. 412 pp.
13. Kimelberg, H. K., Mayhew, E. G. 1978. *CRS Crit. Rev. Toxicol.* 6:25–79
14. Fendler, J. H., Romero, A. 1977. *Life Sci.* 20:1109–20
15. Calvin, M. 1978. *Acc. Chem. Res.* 11:369–74
16. Fendler, J. H. 1980. *J. Phys. Chem.* 84:1485–91
17. Porter, G. 1978. *Proc. R. Soc. London Ser. A* 362:281–303
18. Grätzel, M. 1981. *Acc. Chem. Res.* 14:376–84
19. Berezin, I. V., Martinek, K., Yatsimirskii, A. K. 1973. *Russ. Chem. Rev. Eng. Transl.* 42:787–802
20. Martinek, K., Yatsimirskii, A. K., Levashov, A. V., Berezin, I. V. 1977. See Ref. 45, pp. 489–508
21. Romsted, L. S. 1975. PhD thesis. Indiana Univ.
22. Romsted, L. S. 1977. See Ref. 45, pp. 509–30
23. Chaimovich, H., Bonilha, J. B. S., Politi,

M. J., Quina, F. H. 1979. *J. Phys. Chem. Chem.* 83:1851–54
24. Quina, F. H., Chaimovich, H. 1979. *J. Phys. Chem.* 83:18444–50
25. Quina, F. H., Politi, M. J., Cuccovia, I. M., Baumgarten, E., Martins-Faranchetti, S. M., Chaimovich, H. 1980. *J. Phys. Chem.* 84:361–65
26. Chaimovich, H., Aleixo, R. M. V., Cuccovia, I. M., Zanette, D., Quina, F. H. 1982. See Ref. 46, pp. 949–73
27. Romsted, L. S. 1984. See Ref. 47, pp. 1015–68
28. Atik, S. S., Singer, L. A. 1979. *J. Am. Chem. Soc.* 101:6759–61
29. Infelta, P. P., Grätzel, M. 1979. *J. Chem. Phys.* 70:179–86
30. Almgren, M., Grieser, F., Thomas, J. K. 1979. *J. Am. Chem. Soc.* 101:279–91
31. Yekta, A., Aikawa, M., Turro, N. J. 1979. *Chem. Phys. Lett.* 63:543–48
32. Turro, N. J., Grätzel, M., Braun, A. M. 1980. *Angew. Chem. Int. Ed. Engl.* 19:675–96
33. Hatlee, M. D., Kozak, J. 1980. *J. Chem. Phys.* 72:4358–67
34. Hatlee, M. D., Kozak, J. 1981. *J. Chem. Phys.* 74:1098–1140
35. Hatlee, M. D., Kozak, J. 1981. *J. Chem. Phys.* 74:5627–35
36. Dung, M. H., Knox, D. G., Kozak, J. J. 1980. *Ber. Bunsenges. Phys. Chem.* 84:789–95
37. Dung, M. H., Kozak, J. J. 1982. *J. Chem. Phys.* 76:984–96
38. Hatlee, M. D., Kozak, J. J., Rothenberger, G., Infelta, P. P., Grätzel, M. 1980. *J. Phys. Chem.* 84:1508–19
39. Hatlee, M. D., Kozak, J. J., Grätzel, M. 1982. *Ber. Bunsenges. Phys. Chem.* 86:157–61
40. MacCarthy, J. E., Kozak, J. J. 1982. *J. Chem. Phys.* 77:2214–16
41. Patterson, L. K., MacCarthy, J. E., Kozak, J. J. 1982. *Chem. Phys. Lett.* 89:435–37
42. Musho, M. K., Kozak, J. J. 1984. *J. Chem. Phys.* 80:159–69
43. Lee, P., Kozak, J. J. 1984. *J. Chem. Phys.* In press
44. Mittal, K. L., ed. 1979. *Solution Chemistry of Surfactants.* New York: Plenum. 958 pp.
45. Mittal, K. L., ed. 1977. *Micellization, Solubilization and Microemulsions.* New York: Plenum. 945 pp.
46. Mittal, K. L., Fendler, E. J., eds. 1982. *Solution Behavior of Surfactants, Theoretical and Applied Aspects.* New York: Plenum. 1544 pp.
47. Mittal, K. L., Lindman, B., eds. 1984. *Surfactants in Solution.* New York: Plenum
48. Tanford, C. 1980. *The Hydrophobic Effect, Formation of Micelles and Biological Membranes.* New York: Wiley-Interscience. 233 pp. 2nd ed.
49. Attwood, D., Florence, A. T. 1983. *Surfactant Systems, Their Chemistry, Pharmacy, and Biology.* London: Chapman & Hall. 794 pp.
50. Luisi, P. P. 1983. *Biological and Technical Relevance of Reversed Micelles.* New York: Plenum
51. Fendler, J. H. 1976. *Acc. Chem. Res.* 9:153–61
52. Eicke, H. F. 1980. *Top. Curr. Chem.* 87:85–145
53. Kitahara, A. 1980. *Adv. Colloid Interface Sci.* 12:109–40
54. O'Connor, C. J., Lomax, T. D., Ramage, R. E. 1984. *Adv. Colloid Interface Sci.* 20:21–97
55. Gershfeld, N. C. 1976. *Ann. Rev. Phys. Chem.* 27:349–68
56. Kuhn, H., Mobius, D., Bucher, H. 1972. In *Physical Methods in Chemistry,* ed. A. Weissberger, B. W. Rossiter, Pt. 3B, 1:571–701. New York: Wiley-Interscience
57. Mobius, D. 1981. *Acc. Chem. Res.* 14:63–68
58. Sagiv, J. 1981. *J. Am. Chem. Soc.* 102:92–98
59. Bangham, A. D. 1968. *Prog. Biophys. Mol. Biol.* 18:29–95
60. Fendler, J. H. 1980. *Acc. Chem. Res.* 13:7–13
61. Fendler, J. H., Tundo, P. 1984. *Acc. Chem. Res.* 17:3–8
62. Fendler, J. H. 1984. See Ref. 47, pp. 1947–89
63. Gross, L., Ringsdorf, H., Schupp, H. 1981. *Angew. Chem. Int. Ed. Engl.* 20:305–25
64. Lindman, B., Wennerström, H. 1980. *Top. Curr. Chem.* 87:1–83
65. Wennerström, H., Lindman, B. 1979. *Phys. Rep.* 52:1–86
66. Kosower, E. M. 1968. *An Introduction to Physical Organic Chemistry.* New York: Wiley. 503 pp.
67. Bunton, C. A. 1979. *Catal. Rev. Sci. Eng.* 20:1–56
68. Bunton, C. A., Hong, L. S., Romsted, L. S. 1982. See Ref. 46, pp. 975–91
69. Almgren, M., Rydholm, R. 1979. *J. Phys. Chem.* 83:360–64
70. Funasaki, N. 1979. *J. Phys. Chem.* 83:1998–2003
71. Yalkovski, S. H., Zografi, G. 1980. *J. Pharm. Sci.* 59:798–802
72. Lim, Y. Y., Fendler, J. H. 1978. *J. Am. Chem. Soc.* 100:7490–94
73. Lim, Y. Y., Fendler, J. H. 1979. *J. Am. Chem. Soc.* 101:4023–29

74. Fendler, J. H., Hinze, W. L. 1981. *J. Am. Chem. Soc.* 103:5439–47
75. Mita, K., Kunugi, S., Okubo, T., Ise, N. 1975. *J. Chem. Soc. Faraday Trans. I* 71:936–45
76. Mita, K., Okubo, T., Ise, N. 1976. *J. Chem. Soc. Faraday Trans. I* 72:1033–42
77. Manning, G. S. 1979. *Acc. Chem. Res.* 12:443–49
78. Rothenberger, G., Infelta, P. P., Grätzel, M. 1979. *J. Phys. Chem.* 83:1871–75
79. Maestri, M., Infelta, P. P., Grätzel, M. 1978. *J. Chem. Phys.* 69:1522–26
80. Frank, A. J., Grätzel, M., Kozak, J. J. 1976. *J. Am. Chem. Soc.* 98:3317–21
81. Henglein, A., Proske, Th. 1978. *Ber. Bunsenges. Phys. Chem.* 82:471–76
82. Escabi-Perez, J. R., Fendler, J. H. 1978. *J. Am. Chem. Soc.* 100:2234–36
83. Turro, N. J., Yekta, A. 1978. *J. Am. Chem. Soc.* 100:5951–52
84. Rodgers, M. A. J., Da Silva, E., Wheeler, M. E. 1976. *Chem. Phys. Lett.* 43:587–91
85. Tachiya, M. 1975. *Chem. Phys. Lett.* 33:289–92
86. Tamura, K., Schelly, Z. A. 1981. *J. Am. Chem. Soc.* 103:1013–18
87. Tamura, K., Schelly, Z. A. 1981. *J. Am. Chem. Soc.* 103:1018–22
88. Harada, S., Schelly, Z. A. 1982. *J. Phys. Chem.* 86:2098–2102
89. Adam, G., Delbrüch, M. 1968. In *Structural Chemistry and Molecular Biology*, ed. A. Rich, N. Davidson, pp. 198–203. San Francisco: Freeman
90. Richter, P. H., Eigen, M. 1974. *Biophys. Chem.* 2:255–63
91. Loughran, T., Hatlee, M. D., Patterson, L. K., Kozak, J. J. 1980. *J. Chem. Phys.* 72:5791–97
92. Galla, H. J., Sackmann, E. 1974. *Biochim. Biophys. Acta* 339:103–15
92a. Astumian, R. D., Schelly, Z. A. 1984. *J. Am. Chem. Soc.* 106:304–8
93. Montroll, E. W. 1969. *J. Math. Phys.* 10:753–65
94. Moss, R. A., Bizzigotti, G. O. 1981. *J. Am. Chem. Soc.* 103:6512–14
95. Moss, R. A., Ihara, Y., Bizzigotti, G. O. 1982. *J. Am. Chem. Soc.* 104:7476–78
96. Moss, R. A., Bizzigotti, G. O. 1982. *Tetrahedron Lett.* 23:5235–38
97. Moss, R. A., Schreck, R. P. 1984. *J. Am. Chem. Soc.* 105:6767–68
98. Moss, R. A., Shin, J. J. 1983. *J. Chem. Soc. Chem. Commun.* 1983:1027–28
99. Moss, R. A., Bizzigotti, G. O., Ihara, Y. 1983. In *Biomimetic Chemistry*, ed. Z. Yoshida, N. Ise, pp. 189–205. Tokyo: Kodansha Ltd.
100. Fuhrhop, J. H., Bartsch, H., Fritsch, P. 1981. *Angew. Chem. Int. Ed. Engl.* 20:804–5
101. Fuhrhop, J. H., Mathieu, J. 1983. *J. Chem. Soc. Chem. Commun.* 1983:144–45
102. Murakami, Y., Nakano, A., Akiyoshi, K. 1982. *Bull. Chem. Soc. Jpn.* 55:3004–12
103. Murakami, Y., Aoyama, Y., Kikuchi, J., Nishida, K., Nakano, A. 1982. *J. Am. Chem. Soc.* 104:2937–40
104. Kunitake, T., Sakamoto, T. 1978. *J. Am. Chem. Soc.* 100:4615–17
105. Kunitake, T., Okahata, Y., Ando, R., Shinkai, S., Hirakawa, S. 1980. *J. Am. Chem. Soc.* 102:7877–81
106. Kunitake, T., Sakamoto, T. 1979. *Chem. Lett. Jpn.* 1979:1059–62
107. Kunitake, T. 1983. In *Biomimetic Chemistry*, ed. Z. Yoshida, N. Ise, pp. 147–62. Tokyo: Kodansha Ltd.
108. Kunitake, T., Ihara, H., Okahata, Y. 1983. *J. Am. Chem. Soc.* 105:6070–78
109. Ishiwatari, T., Fendler, J. H. 1984. *J. Am. Chem. Soc.* 106:1908–12
110. Tundo, P., Kurihara, K., Politi, M. J., Fendler, J. H. 1982. *Angew. Chem. Int. Ed. Engl.* 21:81–82
111. Reed, W., Guterman, L., Tundo, P., Fendler, J. H. 1984. *J. Am. Chem. Soc.* 106:1897–1907
112. Shimomura, M., Kunitake, T. 1981. *Chem. Lett. Jpn.* 1981:1001–4

VARIATIONAL TRANSITION STATE THEORY

Donald G. Truhlar

Department of Chemistry, University of Minnesota, Minneapolis, Minnesota 55455

Bruce C. Garrett

Chemical Dynamics Corporation, 1550 W. Henderson Road, Columbus, Ohio 43220

INTRODUCTION

Transition state theory (TST)[1] is the most widely used theory for calculating rates of bimolecular reactions occurring in the gas phase and in condensed phases. TST is also incorporated into the widely used RRKM theory for unimolecular reactions. The popularity of TST is largely due to its simplicity and its usefulness for correlating trends in reaction rates in terms of easily interpreted quantities. Several forms of variational transition state theory (VTST) have been proposed, one as early as 1937; however, until recently, most applications of TST have been limited to the conventional (nonvariational) formulation. In recent years there has been renewed interest in VTST for providing insights into the factors controlling chemical

---

[1] An alphabetical list of the abbreviations used in this article is as follows: CVT, canonical variational transition state theory; DA, dynamical-path vibrational-average tunneling approximation; GTST, generalized transition state theory; ICVT, improved canonical variational transition state theory; LA, least-action tunneling approximation; LAG, least-action ground-state transmission coefficient; LC, large-curvature tunneling approximation; LCG, large-curvature ground-state transmission coefficient; MCP, Marcus-Coltrin path; PA, phase-average tunneling approximation; RRKM, Rice-Ramsperger-Kassel-Marcus; SAG, semiclassical vibrationally(-rotationally) adiabatic ground-state transmission coefficient; SC, small-curvature tunneling approximation; SO, second-order tunneling approximation; SOP, semiclassical optical potential tunneling approximation; TST, transition state theory; US, unified statistical model; VA, vibrational-average tunneling approximation; VTST, variational transition state theory; μVT, microcanonical variational transition state theory.

reaction rates, and VTST has been developed into a practical quantitative tool. The present review is concerned with the most recent developments, and we shall aim to complement rather than duplicate several other recent reviews, which we now summarize.

Chesnavich & Bowers (30, 31) reviewed applications of statistical methods in gas-phase ion chemistry, including detailed discussions of transition-state switching models and applications of VTST to ion-dipole capture. Walker & Light (188) reviewed the theory of reactive collisions, including progress to date on VTST. Truhlar & Garrett (173) provided an introduction to VTST and an overview of their early work on the subject. Two later articles, by Garrett et al (67) and Truhlar et al (181), provided partial reviews of selected aspects of further work of this group; these articles are partly review and partly new material, and they are discussed further below. Pechukas (124, 125) reviewed recent developments in transition state theory, including VTST, and we especially wish to single out his discussions of quantal and semiclassical approaches and of periodic-orbit dividing surfaces. Pollak (138) has also reviewed periodic-orbit dividing surfaces and related topics. Laidler & King (95) reviewed the historical origins of transitions state theory, excluding VTST, with coverage up to about 1938; Hirschfelder (81) provided additional historical comments. Hase (79) reviewed the history and use of variational concepts in unimolecular rate theory. Truhlar et al (177) reviewed the current status of transition state theory, including VTST, with special emphasis on the validity of the equilibrium and dynamical bottleneck assumptions, on localized states in unimolecularly decaying systems, and on frictional effects in solution-phase reactions. Truhlar et al (180) wrote a handbook-type chapter and concentrated on the practical aspects of VTST calculations, with emphasis on reactions of polyatomics, anharmonicity, tunneling, and other corrections. See (200) on TST and VTST, emphasizing organic applications.

The present review of VTST concentrates on work reported since the previous review in this series, by Pechukas (124). Readers are referred to Garrett & Truhlar (56) and to the reviews mentioned above for more extensive references to earlier work. We also restrict the present review to gas-phase reactions.

In addition to variational transition state theory, this chapter briefly considers relevant recent developments in selected aspects of several related subjects: related dynamical theories, the role of tunneling in chemical reactions, the calculation of dynamical bottlenecks and rate constants for state-selected processes, the role of resonances in chemical reactions, and vibrational bonding.

BASIC CONCEPTS

Since VTST is introduced in several of the papers above, we give here only a brief review as background to the later sections.

Transition state theory is a statistical mechanical theory of chemical reaction rates that may be derived from two fundamental assumptions. First one defines a reaction coordinate s leading from reactants (negative s) to products (positive s) and a (generalized) transition state as a system part way between reactants and products with a fixed value for s (thus the transition state is a system with one less degree of freedom than the reactants). The first assumption is that transition state species that originate as reactants are in local equilibrium with reactants. The second assumption is that any system passing through the transition state does so only once (before the next collision or before it is stabilized or thermalized as a reactant or product). These assumptions may be called the *local-equilibrium* and *no-recrossing* assumptions. Early workers were aware that the validity of the no-recrossing assumption depends on the location of the transition state, and that the transition state may be variationally defined as the phase-space hypersurface with the least one-way flux through it (82, 192) or as the hypersurface that yields the smallest free energy of activation (48). But in all conventional formulations of transition state theory, the transition state passes through a saddlepoint on a potential energy surface and the omitted coordinate s is taken as the unbound saddlepoint normal mode (49). Variational transition state theory (VTST) is the name we apply to theories that use the minimum-flux or maximum-free-energy-of-activation criteria to choose the transition state. VTST does not provide exact expressions for rate constants because it still involves the local-equilibrium assumption and because additional approximations are required to translate the flux-through-a-hypersurface argument into practical terms in a quantum mechanical world. Furthermore the variational search for the best transition state is usually carried out with constraints (e.g. one-dimensional search in coordinate space) for practical reasons.

Although the early statements of the classical variational principle by Wigner (192) and Horiuti (82) are correct and clear, the reader should be warned that there has also been some confusion about VTST. Thus Evans (48) incorrectly implied that the maximum free energy of activation corresponds to a minimum-probability rather than a minimum-flux condition and that the minimum-free-energy transition state passes through the saddlepoint. Later workers sometimes confused the minimum-flux condition with a minimum-density-of-states condition. Some workers attempted to discuss variational ideas in terms of free energy surfaces as functions of more than one coordinate, whereas the free energy to be minimized is a function of the location of a transition state surface, not a function of all the coordinates of the system. For another example, Swarc (169) provided a correct description of how to carry out a maximum-free-energy-of-activation calculation with classical reaction-coordinate motion, but Eyring (50) commented incorrectly that it is not possible to define the free energy of activation if the transition-state hypersurface does not pass through a stationary point of the potential energy, i.e. the saddlepoint. Swarc himself was unsure how to compare calculations with different reaction coordinates, but correct appreciation of the variational criterion shows that it applies to arbitrary variations in the transition state, not just to different choices for a given reaction coordinate.

Modern appreciation of VTST concepts includes the discussion of Eliason & Hirschfelder (47) of the relationship of a variational criterion for the transition-state-theory rate constant to collision theory and the applications by Keck (89, see also 90) of variational theory to atom-atom recombination in the presence of a third body. Keck (91) also presented variational theory in a more general context. Bunker & Pattengill (23), Marcus (102, 103), and Wong & Marcus (196) proposed related schemes (not VTST) for unimolecular and bimolecular reactions (see also 55, 56, 171). Tweedale & Laidler (185) provided an example of a free-energy-of-activation curve as a function of reaction coordinate for a collinear atom-diatom reaction, and Quack & Troe (145–149) used VTST and related dynamical schemes for a series of calculations on unimolecular decompositions of triatomics. In the last five years or so there has been considerable activity in elucidating the classical mechanics of variational transition states; general techniques have been proposed for calculating free energy of activation profiles from potential energy surfaces and for performing VTST calculations for arbitrary systems; and variational transition state theory including important quantization and tunneling effects has begun to receive extensive testing as a general practical tool for for the calculation of bimolecular rate constants.

Some comments on notation: Transition state theory (TST) refers to conventional, generalized, or variational transition state theory. When it is necessary to make a distinction, conventional transition state theory refers to placing the transition state at a saddlepoint on the potential energy surface, generalized transition state theory (GTST) refers to arbitrary locations of the transition state, and variational transition state theory refers to GTST when the location of the transition state is determined variationally. The optimum transition states for microcanonical or canonical ensembles correspond to a minimum sum of states or a maximum free energy of activation, respectively. Microcanonical variational theory (μVT) and canonical variational theory (CVT) denote the results obtained making the transition-state theory assumption at the globally best dynamical bottleneck for a microcanonical or canonical ensemble (55–57). Improved canonical variational theory (ICVT) refers to using μVT below the μVT energy threshold and optimizing the variational transition states for the non-zero contributions based on a canonical ensemble truncated from below at the threshold energy (61, 70).

When TST is compared to gas-phase experimental results, one tests both fundamental assumptions as well as the potential energy surface. When classical TST is tested against accurate classical dynamics (trajectories), one makes the same local-equilibrium approximation and uses the same potential energy surface for both calculations; hence only the no-recrossing assumption is tested. When quantized TST is tested against accurate (i.e. converged) quantal dynamical calculations one again uses the local-equilibrium assumption and the same potential energy surface in both cases, but now one tests not only the implicit translation of the classical no-recrossing assumption to a quantum mechanical world but also the accuracy of the incorporation of quantal effects such as tunneling into the TST calculation.

CLASSICAL VARIATIONAL TRANSITION STATE THEORY

The fundamental TST dynamical assumption of no recrossing is inherently a classical approximation, and classical TST can be formulated invoking the fundamental equilibrium and dynamical assumptions without any ambiguity or further approximations. In classical mechanics, TST provides an upper bound on the cross section or the local-equilibrium rate constant, and this bound is the basis for classical VTST, in which the calculated cross section or rate constant is minimized with respect to the location of the transition state. Physically one interprets the generalized transition states as tentative dynamical bottlenecks to the phase-space flow of trajectories from reactants to products. The variational transition state is the best dynamic bottleneck for an equilibrium ensemble.

For collinear atom-diatom reactions, the classical microcanonical variational transition states are periodic trajectories that vibrate between two equipotentials in the interaction region (123, 124, 138). Such trajectories (called pods) may be found numerically. Pechukas (123, 124) and Pollak (128, 135) have discussed the problems with generalizing the pods treatment to reactions in three dimensions. More approximate but more general techniques for variationally optimizing transition states, straightforwardly applicable in any number of dimensions, involve modeling the vibrational and rotational state sums (classical phase space volumes) of generalized transition states by the usual techniques of bound-state theory and searching numerically for the optimum transition states (56, 109).

Before 1979, most tests of the accuracy of the TST no-recrossing assumption were carried out for collinear $H+H_2$ with the conventional transition state location at the saddlepoint (26, 27, 78, 127, 166). These studies showed that this assumption is exact for this system up to about 0.2 eV above the barrier for collinear reactions and in 3D the agreement is better than 10% up to 1 eV above the barrier. More recently, both conventional TST and VTST have been compared to accurate classical calculations for a variety of collinear atom-diatom reactions. These studies show that VTST often provides significant improvements in accuracy as compared to conventional TST (142, 167, 56, 61).

These tests of classical VTST against accurate classical rate constants have been for bimolecular reactions with a single saddlepoint. Although more than one dynamical bottleneck can occur for single-saddlepoint reactions because of the decrease in the bound vibrational frequency in going from reactants toward the saddlepoint (an entropic effect), the presence of the two dynamical bottlenecks is an energetic effect for two-saddlepoint surfaces. Garrett et al (68) tested VTST for a potential energy surface with two identical saddlepoints. The second saddlepoint makes the no-recrossing assumption less valid at the first saddlepoint. In fact, TST and μVT overestimate the exact classical rate constant by a factor of two at total energies infinitesimally above the saddlepoint energy. However, the calculations show that the μVT and conventional TST results overestimate the exact classical one by only 20% at an energy 0.1 kcal/mol above the saddlepoint. Over a temperature range from 100 to 10,000 K, conventional TST rates agree with μVT and CVT ones to within 10%. For the system studied, the worst agreement between any form of TST and the exact classical results is for conventional TST at high temperature; for example at 2400 K conventional TST is too high by 21% and at 10,000 K it is too high by 47%.

Wolf & Hase (194) applied minimum-state-density criteria, which are similar to VTST, to find critical configurations for RRKM calculations on the dissociation of H–C–C model systems. The variational RRKM rate constants were larger than those computed from trajectories, typically by a factor of two for the tighter transition states and by factors of 5–50 for the looser cases. The largeness of the latter is probably due to the use of harmonic, separable approximation for the classical density of states and to an oversimplified treatment of the hindered-rotor degrees of freedom.

J. Miller (111, see also 112) has applied classical generalized TST to a reaction with no intrinsic barrier: $H+O_2$. In these calculations, the harmonic-oscillator, rigid-rotor approximation is used to evaluate the sum of states at the generalized transition state and the density of states of reactants. The vibrational frequencies and moments of inertia are obtained from an ab initio potential energy surface. Although reaction cross sections were computed for several locations of the dividing surface, the location of the dividing surface that gives the minimum reaction cross section for each energy was not found. Using the dividing surface that gives the smallest cross sections, the generalized TST results severely underestimate the quasiclassical trajectory results at all total energies below about 33 kcal/mol. However, we emphasize that these comparisons are for purely classical generalized TST versus trajectories with quantized initial conditions; hence no definite ordering of the resulting cross-sections should be expected.

Martin & Raff (109) have suggested a general procedure for classical variational transition state theory calculations in atom-diatom reactions in three dimensions. The dividing surface is expressed as a linear combination of internal coordinates and the coefficients in this expansion are variationally optimized to minimize the thermal rate constant. Calculations were performed for the $H + H_2$ and $H + I_2$ reactions. For the $H + H_2$ system the variational TST results are within 22% of the exact classical ones over a temperature range 300 to 1100 K. The agreement is not as good for the $H + I_2$ system, in which the VTST result overestimates the classical trajectory rate by a factor of 2.3 at 600 K. By carrying out combined-phase-space-trajectory calculations (4, 85) at their best dividing surface, a factor of 18 reduction in computer time with a decrease of a factor of four in statistical uncertainty was realized for the $H + H_2$ system, as compared to a trajectory calculation with sampling in the reactants' region.

Classical variational transition state theory has also been applied to the calculations of capture rate constants in collisions of ions with polar molecules. Su & Chesnavich (165) have extended earlier calculations (32, 33) to reduce the numerical error. The systems studied corresponded to H^- and H_3^+ reacting with a variety of polar molecules. For these systems the μVT rate constants agree very well with classical trajectory ones.

Swamy & Hase (168) have carried out similar studies for alkali ions recombining with H_2O. In these studies the agreement between classical trajectory rate constants and μVT ones is not as favorable: for the $Li^+ + H_2O$ system errors of 2.3 were found at 300 and 1000 K, and for the $K^+ + H_2O$ system errors of 2.9 and 6 were found at 300 and 1000 K, respectively. The errors are the result of trajectories that form short-lived collision complexes that are not stabilized by a third-body collision, thus leading to recrossing of the dividing surface.

One difficulty in calculating reliable thermal rate constants is the lack of potential energy surface information. In VTST the necessary information is the potential energy in a region about the minimum-energy path, whereas in conventional TST only information about the potential near the saddlepoint and in the reactant region is necessary. Truhlar et al (182) have developed methods of interpolating parameters in the reaction-path Hamiltonian between reactants, the saddlepoint, and products. They compared μVT valcuations based on interpolation to conventional TST, to μVT using the actual potential energy surface information, and to exact classical rate constants for a symmetric, a nearly symmetric, and two asymmetric collinear atom-diatom reactions. In all cases the μVT results computed using the interpolated potential energy surface information are in good agreement with the μVT results obtained using the actual potential. The interpolation schemes provide useful means of obtaining improved estimates of the rate constants for systems with limited potential energy surface information.

VARIATIONAL TRANSITION STATE THEORY IN THE REAL, QUANTIZED WORLD

VTST calculations in the quantum mechanical world have been carried out using the ansatz that if quantum effects on reaction-coordinate motion, which is responsible for movement from the reactants' region of phase space or state space to the products, are temporarily neglected, it still makes physical sense to minimize the rate constant (57). The intermediate-step quantity involved in this step, a rate constant corresponding to classical reaction-coordinate motion but a quantum mechanical treatment of all other degrees of freedom, has been called the *hybrid rate constant*. Minimizing the hybrid rate constant with respect to dividing-surface location is called quantized VTST. (In a quantized calculation there is usually not a large difference between the results of μVT, CVT, and ICVT calculations; in such cases we can just say VTST.) Quantal effects on reaction-coordinate motion and to some extent even quantal nonseparability of the reaction coordinate can be included, if desired, by multiplying the hybrid rate constant by a transmission coefficient. This generally includes both classically forbidden barrier penetration and nonclassical barrier reflection, but since the former usually dominates it is convenient to call this a tunneling correction.

Microcanonical variational theory for the hybrid rate constant is equivalent to making an adiabatic approximation for all degrees of freedom with respect to the reaction coordinate (56, 57). Thus there is a strong connection between VTST and adiabatic collision theories. By use of the adiabatic analogy or diabatic generalizations, the dynamic bottlenecks of

TST can also sometimes be interpreted as dynamical bottlenecks for state-selected reactions or for the decay of quasibound collisional resonance states.

The most important quantum mechanical effect on reaction-coordinate motion is tunneling. Thus the ability to estimate tunneling probabilities accurately is essential to the accurate use of transition state theory for many reactions. In general the tunneling contribution may be estimated by any semiclassical or quantal method; in some cases the adiabatic approximation mentioned above in conjunction with the classical-reaction-coordinate motion part of the calculation is also useful for the tunneling calculations, and the adiabatic derivation of TST makes it clear how to include tunneling consistently (183, 70). Quantized VTST with semiclassical *a*diabatic transmission coefficients based on the *g*round-state *s*-wave reaction probability is abbreviated VTST/SAG. Two kinds of nonadiabatic transmission coefficients have also been applied as corrections to quantized VTST; these have been called the *l*arge-*c*urvature *g*round-state and *l*east-*a*ction *g*round-state methods, and they lead to results abbreviated VTST/LCG and VTST/LAG.

Although it is not a necessary part of VTST, in our own work we have always considered one-dimensional sequences of generalized-transition-state dividing surfaces orthogonal to a gradient-following-path in mass-scaled coordinates. This choice of dividing surfaces is convenient; it eliminates potential coupling between the reaction coordinate and the other degrees of freedom through quadratic terms, and it promotes the dynamic separability of the reaction coordinate, thus tending to minimize local recrossing effects. Furthermore the use of a gradient-following path in mass-scaled coordinates facilitates the inclusion of internal centrifugal effects in tunneling calculations. An excellent discussion of gradient-following paths and the structure of the Hamiltonian in coordinate systems built on such paths has been given by Natanson (120).

Practical VTST calculations for a quantized world have so far been based on variationally optimizing the hybrid rate constant and adding a tunneling correction (70, 181) because more rigorous extensions of VTST to a quantum mechanical world do not provide a useful bound [see the discussions in (123, 172, 43)]. Pollak (131) has presented a new transition state expression with bounding properties and discussed its expansion in a power series in $\hbar$. The $\hbar$ expansion is known to be slowly convergent for tunneling contributions. It would be interesting to see whether practical and accurate bounds for real chemical reactions could be obtained from this formulation or whether the formalism provides a practically advantageous way to choose variational dividing surfaces.

Garrett et al (71) tested VTST/SAG calculations for model collinear reactions with mass combinations $H+FH$ and $D+FD$ and with a low-barrier, twin-saddlepoint potential energy surface by comparison against accurate quantal dynamical calculations. The comparisons showed that VTST/SAG predicts accurate local-equilibrium rate constants within a factor of 1.57 over the 200–7000 K range for the $H+FH$ mass combination and within a factor of 1.42 over the 200–2400 K range for $D+FD$, with the largest errors at the lowest temperature in each case. Employment of fully quantal vibrationally adiabatic tunneling calculations rather than semiclassical ones improved the accuracy. In another series of tests of VTST/SAG calculations against accurate quantal collinear rates, Bondi et al (16) considered the reactions $H+H_2$, $Mu+H_2$, and $Mu+D_2$ at 200–2400 K, using two different potential energy surfaces for two of the mass combinations, for a total of five cases. For $H+H_2$ and $Mu+H_2$ the accuracy was 38% or better over the whole temperature range, but for $Mu+D_2$ the errors were in this range only for $T \gtrsim 500$ K or 800 K, depending on the surface. The accuracy of VTST/SAG calculations for the seven systems discussed in this paragraph is actually slightly worse than typical of that found in 18 previous test cases of VTST/SAG calculations against accurate quantal equilibrium rate constants for collinear systems, as reviewed previously (61, 181, 177). In fact, for 300 K, for the full set of 25 cases the ratio to the accurate quantal equilibrium rate constant of the rate constant calculated by VTST/SAG calculations employing CVT for the VTST part and the small-curvature-tunneling approximation (163) for the SAG part, is in the range 0.49 to 1.54 in 22 cases and in the range 0.62 to 1.30 in 17 cases. In contrast, conventional transition state theory calculations often show large errors in these 25 cases: the ratio of conventional transition state theory rate constants to accurate quantal equilibrium rate constants is in the above two ranges in only seven and five cases, respectively, and even extending the range to 0.4 to 2.5 increases the number of cases to only 13.

For $Mu+D_2$ the VTST/SAG calculation on the most accurate surface decreases the errors from factors of 59, 23, and 1.8 in conventional TST at 200 K, 300 K, and 2400 K to 0.42, 0.49, and 0.92, respectively. These results, like all VTST/SAG calculations mentioned so far, are based on straight-line GTST dividing surfaces with anharmonicity treated by a Morse approximation [the Morse I approximation in the notation of Garrett & Truhlar (56, 57)]. Pollak (138) has shown that more accurate results can be obtained for the $Mu+D_2$ mass combination by semiclassical quantization of pods. Straight-line dividing surfaces for collinear reactions have more straightforward three-dimensional and polyatomic analogs than do pods, so we prefer this treatment to one based on pods; but it is not too impractical to go beyond the Morse approximation for anharmonicity. If the anharmonic energy levels of the straight-line generalized transition states are calculated numerically by the WKB approximation, without the Morse approximation, VTST/SAG agrees with quantal equilibrium rate constants for $Mu + D_2$ on the most accurate surface within 8% over the 200–2400 K temperature range (Garrett & Truhlar 1984, *J. Chem. Phys.* In press); it is encouraging that it is not necessary to use the curved pods as dividing surfaces to achieve this accuracy. Further work (Garrett & Truhlar 1984, *J. Chem. Phys.* In press) shows that although the Morse approximation usually leads to reasonably good agreement with the WKB approximation for the zero point energy at the variational transition state (quantitative differences are largest for reactions with high zero point energies), using the WKB method for stretching vibrations does provide systematic improvement over that achieved in previously reported tests of VTST/SAG calculations against accurate quantal dynamics. The rest of the VTST-plus-tunneling results discussed in the present section were all obtained by the more convenient Morse approximation for stretches and by a mixed quadratic-quartic approximation (62, 84) for bends that have no cubic anharmonicity.

Bondi et al (16) also predicted three-dimensional rate constants for the $Mu + H_2$ reaction based on the most accurate available potential energy surface, the so-called LSTH [Liu (100), Siegbahn & Liu (157), Truhlar & Horowitz (178)] surface. At 600 K, the ratio of the VTST rate to the TST one is 0.11, the ratio of the VTST/SAG rate to one calculated (15) from full quasiclassical trajectory calculations is 0.065, and the kinetic isotope effect (ratio of the rate for $Mu + H_2$ to that for $H + H_2$) is 0.017. The small values for all three ratios are direct consequences of the large zero point effects for this system and the large dependence of the ground-state stretching energy level of the generalized transition state on the value of the reaction coordinate. Despite the size of this effect, the predicted rate constants at 608 K and 875 K are in good agreement with later experimental values [D. M. Garner and D. G. Fleming 1982, unpublished; cited in (15)]. These calculations involve no empirical elements or adjustable parameters and they are believed to be the first totally ab initio reliable, quantitative predictions of chemical reaction rates. With our present confidence in the reliability of our dynamical calculations for a given potential energy surface, the difficulty of making such predictions for other reactions depends more on future advances in electronic-structure calculations of potential energy surfaces than on further advances in treating the dynamics.

Blais et al (13–15), again using the LSTH surface, compared VTST, VTST/SAG, and quasiclassical trajectory calculations of rate constants and activation energies for $H + H_2$, $D + H_2$, and $Mu + H_2$ to each other at 444–2400 K and to experiment at 444–875 K. The VTST/SAG calculations are in best agreement with experiment at the lower temperatures because of the importance of tunneling. The most interesting aspect of these calculations is the temperature dependence of the activation energy. For example the activation energy predicted by the VTST/SAG calculations for $D+H_2$ rises from 6.8 kcal/mol at 300 K to 7.5, 9.0, and 14.0 kcal/mol at 444, 875, and 2400 K. The latter two values, at temperatures at which tunneling effects are less important, are in good agreement with trajectory values of 8.7 and 13.9 kcal/mol. Since recrossing errors increase with temperature in classical tests for this kind of system, the agreement of variational transition state theory with full trajectory calculations for the slope of the rate constant at 2400 K is encouraging, especially for using the simpler theory for the important practical problem of extrapolating rate constants to high temperature for combustion applications.

Clary (38) tested VTST/SAG calculations (72) against presumably accurate quantal calculations (35, 37, 38, see also 40) for the three-dimensional $D+ClH$ exchange reaction and three isotopic analogs, with the same potential energy surface used for both sets of calculations so that the comparison provided a test of the dynamical methods. The quantal calculations were performed by a method (35, 36) combining the energy sudden and centrifugal sudden approximations in a way particularly appropriate for the transfer of a heavy particle between two light ones. The VTST/SAG and quantal rate constants for 295 K differed by only 15%, 5%, 25%, and 12% for the four cases studied. Good (but not as good) agreement, an error of 38% for $T \geqslant 300$ K, had also been obtained for the only previous test (70, 181), for $H+H_2$, of VTST/SAG calculations against presumably accurate three-dimensional rate constants (156) for a given potential energy surface. Note that for $H+H_2$ no sudden approximations were made in the quantal calculations; the accurate calculations are possible in this case because of the lightness of all three atoms, yielding a relatively small number of channels. In the $H+H_2$ case, as for the $H+FH$ case discussed above, the agreement is considerably improved if fully quantal one-dimensional tunneling calculations (60; see also 124, 173) are substituted for the semiclassical tunneling calculations. Using such fully quantal tunneling calculations, VTST-plus-tunneling results have recently been reported (176) for eight tritium-substituted analogs of the $H+H_2$ reaction, using the accurate LSTH potential energy surface that was also used for the $Mu+H_2$ calculations discussed above.

Clary et al (41) tested VTST/SAG calculations against accurate quantal calculations for collinear $H+BrH$ and $D+BrH$ and against energy-sudden-approximation, centrifugal-sudden-approximation calculations for the same reactions in three dimensions. The VTST/SAG and quantal rate constants showed good agreement in all four cases; e.g. for three-dimensional H + BrH they agree to 20% or better for 150–500 K, even though the SAG transmission coefficient is 1050 at 150 K.

Garrett et al (72, 73) applied VTST/SAG calculations to three-dimensional kinetic isotope effects in the reactions $Cl + H_2$, D_2, T_2, HD, DH, HT, and TH. They considered eleven different potential energy surfaces with a goal of finding a surface that was consistent with experiment. They found large differences from conventional TST in many cases, especially for intramolecular HD/DH kinetic isotope effects. None of the VTST/SAG calculations was in completely satisfactory agreement with experiment, perhaps because of errors in all the surfaces but also perhaps because of remaining uncertainties in the tunneling calculations or other errors in the dynamics calculations, such as different amounts of recrossing for the different isotopic combinations.

Isaacson & Truhlar (84) extended the VTST formulation to general nonlinear polyatomic reactions and Skodje et al (162) similarly extended the SAG transmission-coefficient approximation. Both extensions make use of the polyatomic reaction-path Hamiltonian of Miller et al (118) and assume independent generalized normal modes. Extensions of these methods have also been presented (180) for three-dimensional polyatomic reactions with linear generalized transition states. Isaacson & Truhlar (84) and Truhlar et al (181) applied the general polyatomic formalism to the reaction $OH + H_2 \rightarrow H_2O + H$, as well as to reactions of OH with D_2, HD, and DH, using Schatz & Elgersma's (155) fit of Walch & Dunning's (186) ab initio potential energy surface. They found that variational optimization of the generalized transition state lowered the calculated rate by a factor of 1.9 at 298 K, and quantal effects on reaction-coordinate motion increased it by a factory of 17 at the same temperature; both the optimization effect and the tunneling correction are decreasing functions of temperature. The final results agree with the recommended experimental rate constants of Cohen & Westberg (42) within a factor of 1.7 over the 298–2400 K temperature range, over which the rate constant varies by a factor of 2×10^3. The calculated results are a factor of 1.6–1.7 higher than experiment (151, 42) at 298 K, a factor of 0.6–0.8 lower at 600 K, and more accurate at 2400 K. Thus they slightly underestimate the low-temperature activation energy and slightly overestimate the high-temperature activation energy. Nevertheless the agreement with experiment is remarkably good. In general one hopes that an ab initio surface is useful for force constants for bound generalized normal modes but one expects to have to adjust the ad initio barrier height to obtain such good agreement with experiment; in this case good agreement was obtained without adjustment. More important is the insight that the VTST/SAG calculation gives into the detailed dynamics. The variational transition states occur 0.07–0.10 a_0 earlier along the reaction path (in scaled coordinates with a reduced mass of 1.8 amu) than the saddlepoint, curvature of the reaction path increases the SAG transmission coefficient by a factor of 4.3 at 298 K, and the results are sensitive to anharmonicity. The H_2/D_2 kinetic isotope effect is larger than the experimental one (151) at 298–600 K; we may speculate that this and the too-low activation energy at these temperatures occur because the barrier on the potential energy surface is slightly too thin and allows a little too much tunneling, although several other explanations could also be given.

Truhlar et al (174) used a combination of VTST/SAG and trajectory calculations to adjust a new potential energy surface for $F + H_2$ to a variety of experimental data for the reactions $F + H_2$ and $F + D_2$.

Garrett et al (67) presented detailed studies of the trends in variational transition state locations for atom-diatom reactions with a variety of mass combinations and potential energy surfaces, complementing earlier systematic studies (58, 59) of this subject that were limited to rotated-Morse bond-energy-bond-order surfaces. The main conclusions of these studies are as follows. The ratio $k^{\ddagger}/k^{VTST}$ of rate constants calculated by the conventional and variational theories is largest for symmetric or nearly symmetric reactions in which a light particle is transferred between two heavier ones. In these cases the saddlepoint tends to be symmetric or nearly symmetric and small changes in geometry can cause large changes in the zero point energy requirement for a bound stretching coordinate. (There is only one such coordinate for atom-diatom collisions; for polyatomic reactions the analogous stretching coordinate is the one involving atoms participating in the bond changing.) The effect on $k^{\ddagger}/k^{VTST}$ van be very large (up to several orders of magnitude) and decreases with temperature. A second important case is very asymmetric reactions with saddlepoints located well into the reactant or product channel. In these cases the bending effects become dominant. Since potential energy varies more slowly with distance along the reaction coordinate, in these cases the variational transition states may be much farther removed from the saddlepoint and more temperature dependent. The ratio $k^{\ddagger}/k^{VTST}$ increases with temperature in these cases but is usually only a factor of two to three.

In order to provide further evidence that the large $k^{\ddagger}/k^{VTST}$ ratio for symmetric heavy-light-heavy systems are not artifacts of the potential energy surfaces considered, Garrett et al (74) performed VTST calculations for an ab initio potential energy surface for the reactions $^{37}Cl + H^{35}Cl$ and $^{37}Cl + D^{35}Cl$. For three-dimensional $^{37}Cl + H^{35}Cl$ they obtained $k^{\ddagger}/k^{VTST}$ values of 110, 28, and 9 at 200 K, 300 K, and 600 K, confirming the large effect. The saddlepoint on the ab initio surface is symmetric with nearest-neighbor distances at 1.47 Å, potential energy 6.3 kcal/mol, and a bound stretching frequency of 337 cm^{-1}. At the 300 K twin asymmetric variational transition states, these values are 1.60 Å, 1.35 Å, 5.8 kcal/mol, and 1682 cm^{-1}. Tunneling calculations based on vibrational adiabaticity along the minimum-energy path are not valid for these mass combinations because the minimum-energy reaction path has large curvature in mass-scaled coordinates. Instead, the authors used a new large-curvature tunneling method, leading to the LCG transmission coefficient mentioned above. In order to compare to experiment (92, 93), the surface was scaled along the minimum-energy path (but not orthogonal to it, see the discussion above for the OH+H$_2$ reaction) so that the VTST/LCG rate constant for ^{37}Cl +D^{35}Cl agreed with experiment at 368.2 K; this yielded a collinear saddlepoint potential energy of 9.0 kcal/mol and a noncollinear saddle-point potential energy about 1.5 kcal/mol lower. The kinetic isotope effects calculated for the scaled surface, as calculated by the conventional TST, VTST, and VTST/LCG methods, are compared to experiment (92) in Table 1. Although both the conventional TST results and the VTST/LCG results are in qualitative agreement with experiment, the physical factor controlling the kinetic isotope effect is entirely different in the two calculations. In conventional TST this kind of isotope effect is determined by the saddlepoint stretching force constants, as in the widely used Melander-Westheimer model (110, 190). In the VTST calculations without tunneling, the kinetic isotope effect at the temperatures of Table 1 is less than 1.03, so essentially the entire effect in the VTST/LCG calculations is due to quantal effects on reaction coordinate motion. Furthermore, in the LCG model the tunneling for these reactions occurs by rapid light-atom motions at fixed Cl–Cl distance, and most of it occurs for Cl–Cl distances much larger than the Cl–Cl distance at the saddlepoint or even at the outer turning-point distance of the Cl–Cl symmetric-stretch zero-point motion of the conventional transition state. These results cast strong doubts on the validity of the common practice in physical organic chemistry of interpreting this kind of isotope effect for H or H$^+$ transfer in terms of transition state force constants. In a more general context, our VTST calculations call for a critical reexamination of conventional TST interpretations (110) of kinetic isotope effects even in cases when tunneling effects on the kinetic isotope

Table 1 H/D kinetic isotope effects for ^{35}Cl+H^{37}Cl on scaled ab initio surface

T (K)	‡	CVT	CVT/LCG	Experimental
368	2.8	1.0	4.2	5.0±0.7
423	2.5	1.0	3.4	4.1±0.4

effect are small because we find in many cases that variational transition states for different isotopic versions of a given reaction are different whereas the basic principle of the conventional analysis is that they are not.

Bondi et al (17) tested the VTST/LCG method against accurate quantal equilibrium rate constants for collinear Cl+HCl, Cl+DCl, and Cl +MuCl, using not the ab initio or scaled ab initio surface but a similar semiempirical surface with a collinear saddlepoint potential energy of 8.55 kcal/mol. The comparison for the Cl + MuCl case is given in Table 2. Both the variational effect, as measured by $k^{\ddagger}/k^{\text{VTST}}$, and the quantal effect on reaction-coordinate motion, as measured by $k^{\text{VTST/LCG}}/k^{\text{VTST}}$, are very large, but the final results are accurate within 39% over a factor of five in temperature. The success of the VTST/LCG method in this case is a consequence of the success of the tunneling calculations. To verify that the VTST part of the calculation is also meaningful, Truhlar, Garrett, Hipes & Kuppermann (1984, *J. Chem. Phys.* In press) tested the same methods against accurate quantal equilibrium rate constants for the reaction I+HI on a low-barrier surface for which tunneling effects are negligible. The results are shown in Table 2, and they verify that the VTST and VTST/LCG methods are reliable for heavy-light-heavy reactions in the low-barrier, no-tunneling limit as well as the high-barrier, tunneling-dominated limit.

So far in this article we have considered primarily tight transition states in which two bonds are simultaneously appreciably partially broken or newly made. Variational transition state theory is also applicable to loose and nearly loose transition states, and we now consider recent papers on that subject.

Cates et al (24) considered the reactions $Cl^+ + H_2 \rightarrow HCl^+ + H$ and $HCl^+ + H_2 \rightarrow H_2Cl^+ + H$. Both reactions are exoergic but the authors found a positive temperature dependence for the former and a negative

Table 2 Ratio of approximate rate constants to accurate quantal equilibrium ones for collinear reactions on semiempirical surfaces

Reaction	T (K)	$\ddagger$	ICVT	ICVT/LCG
Cl+MuCl	200	92300	0.003	0.68
	400	581	0.081	1.02
	1000	44	0.56	1.39
I+HI	100	17500	0.77	0.77
	200	214	0.99	0.99
	400	96	1.1	1.1
	1000	19	1.4	1.4

temperature dependence for the latter. They interpreted this in terms of an early barrier for the former and, following Farneth & Brauman (51), Olmstead & Brauman (122), Asubiojo & Brauman (5), Jasinski & Brauman (88), and Pellerite & Brauman (129), in terms of a tight variational transition state for the latter. It would be interesting to see whether the latter interpretation could be supported by actual VTST calculations on a full potential energy surface.

Troe (170) provided a simplified version of the statistical adiabatic channel model (145), which is similar to VTST, for unimolecular bond fission reactions and the reverse radical association schemes. This work addresses the difficult question of the correlation of vibrational, rotational, and orbital (centrifugal) energies between the two limits of tight and loose generalized transition states. The properties of the potential energy surface are interpolated by a scheme similar to that originally applied by Quack & Troe (145).

In the section on classical VTST we discussed the work of J. Miller (111) on $H + O_2 \rightarrow HO + O$. This is an endoergic reaction whose dynamical bottleneck lies in the exit channel; it is equivalent but more straightforward to consider the early generalized transition states of the reaction $O + OH$. Rai & Truhlar (150) applied quantized VTST to this case using the ab initio potential energy surface of Melius & Blint (201), and the semiempirical reaction-path correlation scheme of Quack & Troe (148). For the Melius-Blint surface the variational transition state for 300 K occurs at an O-to-OH distance of 5.4 Å, which is much larger than the range of O-to-OH distances for which most of the electronic structure calculations were performed. One advantage of VTST calculations over collision theory calculations is that this kind of information about critical geometries is available and may serve as a guide to future electronic structure calculations so that they may be carried out at the dynamically most important geometries. For 300 K the VTST rate constants, as well as the trajectory calculations of Miller (111), are larger than the experimental rate constants (42), presumably because the ab initio surface is too attractive. It is not clear whether this is a fault of the electronic structure calculations or the extrapolation to large O-to-OH distances. The calculations based on the Quack-Troe scheme were more successful, but it is not known whether this is fortuitous or meaningful, especially since the calculations are sensitive to how the rotational-orbital-motion correlations are treated, and this is quite uncertain.

Clearly further progress on the transition state theory of systems with loose and nearly loose transition states will require better knowledge of potential energy surfaces for such systems. Duchovic et al (45) have recently performed state-of-the-art electronic structure calculations for the potential energy along the dissociation coordinate in $CH_4 \to CH_3 + H$. Further work along this line is sorely needed.

VTST concepts have also been applied to a few reactions involving more than four atoms. See the work of Brauman and co-workers mentioned above and also Agmon (1, 2), Chesnavich et al (29), and Jarrold et al (86). Bowers and co-workers proposed a transition-state switching model for ion-molecule reactions involving tight and orbiting generalized transition states. In this model the existence of tight generalized transition states is postulated even for reactions without a saddlepoint; for such reactions there may be a local maximum in the free-energy of activation for tight geometries because, as the system moves along the reaction coordinate in the exoergic direction, rotational-orbital motions of the reactants are converted to vibrations. Such dynamical bottlenecks were found by Garrett & Truhlar (58) for neutral reactions with very small barrier heights in the entrance channel and by Rai & Truhlar (150) for the no-saddlepoint $O + OH$ radical-radical reaction discussed above. As the temperature increases, the canonical variational transition state becomes tighter in such systems; in some cases there may be a tighter and a looser bottleneck even at a single temperature.

The central barriers in long-lived ion-molecule complexes have been further characterized by Wolfe et al (195) and Squires et al (164).

In attempting to use VTST concepts in a qualitative sense one should be careful to distinguish free energies of activation from free energies of formation. Thus, as the generalized transition state tends to reactants or products, it should not be assumed that the free energy of activation tends to zero and to the free energy of reaction, respectively. The difference arises because the free energy of activation is a quasithermodynamic quantity referring to transition states, which are missing one degree of freedom, whereas free energies of reaction and formation include all degrees of freedom.

RELATED TOPICS

Above we have reviewed recent developments in transition state theory. We now briefly consider recent developments in a few closely related subjects. We do not attempt to present self-contained discussions of these subjects in their own context but rather discuss them in relation to VTST concepts.

Related Dynamical Approximations

The unified statistical (US) theory provides a generalization of VTST to the case of two (113) or more (114) dynamical bottlenecks. Unlike VTST, the unified statistical theory does not give a bound even in classical mechanics.

Pollak & Levine [(140, 141), see also Davis (44)] have emphasized that the US theory for classical systems can be derived by information theory where the average number of crossings of a critical surface is imposed as a constraint, and they have also proposed a generalization involving a second constraint, which may be computed from the entropies of the reactive and nonreactive state-to-state probability matrices. They (141) also suggested a canonical generalization by replacing microcanonical fluxes by canonical ones. A canonical unified statistical theory was also suggested by Garrett & Truhlar (63), who tested its predictions against accurate classical dynamics for several collinear atom-transfer reactions. Garrett et al (68) tested the original classical unified statistical theory against accurate dynamics for a reaction with two saddlepoints and found that it overestimates the extent of recrossing and hence underestimates the rate constant. Truhlar et al (180) discussed the incorporation of quantization and tunneling effects in the original and canonical unified statistical theories. They also reported that the quantized canonical unified statistical theory does not systematically improve on the canonical variational theory in accuracy tests against accurate quantal equilibrium rate constants for collinear reactions, although it can change predicted kinetic isotope effects by a non-negligible amount. In the limit of a strongly bound intermediate between the dynamical bottlenecks, the unified statistical theory reduces to the statistical theory of Pechukas & Light (126) and Nikitin (121). That statistical theory was originally formulated for loose dynamical bottlenecks so that the fluxes through the dividing surfaces were proportional to the asymptotically available phase space, but it was generalized to tight generalized transition states by Lin & Light (99). Webb & Chesnavich (189) have used models involving both tight and orbiting transition states in generalized statistical phase-space theory calculations on the energy dependence of the cross sections for the reaction $C^+ + D_2$.

Chesnavich (28) proposed a theory related to VTST in which, rather than varying the dividing surface location, he fixed its location in the entrance channel and varied its boundary. He obtained upper bounds on cross sections for atom-diatom exchange reactions.

An important remaining problem in generalized transition state theory is to estimate recrossing corrections. It would be very convenient if these could be estimated from local properties of the potential surface. Miller (114a) attempted to do this using the curvature of the minimum-energy reaction path at the saddlepoint or the point of maximum curvature; unfortunately, as discussed elsewhere (180), the predictions of his formulas do not correlate well with accurate classical dynamics. Global trajectories provide a more reliable, but more expensive, guide to recrossing effects. Bowman et al (19) have evaluated transmission coefficients from trajectories starting in asymptotic regions, and Truhlar & Garrett [as discussed in (180, 181)] have evaluated them from trajectories beginning at variational transition states. Further work using this approach is in progress.

Lee, Bowman, and colleagues (18, 97) suggested using reduced-dimensionality accurate quantal calculations to obtain transmission coefficients for full-dimensional TST calculations. In further work, Walker & Hayes (187) and Bowman et al (21) presented reduced-dimensionality calculations for reaction of H with vibrationally excited H_2. The reduction in dimensionality was achieved by treating bending degrees of freedom adiabatically, and it corresponds to using generalized transition state theory for bending and rotational degrees of freedom and full dynamics for the two most strongly coupled degrees of freedom. Miller & Schwartz (119) and Skodje & Truhlar (161) have presented improved system-bath decompositions of reaction-path Hamiltonians that might be used for this kind of approximation.

Kuppermann (94), Christov (34), and Truhlar et al (180) have provided discussions of the relation of transition state theory to accurate collision theory. The goal of this kind of analysis is to provide further insight into the dynamical corrections to TST, such as those considered in the previous paragraphs.

In microcanonical transition state theory one calculates a rate constant for each total energy of the transition state. Miller (115–117) has pointed out that one should calculate a distinct microcanonical rate constant for each irreducible representation of the transition state in the symmetry group that applies along the reaction path, since states of different symmetry are decoupled. The difference between the rate constants for different symmetries is largest for energies near threshold. A related practical point is that it is sometimes better to base transition state theory on a reference path through a saddlepoint with two imaginary frequencies. This kind of reference path has been used for the unimolecular decomposition of H_2CO (116) and for $^{37}Cl + H^{35}Cl$ (74). Celli et al (202) and Sakimoto (203) have calculated ion-dipole capture rate constants using an average-free-energy-function method and an adiabatic method, respectively; both methods are closely related to μVT (31).

Tunneling

As discussed above, accurate transmission coefficients that account for tunneling contributions are an important ingredient in transition state theory calculations for many cases. We have already mentioned some aspects of new developments in the theory of tunneling in conjunction with variational transition state theory calculations.

The most significant qualitative points to emerge from recent work on tunneling in chemical reactions are: 1. The tunneling contributions are usually larger than would be expected by most workers. 2. Reaction-path curvature effects are often very large, i.e., accurate transmission coefficients can be calculated only by using dominant tunneling paths systematically displaced from the minimum-energy reaction path. 3. Tunneling probabilities for multidimensional systems can nevertheless be calculated reliably in most or all cases by reduced-dimensionality semi-classical methods.

The simplest and most commonly used methods for approximating tunneling effects in conventional TST are the methods of Wigner (191) and Bell (10). In both these methods the potential along the minimum energy path in the vicinity of the saddlepoint is approximated as a truncated parabola. The transmission coefficient is obtained by Boltzmann-averaging the semiclassical barrier penetration probabilities. Wigner's tunneling correction is a semiclassical approximation to lowest order in $\hbar$. This correction factor is valid only when the correction is small, typically less than a factor of 2. Bell's method has a larger region of validity but the expressions are discontinuous and contain divergences. Recently, Skodje & Truhlar (160) have presented a continuous, divergence-free analytic expression for the transmission coefficient for a truncated parabolic barrier that approximates the accurate uniform semiclassical transmission coefficients over a wide range of parameters (see also 11). The method is also applicable to unsymmetric barriers and is shown to be useful for barriers with shapes other than parabolic. They also found that it is best to fit the barrier to a parabola using the effective parabolic width of the nonparabolic barrier at energies that contribute appreciably to the transmission coefficient.

The first successful approximation for tunneling in systems with significant reaction-path curvature was developed by Marcus & Coltrin (108) and extended by Garrett & Truhlar (54, 57, 60, 62). This method calculates the tunneling action integral along the caustic envelope of a family of unbound trajectories with quantized adiabatic vibrations; this is called the Marcus-Coltrin path (MCP). More recently, Gray and coworkers (76) developed a semiclassical adiabatic model using the reaction-path Hamiltonian (118) and treating the kinetic energy terms containing curvature coupling by second-order classical perturbation theory. They used this second-order (SO) tunneling method involving the adiabatic barrier to study the unimolecular isomerization of HNC to HCN (76) and the unimolecular decomposition of formaldehyde (77). Forst (52) also treated tunneling in formaldehyde decomposition; however, he used the classical barrier and neglected reaction-path curvature. Cerjan et al (25) unified the semiclassical perturbation approximation with the infinite order sudden approximation applied to the reaction-path Hamiltonian to obtain an expression for the total reaction probability that takes the form of a zero-curvature adiabatic reaction probability times a curvature-dependent correction factor.

Skodje et al (162, 163) developed a semiclassical adiabatic model that is valid for systems with small reaction-path curvature, and they derived a criterion for the validity of the adiabatic approximation in curvilinear natural collision coordinates. The small-curvature (SC) tunneling method is similar to the collinear-reaction method of Marcus & Coltrin (108) but can be applied without singularities to systems with large reaction-path curvature, and it is expected to be more accurate for small-curvature systems. In addition it is applicable to noncollinear systems with reaction-path curvature components in more than one degree of freedom, for example it has been applied to calculate large tunneling corrections for the reactions $OH + H_2$ and isotopic analogs, which have curvature components in four of the five vibrational coordinates (84, 181). Skodje et al also compared, both formally and numerically, the MCP, SC, and SO tunneling methods, a previously suggested method [the phase average (PA) method (118)], and three new methods [the vibrational average (VA), the dynamical-path vibrational average (DA), and the semiclassical optical potential (SOP) methods]. These adiabatic methods may be classified into two general groups, depending upon the method used to remove from the kinetic energy term the dependence upon the coordinates of the bound degrees of freedom orthogonal to the reaction coordinate. One class of models (PA, SO, VA, DA, and SOP) accomplishes this by "averaging" the reaction-path Hamiltonian over the vibrational coordinates either classically or quantally. The other class (MCP and SC), which is systematically more successful, defines single values of the vibrational coordinates for each value along the reaction coordinate. These "vibrational-collapse" models have the physical interpretation that the tunneling is forced to occur along a specified path through the interaction region.

For reactions with large reaction-path curvature, the adiabatic approximation breaks down. Large reaction-path curvature occurs, for example, in systems in which a light atom is transferred between two heavy atoms or molecules. Babamov, Marcus, and Lopez (6–8) developed a method for computing the reaction probability for this type of system, and they applied it to study tunneling probabilities in the threshold region as well as the oscillations of the reaction probability as a function of energy for energies above threshold. Garrett, and co-workers (74) developed a similar method, which they called the large-curvature (LC) method, and they used it to calculate thermally averaged tunneling correction factors for VTST. The physical model for a collinear atom-diatom reaction is that the tunneling occurs by the most direct path (a straight line) connecting the reactant and product regions. Motion in the bound vibrational coordinate (rather than translational motion along the reaction coordinate) promotes tunneling, and for a fixed total energy, tunneling can begin at a wide range of geometries along the caustic parallel to the reaction coordinate from the asymptotic reactant region to the turning point in the adiabatic potential. This method was demonstrated to work well for the Cl + HCl reaction and isotopic variants in both 1D (17) and 3D (74).

Garrett & Truhlar (65) unified the LC method with the vibrational-collapse adiabatic models by developing a least-action (LA) tunneling method. In this method the optimum tunneling path is chosen from a set of parameterized paths by requiring it to be the one that accumulates the least imaginary action along the tunneling path. This method was found to be extremely successful for a system with small-to-large reaction-path curvature. In practice, transmission coefficients based on the LC and LA approximations are based on the ground state and are called LCG and LAG, respectively. See (180) for formulas for applying these methods to general polyatomic systems.

The methods described above have been applied to reactions with barriers in the regions of large reaction-path curvature, and reaction-path curvature effects on tunneling probabilities have been found to be very important in many cases. For reactions with no barrier, or barriers far into the reactant and product regions, it is possible to simplify the treatment of reaction-path curvature. Illies, Jarrold, and Bowers (83, 87) proposed a tunneling model to describe the unimolecular fragmentation of CH_4^+ and NH_3^+, which are reactions with loose transition states. They approximated the potential in the tunneling region by a dipole term plus a rotational barrier from free internal rotation of the molecular fragments and orbital rotation. Using this model they obtained good agreement between calculated rate constants and experimental ones.

Heller & Brown (80) presented a method to estimate surface-hopping probabilities from a bound state on an upper surface to a bound state on a steep lower surface that does not cross (or avoids crossing) the upper surface in the classically allowed region, for the case in which a single path dominates the tunneling. Although the problem is formally quite different from the problem of single-surface reactive scattering for which the LA method was developed, the semiclassical solution has some points in common, especially with our small-curvature limit. Cross-fertilization of the two methods may provide clues as to how to extend both to a wider range of problems.

Transition state theory with tunneling has also been used to examine intramolecular hydrogen-transfer reactions. LeRoy (98) used a phenomenological model to calculate the rate of transfer of hydrogen atoms between two nonequivalent sites in large polyatomic molecules. The physical model is that vibrational stretching of the bond being broken initiates the reaction, although it is not necessary that 100% of the energy in the vibration is available for promoting the reaction. The effects of the degrees of freedom orthogonal to the reaction coordinate enter the rate expression through a steric factor. Adjusting the steric factor, the percentage of vibrational energy available to the reaction coordinate, and the effective one-dimensional potential along the tunneling path, LeRoy found he could reproduce experimental data for several intramolecular H atom transfer reactions. Bicerano et al (12) studied a similar problem, the transfer of a hydrogen atom between equivalent sites in malonaldehyde. For this symmetric system the potential along the minimum energy path between the two equilibrium geometries is a symmetric double well potential. Tunneling was included in the vibrationally adiabatic approximation with potential parameters taken from ab initio electronic structure calculations. Instead of calculating the rate of transfer from one well to another, they computed the effect of tunneling upon the energy level splitting, obtaining a result within a factor of two of the experimental one. The polyatomic VTST formalism discussed above can also be applied to multidimensional unimolecular isomerizations, with SC, LC, or LA tunneling corrections (180; F. B. Brown and D. G. Truhlar 1984, unpublished).

Vibrationally Adiabatic Barriers

The free energy of activation curve as a function of reaction coordinate reduces at 0 K to the vibrationally-rotationally adiabatic ground-state, s-wave potential curve, or, for short, the vibrationally adiabatic ground-state potential curve $V_a^G(s)$. When the shape of this curve is dominated by the s dependence of high-frequency modes, then the barriers of $V_a^G(s)$ provide a guide to the location of dynamical bottlenecks at nonzero temperature or nonzero microcanonical energy. Similarly, the barriers of vibrationally adiabatic excited-state curves may provide dynamical bottlenecks for reactions of vibrationally excited species.

Agmon (3) suggested using an alternative coordinate system to calculate $V_a^G(s)$, with the goal of improving the accuracy of the separability of the reaction coordinate that must be assumed in TST or VTST. A difficulty with Agmon's coordinate system is that the kinetic energy operator of the generalized transition state is complicated because the vibrational coordinates are curved. Reaction-path Hamiltonians based on the minimum-energy path and non-curved vibrational coordinates (56, 57, 70, 84, 104–107, 118) allow for more convenient calculations of the vibrational energies of the generalized transition states; yet, in a quadratic expansion about any point on the reaction path, the potential energy contains no cross-term coupling the reaction coordinate to the vibrational coordinates.

Pollak (130, 132–134) calculated vibrationally adiabatic potential curves and transmission probabilities for the vibrationally excited collinear $H+H_2$ reaction and isotopic analogs by quantizing pods, and also, in

Jacobi coordinates, by treating the one-dimensional bound vibrational motion quantum mechanically and the reaction-coordinate motion by a parabolically uniformized semiclassical approximation. He obtained qualitatively similar results to earlier state-selected vibrationally adiabatic calculations (57), but better agreement with accurate quantal results for reaction probabilities of vibrationally excited species. The earlier calculations had been carried out by the Morse I approximation applied to locally straight dividing surfaces in coordinates based on the minimum-energy reaction path. The quantitative differences were attributed to the Morse I approximation and to the neglect of important curvature corrections that are contained in pods. Garrett & Truhlar (1984, *J. Chem. Phys.* In press) have performed calculations employing straight-line dividing surfaces perpendicular to the minimum-energy path and using the WKB approximation for vibrational energies; the new calculations yield excellent agreement with adiabatic barrier heights obtained by quantizing pods and also with accurate quantal rate constants for the vibrationally excited case. This shows that curved generalized transition states are not necessary for high accuracy. As mentioned in a previous section, the Morse I approximation, which is very convenient, is usually adequate for thermal reactions, but WKB or quantal vibrational eigenvalues may be required for good accuracy for excited states.

Pollak (130, 132), Lee et al (96), and Ron et al (153) also used pods and vibrational energy calculations in Jacobi coordinates to evaluate adiabatic barriers for collinear and reduced-dimensionality calculations on the reactions $F+H_2$ and isotopic analogues and $O+H_2$. A disadvantage of Jacobi coordinates is that they yield accurate adiabatic barriers only relatively far out in the reactants and products channels; and a disadvantage of pods is that they exist only for collinear atom-diatom reactions. Methods based on minimum-energy reaction paths are more general, although they may be inappropriate in regions of very large reaction-path curvature; fortunately we have found in applications that this is not a problem because the variational transition state tends not to be located in such regions.

Garrett & Truhlar (57) and Pollak (132) also used adiabatic transmission probabilities to calculate the cumulative reaction probability, which has a step-like character due to channel openings; these steps should not be confused with oscillations in the state-selected reaction probabilities, which are due to interference effects such as resonances, but may sometimes be explained (46) by invoking only quantal discreteness.

The main reason that quantized VTST is more reliable than standard trajectory calculations for thermal rate constants is that it incorporates quantized energy requirements at dynamical bottlenecks, i.e. it incorporates the constraints of quantized adiabatic barriers. Schatz (154) suggested incorporating such energy constraints for low-frequency, classically non-adiabatic bending modes as ad hoc additions to the potential energy surface for three-dimensional trajectory calculations.

State-selected Reactions

The discussion in the main part of this review is centered on thermal rate constants, which are the traditional domain for transition state theory. The methods of variational transition state theory and related methods are also useful for understanding excited-state reactivity in certain cases. For example, adiabatic barriers, as discussed above, may be used to interpret excited-state reactivity and product-state distributions (130, 133, 197). Full rate constant calculations for vibrationally excited species may also be performed by invoking the vibrationally adiabatic or diabatic approximation for one degree of freedom and variational transition state theory for others; such calculations have been performed for several collinear reactions (57, Garrett & Truhlar 1984, *J. Chem. Phys.* In press), for three-dimensional $H+H_2$ and $D+H_2$ (B. C. Garrett and D. G. Truhlar 1984, unpublished), and for three-dimensional $OH+H_2$ (179). For $OH+H_2$ ($n = 1$), conventional TST predicts a vibrational rate enhancement of $> 10^4$, whereas state-selected VTST predicts 10^2, which is in good agreement with experiment (75, 199). State-selected VTST calculations (179) for $OH+H_2$ also imply that the large-non-Arrhenius behavior for $OH+H_2$ is not a consequence of the rate enhancement for vibrationally excited H_2, as had been suggested (198).

Pollak & Pechukas (143) showed that one may map out the reactant and product classical vibrational energy distributions by studying trajectories initiated in the immediate vicinity of variational transition states.

Resonances

The vibrationally adiabatic potential curves of variational transition state theory are also very useful for predicting and classifying collisional resonances in many chemical reactions, especially for thermoneutral and nearly thermoneutral reactions for which reaction-path curvature is small or intermediate (9, 64, 71, 158, 159). Variational transition states provide the barriers to decay of the resonance in the one-dimensional vibrationally adiabatic model. For large reaction-path curvature or strongly exothermic reactions, approaches based on resonant periodic orbits or on adiabaticity in hyperspherical coordinates appear more useful [see, for example, (139, 144, 152) and references therein].

Vibrational Bonding

Vibrational bonding has received considerable attention in the last couple of years, and it is interesting to point out how, like resonance phenomena, it can often be predicted and understood in terms of the same concepts and quantities as developed for variational transition state theory. In particular, vibrational bonding may be considered as the extreme of a pre-threshold resonance. For example, we observed pre-threshold resonances for collinear reactions with mass combinations H+FH and D+FD on a low-barrier potential energy surface, and these can be understood vibrationally adiabatically (71, 159). If the mass combination is changed to heavy-light-heavy, for which generalized-transition-state vibrational energy requirements show the most pronounced minimum in the interaction region (58), the vibrational energy of the resonance will decrease, and the resonance energy may drop below the zero point energy of the atom-diatom reactants and thus become a true bound state, even though the lowest point on the potential surface still occurs for the asymptotic atom-diatom reactants. This is called vibrational bonding, as opposed to ordinary bonding, with an equilibrium geometry corresponding to the minimum in the potential energy surface. A vibrational bonding state was first reported for collinear IHI (101), and shortly thereafter a vibrational bonding state for three-dimensional IHI was calculated (39, see also 101, 137). Variational transition states may serve as effective barriers that contribute to localizing a vibrational-bonding state in the strong interaction region. Adiabatic bonding is expected more generally in excited-state vibrationally adiabatic curves than in ground-state ones; if the adiabatically bound state in an excited-state vibrationally adiabatic potential curve lies below the asymptote, it may still decay nonadiabatically (158, 159), and thus the state is only quasibound. Only when vibrational effects are largest does one expect to find states below the asymptote of $V_a^G(s)$, and hence vibrational bonding will occur far less frequently than the similar resonance effect.

CONCLUDING REMARKS

In the last few years it has been shown that variational transition state theory can be implemented usefully for practical calculations of chemical reaction rates from potential energy surfaces. When combined with accurate semiclassical tunneling calculations, VTST is the most accurate practical method available for such calculations. Variational-transition-state constructs are also useful for quantitative interpretations of excited-state reactivity and resonances.

Acknowledgment

The work at the University of Minnesota was supported in part by the US Department of Energy, Office of Basic Energy Sciences, under contract no. DE-AC02-79ER10425. The work at Chemical Dynamics Corporation was supported by the US Army Research Office under contract no. DAAG-29-81-C-0015.

Literature Cited

1. Agmon, N. 1980. *J. Am. Chem. Soc.* 102: 2164–67
2. Agmon, N. 1981. *Int. J. Chem. Kinet.* 13: 333–65
3. Agmon, N. 1983. *Chem. Phys.* 76: 203–18
4. Anderson, J. B. 1973. *J. Chem. Phys.* 58: 4684–92
5. Asubiojo, O. I., Brauman, J. I. 1979. *J. Am. Chem. Soc.* 101: 3715–24
6. Babamov, V. K., Lopez, V., Marcus, R. A. 1983. *J. Chem. Phys.* 78: 5621–28
7. Babamov, V. K., Lopez, V., Marcus, R. A. 1983. *Chem. Phys. Lett.* 101: 507–11
8. Babamov, V. K., Marcus, R. A. 1981. *J. Chem. Phys.* 74: 1790–1803
9. Basilevsky, M. V., Ryaboy, V. M. 1981. *Int. J. Quantum Chem.* 19: 611–35
10. Bell, R. P. 1959. *Trans. Faraday Soc.* 55: 1–4
11. Bell, R. P. 1980. *The Tunnel Effect in Chemistry*, appendix C. London: Chapman & Hall
12. Bicerano, J., Schaefer, H. F. III, Miller, W. H. 1983. *J. Am. Chem. Soc.* 105: 2550–53
13. Blais, N. C., Truhlar, D. G., Garrett, B. C. 1981. *J. Phys. Chem.* 85: 1094–96
14. Blais, N. C., Truhlar, D. G., Garrett, B. C. 1982. *J. Chem. Phys.* 76: 2768–70
15. Blais, N. C., Truhlar, D. G., Garrett, B. C. 1983. *J. Chem. Phys.* 78: 2363–67
16. Bondi, D. K., Clary, D. C., Connor, J. N. L., Garrett, B. C., Truhlar, D. G. 1982. *J. Chem. Phys.* 76: 4986–95
17. Bondi, D. K., Connor, J. N. L., Garrett, B. C., Truhlar, D. G. 1983. *J. Chem. Phys.* 78: 5981–89
18. Bowman, J. M., Ju, G.-Z., Lee, K. T. 1982. *J. Phys. Chem.* 86: 2232–39
19. Bowman, J. M., Ju, G.-Z., Lee, K. T., Wagner, A. F., Schatz, G. C. 1981. *J. Chem. Phys.* 75: 141–47
20. Deleted in proof
21. Bowman, J. M., Lee, K. T., Walker, R. B. 1983. *J. Chem. Phys.* 79: 3742–45
22. Deleted in proof
23. Bunker, D. L., Pattengill, M. 1968. *J. Chem. Phys.* 48: 772–76
24. Cates, R. D., Bowers, M. T., Huntress, W. T. Jr. 1981. *J. Phys. Chem.* 85: 313–15
25. Cerjan, C. J., Shi, S.-H., Miller, W. H. 1982. *J. Phys. Chem.* 86: 2244–51
26. Chapman, S., Hornstein, S. M., Miller, W. H. 1975. *J. Am. Chem. Soc.* 97: 892–94
27. Chesnavich, W. J. 1978. *Chem. Phys. Lett.* 53: 300–3
28. Chesnavich, W. J. 1982. *J. Chem. Phys.* 77: 2988–95
29. Chesnavich, W. J., Bass, L., Su, T., Bowers, M. T. 1981. *J. Chem. Phys.* 74: 2228–46
30. Chesnavich, W. J., Bowers, M. T. 1979. Statistical methods in reaction dynamics. In *Gas-Phase Ion Chemistry*, ed. M. T. Bowers, pp. 119–51. New York: Academic
31. Chesnavich, W. J., Bowers, M. T. 1982. *Prog. React. Kinet.* 11: 137–267
32. Chesnavich, W. J., Su, T., Bowers, M. T. 1979. In *Kinetics of Ion-Molecule Reactions*, ed. P. Ausloos, pp. 31–53. New York: Plenum. 508 pp.
33. Chesnavich, W. J., Su, T., Bowers, M. T. 1980. *J. Chem. Phys.* 72: 2741–55
34. Christov, S. G. 1980. *J. Res. Inst. Catal. Hokkaido Univ.* 28: 119–36
35. Clary, D. C. 1981. *Chem. Phys. Lett.* 80: 271–74
36. Clary, D. C. 1981. *Mol. Phys.* 44: 1067–81
37. Clary, D. C. 1981. *Mol. Phys.* 44: 1083–97
38. Clary, D. C. 1982. *Chem. Phys.* 71: 117–25
39. Clary, D. C., Connor, J. N. L. 1983. *Chem. Phys. Lett.* 94: 81–84
40. Clary, D. C., Drolshagen, G. 1982. *J. Chem. Phys.* 76: 5027–33
41. Clary, D. C., Garrett, B. C., Truhlar, D. G. 1983. *J. Chem. Phys.* 78: 777–82
42. Cohen, N., Westberg, K. R. 1982. *Aerospace Report ATR-82(7888)-3*, pp. 39–44. El Segundo: Aerospace Corp.
43. Costley, J., Pechukas, P. 1981. *Chem. Phys. Lett.* 83: 139–44
44. Davis, J. P. 1981. *J. Chem. Phys.* 75: 2011–12. Erratum: 1982. 76: 753
45. Duchovic, R. J., Hase, W. L., Schlegel, H. B., Frisch, M. J., Raghavachari, K. 1982. *Chem. Phys. Lett.* 89: 120–25
46. Duff, J. W., Truhlar, D. G. 1975. *Chem. Phys. Lett.* 36: 551–54
47. Eliason, M. A., Hirschfelder, J. O. 1959. *J. Chem. Phys.* 30: 1426–36
48. Evans, M. G. 1938. *Trans. Faraday Soc.* 34: 49–57, 73
49. Eyring, H. 1935. *J. Chem. Phys.* 3: 107–15
50. Eyring, H. 1962. Discussion. In *The Transition State, Chem. Soc. Special Publ. 16*, p. 27. London: Chem. Soc.
51. Farneth, W. E., Brauman, J. I. 1976. *J. Am. Chem. Soc.* 98: 7891–98
52. Forst, W. 1983. *J. Phys. Chem.* 87: 4489–94
53. Deleted in proof
54. Garrett, B. C., Truhlar, D. G. 1979. *J. Phys. Chem.* 83: 200–3; Erratum 83: 3058

55. Garrett, B. C., Truhlar, D. G. 1979. *J. Chem. Phys.* 70:1593–98
56. Garrett, B. C., Truhlar, D. G. 1979. *J. Phys. Chem.* 83:1052–79; Errata 83:3058, 87:4553
57. Garrett, B. C., Truhlar, D. G. 1979. *J. Phys. Chem.* 83:1079–1112; Errata 84:692–86, 87:4553–54
58. Garrett, B. C., Truhlar, D. G. 1979. *J. Am. Chem. Soc.* 101:4534–48
59. Garrett, B. C., Truhlar, D. G. 1979. *J. Am. Chem. Soc.* 101:5207–17
60. Garrett, B. C., Truhlar, D. G. 1979. *Proc. Natl. Acad. Sci. USA* 76:4755–59
61. Garrett, B. C., Truhlar, D. G. 1980. *J. Phys. Chem.* 84:805–12
62. Garrett, B. C., Truhlar, D. G. 1980. *J. Chem. Phys.* 72:3460–71
63. Garrett, B. C., Truhlar, D. G. 1982. *J. Chem. Phys.* 76:1853–58
64. Garrett, B. C., Truhlar, D. G. 1982. *J. Phys. Chem.* 86:1136–41; Erratum: 1983. 87:4554
65. Garrett, B. C., Truhlar, D. G. 1983. *J. Chem. Phys.* 79:4931–38
66. Deleted in proof
67. Garrett, B. C., Truhlar, D. G., Grev, R. S. 1981. Determination of the bottleneck regions of potential energy surfaces for atom transfer reactions by variational transition state theory. In *Potential Energy Surfaces and Dynamics Calculations*, ed. D. G. Truhlar, pp. 587–637. New York: Plenum. 866 pp.
68. Garrett, B. C., Truhlar, D. G., Grev, R. S. 1981. *J. Phys. Chem.* 85:1569–72
69. Deleted in proof
70. Garrett, B. C., Truhlar, D. G., Grev, R. S., Magnuson, A. W. 1980. *J. Phys. Chem.* 84:1730–48; Erratum: 1983. 87:4554
71. Garrett, B. C., Truhlar, D. G., Grev, R. S., Schatz, G. C., Walker, R. B. 1981. *J. Phys. Chem.* 85:3806–17
72. Garrett, B. C., Truhlar, D. G., Magnuson, A. W. 1981. *J. Chem. Phys.* 74:1029–43
73. Garrett, B. C., Truhlar, D. G., Magnuson, A. W. 1982. *J. Chem. Phys.* 76:2321–31
74. Garrett, B. C., Truhlar, D. G., Wagner, A. F., Dunning, T. H. Jr. 1983. *J. Chem. Phys.* 78:4400–13
75. Glass, G. P., Chaturvedi, B. K. 1981. *J. Chem. Phys.* 75:2749–52
76. Gray, S. K., Miller, W. H., Yamaguchi, Y., Schaefer, H. F. III. 1980. *J. Chem. Phys.* 73:2733–39
77. Gray, S. K., Miller, W. H., Yamaguchi, Y., Schaefer, H. F. III. 1981. *J. Am. Chem. Soc.* 103:1900–4
78. Grimmelmann, E. K., Lohr, L. L. 1977. *Chem. Phys. Lett.* 48:487–90
79. Hase, W. L. 1983. *Acc. Chem. Res.* 16:258–64
80. Heller, E. J., Brown, R. C. 1983. *J. Chem. Phys.* 79:3336–51
81. Hirschfelder, J. O. 1983. *Ann. Rev. Phys. Chem.* 34:1–29
82. Horiuti, J. 1938. *Bull. Chem. Soc. Jpn.* 13:210–16
83. Illies, A. J., Jarrold, M. F., Bowers, M. T. 1982. *J. Am. Chem. Soc.* 104:3587–93
84. Isaacson, A. D., Truhlar, D. G. 1982. *J. Chem. Phys.* 76:1380–91
85. Jaffe, R. L., Henry, J. M., Anderson, J. B. 1973. *J. Chem. Phys.* 59:1128–41
86. Jarrold, M. F., Bass, L. M., Kemper, P. R., van Koppen, A. M., Bowers, M. T. 1983. *J. Chem. Phys.* 78:3756–66
87. Jarrold, M. F., Illies, A. J., Bowers, M. T. 1982. *Chem. Phys. Lett.* 92:653–58
88. Jasinski, J. M., Brauman, J. I. 1980. *J. Am. Chem. Soc.* 102:2906–13
89. Keck, J. C. 1960. *J. Chem. Phys.* 32:1035–50
90. Keck, J. C. 1962. *Discuss. Faraday Soc.* 33:173–82, 291–93
91. Keck, J. C. 1967. *Adv. Chem. Phys.* 13:85–121
92. Klein, F. S., Persky, A., Weston, R. E. Jr. 1964. *J. Chem. Phys.* 41:1799–1807
93. Kneba, M., Wolfrum, J. 1979. *J. Phys. Chem.* 83:69–73
94. Kuppermann, A. 1979. *J. Phys. Chem.* 83:171–87
95. Laidler, K. J., King, M. C. 1983. *J. Phys. Chem.* 87:2657–64
96. Lee, K. T., Bowman, J. M., Wagner, A. F., Schatz, G. C. 1982. *J. Chem. Phys.* 76:3563–82
97. Lee, K. T., Bowman, J. M., Wagner, A. F., Schatz, G. C. 1982. *J. Chem. Phys.* 76:3583–96
98. LeRoy, R. J. 1980. *J. Phys. Chem.* 84:3508–16
99. Lin, J., Light, J. C. 1966. *J. Phys. Chem.* 45:2545–59
100. Liu, B. 1973. *J. Chem. Phys.* 58:1925–37
101. Manz, J., Meyer, R., Römelt, J. 1983. *Chem. Phys. Lett.* 96:607–12
102. Marcus, R. A. 1965. *J. Chem. Phys.* 43:1598–1605
103. Marcus, R. A. 1966. *J. Chem. Phys.* 45:2630–38
104. Marcus, R. A. 1966. *J. Chem. Phys.* 45:4493–99
105. Marcus, R. A. 1966. *J. Chem. Phys.* 45:4500–4
106. Marcus, R. A. 1968. *Discuss. Faraday Soc.* 44:7–13
107. Marcus, R. A. 1974. Activated-complex theory: Current status, extensions, and applications. In *Investigation of Rates and Mechanisms of Reaction, Techniques of Chemistry*, ed. E. S. Lewis,

6(Pt. 1): 13–46. New York: Wiley-Interscience
108. Marcus R. A., Coltrin, M. E. 1977. *J. Chem. Phys.* 67:2609–13
109. Martin, D. L., Raff, L. M. 1982. *J. Chem. Phys.* 77:1235–47
110. Melander, L., Saunders, W. H. Jr. 1980. *Reaction Rates of Isotopic Molecules*, pp. 29–36. New York: Wiley. 2nd ed.
111. Miller, J. A. 1981. *J. Chem. Phys.* 74:5120–32
112. Miller, J. A. 1981. *J. Chem. Phys.* 75:5349–54
113. Miller, W. H. 1976. *J. Chem. Phys.* 65:2216–23
114. Miller, W. H. 1981. Reaction path Hamiltonian for polyatomic systems: Further developments and applications. In *Potential Energy Surfaces and Dynamics Calculations*, ed. D. G. Truhlar, pp. 265–86. New York: Plenum. 866 pp.
114a. Miller, W. H. 1982. *J. Chem. Phys.* 76:4904–8
115. Miller, W. H. 1983. *J. Am. Chem. Soc.* 105:216–20
116. Miller, W. H. 1983. *J. Phys. Chem.* 87:21–22
117. Miller, W. H. 1983. *J. Phys. Chem.* 87:2731–33
118. Miller, W. H., Handy, N. C., Adams, J. E. 1980. *J. Chem. Phys.* 72:99–112
119. Miller, W. H., Schwartz, S. 1982. *J. Chem. Phys.* 77:2378–82
120. Natanson, G. 1982. *Mol. Phys.* 46:481–512
121. Nikitin, E. E. 1965. *Teor. Eksp. Khim.* 1:135–43
122. Olmstead, W. N., Brauman, J. I. 1977. *J. Am. Chem. Soc.* 99:4219–28
123. Pechukas, P. 1976. Statistical approximations in collision theory. In *Dynamics of Molecular Collisions, Part B, Modern Theoretical Chemistry*, ed. W. H. Miller, 2:269–322. New York: Plenum. 380 pp.
124. Pechukas, P. 1981. *Ann. Rev. Phys. Chem.* 32:159–77
125. Pechukas, P. 1982. *Ber. Bunsenges. Phys. Chem.* 86:372–98
126. Pechukas, P., Light, J. C. 1965. *J. Chem. Phys.* 42:3281–91
127. Pechukas, P., McLafferty, F. J. 1973. *J. Chem. Phys.* 58:1622–25
128. Pechukas, P., Pollak, E. 1979. *J. Chem. Phys.* 71:2062–67
129. Pellerite, M. J., Brauman, J. I. 1982. Nucleophilic substitution. In *Mechanistic Aspects of Inorganic Reactions, Am. Chem. Soc. Symp. Ser.* 198, ed. D. B. Rorabacher, J. F. Endicott, pp. 81–95. Washington: Am. Chem. Soc.
130. Pollak, E. 1981. *J. Chem. Phys.* 74:5586–94
131. Pollak, E. 1981. *J. Chem. Phys.* 74:6765–70
132. Pollak, E. 1981. *J. Chem. Phys.* 75:4435–40
133. Pollak, E. 1981. *Chem. Phys. Lett.* 80:45–54
134. Pollak, E. 1981. *Chem. Phys.* 61:305–16
135. Pollak, E. 1982. *J. Chem. Phys.* 78:1228–36
136. Pollak, E. 1982. *Chem. Phys. Lett.* 91:27–33
137. Pollak, E. 1983. *Chem. Phys. Lett.* 94:85–89
138. Pollak, E. 1984. Periodic orbits and the theory of reactive scattering. In *The Theory of Chemical Reaction Dynamics*, ed. M. Baer. Boca Raton, FL: CRC Press. In press
139. Pollak, E., Child, M. S. 1981. *Chem. Phys.* 60:23–32
140. Pollak, E., Levine, R. D. 1982. *Ber. Bunsenges. Phys. Chem.* 86:458–64
141. Pollak, E., Levine, R. D. 1982. *J. Phys. Chem.* 86:4931–37
142. Pollak, E., Pechukas, P. 1978. *J. Chem. Phys.* 69:1218–26
143. Pollak, E., Pechukas, P. 1983. *J. Chem. Phys.* 79:2814–21
144. Pollak, E., Wyatt, R. E. 1982. *J. Chem. Phys.* 77:2689–91
145. Quack, M., Troe, J. 1974. *Ber. Bunsenges. Phys. Chem.* 78:240–52
146. Quack, M., Troe, J. 1975. *Ber. Bunsenges. Phys. Chem.* 79:170–83
147. Quack, M., Troe, J. 1975. *Ber. Bunsenges. Phys. Chem.* 79:469–75
148. Quack, M., Troe, J. 1977. *Ber. Bunsenges. Phys. Chem.* 81:329–37
149. Quack, M., Troe, J. 1977. *Gas Kinetics Energy Transfer: Specialist Periodical Report*, 2:175–238. London: Chemical Soc.
150. Rai, S. N., Truhlar, D. G. 1983. *J. Chem. Phys.* 79:6046–59
151. Ravishankara, A. R., Nicovich, J. M., Thompson, R. L., Tully, F. P. 1981. *J. Phys. Chem.* 85:2498–2503
152. Römelt, J. 1983. *Chem. Phys.* 79:197–209
153. Ron, S., Baer, M., Pollak, E. 1983. *J. Chem. Phys.* 78:4414–22
154. Schatz, G. C. 1983. *J. Chem. Phys.* 79:5386–91
155. Schatz, G. C., Elgersma, H. 1980. *Chem. Phys. Lett.* 73:21–25
156. Schatz, G. C., Kuppermann, A. 1976. *J. Chem. Phys.* 65:4668–92
157. Siegbahn, P., Liu, B. 1978. *J. Chem. Phys.* 68:2457–65
158. Skodje, R. T., Schwenke, D. W.,

Truhlar, D. G., Garrett, B. C. 1984. *J. Phys. Chem.* 88:628–36
159. Skodje, R. T., Schwenke, D. W., Truhlar, D. G., Garrett, B. C. 1984. *J. Chem. Phys.* 80:3569–73
160. Skodje, R. T., Truhlar, D. G. 1981. *J. Phys. Chem.* 85:624–28
161. Skodje, R. T., Truhlar, D. G. 1983. *J. Chem. Phys.* 79:4882–88
162. Skodje, R. T., Truhlar, D. G., Garrett, B. C. 1981. *J. Phys. Chem.* 85:3019–23
163. Skodje, R. T., Truhlar, D. G., Garrett, B. C. 1982. *J. Chem. Phys.* 77:5955–76
164. Squires, R. R., Bierbaum, V. M., Grabowski, J. J., dePuy, C. H. 1983. *J. Am. Chem. Soc.* 105:5185–92
165. Su, T., Chesnavich, W. J. 1982. *J. Chem. Phys.* 76:5183–85
166. Sverdlik, D. I., Koeppl, G. W. 1978. *Chem. Phys. Lett.* 59:449–53
167. Sverdlik, D. I., Stein, G. P., Koeppl, G. W. 1979. *Chem. Phys. Lett.* 67:87–92
168. Swamy, K. N., Hase, W. L. 1982. *J. Chem. Phys.* 77:3011–21
169. Swarc, M. 1962. Discussion. In *The Transition State, Chem. Soc. Spec. Publ.* 16, pp. 25–27. London: Chem. Soc.
170. Troe, J. 1981. *J. Chem. Phys.* 75:226–37
171. Truhlar, D. G. 1970. *J. Chem. Phys.* 53:2041–44
172. Truhlar, D. G. 1979. *J. Phys. Chem.* 83:199
173. Truhlar, D. G., Garrett, B. C. 1980. *Acc. Chem. Res.* 13:440–48
174. Truhlar, D. G., Garrett, B. C., Blais, N. C. 1984. *J. Chem. Phys.* 80:232–40
175. Deleted in proof
176. Truhlar, D. G., Grev, R. S., Garrett, B. C. 1983. *J. Phys. Chem.* 87:3415–19
177. Truhlar, D. G., Hase, W. L., Hynes, J. T. 1983. *J. Phys. Chem.* 87:2664–82; 1983. Erratum: 87:5523
178. Truhlar, D. G., Horowitz, C. J. 1978. *J. Chem. Phys.* 58:2466–76. Erratum 71:1514
179. Truhlar, D. G., Isaacson, A. D. 1982. *J. Chem. Phys.* 77:3516–22
180. Truhlar, D. G., Isaacson, A. D., Garrett, B. C. 1984. Generalized transition state theory. See Ref. 138, In press
181. Truhlar, D. G., Isaacson, A. D., Skodje, R. T., Garrett, B. C. 1982. *J. Phys. Chem.* 86:2252–61; Erratum: 1983. 87:4554
182. Truhlar, D. G., Kilpatrick, N. J., Garrett, B. C. 1983. *J. Chem. Phys.* 78:2438–42
183. Truhlar, D. G., Kuppermann, A. 1971. *J. Am. Chem. Soc.* 93:1840–51
184. Truhlar, D. G., Kupperman, A. 1972. *J. Chem. Phys.* 56:2232–52
185. Tweedale, A., Laidler, K. J. 1970. *J. Chem. Phys.* 53:2041–44
186. Walch, S. P., Dunning, T. H. 1980. *J. Chem. Phys.* 72:1303–11
187. Walker, R. B., Hayes, E. F. 1983. *J. Phys. Chem.* 87:1255–63
188. Walker, R. B., Light, J. C. 1980. *Ann. Rev. Phys. Chem.* 31:401–33
189. Webb, D. A., Chesnavich, W. J. 1983. *J. Phys. Chem.* 87:3791–98
190. Westheimer, F. H. 1961. *Chem. Rev.* 61:265–73
191. Wigner, E. 1932. *Z. Phys. Chem.* B 19:203–16
192. Wigner, E. 1937. *J. Chem. Phys.* 5:720–25
193. Wigner, E. 1938. *Trans. Faraday Soc.* 34:29–41
194. Wolf, R. J., Hase, W. L. 1980. *J. Chem. Phys.* 72:316–31
195. Wolfe, S., Mitchell, D. J., Schlegel, H. B. 1981. *J. Am. Chem. Soc.* 103:7694–96
196. Wong, W. H., Marcus, R. A. 1971. *J. Chem. Phys.* 55:5625–29
197. Zeiri, Y., Shapiro, M., Pollak, E. 1981. *Chem. Phys.* 60:239–47
198. Zellner, R. 1979. *J. Phys. Chem.* 83:18–23
199. Zellner, R., Steinert, W. 1981. *Chem. Phys. Lett.* 81:568–72

References added in proof:

200. Kreevoy, M. M., Truhlar, D. G. 1984. Transition state theory. In *Investigation of Rates and Mechanisms of Reactions*, ed. C. F. Bernasconi. New York: Wiley. 4th ed. In press
201. Melius, C. F., Blint, R. J. 1979. *Chem. Phys. Lett.* 64:183–89
202. Celli, F., Weddle, G., Ridge, D. P. 1980. *J. Chem. Phys.* 73:801–12
203. Sakimoto, K. 1984. *Chem. Phys.* 85:273–78

AB INITIO VIBRATIONAL FORCE FIELDS

Géza Fogarasi

Department of General and Inorganic Chemistry, Eötvös L. University, Budapest, H-1088, Hungary

Péter Pulay

Department of Chemistry, The University of Arkansas, Fayetteville, Arkansas 72701

INTRODUCTION

Vibrations of a molecule take place on a multidimensional potential surface. For not too large displacements around a reference geometry, the potential energy can be expanded in a power series, the coefficients of which are the energy derivatives with respect to the nuclear displacement coordinates. The latter are the force constants of various order. Frequently, the term *force constant* is used specifically for the quadratic (harmonic) constants. The negative first derivatives are generally referred to as the *forces*.

The purely experimental determination of force constants is a very difficult task. The major problem is the large number of unknown constants as compared to the amount of easily accessible data. This difficulty is already pronounced at the harmonic level, except for the simplest molecules. The lack of reliable force fields for prototype molecules prevented the development of accurate transferable harmonic force fields for molecules more complex than simple alkanes. The difficulties are compounded at the anharmonic level.

The potential surface governing the nuclear movement is given, in the Born-Oppenheimer approximation, by the electronic energy of the molecule (including the nuclear repulsion) as a function of the displacement coordinates. Theoretical calculation of force constants is thus equivalent to the calculation of derivatives of the molecular energy as obtained from the electronic Schrödinger equation.

The simplest method for evaluating derivatives is numerical differentiation, using finite differences or least-squares fitting of energy points. For larger systems with many degrees of freedom the numerical approach becomes increasingly inefficient because the complete wave function calculation has to be repeated for a great number of points. In addition, this procedure is very sensitive to numerical errors.

The rapid, recent progress in the theoretical investigation of potential surfaces is partly due to the development of analytical derivative techniques. Obviously, the direct evaluation of derivatives from the wave function is fundamentally more economical than the numerical approach, provided that the theory is tractable and efficient algorithms can be constructed. The first major breakthrough was the introduction of the analytic gradient formalism for Hartree-Fock wave functions. As the result of recent developments, it has now become possible to calculate first derivatives for several types of wave functions that include electron correlation. In addition, economical techniques are emerging for the calculation of higher derivatives.

Recent progress in computational techniques and machinery has made force field calculations at the ab initio level increasingly accessible to spectroscopists. The purpose of this review is to summarize the techniques and the main results of such calculations. Our main conclusion is that quantum chemical calculations can significantly improve our knowledge of molecular force fields. Nevertheless, for the dominant (quadratic diagonal) force constants, the ab initio results cannot compete with the high accuracy of spectroscopic measurements. Less dominant force constants (quadratic coupling and higher terms) are, however, usually more accurately determined from theory than from experiment. Because of their complementary nature, the most reliable force fields can be obtained at present from the combined use of theoretical and experimental information.

Force constants contain all the information necessary for the calculation of vibrational frequencies. Reliable prediction of the latter would greatly enhance the usefulness of vibrational spectroscopy in structural studies. Correct identification of the observed bands in experimental work is a difficult task, and misassignments often occur even in structurally simple organic molecules. Reliable predictions would be of especially great value for systems not easily accessible to experiment (unstable molecules, radicals, conformers, transition states, and excited states). Good force fields are also needed for the evaluation of vibrational averaging effects in, for example, microwave rotational spectroscopy.

Most vibrational studies are based on the harmonic approximation because higher force constants are not easily accessible, and the vibrational problem itself is easily tractable at the harmonic level only. Although this is satisfactory for many purposes, there is increasing interest also in anharmonic vibrational levels, due to developments in infrared laser technology, laser isotope separation, and high resolution vibrational spectroscopy. Thus the calculation of higher derivatives will also be considered.

Because developments of the past 15 years have been reviewed elsewhere (1, 2) we concentrate here mainly on more recent results and expected future trends. The emphasis is on small and medium sized polyatomic molecules; the simple case of diatomics is not treated. Vibrational intensities (dipole moment and polarizability derivatives), although closely related both in subject and in the computational technique, cannot be discussed here due to space limitations.

THEORETICAL ASPECTS

In this section we outline briefly the theoretical background of the direct calculation of energy derivatives. A more detailed exposition of the quantum chemical aspects is given elsewhere (2, 3; P. Pulay, *Adv. Quantum. Chem.*, in preparation). The presentation follows a recent formulation of second and third derivatives (4).

General Considerations

Let us consider the most important type of correlated wave function, written as

$$\Psi = \Sigma_K A_K \Phi_K \qquad 1.$$

where Φ_K denotes an electronic configuration described by a Slater determinant or by a fixed (spin-adapted) linear combination of several such determinants. The latter are constructed from n orthogonal molecular orbitals (MO) which, in turn, are linear combinations of m basis functions:

$$\phi_i = \Sigma_r^m C_{ri} \chi_r. \qquad 2.$$

The dependence of the energy, $E = \langle \Psi | \hat{H} \Psi \rangle$, on the nuclear coordinates **R** comes from different sources. It depends most directly on **R** through the Hamiltonian; it depends indirectly on **R** through the dependence of (a) the configuration interaction (CI) coefficients A_K, (b) the MO coefficients C_{ri}, and (c) the basis functions χ_r, on the nuclear configuration. The last point, the dependence of the basis set, is a matter of definition. For the usual limited basis sets, there is only one reasonable definition: The positions of the basis functions are anchored rigidly to the nuclei, and their exponents are kept constant (2).

From the computational point of view, it is convenient to distinguish between only two groups, variational versus nonvariational parameters. The dependence of the energy on the latter is then included in the direct dependence on **R**. Denoting the total set of variational parameters by **c**, the energy is formally:

$$e = e(\mathbf{c}, \mathbf{R}). \qquad 3.$$

The idea behind this separation of the parameters, as shown below, is that variational parameters allow the simplification of derivative formulas.

Variational parameters are determined from the condition of making Eq. 3 stationary, subject usually to a set of constraint equations:

$$f_m(\mathbf{c}, \mathbf{R}) = 0, \quad m = 1, \ldots M. \qquad 4.$$

Normally, the constraints are orthonormality conditions for the molecular orbitals and for the electronic configurations. Defining the Lagrangian function as $F(\mathbf{c}, \lambda, \mathbf{R}) = e - \Sigma_m \lambda_m f_m$, constrained optimization leads to the equations

$$\partial F/\partial c_i = \partial e/\partial c_i - \Sigma_m \lambda_m (\partial f_m/\partial c_i) = 0. \qquad 5.$$

Let us consider two important examples: (a) in the multiconfiguration self-consistent field (MC-SCF) scheme both the CI and the MO coefficients, A_K and C_{ri} in Eqs. 1, 2, are optimized and thus belong to set **c**; (b) in the CI method, only the CI coefficients are optimized and qualify as variational parameters, whereas the dependence through the MOs is nonvariational and must be included in the direct dependence on R.

First Derivatives

The advantage of distinguishing variational and nonvariational parameters becomes apparent when we consider the first derivatives. Solution of Eqs. 4 and 5 gives the functions $\mathbf{c}(\mathbf{R})$. With this, the final energy formula becomes

$$E(\mathbf{R}) = e[\mathbf{c}(\mathbf{R}), \mathbf{R}] \qquad 6.$$

and the gradient of E can be written as

$$dE/dR_a \equiv E^a = \Sigma_i (\partial e/\partial c_i) c_i^a + \partial e/\partial R_a \qquad 7.$$

where we introduced the superscript ^{a} to denote differentiation with respect to the nuclear coordinate R_a. The troublesome first term in Eq. 7 can be eliminated. Multiplying Eq. 5 by c_i^a and summing over i, we obtain

$$\Sigma_i (\partial e/\partial c_i) c_i^a - \Sigma_{im} \lambda_m (\partial f_m/\partial c_i) c_i^a = 0. \qquad 8.$$

Differentiation of the constraint Eq. 4 yields:

$$\Sigma_i (\partial f_m/\partial c_i) c_i^a + (\partial f_m/\partial R_a) = 0, \quad m = 1, \ldots M. \qquad 9.$$

Substitution of Eqs. 8 and 9 into Eq. 7 gives finally

$$E^a = (\partial e/\partial R_a) - \Sigma_m \lambda_m (\partial f_m/\partial R_a) = (\partial F/\partial R_a). \qquad 10.$$

The final result shows the important fact that the derivatives of the variational parameters do not enter the gradient formula. To take the simplest case of a Hartree-Fock (HF) wave function, evaluation of the derivatives of the MO coefficients can be avoided. In the more general MC-SCF case, the same is true also for the CI coefficients. In these schemes the only major computational task is the calculation of the derivatives of the basis function integrals appearing in the energy formula. In a CI calculation, however, the MO coefficients are not optimized and their derivatives must also be evaluated (see below).

As an example, we give here the gradient formula for the MC-SCF case, as derived by W. Meyer [published in Ref. (2)]. Based on the energy expression for an MC-SCF wave function [e.g. see (5)] and using the orthonormality conditions for the configurations and the orbitals, the gradient can be written in the following form:

$$E^a = (\partial F/\partial R_a)$$
$$= \Sigma_{rs} \Sigma_{ij} \gamma_{ij} C_{ri} C_{sj} h^a_{rs} + 1/2 \Sigma_{rstu} \Sigma_{ijkl} \Gamma_{ijkl}$$
$$\times C_{ri} C_{sj} C_{tk} C_{ul} (rs|tu)^a - \text{Tr}(S^a C \varepsilon C^+). \qquad 11.$$

In this formula, γ_{ij} and Γ_{ijkl} are the orbital density matrices, $h_{rs} = \langle \chi_r | \hat{h} \chi_s \rangle$ with $\hat{h}$ denoting the one-electron Hamiltonian, $(rs|tu) = \langle \chi_r(1)\chi_s(1)|1/r_{12}|\chi_t(2)\chi_u(2)\rangle$, S is the overlap matrix, $S_{rs} = \langle \chi_r | \chi_s \rangle$ and ε_{ij} are Lagrangian multipliers, obtained from the MC-SCF equations.

The first two terms in Eq. 11 include the Hellmann-Feynman force and the integral force (6), the latter arising from the change of the basis set with nuclear geometry. Note that the integral force is zero in the limiting case of a complete basis set but is very important for all practical, even large, basis sets. The last term is the *density force* (6) or, better, the *normalization force* (3). For a short MC-SCF expansion, it is practical to evaluate the eightfold summation in Eq. 11. However, for a longer expansion the two-electron density matrix must be transformed to the AO basis prior to the evaluation of the gradient. Transforming one subscript at a time, this can be accomplished in $O(N^5)$ operations instead of $O(N^8)$.

When evaluating Eq. 11, the major computational task is the calculation of the integral derivatives; this, although simple in principle, needs considerable extra computer time. In the Hartree-Fock (HF) case, a complete calculation including the gradient costs roughly twice as much as the traditional calculation of the wave function (and energy) only, with the timing data depending strongly on the contraction of the basis set. In spite of the additional computer time, the analytic gradient method, which gives the complete gradient vector in one calculation, is obviously far superior to numerical differentiation.

Calculations of the first derivatives is thus relatively simple for SCF type wave functions. Realization of this fact led to the introduction of gradient techniques, implemented first for atom-anchored basis sets within the HF closed shell scheme (6). The method, often referred to as the *force* or *gradient method*, was devised primarily for the calculation of (harmonic) force constants and also for efficient geometry optimization by force relaxation (6). The force constants are calculated numerically from the analytic gradients evaluated at appropriately distorted nuclear configurations around a reference geometry.

The force method has found wide applications in the past 15 years (see 1–3, 7 and subsequent sections of this review). There are now several SCF gradient programs available. In our group, a number of calculations have been carried out by MOLPRO (8), the first general gradient program, and more recently by TEXAS (9). A new efficient algorithm has been published recently by Schlegel (10). The latter paper gives also a list of several other gradient programs.

The force method was soon generalized also to open-shell wave functions by Meyer & Pulay (11). It was implemented for both the unrestricted and restricted HF scheme in MOLPRO and several molecules were treated with it, for example triplet N_2H_2 (11), CH_4^+ (12), CHO (13), and $C_2H_4^+$ (2).

MC-SCF gradient calculations have been delayed by convergence difficulties in the MC-SCF procedure itself. There has been significant progress in this field recently and, although limited to certain simple MC-SCF wave functions, several gradient implementations have been reported (14–16).

The MC-SCF scheme is limited to small CI expansions. Large scale correlation calculations presently apply three principal techniques: (*a*) variational CI, (*b*) coupled-cluster methods, and (*c*) perturbation theory. The advantage of variational CI is that the CI expansion coefficients, being variational parameters, do not appear in the gradient formula. We do need the derivatives of the MO coefficients with respect to the nuclear distortions. These are obtained from the coupled-perturbed-Hartree-Fock (CPHF) equations (17). The first computationally viable solution of the latter was only recently given by Pople et al (18) in connection with HF second derivatives (see next subsection). Another difficulty is the evaluation of the two-electron integral forces, which requires the two-electron density matrix in AO basis, as pointed out first by Meyer (19). The first CI gradient algorithms have been developed simultaneously by the Schaefer (20) and the Pople (21) groups. The availability of the CI gradient is a significant break-through toward more efficient potential surface studies using correlated wave functions. The CI gradient has also been formulated for the most common types of open-shell CI wave functions (22), as well as for CI wave functions based on an MC-SCF reference configuration (23).

In coupled-cluster methods (see 24 for a review), although the computational procedure is similar to CI, the expansion coefficients are not determined from a variational condition on the total energy. The gradient thus contains the derivatives of the coefficients; this makes its analytic evaluation too expensive. A possible solution is to reformulate the coupled-cluster equations in a form equivalent to a variational condition on the CI coefficients, although not to a strict energy expectation value. This has been done recently for CEPA-2 (25); the modified version CEPA-2V (V for variational) is, for all practical purposes, identical to CEPA-2 but allows the efficient evaluation of the gradient (26).

Methods based on perturbation theory allow a more efficient treatment of electron correlation than do variational methods, as a result of the lack of iterative cycles. Unfortunately, the energy expression of perturbative methods is not variational in the CI coefficients. As a consequence, the energy derivatives contain the derivatives of the CI coefficients. Evaluation of the latter is straightforward in principle, but expensive numerically. The first analytic gradient method for correlated wave functions was developed by Pople et al (18) for the second-order Møller-Plesset perturbation theory (27). This formulation does not seem to be competitive computationally; recent extensive geometry (28) and vibrational frequency (29) calculations have been carried out by numerical differentiation.

Second and Higher Derivatives

Analytic calculation of the force constants themselves (second derivatives) is complicated by the fact that the evaluation of the (first) derivatives of the variational parameters, even in the SCF case, can no longer be avoided. Continuing the general formulation given above, the derivatives c_i^a are obtained from Eq. 9 and from the first derivatives of the variational Eq. 5:

$$\Sigma_j(\partial^2 F/\partial c_i \partial c_j)c_j^a + (\partial^2 F/\partial c_i \partial R_a) - \Sigma_m \lambda_m^a(\partial f_m/\partial c_i) = 0, \quad i = 1,\ldots N. \quad 12.$$

Eqs. 9 and 12, a very large system of $N \times M$ linear equations, form the coupled-perturbed Hartree-Fock (CPHF) equations. Once they are solved, the second derivatives of the energy can be expressed in a straightforward way. For HF wave functions, explicit formulas were given long ago by Bratož (30) and by Gerratt & Mills (17). The generalization for MC-SCF case has been worked out only recently (4, 31, 32, 41).

Although the theory had been available, the analytic calculation of second derivatives seemed impractical for a long time. Only recently did

Pople et al (18) find a computationally efficient solution, based on two important innovations. First, they used the Rys polynomial integral evaluation technique of Dupuis et al (33, 34), which speeds up the calculation of higher integral derivatives.[1] Second, they applied an efficient combination of direct and iterative methods for the solution of large systems of linear equations, in the spirit of the matrix diagonalization techniques of Roos & Siegbahn (35, 36) and Davidson (37). We call this type of method DIIS, direct inversion in the iterative subspace (38). It may be noted that DIIS has also proven very useful in accelerating the convergence of conventional SCF iteration (38, 39) and even geometry optimization (40).

The second derivative method (18) was worked out explicitly for one-determinant SCF wave functions within the unrestricted HF scheme. Recently, Saxe et al (41) also worked it out for the simplest (high-spin) open shell case with restricted HF wave functions. Yamaguchi et al (42) first implemented it for the simplest MC-SCF scheme, a two-configuration SCF wave function.

Analytic second derivatives represent a further significant breakthrough in the investigation of potential surfaces. According to the available timing data (18, 41), a complete calculation takes about four to five times the computer time needed for the gradient. Considering that at least $n+1$ points are necessary to obtain the complete force field by numerically differentiating the gradient, the fully analytic evaluation is clearly more efficient if the number of degrees of freedom, n, is larger than 4 or 5.

Direct evaluation of the third derivatives (4) offers further substantial advantages for the study of potential surfaces. This method has not yet been implemented.

Analytical second derivatives for general configuration interaction wave functions have been derived and implemented recently (41a). In principle, this method is not expected to bring a major advance in economy as compared with the numerical differentiation of the gradients (4). There are, however, substantial savings in integral evaluation and in logic, and even the first implementation is economical (41a).

PRACTICAL ASPECTS

Theoretical calculations, at any level, inevitably give approximate results. As long as this is the situation, some practical aspects of the theoretical

---

[1] Pople et al (18) claim that as a result of the Rys polynomial technique gradient evaluation in their program is less costly than integral calculation, in contrast to other programs in which the ratio is about 4 : 1. However, this comparison is invalid, as they do not distinguish between integral time and total (SCF + integral) energy calculation time, and because of the selective use of symmetry in the gradient calculation (see 1).

approach are not negligible. We have explained in detail our recommendations for a strategy for systematic calculations (7, 43). Three main points are to be considered: (a) selection of the basis set, (b) choice of the reference geometry, and (c) representation of the force constants in a suitable coordinate system.

For larger molecules one is necessarily restricted to the Hartree-Fock (HF) approximation and to relatively modest basis sets. The latter still must be flexible enough to give a reasonable description of the wave function. Basis sets of double-zeta (DZ) quality in the valence shell (split valence basis) are the minimum requirement. The 4-31G (44) and 4-21G (7) Gaussian basis sets perform well; these have been used most widely in force constant calculations (see 1 for a review and the next section). Note that the latter is significantly more economical in analytic derivative calculations. The 3-21G set (45) is practically equivalent to the 4-21G one, both in quality and in economy. Inclusion of polarization functions may be necessary for molecules with lone electron pairs, cumulated double bonds, etc. For second row atoms the polarized 3-21G(*) basis set (46) is the one of comparable quality with the above basis sets for first row atoms. Large scale CI calculations need matching large basis sets. Too small basis sets yield exaggerated correlation effects, and even DZ plus polarization basis sets contain significant basis truncation errors (47, and the section below on CI results).

Selection of the reference geometry around which the derivatives are evaluated has quite a significant effect on the (harmonic) force constants, especially for the stretchings, due to anharmonicity. The most consistent procedure would be to take the minimum energy geometry determined within the same theoretical scheme to be used for the evaluation of the force field. This procedure has, however, the disadvantage that the error in the geometry appears as a considerable, often dominant, contribution to the errors in the force constants [for a more detailed exposition of this argument see e.g. (2, 47)]. To eliminate this effect, we have therefore advocated calculating force constants around the exact equilibrium geometry (7, 43). The latter could be taken from experimental results, but this introduces random errors into the procedure; also, most experimental structures are vibrational averages, rather than equilibrium data. For simple organic molecules, systematic corrections can be applied on the theoretical geometries obtained from SCF calculations to estimate the equilibrium geometries (48–50, 7). For molecules accessible to high-level CI calculations, it is quite practical to determine the geometry from the CI wave function, while using the less expensive SCF method for the much more demanding force constant calculations. Correlation effects on the latter become much smaller with this procedure (1, 47).

The direct quantum mechanical evaluation of energy derivatives is efficient only if Cartesian nuclear coordinates are used. Cartesian force constants are perfectly adequate for the calculation of vibrational frequencies. However, force constants expressed in internal valence coordinates are superior in three respects:

1. They are more transferable than Cartesian constants.
2. They allow a more compact representation of the force fields of large molecules, particularly if anharmonic terms are retained.
3. They are naturally adapted to simple physical models.

A selective empirical scaling of the theoretical force constants can be done in valence coordinates only (see next section). A slight disadvantage of internal valence coordinates is the multitude of coordinates in use, necessitating awkward transformations. A standardized set of local symmetry coordinates has been recommended recently (7). This set is applicable for most organic molecules, avoids the problem of redundancy, and is well transferable. The transformation of the gradients and quadratic force constants from Cartesian to internal valence coordinates is a simple procedure (2). Cartesian coordinates can also be used as intermediate quantities in the transformation from one internal coordinate system to another (1).

Analysis of the ab initio results and comparison with experiment can be carried out either at the level of force constants or the frequencies. For reasons listed below, we greatly prefer the former. First, one can determine the frequencies from the force constants but not vice versa. Second, the frequencies depend on many physically different types of force constants simultaneously. Some of these constants are more sensitive to deficiencies in the theory than others. Comparison of the frequencies with experiment reflects primarily the accuracy of the dominant diagonal terms in the potential and does not allow the utilization of theoretical predictions for the nondominant force constants, which may be substantially more accurate than the corresponding experimental quantities. An empirical scaling of the force constants is obviously a much sounder procedure than that of the frequencies.

HARTREE-FOCK FORCE CONSTANTS AND THEIR COMBINATION WITH EXPERIMENT

Systematic force constant calculations at the Hartree-Fock (HF) level were started in the early 1970s on small polyatomic molecules [(51–53; see (7) for a review for further early works by Pulay & Meyer, Pulay, Fogarasi & Sawodny, Botschwina et al, Schlegel et al, and Blom & Altona]. By now, it has become possible to calculate fairly large molecules like, for example, benzene (54) or naphthalene (H. Sellers, J. E. Boggs, P. Pulay in preparation). For a number of basic prototypical molecules, complete HF ab initio harmonic force fields are available (see 1 for a review of recent work by Bock et al, Ha et al, Pouchan et al, Wiberg & Wendoloski, Pacansky & Dupuis, Morokuma and co-workers, Komornicki et al, and Steele et al).

The accuracy of the results at this moderate level of theory is obviously limited. Experience shows, however, that the errors are largely systematic. In general, the following conclusions can be drawn: (a) diagonal stretching force constants are overestimated by 10–15%; (b) overestimation of diagonal bending force constants is usually somewhat higher, about 20–30%; (c) for the wide variety of off-diagonal (coupling) constants, the results are somewhat less systematic; still, large values are obtained within comparable accuracy of about 20%, while small values, for which an absolute error is more appropriate, are obtained within 0.05–0.10 aJ/$Å^n$.

The above accuracy in the force constants leads to an overestimation of the vibrational frequencies by roughly 10%. This conclusion was also demonstrated by a recent extensive study in which the frequencies of 37 molecules were calculated (55; valence force constants were not evaluated). Considering the modest level of theory, this accuracy is quite remarkable. Still, a purely theoretical approach at this level does not allow quantitative predictions.

The significance of HF force constants lies in their combination with experimental data. The idea of combining the theoretical and experimental information was suggested very early (6, 53), and several slightly different procedures have been used since. The basic feature of all these simple schemes is the scaling down of the systematically overestimated diagonal force constants. Handling of the coupling force constants is less clear-cut; they may be left unchanged or may also be included in the scaling. With either method, high accuracy for these, generally small, quantities is not required. The important point is that theoretical calculations give good approximations for these terms, which are poorly determined experimentally.

The usefulness of the above idea was first demonstrated in an early study on C_2 hydrocarbons (56), formaldehyde (56a), and methylamine (56b). A simple adjustment on the diagonal force constants, leaving the coupling terms unchanged, gave already remarkably good frequencies. In some other studies (57–60) the off-diagonal terms were again taken fixed from the calculations, while the diagonal force constants were determined from the experimental frequencies. The complete theoretical information is preserved in two slightly different scaling schemes introduced by Blom & Altona (48) and by our group [first in semiempirical calculations (61, 62)]. A few scale factors are assigned to chemically different types of force constants, and the complete force field is scaled down by fitting it, through the scale parameters, to the experimental frequencies. Details of these schemes and their applications are given elsewhere (43).

Empirically adjusted HF level force constants are now available for a number of prototypical organic molecules. Exact comparisons are, however, hampered by the differences, even if slight, in the procedures. We have therefore suggested a uniform procedure recently, calling the resulting force constants the Scaled Quantum Mechanical (SQM) force field (43). The recommendations include not only the scaling scheme, but also the selection of basis set, reference geometry, and internal coordinates, as was outlined in the previous section.

To illustrate the accuracy of SQM type force fields we discuss here ethylene at some length. Ethylene is one of the few polyatomic molecules for which very precise experimental results are available (63, 64) to compare with the theoretical values (65, 66). It is interesting to note that ethylene was the first example of a successful application of quantum chemical force constant calculations. Until 1971 the numerous experimental force fields fell into two main categories, both of which reproduced all isotope frequencies available at that time. The ab initio calculations of Pulay & Meyer (53) selected clearly the right solution, and this was simultaneously confirmed by an experimental study, based on more data, of McKean & Duncan (67).

Alternative solutions to the experimental force field problem seem to be quite frequent; other early examples include ONF, NF_3 (68), and O_2NF (69).

Turning to the force field of ethylene given in Table 1, the overall agreement of the various results is excellent. Nevertheless, the earlier experimental force field of Duncan et al (DMM in Table 1) still has some discrepancies as compared to theory. It is thus reassuring that most of these are eliminated in the improved experimental force field (D-H in Table 1), based on 116 (!) pieces of data. In particular, the CC str/CH_2 sym str constant $F_{1,2}$, which was too high by a factor of about 2 in the DMM field, is now in perfect accord with the theoretical results. The only remaining discrepancy concerns the as str/rock interactions $F_{4,8}$ and $F_{4,9}$ given also in symmetry coordinates at the bottom of Table 1. As can be seen more clearly in terms of these latter, the disagreement is basically in the B_{2u} symmetry species. Considering that the sensitivity of the individual force constants to the experimental data is quite different, whereas fairly balanced results can be expected from the calculations, the theoretical result is probably more reliable. One may add that the reproduction of the experimental data (not given here) was as good with the SQM type force fields as with those derived directly from the experimental data. The frequencies of all isotopomers were reproduced within ~ 10 cm^{-1}, other data were also calculated within their experimental error limits (66).

Although ethylene may be an ideal case, the results on a number of molecules are of comparable quality. Scaled ab initio force fields have been systematically developed by Blom & Altona [ethane, propane, cyclopropane (48); methanol (70); propene (71); dimethylether (72)], by our group [benzene (54); cyclobutane (73); oxetane (74); CH_3POF_2 (75); a series of conjugated molecules (43); cubane (75a); papers are in preparation on

Table 1 Scaled quantum mechanical and experimental force constants of ethylene[a]

Force constant	$F_{i,j}$	Ab initio[b]		Experimental	
		4-21G	6-31G**	DMM[c]	DH[d]
CC str	1,1	9.351	9.152	9.395	9.418
CH_2 sym str	2,2	5.606	5.601	5.620	5.604
CH_2 as str	4,4	5.547	5.537	5.575	5.520
CH_2 sym def	6,6	0.467	0.470	0.470	0.470
CH_2 rock	8,8	0.560(0.570)[e]	0.558	0.572	0.567
CH_2 wag	10,10	0.267(0.258)[e]	0.268	0.257	0.257
torsion	12,12	0.135(0.139)[e]	0.135	0.139	0.139
CC str/CH_2 sym str	1,2	0.112	0.123	0.257	0.113
CC str/CH_2 sym def	1,6	−0.220	−0.190	−0.222	−0.234
CH_2 sym str/sym str	2,3	0.011	0.009	0.017	0.025
CH_2 sym str/sym def	2,6	0.093	0.087	0.018	0.083
CH_2 sym str/sym def	2,7	−0.013	−0.016	−0.074	−0.005
CH_2 sym def/sym def	6,7	0.017	0.018	0.018	0.018
CH_2 as str/as str	4,5	0.022	0.019	−0.082	0.005
CH_2 as str/rock	4,8	0.161	0.123	0.111	0.056
CH_2 as str/rock	4,9	−0.055	−0.060	−0.285	−0.182
CH_2 rock/rock	8,9	−0.086	−0.093	−0.087	−0.082
CH_2 wag/wag	10,11	0.035	0.039	0.039	0.040
CH_2 as str/rock interactions in symmetry coordinates					
$F_{4,8} - F_{4,9} = F^{sym}_{5,6}(B_{1g})$		0.216	0.182	0.396	0.258
$F_{4,8} + F_{4,9} = F^{sym}_{9,10}(B_{2u})$		0.106	0.062	−0.174	−0.136

[a] For details, especially the definition of local symmetry coordinates, see Ref. (66).
[b] Scaled ab initio results with two basis sets (66). Cf the results of Blom & Altona (65), which are very similar.
[c] Experimental results based on 55 pieces of data (63).
[d] Experimental results based on 116 pieces of data (64).
[e] The deviations of the rocking, wagging, and torsional force constants from the experimental results are mainly due to the different assumptions about the equilibrium geometry. To show this, the scaling of the 4–21G force constants was also carried out using the G matrix evaluated at the reference geometry of DMM (63). These values are given in parentheses.

naphthalene and hexatriene], and by Bock et al [fluoroethylenes (76); glyoxdl and acrolein (77); butadiene (78)]. Important systematic work, without scaling, has been performed by Schlegel et al (79–82), including anharmonic force constants.

In general, SQM type force fields reproduce the experimental frequencies within the excellent accuracy of approximately 2%. At least part of the residual errors can be attributed to the limitations inherent in the harmonic approximation. One must of course not forget that the theoretical force field is fitted to these frequencies. The fact, however, that this accuracy is achieved by using only a few variable parameters (the scale factors) proves that the theoretical information is basically correct. For medium sized molecules this combined approach is certainly the best method to obtain reliable force fields.

The final goal is of course the prediction of vibrational frequencies. For this purpose, one should calculate the theoretical force field and apply on it scale factors taken over from similar molecules. The success of this procedure depends on the transferability of scale factors. Estimated scale factors have proven good enough to check experimental assignments in several cases. For example, we were able to reassign some rocking frequencies of cyclobutane (73) and oxetane (74), showing that they must be placed at much higher values than expected from spectroscopic experience. An earlier example is the correct assignment of the fundamental frequencies of methylamine (56b). The successful reproduction of the fundamental frequencies in a series of analogous molecules by a common, small set of scale factors (43, 48) is a proof of the transferability of the scale factors. Recently, we have tested this transferability in a more direct way (82a): the scale factors of benzene were used to correct the theoretical force field of pyridine. The harmonic frequencies calculated this way are reproduced in Table 2. Except for the CH stretching frequencies, which are strongly influenced by anharmonicity, the mean deviation is 5.7 cm^{-1}. Obviously, this is an ideal example because of the close similarity of benzene and pyridine. Still, this successful a priori prediction of the vibrational spectrum of a moderately complex molecule offers hope that unknown spectra can be computed to harmonic oscillator accuracy.

It should be mentioned finally that the concept of combining theoretical and experimental information has been used with success to calculate also anharmonic frequencies. In a series of papers by Botschwina (83–86), calculated HF values are used for the higher order terms and usually also for the harmonic coupling terms, while the geometry is taken from experiment and the diagonal quadratic force constants are fitted to the experimental frequencies.

Although the best results can be obtained from combined approaches

Table 2 Calculated and experimental in-plane fundamental frequencies of isotopic pyridines[a]

d_0		$4\text{-}d_1$		$2,6\text{-}d_2$		$3,5\text{-}d_2$	
Exp	Calc	Exp	Calc	Exp	Calc	Exp	Calc
A_1		A_1		A_1		A_1	
603	604	596	597	595	596	595	597
991	980	989	980	891	881	827	824
1030	1025	1010	1005	985	980	965	955
1069	1073	1067	1073	1019	1016	1033	1032
1217	1214	1214	1214	1089	1087	1174	1171
1483	1475	1475	1470	1414	1411	1434	1422
1581	1585	1574	1578	1577	1573	1573	1569
3042	3065	2286	2276	2257	2285	2294	2286
3065	3078	3050	3073	3063	3067	3044	3075
3077	3102	3050	3098	3063	3097	3044	3082
B_2		B_2		B_2		B_2	
654	654	648	649	641	644	641	642
1052	1053	861	861	906	900	866	870
1146	1156	1085	1090	1131	1134	1086	1076
1227	1240	1223	1231	1184	1183	1224	1233
1355	1353	1330	1333	1245	1240	1323	1318
1437	1436	1412	1403	1418	1419	1408	1405
1574	1582	1559	1570	1567	1574	1558	1568
3034	3071	3035	3071	2249	2274	2272	2286
3079	3093	3072	3093	3063	3088	3035	3076

[a] Ref. (82a).

discussed above, even the unadjusted theoretical force field and frequencies can be used as a guide in interpreting experimental data. This is especially true for unstable systems for which the spectroscopic data are usually incomplete. To mention a few examples of applications, radicals (87, 88, 88a), excited electronic states (89), ionized core-hole states [important in ESCA (90, 91)], conformers (77, 78, 92), transition states (93), and unstable molecules (94) have been treated. For most ground-state molecules, even for those with unusual structures, the predictions can be improved significantly by using empirical scale factors transferred from other molecules.

FORCE CONSTANTS CALCULATED FROM CORRELATED WAVE FUNCTIONS

Force constants calculated with good-quality correlated wave functions are still very expensive at the present time, and results are available for small molecules only. However, there is rapid development in this field, as shown for example by the recent introduction of analytical CI gradient techniques (20, 21, 41a).

All CI methods are suitable for force field calculations, provided that they yield smooth potential surfaces. In general, this excludes methods with extensive individual configuration selection. The leading techniques consider explicitly all double (and in most cases also all single) substitutions relative to a reference configuration. They include variational CI, possibly with Davidson's correction (95), coupled-cluster methods like CEPA [12; for a review see (24)], and perturbative techniques. Only dynamical electron correlation is considered in this review.

We begin with a brief survey of high-quality representative calculations. The molecules best studied are water (96–101), hydrogen cyanide (47, 85, 101–106), acetylene (85, 108), methane (101, 109), formaldehyde (101, 110), ethylene (18), and carbon dioxide (111). Systematic calculations, with several basis sets, have been performed for the frequencies of HCN, H_2CO, H_2O, CH_4, and NH_4 (101), and for the force constants of HF, HCN, and NH_3 (47). The fundamental frequencies of 36 molecules, calculated at the second-order Møller-Plesset level of perturbation theory (MP2), have been published recently (29).

A particularly thorough comparison of different correlation techniques has been carried out for the water molecule (98, 99). This study shows that the inclusion of dynamic electron correlation leads to significant improvement, although the final results still fall somewhat short of quantitative accuracy. Surprisingly, the major difference between various many-body techniques appears in the linear force constants (forces), i.e. in the molecular geometry (4).

A more recent study of water (100) uses MC-SCF based wave functions to generate the ground-state potential surface for water. The results of the state-of-the-art calculations by Bartlett et al (99) and Kraemer et al (100) disagree for the anharmonic force constants. Most of this puzzling behavior can be attributed to numerical difficulties in the fitting of the potential surface (1). The difficulty is caused by the fitting of the total (SCF plus correlation) potential energy surface with a considerable number of force constants. As pointed out for example in (109), the SCF and the correlation energy surfaces are very unlike, and require different treatment. The SCF energy is generally very anharmonic, and requires a considerable number of fitting constants. Because it is not too expensive to generate, and the energies are highly accurate numerically, a large number of closely spaced grid points may be used. The correlation energy, on the other hand, is generally a smooth function of the nuclear coordinates, and can be fitted with a low-degree polynomial. Because of the high cost and the lower numerical accuracy of the energies, a smaller number of widely spaced grid points should be used.

If only the correlation energy contributions are considered, the differences between the two calculations (99, 100) indeed diminish, as shown in part in Table 3. It is noteworthy that only 14 constants are needed to reproduce the correlation energy surfaces of Bartlett et al (99) with a root-mean-square deviation of $1-3 \times 10^{-8} E_h$ (roughly the accuracy of the points), showing that only 11 force constants are well determined.

Because it costs so much to achieve even slight improvements in the theory, it is of much interest to determine the sensitivity of different force constants to the main variables in the theory: basis set quality and the level of treatment of electron correlation. The representative calculations cited above allow a number of conclusions. We shall particularly rely on a recent study (47) that investigated these questions for the small prototype molecules HF, HCN, and NH_3. The valence force constants up to the dominant quartic terms were evaluated at three correlation levels: SCF, CI with all single and double substitutions (SD-CI), and the latter augmented with the unlinked cluster contribution, estimated from Davidson's correction (DC) (95). At each level, calculations were carried out with five basis sets, ranging from small split-valence (DZ), through DZ plus polarization (DZ+P) and triple-zeta plus polarization (TZ+P), to TZ plus two sets of polarization functions (TZ+2P). As an illustration, part of the results for ammonia are reproduced in Table 4.

The main conclusions pertaining to dynamic correlation effects on force constants are the following.

1. The dominant correlation contribution to potential surfaces near the equilibrium geometry is the linear (gradient) term. Unlinked cluster effects

Table 3 Comparison of the cubic force constants of water, obtained by Bartlett et al (99) and Kraemer et al (100)[a]

Constant	Total		Correlation contribution	
	Barlett CCD	Kraemer MR-CI	Bartlett CCD	Kraemer MR-CI
F_{rrr}	−57.3	−65.0	0.23	−0.02
F_{aaa}	−0.678	−0.71	0.04	0.00
$F_{rrr'}$	−0.06	−0.55	−0.05	−0.08
F_{rra}	−0.08	−0.02	−0.01	−0.05
$F_{rr'a}$	−0.48	−0.50	0.04	0.05
F_{raa}	−0.286	−0.24	0.012	0.017

[a] P. Pulay, unpublished. Only the first 33 points of Ref. (100) were used in the fitting.

Table 4 Test of the basis set and electron correlation effects on selected force constants of ammonia[a]

Constant	4-21G SCF	4-21G SD-CI	5-31G** SCF	5-31G** SD-CI	6-311G** SCF	6-311G** SD-CI	6-311G** SCF	6-311G** SD-CI	6-311G** DC
g_r	0.0203	−0.1057	0.0760	−0.0643	0.0740	−0.0756	0.0827	−0.0703	−0.0266
g_s	0.04040	−0.01007	0.00113	−0.00781	0.00221	−0.00722	−0.00252	−0.00554	−0.00147
F_{rr}	7.231	−0.110	7.460	−0.020	7.385	−0.025	7.343	−0.109	−0.052
F_{ss}	0.1553	−0.0087	0.1293	−0.0128	0.1285	−0.0109	0.1263	−0.0133	−0.0023
F_{aa}	0.9003	−0.0546	0.7683	−0.0507	0.7584	−0.0752	0.7510	−0.0520	−0.0101
F_{rs}	−0.181	−0.016	−0.133	−0.010	−0.128	−0.008	−0.129	−0.007	
F_{ra}	−0.220	−0.012	−0.156	−0.004	−0.152	0.006	−0.156	−0.002	
F_{rrr}	−47.3	−0.5	−47.6	−0.6	−46.5	−0.3	−46.1	0.0	

[a] From Ref. (47). Gradients are denoted by g, force constants by F. The definition of the coordinates is as follows: r is the NH stretching; $s = 3^{-1/2}(t_1 + t_2 + t_3)$ where e.g. t_1 is the out-of-plane angle of N-H$_1$ with the H$_2$NH$_3$ plane; $a = 6^{-1/2}(2\alpha_1 - \alpha_2 - \alpha_3)$ where α_1 is the H$_2$NH$_3$ angle. All force constants have been evaluated at the reference geometry, $r = 1.0116$ Å, $\alpha = 106.67°$. Values in the SD-CI and DC columns are contributions from singles and doubles CI and from Davidson's correction, which have to be added to the SCF values to obtain the total.

for the gradient are significant. For example, in ammonia (TZ+2P basis), the SD-CI contribution contributes about 0.010 Å to the lengthening of the N-H bond; the contribution from Davidson's correction is 0.003 Å (see Table 4). Correlation effects are larger in multiple bonds. For the CN bond in HCN, the optimized SCF, SD-CI, and DC lengths are, in this order, 1.1236, 1.1451, and 1.1538 Å (cf the experimental $R_e = 1.1532$ Å), using the TZ+2P basis. The correlation effect on the valence angles of ammonia and water is 1–2°.

2. Correlation contributions to quadratic stretching force constants are not large if the force constants are evaluated at the accurate equilibrium geometry. For example, as Table 4 shows, the combined SD-CI and DC contributions to the N-H stretching force constant in ammonia is only 2% (TZ+2P basis). Part of the deviation in earlier calculations obviously was due to the insufficient basis set, and was erroneously attributed to electron correlation. The effect is again larger for multiple bonds. For the CN bond, the combined SD-CI and DC contributions are 3.8% of the SCF (TZ+2P basis) (47). It must be pointed out, however, that the best final value, 19.42 aJ/Å, still exceeds the accurate experimental value, 18.70 aJ/Å, by another 3.8% (47). The origin of this deviation is not clear at this time. If, as it seems likely, it is caused by higher-order correlation, then the total correlation contribution to the triple-bond stretching force constant is 7.9%. This is in contrast with the case of single bonds, particularly X-H bonds, where the TZ+2P basis, SD-CI+DC results are generally accurate to within 2% (47).

3. Correlation contributions to the diagonal deformation force constants amount to about 10% for single bonds, and up to 20% for multiple bonds at the basis set limit.

4. The effect of electron correlation on the coupling force constants in valence coordinates is generally within 10%. Considering that most of these quantities are small, the correlation contribution is not very significant. An exceptional case is carbon dioxide, in which the CO/CO coupling is significantly overestimated at the Hartree-Fock level (111).

5. In general, the effect of electron correlation on the quadratic force field can be reproduced quite well by scaling, providing justification to the empirical scale factor methods.

6. An important, and quite counterintuitive, result is that cubic force constants, particularly the important cubic diagonal stretching ones, are reproduced to very high accuracy at the Hartree-Fock level (47, 109); correlation contributions amount to only 1–2% in general at the correct r_e geometry.

7. Basis set effects are significant, even beyond the usual double-zeta plus polarization level. The transferability of correlation contributions between

different basis sets has been investigated recently (47). If the correlation contributions were well transferable then the high cost of CI calculations with large basis sets could be avoided by evaluating the correlation contributions with smaller basis sets. The results, as illustrated in Table 4, are rather disappointing in this respect: the SD-CI contributions do not show better convergence than do the SCF values (see particularly the variation in F_{rr}).

It is interesting to compare the best theoretical values with experiment (47). For the XH stretching force constants the following deviations were found: 2.3% in HF, 0.8% in HCN, and 1.8% in NH_3. The discrepancy for the CN stretching force constant is 3.8%. Deviations in the final values of the deformational force constants are 4.8% in HCN and 1.2% for F_{aa} in NH_3; a seemingly perfect agreement found for F_{ss} in ammonia may be partly fortuitous because of the uncertainty of the experimental value. The quadratic coupling terms and the anharmonic force constants are calculated within their experimental uncertainty. While these results show the potential of the theoretical calculation of force constants, they also demonstrate that even present high-quality calculations have nonnegligible residual errors. These observations are also supported by results for larger molecules, e.g. methyl fluoride (G. Fogarasi, J. E. Boggs, P. Pulay, in preparation).

CONCLUSION

The ab initio calculation of force constants can significantly contribute to our knowledge of molecular force fields. For the small coupling and anharmonic constants, good theoretical values are generally more accurate than all but the most exacting experimental data. Modest calculations at the Hartree-Fock level, using small split-valence basis sets, already yield fairly accurate results, particularly in conjunction with empirical scaling. However, for the dominant quadratic force constants, theory is not yet competitive with the high accuracy of spectroscopic measurements. The present review therefore emphasizes a pragmatic approach, using both theoretical and experimental information. The economy of the theoretical methods has greatly improved in the past decade, mainly because of the introduction of direct derivative techniques.

Literature Cited

1. Fogarasi, G., Pulay, P. 1984. *Vibrational Spectra and Structure*, ed. J. R. Durig, Vol. 14. Amsterdam/Oxford/New York: Elsevier. In press

2. Pulay, P. 1977. *Applications of Electronic Structure Theory*, ed. H. F. Schaefer III, pp. 153–85. New York/London: Plenum

3. Pulay, P. 1981. *The Force Concept in Chemistry*, ed. B. M. Deb, pp. 449–80. New York: Van Nostrand Reinhold
4. Pulay, P. 1983. *J. Chem. Phys.* 78:5043–51
5. Hinze, J. 1974. *Adv. Chem. Phys.* 26:213–63
6. Pulay, P. 1969. *Mol. Phys.* 17:197–204
7. Pulay, P., Fogarasi, G., Pang, F., Boggs, J. E. 1979. *J. Am. Chem. Soc.* 101:2550–60
8. Meyer, W., Pulay, P. 1969. *MOLPRO Description*. München/Stuttgart
9. Pulay, P. 1979. *Theor. Chim. Acta* 50:299–312
10. Schlegel, H. B. 1982. *J. Chem. Phys.* 77:3676–81
11. Meyer, W., Pulay, P. 1972. *Proc. 2nd Seminar on Computational Problems in Quantum Chem., Strasbourg, France*, ed. A. Veillard, G. H. R. Diercksen, pp. 44–48. Munich: Max-Planck Inst.
12. Meyer, W. 1973. *J. Chem. Phys.* 58:1017–35
13. Botschwina, P. 1974. *Chem. Phys. Lett.* 29:98–101
14. Goddard, J. D., Handy, N. C., Schaefer, H. F. III. 1979. *J. Chem. Phys.* 71:1525–30
15. Kato, S., Morokuma, K. 1979. *Chem. Phys. Lett.* 65:19–25
16. Wendoloski, J. J. 1981. Quoted in Ref. 32
17. Gerratt, J., Mills, I. M. 1968. *J. Chem. Phys.* 49:1719–29
18. Pople, J. A., Krishnan, R., Schlegel, H. B., Binkley, J. S. 1979. *Int. J. Quantum Chem. Symp.* 13:225–41
19. Meyer, W. 1976. *J. Chem. Phys.* 64:2901–7
20. Brooks, B. R., Laidig, W. D., Saxe, P., Goddard, J. D., Yamaguchi, Y., Schaefer, H. F. III. 1980. *J. Chem. Phys.* 72:4652–53
21. Krishnan, R., Schlegel, H. B., Pople, J. A. 1980. *J. Chem. Phys.* 72:4654–55
22. Osamura, Y., Yamaguchi, Y., Schaefer, H. F. III. 1981. *J. Chem. Phys.* 75:2919–22
23. Osamura, Y., Yamaguchi, Y., Schaefer, H. F. III. 1982. *J. Chem. Phys.* 77:385–90
24. Kutzelnigg, W. 1977. *Methods of Electronic Structure Theory*, ed. H. F. Schaefer III, pp. 129–88. New York/London: Plenum
25. Pulay, P. 1983. *J. Mol. Struct. THEOCHEM.* 103:57–66
26. Pulay, P. 1983. *Int. J. Quantum Chem. Symp.* 17:257–63
27. Pople, J. A., Binkley, J. S., Seeger, R. 1976. *Int. J. Quantum Chem. Symp.* 10:1–19
28. DeFrees, D. J., Raghavachar, K., Schlegel, H. B., Pople, J. A. 1982. *J. Am. Chem. Soc.* 104:5576–80
29. Hout, R. F. Jr., Levi, B. A., Hehre, W. J. 1982. *J. Comp. Chem.* 3:234–50
30. Bratož, S. 1958. *Colloq. Int. CNRS* 82:287–301
31. Jaszunski, M., Sadlej, A. J. 1977. *Int. J. Quantum Chem.* 11:233–45
32. Dupuis, M. 1981. *J. Chem. Phys.* 74:5758–65
33. Dupuis, M., Rys, J., King, H. F. 1976. *J. Chem. Phys.* 65:111–16
34. Dupuis, M., King, H. F. 1978. *J. Chem. Phys.* 68:3998–4004
35. Roos, B. O. 1972. *Chem. Phys. Lett.* 15:153–59
36. Roos, B. O., Siegbahn, P. E. M. 1977. *Methods of Electronic Structure Theory*, ed. H. F. Schaefer III, pp. 277–318. New York/London: Plenum
37. Davidson, E. R. 1975. *J. Comp. Phys.* 17:87–94
38. Pulay, P. 1980. *Chem. Phys. Lett.* 73:393–98
39. Pulay, P. 1982. *J. Comput. Chem.* 3:556–60
40. Császár, P., Pulay, P. 1984. *J. Mol. Struct.* In press
41. Saxe, P., Yamaguchi, Y., Schaefer, H. F. III. 1982. *J. Chem. Phys.* 77:5647–54
41a. Fox, D. J., Osamura, Y., Hoffmann, M. R., Gaw, J. F., Fitzgerald, G., et al. 1983. *Chem. Phys. Lett.* 102:17–19
42. Yamaguchi, Y., Osamura, Y., Fitzgerald, G., Schaefer, H. F. III. 1983. *J. Chem. Phys.* 78:1607–8
43. Pulay, P., Fogarasi, G., Pongor, G., Boggs, J. E., Vargha, A. 1983. *J. Am. Chem. Soc.* 105:7037–47
44. Ditchfield, R., Hehre, W. J., Pople, J. A. 1971. *J. Chem. Phys.* 54:724–28
45. Binkley, J. S., Pople, J. A., Hehre, W. J. 1980. *J. Am. Chem. Soc.* 102:939–47
46. Pietro, W. J., Francl, M. M., Hehre, W. J., DeFrees, D. J., Pople, J. A., Binkley, J. S. 1982. *J. Am. Chem. Soc.* 104:5039–48
47. Pulay, P., Lee, J.-G., Boggs, J. E. 1983. *J. Chem. Phys.* 79:3382–91
48. Blom, C. E., Altona, C. 1976. *Mol. Phys.* 31:1377–91
49. Blom, C. E., Slingerland, P. J., Altona, C. 1976. *Mol. Phys.* 31:1359–76
50. Schäfer, L., van Alsenoy, C., Scarsdale, J. N. 1982. *J. Mol. Struct. THEOCHEM.* 86:349–64
51. Pulay, P. 1970. *Mol. Phys.* 18:473–80
52. Pulay, P. 1971. *Mol. Phys.* 21:329–39
53. Pulay, P., Meyer, W. 1971. *J. Mol. Spectrosc.* 40:59–70
54. Pulay, P., Fogarasi, G., Boggs, J. E. 1981. *J. Chem. Phys.* 74:3999–4014

55. Pople, J. A., Schlegel, H. B., Krishnan, R., DeFrees, D. J., Binkley, J. S., et al. 1981. *Int. J. Quantum Chem. Symp.* 15:269–78
56. Pulay, P., Meyer, W. 1974. *Mol. Phys.* 27:473–90
56a. Meyer, W., Pulay, P. 1975. *Theor. Chim. Acta* 32:253–64
56b. Pulay, P., Török, F. 1975. *J. Mol. Struct.* 29:239–46
57. Bleicher, W., Botschwina, P. 1975. *Mol. Phys.* 30:1029–36
58. Botschwina, P., Meyer, W., Semkov, A. M. 1976. *Chem. Phys.* 15:25–34
59. Fogarasi, G., Pulay, P., Molt, K., Sawodny, W. 1977. *Mol. Phys.* 33:1565–70
60. Christe, K. O., Curtis, E. C., Sawodny, W., Härtner, H., Fogarasi, G. 1981. *Spectrochim. Acta Part A* 37A:549–56
61. Török, F., Hegedüs, A., Kósa, K., Pulay, P. 1976. *J. Mol. Struct.* 32:93–99
62. Fogarasi, G., Pulay, P. 1977. *J. Mol. Struct.* 39:275–80
63. Duncan, J. L., McKean, D. C., Mallinson, P. D. 1973. *J. Mol. Spectrosc.* 45:221–46
64. Duncan, J. L., Hamilton, E. 1981. *J. Mol. Struct.* 76:65–80
65. Blom, C. E., Altona, C. 1977. *Mol. Phys.* 34:177–92
66. Fogarasi, G., Pulay, P. 1981. *Acta Chim. Acad. Sci. Hung.* 108:55–73
67. McKean, D. C., Duncan, J. L. 1971. *Spectrochim. Acta Part A* 27A:1879–91
68. Sawodny, W., Pulay, P. 1974. *J. Mol. Spectrosc.* 51:135–41
69. Molt, K., Sawodny, W., Pulay, P., Fogarasi, G. 1976. *Mol. Phys.* 32:169–76
70. Blom, C. E., Otto, L. P., Altona, C. 1976. *Mol. Phys.* 32:1137–49
71. Blom, C. E., Altona, C. 1977. *Mol. Phys.* 33:875–85
72. Blom, C. E., Oskam, A., Altona, C. 1977. *Mol. Phys.* 34:557–71
73. Bánhegyi, Gy., Fogarasi, G., Pulay, P. 1982. *J. Mol. Struct. THEOCHEM.* 89:1–13
74. Bánhegyi, Gy., Pulay, P., Fogarasi, G. 1983. *Spectrochim. Acta Part A* 39A:761–69
75. von Carlowitz, S., Zeil, W., Pulay, P., Boggs, J. E. 1982. *J. Mol. Struct.* 87:113–24
75a. Dunn, K. M., Pulay, P., Van Alsenoy, Ch., Boggs, J. E. 1984. *J. Mol. Spectrosc.* 103:268–80
76. Bock, C. W., George, P., Trachtman, M. 1979. *J. Mol. Spectrosc.* 76:191–203
77. Bock, C. W., George, P., Trachtman, M. 1979. *J. Mol. Spectrosc.* 78:248–56, 298–308
78. Bock, C. W., Trachtman, M., George, P. 1980. *J. Mol. Spectrosc.* 84:243–55
79. Schlegel, H. B., Wolfe, S., Bernardi, F. 1975. *J. Chem. Phys.* 63:3632–38
80. Schlegel, H. B., Wolfe, S., Mislow, K. 1975. *J. Chem. Soc. Chem. Commun.* 246–47
81. Schlegel, H. B., Wolfe, S., Bernardi, F. 1977. *J. Chem. Phys.* 67:4181–93
82. Schlegel, H. B., Wolfe, S., Bernardi, F. 1977. *J. Chem. Phys.* 67:4194–98
82a. Pongor, G., Pulay, P., Fogarasi, G., Boggs, J. E. 1984. *J. Am. Chem. Soc.* In press
83. Botschwina, P. 1979. *Chem. Phys.* 40:33–44
84. Botschwina, P., Härtner, H., Sawodny, W. 1980. *Chem. Phys. Lett.* 74:156–59
85. Botschwina, P. 1982. *Chem. Phys.* 68:41–63
86. Botschwina, P. 1982. *Mol. Phys.* 47:241–49
87. Pacansky, J., Dupuis, M. 1982. *J. Am. Chem. Soc.* 104:415–21
88. Dupuis, M., Pacansky, J. 1982. *J. Chem. Phys.* 76:2511–15
88a. Komornicki, A., Jaffe, R. L. 1979. *J. Chem. Phys.* 71:2150–55
89. Gregory, A. R., Kidd, K. G. 1980. *Chem. Phys. Lett.* 72:385–87
90. Clark, D. T., Müller, J. 1977. *Chem. Phys.* 23:429–36
91. Clark, D. T., Smagellotti, A., Tarantelli, F. 1980. *Chem. Phys.* 52:1–9
92. Čarsky, P., Zahradnik, R. 1979. *J. Mol. Struct.* 54:247–55
93. Saxe, P., Yamaguchi, Y., Pulay, P., Schaefer, H. F. III. 1980. *J. Am. Chem. Soc.* 102:3718–23
94. Schaad, L. J., Hess, B. D., Ewig, C. S. 1979. *J. Am. Chem. Soc.* 101:2281–83
95. Langhoff, S. R., Davidson, E. R. 1974. *Int. J. Quantum Chem.* 8:61–72
96. Meyer, W. 1975. *Proc. of the SCR Atlas Symp. No. 4*, ed. V. R. Saunders, J. Brown. pp. 97–104. Chilton, U.K.: Atlas Computer Lab.
97. Hennig, P., Kraemer, W. P., Diercksen, G. H., Strey, G. 1978. *Theor. Chim. Acta* 47:233–48
98. Rosenberg, B. J., Ermler, W. C., Shavitt, I. 1976. *J. Chem. Phys.* 65:4072–80
99. Bartlett, R. J., Shavitt, I., Purvis, G. D. III. 1979. *J. Chem. Phys.* 71:281–91
100. Kraemer, W. P., Roos, B. O., Siegbahn, P. E. M. 1982. *Chem. Phys.* 69:305–21
101. Yamaguchi, Y., Schaefer, H. F. III. 1980. *J. Chem. Phys.* 73:2310–18
102. Taylor, P. R., Bacskay, G. B., Hush, N. S., Hurley, A. C. 1978. *J. Chem. Phys.* 69:1971–79
103. Wahlgren, U., Pacansky, J., Bagus, P. S. 1975. *J. Chem. Phys.* 63:2874–81

104. Liu, B., Sando, K. M., North, C. S., Friedrich, H. B., Chipman, D. M. 1978. *J. Chem. Phys.* 69:1425-28
105. Lie, G. C., Peyerimhoff, S. D., Buenker, R. J. 1981. *J. Chem. Phys.* 75:2892-98
106. Dykstra, C. E., Secrest, D. 1981. *J. Chem. Phys.* 75:3967-72
107. Deleted in proof
108. Bagus, P. S., Pacansky, J., Wahlgren, U. 1977. *J. Chem. Phys.* 67:618-23
109. Pulay, P., Meyer, W., Boggs, J. E. 1978. *J. Chem. Phys.* 68:5077-85
110. Jaquet, R., Kutzelnigg, W., Staemmler, V. 1980. *Theor. Chim. Acta* 54:205-27
111. Ha, T. K., Blom, C. E., Günthard, H. H. 1980. *Chem. Phys. Lett.* 70:473-76

CHARACTERIZATION OF SURFACES THROUGH ELECTRON AND PHOTON STIMULATED DESORPTION[1]

Theodore E. Madey, David E. Ramaker, and Roger Stockbauer

Surface Science Division, National Bureau of Standards, Washington DC 20234

INTRODUCTION

Electron stimulated desorption (ESD) and photon stimulated desorption (PSD) are the surface analogues of electron- and photon-induced dissociation of gaseous molecules. In ESD and PSD, beams of energetic electrons or photons (typical energies range from ~ 10 eV to > 1000 eV) incident on a surface containing terminal bulk atoms or an adsorbed monolayer of molecules or atoms will cause electronic excitations in the surface species; these excitations can cause desorption of ions, ground state neutrals, or metastables from the surface. A basic difference between surface and gas phase dissociation processes is that a solid surface provides extra pathways for deexcitation that are not possible in the gas phase.

There has been a great deal of interest during the last few years in the experimental and theoretical aspects of ESD and PSD (1). Recent theoretical work (2a–e) has provided new understanding of processes involving ion desorption from the surfaces of ionic solids, as well as of mechanisms for ion formation and desorption from covalently-bonded species at surfaces. Angle-resolved ESD and PSD measurements have been established as valuable tools for characterizing the geometry of surface

---

[1] The US Government has the right to retain a nonexclusive, royalty-free license in and to any copyright covering this paper.

molecules (3a–c): ion desorption directions are related to surface bond angles. In addition, the use of synchrotron radiation for the accurate measurements of desorption thresholds and yield vs photon energy for both ions and neutrals provides new data to test conflicting theoretical models (4). An ubiquitous practical problem in surface chemical and topographical analysis using electron and photon beams (Auger spectroscopy, x-ray photoelectron spectroscopy, electron microscopy) is the beam damage that can impede the analysis (5). Thus, ESD and PSD studies of surfaces provide insights into fundamental electronic excitations at surfaces, the structure and bonding of atoms and molecules at surfaces, and a basic understanding of radiation damage processes at surfaces.

The plan of the present review is as follows: the next section contains a summary of experimental observations that typify ESD/PSD observations, and the following section is a description of experimental methods. We then provide a detailed examination of the excitation mechanisms in ESD and PSD, and of the similarities and differences between electron and photon excitation. We examine the utility of ESD ion angular distributions (ESDIAD) in determining local molecular structure on surfaces, and, finally, we look at future challenges.

A recent book (1) and several recent review articles (3–10) provide overviews of different aspects of the physics and chemistry of ESD/PSD, and the interested reader is referred to these for additional information.

GENERAL EXPERIMENTAL OBSERVATIONS IN ELECTRON AND PHOTON STIMULATED DESORPTION

Some of the basic experimental observations (1–10) that characterize electron and photon stimulated desorption of surface species are as follows:

1. As indicated schematically in Figure 1a, bombardment of solid surfaces containing adsorbed layers with electron or photon beams of sufficient energy (e.g. ~ 10 to > 1000 eV) can cause desorption of ground state neutral atomic and molecular fragments, positive and negative ions, and metastable species. In addition, dissociation or "cracking" of surface molecules and surface diffusion of adsorbed species can be induced by electron bombardment. Because of the ease of detection of ions, most ESD/PSD studies have concentrated on measurements of ion desorption.

2. Most ESD/PSD processes have cross sections that are smaller than those for electron- and photon-induced dissociation and ionization of gaseous molecules. For 100 eV electrons, typical cross sections for gas phase dissociative ionization are $\sim 10^{-16}$ cm^2. Typical cross sections for ESD of

adsorbed molecules are in the range 10^{-18} to 10^{-23} cm^2. In general, cross sections for desorption of neutral species are much larger than cross sections for ions, by factors of 10–100. For ESD, an ion yield of 10^{-6} ions per incident electron is considered large.

3. Most ESD/PSD ions are atomic, with H^+, O^+, F^+, and Cl^+ the species having the highest yields. Negative ions of each of these species are also seen (11), but their yields are generally smaller than for positive ions. The dominant molecular ions observed in ESD/PSD of various adsorbed monolayers include CO^+ and $OH^\pm$. ESD/PSD of condensed multilayers often result in larger, more complex ions also [e.g. $C_xH_y^+$ from condensed C_6H_{12} (12a, b) and N_3^+ from condensed N_2 (13)]. Metastable states of CO, Li, and Na have been observed (14a–c).

4. Whereas energy thresholds for the ESD of neutral species are as low as 5 eV for CO from W, thresholds for desorption of ions are higher. Thresholds for ESD/PSD of CO^+ from W and H^+ from condensed CH_3OH are 15–20 eV, while O^+ from adsorbed CO has an onset in the range 25–30 eV. The desorption of O^+ from the surfaces of certain maximal valency metal oxides are related to core hole ionization in the metal cation, typically 25–45 eV for 3d transition metal oxides. New channels for desorption are seen to appear at energies above deep core ionization thresholds, e.g. O^+ from CO and NO above O (1s) at 530 eV. Double positive ions are also observed above core hole thresholds.

5. Ion kinetic energies are generally in the range 0–10 eV.

6. ESD/PSD cross sections are very sensitive to the mode of bonding of a molecule to the substrate surface. In general, cross sections for rupture of internal bonds in adsorbed molecules (e.g. the C–O bond in adsorbed CO bound via the C atom to the substrate, or, the H–O bond in adsorbed H_2O) are higher than cross sections for rupture of metal-atom bonds (e.g. for fractional monolayers of adsorbed H or O on a metal surface). ESD cross sections for desorption of O^+ from metal surfaces are higher for adsorption of O on step and defect sites than on atomically smooth surfaces.

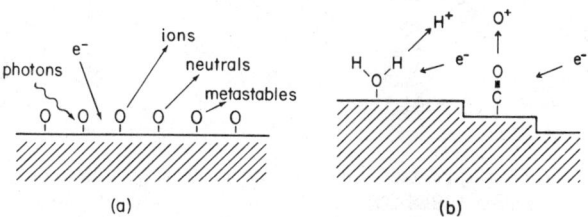

Figure 1 (a) Schematic of electron and photon stimulated desorption processes (ESD/PSD). (b) Schematic bonding configurations for adsorbed molecules on surfaces, showing relationship between surface bond angle and initial ion desorption angle.

7. ESD/PSD ions do not generally desorb in isotropic distributions. Rather, they desorb in discrete cones of emission, in directions determined by the orientations of the surface molecular bonds that are "ruptured" by the excitation (3a–c). As shown in Figure 1b, ESD of adsorbed H_2O produces H^+ ions that desorb in directions determined by the OH bond angle. The resulting ESD ion angular distribution (ESDIAD) patterns provide a direct display of the geometrical structure of surface molecules in the adsorbed monolayer.

In the following sections, we examine the measurement methods and the physical models that provide a theoretical framework for understanding most of the above observations. Details and utility of the ESDIAD method are then elucidated.

EXPERIMENTAL METHODS

The majority of ESD/PSD measurements have involved detection of desorbed ions, for which the basic quantities to be measured are ion mass/charge, yield, kinetic energy distribution, and angular distribution. The instruments used for these measurements have been reviewed in two recent articles (15, 16), and only those more frequently used are mentioned here.

Much of the recent work in ESD has involved the measurement of ion angular distributions, ESDIAD. In the display-type apparatus developed at the National Bureau of Standards (3a–c) and shown in Figure 2a, a

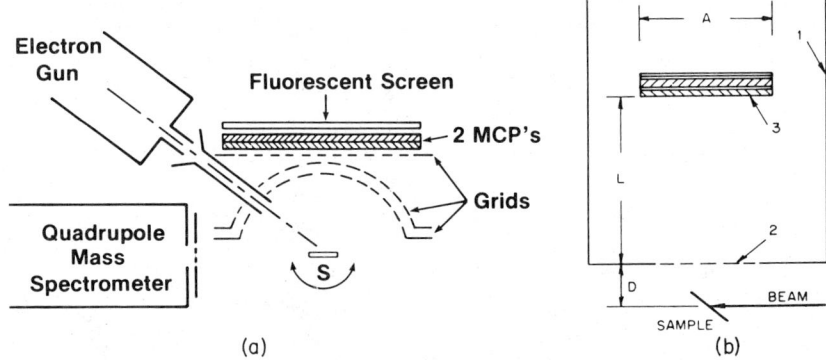

Figure 2 (a) Drawing of apparatus used for measurements of electron stimulated desorption ion angular distributions (ESDIAD). ESDIAD patterns are displayed on the fluorescent screen using the grid-microchannel plate fluorescent screen detector array. The mass spectrometer is used for mass analysis of desorbing ion beams. (b) Time-of-flight mass analyzer used for photon stimulated desorption (PSD). The synchrotron light pulse provides the "start" pulse for the ion flight times (see text). From Ref. (19).

focused electron beam bombards a single crystal sample. The ion beams that desorb via ESD pass through a hemispherical grid and impinge on the front surface of a double microchannel plate (MCP) assembly. The output electron signal from the MCP assembly is accelerated to a fluorescent screen where it is displayed visually (the ESDIAD pattern) and photographed. By changing potentials, the elastic low energy electron diffraction (LEED) pattern from the sample can be studied. Mass identification of ESD ions is accomplished using a quadrupole mass spectrometer, which is also used as a detector in thermal desorption studies from the adsorbed layer.

Niehus (8, 17) uses a channeltron multiplier mounted on a computer-driven goniometer as a moveable ion detector for ESDIAD. His data are in the form of computer-generated plots of ion intensity vs desorption angle; ion mass is determined using a time-of-flight method.

PSD studies have concentrated on determining the ion desorption mechanism by correlating structure in the ion yield curves with specific excitations. Due to the relatively high photon energies involved (>20 eV) and the need for photon tunability, these experiments are done exclusively with synchrotron radiation.

The simplest apparatus for these measurements is a time-of-flight (TOF) analyzer, which is used at the large electron storage rings where the time between light pulses is sufficiently long. The analyzer (Figure 2b) consists of 1. an electrostatic shield; 2. a meshed entrance aperture; and 3. a microchannel plate array detector (18, 19).

At smaller storage rings where the interval between light pulses is not long enough for TOF ion analysis, electrostatic deflection analyzers are employed. A schematic of the instrument in use at NBS (20) is shown in Figure 3. It consists of a commercially available double pass cylindrical mirror analyzer with voltages suitable for ion analysis. Mass analysis is

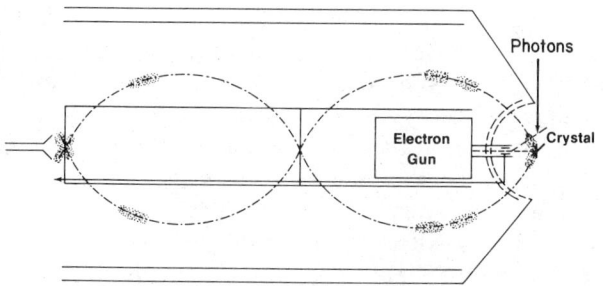

Figure 3 Double pass cylindrical mirror analyzer used for time of flight mass analysis of ESD ions, and for energy analysis of both ESD and PSD ions. The "start" pulse is provided by the electron beam. From Refs. (15, 20).

accomplished under electron impact by pulsing the electron beam onto the sample and measuring the TOF of the resulting ion burst.

Recent techniques for detection of ground state neutrals and metastables are described in Refs. (14a–c, 15).

MECHANISMS

Electron vs Photon Stimulated Desorption

The equivalence of the electron and photon excitation processes has been well documented by the agreement of desorption thresholds, ion energy and angular distributions, and the nature of the surface species from which desorption occurs (4). These data indicate that PSD and ESD are initiated by very similar elementary excitations at the surface. Nevertheless, differences do appear in the spectral shape of the ion yield vs excitation energy under photon or electron excitation. These different spectral dependences reflect the differences in the physics of the initial excitation process itself.

Excitation of a core level or even a deep valence level can be thought of as a fundamentally atomic excitation process, with the neighboring atomic interaction effects superimposed on this atomic excitation envelope. Figure 4 compares the electron and photon excitation functions of the He(1s) and Ne(2p) levels (4). The He(1s) and Ne(2p) photon excitation curves show relatively sharp thresholds followed by sharp decay with an energy dependence of the order of $1/E^n$ ($2 < n < 3$). The Ne(2p) curve shows a delayed onset with the maximum well above the threshold energy, resulting from the centrifugal barrier present in all states having high angular momentum. The electron excitation curves show a much weaker rise at threshold relative to the photon excitation curves. The He $^1S \to {}^1P$ excitation rises very weakly from threshold, peaking at 3–4 times the threshold energy and decaying as $\ln(E)/E$; indeed the spectrum resembles more of a step function than a peaked lineshape. The electron excitation curves reveal large differences between dipole-allowed and forbidden excitations. The He $^1S \to {}^3P$ excitation cross section peaks at 1.5 times the threshold energy and decays as $1/E$.

The much slower rise in the electron excitation cross section just above threshold relative to the photon cross section results because the primary electron must have an empty final state to occupy in the solid. In addition, an electron beam acts essentially as a *white radiation* source for exciting both dipole allowed and dipole forbidden excitations over a wide spectral range. Thus PSD excited by monochromatic photons exhibits much sharper thresholds than ESD, and the spectral lineshape may in some instances be expected to exhibit directly the final state density information.

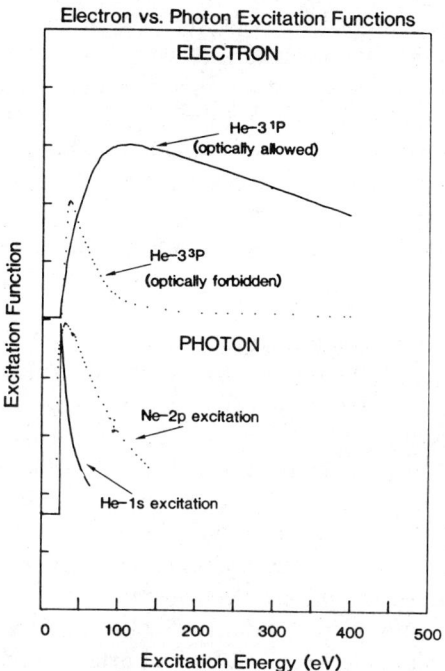

Figure 4 Electron and photon excitation functions of the He(1s) and Ne(2p) levels displaying atomic effects in absorption spectra. From Ref. (4).

ESD has much weaker thresholds, with the spectrum spread out over a much larger energy and with the small dipole-forbidden contributions superimposed on the larger dipole-allowed "step-function" contributions. This makes it more difficult to interpret and extract quantitative information from ESD spectra than from PSD spectra.

Competitive Decay Processes

The stimulated desorption process can be described approximately as a sequence of three steps (2b):

1. a fast initial electronic excitation (10^{-16} sec);
2. decay of the excited state by displacement of the atomic positions but in competition with other electronic decay mechanisms, which redistribute the electronic energy (10^{-15}–10^{-14} sec);
3. a modification (e.g. energy, charge state, etc) of the desorbing species as it recedes from the surface (10^{-14}–10^{-13} sec).

Our knowledge of desorption processes is based, in large part, on examination of the desorbed species (2b): their identities, angular and energy distributions, charge states, and electronic and vibrational energy distributions. All of these quantities can provide clues to the desorption mechanisms, but modifications of the desorbing species (step 3) can obscure the interpretation of the interesting dynamics that occurs in step 2. In the following paragraphs, we describe measurements that provide insights into step 2—this is the step that differentiates between theoretical mechanisms or models for desorption. We then explore the impact of the step 3 modification processes (e.g. reneutralization, image forces) on ion trajectories and angular distributions.

Most helpful to an elucidation of step 2 has been the careful comparison of PSD spectral data with photoelectron CIS (constant initial state) data or photoabsorption data. This comparison provides insight into the dynamics of step 2 by revealing which initial states eventually result in desorption, and which do not. In this context, the possible competitive decay mechanisms are noted, such as Auger decay, autoionization, resonant photoemission, hole delocalization, etc. The resultant intermediate excited states can be categorized by the number of particles, such as one-hole ($1h$), two-hole ($2h$), or two-hole one-electron ($2h1e$) type states; indeed the proposed mechanisms to be summarized below can best be described and differentiated in this context (2d, 21). It is important to emphasize that these decay mechanisms are occurring in competition with the atom displacement, but only those decay processes which occur in the time t_c that the receding atom can still be trapped [i.e. which occur when the atom is within a critical internuclear distance, R_c, within which reneutralization and recapture is highly probable (22)] can alter the total desorption cross-section. The processes occurring after this period are included in step 3. Since, generally, the distance R_c is $\lesssim$ a bond length, one can discuss the decay mechanisms in terms of states appropriate for the equilibrium atom configurations, and hence they can be probed by normal spectroscopic techniques. A summary of these models is given in Table 1.

Ionic desorption from the surface is usually discussed in the context of several different models. The first of these is known as the Menzel-Gomer-Redhead (MGR) model (22, 23); it provided the only conceptual framework for ESD/PSD processes for nearly 15 years after it appeared in 1964, and it is illustrated in Figure 5a. In this model the primary process is a Franck-Condon excitation or ionization to a repulsive neutral or ionic state. Although the actual nature of the repulsive state is not usually specified, the initial proposal was in the context of a valence level excitation followed by a recapture or neutralization step that involves transfer of the excitation energy into the bulk. As such, this view is supported by desorption thresholds corresponding to valence electron excitations and by the dominance of the neutral yield over the ion yield. Since the detailed nature of the electronic excitations is not specified, this model does not have predictive capabilities concerning the initial excited state or about the surface atomic arrangements required for desorption to occur. To be more specific and to contrast with other many-particle mechanisms, some recent authors have limited (perhaps unfairly) the MGR model to a $1h$ or $1h1e$ excitation similar to that which occurs in some gas phase dissociation or predissociation processes (2d, 21). The MGR model is similarly classified in this work. On the other hand, the general definition of the MGR model based on a Franck-Condon excitation to a repulsive final state is sufficiently broad as to encompass all of the following more specific excitation processes.

An elaboration of the MGR desorption model has recently been suggested by Antoniewicz (24). In this model the ion begins its trajectory by moving toward the surface, where it is neutralized (see Figure 5b). After neutralization, the atom exists on a strongly repulsive potential curve leading to probable desorption as a neutral (or perhaps even an ion).

The second basic model is the Knotek-Feibelman (KF) model (25a–c), which is particularly applicable to highly ionic systems (i.e. maximal valency systems). This model for ion desorption is based on the ionization of a core level as the primary process. The interatomic Auger decay of the core hole creates a two hole $(2h)$ positive anion at an initially negative ion site (see Figure 5c). The expulsion of the positive ion results from the reversal of the Madelung potential.

The Auger process has recently been demonstrated to be of importance in

Table 1 Summary of proposed models for desorption

Model	Excitation level	Final state	Applicability
Menzel-Gomer-Redhead	valence	$1h$ or $1h1e$	see below[a]
Antoniewicz	valence	$1h$ or $1h1e$	physisorbed inert gases
Knotek-Feibelmann	core	$2h$	ionic, maximal valency systems
Auger stimulated desorption	core	$2h$, $2h1e$	covalent systems, adsorbates
Many particle CI excitations	valence	$2h1e$, $2h$	covalent systems, adsorbates

[a] As narrowly defined in the present text, this model may be applicable to some situations involving ESD/PSD of neutral species; as broadly defined, it includes all the other models and is applicable anywhere.

desorption of ions from some covalent systems and for non-maximal valency systems, but its role is somewhat different from the KF model. In covalent systems the Auger process must create two holes localized in a bonding orbital (26a, b). The expulsion of the positive ion still results from the unshielded nuclear-nuclear repulsion (also called hole-hole repulsion), but an additional condition must exist; the holes must remain localized for a sufficiently long time in the bonding orbital that initially held the atom to the surface. This mechanism is more appropriately called the Auger stimulated desorption (ASD) model. It has recently also been shown to describe desorption from molecular adsorbates (i.e. dissociation of OH on Ti, Cr, and Cu) as a result of metal core level excitation (27).

Critical to the $2h$ desorption models is the localization of the two holes, a condition necessary to provide the Coulomb repulsion for expulsion of the ion. General localization criteria have been proposed recently in the context of Auger Spectroscopy. According to the Cini-Sawatzky (CS) theory or configuration interaction (CI) theory, localization of the valence holes results only if the effective hole-hole repulsion U^e is greater than some appropriate covalent interaction or bandwidth V (i.e. $U^e > V$) (9, 22, 28). Core-valence-valence (CVV) Auger lineshapes provide direct experimental evidence for localization or delocalization of the valence holes.

Results from electron-electron and electron-ion coincidence data, photo-dissociation studies, and from theoretical calculations on small molecules such as N_2, CO, and H_2O, indicate that $2h1e$ "multi-excited" states may also be instrumental in initiating dissociation (21, 29). These states appear

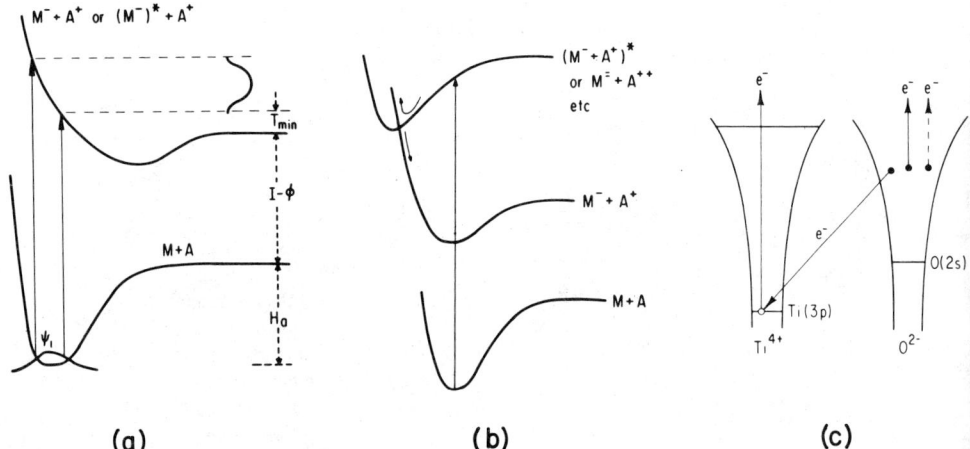

Figure 5 Illustration of mechanisms of ESD and PSD. (*a*) Menzel-Gomer-Redhead, (*b*) Antoniewicz, and (*c*) Knotek-Feibelman. From R. Gomer, in Ref. (2c).

as satellites in photoelectron energy distributions. They derive their intensity either from correlation mixing with the one electron states in the deep valence region or from core hole shakeup. Although they are of relatively minor importance in the gas phase because of small branching ratios, recent work indicates that they are more significant for dissociation of chemisorbed molecules (21, 29). Experimental evidence indicates that molecular desorption from chemisorbed molecules on metals is also dominated by $2h1e$ excitations. Finally, recent data showing ESD of molecular ions from condensed phase molecules such as neopentane, cyclohexane, benzene, and fluorinated analogs suggest that "multi-electron (hole) states" may also be involved in the photo-dissociation of large covalent molecules (30a–c). In these $2h1e$ states, the expulsion of the ion usually results from occupation of a strongly antibonding orbital (i.e. the $1e$ part) but may also result from the emptying of a bonding orbital (i.e. the $2h$ part). In any event, the many particle $2h1e$ state is expected to be long lived.

As a result of recent detailed examinations on several systems, some conclusions have been made on the importance of $1h$, $2h$, and $2h1e$ mechanisms (22). They are as follows:

1. At core excitation energies, the Auger induced $2h$ states are the most important for desorption if $U^e > V$. If $U^e < V$ the $2h$ states are ineffective and the $2h1e$ states become more important.

2. At valence excitation energies, the $2h$ states are excited with low intensity so that, in most instances, the $2h1e$ states have comparable or even greater intensity. In either case, the $2h1e$ states generally have lower excitation energies than the $2h$ states (especially true for large U^e) so that the initial desorption thresholds on the surface generally result from $2h1e$ excitations.

3. In general, the lifetimes, τ, of the $2h1e$ states are greater or of the order of the $2h$ states, with the $1h$ states having much shorter lifetimes. This is particularly true if $U^e > V$. If $U^e < V$, the $2h$ states have lifetimes of the order of the $1h$ states. This can be summarized as follows:

$U^e > V: \tau(2h1e) \sim \tau(2h) \gg \tau(1h)$ 1.

$U^e < V: \tau(2h1e) \gg \tau(2h) \sim \tau(1h)$. 2.

4. The above relationships indicate that the $1h$ states are rarely effective for desorption from the surface; however, the deep valence $1h$ states such as the 3σ state in CO may be sufficiently localized and narrow that they can initiate desorption; $1h$ states are also believed to be effective for desorption of inert gases from metals.

5. Important alternative decay mechanisms of the excited states include

resonant decay (RD), valence one hole and two hole hopping (this may occur via a valence VVV Auger process), and core hole Auger decay (AD). Core hole Auger decay generally initiates desorption by creating a $2h$ state; the other decay mechanisms abort the desorption process. Typical decay rates, R, may be summarized as follows: $R(RD) \sim 10^{16}$ sec^{-1}, $R(1h) \sim 10^{15}$ sec^{-1}, $R(2h) \sim 10^{14}$ sec^{-1} ($U^e > V$), $R(AD) \sim 10^{14}$–10^{15} sec^{-1}. Typical desorption rates are 10^{13}–10^{14} sec^{-1}.

6. Covalent interaction effects can dramatically alter the decay rates and hence the ion yields. Bond polarity effects, adsorbate-substrate effects (in the form of metal to adsorbate screening charge transfer), adsorbate-adsorbate interaction effects, and hydrogen bonding effects have been observed in recent studies.

Specific Examples

In recent years, the mechanisms of ESD and PSD have been explored by many authors for a variety of systems: desorption of substrate and adsorbate atoms and ions from surfaces of oxides and alkali halides (4, 14, 25a–c), desorption of atomic and molecular fragments from fractional monolayers adsorbed on metal and semiconductor surfaces (1, 3a–c, 7a–c, 10), desorption of atoms from physically adsorbed layers (1), and desorption of atomic and molecular ions from condensed multilayers (13). In the next paragraphs, we examine a few illustrative examples of ESD/PSD mechanisms based in large part on our own work.

TRANSITION METAL OXIDES   The ideal system for application of the Knotek-Feibelman $2h$ mechanism of ion desorption from maximal valency ionic solids appears to be the surface of the ionic oxide TiO_2. Indeed, evidence for the relevance of the KF mechanism to ESD/PSD of O^+ from both TiO_2 and an oxidized Ti(0001) surface has been reported (20, 25a–c). On the other hand, O^+ PSD data have been reported for Cr(110) oxide films, in which UPS data indicate that the stoichiometry of the stable oxide is Cr_2O_3, a nonmaximal valency oxide (31). Here the ASD model must be applied.

The PSD of O^+ ions having a most probable kinetic energy of ~ 4 eV is observed from Cr(110) for oxygen exposures $\gtrsim 1$ Langmuir. ESDIAD data indicate that O^+ desorbs from Cr(110) in a normal direction, for a range of oxygen exposures (0–50 L) and adsorption temperatures (80–700 K). For all of these oxygen exposures and adsorption temperatures, the O^+ ion yield vs photon energy is similar to Figure 6a for the lightly-oxidized Cr(110) surface. Figure 6b is the 2.5 eV constant final state (CFS) spectrum corresponding to the same thin oxide layer as Figure 6a; the structure in the CFS curve is similar to that of the Cr $3p$ photoabsorption spectrum. The CFS spectrum exhibits a pronounced shoulder at ~ 45 eV, which is absent in the PSD O^+ yield curve. The major peaks at ~ 51 eV are similar for both curves. Structures above 55 eV are not distinguishable from the noise (31).

Based on O K ELS (electron energy loss spectroscopy) and Cr K XAS (x-ray absorption spectroscopy) data for Cr_2O_3, SXA (soft x-ray absorption) data for atomic Cr, and theoretical calculations, it is believed that the shoulder at 45 eV in Figure 6b is due to excitation to $2t_{2g}$ and $3e_g$ molecular orbitals (which are primarily localized Cr $3d$ orbitals), and the peak at 51 eV corresponds to excitations to a $3d$ state mixed with the εd

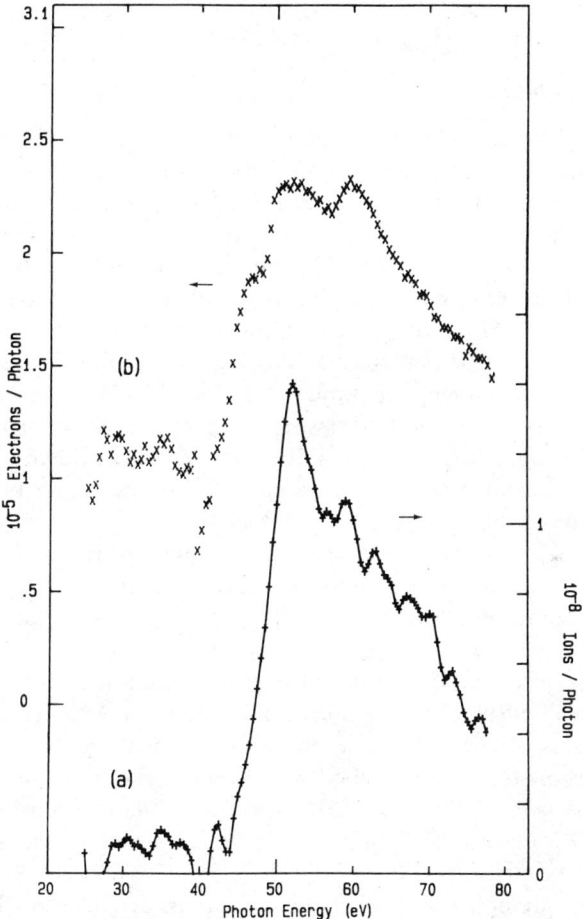

Figure 6 (a) O^+ PSD ion yield for lightly oxidized Cr(110), corrected for monochromator transmission and second order radiation (31). (b) Constant final state secondary electron yield curve for lightly oxidized Cr(110), corrected as above (31).

continuum, thus designated as more delocalized $\overline{3\varepsilon}d$ state. The $3d$ orbitals are dominated by resonant photoemission decay; this leaves a $1h$ state that does not initiate desorption. In contrast, the $\overline{3\varepsilon}d$ state decays with significant probability by Auger processes [i.e. where the excited electron either escapes before the Auger decay, or remains a spectator (satellite) leaving a $2h$ or $2h1e$ state], and hence can produce desorption (31).

Just below the Fermi level, UPS (ultraviolet photoemission spectroscopy) data reveal two main valence bands in Cr oxide, which can be identified as the Cr $3d$ and O valence bands (v_0 and v_1) with electron density (31)

$$v_0 = c\ 3d$$
$$v_1 = a\ 3d + b\ O2p.$$
3.

Normal Auger decay of the core hole creates the valence band holes v_0^{-2}, $v_0^{-1}v_1^{-1}$, and v_1^{-2} with relative intensities c^2, ac, and a^2. These relative intensities are based on the assumption that the Cr intra-atomic Auger process dominates the Auger decay. Only the v_1^{-2} final state (or the $v_1^{-2}\ 2t'_{2g}$ state in the case of the satellite) is thought to be responsible for the O^+ desorption, because v_1 is a bonding orbital, albeit localized strongly on the O atom ($b \gg a$). The strong localization on the O atom decreases the covalent interaction between neighboring v_1 bond orbitals and consequently increases the v_1^{-2} lifetime for the desorption process (26a, b) (i.e. $U^e > V$). In this Auger induced desorption picture, the desorption results from the Coulomb correlation of a localized 2-hole covalent antibonding state that was created by a metal atom *intra*-atomic Auger process. This appears to be in sharp contrast to the Knotek-Feibelman model for ionic systems, which assumes an *inter*-atomic Auger process in a maximal valency system. However as a and c approach zero in Eq. 3 (such as in TiO_2), the interatomic Auger process becomes more important and the v_1 bond orbital polarizes to an atomic $O2p$ orbital. Thus the KF and ASD mechanisms become indistinguishable at this point and one can consider the ASD mechanism as a generalization of the KF mechanism.

Figure 6 is a clear example of how information on the dynamic mechanisms of step 2 as described in a preceding section can be obtained from comparison of PSD and UPS spectral data. The sensitivity to the electronic configuration of the metal $3p$ core hole state seen in Cr_2O_3 is believed to be present also in TiO_2 (32a, b). This merely indicates that competitive core hole decay mechanisms may be active even when the KF mechanism is applicable, as one would expect. This sensitivity to the electron configuration of the core hole state suggests the possibility of selectively stimulating surface chemical reactions. On the other hand, the deeper core levels [e.g. O $1s$ in TiO_2 (33), Mo $3d$, $3p$, $1s$ in Mo oxides (19) etc] undergo an Auger cascade decay process, which generally results in a PSD ion yield varying linearly with the core hole creation rate. In the latter situation, the PSD ion yield reflects a local probe of the near edge or extended x-ray absorption fine structure (NEXAFS or EXAFS).

ADSORBATES—CO, H_2O, AND OH Figure 7 compares the O^+ ion yield obtained by PSD from CO/Ru(001) (34) with that obtained by $(e, 2e)$ or photodissociation from CO(g). The contribution around 40 eV has been attributed to the $5\sigma^{-2} 6\sigma$ excitation; the major peak has been attributed to the $3\sigma^{-1}$ excitation, which is heavily mixed with other $2h1e$ ionic states (28). The 6σ orbital is so strongly antibonding that it lies above the $k\sigma$ threshold; thus excitation to this state is a resonance allowing for a fast $6\sigma \rightarrow k\sigma$ resonant decay. However, in the presence of two 5σ holes it becomes stabilized as an excitonic-like state. The much narrower $5\sigma^{-2} 6\sigma$ peak in CO/Ru compared with CO(g) has been attributed to the metal

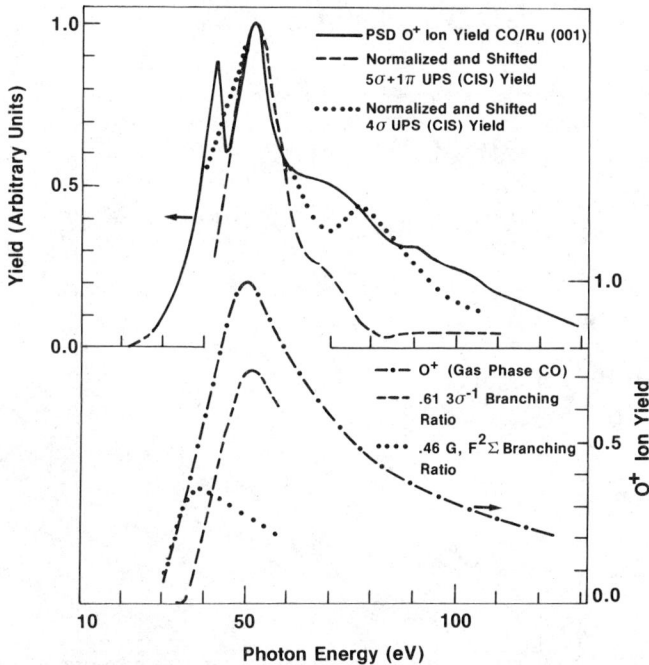

Figure 7 Top: Comparison of the PSD O^+ ion yield for CO/Ru(001) with the $(5\sigma + 1\pi)$ and 4σ UPS (CIS) yield. Bottom: The O^+ ion yield from photodissociation of CO gas. Also shown is the $3\sigma^{-1}$ and the G, F $^2\Sigma$ branching ratios multiplied by the fragmentation ratios for O^+ formation as determined by $(e, 2e)$ and $(e, e+\text{ion})$ measurements. From Refs. (34, 29).

$d\pi \to CO(2\pi)$ charge transfer that occurs on the surface, particularly in the presence of a core or valence hole. This charge screens the two 5σ holes, causing the 6σ orbital again to rise above the $k\sigma$ threshold and allow the resonant $6\sigma \to k\sigma$ decay. This decay aborts the desorption process since the remaining $5\sigma^{-2}$ state alone will not dissociate; i.e. an electron in the strongly antibonding 6σ orbital is necessary to break the CO bond. The $5\sigma^{-2} 6\sigma^2$ resonance, on the other hand, must undergo a relatively slow (10^{14} sec^{-1}) autoionization process ($5\sigma^{-2} 6\sigma^2 \to 5\sigma^{-1} k\sigma$) to abort the desorption: consequently this excitation has an excellent chance of leading to desorption. Therefore in CO(g), the $5\sigma^{-2} 6\sigma k\sigma$ excitation (including the $5\sigma^{-2} 6\sigma^2$ resonance) dissociates; in CO/Ru only the relatively narrow $5\sigma^{-2} 6\sigma^2$ resonance leads to dissociation.

The similarity of the $(5\sigma + 1\pi)$ or 4σ UPS (CIS) lineshape for CO/Ru and the O$^+$ ion yield peak around 50 eV (Figure 7) provides good evidence that this peak indeed arises from the $3\sigma^{-1}$ excitation. The structure above 60 eV in the PSD O$^+$ yield (Figure 7) has been attributed to the $5\sigma^{-2} 7\sigma$ excitation and other $2h1e$ states involving the 4σ and 1π orbitals (28).

Figure 8 shows the PSD O$^+$ yield from CO/Ni(100) at excitation energies

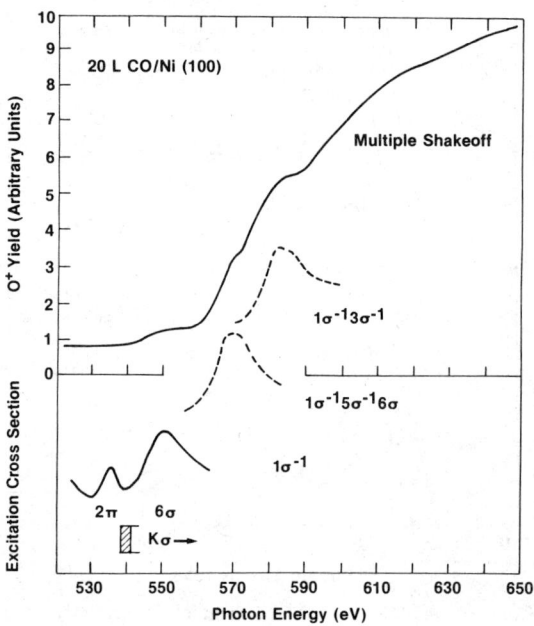

Figure 8 Comparison of the O$^+$ PSD ion yield for 29 L exposure of CO/Ni(100) with the 1σ excitation cross-section as determined by the O KVV Auger yield. The $1\sigma^{-1} 5\sigma^{-1} 6\sigma$ and $1\sigma^{-1} 3\sigma^{-1}$ excitation cross-sections are also schematically indicated. The ion yield above 600 eV has been attributed to multiple shake-off (35, 29).

near the O K core level (35, 36, 28). The excitation cross sections of the $1\sigma^{-1}$ $(2\pi, 6\sigma)$, $1\sigma^{-1} 5\sigma^{-1} 6\sigma$, and $1\sigma^{-1} 3\sigma$ states are also indicated. Although the $1\sigma^{-1}$ $(2\pi, 6\sigma)$ excitation is expected to dominate the CO(g) dissociation spectrum, for CO/Ni it is barely visible. This can be understood by examination of the various decay mechanisms of the levels involving the O K levels. In general, resonant decay is faster (10^{16} sec^{-1}) than Auger decay (10^{14} sec^{-1}), which is faster than the desorption process (10^{14}–10^{13} sec^{-1}). O$^+$ desorption results only if the final electronic state has a 6σ electron or a 3σ hole, since the $2h$ mechanism is not effective on the surface. The ineffectiveness of the simple core level ASD process in this case causes the PSD spectrum to exhibit a delayed onset: this apparently occurs quite often in chemisorbed systems (37).

The situation with H_2O is similar to CO but with some significant differences (29). A strongly antibonding $4a_1$ orbital in H_2O plays the same role as the 6σ orbital in CO; however, the $4a_1$ orbital lies below the ionization threshold so that resonant decay of the $4a_1$ electron is not important in H_2O. A deep valence $2a_1$ level is comparable to the 3σ level in CO. Both undergo a large mixing with the $2h1e$ states so that the $2a_1$ or 3σ orbitals are not strictly $1h$ in character (29). Dissociation of $H_2O(g)$ results from $2h1e$ excitations (e.g. $1b_1^{-2} 4a_1$), $1h$ excitations ($1b_2^{-1}$) via predissociation, $2h$ excitations and via the deep valence $2a_1^{-1}$ excitations. Condensed H_2O and molecularly chemisorbed H_2O reveal only the $2h1e$ and $2h$ contributions to the PSD H$^+$ ion yield (38). This clearly reveals the effects of competitive core hole decay processes on the $1h$ desorption process in the solid. Chemisorbed OH on Ti, Cr, and Cu exhibits both the intramolecular $2h1e$ and $2h$ contributions and intermolecular Auger stimulated desorption contributions as a result of core hole excitation of the substrate metal atoms (27). The latter ASD contributions reveal features remarkably similar to the O$^+$ ASD contributions from Cr oxide and represent clear evidence of molecular adsorbate dissociation as a result of a substrate metal atom Auger process (31).

NEUTRAL DESORPTION   Recently, two widely different examples of neutral desorption have been reported; these involve PSD of excited neutral alkali atoms from an alkali halide surface (14a–c), and ESD of physisorbed rare gases from a metal surface (39a, b). These experiments are important because they are among the first detailed experiments to characterize the majority of the desorbed species. Clearly in these two instances, neutral atoms dominate over ions; it is generally believed that this is true for most systems. The proposed mechanisms are discussed in the context of those described above.

Both ESD and PSD data of excited alkali atoms desorbed from alkali halide surfaces have been reported (14a–c). The first-resonance-line photon yields of Na, Li, and K due to ESD from NaCl, LiF, and KCl indicate thresholds at both the alkali and anion core levels. PSD data for Li* from LiF indicates a large resonant enhancement of the desorption cross-section at the Li 1s core exciton level. This strongly suggests that a $2h1e$ type final state is involved, arising from Auger decay of the 1s core hole with the excitonic electron remaining as a spectator on the Li atom (hence the neutral excited Li atom). The Li* yield is very large, $\sim 7 \times 10^{-3}$ atoms/incident photon. Li^+ ions appear over a wide energy range, perhaps as a result of the $2h$ KF mechanism in these ionic systems.

Zhang & Gomer have recently reported that Xe and Kr physisorbed on W yield only neutral atoms in ESD (39a, b). At least for Kr, the desorption threshold appears just at the threshold energy for production of the ion. These authors present arguments that this is one instance in which the Antoniewicz "bounce" mechanism is active (24). It is believed that initial ionization causes the ion to move inward where it makes a transition to a neutral, weakly bonding potential curve, resulting in desorption of a neutral atom. The exact nature of the initial ionic state here is relatively unimportant, and since the cross sections are huge (0.013 desorbed atoms/ion formed), a $1h$ state is probably involved.

Factors Influencing Energy and Angular Distributions of Electron and Photon Stimulated Desorption Ions: The Image Force and Neutralization Processes

We now address step 3 in the desorption process, viz. modification of the desorbing species as it recedes from the surface. The final measured yields of ions, neutrals, and metastables and the energy and angular distributions of desorbing species can all be influenced by processes occurring on time scales of the order 10^{-14} to 10^{-13} s, i.e. beyond the time t_c during which the desorbing atom can be trapped by rapid electronic decay processes (cf section on *Competitive Decay Processes*, above).

Consider the angular distributions of desorbing species. Although there have been no general calculations of the expected angular distribution of PSD/ESD products, there are good physical reasons to expect that initial desorption angles are directly related to the direction of the bond ruptured by the excitation; for ion desorption this view is substantiated by experiment, as is below. First, typical ion desorption times (10^{-14} to 10^{-15} s) are much faster than molecular vibration and rotation times (10^{-13} to 10^{-12} s). For example, an H^+ ion desorbing with 2 eV kinetic energy will travel 1 Å in 5×10^{-15} s and an 8 eV O^+ ion will travel 1 Å in 1×10^{-14} s (3a–c). After an ion has moved ~ 1 Å from the equilibrium bond length, the probability of recapture is small. Thus, ion desorption times are sufficiently rapid with respect to vibration times that significant molecular arrangements are unlikely to occur prior to desorption. Second, it appears clear that ion desorption processes generally involve localization of 2 holes in a bonding orbital, with conversion of electronic (Coulomb) repulsion to nuclear motion (9). The repulsive force is expected to be along the line of centers between the atoms in the bond being ruptured, i.e. along the bond direction. Third, gas phase photo-dissociation can often be described by impulsive models in which the dissociation fragments fly away along the direction of the broken bond (40a–c). Thus, it is expected that *initial* ESD/PSD ion desorption angles are determined by surface bond angles. However, as is shown below, the *measured* energy and angular distributions of desorbing ions can be strongly influenced by different final state effects, including distortions of trajectories by the image potential, and by neutralization processes.

It can easily be shown that an ESD/PSD ion leaving a surface with initial polar angle θ_0 will arrive at a detector with apparent angle θ_f, where θ_f is invariably greater than θ_0 due to the attractive image potential between the ion and the conducting surface (3a–b, 41) ($\theta = \tan^{-1}(dz/dx)$ in Figure 9a). For values of θ_0 greater than a "critical angle" θ_c (i.e. when $p_\perp^2/2m < V_I$ where $p_\perp$ is the normal ion momentum at the instant the ion is formed, m is ion mass, and V_I is the image potential at the surface), the ion cannot leave the surface and is recaptured. Several calculated ion trajectories for different values of θ_0, for $\theta_c = 45°$, are shown in Figure 9a. These curves are based on an impulsive model of ion desorption, in which the ion kinetic energy is a maximum at the instant the ion is formed (42). Note that distortions of the *azimuthal* angles of ion desorption are not expected due to the image field.

As shown by Tully (2b), the situation is further complicated by neutralization effects: the translational energy and angular distributions of ions or of electronically excited species may be strongly affected by neutralization or de-excitation processes during their escape. In the Hagstrum (43) picture of ion neutralization, faster particles have higher survival probabilities P_s, given by

$$P_s \propto \exp(-c/v_1),$$

where v_1 is the velocity component perpendicular to the surface and c is a constant. For c having a typical value of 2×10^6 cm/s (44), the ion survival probability is proportional to the *dot-dash curve* of Figure 9b. Multiplying this by a Maxwell-Boltzmann distribution at 1000 K (*dashed curve*) produces the velocity distribution of the surviving species (*solid curve*). The translational energies of surviving ions are shifted to higher energies because the slower ions (or excited states) are preferentially destroyed. Figure 9c shows the angular distribution of surviving desorbed ions calculated using the same assumptions as in Figure 9b, with the additional assumption that the neutralization rate is not dependent on the tangential velocity component. The surviving ion distribution is forced considerably toward the surface normal (*solid curve*), compared to the statistical cos θ distribution (*dashed curve*).

To summarize: Surface bond angle and initial ion desorption angle are related, but it is not always possible to extract a quantitative value for the bond angle from the measured ion desorption angle, due to final state effects. However, despite the problems of quantitative ESDIAD and angle-resolved PSD implied by the above discussion, there are many useful semiquantitative and qualitative conclusions to be drawn from angle-resolved ion desorption measurements at large polar angle. First, the differences between perpendicular and "inclined" molecules can be observed readily: The ion emission is either centered about the surface

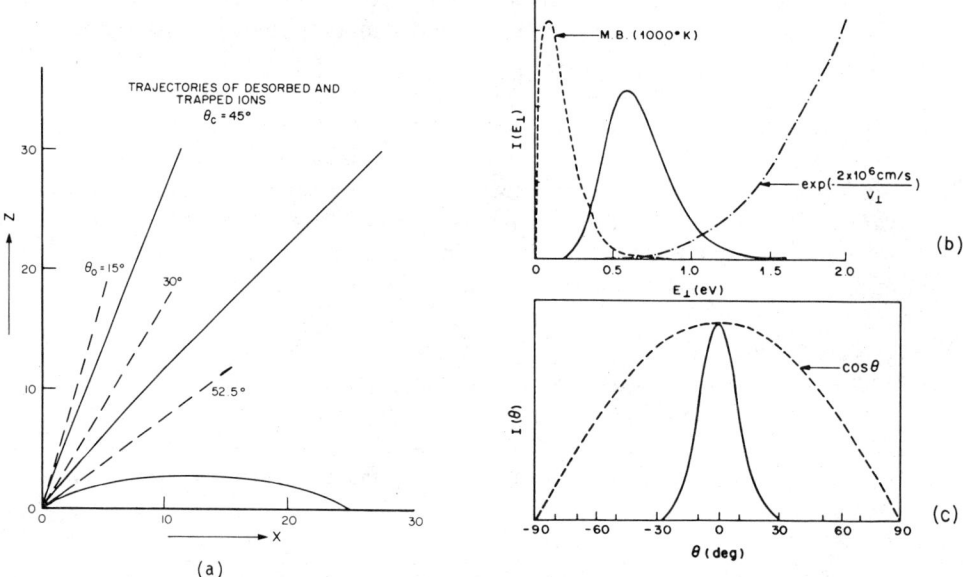

Figure 9 Illustration of final-state effects in ESD/PSD. (*a*) Ion trajectories in the presence of the surface image field are plotted as vertical distance z above the surface vs horizontal distance x along the surface. *Solid lines* correspond to desorption with initial angle θ in the absence of the image field; *dashed lines* are the trajectories in the image field, for critical recapture angle $\theta_c = 45°$ (42). (*b*) Kinetic energy distribution of desorbed ions (*solid curve*), assuming that a Maxwell-Boltzmann distribution initially leaves the surface at 1000 K (*dashed curve*); ion survival probability given by *dot-dash curve* (2b). (*c*) Angular distribution of ejected ions (*solid curve*) using assumptions as in (*b*), and neglecting image distortion of trajectories. The strong attenuation at large polar angles is due, in part, to the assumption regarding low kinetic energies (2b).

normal, or in an off-normal direction. Second, the presence or absence of azimuthal ordering in adsorbed molecules can be seen easily in ESDIAD patterns because there is little or no distortion of the azimuthal angle in the image field. Finally, ESDIAD and PSD measurements at small polar angle $\theta_0 \ll \theta_c$ are less strongly perturbed by neutralization anisotropies and the image force (Figure 9a), so that quantitative determinations of the polar bond angle are possible.

In the next section, several striking examples of the use of ESDIAD to determine the local molecular structure of surface molecules are presented.

ANGLE RESOLVED ESD/PSD: RELATION TO MOLECULAR STRUCTURE AT SURFACES

Experimental Basis of Electron Stimulated Desorption Ion Angular Distributions (ESIAD)

The physical basis for relating ESD/PSD ion desorption directions to surface bond angles has already been discussed, but the strongest evidence for this relation has been found in experimental ESDIAD measurements. To "calibrate" the method, ESDIAD has been applied to a variety of adsorbate-substrate systems whose surface geometry had been predicted or determined using other surface-sensitive spectroscopies (3). In every case studied to date, the ESDIAD results are consistent with the other methods. For example, for "standing up" CO on Ni(111) and Ru(0001), the molecule is known to be bound via the C atom to the metal surface, with its molecular axis perpendicular to the surface. ESD O^+ (and CO^+) ions desorb in the direction perpendicular to the surface (45, 46). For inclined CO on Pd(210), ESD of O^+ occurs in the predicted off-normal direction (47). For additional systems, including NH_3 on Ru(0001) and Ni(111) (48a, b, 49), NO on Ni(111), H_2O on Ru(0001) (50), and C_6H_{12} on Ru(0001), the structural assignments based on ESDIAD measurements are consistent with bonding geometries determined using other surface methods (3a–c).

Application of ESDIAD to Short-range Local Ordering in Adsorbed Molecules

There is an important area in which ESDIAD has provided new information about local structures of surface molecules—information not readily available using other techniques. Netzer & Madey (48) found that traces of preadsorbed oxygen on various surfaces will induce a high degree of azimuthal order in adsorbed molecules that are disordered azimuthally on the clean surface. This surprising oxygen-stabilized azimuthal ordering has been studied for NH_3 and H_2O on Ni(111) (48a, b), Al(111) (51),

Ru(0001) (52), and Ni(110). In all cases, the short-range local ordering has been found to occur in the absence of long range two-dimensional ordering observed using low energy electron diffraction (LEED).

Figure 10 contains a sequence of ESDIAD patterns for the adsorption of NH_3 on Ni(111) and Ru(0001), along with bonding models (48a, b, 52). Figure 10a is the H^+ ESDIAD pattern characteristic of a fractional monolayer ($\theta_{NH_3} < 0.15$ monolayers) of NH_3 adsorbed at $T < 150$ K on both Ru(0001) and Ni(111) surfaces. The center of the pattern is dim, with the H^+ emission directed into a "halo," away from the surface normal. This pattern is consistent with the model of Figure 10d, in which NH_3 is bonded via the N atom, with H atoms oriented away from the surfaces. The "halo" arises from a random azimuthal distribution of NH-ligands among the adsorbed NH_3 species.

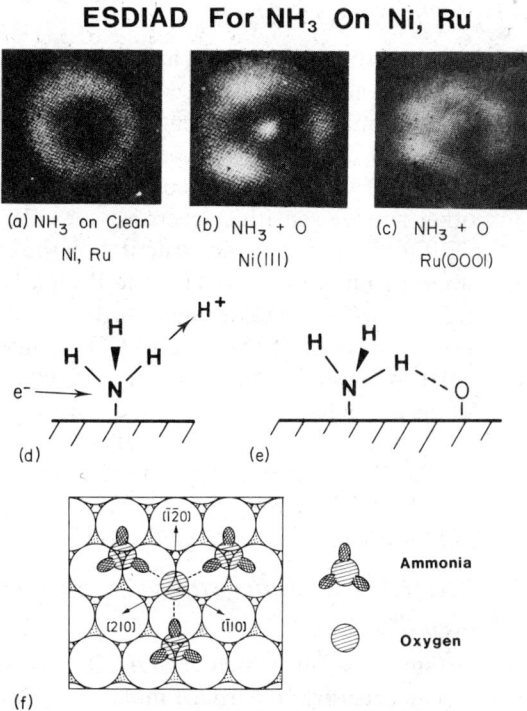

Figure 10 The structure of adsorbed NH_3 as determined using ESDIAD. (a) H^+ ESDIAD pattern for <0.15 monolayers of NH_3 on Ru(0001) and Ni(111); (b) H^+ ESDIAD pattern for $NH_3 + O$ on Ni(111); (c) H^+ ESDIAD pattern for $NH_3 + O$ on Ru(0001); (d, e) schematics of NH_3 structure on clean and oxygen-dosed surfaces; (f) bonding model for azimuthal stabilization of NH_3 by O on Ru(0001). Data from Refs. (48a, b, 49, 52).

In contrast, the adsorption of low coverages of NH_3 at 80 K onto the oxygen pre-dosed and annealed Ni(111) surface produces a highly ordered threefold H^+ ESDIAD pattern (Figure 10b) with intense ion emission along [$\bar{1}\bar{1}2$] azimuths. Even at the lowest oxygen coverages used in this work ($\theta_0 \leq 0.03$ monolayers), evidence was found for this azimuthal ordering induced by oxygen (48a, b). NH_3 adsorbed on the oxygen pre-dosed Ru(0001) surface results in a sixfold H^+ ESDIAD pattern (Figure 10c) (52).

In Figure 10e, f structural models are proposed. From low energy electron diffraction (LEED), it is believed that oxygen is adsorbed in threefold hollow sites on Ni(111). For NH_3 adsorbed in neighboring threefold hollow sites the O–N distance is 2.87 Å, a typical distance for hydrogen bonding interactions. We suggest that the strength of the (hydrogen-bonding) NH_3–O interaction is sufficient to orient the molecules azimuthally on the surface. The model of Figure 10e is consistent with the threefold ESDIAD pattern of Figure 10b for NH_3 on fcc Ni(111). The sixfold hexagonal pattern of Figure 10c is believed to be generated because structures similar to those of Figure 10e are rotated by 60° upon crossing a monatomic step on the hcp Ru(0001) surface (52).

In recent work, it has been found that an electropositive additive, Na, also induced molecular reorientations in coadsorption with NH_3, H_2O, and CO on metal surfaces (52, 53a, b).

Note that these observations of steric effects in coadsorption have implications for our understanding of the mechanisms by which promoters and poisons function in heterogeneous catalysis. It may be that a catalyst promoter or poison can alter a reaction pathway by inducing the formation of a surface structure that has a low probability of formation in the absence of the promoter or poison.

Another example of the use of ESDIAD in surface chemistry involves H_2O on Ni(110) (54). H_2O has a very low reaction probability on clean Ni(110) at 300 K, but the presence of a fractional monolayer of oxygen ($\theta_0 < 0.5$ monolayers) promotes the adsorption and decomposition of H_2O. ESDIAD, angle resolved UPS (ultraviolet photoemission spectroscopy), and isotopic mixing experiments all indicate that OH(ad) is formed on the surface, presumably via a hydrogen abstraction reaction, $H_2O + O(ad) \rightarrow 2OH(ad)$. Figure 11a is a LEED pattern of the clean Ni(110) surface, and Figure 11b is an H^+ ESDIAD pattern characteristic of adsorbed OH. The two H^+ beams of Figure 11b indicate that the molecular axes of the OH(ad) species are inclined with respect to the surface normal, and are oriented along [001] and [00$\bar{1}$] azimuths. This assignment, illustrated in Figure 11c, d is consistent with angle resolved UPS data (54).

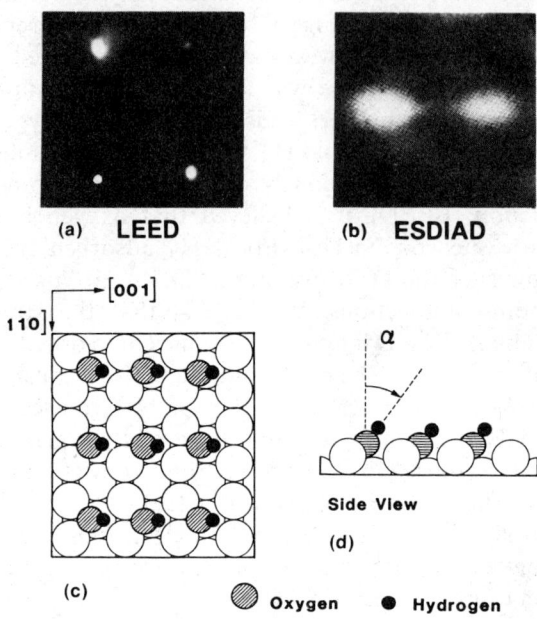

Figure 11 The structure of OH on Ni(110). (a) Clean LEED pattern for Ni(110); (b) H^+ ESDIAD pattern for inclined OH on Ni(110); (c, d) bonding models for OH on Ni(110). From Ref. (54).

SUMMARY: THE FUTURE

In summary, we have seen that electron and photon stimulated desorption processes arise from similar electronic excitations, and that much progress has been made in the theoretical descriptions of desorption mechanisms for different classes of adsorbate-substrate systems (ionic surfaces; covalent molecules on metal surfaces; physical adsorption of molecules on metal surfaces). Despite perturbations of ion trajectories by final state effects (image force; reneutralization), the electron stimulated desorption ion angular distributions method (ESDIAD) is seen to be a valuable tool for probing the local structure of surface molecules: new insights into impurity-stabilized structures have been revealed.

In the future, we anticipate that ESDIAD and angle resolved PSD will find further use as techniques to complement and extend other surface-sensitive structural probes, particularly in studies of local structure in the absence of long range order. Most ESDIAD studies to date have concentrated on adsorbed molecules; the method is also useful for characterizing the structures of certain oxide surfaces. The angle resolved ESD of neutrals recently reported by Menzel et al (55a, b) may be particularly useful in structural studies since there are no image force effects to alter trajectories. Laser techniques (laser-induced fluorescence, multiphoton ionization) will be used to determine desorption yields, energy partition, and angular distribution of both ground state and metastable species. ESD/PSD of negative ions will continue to attract more attention. The availability of intense sources of synchrotron radiation and high flux monochromators at photon energies $\gg 500$ eV makes possible PSD studies for deep core levels in both the adsorbate and substrate. Much new chemical physics is anticipated in studies of bond specificity in desorption from complex molecules, as well as studies of branching ratios and angular distributions of desorption fragments following core level excitations. Comparison of angle resolved ion desorption yields with electronic excitations at core hole thresholds [NEXAFS-near edge x-ray adsorption fine structure (56)] will be illuminating, particularly in studies of the hv-polarization-dependence of PSD. There is a multitude of opportunities for exciting work in electron and photon induced surface processes!

ACKNOWLEDGMENTS

This work was supported in part by the Office of Naval Research, and by the Office of Basic Energy Sciences, Department of Energy.

Literature Cited

1. Tolk, N. H., Traum, M. M., Tully, J. C., Madey, T. E., eds. 1983. *Desorption Induced by Electronic Transitions*, Springer Ser. in Chem. Phys., Vol. 24. Heidelberg: Springer-Verlag
2a. Jennison, D. R. 1983. Ref. 1, p. 26
2b. Tully, J. C. 1983. Ref. 1, p. 31
2c. Gomer, R. 1983. Ref. 1, p. 40
2d. Feibelman, P. J. 1983. Ref. 1, p. 61
2e. Ramaker, D. E. 1983. Ref. 1, p. 70
3a. Madey, T. E. 1981. In *Inelastic Particle-Surface Collisions*, ed. W. Heiland, E. Taglauer, p. 80. Springer Ser. in Chem. Phys., Vol. 17. Heidelberg: Springer-Verlag
3b. Madey, T. E., Netzer, F. P., Houston, J. E., Hanson, D. M., Stockbauer, R. 1983. Ref. 1, p. 120
3c. Madey, T. E., Yates, J. T. Jr. 1977. *Surf. Sci.* 63:203
4. Knotek, M. L. 1983. Ref. 1, p. 139
5. Pantano, C. G., Madey, T. E. 1981. *Appl. Surf. Sci.* 13:115
6. Knotek, M. L. 1983. *Phys. Scripta* T6:94
7a. Menzel, D. 1982. *J. Vac. Sci. Technol.* 20:538
7b. Menzel, D. 1975. In *Interactions on Metal Surfaces*, ed. R. Gomer, p. 124. Berlin: Springer
7c. Menzel, D. 1975. *Surf. Sci.* 47:370
8. Niehus, H. 1982. *Appl. Surf. Sci.* 13:292
9. Ramaker, D. E. 1983. *J. Vac. Sci. Technol.* A1:1137
10. Madey, T. E., Doering, D. L., Bertel, E., Stockbauer, R. 1983. *Ultramicroscopy* 11:187
11. Lichtman, D. 1983. Ref. 1, p. 117
12a. Stockbauer, R., Bertel, E., Madey, T. E. 1983. *J. Vac. Sci. Technol.* A1:1162
12b. Stockbauer, R., Bertel, E., Madey, T. E. 1983. Ref. 1, p. 267
13. Rosenberg, R. A., Rehn, V., Green, A. K., LaRoe, P. R., Parks, C. C. 1983. Ref. 1, p. 247
14a. Tolk, N. H., Traum, M. M., Kraus, J. S., Pian, T. R., Collins, W. E., Stoffel, N. G.,

Margaritondo, G. 1982. *Phys. Rev. Lett.* 49:812
14b. Pian, T. R., Traum, M. M., Kraus, J. S., Collins, W. E. 1983. *Surf. Sci.* 129:573
14c. Pian, T. R., Traum, M. M., Kraus, J. S., Tolk, N. H., Stoffel, N. G., Margaritondo, G. 1983. *Surf. Sci.* 128:13
15. Madey, T. E., Stockbauer, R. 1984. *Methods of Experimental Surface Science*, ed. R. L. Park, M. G. Lagally. New York: Academic. In press
16. Stockbauer, R. 1984. *J. Nucl. Inst. Methods*. In press
17. Niehus, H., Krahl-Urban, B. 1981. *Rev. Sci. Instr.* 52:68
18. Knotek, M. L., Jones, V. O., Rehn, V. 1979. *Phys. Rev. Lett.* 43:300
19. Jaeger, R., Stohr, J., Feldhaus, J., Brennan, S., Menzel, D. 1981. *Phys. Rev. B* 23:2102
20. Hanson, D. M., Stockbauer, R., Madey, T. E. 1981. *Phys. Rev. B* 24:5513
21. Ramaker, D. E. 1983. *Chem. Phys.* 80:183
22. Menzel, D., Gomer, R. 1964. *J. Chem. Phys.* 41:3311
23. Redhead, P. A. 1964. *Can. J. Phys.* 42:886
24. Antoniewicz, P. 1980. *Phys. Rev. B* 21:3811
25a. Knotek, M. L., Feibelman, P. J. 1978. *Phys. Rev. Lett.* 40:964
25b. Knotek, M. L., Feibelman, P. J. 1979. *Surf. Sci.* 90:78
25c. Feibelman, P. J., Knotek, M. L. 1978. *Phys. Rev. B* 18:6531
26a. Ramaker, D. E., White, C. T., Murday, J. S. 1981. *J. Vac. Sci. Technol.* 18:748
26b. Ramaker, D. E., White, C. T., Murday, J. S. 1982. *Phys. Lett. A* 89:211
27. Bertel, E., Ramaker, D. E., Kurtz, R. L., Stockbauer, R., Madey, T. E. 1984. Submitted for publication
28. Jennison, D. R., Emin, D. 1983. *Phys. Rev. Lett.* 51:1390
29. Ramaker, D. E. 1983. *J. Chem. Phys.* 78:2998
30a. Kelber, J. A., Knotek, M. L. 1982. *Surf. Sci.* 121:L499
30b. Kelber, J. A., Knotek, M. L. 1983. *J. Vac. Sci. Technol.* A1:1149
30c. Kelber, J. A., Jennison, D. R. 1982. *J. Vac. Sci. Technol.* 20:848
31. Bertel, E., Stockbauer, R., Kurtz, R. L., Ramaker, D. E., Madey, T. E. 1984. *J. Vac. Sci. Technol.* In press
32a. Bertel, E., Stockbauer, R., Madey, T. E. 1983. *Phys. Rev. B* 27:1939
32b. Bertel, E., Stockbauer, R., Madey, T. E. 1984. *Surf. Sci.* 142: In press
33. Parks, C. C., Shirley, D. A., Knotek, M. L., Loubriel, G., Stuhlen et al, *Surf. Sci.* In press
34. Madey, T. E., Stockbauer, R., Flodstrom, S. A., Van der Veen, J. F., Himpsel, F. J., Eastman, D. 1981. *Phys. Rev. B* 23:6847
35. Jaeger, R., Stohr, J., Treichler, R., Baberschke, K. 1981. *Phys. Rev. Lett.* 47:1300
36. Jaeger, R., Stohr, J., Treichler, R. 1982. *Surf. Sci.* 117:533
37. Franchy, R., Menzel, D. 1979. *Phys. Rev. Lett.* 43:855
38. Stockbauer, R. L., Hanson, D. M., Flodstrom, S. A., Madey, T. E. 1982. *Phys. Rev. B* 26:1885
39a. Zhang, Q. J., Gomer, R. 1981. *Surf. Sci.* 109:567
39b. Zhang, Q. J., Gomer, R., Bowman, D. R. 1983. *Surf. Sci.* 129:535
40a. Busch, G. E., Wilson, K. R. 1972. *J. Chem. Phys.* 56:3626
40b. Busch, G. E., Wilson, K. R. 1972. *J. Chem. Phys.* 56:3638
40c. Busch, G. E., Wilson, K. R. 1972. *J. Chem. Phys.* 56:3655
41. Clinton, W. L. 1981. *Surf. Sci.* 112:L791
42. Miskovic, Z., Vukanic, J., Madey, T. E. 1984. *Surf. Sci.* 142: In press
43. Hagstrum, H. D. 1977. *Inelastic Ion-Surface Collisions*, ed. N. H. Tolk, p. 1. New York: Academic
44. Vasile, M. J. 1982. *Surf. Sci.* 115:L141
45. Madey, T. E. 1979. *Surf. Sci.* 79:575
46. Netzer, F. P., Madey, T. E. 1982. *J. Chem. Phys.* 76:710
47. Madey, T. E., Yates, J. T. Jr., Bradshaw, A. M., Hoffmann, F. M. 1979. *Surf. Sci.* 89:370
48a. Netzer, F. P., Madey, T. E. 1982. *Surf. Sci.* 119:422
48b. Netzer, F. P., Madey, T. E. 1981. *Phys. Rev. Lett.* 47:928
49. Benndorf, C., Madey, T. E. 1983. *Surf. Sci.* 135:164
50. Doering, D. L., Madey, T. E. 1982. *Surf. Sci.* 123:305
51. Netzer, F. P., Madey, T. E. 1982. *Chem. Phys. Lett.* 88:315
52. Benndorf, C., Madey, T. E. 1983. *Chem. Phys. Lett.* 101:59
53a. Doering, D. L., Semancik, S., Madey, T. E. 1983. *Surf. Sci.* 133:49
53b. Netzer, F. P., Doering, D. L., Madey, T. E. 1984. *Surf. Sci.* 143: In press
54. Benndorf, C., Nobl, C., Madey, T. E. 1984. *Surf. Sci.* 138:292
55a. Feulner, P., Riedl, W., Menzel, D. 1983. *Phys. Rev. Lett.* 50:586
55b. Feulner, P., Treichler, R., Menzel, D. 1981. *Phys. Rev. B* 24:7427
56. Stohr, J., Jaeger, R. 1982. *Phys. Rev. B* 26:4111

METALLIC GLASSES

Frans Spaepen and David Turnbull

Division of Applied Sciences, Harvard University,
Cambridge, Massachusetts 02138

INTRODUCTION

Until quite recently it was generally doubted that metals could exist in a glassy form. There were reports as early as three to four decades ago that certain electro- or vapor-deposited metal films exhibited diffraction patterns characteristic of amorphous structures, but this evidence was considered inconclusive. It was thought that there could be little kinetic resistance to the crystallization of metal melts because of the relatively small attending changes in volume, coordination number, and nearest neighbor spacings. However, the demonstration (1, 2) that pure metal melts could sustain undercoolings of ~ 0.2 or more of their thermodynamic crystallization temperature, T_m, without measurable crystal nucleation suggested that crystallization must be attended by some essential reconstruction of the atomic short-range order (SRO). Actually, to characterize SRO fully, the angular distribution or "topology," as well as the number and spacing, of near neighbors about a reference atom must be specified. Frank (3) presented arguments that the energetically preferred topology of the SRO in monatomic liquids should be icosahedral, and so would have to be reconstructed in transforming to a close-packed crystal. He associated the high kinetic resistance to nucleation of such crystals in metallic liquids to the difficulty of this reconstruction. Later, the model studies of Bernal and his co-workers (4, 5) showed that uniform unattracting hard spheres, when randomly packed to maximum density without constraint by flat bounding surfaces, formed an amorphous structure (dense random packing, DRP), which apparently could not be collapsed to a crystalline structure. The SRO in this DRP structure was formed predominantly by polytetrahedral, including icosahedral, configurations, whereas the occurrence of octahedral configurations was far less than in close-packed crystal structures.

These studies suggested that monatomic materials could exist in metastable amorphous solid form but that physical realization of such a form would require either quenching through a highly labile undercooled liquid regime or condensation onto a substrate at a temperature well below the glass, or "configurational freezing," temperature, T_g. Efforts to melt-quench pure metals to glass have failed, but in 1960 Duwez, Klement & Willens (6) reported quenching a molten Au_4Si based alloy to an apparently amorphous structure. Further diffraction studies (7) confirmed the amorphous nature of the alloy structure, and Chen & Turnbull (8) showed that when about 1/2 of the Si in the alloy was replaced by Ge, the resulting mixture melt-quenched to an amorphous form that exhibited both the thermal and rheological characteristics of the melt ↔ glass transition upon thermal cycling about a temperature, T_g. The volumetric manifestation of the transition in other alloys was observed later by Chen et al (9).

Subsequent to the discovery of Duwez et al, numerous alloys—generally of the metal-metalloid, early-late transition metal (10) or even simple metal (e.g. Ca-Al) type (11)—have been melt-quenched to glasses. Some of the glasses have exhibited quite unique and possibly technically important properties such as (a) exceptionally high mechanical strength attended by considerable ductility, (b) very low magnetic hysteresis losses, if Fe or Co based, and (c) high corrosion resistance relative to crystalline alloys with the same composition.

As knowledge developed it became apparent that the main issues concerning the flow, relaxation, and structure of metallic glasses are remarkably parallel to those which emerged much earlier in the studies of the common nonmetallic glasses. These common glasses, usually formed in bulk quantities at low cooling rates, generally exhibit reduced glass temperatures, T_{rg}'s ($T_{rg} = T_g/T_l$, where T_l is the liquidus temperature) ranging from 0.84 to 0.60 (12). In contrast, the labile regime of the metal alloy glass formers, so far identified, is usually much wider, as indicated by relatively low T_{rg} values, which have ranged from 0.68 to as low as ~ 0.4. It is because of these extended labile regimes that exposure of glass formation by metal alloys has usually required extraordinarily high melt quench rates. Also, the essential role of impurity admixture in metal glass formation has presented problems not encountered in the formation of other types of glasses.

The literature on metallic glasses has been burgeoning for the past several years. Numerous reviews have appeared; those listed in references (13–22) are among the more recent of them. In the brief space of this article, we attempt to survey the current status of our understanding of the formation, structure, and atomic transport properties of metallic glasses.

MELT-GLASS TRANSITION

When a melt is cooled below the thermodynamic crystallization temperature, T_l, and formation of crystals is kinetically suppressed, the atomic transport rates decrease continuously with decreasing temperature. This is illustrated by Figure 1, which shows the temperature dependence of the shear viscosity, η, and the corresponding time constant for atomic rearrangement $\tau = \eta \bar{V}/RT$ ($\bar{V}$: molar volume). Curve (a) corresponds to the viscosity of the melt in (metastable) internal equilibrium. Since the molecular rearrangements required to reach this equilibrium state take an increasing amount of time with decreasing temperature, below a certain temperature, T_g, metastable equilibrium can no longer be reached on the experimental time scale. The nonequilibrium system is then called a "glass," and T_g is the glass transition temperature. Conventionally, T_g is chosen as the temperature at which the metastable equilibrium viscosity is 10^{13} poise, which corresponds to a time constant τ of about one hour (23). Lines (b) and (c) in Figure 1 represent the temperature dependence of the viscosity of the glass in a particular nonequilibrium configuration, and are therefore designated as "isoconfigurational" lines. The state corresponding to (c) can be obtained either by cooling more slowly than for (b), and thus allowing more time for equilibration, or by so-called "structural relaxation" of state (b) toward its metastable equilibrium, as marked by the arrows in the figure.

Measurements of the equilibrium viscosity near T_g are usually performed

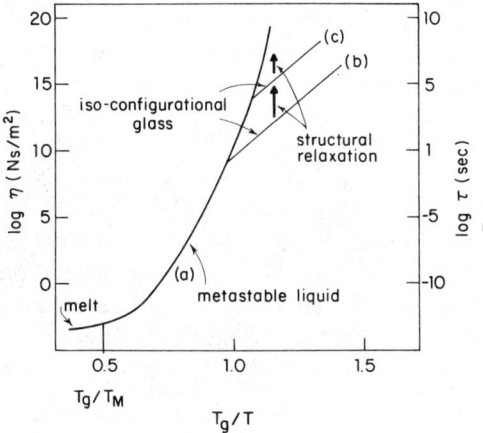

Figure 1 Schematic representation of the temperature dependence of the shear viscosity (η) and the time constant for molecular rearrangement (τ) in a metallic liquid and glass.

in a creep experiment. In order to ascertain that equilibrium has been reached, it is crucial that possible effects of structural relaxation be taken into account by establishing that the asymptotic values of $\eta(t)$ for positive and negative isothermal relaxation are the same. So far, this has only been established for $Au_{77}Ge_{13.6}Si_{9.4}$ (8) and $Pd_{77.5}Cu_6Si_{16.5}$ (24). Earlier measurements on the latter system (25) had resulted in lower values for η due to insufficient relaxation.

The temperature dependence of the equilibrium viscosity for these two metallic glasses obeys the Fulcher-Vogel law (26, 27):

$$\eta(T) = \eta_0 \exp\left(\frac{B}{T-T_0}\right). \qquad 1.$$

In some organic and ionic glasses (28, 29) at high viscosities a deviation from this law toward the Arrhenius temperature dependence

$$\eta(T) = \eta_0' \exp\left(\frac{B'}{T}\right) \qquad 2.$$

has been observed. In the Pd-Cu-Si glass, measurements have been made up to 10^{15} poise but no such deviation has been observed (24).

The Fulcher-Vogel law has been accounted for by two microscopic models. The configurational entropy model of Adam & Gibbs (30) and Gibbs & DiMarzio (31) leads to an expression for the viscosity

$$\eta = \eta_0 \exp\left(\frac{C}{TS_c}\right) \qquad 3.$$

where S_c is the configurational entropy of the melt, and C is a constant containing the minimum configurational entropy of a rearrangeable subsystem, Δs_c^*, and the free energy of activation, $\Delta\mu$. Specific heat measurements on undercooled melts show that S_c decreases rapidly with temperature, and, if equilibrium could be maintained, would become zero at a finite temperature T_0 (32). Linearization of $TS_c(T)$ near T_0 then converts Eq. 3 into the Fulcher-Vogel law, Eq. 1.

The second model, developed by Cohen & Turnbull (33–35), is based on the concept of the free volume, which is defined as that part of the atomic volume that can be redistributed throughout the system without change in energy. From the probability that a particular atom has a minimum free volume v^*, required for local rearrangement, an expression for the viscosity is obtained

$$\eta = \eta_0 \exp\left(\frac{\gamma v^*}{v_f}\right) \qquad 4.$$

where v_f is the average free volume per atom and γ is a geometric factor, between 0.5 and 1, that takes into account overlap between neighboring atomic volumes. Since the free volume vanishes in an ideally ordered liquid at T_0, $v_f(T)$ can be approximated as $\alpha(T-T_0)v_0$, where α is the thermal expansion, which makes Eq. 4 then equivalent to the Fulcher-Vogel Eq. 1.

An advantage of the configurational entropy model is that it allows use of thermodynamic parameters, obtained either from experiment or from statistical mechanics calculations [as performed by Gibbs & DiMarzio (31) for polymer glasses, for example]. The free volume model, on the other hand, is a more microscopic one, in that it describes the detailed atomic rearrangement required to get diffusion or flow. It should be emphasized that the two models are conceptually equivalent; in fact, Chen & Turnbull (8) formally demonstrated this in the limit of small free volume.

Recently, Cohen & Grest (36–38) have proposed an extension of the free volume theory, and one of their objectives has been to develop the statistical mechanics of the free volume distribution, and hence to combine the advantages of the earlier free volume and configurational entropy models over the entire range of free volume values.

To this end, they divide the local atomic volumes or "cells" into two categories: (a) "liquid-like" cells, with an atomic volume, v, greater than a critical value, v_c; each cell then has a free volume $v_f = v - v_c$; (b) "solid-like" cells, with $v < v_c$, which contain no free volume and in which the atoms simply vibrate in a harmonic potential. The free volume is again assumed to be redistributable among the liquid-like cells without any change in free energy. The statistical mechanics of a mixture of these two types of cells is then developed by calculating the distribution of liquid-like clusters from percolation theory. A communal entropy is assigned to the liquid clusters larger than a critical size that allows diffusive motion throughout the cluster; this leads to a sharp increase in the total communal entropy at the percolation threshold fraction of liquid-like cells, p, and a first order transition from the liquid to the glassy state, manifested by a discontinuous jump in p. That this first order transition is not observed experimentally is attributed to its preemption by the kinetic glass transition at T_g. Use of the communal entropy in this theory, however, can be criticized on two grounds: (a) the communal entropy of a liquid, even at the melting point of the crystal, is much less than $Nk \ln V$ (39), so that its contribution to the entropy of the liquid clusters may in fact be negligibly small; (b) the communal entropy used here is based on the accessibility of the volume by diffusive motion of atoms, which is a *kinetic* and not a thermodynamic equilibrium property. It is unclear to what extent the results of the Cohen & Grest theory depend on the treatment of the entropy problem. Details, such as the first order nature of the "equilibrium glass transition," are likely to be affected; the temperature dependence of the equilibrium viscosity, discussed below, may be less sensitive to the choice of entropy treatment. As the authors themselves state: "The problem of how to incorporate communal entropy into the [configurational] entropy theory or internal configurational entropy into the free volume theory remains unsolved."

Based on the above calculation of the equilibrium value of the fraction of liquid-like cells, Cohen & Grest calculate the temperature dependence of the viscosity:

$$\eta = \eta_0 \exp\left[\frac{B}{T - T_0 + [(T - T_0)^2 + CT]^{1/2}}\right]. \qquad 5.$$

This expression reduces to the Fulcher-Vogel form at high temperatures, and due to a fourth fitting parameter, it can accurately describe the deviations toward Arrhenius behavior at lower temperatures, observed in many nonmetallic glasses (28, 29). Predictions of other equilibrium physical properties, such as the thermal expansion, have been less successful.

The model has also been applied to nonequilibrium phenomena, such as the rate dependence of the thermal effects accompanying the kinetic glass transition. The fraction of liquid-like cells, p, is used as the governing parameter in a numerical simulation, in which p is calculated incrementally in a series of time steps Δt:

$$p = p_{eq} + [p' - p_{eq}]e^{\Delta t/\tau} \qquad 6.$$

where p_{eq} is the equilibrium value, and p' the value at the end of the previous time step; the time constant for relaxation, τ, is taken proportional to the instantaneous viscosity. Although this procedure is qualitatively successful, the relaxation effects have been equally well described by using a different parameter, such as the fictive temperature (40), the enthalpy (41), or the volume, in Eq. 6.

STRUCTURE

The diffraction pattern of a metallic glass consists of broad peaks, demonstrating the absence of long-range translational symmetry in these materials. Fourier transformation of the diffracted intensity yields the pair correlation function, g (which is, for these homogeneous and isotropic materials, equivalent to the radial distribution function); the experimental techniques and inversion procedures have been summarized in several extensive reviews (15, 42, 43). Since the atom positions cannot be determined unambiguously from the radial distribution function, testing of structural models must be extended to other physical properties, such as the density, and to the various spectroscopic measurements, such as NMR and Mössbauer. For multicomponent systems, which all metallic glasses are, full structural characterization requires the determination of partial pair correlation functions (e.g. in a binary system AB there are three: g_{AA}, g_{AB}, and g_{BB}); they can be obtained from a combination of diffraction experiments using different types of radiation [X-rays, neutrons, polarized neutrons (44)], by exploiting the X-ray anomalous dispersion (42, 45) or, less directly, from isotopic substitution. The nearest neighbor environment of specific atom species can be obtained from the extended X-ray absorption fine structure (EXAFS) (46), or from spectroscopic measurements [QMR (47), Mössbauer (48)].

Although an assembly of randomly oriented microcrystals does not have long-range translational symmetry, it was shown in the earliest structural studies of amorphous metals that their radial distribution functions could not be accounted for by a microcrystalline model, even if the crystals were allowed to be very small or strained (15); in fact, the inadequacy of the microcrystalline model reflects the fundamental difference in short-range order of the glassy and crystalline phases discussed in the introduction.

Cargill showed that the radial distribution function of amorphous Ni–P alloys could most satisfactorily be modeled by the dense random packing (DRP) of hard spheres, developed earlier by Bernal and co-workers (4, 5). Discrepancies in the details of the fit were due to the hard sphere potential and the monatomic composition of the DRP; indeed, as was shown later (49, 50), the DRP is a fully adequate model for the structure of nearly pure amorphous metals produced by vapor deposition on a cold substrate.

The first attempt to deal with the problem of modeling the alloy was made by Polk (51, 52), who proposed that metal-metalloid glasses could be modeled by a DRP skeleton of metal atoms, with the metalloid atoms occupying the largest interstices. The number of large interstices in the DRP was later shown to be insufficient to incorporate typically 20% metalloids (15, 53). Polk's proposal nevertheless brought out an important qualitative aspect of the alloy structure: the coordination shell of the metalloid atom consists of metal atoms only, and is similar to that in metal-rich intermetallic compounds. For example, in crystalline Ni_3P, each phosphorous atom is surrounded by nine nickel atoms forming a capped trigonal prism; this is similar to the metalloid coordination found experimentally in the glass (44, 54).

Later attempts to construct specific models for binary glasses fall into two categories:

1. Computer-built dense random packings of hard spheres, relaxed

afterwards in appropriate potentials V_{AA}, V_{BB}, and V_{AB}, can account for most of the features of the total radial distribution function (55, 56). Although the construction algorithm for metal-metalloid models is designed to minimize the number of metalloid-metalloid neighbors, the resulting number is still appreciably greater than zero (57).

2. "Stereochemically defined" models, introduced by Gaskell (54, 58), in which clusters of a metalloid surrounded by a well-defined coordination shell of metal atoms (e.g. in a trigonal prism arrangement) are packed together, and finally computer relaxed: So far the size of these models has been limited, so that a detailed comparison with the partial radial distribution functions at intermediate interatomic distances cannot yet be made.

Although the very strong chemical short-range order in the metal-metalloid glasses is clearly recognized, its characterization in the metal-metal glasses remains more of a problem. Cargill & Spaepen (59) have proposed an order parameter, η_{AB}, based on the average partial coordination numbers Z_{ij}, with $Z_A = Z_{AA} + Z_{AB} \neq Z_B = Z_{BB} + Z_{BA}$ (with $Z_{AB} = (x_B/x_A)Z_{BA}$; x: atom fraction)

$$\eta_{AB} = \frac{Z_{AB}(x_A Z_z + x_B Z_B)}{x_B Z_A Z_B} - 1. \qquad 7.$$

For a fully disordered alloy $\eta = 0$; $\eta < 0$ or > 0 correspond to a clustering or ordering tendency, respectively. The maximum value of η_{AB} in a concentrated ordered alloy remains an unresolved problem, however. Calculation of this order parameter for experimental systems and models for which partial coordination numbers are available shows that some of them are quite ordered (56, 59).

The question of topological order in amorphous systems is a much more difficult one, since the choice of an order parameter is not obvious. In the introduction, we have remarked that the short-range order in the glass is essentially polytetrahedral; such SRO is incompatible with space-filling in three dimensions. It has been shown, however, that in curved three dimensional space (i.e. on the surface of a four-dimensional polytope) perfect polytetrahedral packing becomes possible (60, 61). To map this structure onto three-dimensional flat space, "cuts" or defects have to be introduced. For example, to this end Nelson (62, 63) has used an array of disclination lines; he demonstrates that an ordered array describes the polytetrahedral Frank-Kasper phases, and postulates that a glass may be described by a disordered array.

Although *translational* order is entirely absent in amorphous systems, the degree of *orientational* order has only recently begun to be studied. Nelson et al (64) showed that in a two-dimensional amorphous model, consisting of a mixture of two sizes of spheres, the correlations of hexagonal orientational order (defined by angular Fourier analysis of the local bond directions) extended over a 10 sphere diameter distance, whereas the translational correlations dropped off at very close range. Similarly, Steinhardt et al (65, 66) showed that in a very undercooled argon molecular dynamics model, icosahedral orientational order (defined from the characteristic spherical harmonic components Y_{6m} of all the bond directions in the system) extended over several interatomic distances. Experimental confirmation of these correlations will require quantitative microdiffraction on ultra-thin foils of amorphous metals.

FORMATION AND KINETIC STABILITY

Experience indicates that all materials are less stable in an amorphous solid than in some crystalline form. In some instances, realization of the stable crystalline form may necessitate the breaking of covalent bonds, as in the crystallization of an atactic polymer, or phase separation, as in the crystallization of those alloys that are more stable in an amorphous solid than in *homogeneous* crystallized form. Thus, for an amorphous solid to form and persist, a thermodynamically favored crystallization process must be bypassed and then delayed (23).

Crystallization may be bypassed by a procedure where the material is energized, by melting, irradiation, or dissolution, and then deenergized by quenching or condensation to a temperature well below T_g. Deenergization exposes the possibility of amorphous solid formation, as well as crystallization, and its end stages may trap the amorphous state, when formed (12). Our discussion is concentrated mainly on the conditions for glass formation in melt cooling.

These conditions are described in earlier publications (67–70) and are reviewed here only briefly. To solidify to a glass the melt must be undercooled through the labile regime of temperature between T_l and T_g. The extent of this labile regime, $(T_l - T_g)/T_l = 1 - T_{rg}$, ranges from 0.2 to 0.4 for easy glass formers such as SiO_2, and from 0.3 to 0.6 for the metal alloy glass formers. Within the labile regimes the atomic mobility is quite high and the crystallization front, once formed, generally progresses rapidly. This progression occurs in a sequence of two steps: an interfacial rearrangement process and the transport of the crystallization heat, and possibly impurity, away from the interface. Where only the heat transport need be considered, at steady state the rate of advance, u, of a planar crystal-melt interface may be expressed as (71):

$$u \cong fk_i\lambda(1 - e^{-\Delta S_m \Delta T_i/kT_i}) = \frac{\kappa \bar{V}}{\Delta H_m}(\text{grad } T)_i \qquad 8.$$

where k_i = frequency of interfacial rearrangement; f = fraction of interfacial sites at which rearrangement can occur; λ = interface displacement/rearrangement; ΔS_m = molar entropy of crystal melting, here assumed to be temperature independent; $\Delta T_i = T_l - T_i$ where T_i is the interface temperature; ΔH_m = enthalpy of crystal melting; κ = thermal conductivity of phase through which the heat is extracted; $(\text{grad } T)_i$ = thermal gradient at the interface; $\bar{V}$ = molar volume. In the linear kinetic regime $(RT_i \gg \Delta S_m \Delta T_i)$, the interfacial undercooling becomes

$$\Delta T_i \cong \left(\frac{\kappa (\text{grad } T)_i}{fk_i}\right) \frac{\bar{V} R T_i}{\lambda \Delta H_m \Delta S_m}. \qquad 9.$$

For normal values of $(\text{grad } T)_i$, and when the interfacial rearrangement is very rapid, so that $fk_i \gg \kappa (\text{grad } T)_i$, the interface temperature, T_i, is only a little displaced from equilibrium and growth is then said to be "transport limited." Alternatively when, because of sluggish interface rearrangement or extremely high thermal gradients, $\kappa (\text{grad } T)_i \gg fk_i T_i$ may be displaced to levels far below equilibrium and approaching the temperature, T_s, of the heat sink; under these conditions growth is said to be "interface limited."

It follows from these considerations that for a melt in contact with its own crystal to form a glass, by heat extraction through the crystal, T_s must be well below T_g and $(\text{grad } T)_i$ must be sufficient to depress T_i to a level below T_g. To form metal glass under these conditions the imposed $(\text{grad } T)_i$ and the resultant quench rate, $\dot{T}$, in general, must be very high, owing to the rapidity of interfacial rearrangement and atomic transport in metal melts.

Even with $(\text{grad } T)_i$ sufficient to depress T_i to a level below T_g, a melt could not be quenched to a glass unless it were highly resistant to homogeneous crystal nucleation over the entire labile range. This resistance does not correspond directly to that opposing crystal growth. In simple ("classical") nucleation theory it is the product of an atomic transport resistance factor that may, indeed, correspond to that opposing crystal growth, and of a factor— which does not correlate with the transport factor—associated with the considerable work, σ, of forming the crystal-melt interface. More particularly, in the simple theory the *steady state* frequency, I, of homogeneous nucleation may be expressed, in terms of scaled variables, as follows (23, 69, 70):

$$I \cong nk_i \exp\left[\frac{-16\pi\alpha^3\beta}{3(\Delta T_r)^2 T_r}\right] \qquad 10.$$

where n = number density of melt atoms or molecules; $\beta = \Delta S_m/R$; $T_r = T/T_l$; $\Delta T_r = (T_l - T)/T_l$; T = absolute temperature of the undercooled melt; $\alpha = (\bar{N}\bar{V}^2)^{1/3}\sigma/\Delta H_m$, assumed temperature independent and isotropic; $\bar{N}$ = Avogadro's number; $\bar{V}$ = molar volume of the melt. I, calculated from this expression, is immeasurably low at small undercooling, but it rises sharply as ΔT_r increases and reaches measurable levels at some onset temperature $T_r''(\dot{T})$, which depends on quench rate, but only weakly in the labile range. Reflecting this weak dependence, at the lowest realizable cooling rate $T_r''(\dot{T})$ approaches some value T_{ro}'', the temperature at the onset of measurable nucleation, which generally lies well below unity. In some melts (e.g. SiO$_2$) I does not reach measurable levels at any temperature so that T_{ro}'' does not exist. To put such melts into glass form it is only necessary to eliminate seeds and other heterophase nucleants, or greatly reduce their number density, and then cool at any convenient rate.

When T_{ro}'' lies in the labile range and heterophase nucleants are absent, whether or not a glass forms will be determined by the quench rate between T_{ro}'' and T_{rg}, almost independently of its value from 1 to T_{ro}''. From I and u and their time and ΔT_r dependences, the minimum rate, $-\dot{T}_{min}$, for quenching a melt to glass can be calculated. The amorphous form will remain trapped at $T < T_g$ provided reconstruction of the SRO in the interfacial layers must attend crystal growth. Metal glass formation under these conditions is considered first.

The minimum scaled undercooling, $\Delta T_{ro}''$, at nucleation onset in pure metal melts is quite high, typically $\gtrsim 0.2$ to 0.3 corresponding to $\alpha \sim 0.6$ (23). Using this value of α in Eq. 10 and assuming that k_i scales with the reciprocal shear viscosity as described by the Vogel-Fulcher equation, we have computed the steady nucleation frequency, $I(\Delta T_r)$, at various T_{rg} (70). At given T_{rg}, I increases with ΔT_r to a maximum, $I(\Delta T_r)_{max}$, and then falls sharply as $\Delta T_r \to \Delta T_{rg}$. As T_{rg} increases, $I(\Delta T_r)_{max}$ shifts to lower ΔT_r and decreases in magnitude, becoming immeasurably small at $T_{rg} > 2/3$. Such suppression of $I(\Delta T_r)_{max}$ would be a necessary condition for bulk glass formation at low quench rates. By extending this analysis, $-\dot{T}_{min}(T_{rg})$ for bypassing homogeneous nucleation was computed. It falls sharply with increasing T_{rg} and is $\sim 10^6$ deg/sec at $T_{rg} \sim 1/2$ and ~ 1 deg/sec at $T_{rg} \sim 2/3$.

Actually, considerable homophase impurity admixture seems always required, even at the highest realizable quench rates, $\sim 10^{12}$ deg/sec, for metal glass formation in melt quenching. Such admixture might promote glass formation by kinetically restraining crystal growth (67) and by increasing the resistance to homogeneous nucleation. First the effects of homophase impurities on nucleation resistance are considered.

These effects should be determined principally by the dependence of the nucleation onset temperature, $T_0''(x)$ and glass temperature $T_g(x)$ on solute concentration, x. Experience indicates (72) that at temperatures high in the labile range $T_0''(x)$ tends (see Figure 2) roughly to parallel $T_l(x)$. Thompson & Spaepen (73) showed that such parallel behavior should be expected, provided:

1. α is, as in Spaepen's model (74, 75) for the interface, mainly topological in origin and so only weakly dependent on x;
2. the crystal nucleus is always in interaction with melt of average composition.

If T_{rg} increased with x, the extent of the labile range, $1-T_{rg}$, would decrease with x, and $T_0''(x)$ on a parallel course with $T_l(x)$ would approach $T_g(x)$. Actually, calculations based on simple nucleation theory with provisos 1 and 2 above indicate that $T_0''(x)$ should fall away from this parallel course and then drop precipitously to 0 as it nears T_g. Increasing $T_{rg}(x)$ could result from increasing $T_g(x)$ or decreasing $T_l(x)$, but it seems that the latter effect generally predominates since present experience indicates that $T_g(x)$ usually varies rather slowly with x. Thus a falling $T_l(x)$ would approach $T_g(x)$ more closely as x increases with attendant sharply increasing nucleation resistance and decreasing $-\dot{T}_{min}$. It was these considerations that led Cohen (76), Marcus & Turnbull (77) to their association of easiest glass forming compositions with those around deep eutectic points.

The T_{rg} values reported for alloy glasses (16) have ranged from ~ 0.7 to as low as ~ 0.4. Generally $-\dot{T}_{min}$ increases sharply with decreasing (78) T_{rg} in rough accord with theory. Thus, quench rates of 10^6 deg/sec or higher are, indeed, required to expose glass formation by alloys with T_{rg} near the lower end of the range (e.g. Au_4Si, $T_{rg} \sim 0.45$; Fe_4B, $T_{rg} \sim 0.5$). An extreme example is (79) glass formation by $Fe_{95}B_5$, $T_{rg} \sim 0.4$, requiring quench rates $\sim 10^{12}$ deg/sec. Certain Pd based alloys with T_{rg} in the higher part of the range can be melt quenched to glasses at much lower rates. Examples are $Pd_{82}Si_{18}$, $T_{rg} \sim 0.60$, $-\dot{T}_{min} < 760$ deg/sec in drop tube quenches; $Pd_{77}Cu_6Si_{17}$, $T_{rg} \sim 0.63$, $-\dot{T}_{min} \sim 100$ deg/sec; $Pd_{82}Si_{15}Sb_3$, $T_{rg} \sim 0.6$, $\dot{T}_{min} \sim 100$ deg/sec. Recently Drehman et al (80) found that when special care was taken to eliminate surface nucleants, bulk melts (minimum dimensions 6 mm) of $Pd_{40}Ni_{40}P_{20}$ alloy, $T_{rg} \sim 0.67$, formed glass when cooled at rates of only 1 deg/sec. These results suggest that methods designed to eliminate heterophase nucleants prior to quenching (e.g. by dispersion), containerless processing (as in drop tube experiments) (81), and fluxing, could be applied more widely in forming glasses from alloy melts with relatively high T_{rg}.

While there are apparent exceptions to the $-\dot{T}_{min}$, T_{rg} correlation it should be recognized that the correlation can be tested conclusively only when it is demonstrated that heterogeneous nucleation is negligible.

Selective oxidation of some readily oxidizable component of an alloy glass often results in easily crystallizable residues of other components at the external surface of the glass. Also an easily crystallizable phase may sometimes form internally by phase separation of the melt in its highly undercooled state.

Although pure metal melts have not been melt quenched to glass, it might seem that they would form amorphous solids when condensed on chilled substrates. Actually, Buckel & Hilsch (82, 83) found that pure metal films formed by condensation at 4–5 K were either crystalline as formed or when amorphous, as were Ga and Bi, crystallized upon warming, at $T < 20$ K. However, with a few percent impurity admixture some metals condensed to amorphous films that resisted crystallization to temperatures well in excess of 300 K. These experiments suggested to us (23, 67, 84, 85) that growth of a metal crystal into its pure amorphous state, liquid or solid, is a non-reconstructive process requiring no thermal activation and so would not be suppressed by configurational freezing. This view is further supported by measurements showing the extreme rapidity of growth of metal crystals into their pure melts at high undercooling. For example, Lin & Spaepen (86) estimate growth rates of Fe crystals into their undercooled melts as high as 500 meters/sec. To interpret these results we and our colleagues suggested that in pure metals the crystal-melt interfacial rearrangement rate is limited only by the collision rate of atoms from the melt onto the interface. In this event $fk_i\lambda$ in Eq. 1 should, with $f = 1$, approximate the sound speed, u_s, in the melt (71, 84, 85). This kinetic model is supported by Coriell & Turnbull's analysis (87), in which due account was taken of heat transport and shape stability, of the ultra rapid growth of Ni dendrites in their undercooled melts reported by Walker (88a, see also 88b) and Colligan & Bayles (89).

This rapid and presumably non-reconstructive crystal growth in monatomic systems may relate to its two-dimensional nature, in contrast with the reconstructive three-dimensional nucleation step. Such a relation may be understood in terms of the Spaepen-Meyer (74, 75) model of the crystal-melt interface or the model of Nelson & Halperin (90) for the crystallization of two-dimensional liquids. In the Spaepen-Meyer model, atoms from the melt are localized in a liquid monolayer on the interface without an attendant density deficit. It was demonstrated that this layer could crystallize without reconstruction by small atom displacements. In the Halperin-Nelson treatment there is continuity of states between liquid and crystal in two dimensions, and crystallization may occur continuously by motion of dislocations in an array that can describe the structure of two-dimensional monatomic liquids.

From the foregoing considerations it appears that a pure metal amorphous film would be susceptible at all temperatures below T_l to crystallization by movement of crystal-amorphous phase interfaces from any heterophase nucleants, e.g. flat solid surfaces to which the metal was well bonded, present in the system. However, if homophase impurities are incorporated in the metal, any redistribution of these impurities (whether by partition or short-range ordering) thermodynamically demanded for crystal growth would be reconstructive and so would impose serious kinetic restraint on growth. In this event k_i in Eq. 8 would scale with the impurity diffusivity, D_i, rather than with u_s, and would, with the attending interface motion, become negligible at T well below T_g. It follows that presence of such impurities would be a necessary condition for amorphous solid formation by any metal in contact with crystal nucleants.

The importance of kinetic resistance to partitioning in alloy glass formation has been demonstrated by Boettinger (91). For partitionless crystallization to occur the undercooling must exceed that at the temperature $T_0(x)$, where the free energies of crystal and melt alloy phases of identical composition, x, are equal. The course of $T_0(x)$ for an alloy with a falling liquidus, $T_l(x)$, is shown schematically in Figure 2. Calculations based on the regular solution model indicate that in many binary systems $T_0(x)$ should be continuous across the phase diagram with a minimum at

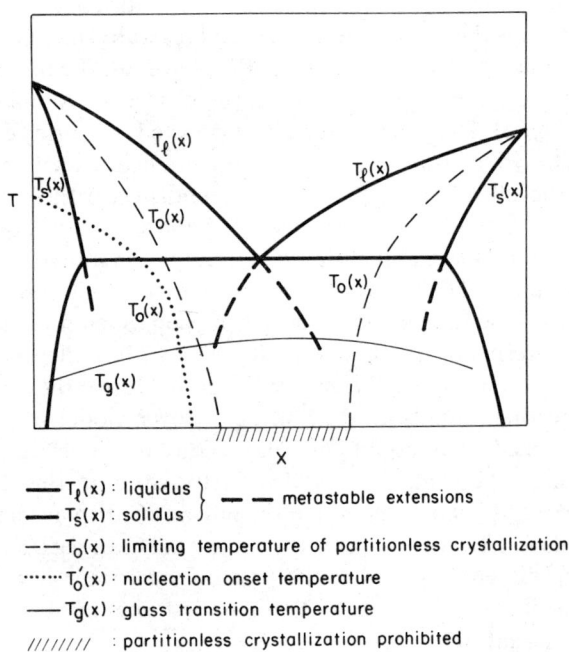

——— $T_l(x)$: liquidus
——— $T_s(x)$: solidus   } — — metastable extensions

— — $T_0(x)$: limiting temperature of partitionless crystallization

······ $T_0'(x)$: nucleation onset temperature

——— $T_g(x)$: glass transition temperature

/////// : partitionless crystallization prohibited

Figure 2 Schematic phase diagram, showing characteristic temperatures for crystallization and glass formation.

some intermediate concentration (93). However, there are systems in which $T_0(x)$ may go to 0 at compositions x_0 and x'_0 on opposite sides of the diagram. Consequently, over the composition range x_0 to x'_0 partitionless crystallization is excluded at all temperatures. As Boettinger (92) has emphasized, alloys with compositions in this range should be relatively good glass formers. Such alloys also should exhibit low scaled liquidus temperatures, $T_l(x)/T_l(0)$ reflecting an increasing stability of the amorphous relative to the crystalline state of the homogeneous alloy with increasing x. Thus, Boettinger's conditions for easy glass formation in melt quenching are qualitatively similar to the low $T_l(x)/T_l(0)$ requirement pointed out earlier.

In partitionless growth, when it can occur without local reordering of impurity, the rate of interfacial rearrangement should be virtually equal to that in pure metals, i.e. $k_i\lambda \rightarrow u_s$, so that amorphous solid formation by alloys in which such growth is permitted would hardly be possible. However, Lin & Spaepen (79, 86, 94) found that in rapid melt quenches, following high energy 30 picosecond pulses, alloys of Fe–B and Ni–Mo formed glasses at compositions well into the range in which partitionless crystallization should have been thermodynamically permissible even at temperatures well above T_g.

It appears that in these alloys, local impurity reordering may have been necessary for crystal growth and that this reconstructive process slowed growth sufficiently for trapping the glass forms of the alloys.

In condensation processes, the deenergization times are typically one picosecond or less. These are orders of magnitude smaller than achieved in melt quenching processes, excepting those following high energy pulses of a picosecond or less duration. Consequently, the range and variety of alloy compositions that have been condensed in amorphous solid form are much wider than those exhibited by glasses formed by the more common melt quenching procedures (16). The ultra-fast conditions of picosecond pulsed laser melt quenching match those of condensation, as has been demonstrated for the Ni–Nb system (95). The essential condition that a condensed alloy, as well as a melt-quenched one, persist in an amorphous solid form, when in contact with crystal nucleants, is that crystallization front movements must be thermally activated.

An important recent development is the demonstration by Johnson and co-workers (96) that amorphous solid alloys can form by solid state reaction. For example, thin (100–600 Å thick) contacting layers of polycrystalline Au and La transformed to amorphous alloys, with compositions ranging from 30–50 at.% La, when annealed at temperatures in the range 50–100°C. Their thermodynamic analysis indicated that although these alloys would have been most stable as phase separated mixtures of crystalline intermetallic phases, their amorphous solid forms should be more stable than phase separated mixtures of the crystalline elements as well as homogeneous f.c.c. alloy forms. The conditions that they thought essential for amorphous solid formation in this way are (a) prohibitively high kinetic resistance to the nucleation and growth of the crystalline intermetallic phases, and (b) "fast" diffusion (97) of the constituent, A, in the crystalline phase of the other, B (in the Au–La example, Au → La), so permitting rapid growth of the amorphous phase by transport of A through B to the amorphous-B interface. However, it seems that the growth could have occurred by rapid ("short-circuit") transport of the elements along the amorphous-crystalline interfaces. Koch et al (98) reported a further interesting example of formation of an amorphous alloy ($Ni_{60}Nb_{40}$) by solid state reaction that occurred when mixtures of the elemental powders were ball milled at room temperature. According to the general criteria for fast diffusion it is expected that Ni impurity should be a fast diffuser in crystalline Nb. It seems likely that the predominant mechanism in forming amorphous Ni–Nb in these experiments would be similar to that operative in the solid state formation of amorphous Au–La alloys.

PROPERTIES

Atomic Transport

The coefficients that describe the kinetics of the principal atomic transport processes in glasses are:

1. The shear viscosity, η;
2. The diffusivities, D, of the various constituents;
3. The frequency, k_s, of structural relaxation unattended by crystallization. This frequency may be indicated by the rates of irreversible changes in D, $1/\eta$, or specific volume with thermal treatment;
4. The frequency, k_i, discussed above, of interfacial rearrangement in crystal growth.

D, k_i, and k_s of nonmetallic glasses often all scale roughly with reciprocal shear viscosity, $1/\eta$. However, we noted that in the crystallization of liquid metals it seems clear, where there is no accompanying impurity redistribution, that k_i does not scale with $1/\eta$.

The now fairly extensive measurements of diffusivities in metallic glasses, and their relation to structural relaxation rates and shear viscosity, are reviewed briefly here. More comprehensive reviews have been presented by Greer and others (99, 100).

The earliest measurements of diffusivities in metallic glasses indicated that the values for the metal elements are extremely low, $< 10^{-16}$ cm^2/sec, and thus difficult to measure even at the onset temperatures of rapid crystallization. A number of high resolution concentration profiling techniques have been applied to the determinations of these low diffusivities. The resolution limit of most of these techniques is ~ 30 Å, which would permit evaluations of D as low as 10^{-20} cm^2/sec (101–103).

Apparently, the most sensitive technique for determining D is to monitor the decrease in amplitude, as indicated by the intensity of x-ray satellite peaks, of the composition modulation in artificially layered structures. This method was first used by Dumond & Youtz (104) and refined and applied extensively to studies of interdiffusion in crystalline alloys by Hilliard and co-workers (105, 106). It was first applied to glasses by Rosenblum et al (107). Diffusivities as low as 10^{-23} cm^2/sec, corresponding to only 0.2 jump/atom in the time of the anneal, can be measured accurately. It is the interdiffusivity, $\tilde{D}_\lambda$, where λ is the wavelength of the modulation, of the constituents of the two layers, rather than self or single constituent D, which is measured. The bulk interdiffusivity, $\tilde{D}$, may be obtained by extrapolating the $\tilde{D}_\lambda$ for different λ, to $\lambda = \infty$. The method permits continuous monitoring of $\tilde{D}_\lambda$ exhibited by a single specimen with differing annealing treatments. Thus, it is especially well suited to the characterization of the structural relaxation and isoconfigurational temperature dependence of $\tilde{D}$. $\tilde{D}_{ic}(T)$ is determined by monitoring $\tilde{D}$ continuously during cyclic temperature annealing of a single specimen. When conditions are reached such that $\tilde{D}$ at a given temperature remains the same after thermal cycling it is concluded that the configuration stayed essentially fixed during the cycles and anneals. This technique was adapted from that used by Taub & Spaepen (108) to determine the isoconfigurational temperature dependence of creep rates (see also following section, *Mechanical*).

Greer et al (109) and Cammarata & Greer (110) determined $\tilde{D}_\lambda$ of Pd and Fe in artificially layered amorphous $Pd_{85}Si_{15}/Fe_{85}B_{15}$ films at λ ranging from 2 to 70 Å. At $\lambda > 30$ Å, $\tilde{D}$ depended only weakly on λ, but dropped sharply in the early stages of isothermal relaxation anneals. For $\lambda = 32.3$ Å, $1/\tilde{D}_\lambda$ increased linearly with time, similarly to the isothermal shear viscosity of Pd based glassy metals. The temperature dependence of $\tilde{D}$ of an isoconfigurational state was described by the Arrhenius equation with activation energy 195 kJ/mole, very close to that for viscous flow of $Pd_{82}Si_{18}$, and a prefactor ~ 0.1 cm^2/sec.

The Cammarata-Greer $\tilde{D}_{ic}(T)$ relation falls, when extrapolated to higher temperatures, within the range of metal diffusivities in Pd and Fe–Ni based alloy glasses determined by composition profiling (CP) methods. Excepting the experiments of Chen et al (103), there were in the CP studies no clear indications of structural relaxation effects on either the metal or metalloid diffusivities. However, the Cammarata-Greer results indicate that most of the observable relaxation effect occurs early in the isothermal diffusion anneal, and so might not be detected easily after the relatively much longer anneals required in the CP experiments. There was no demonstration in any of the latter experiments that the observed diffusion was isoconfigurational. In no case, with the possible exception of the study by Chen et al, was the diffusivity measured when the glasses were in states approaching full structural relaxation.

Glassy metal diffusivities measured in different laboratories, or even in a single laboratory, apparently scatter by a factor of 2 to 3. However, taken as a whole, the results on Fe–Ni and Pd based metal-metalloid glasses in the temperature range 0.75 T_g to 1.0 T_g, in which the glasses could not have been fully relaxed, exhibit some general trends as follows (see Figure 3):

1. All of the diffusivities, metal and metalloid, fall, on a reciprocal temperature plot, within a band which, at a given temperature, is no more than two to three orders of magnitude wide.
2. The metal diffusivities are three to five orders of magnitude higher than in their pure crystal but about three orders of magnitude lower than along grain boundaries in these crystals. Their isoconfigurational diffusivity and its temperature dependence is described by a normal

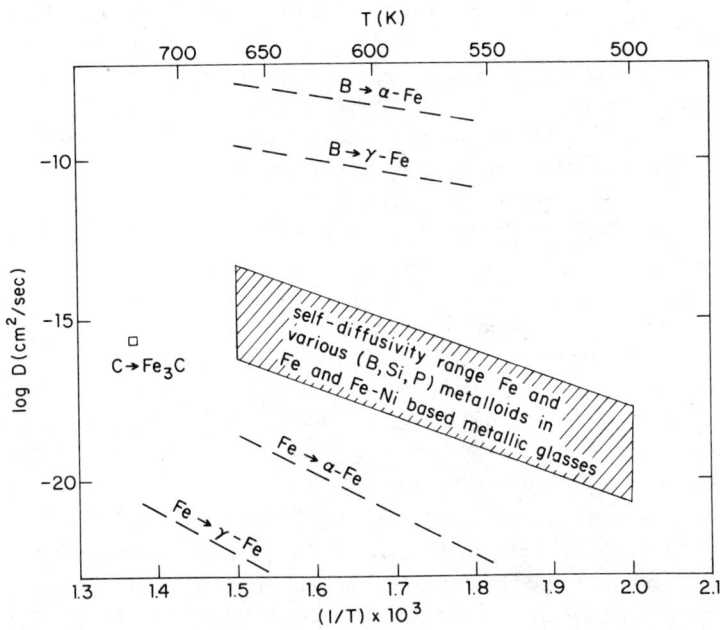

Figure 3 Comparison of the diffusivity in crystalline and amorphous Fe-based alloys. Self-diffusivity data in metallic glasses from Cantor & Cahn (100); B in α-Fe from Thomas & Leak (112); B in γ-Fe from Busby et al (111); C in Fe$_3$C from Ozturk et al (113); Fe in α- and γ-Fe from Fridberg et al (134).

Arrhenius prefactor and an activation energy about 2/3 of that for self-diffusion in the pure crystal.
3. The metalloid (Si,P,B) diffusivities are often equal, within the large experimental uncertainties, to those of the metal. Even the B diffusivity is in Fe–Ni based glasses no more than 1/2 order of magnitude greater than the metal diffusivity, and it is eight to ten orders of magnitude below that in its highly dilute state in Fe crystals (111, 112a, see also 112b).

Diffusivities of the elements in crystalline intermetallic phases with composition near those of the glasses would provide more significant standards for comparison with glass diffusivities than those in dilute crystalline solutions of metalloid in metal. Unfortunately, there have been few measurements of D in crystalline intermetallic phases. However, it is interesting that the one reported D of C in crystalline Fe_3C is about six to eight orders of magnitude below that in the crystalline phases of Fe and about two orders of magnitude below the metalloid D's in Fe–Ni based glasses (113).

The near equivalence between metal and metalloid atom diffusivities in these glasses suggests that some type of metal-metalloid exchange might be the predominant diffusion mechanism. One possibility might be an atomic exciton or interstitial-vacancy bound pair mechanism analogues to that proposed for the transport of Cd impurity in crystalline Pb (114). In this process, a metalloid atom might move to an adjacent vacant interstitial site so that the metal atom could shift to the metalloid position left vacant. This exchange would create a defect in the SRO of the glass (i.e. a metal at a metalloid site and vice versa), which would be removed by a reverse exchange involving a different metalloid atom. The process could be enhanced at positions of high free volume concentration.

Mechanical

Metallic glasses deform plastically in two qualitatively different modes (115): (a) at high temperatures ($T > 0.6T_g$) and low stresses ($\tau < 10^{-2}\mu$; μ shear modulus), they deform by *homogeneous* flow, i.e. each volume element in the specimen undergoes the same strain; (b) at lower temperatures and higher stresses, they deform by *inhomogenous* flow, in which plastic deformation is localized in a few, very narrow shear bands.

Homogeneous flow is measured in creep experiments. Special care must be taken to eliminate, or at least control, the effects of structural relaxation, which can change the isothermal viscosity by many orders of magnitude (see Figure 1). The kinetics of structural relaxation in metallic glasses have been shown to be bimolecular in nature (24), similar to what is observed, and expected on the basis of broken bond annihilation, in silicate glasses (116, 117). Below T_g, bimolecular kinetics are manifested by a *linear* increase of the viscosity with time (118). Therefore, the *relative* increase in viscosity, $d \ln \eta/dt$, decreases with increasing annealing time; sufficient preannealing, therefore, permits measurements at several temperatures, without appreciable further relaxation. To ascertain that measurements of the temperature dependence of the viscosity are truly iso-configurational, i.e. no relaxation has occurred [see lines (b) and (c) on Figure 1], the data must be reproducible upon cycling of the temperature.

Homogeneous flow is Newtonian for shear stresses below a few hundred MPa. The stress-strain rate dependence up to the fracture stress can be described satisfactorily by simple transition state kinetic theory:

$$\dot{\gamma} = \gamma_0 \sinh\left[\frac{\gamma_0 v_0 (\tau - \tau_0)}{kT}\right] \qquad 11.$$

where the strain product, $\gamma_0 v_0$, is on the order of a few atomic volumes and τ_0 is a small threshold stress (119, 120).

It is worth noting that the iso-configurational activation energy for flow is very similar to that for diffusion, and that the linear increase in η correlates with the linear increase in $(1/\tilde{D}_\lambda)$ observed in modulated film experiments (see preceding section, *Atomic Transport*). The magnitude of the viscosity, however, is two orders of magnitude below that predicted from the diffusivity by the Stokes-Einstein relation (99). All this indicates that one microscopic mechanism may be governing both flow and diffusion, but that only a fraction of the diffusional rearrangements is accompanied by flow.

Inhomogeneous flow in metallic glasses occurs in thin, shear bands, the position of which is determined by the planes of maximum shear stress. Although the plastic strain inside the bands is very large, the overall strain in a tension experiment is small ($\sim 0.2\%$) due to the small number of shear bands that form prior to failure. In compression, bending, wire drawing, or rolling, however, large macroscopic plastic strains can be achieved. The localization of shear in narrow bands is a macroscopic phenomenon that can be explained on the basis of continuum mechanics using a constitutive law that incorporates softening of the metallic glass at high stress, due to strain disordering (121, 122).

Ductile fracture of a metallic glass is preceded by inhomogeneous flow, and occurs *along* the plane of the shear band. The fracture surface morphology has a characteristic vein pattern (123), caused by the softening of the material in the band and its subsequent failure by the Taylor instability (124–126). The high strength of the metallic glasses is a result of their local ductility, which allows relaxation of the stresses at concentrators. The microscopic basis of this local ductility, as explained above, is a result of strain disordering that is possible in an amorphous material with metallic bonding. In the covalently bonded silicate glasses, on the other hand, this mechanism is not available, and their strength is therefore limited by brittle fracture.

Under certain conditions, some metallic glasses also fail by brittle fracture (127a, see also 127b). The Fe-based glasses, in particular, are sensitive to this (128). They exhibit a ductile-to-brittle transition temperature, which increases with increasing annealing temperature or time (T. W. Wu, F. Spaepen, unpublished observations). The microscopic basis of this transition is as yet poorly understood. Some authors claim that it is due to changes in microstructure, such as liquid-liquid phase separation (130–132), others have attempted to explain it with fracture mechanics, taking into account the viscosity increase upon annealing, and the time dependence of the strain softening mechanism (133).

ACKNOWLEDGMENTS

The authors' research in this area has been supported by the Office of Naval Research, under Contract No. N00014-77-C-0002.

Literature Cited

1. Turnbull, D., Cech, R. E. 1950. *J. Appl. Phys.* 21:804
2. Turnbull, D. 1952. *J. Chem. Phys.* 20:411
3. Frank, F. C. 1952. *Proc. R. Soc. Ser. A* 215:43
4. Bernal, J. D. 1960. *Nature* 185:68
5. Finney, J. L. 1970. *Proc. R. Soc. Ser. A* 319:497
6. Klement, W., Willens, R. H., Duwez, P. 1960. *Nature* 187:869
7. Duwez, P. 1967. *Trans. Am. Soc. Met.* 60:607 (for review)
8. Chen, H. S., Turnbull, D. 1968. *J. Chem. Phys.* 48:2560
9. Chen, H. S., Krause, J. T., Sigety, E. A. 1973/1974. *J. Non-Cryst. Solids* 13:321
10. Giessen, B. C., Madhava, M., Polk, D. E., VanderSande, J. 1976. *Mat. Sci. Eng.* 23:145 (for review)
11. St. Amand, R., Giessen, B. C. 1978. *Scripta Met.* 12:1021
12. Turnbull, D. 1981. *Metall. Trans.* 12A:695 (for summary and original references)
13. Giessen, B. C., Wagner, C. N. J. 1972. *Liquid Metals: Chemistry and Physics*, ed. S. Z. Beer, pp. 633–95. New York: Dekker
14. Duwez, P. 1976. *Ann. Rev. Mat. Sci.* 6:83
15. Cargill, G. S. III. 1975. *Solid State Phys.* 30:227
16. Chaudhari, P., Turnbull, D. 1978. *Science* 199:11
17. Chen, H. S. 1980. *Rep. Prog. Phys.* 43:353
18. Nagel, S. R. 1982. *Adv. Chem. Phys.* 51:227
19. Cahn, R. W. 1980. *Contemp. Phys.* 21:43–75
20. Luborsky, F. E., ed. 1983. *Amorphous Metallic Alloys*. London: Butterworths
21. Guntherodt, H. J., Beck, H., eds. 1981. *Glassy Metals, I, Top. Appl. Phys.* Vol. 46. Berlin: Springer
22. Beck, H., Guntherodt, H. J., eds. 1983. *Glassy Metals, II, Top. Appl. Phys.* Vol. 53. Berlin: Springer
23. Turnbull, D. 1969. *Contemp. Phys.* 10:473
24. Tsao, S. S., Spaepen, F. 1983. *Amorphous Metals, Modeling of Structure and Properties*, ed. V. Vitek, p. 323. New York: TMS-AIME
25. Chen, H. S., Goldstein, M. 1971. *J. Appl. Phys.* 43:174
26. Vogel, H. 1921. *Phys. Z.* 22:645
27. Fulcher, G. 1925. *J. Am. Ceram. Soc.* 6:339
28. Laughlin, W. T., Uhlmann, D. R. 1972. *J. Phys. Chem.* 76:2317
29. Weiler, R., Blaser, S., Macedo, P. B. 1969. *J. Phys. Chem.* 73:4147
30. Adam, G., Gibbs, J. H. 1965. *J. Chem. Phys.* 43:139
31. Gibbs, J. H., DiMarzio, E. A. 1958. *J. Chem. Phys.* 28:373

32. Kauzmann, W. 1948. *Chem. Rev.* 42: 219
33. Cohen, M. H., Turnbull, D. 1959. *J. Chem. Phys.* 31:1164
34. Turnbull, D., Cohen, M. H. 1961. *J. Chem. Phys.* 34:120
35. Turnbull, D., Cohen, M. H. 1970. *J. Chem. Phys.* 52:3088
36. Cohen, M. H., Grest, G. S. 1979. *Phys. Rev. B* 20:1077
37. Grest, G. S., Cohen, M. H. 1980. *Phys. Rev. B* 21:4113
38. Grest, G. S., Cohen, M. H. 1981. *Adv. Chem. Phys.* 48:455
39. Hoover, W. G., Ree, F. H. 1968. *J. Chem. Phys.* 49:3609
40. Narayanaswamy, O. S. 1971. *J. Am. Cer. Soc.* 54:491
41. Stephens, R. B. 1978. *J. Appl. Phys.* 49:5855
42. Waseda, Y. 1981. *Prog. Mat. Sci.* 26:1
43. Wagner, C. N. J. 1983. See Ref. 20, p. 58
44. Sadoc, J. F., Dixmier, J. 1976. *Mat. Sci. Eng.* 23:197
45. Fuoss, P. H., Warburton, W. K., Bienenstock, A. 1980. *J. Non-Cryst. Solids* 35/36:1233
46. Hayes, T. M., Boyce, J. B. 1982. *Solid State Phys.* 37:173
47. Panissod, P., Aliaga Guerra, D., Amamou, A., Durand, J., Johnson, W. L., Carter, W. L., Poon, S. J. 1980. *Phys. Rev. Lett.* 44:1465
48. Gonser, U., Preston, R. 1983. See Ref. 22, p. 93
49. Ichikawa, T. 1973. *Phys. Stat. Sol. A* 19:707
50. Davies, L. B., Grundy, P. J. 1972. *J. Non-Cryst. Sol.* 11:179
51. Polk, D. E. 1970. *Scripta Met.* 4:117
52. Polk, D. E. 1972. *Acta Met.* 20:485
53. Frost, H. J. 1982. *Acta Met.* 30:889
54. Gaskell, P. H. 1983. See Ref. 22, p. 5
55. Boudreaux, D. S., Gregor, J. M. 1977. *J. Appl. Phys.* 48:156, 5057
56. Cargill, G. S. III. 1983. *Amorphous Materials: Modeling of Structure and Properties*, ed. V. Vitek, p. 15. New York: TMS-AIME
57. Boudreaux, D. S., Frost, H. J. 1981. *Phys. Rev. B* 23:1506
58. Gaskell, P. H. 1979. *J. Non-Cryst. Solids* 32:207
59. Cargill, G. S. III, Spaepen, F. 1981. *J. Non-Cryst. Solids* 43:91
60. Sadoc, J. F. 1981. *J. Non-Cryst. Solids* 44:1
61. Kleman, M., Sadoc, J. F. 1979. *J. Phys.* 40:L569
62. Nelson, D. R. 1984. *J. Non-Cryst. Sol.* 61/62:475
63. Nelson, D. R. 1983. *Phys. Rev. B* 28:5515
64. Nelson, D. R., Rubinstein, M., Spaepen, F. 1982. *Philos. Mag. A* 46:105
65. Steinhardt, P. J., Nelson, D. R., Ronchetti, M. 1981. *Phys. Rev. Lett.* 47:1297
66. Steinhardt, P. J., Nelson, D. R., Ronchetti, M. 1983. *Phys. Rev. B* 28:784
67. Turnbull, D. 1971. *Solidification*, pp. 1–22. Metals Park, Ohio: Am. Soc. Metals
68. Uhlmann, D. R. 1972. *J. Non-Cryst. Solids* 7:337
69. Turnbull, D. 1964. *Physics of Non-Crystalline Solids*, ed. J. W. Prins, pp. 41–56. Amsterdam: North Holland
70. Spaepen, F., Turnbull, D. 1976. *Rapidly Quenched Metals*, ed. N. J. Grant, B. C. Giessen, pp. 205–9. Cambridge, Mass: MIT Press
71. Spaepen, F., Turnbull, D. 1982. *Laser Annealing of Semi-Conductors*, ed. J. M. Poate, J. W. Mayer, pp. 15–42. New York: Academic
72. Perepezko, J. H. 1980. *Proc. 2nd Int. Conf. Rapid Solidification Processing*, ed. R. Mehrabian, B. H. Kear, M. Cohen, p. 56. Baton Rouge, LA: Claitor's
73. Thompson, C. V., Spaepen, F. 1983. *Acta Met.* 31:2021
74. Spaepen, F. 1975. *Acta Met.* 23:729
75. Spaepen, F., Meyer, R. B. 1976. *Scripta Met.* 10:257
76. Cohen, M. H., Turnbull, D. 1961. *Nature* 189:131
77. Marcus, M., Turnbull, D. 1976. *Mat. Sci. Eng.* 23:211
78. Davies, H. A. 1983. See Ref. 20, pp. 8–25
79. Lin, C. J., Spaepen, F. 1982. *Appl. Phys. Lett.* 41:716
80. Drehman, A. J., Greer, A. L., Turnbull, D. 1982. *Appl. Phys. Lett.* 41:716
81. Drehman, A. J., Turnbull, D. 1981. *Scripta Met.* 15:543
82. Hilsch, R. 1960. *Non-Cryst. Solids*, ed. V. C. Frechette, pp. 348–73. New York: Wiley
83. Buckel, W., Hilsch, R. 1954. *Z. Phys.* 138:109
84. Turnbull, D. 1974. *J. Phys.* 35-C-4:1–9
85. Turnbull, D., Bagley, B. G. 1975. *Treatise on Solid State Chemistry*, ed. N. B. Hannay, 5:513–54. New York: Plenum
86. Lin, C. J., Spaepen, F. 1983. *Chemistry and Physics of Rapidly Solidified Materials*, ed. B. J. Berkowitz, R. O. Scattergood, pp. 273–80. New York: TMS-AIME
87. Coriell, S. C., Turnbull, D. 1982. *Acta Met.* 30:2133
88a. Walker, J. L. 1961. *Phys. Chem. of Process Metallurgy*, Pt. 2, p. 845. New York: Interscience

88b. Chalmers, B. 1964. *Principles of Solidification*, Ch. 4. New York: Wiley
89. Bayles, B. J., Colligan, G. A. 1962. *Acta Met.* 10:895
90. Nelson, D. R., Halperin, B. I. 1979. *Phys. Rev. B* 19:2457
91. Boettinger, W. J. 1982. *Proc. 4th Int. Conf. on Rapidly Quenched Metals*, ed. T. Masumoto, K. Suzuki, 1:99–102. Sendai: Jap. Inst. Metals
92. Boettinger, W. J., Coriell, S. C., Sekerka, R. F. 1984. *Mat. Sci. Eng.* In press
93. Baker, J. C., Cahn, J. W. 1970. *Solidification*, pp. 23–58. Metals Park, Ohio: Am. Soc. Metals. See also 1969. *Acta Met.* 17:573
94. Lin, C. J., Spaepen, F., Turnbull, D. 1984. *J. Non-Cryst. Solids* 61/62:767
95. Lin, C. J., Spaepen, F. 1984. *Rapidly Solidified Metastable Materials*, ed. B. H. Kear, B. C. Giessen. *MRS Symp. Proc.* Amsterdam: North-Holland. In press
96. Schwarz, R. B., Johnson, W. L. 1983. *Phys. Rev. Lett.* 51:415
97. Warburton, W. K., Turnbull, D. 1975. *Diffusion in Solids: Recent Developments*, ed. A. S. Nowick, J. J. Burton, pp. 171–229. New York: Academic
98. Koch, C. C., Cavin, O. B., McKamey, C. G., Scarbrough, J. O. 1983. *Appl. Phys. Lett.* 43:1017
99. Greer, A. L. 1984. *J. Non-Cryst. Solids* 61/62:737
100. Cantor, B., Cahn, R. W. 1983. *Amorphous Metallic Alloys*, ed. F. E. Luborsky, pp. 487–505. London: Butterworths
101. Gupta, D., Tu, K. N., Asai, K. W. 1975. *Phys. Rev. Lett.* 35:796
102. Birac, C., Lesueur, D. 1976. *Phys. Stat. Solidi A* 36:247
103. Chen, H. S., Kimerling, L. C., Poate, J. M., Brown, W. L. 1978. *Appl. Phys. Lett.* 32:461
104. Dumond, J., Youtz, J. P. 1940. *J. Appl. Phys.* 11:357
105. Cook, H. E., Hilliard, J. E. 1969. *J. Appl. Phys* 40:2191
106. Greer, A. L., Spaepen, F. 1984. *Synthetic Modulated Structure Materials*, ed. L. Chang, B. C. Giessen (for a review). New York: Academic. In press
107. Rosenblum, M. P., Spaepen, F., Turnbull, D. 1980. *Appl. Phys. Lett.* 37:184
108. Taub, A. I., Spaepen, F. 1979. *Scripta Met.* 13:195
109. Greer, A. L., Lin, C. J., Spaepen, F. 1982. *Proc. 4th Int. Conf. Rapidly Quenched Metals*, ed. T. Masumoto, K. Suzuki, p. 567. Sendai: Jap. Inst. Metals
110. Cammarata, R. C., Greer, A. L. 1984. *J. Non-Cryst. Solids* 61/62:889
111. Busby, P. E., Warga, M. E., Wells, C. 1953. *Trans. AIME (J. Met.)* 5:1463
112a. Thomas, W. R., Leak, G. M. 1955. *Nature* 176:29
112b. Busby, P. E., Wells, C. 1954. *Trans. AIME (J. Met)* 6:972
113. Ozturk, B., Fearing, V. L., Ruth, J. A. Jr., Simkovich, G. 1982. *Met. Trans.* 13A:1871
114. Miller, J. W. 1969. *Phys. Rev.* 181:1095; 1969. 188:1074
115. Spaepen, F. 1981. *Physics of Defects*, ed. R. Balian, et al. Les Houches Lectures XXXV, pp. 133–74. Amsterdam: North-Holland
116. Lillie, H. R. 1933. *J. Am. Ceram. Soc.* 16:619
117. Roberts, G. J., Roberts, J. P. 1965. *Proc. 7th Int. Conf. Glass*, Brussels, p. 31. London: Gordon & Breach
118. Taub, A. I., Spaepen, F. 1980. *Acta Met.* 28:1781
119. Taub, A. I. 1980. *Acta Met.* 28:633
120. Taub, A. I. 1982. *Acta Met.* 30:2117
121. Steif, P. S., Spaepen, F., Hutchinson, J. W. 1982. *Acta Met.* 30:447
122. Spaepen, F. 1977. *Acta Met.* 25:407
123. Spaepen, F., Turnbull, D. 1974. *Scripta Met.* 8:563
124. Saffman, P. G., Taylor, G. I. 1958. *Proc. R. Soc. London Ser. A* 245:312
125. Spaepen, F. 1975. *Acta Met.* 23:615
126. Argon, A. S., Salama, M. 1976. *Mat. Sci. Eng.* 23:219
127a. Davis, L. A. 1978. *Metallic Glasses*, ed. J. J. Gilman, H. J. Leamy, p. 180 (for a review). Metals Park, Ohio: Am. Soc. Metals
127b. Kimura, H., Masumoto, T. 1983. *Amorphous Metallic Alloys*, ed. F. E. Luborsky, p. 187. London: Butterworths
128. Luborsky, F. E., Walter, J. L. 1976. *J. Appl. Phys.* 47:3648
129. Deleted in proof
130. Chen, H. S. 1977. *Scripta Met.* 11:367
131. Walter, J. L., Bacon, F., Luborsky, F. 1976. *Mat. Sci. Eng.* 24:239
132. Piller, J., Haasen, P. 1983. *Acta Met.* 30:1
133. Steif, P. S. 1983. *J. Mech. Phys. Solids* 31:359
134. Fridberg, J., Torndahl, L. E., Hillert, M. 1969. *Jernkont. Ann.* 153:263

EFFECTS OF SATURATION ON LASER-INDUCED FLUORESCENCE MEASUREMENTS OF POPULATION AND POLARIZATION

Robert Altkorn* and Richard N. Zare

Department of Chemistry, Stanford University, Stanford, California 94305

INTRODUCTION

This article addresses the effects of saturation in laser-induced fluorescence on both population and polarization measurements under certain conditions. Specifically, we examine low-pressure experiments in which collisional relaxation does not occur between excitation and detection. We use a rate equation approach to model the dynamics of spectroscopic transitions, and introduce directional Einstein coefficients to treat the interaction of a pulsed beam of polarized light with an anisotropic molecular distribution. There is an extensive body of literature related to saturation effects most of which concerns (a) optical pumping experiments, (b) laser physics, (c) high pressure studies, and (d) atomic systems. No attempt is made here to present a complete review of this field. Instead we cite mainly references that we feel would be of greatest relevance to workers applying LIF techniques to reaction dynamics.

Background

The laser-induced fluorescence (LIF) technique has become a workhorse for detecting molecular species and determining their internal state

---

* Present address: Department of Chemistry, Northwestern University, Evanston, Illinois 60201.

distributions (1–5). It has been applied with advantage in various environments, from the ultra low pressure regimes of molecular beams and beam-surface scattering to the relatively high pressures of discharges (6) and flames (7). A great deal of attention has been concentrated on extracting relative populations from the variation of the fluorescence signal with wavelength (fluorescence excitation spectrum). Increasingly it has been realized that the accurate determination of relative populations requires an understanding of how the fluorescence signal varies with the polarization of the incident and detected photons (8, 9). Some polarization measurements are also interesting in their own right because they contain information about the anisotropy of the distribution of angular momentum vectors in the species being studied. In the past, workers using LIF to study reaction dynamics or molecular scattering assumed rather idealized conditions, in particular that the excited state population was strictly proportional to the laser intensity. This assumption fails as the laser power increases, and we refer to such nonlinear intensity-dependent behavior as *saturation*. Although the effects of saturation can be avoided in the limit of low laser power, in practice saturation effects are expected to be important, particularly when pulsed lasers are used.

Saturation effects were first observed in the radiofrequency (10, 11) and microwave (12–14) regimes, where excited state relaxation times are long and radiation sources of very high spectral brightness are common. As lasers were developed, saturation of molecular transitions with infrared radiation became possible (15, 16). With the advent of visible lasers, saturation effects were extended to include electronic transitions in atoms and molecules.

Nonlinear effects in optical pumping experiments have been treated in considerable generality and detail. Usually these experiments are not designed to measure the anisotropy or relative populations in a molecular distribution, but rather to determine atomic or molecular parameters such as Landé g factors, excited state lifetimes, cross sections for energy transfer or depolarization, or various spectroscopic constants. Nevertheless, the theory developed for these experiments can also be applied to LIF measurements of the products of chemical reactions.

In a series of elegant papers (17–20), Ducloy discussed nonlinear behavior associated with the laser excitation of atoms or molecules in states of low (17, 18) or very large (classical limit) (19, 20) angular momenta. Ducloy's treatment assumes continuous-wave excitation and requires that a "broad line approximation" apply, in which the excitation probability is constant over the Doppler profile of the absorber. The excitation process is described in a density matrix formalism and the equations are solved nonperturbatively. Ducloy allowed for the presence of an external field, arbitrary anisotropy in the molecular distribution, and excitation by linearly or circularly polarized light. This and other work on saturation effects in optical pumping experiments has been reviewed by Lehmann (21), Broyer et al (22), and Decomps, Dumont & Ducloy (23).

At the opposite extreme, a large amount of work has been done on the interaction of intense radiation with matter in the limit that the light source has infinitely narrow bandwidth, predominantly in connection with the theory of lasers (24–34). However, it is unlikely that this limit is reached in experiments using most pulsed tunable dye lasers. Behavior intermediate between the broad-line and infinitely narrow bandwidth limits is more complicated and has been discussed by Avan & Cohen-Tannoudji (35, 36).

The effect of saturation on LIF measurements conducted in the relatively high pressure regime has also been the subject of considerable effort. Most of the LIF measurements performed in this regime are designed to measure the temperature or concentration of various species in given portions of a flame or plasma. Piepmeier first suggested that saturation may actually prove beneficial in these measurements, pointing out that it could reduce the dependence of the fluorescence intensity on quenching rate (37, 38). This method was first applied to the detection of atoms in flames by Omenetto et al (39). Subsequently, a large number of saturated LIF measurements have been performed in the high pressure regime, the majority of which have involved the detection of atomic species. This work has been the subject of a number of reviews (7, 40–45). Saturated LIF experiments in flames have also been conducted with molecular species such as C_2 (46–48), CH (48, 49), CN (49), OH (43, 50–52), and MgO (53). This work has also been discussed in several reviews (7, 42–45).

Although the need for quenching rate data is minimized in saturated LIF measurements in flames, these experiments do have a number of complications. For example, the possibility of laser-induced chemistry (54–56) or ionization must be considered, complicated models must be used to take into account partial vibrational and rotational relaxation during the laser pulse (43, 57–60), and extrapolations to complete saturation must be carried out if only partial saturation can be achieved (47, 48). In addition, most models of saturation in flames have assumed that the collisional environment of atmospheric flames causes emission to be unpolarized. However, recent studies by Doherty & Crosley (61) have shown that this assumption need not hold.

Several groups have considered the effects of saturation on population measurements performed in the low pressure regime. These are inherently low-signal experiments because of the small concentrations of detected species. Thus it might be desirable, or even necessary, to saturate the molecular transitions in order to obtain an acceptable fluorescence intensity. The importance of taking saturation into account has been emphasized by Allison, Johnson & Zare (62), who showed that saturation effects had caused a previous study of the Ba + CF_3I reaction system (63) to be misinterpreted. Liu & Parson (64) and Córdova, Rettner & Kinsey (65) used rate equation models of the saturation process to aid in their analyses of data taken using tunable pulsed dye lasers. Neither group included polarization effects, although both groups, as well as Geraedts et al (66), pointed out that such effects should be taken into account. More recently, Guyer, Hüwel & Leone (67) reported LIF studies on CO^+ in which a strong saturation limit was achieved.

In addition to the previously mentioned work concerning optical pumping experiments and LIF studies in the low- and high-pressure regimes, the advent of the laser has motivated the study of saturation phenomena in a number of different areas. Just a few of these include the use of saturable absorbers as passive mode-locking devices in lasers (16, 68–71), work on saturation in semiconductors (72), and studies of saturation in inhomogeneously broadened laser gain media, especially concerning the Lamb dip (24, 73, 74). The Lamb dip provided perhaps the first of a large number of high-resolution spectroscopic methods falling under the general heading of saturation spectroscopy. The field of saturation spectroscopy has been reviewed in a number of books and articles (30, 75–79). A number of applications concern atomic systems, but saturation techniques can also be used to unravel complicated molecular spectra (80, 81).

MODELING TRANSITION DYNAMICS

We use a rate equation approach to model the dynamics of spectroscopic transitions. Such an approach is justified when coherent transients (82–84) can be ignored. The suitability of rate equations for describing the interaction of laser light with atoms or molecules in the low pressure (collisionless) regime has been discussed by Cohen-Tannoudji (85), Avan & Cohen-Tannoudji (35), and Hertel & Stoll (86). These authors show that coherent effects can be neglected (a) if the coherence time of the laser (35, 85) is short compared to the pumping time (reciprocal of absorption or stimulated emission probability), or (b) the laser pulse is long compared to the excited-state radiative lifetime. How well these conditions are met depends on the type of laser used and the molecule being studied.

The second condition should easily be fulfilled if long-pulse (such as flashlamp-pumped) dye lasers are used to probe molecules with short lifetimes. However, a situation in which the radiative lifetime is much longer than the laser pulse can occur if short-pulse (such as YAG-pumped) dye lasers are used to study molecules having long lifetimes (such as OH $A^2\Sigma^+$, which has a radiative lifetime of ~ 700 ns, or NO $A^2\Sigma^+$ with a radiative lifetime of ~ 180 ns).

Whether or not the first condition is met depends on the distribution of frequencies in the laser pulse. If the laser output consists of a very large number of closely spaced (overlapping) modes, the coherence time can be quite short. A long coherence time is obtained if the laser is run single mode and there are negligible fluctuations in the amplitude or phase of the laser output during the pulse. (This is probably best approximated by a pulse-amplified CW single-mode laser.)

Even if the experimenter is working in the limit at which coherent effects might manifest themselves, he/she must at least consider that:

1. The spatial profile of pulsed dye laser beams is usually quite inhomogeneous. Thus, there are regions of widely different intensity whose interactions with the molecules being probed must be averaged.
2. Dye lasers often have significant pulse-to-pulse intensity fluctuations. Because LIF measurements usually represent the average of a number of laser shots, again coherent transients are masked.

Thus it is likely that rate equations are adequate in most cases to calculate the effects of saturation on population and polarization measurements. When these assumptions break down, the optical Bloch equations (28, 29) still apply, but such studies commonly become more an investigation of the characteristics of the laser light source than of the atomic or molecular system under investigation (35).

DIRECTIONAL EINSTEIN COEFFICIENTS

Rate equation models of spectroscopic transitions generally involve the Einstein A and B coefficients (87, 88). Implicit in the derivation of these quantities is an average over light of all directions and two orthogonal polarizations, as well as molecules of all orientations. Hence they are sometimes called the integrated Einstein coefficients. In LIF experiments, we are concerned with directional (polarized) radiation and molecular distributions that are often anisotropic. In order to treat the LIF process, we find it convenient to introduce the directional (or differential) Einstein coefficients, a and b, which are used to model the interaction of directional, polarized radiation with molecules having a specific orientation. Although the directional Einstein coefficients are not widely known, they have been discussed in detail by Stepanov & Gribkovskii (89). They may be defined in terms of the more familiar integrated Einstein coefficients by

$$a_{21} = \frac{8\pi^3 v^3}{hc^3} |\mu_{12}|^2 (\hat{\varepsilon} \cdot \hat{\mu})^2 = \frac{3A_{21}}{8\pi} (\hat{\varepsilon} \cdot \hat{\mu})^2 \qquad 1.$$

and

$$b_{12} = b_{21} = \frac{8\pi^2}{h^2} |\mu_{12}|^2 (\hat{\varepsilon} \cdot \hat{\mu})^2 = 3B_{12}(\hat{\varepsilon} \cdot \hat{\mu})^2, \qquad 2.$$

where $(\hat{\varepsilon} \cdot \hat{\mu})$ is the cosine of the angle between the electric field vector of the light and the transition dipole moment of the molecule and $|\mu_{12}|^2$ is the absolute square of the transition dipole between states 1 and 2 separated by the energy hv. It has also been assumed that states 1 and 2 have equal degeneracies. The factor of 1/3 in the relation between b_{21} and B_{21} arises from averaging $(\hat{\varepsilon} \cdot \hat{\mu})^2$ over light of all directions and two orthogonal polarizations. The factor of $8\pi/3$ in the relation between a_{21} and A_{21} comes from integrating $(\hat{\varepsilon} \cdot \hat{\mu})^2$ over an isotropic distribution of emitted light. We introduce the quantity $\rho(v, \Omega, \hat{\varepsilon})$ as the density of radiation (energy per unit volume per unit frequency interval) directed into the solid angle $d\Omega$ whose electric field vector is specified by $\hat{\varepsilon}$. Then, as shown in Figure 1, the rate of absorption of radiation characterized by $\rho(v, \Omega, \hat{\varepsilon})d\Omega$ is given by

$$b_{12}\rho(v, \Omega, \hat{\varepsilon})d\Omega N_1 \qquad 3.$$

and the rate of emission of radiation having the same properties is given by

$$[b_{21}\rho(v, \Omega, \hat{\varepsilon})d\Omega + a_{21}d\Omega]N_2. \qquad 4.$$

Here b_{12}, b_{21}, and a_{21} are the directional Einstein coefficients for absorption, stimulated emission, and spontaneous emission, respectively;

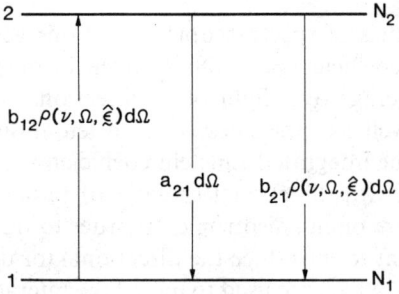

Figure 1 Schematic diagram of a two-level system. The probability of absorption of light of frequency v and polarization $\hat{\varepsilon}$ directed into the solid angle $d\Omega$ is given by $b_{12}\rho(v, \Omega, \hat{\varepsilon})d\Omega$. The probability of emission of light having the same properties is given by $b_{21}\rho(v, \Omega, \hat{\varepsilon})d\Omega + a_{21}d\Omega$.

N_1 is the density of molecules in the lower state; and N_2 is the density of molecules in the excited state.

Before proceeding to describe LIF measurements in terms of the directional Einstein coefficients, we note the following points:

1. Molecular transitions have finite bandwidths and lasers often have sharp spectral mode structure. Implicit in the use of $b_{12}\rho(v, \Omega, \hat{\varepsilon})d\Omega$ for an absorption probability is the requirement that the laser output be effectively constant over the Doppler profile of the transition. If this is not the case, one must obtain an average absorption probability by integrating the mode structure of the laser over the bandwidth of the transition. As pointed out by Kinsey (2), if each of the modes is saturating a subset of molecules in a given velocity range, failing to perform this integration can introduce a significant error into the calculated absorption probability. Killinger, Wang & Hanabusa (51) have considered in some detail the effects of laser spectral profile on saturation behavior in the rate equation limit.

2. We have omitted M sublevel degeneracy in discussing the directional Einstein coefficients. This is appropriate because the directional Einstein coefficients depend on the position of the transition dipole and therefore the M sublevel. In practice, then, the experimenter must average the absorption probability over the anisotropy of the distribution. Our treatment applies to the large angular momentum limit when the M level distribution can be approximated by a continuous function of the angle between the $\mathbf{J}$ vector and the axis of quantization.

3. We have chosen to describe the laser output in terms of its radiation density and have employed the cgs unit system. If one chooses for example to use the MKS system or to replace radiation density by intensity, slightly different formulae must be used for the Einstein coefficients. This has been discussed in detail by Hilborn (90).

LASER-INDUCED FLUORESCENCE INTENSITY

Consider the experimental arrangement shown in Figure 2. This is the traditional excitation-detection geometry in which the laser is incident along the X axis and linearly polarized along $\hat{\varepsilon}_a$ and fluorescence of linear polarization $\hat{\varepsilon}_d$ is collected in the Y direction. For simplicity, we suppose that the molecular distribution is cylindrically symmetric about the Z axis. This is the case, for example, in a beam-gas collision geometry when the beam is directed along the Z axis or in a photodissociation experiment when the electric vector of the polarized photolysis beam is along the Z axis.

The LIF measurement consists of two parts: We must first calculate the number of molecules excited by the laser and then the amount of fluorescence incident on the detector. Initially, we consider only those molecules oriented such that $(\hat{\varepsilon}_a \cdot \hat{\mu})$ specifies the cosine of the angle between the transition dipole and the electric field vector of a laser. For these molecules, the excited state population is governed by the differential equation

$$\frac{dN_2}{dt} = N_1 b_{12} \rho(v, \Omega_a, \hat{\varepsilon}_a) d\Omega_a - N_2 [A_{21} + b_{21} \rho(v, \Omega_a, \hat{\varepsilon}_a) d\Omega_a] \qquad 5.$$

which is easily solved for a rectangular pulse of duration Δt_L to yield

$$N_2 = \frac{N b_{12} \rho(v, \Omega_a, \hat{\varepsilon}_a) d\Omega_a}{2 b_{12} \rho(v, \Omega_a, \hat{\varepsilon}_a) d\Omega_a + A_{21}}$$
$$\times \{1 - \exp[-(2 b_{12} \rho(v, \Omega_a, \hat{\varepsilon}_a) d\Omega_a + A_{21}) \Delta t_L]\}, \qquad 6.$$

where $N = N_1 + N_2$ is the total concentration of the molecule in the state being probed. We have used A_{21} rather than a_{21} because the total rate at which the ground state is repopulated is of interest. We note that Eq. 5 is equivalent to that derived in the density matrix treatment in the limits of large angular momentum, broadband excitation, and no external fields (20–23).

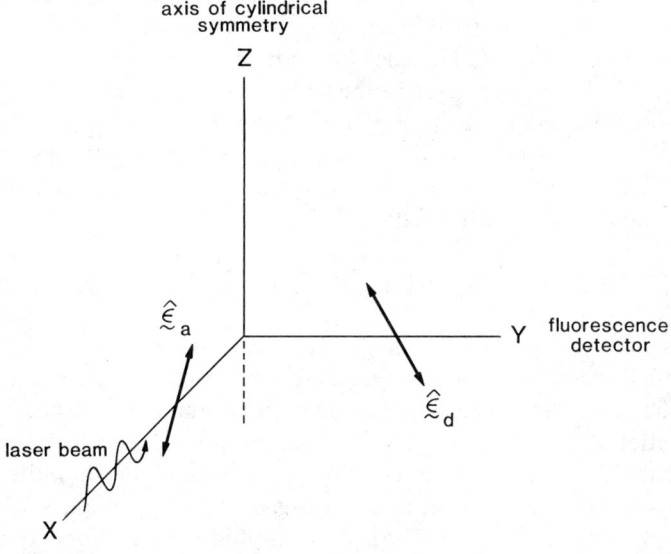

Figure 2 Traditional right angle excitation-detection geometry. Molecules are excited by a laser of linear polarization $\hat{\varepsilon}_a$ incident along the X axis. Fluorescence of linear polarization $\hat{\varepsilon}_d$ is detected along the Y axis. The Z axis is chosen to lie along the axis of cylindrical symmetry.

Having determined the number of molecules excited by the laser, we now consider the amount of fluorescence incident on the detector. The density of fluorescent radiation of polarization $\hat{\varepsilon}_d$ incident into the element of solid $d\Omega_d$ is given simply by $N_2 a_{21} d\Omega_d$. The intensity of detected fluorescence may be written as

$$I = K a_{21} N_2 \qquad \qquad 7.$$

where K is a constant depending on the details of the detection system and N_2 is given by Eq. 6. Of course Eq. 7 applies only to those molecules oriented with a specific value of the cosine of the angle between the transition dipole and the electric field vector of the laser, given by $(\hat{\varepsilon}_a \cdot \hat{\mu})$ and a specific value of the cosine of the angle between the transition dipole and the electric vector of the detected fluorescence, given by $(\hat{\varepsilon}_d \cdot \hat{\mu})$. To obtain the total fluorescence intensity we must average over the entire distribution of molecules. In the classical limit, the molecular distribution is specified as $n(\theta)$ where θ is the angle between the **J** vector of the molecule and the Z axis.

The position of the **J** vector can be related (8) to the position of the transition dipole μ through

$$(\hat{\mu} \cdot \hat{\varepsilon})^2 = (\hat{\mathbf{J}} \cdot \hat{\varepsilon})^2 \qquad (Q \text{ branch}) \qquad \qquad 8.$$

and

$$(\hat{\mu} \cdot \hat{\varepsilon})^2 = \tfrac{1}{2}[1 - (\hat{\mathbf{J}} \cdot \hat{\varepsilon})^2] \qquad (P \text{ or } R \text{ branch}) \qquad \qquad 9.$$

for linearly polarized light. Substituting $n(\theta)$ for N and using Eqs. 6 and 7, we obtain for the fluorescence intensity

$$I = K \int \frac{n(\theta) a_{21} b_{12} \rho}{2 b_{12} \rho + A_{21}}$$
$$\times \{1 - \exp[-(2 b_{12} \rho + A_{21})\Delta t_\text{L}]\} d\Omega \qquad \qquad 10.$$

where for convenience we write $\rho = \rho(v, \Omega_a, \hat{\varepsilon}_a) d\Omega_a$, $d\Omega$ is the solid angle element about **J**, and Eq. 8 or Eq. 9 must be substituted into Eq. 1 and Eq. 2 for a_{21} and b_{12}. The angular dependence of $n(\theta)$ is commonly expressed through an expansion in Legendre polynomials:

$$n(\theta) = n \sum_{l=0}^{\infty} a_{2l} P_{2l}(\cos \theta) \qquad \qquad 11.$$

where n represents the total concentration of molecules in the state of interest (assuming the normalization of $a_0 = 1$) and all odd moments are zero for a cylindrically symmetric distribution of **J** without handedness. (In any case only the even moments can be detected with linearly polarized light.) Substituting Eq. 11 into Eq. 10, we obtain the grand result:

$$I = Kn \int \frac{\left(\sum_{l=0}^{\infty} a_{2l} P_{2l}(\cos\theta)\right) a_{21} b_{12} \rho}{2b_{12}\rho + A_{21}}$$

$$\times \{1 - \exp[-(2b_{12}\rho + A_{21})\Delta t_L]\} d\Omega. \qquad 12.$$

The above derivation is predicated on the following assumptions concerning the LIF experiment:

1. The experiment involves a pulsed laser and fluorescence is collected after the laser is off.
2. The laser pulse is perfectly uniform (although any coherent effects are assumed to be absent).
3. The solid angle subtended is sufficiently small so that a_{21} is constant over the detector.
4. All states being studied have the same lifetime and quantum yield for fluorescence.
5. The LIF process in the atom or molecule can be accurately modeled using a two-level system (extension to a three-level system is considered in the Appendix).

Assuming that the laser pulse is perfectly uniform allows us to neglect any "hot spots" that might be present in the beam profile, the low-intensity wings of a Gaussian beam, and any intensity variations between the beginning and end of a pulse. Clearly, the experimenter may face the problem of saturating molecules with the more intense portions of the beam but obtaining linear behavior in regions of lower intensity. Thus Liu & Parson (64) and Córdova, Rettner & Kinsey (65) chose to measure a "saturation parameter" rather than use calculated values of the laser power and Einstein coefficients. We believe this procedure has much practical merit.

Effects due to a nonuniform beam profile have been considered by a number of workers, mostly in the field of flame and plasma diagnostics. Rodrigo & Measures (91) first pointed out that spatial inhomogeneity in the laser profile could cause the intensity actually needed to saturate a transition to be considerably higher than that calculated under the assumption of a homogeneous beam. Daily (92) analyzed the problem of saturating atoms in a flame with a Gaussian beam and came to a similar conclusion. Mailänder (48) discussed a method for determining a "saturated volume" throughout which the number of molecules in the excited state is constant. Van Calcar et al (93) have considered the importance of spatial beam profile in some detail. Using diaphragms and a diffuser to improve the homogeneity of the laser beam, they report actual saturation behavior that agrees with their calculations to within 25% in studies of sodium atoms in flames. Van Calcar et al (94) also reported detailed studies of the spatial properties of a flashlamp-pumped dye laser beam.

In practice, the detector subtends a finite solid angle over which the fluorescence must be integrated. Usually the assumption is made that a_{21} is constant over the detector. If this is not the case for a polarization experiment, a correction for finite solid angle must be included (95).

We also assume that all states being studied have the same lifetime. If this is not the case, it is important when interpreting the data to recognize that the populations of different excited states are decaying exponentially at different rates, and to integrate the expression for fluorescence intensity over the collection time. Alternatively, the experimenter may choose to collect fluorescence for a time long compared to the longest radiative lifetime, provided that the quantum yield is the same for all states. Then the fluorescence signal is once more proportional to the excited state population if the detector response is not biased by the rate of arrival of the fluorescent photons.

Finally, we turn to the assumption that the LIF process in the atom or molecule can be treated as a two-level system. If a short-pulse dye laser is being used, the laser is often on for a time Δt_L much shorter than the excited state radiative lifetime τ. Thus, spontaneous emission may be neglected during the excitation process, and a two-level system becomes appropriate. If the dye laser pulse is longer than the excited state radiation lifetime, a three-level system must be used to model the excitation process. A three-level system is straightforward but somewhat more complicated algebraically. We discuss such a treatment in the Appendix. The fluorescence process (after the laser is off) cannot in general be modeled by a two-level system as the atom or molecule usually radiates to a number of lower states. However, corrections for a three-level system are simple and also discussed in the Appendix.

EFFECTS OF SATURATION ON POPULATION MEASUREMENTS

In the limit at which the exponential in Eq. 12 may be replaced by the first two terms in its Taylor series expansion

$$I = Kn\rho \int [1 + a_2 P_2 (\cos \theta) + a_4 P_4 (\cos \theta)] a_{21} b_{12} \Delta t_L d\Omega \qquad 13.$$

where terms containing Legendre polynomials of order greater than 4 have not been included because they vanish upon integration. This approximation is expected to be valid when the pumping rate is low (no saturation).

Using Eq. 1 and Eq. 2, we substitute A_{21} and B_{12} for a_{21} and b_{12} in Eq. 13 and obtain

$$I = \frac{9}{8\pi} Kn\rho A_{21} B_{12} \Delta t_L \int [1 + a_2 P_2(\cos\theta) + a_4 P_4(\cos\theta)]$$
$$\times [\hat{\varepsilon}_a \cdot \hat{\mu}]^2 [\hat{\varepsilon}_d \cdot \hat{\mu}]^2 d\Omega. \qquad 14.$$

Equation 14 can be further simplified if the following three conditions are satisfied:

1. The polarization of the laser and the detected fluorescence are unchanged during the experiment.
2. The distributions of molecules in all states probed by the laser are characterized by the same values of a_2 and a_4.
3. The fractions of collected fluorescence due to Q branch and P or R branch transitions are the same for all states studied.

In this limit, the integral in Eq. 14 may be incorporated into the proportionality constant to give the familiar expression (2)

$$I = K'nB_{12}\rho\Delta t_L \qquad 15.$$

where

$$K' = \frac{9}{8\pi} KA_{21} \int (1 + a_2 P_2(\cos\theta) + a_4 P_4(\cos\theta)) [\hat{\varepsilon}_a \cdot \hat{\mu}]^2 [\hat{\varepsilon}_d \cdot \hat{\mu}]^2 d\Omega. \qquad 16.$$

It is Eq. 15 that is almost always assumed to be valid in extracting populations from LIF measurements. In order to compare the concentration of molecules in states i and j, it is common (1) to form the ratio

$$\frac{n_i}{n_j} = \frac{I_i}{I_j} \frac{B_{12}^j \rho^j}{B_{12}^i \rho^i}. \qquad 17.$$

(Here we have neglected any corrections due to changes in detector response with frequency.) Equation 15 is valid only in the limit of low laser power. As the laser power is increased, nonlinear behavior becomes apparent, and we must use the following expression, obtained from Eq. 12, to find n_i/n_j:

$$\frac{n_i}{n_j} = \frac{I_i \int \frac{\left[\sum_{l=0}^{\infty} a_{2l}^j P_{2l}(\cos\theta)\right] a_{21}^j b_{12}^j \rho^j \{1 - \exp[-(2b_{12}^j \rho^j + A_{21}^j)\Delta t_L]\}}{2b_{12}^j \rho^j + A_{21}^j} d\Omega}{I_j \int \frac{\left[\sum_{l=0}^{\infty} a_{2l}^i P_{2l}(\cos\theta)\right] a_{21}^i b_{12}^i \rho^i \{1 - \exp[-(2b_{12}^i \rho^i + A_{21}^i)\Delta t_L]\}}{2b_{12}^i \rho^i + A_{21}^i} d\Omega}. \qquad 18.$$

There exists a limit in which Eq. 17 remains accurate for interpreting even saturated spectra. If the b coefficients and the radiation density—as well as all angular factors—are the same for states i and j, then both Eq. 17 and Eq. 18 reduce to

$$\frac{n_i}{n_j} = \frac{I_i}{I_j}. \qquad 19.$$

Thus in many practical cases, saturation should not change the appearance of LIF spectra. For example, such behavior might be expected for the members of a band progression (Δv constant) in which the Franck-Condon factors are roughly all the same. In fact, saturation might be desirable in this case as it will increase the fluorescence signal. However, in probing features whose transition strengths differ significantly, saturation can seriously alter the appearance of the spectra. If this occurs, careful analysis is required to extract relative populations from the experimental data.

A dramatic example of the change in the appearance of a spectrum due to saturation is shown in Figure 3, taken by Allison, Johnson & Zare (62) with a maximum power of 5 MW/cm² in a 0.3 Å bandwidth. This spectrum includes $\Delta v = 0$ and $\Delta v = -1$ progressions of the $C^2\Pi_{3/2} - X^2\Sigma^+$ transition in BaI. It is seen that the $\Delta v = 0$ progression having large Franck-Condon factors "saturates first" and, as the laser power is increased, the members of the $\Delta v = -1$ progression increase in relative intensity. A change of about a factor of two occurs in the ratio of the $\Delta v = 0$ to $\Delta v = -1$ features when the laser power is attenuated by a factor of 100.

It is tempting to imagine that the experimenter can work in the limit of very strong saturation where simple expressions for the fluorescence intensity are recovered. In this case, conditions are such that $b_{12}\rho\Delta t_L \gg 1$ (and $b_{12}\rho \gg A_{21}$); hence Eq. 12 can be reduced to

$$I = \frac{Kn}{2} \int [1 + a_2 P_2 (\cos\theta)] a_{21} d\Omega. \qquad 20.$$

(Here terms containing Legendre polynomials of order greater than 2 integrate to zero.) We note that Eq. 20 is independent of the Einstein B coefficients and laser power, because half of the molecules in the probed state are excited, regardless of their line strengths. Thus it becomes possible in principle to obtain relative populations or concentrations without knowledge of absorption line strengths! Unfortunately, this limit is difficult to achieve in practice, particularly with molecules. Some of the problems that must be confronted include (*a*) low-intensity wings of the laser beam and inhomogeneities in the beam profile, (*b*) the possibility of multiphoton processes as the laser intensity is increased, and (*c*) power restrictions of available dye lasers.

Nevertheless, some workers appear to have come very close to reaching this limit. For example, in a particularly careful study, Van Calcar et al (93) were able to saturate sodium atoms in flames strongly with a flashlamp-pumped dye laser. They used a series of diaphragms to ensure that the flame was uniformly irradiated, and collected fluorescence from the uniformly irradiated volume during the high intensity portion of the laser pulse. Under these conditions, the fluorescence intensity increased very slowly as the laser power was raised. However, if the fluorescence was integrated over the low-intensity tail of the laser pulse, or if the beam was focused, it was much more difficult to reach the strong saturation limit.

Molecular transitions are more difficult to saturate than atomic transitions because they generally have considerably weaker absorption strengths. However, it appears possible to approach the strong saturation limit even if molecules are being studied. For example, in a study of OH in flames, Lucht, Sweeney & Laurendeau (43) found approximately equal

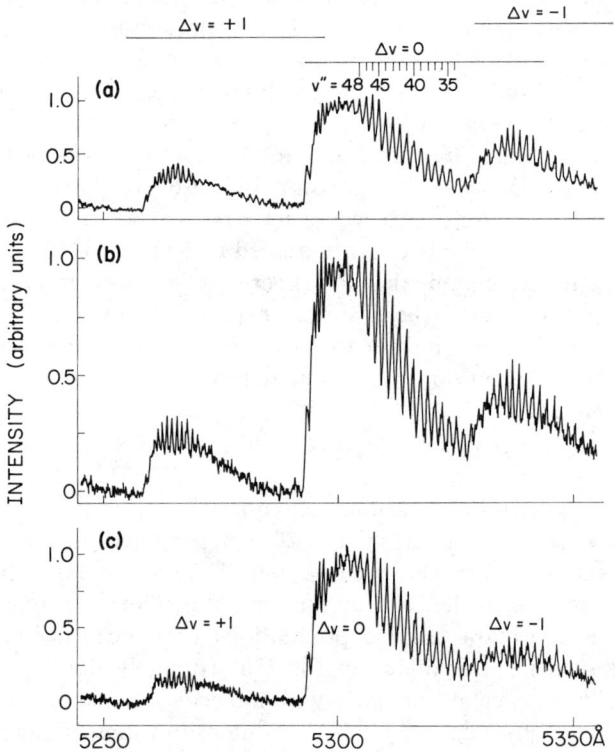

Figure 3 Excitation spectra of BaI ($C^2\Pi_{3/2} - X^2\Sigma^+$) from the reaction Ba+CF$_3$I obtained using (*a*) full laser power (5 MW/cm^2 over a 0.3 Å bandwidth), (*b*) one-tenth laser power, and (*c*) one-hundredth laser power. From Allison, Johnson & Zare (62).

fluorescence intensity from excitation by two peaks in the temporal output of a frequency-doubled YAG-pumped dye laser, even though the laser power in the two peaks differed by approximately a factor of two. (The fluorescence intensity dropped between the peaks, apparently due to collisional deexcitation of OH.)

EFFECT OF SATURATION ON POLARIZATION MEASUREMENTS

Polarization measurements are designed to obtain information about the distribution of angular momentum vectors in an atomic or molecular ensemble. When discussing polarization measurements in the absence of saturation, we are again interested in Eq. 14, which we rewrite (to consolidate as many constants as possible) as

$$I = K'' \int (1 + a_2 P_2(\cos\theta) + a_4 P_4(\cos\theta))[\hat{\varepsilon}_a \cdot \hat{\mu}]^2 [\hat{\varepsilon}_d \cdot \hat{\mu}]^2 d\Omega \qquad 21.$$

where

$$K'' = \frac{9}{8\pi} K n \rho A_{21} B_{12} \Delta t_L. \qquad 22.$$

In principle, it is possible to obtain a_2 and a_4 by performing four LIF measurements on the molecular ensemble (8, 9, 96). The four measurements usually involve polarizing the laser along each of the Z and Y axes and collecting fluorescence polarized along each of the Z and X axes. The fluorescence intensities for each of the four excitation geometries and the four possible combinations of Q and P or R branch transitions are listed in Table 1.

Table 1 The fluorescence signal $I(a_2, a_4)$ obtained from a cylindrical molecular distribution[a] $n(\theta)$ for different resonance fluorescence branches

Excitation-detection polarization		Branch type[b]			
$\hat{\varepsilon}_a$	$\hat{\varepsilon}_d$	(P or $R\uparrow$, P or $R\downarrow$)	($Q\uparrow, Q\downarrow$)	(P or $R\uparrow$, $Q\downarrow$)	($Q\uparrow$, P or $R\downarrow$)
$\hat{Z}$	$\hat{Z}$	$168 - 48a_2 + 8a_4$	$252 + 144a_2 + 32a_4$	$84 + 12a_2 - 16a_4$	$84 + 12a_2 - 16a_4$
$\hat{Z}$	$\hat{X}$	$126 - 18a_2 - 4a_2$	$84 + 12a_2 - 16a_2$	$168 - 48a_2 + 8a_4$	$168 + 78a_2 + 8a_4$
$\hat{Y}$	$\hat{Z}$	$126 - 18a_2 - 4a_4$	$84 + 12a_2 - 16a_4$	$168 + 78a_2 + 8a_4$	$168 - 48a_2 + 8a_4$
$\hat{Y}$	$\hat{X}$	$126 + 36a_2 + a_4$	$84 - 24a_2 + 4a_4$	$168 - 30a_2 - 2a_4$	$168 - 30a_2 - 2a_4$

[a] $n(\theta)$ has the form $\sum_l a_{2l} P_{2l}(\cos\theta)$ where $a_0 = 1$; only the a_2 and a_4 coefficients can be determined by the measurements.
[b] All entries should be multiplied by $\pi K''/315$.

In the limit of very strong saturation, Eq. 20 again becomes applicable. In this case the probability of exciting a molecule is independent of its orientation or the polarization of the laser. The analysis of the LIF experiment then becomes equivalent to that of an excited-state emission experiment (97, 98), only permitting the determination of a_2. The intensities of Z and X polarized fluorescence due to Q and P or R branch transitions in the strong saturation limit are shown in Table 2.

Between the unsaturated and strongly saturated limits, Eq. 12 relates LIF measurements to moments of the molecular distribution. We have written a computer program to evaluate Eq. 12. In Figure 4 we show calculated fluorescence polarizations ($P = [I_Z - I_X]/[I_Z + I_X]$) for measurements involving (P or $R\uparrow$, P or $R\downarrow$) and ($Q\uparrow, Q\downarrow$) transitions, respectively. The fluorescence polarization is displayed as a function of the reduced absorption probability, D, where

$$D = b_{12}\rho\Delta t_L/[\hat{\varepsilon}_a \cdot \hat{\mu}]^2. \qquad 23.$$

(The absorption probability is given by $D[\hat{\varepsilon}_a \cdot \hat{\mu}]^2$.) We have assumed the laser to be polarized along the Z axis and have evaluated Eq. 12 for the following molecular distributions:

1. isotropic ($a_0 = 1$);
2. $\sin^2 \theta$ distribution ($a_0 = 1$, $a_2 = -1$);
3. $\cos^2 \theta$ distribution ($a_0 = 1$, $a_2 = 2$);
4. pure quadrupole distribution ($a_0 = 1$, $a_4 = -1$); and
5. pure hexadecapole distribution ($a_0 = 1$, $a_6 = -1$). (In distributions 1–5, $a_n = 0$ for all a_n not specified.)

It is seen that between the unsaturated and strongly saturated limits, the fluorescence intensity is dependent on a_6 for the distributions considered in Figures 4a and 4b. In general the fluorescence intensity in the moderately

Table 2 The fluorescence signal $I(a_2)$ obtained from a cylindrical molecular distribution[a] $n(\theta)$ for different fluorescence branches in the strong saturation limit

Detection polarization	Branch type[b]	
$\hat{\varepsilon}_d$	P or R	Q
$\hat{Z}$	$10 - 2a_2$	$10 + 4a_2$
$\hat{X}$	$10 + a_2$	$10 - 2a_2$

[a] $n(\theta)$ has the form $\sum_l a_{2l} P_{2l}(\cos \theta)$ where $a_0 = 1$; only the a_2 coefficients can be determined by the measurements.
[b] All entries should be multiplied by $KnA_{21}/40$.

saturated regime depends on all higher-order even moments of the distribution. This opens in principle the possibility of detecting these higher-order moments, although their quantitative determination is at best problematic.

It is also evident from comparing Figures 4a and 4b that the polarizations involving $(Q\uparrow, Q\downarrow)$ transitions seem to converge to the strongly

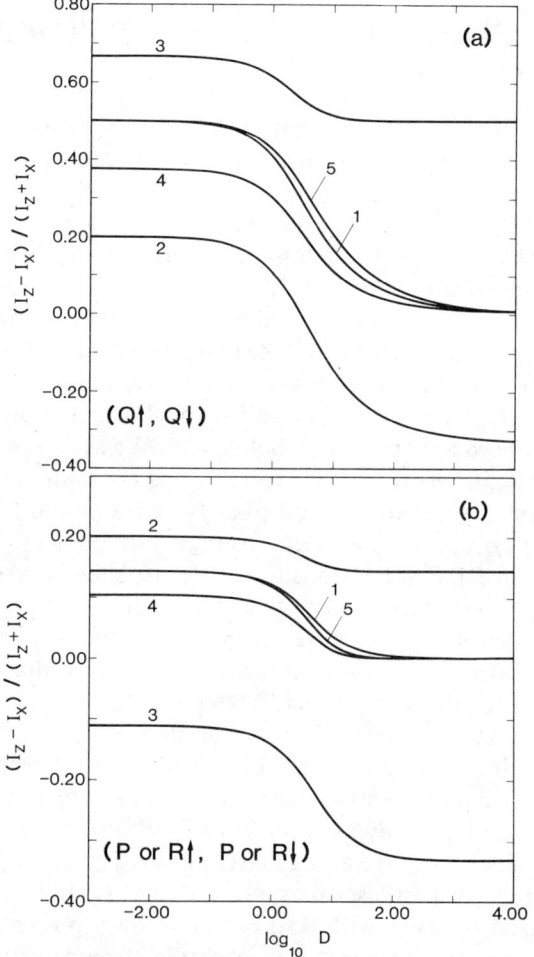

Figure 4 Calculated change in polarization vs the logarithm of the reduced absorption probability D for five molecular **J** vector distributions. The distributions are as follows: 1. isotropic $(a_0 = 1)$; 2. $\sin^2 \theta$ $(a_0 = 1, a_2 = -1)$; 3. $\cos^2 \theta$ $(a_0 = 1, a_2 = 2)$; 4. pure quadrupole $(a_0 = 1, a_4 = -1)$; 5. pure hexadecapole $(a_0 = 1, a_6 = -1)$. Here (a) refers to excitation and fluorescence via Q branches, and (b) via P or R branches.

saturated limit much more slowly than do those involving (P or $R\uparrow$, P or $R\downarrow$) transitions. This follows from the fact that Q branch transition dipoles lie along the **J** vector and undergo no spatial averaging as the molecule rotates. Thus, exciting with a Z-polarized laser and collecting X-polarized fluorescence sets up a situation in which the molecules that contribute the most fluorescence intensity are the most difficult to saturate because of the angular term in the b coefficient.

ROLE OF ANGULAR FACTORS IN SATURATION BEHAVIOR

Our treatment of saturation differs from previous work (64, 65) in that we include explicitly the dependence of fluorescence intensity on the excitation-detection geometry, including the type (P, R or Q) of resonance fluorescence branches, the polarization of the absorbed and detected photons, and the (possible) anisotropy of the ensemble of absorbing species. It is useful to determine whether this treatment involving angular factors, such as directional Einstein coefficients, may be replaced by a treatment involving simply the integrated Einstein coefficients. To explore the adequacy of this simpler approximation, we have plotted in Figure 5 the fluorescence intensity arising from an isotropic distribution of absorbers against the reduced absorption probability, D. We assume the traditional excitation-detection geometry of Figure 2, and we consider the incident light to be linearly polarized. We plot $I_\parallel$ corresponding to $\hat{\varepsilon}_a \parallel \hat{\varepsilon}_d$, $I_\perp$ corresponding to $\hat{\varepsilon}_a \perp \hat{\varepsilon}_d$, and their average value, $(I_\parallel + I_\perp)/2$, for both ($Q\uparrow,Q\downarrow$) and ($P$ or $R\uparrow$, P or $R\downarrow$) transitions. We also show by a dotted line the result when $[\hat{\varepsilon}\cdot\hat{\mu}]^2$ is replaced by its average value of one-third, and the fluorescence intensity is normalized to have the same maximum value at infinite laser power as in the other three cases. This procedure is equivalent to using integrated Einstein coefficients (see Eq. 2).

Figure 5 shows the existence of three intensity regimes. The first is a linear dependence at low laser power (see insets); the second regime occurs at intermediate laser power and is characterized by strong nonlinear dependence of the fluorescence intensity on D; and the third regime is once again linear with almost zero slope at high laser power. The onsets of these three regimes occur at different laser powers for $I_\parallel$ and $I_\perp$. Indeed, $I_\parallel$ begins to exhibit nonlinear behavior with laser power at lower power levels than $I_\perp$. Moreover, $I_\parallel$ much more rapidly reaches the strongly saturated regime than $I_\perp$. It is a matter of much practical concern to test for saturation effects. Clearly, the (linear) variation of fluorescence signal with laser power is the most straightforward diagnostic. Figure 5 shows that the measurement of $I_\parallel$ is the most sensitive way of carrying out this test. Conversely, if the experimenter wishes to achieve the strongly saturated limit, the observation of $I_\parallel$ is again to be commended.

Figure 5 also shows that $(I_\parallel + I_\perp)/2$ is extremely well approximated by the calculation of saturation effects based on the integrated Einstein coefficients (46, 47), at least for low and intermediate laser power levels. This conclusion justifies the previous neglect of angular factors in calculating the behavior of fluorescence intensity with laser power for isotropic distributions, but cannot be expected to hold for nonisotropic distributions.

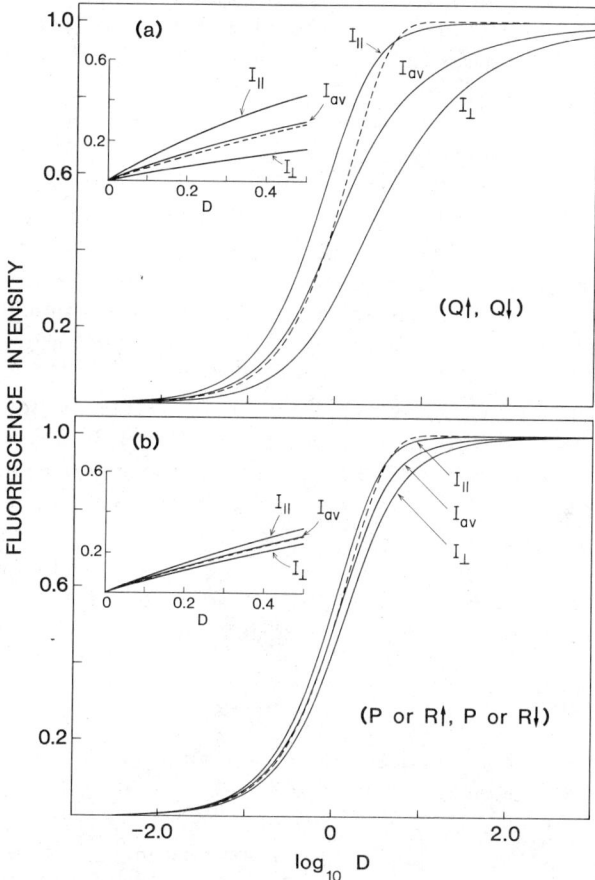

Figure 5 Fluorescence intensity versus reduced absorption probability: (*a*) for $(Q\uparrow, Q\downarrow)$ transitions, and (*b*) for $(P \text{ or } R\uparrow, P \text{ or } R\downarrow)$ transitions. The *solid lines* are $I_\parallel$, $I_\perp$, and $I_{av} = (I_\parallel + I_\perp)/2$; the *dotted line* is a calculation based on integrated Einstein coefficients. All curves are normalized to unit value at infinite laser power.

USING SATURATION TO MEASURE FLUX—A SPECIAL CASE

The measurement of reaction rates or cross sections requires a knowledge of product flux (the number of molecules crossing a unit area in a unit time, or equivalently, the density of molecules multiplied by their velocity). However, LIF usually functions as a density detector (99). An important question is whether saturation can transform the LIF technique from a density to a flux detector. Unfortunately the answer is seldom in the affirmative.

Using the first definition of flux given above, we see that in order for a laser to detect flux it must effectively define a surface and excite molecules that cross that surface with a probability that is independent of molecular velocity. For this condition to be fulfilled, the number of molecules that pass through the beam while it is on must be much greater than the number of molecules present in the beam at any instant. This is clearly not the case in most experiments involving short-pulse dye lasers. The diameter of an average pulsed laser beam is much greater than the distance moved by most chemical reaction products in the time during which a short-pulse laser is on. Thus the molecules are stationary on the time scale of the laser pulse and the laser must rigorously function as a density detector (although in special circumstances, density and flux detection could be equivalent). On the other hand, a CW laser might be used as a flux detector (100) if the strong saturation limit were achieved and the excited molecules did not return to a state that can reabsorb the laser radiation. Then each molecule passing through the laser beam can be detected once and only once in principle. Often the situation is much more complicated and something between flux and density is measured. In a different scheme, a laser might be used under some conditions to measure the density and velocity simultaneously, for example through the Doppler shift (101–104) or through "hole burning" (104, 105), and thereby allow flux to be determined.

SUMMARY

Experimental and theoretical studies of saturation effects on resonance fluorescence have been briefly reviewed. We have restricted our attention to molecules in a collisionless environment excited by pulsed laser sources. In the limit of large rotational angular momentum, directional Einstein coefficients were introduced and rate equations developed to describe the fluorescence intensity as a function of both laser and fluorescence polarization for linearly polarized light interacting with a cylindrically symmetric molecular distribution.

Saturation may actually prove beneficial in some population measurements. This can occur if the strengths of all the transitions composing the spectrum are approximately the same. In this case saturation will not significantly alter the appearance of the spectrum but it may improve signal to noise. If the transition strengths differ greatly, however, saturation can significantly distort the appearance of the spectrum. On the other hand, saturation is rarely advantageous in polarization measurements. In these experiments a "spectrum" is obtained not by scanning wavelength but by varying the polarization of the incident and/or detected light. As the polarization is varied, molecules having different orientations and therefore different transition strengths due to angular factors are selectively detected, and the "spectrum" must change with saturation. In moderately saturated regimes, the fluorescence intensity depends on higher orders of the molecular distribution, making polarization data difficult to invert. In the strongly saturated limit, only the second Legendre moment of the angular momentum distribution is obtainable, whereas in unsaturated measurements both the second and fourth moments are accessible to study.

Acknowledgments

We thank the many people who read earlier drafts of this manuscript and offered suggestions for its improvement. In particular, we are especially grateful to Klaas Bergmann, Paul J. Dagdigian, James DeHaven, Mark A. Johnson, Stephen R. Leone, Guang Hai Lin, Kopin Liu, John McKillop, John M. Parson, Charles T. Rettner, Thomas Trickl, and Jochen Wanner for their assistance. This work was supported by the National Science Foundation. R. N. Z. gratefully acknowledges support through the Shell Distinguished Chairs Program, funded by Shell Companies Foundation, Inc.

Appendix

We consider here the use of a three-level system, shown in Figure 6, for modeling the LIF process. We assume that the laser is resonant on the transition between states 1 and 2. The excited state can radiate to a number of states other than state 1. We call these collectively state 3 and assume that they are not connected to the ground state on the time scale of the laser pulse, Δt_L. The populations evolve according to the coupled differential equations

$$\frac{dN_1}{dt} = -b_{12}\rho N_1 + (b_{21}\rho + A_{21})N_2 \qquad \text{A-1.}$$

$$\frac{dN_2}{dt} = b_{12}\rho N_1 - (b_{21}\rho + \tau^{-1})N_2 \qquad \text{A-2.}$$

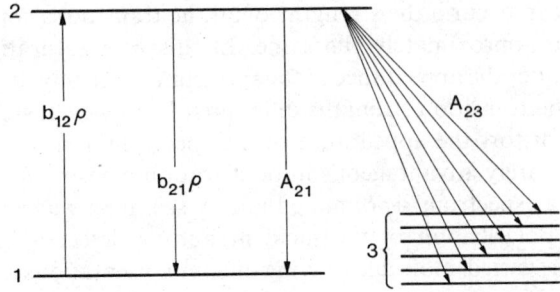

Figure 6 Schematic diagram of a three-level system. All ground state levels other than that being pumped are grouped together as level 3. The probability of exciting a molecule oriented such that $(\hat{\varepsilon}\cdot\hat{\mu})$ is the cosine of the angle between the electric vector of the light and the transition dipole is given by $b_{12}\rho$. The total probability of emission from excited molecules having the same orientation is given by $b_{21}\rho + A_{21} + A_{23}$.

where τ is the lifetime of state 2. The population of state 3 follows by subtraction from the total concentration of molecules, N. Equations A-1 and A-2 may be solved to yield

$$N_2 = \frac{b_{12}\rho N(e^{\lambda_1 \Delta t_L} - e^{\lambda_2 \Delta t_L})}{\lambda_1 - \lambda_2} \qquad \text{A-3.}$$

and

$$N_1 = \frac{N[(b_{12}\rho + \tau^{-1} + \lambda_1)e^{\lambda_1 \Delta t_L} - (b_{12}\rho + \lambda_2 + \tau^{-1})e^{\lambda_2 \Delta t_L}]}{\lambda_1 - \lambda_2} \qquad \text{A-4.}$$

where N is the total number of atoms or molecules, and λ_1 and λ_2 are given by

$$\lambda_{1,2} = \tfrac{1}{2}\{-2b_{12}\rho - \tau^{-1} \pm (\tau^{-2} + 4b_{12}\rho A_{21} + 4b_{12}{}^2\rho^2)^{1/2}\}. \qquad \text{A-5.}$$

The three-level system described here has been used previously by Córdova, Rettner & Kinsey (65) (neglecting polarization effects) in their study of OH. A more general description of three-level systems as well as a discussion of systems with arbitrary numbers of energy levels is given by Stepanov & Gribkovskii (89).

If a three-level system must be used to describe the excitation process, Eq. 12 should be replaced by

$$I = Kn \int \frac{\left[\sum_{l=0}^{\infty} a_{2l}P_{2l}(\cos\theta)\right] b_{12}\rho(e^{\lambda_1 \Delta t_L} - e^{\lambda_2 \Delta t_L})(a_{21} + a_{23}) d\Omega}{\lambda_1 - \lambda_2}. \qquad \text{A-6.}$$

If a two-level system is adequate to model the excitation process, but a three-level system is necessary to describe fluorescence, $a_{21} + a_{23}$ should be substituted for a_{21} in Eq. 12.

Literature Cited

1. Zare, R. N., Dagdigian, P. J. 1974. *Science* 185:739-47
2. Kinsey, J. L. 1977. *Ann. Rev. Phys. Chem.* 28:349-72
3. Levy, M. R. 1979. *Progr. React. Kinet.* 10:1-252
4. Clyne, M. A. A., McDermid, I. S. 1982. *Dynamics of the Excited State, Adv. Chem. Phys.*, ed. K. P. Lawley, 50:1-104
5. Bernstein, R. B. 1982. *Chemical Dynamics via Molecular Beam and Laser Techniques.* Oxford: Clarendon. 262 pp.
6. Gottscho, R. A., Miller, T. A. 1984. *Pure Appl. Chem.* 56:189-208
7. Crosley, D. R., ed. 1980. *Laser Probes for Combustion Chemistry*, ACS Symp. Ser. 134. Washington DC: Am. Chem. Soc. 495 pp.
8. Case, D. A., McClelland, G. M., Herschbach, D. R. 1978. *Mol. Phys.* 35:541-73
9. Greene, C. H., Zare, R. N. 1983. *J. Chem. Phys.* 78:6741-53
10. Abragam, A. 1961. *The Principles of Nuclear Magnetism.* Oxford: Clarendon. 599 pp.
11. Slichter, C. P. 1978. *Principles of Magnetic Resonance.* New York: Springer. 397 pp. 2nd ed.
12. Townes, C. H. 1946. *Phys. Rev.* 70:665-71
13. Karplus, R., Schwinger, J. 1948. *Phys. Rev.* 73:1020-26
14. Karplus, R. 1948. *Phys. Rev.* 74:223-24
15. Burak, I., Steinfeld, J. I., Sutton, D. G. 1969. *J. Quant. Spectrosc. Radiat. Transfer* 9:959-80
16. Giuliano, C. R., Hess, L. D. 1967. *IEEE J. Quant. Elec.* QE-3:358-67
17. Ducloy, M. 1973. *Phys. Rev. A* 8:1844-59
18. Ducloy, M. 1974. *Phys. Rev. A* 9:1319-42
19. Ducloy, M. 1975. *J. Phys.* 36:927-41
20. Ducloy, M. 1976. *J. Phys. B* 9:357-81
21. Lehmann, J. C. 1977. *Les Houches Session XXVII, Frontiers in Laser Spectroscopy*, ed. R. Balian, S. Haroche, S. Liberman, pp. 475-527. Amsterdam: North Holland
22. Broyer, M., Gouedard, G., Lehmann, J. C., Vigué, J. 1976. *Adv. At. Mol. Phys.* 12:165-213
23. Decomps, B., Dumont, M., Ducloy, M. 1976. *Topics in Applied Physics, Vol. 2, Laser Spectroscopy of Atoms and Molecules*, ed. H. Walther, pp. 283-347. New York: Springer
24. Lamb, W. E. Jr. 1964. *Phys. Rev.* 134:A1429-50
25. Stenholm, S., Lamb, W. E. Jr. 1969. *Phys. Rev.* 181:618-35
26. Feldman, B. J., Feld, M. S. 1970. *Phys. Rev. A* 1:1375-96
27. Holt, H. K. 1970. *Phys. Rev. A* 2:233-49
28. Sargent, M. III, Scully, M. O., Lamb, W. E. Jr. 1974. *Laser Physics.* Reading, Mass: Addison-Wesley. 432 pp.
29. Allen, L., Eberly, J. H. 1975. *Optical Resonance and Two Level Atoms.* New York: Wiley-Interscience. 233 pp.
30. Letokhov, V. S., Chebotayev, V. P. 1977. *Nonlinear Laser Spectroscopy.* Berlin: Springer-Verlag. 466 pp.
31. Smith, R. A. 1978. *Proc. R. Soc. London Ser. A* 362:1-12
32. Smith, R. A. 1978. *Proc. R. Soc. London Ser. A* 362:13-25
33. Smith, R. A. 1979. *Proc. R. Soc. London Ser. A* 368:163-75
34. Smith, R. A. 1980. *Proc. R. Soc. London Ser. A* 371:319-29
35. Avan, P., Cohen-Tannoudji, C. 1977. *J. Phys. B* 10:155-70
36. Avan, P., Cohen-Tannoudji, C. 1977. *J. Phys. B* 10:171-85
37. Piepmeier, E. H. 1972. *Spectrochim. Acta B* 27:431-43
38. Piepmeier, E. H. 1972. *Spectrochim. Acta B* 27:445-52
39. Omenetto, N., Bennetti, P., Hart, L. P., Winefordner, J. D., Alkemade, C. Th. J. 1973. *Spectrochim. Acta B* 28:289-300
40. Winefordner, J. D., Omenetto, N. 1979. *Analytical Laser Spectroscopy*, ed. N. Omenetto, pp. 167-218. New York: Wiley
41. Winefordner, J. D. 1978. *New Applications of Lasers to Chemistry*, ACS Symp. Ser. 85, ed. G. M. Hieftje, pp. 50-79. Washington DC: Am. Chem. Soc.
42. Eckbreth, A. C., Bonczyk, P. A., Verdieck, J. F. 1977. *Appl. Spectrosc. Rev.* 13:15-164
43. Lucht, R. P., Sweeney, D. W., Lauren-

deau, N. M. 1983. *Combust. Flame* 50:189–205
44. Schofield, K., Steinberg, M. 1981. *Opt. Eng.* 20:501–10
45. Crosley, D. R. 1981. *Opt. Eng.* 20:511–21
46. Baronavski, A. P., McDonald, J. R. 1977. *J. Chem. Phys.* 66:3300–1
47. Baronavski, A. P., McDonald, J. R. 1977. *Appl. Opt.* 16:1897–1901
48. Mailänder, M. 1978. *J. Appl. Phys.* 49:1256–59
49. Bonczyk, P. A., Shirley, J. A. 1979. *Combust. Flame* 34:253–64
50. Takubo, Y., Yano, H., Matsuoka, H., Shimazu, M. 1983. *J. Quant. Spectrosc. Radiat. Transfer* 30:163–68
51. Killinger, D. K., Wang, C. C., Hanabusa, M. 1976. *Phys. Rev. A* 13:2145–52
52a. Lucht, R. P., Sweeney, D. W., Laurendeau, N. M. 1980. See Ref. 7, pp. 145–51
52b. Lucht, R. P., Laurendeau, N. M., Sweeney, D. W. 1982. *Appl. Opt.* 21:3729–35
52c. Lucht, R. P., Sweeney, D. W., Laurendeau, N. M., Drake, M. C., Lapp, M., Pitz, R. W. 1984. *Opt. Lett.* 9:90–92
53. Pasternack, L., Baronavski, A. P., McDonald, J. R. 1978. *J. Chem. Phys.* 69:4830–37
54. Muller, C. H. III, Schofield, K., Steinberg, M. 1978. *Chem. Phys. Lett.* 57:364–68
55. Muller, C. H. III, Schofield, K., Steinberg, M. 1979. *Chem. Phys. Lett.* 61:212
56. Muller, C. H. III, Schofield, K., Steinberg, M. 1980. *J. Chem. Phys.* 72:6620–31
57. Lucht, R. P., Laurendeau, N. M. 1979. *Appl. Opt.* 18:856–61
58. Lucht, R. P., Sweeney, D. W., Laurendeau, N. M. 1980. *Appl. Opt.* 19:3295–3300
59. Berg, J. O., Shackleford, W. L. 1979. *Appl. Opt.* 18:2093–94
60. Kotlar, A. J., Gelb, A., Crosley, D. R. 1980. See Ref. 7, pp. 137–44
61. Doherty, P. M., Crosley, D. R. 1984. *Appl. Opt.* 23:713–21
62. Allison, J., Johnson, M. A., Zare, R. N. 1979. *Faraday Discuss. Chem. Soc.* 67:124–26
63. Smith, G. P., Whitehead, J. C., Zare, R. N. 1977. *J. Chem. Phys.* 67:4912–16
64. Liu, K., Parson, J. M. 1977. *J. Chem. Phys.* 67:1814–28
65. Córdova, J. F., Rettner, C. T., Kinsey, J. L. 1981. *J. Chem. Phys.* 75:2742–48
66. Geraedts, J., Waayer, M., Stolte, S., Reuss, J. 1982. *Faraday Discuss. Chem. Soc.* 73:375–86
67. Guyer, D. R., Hüwel, L., Leone, S. R. 1983. *J. Chem. Phys.* 79:1259–71
68. Mocker, H. W., Collins, R. J. 1965. *Appl. Phys. Lett.* 7:270–73
69. Garmire, E. M., Yariv, A. 1967. *IEEE J. Quant. Elec.* QE-3:222–26
70. Garmire, E. M., Yariv, A. 1967. *IEEE J. Quant. Elec.* QE-3:377
71. Muller, D. F., Rothschild, M., Boyer, K., Rhodes, C. K. 1982. *IEEE J. Quant. Elec.* QE-18:1865–71
72. James, R. B., Smith, D. L. 1982. *IEEE J. Quant. Elec.* QE-18:1841–64
73. McFarlane, R. A., Bennett, W. R. Jr., Lamb, W. E. Jr. 1963. *Appl. Phys. Lett.* 2:189–90
74. Szöke, A., Javan, A. 1963. *Phys. Rev. Lett.* 10:521–24
75. Chebotayev, V. P., Letokhov, V. S. 1975. *Prog. Quant. Elect.* 4:111–206
76. Shimoda, K., ed. 1976. *Topics in Applied Physics, Vol. 13, High Resolution Laser Spectroscopy*. Berlin: Springer. 378 pp.
77. Hänsch, T. W. 1977. *Proc. Int. School Phys., Enrico Fermi: Nonlinear Spectroscopy. Course LXIV*, ed. N. Bloembergen, pp. 17–86. Amsterdam: North Holland
78. Demtröder, W. 1981. *Laser Spectroscopy*. New York: Springer. 694 pp.
79. Levenson, M. D. 1982. *Introduction to Nonlinear Laser Spectroscopy*. New York: Academic. 256 pp.
80. Johnson, M. A., Webster, C. R., Zare, R. N. 1981. *J. Chem. Phys.* 75:5575–77
81. Johnson, M. A., Rostas, J., Zare, R. N. 1982. *Chem. Phys. Lett.* 92:225–31
82. Macomber, J. D. 1968. *IEEE J. Quant. Elec.* QE-4:1–10
83. Brewer, R. G. 1978. *Coherence in Spectroscopy and Modern Physics*, ed. F. T. Arecchi, R. Bonifacio, M. O. Scully, pp. 41–84. New York: Plenum
84. Shoemaker, R. L. 1979. *Ann. Rev. Phys. Chem.* 30:239–70
85. Cohen-Tannoudji, C. 1975. *Atomic Physics*, ed. G. Zu Putlitz, E. W. Weber, A. Winnacker, 4:589–614. New York: Plenum
86. Hertel, I. V., Stoll, W. 1977. *Adv. At. Mol. Phys.* 13:113–228
87. Einstein, A. 1917. *Phys. Z.* 18:121–28
88. Loudon, R. 1983. *The Quantum Theory of Light*. Oxford: Clarendon. 393 pp. 2nd ed.
89. Stepanov, B. I., Gribkovskii, V. P. 1968. *Theory of Luminescence*. London: Iliffe. 497 pp.
90. Hilborn, R. C. 1982. *Am. J. Phys.* 50:982–86
91. Rodrigo, A. B., Measures, R. M. 1973. *IEEE J. Quantum Electron.* QE-9:972–78

92. Daily, J. W. 1978. *Appl. Opt.* 17:225–29
93. Van Calcar, R. A., Van de Ven, M. J. M., Van Uitert, B. K., Biewenga, K. J., Hollander, Tj., Alkemade, C. Th. J. 1979. *J. Quant. Spectrosc. Radiat. Transfer* 21:11–18
94. Van Calcar, R. A., Heuts, M. J. G., Van Uitert, B. K., Meijer, H. A. J., Hollander, Tj., Alkemade, C. Th. J. 1981. *J. Quant. Spectrosc. Radiat. Transfer* 26:495–502
95. Zinsli, P. E. 1978. *J. Phys. E Sci. Instrum.* 11:17–19
96. Sinha, M. P., Caldwell, C. D., Zare, R. N. 1974. *J. Chem. Phys.* 61:491–503
97. Fano, U., Macek, J. H. 1973. *Rev. Mod. Phys.* 45:553–73
98. Greene, C. H., Zare, R. N. 1982. *Ann. Rev. Phys. Chem.* 33:119–50
99. Cruse, H. W., Dagdigian, P. J., Zare, R. N. 1973. *Faraday Discuss. Chem. Soc.* 55:277–92
100. Bergmann, K., Engelhardt, R., Hefter, U., Witt, J. 1979. *J. Phys. E* 12:507–14
101. Kinsey, J. L. 1977. *J. Chem. Phys.* 66:2560–65
102. Murphy, E. J., Brophy, J. H., Arnold, G. S., Dimpfl, W. L., Kinsey, J. L. 1979. *J. Chem. Phys.* 70:5910–11
103. Murphy, E. J., Brophy, J. H., Kinsey, J. L. 1981. *J. Chem. Phys.* 74:331–36
104. Bergmann, K., Hefter, U., Hering, P. 1978. *Chem. Phys.* 32:329–48
105. Bergmann, K., Hefter, U., Witt, J. 1980. *J. Chem. Phys.* 72:4777–90

TRANSITION METAL MOLECULES

William Weltner, Jr. and Richard J. Van Zee

Department of Chemistry, University of Florida, Gainesville, Florida 32611

INTRODUCTION

What is our present experimental knowledge of the electronic, magnetic, and structural properties of transition-metal molecules and clusters? How do these properties compare with theoretically-computed ones? These are the questions that we try to answer here.[1] Although the distinction between a molecule and a cluster remains undefined, in this context the word "cluster" hardly seems appropriate since all of the presently characterized species, with only one exception, contain less than six metal atoms.[2]

The enthusiastic and welcome interest of theoreticians has produced a literature that has more than compensated for the dearth of experimental data on transition-metal molecules. They have not been daunted by the difficulties involved in making ab initio all-electron calculations on these heavy-atomed molecules, and even on clusters as large as Cu_{13}. The numerous semiempirical and ab initio methods continually being introduced by quantum chemists are now customarily abbreviated to initials which are proliferating beyond perhaps even the theorists' memories. A glossary of such abbreviations used throughout this chapter is given in the footnote below.[3]

---

[1] This literature survey extended up to January 1, 1984.

[2] "Metal cluster" as generally used by the inorganic chemist implies a core of metal atoms stabilized by attached ligands, not the "naked" metal clusters referred to here. The former is well covered in a recent book on metal-metal bonding by Cotton & Walton (1).

[3] Abbreviations used: AREP, averaged relativistic core potential; CAS-SCF, complete active space–self-consistent field; CI, configuration interaction; CNDO, complete neglect of differential overlap; CVP, core-valence polarization (pseudopotential); DIM, diatomics-in-molecules; DVM, discrete variational method; EH, extended Hückel; ECP, effective core

The exploration of this interesting intermediate state of matter lying between an atom and bulk metal is approachable from either direction. Some recent reviews covering both approaches are the following: a *Faraday Symposium* of the Chemical Society in 1980 (2), *Growth and Properties of Metal Clusters* (Bourdon, editor) (3), Perenboom et al in *Physics Reports* (4), Davis & Klabunde in *Chemical Reviews* (5), *Metal Bonding and Interactions in High Temperature Systems* (Gole & Stwalley, editors) (6), Baetzold & Hamilton in *Progress in Solid State Chemistry* (7), Ozin & Mitchell in *Angewandte Chemie* (7a), and Sattler in *Festkörperprobleme* (7b).

HOMONUCLEAR DIATOMICS

Table 1 summarizes succinctly the experimental and theoretical information available on these molecules. For completeness, and contrary to the theme of this review, a few molecules are included for which the information available is derived solely from theory.

From First-Row Elements

Sc_2: Other than the mass spectrometry of Verhaegen et al, yielding $D_e = 1.1 \pm 0.2$ eV (8), there are no gas-phase experimental data on Sc_2. Apparently the molecule was first detected spectroscopically by Busby et al (9), in an argon matrix where three electronic transitions (no vibrational structure) were observed. Its vibrational frequency in the ground state has been determined by Moskovits et al (10) to be 238.9 cm^{-1} from resonance Raman spectra in solid argon. Knight et al (11), measured the ESR spectrum of Sc_2 at 4 K in neon and argon matrices. The ^{45}Sc nucleus has spin $I = 7/2$ so that the observed hyperfine structure identified the spectrum as due to discandium. Analysis of two fine-structure transitions established the ground state as $^5\Sigma$ with a zero-field splitting $|D| = 0.1112$ cm^{-1}. The correct electron configuration is then essentially $(4s\sigma_g)^2(4s\sigma_u)^1(3d\sigma_g)^1(3d\pi_u)^2$ and the hyperfine splittings are approximately

---

potential (pseudopotential); GMO, generalized molecular orbital; GVB-vdw, generalized valence bond–van der Waals; HFS, Hartree-Fock-Slater; LCGTO, linear combination of Gaussian-type orbitals; LD, local density; LSD, local spin density; MC-SCF, multiconfigurational self-consistent field; MEH, modified extended Hückel; MINDO, modified intermediate neglect of differential overlap; MP, model potential; MRD, multi-reference double excitation; NRMP, nonrelativistic model potential; POL-CI, polarized configuration interaction; QRMP, quasirelativistic model potential; RECP, relativistic effective core potential; RHF, restricted Hartree-Fock; SC-CMS, self-consistent cellular multiple scattering; SCF-Xα-SW, self-consistent field-Xα-scattered wave; SC-LSD, self-consistent local spin density; SH, simple Hückel; Xα, a local density method, α is a parameter.

in accord with that spin distribution. A valid chemical bond, and not van der Waals bonding, is therefore implied, in agreement with the mass spectrometric D_e.

Although having the fewest electrons of any transition-metal diatomic, Sc_2 has been somewhat troublesome theoretically. EH theory incorrectly predicted a $^5\Delta_g$ or $^1\Sigma_g$ ground state (9, 12) and recent ab initio calculations originally indicated a long interatomic distance and van der Waals bonding (13, 14). The error in the latter calculations arose because only bound states arising from the $^2D + ^2D$ atomic limits were considered. When the $^2D + ^4F$ atomic limits were included a $^5\Sigma_u^-$ state was found to be strongly bound with respect to the $^2D + ^2D$ limit (in agreement with the ESR results) with three one-electron $3d$ bonds (15).

Ti_2: The D_e determined by Kant & Lin (19) of about 1.3 eV suggests a single bonded molecule, but its vibrational frequency of 407.9 cm^{-1} obtained from resonance Raman spectra in argon matrices (20, 21) indicates a stronger bond. Theory has yielded a variety of ground states (see Table 1). Among them the most recent ab initio calculation by Walch & Bauschlicher (22) tentatively suggests that the ground state is $^1\Sigma_g^+$ with a triple $3d$ bond; however, a $^7\Sigma_u^+$ state is apparently also possible.

V_2: Mass selective resonant two-photon ionization spectroscopy of the expansion-cooled metal beam (see Cr_2 and Mo_2) led to the observation of a strong band near 700 nm (27). When rotationally analyzed, it proved to be a $^3\Pi_u \leftarrow X^3\Sigma_g^-$ transition with $r_e'' = 1.76$ Å and $\omega_e'' = 535$ cm^{-1}. Resonance Raman of V_2 trapped in matrices had previously yielded $\omega_e'' = 537.5$ cm^{-1} (20, 21). Unusual case (a) coupling and large second-order spin-orbit splitting (zero-field splitting, $2\lambda = D = 75$ cm^{-1}) in the $^3\Sigma_g^-$ state implies an electronic configuration containing a half-filled $3d\pi$ or $3d\delta$ orbital and a very low-lying $^1\Sigma_g^+$ excited state. In contrast to the mass spectrometry value of 2.5 eV (19, 23), Smalley, Merer et al estimate that the dissociation energy of V_2 is closer to 1.85 eV from apparent predissociation of higher vibrational levels of the $^3\Pi_u$ state. The short bond distance and high vibrational frequency show that the molecule is bonded strongly by $3d$ electrons. The bond length is only two-thirds of the nearest-neighbor distance in the bulk metal, a larger contraction than any other element except Cr_2.

The ground state and short bond distance were predicted by the ab initio calculations of Walch et al (22a). They show that the dominant configuration is $s\sigma_g^2 d\sigma_g^2 d\pi_u^4 d\delta^2$ involving $d\sigma$ and $d\pi$ two-electron bonding and $d\delta$ one-electron bonds.

Cr_2 (see also Mo_2): Gas phase spectroscopy of photolyzed $Cr(CO)_6$ by Efremov et al (28, 29) first suggested that Cr_2 had a very short bond distance

Table 1 Summary of experimental data and theoretical studies of homonuclear diatomic transition metal molecules

	r_e (Å)	ω_e (cm^{-1})	D_e (eV)	Ground electronic state
Sc_2				
Experiment				
Bulk metal	3.20		3.93/at.	
Gas				
(8) mass spec.			1.1(2)	
Matrix				
(9) UV-vis. absorption				
(10) reson. Raman		238.9		
(11) ESR, zfs $|D| = 0.1112(2)$ cm^{-1}				$^5\Sigma$
Theory (approx. chronological)				
Semiempirical				
EH (12)	2.20	250	1.25	$^5\Delta_g$
EH (9) compared with electronic transitions				
Nonempirical				
LSD (14)	3.25	235	1.00	$^5\Sigma_u^-$
MC-SCF/CI (17); see, however (14)		210	1.12	$^1\Sigma$
RHF (18)	3.05		0.2	$^1\Sigma_g^+$
MC-SCF/ECP (13)	5		0.06	
CAS-SCF/CI (14)			0.44	$^5\Sigma_u^-$
CAS-SCF/CI (15)	2.79	184		$^1\Sigma_g^+$
				$^3\Sigma_g^-$
				$^{1,3}\Sigma$
Ti_2				
Experiment				
Bulk metal	2.915		4.86/at.	
Gas				
(19, 23) mass spec.			1.3(1)	

Matrix				
(9) UV-vis. absorption				
(20, 21) reson. Raman				
Theory				
Semiempirical				
EH (12)	2.30			$^1\Sigma_g^+$
MEH (22)	3.0	407.9		$^3\Delta_g$
EH (9)		250	1.88 0.72	$^3\Sigma$
Nonempirical				
LSD (16)	2.52	220	2.30	$^1\Sigma_g^+$ / $^7\Sigma_u^+$
RHF (18)	1.87	580		$^1\Sigma_g^+$
CAS-SCF/CI (22)				

V_2

Experiment				
Bulk metal	2.627		5.30/at.	
Gas				
(19, 23) mass spec.			2.5(2)	
(27) superson. beam	1.76	$\Delta G_{1/2} = 529.5$ $\lambda = 37.5\ \text{cm}^{-1},\ IP < 6.4\ 3V$	1.85	$^3\Sigma_g^-$
Matrix				
(25) UV-vis. absorption		537.5		
(20, 21) reson. Raman		$\Delta G_{1/2} = 529.0$		
Theory				
Semiempirical				
EH (12)	2.10	300	3.05	$^1\Sigma_g^+$
EH (25, 26)	1.9		0.99	
Nonempirical				
LSD (16)	2.65	230	100	$^9\Sigma_u^-$
RHF (18)	1.96	420		
CAS-SCF + 4f (22)	1.77	593.6	0.33	$^3\Sigma_g^-$

Table 1 (*continued*)

	r_e (Å)	ω_e (cm^{-1})	D_e (eV)	Ground electronic state
Cr$_2$				
Experiment				
Bulk metal	2.493		4.10/at.	
Gas				
(30) mass spec.			1.6(3)	
(29) opt. spec.	1.6847(8)			
(33, 34) superson. beam	1.6788	$\Delta G_{1/2} = 452.34(4)$		$^1\Sigma_g^+$
(32) fluoresc.	1.6858	~470		
Matrix		$\Delta G_{1/2} = 452.34(2)$		$^1\Sigma_g^+$
(35) UV-vis. absorption				
(35a) UV-vis. absorpt. and emission		$\Delta G_{1/2} = 396$		
(31, see also 32) reson. Raman				
Theory (approx. chronological)				
Semiempirical				
EH (12)	1.90		2.6	$^3\Sigma_g^+$
MEH (24)	2.5		1.0	assumed low spin
EH (26, 36)	1.8, 1.7		2.65, 1.6	assumed closed shell
Nonempirical				
SCF-Xα-SW (36)	(1.90)	assign. electron. transitions		
(37)	(1.685)	bonding examined		
LSD (16)	3.66	55	0.20	$^{13}\Sigma_g^+$
MC-SCF (17)	1.9		1.48	$^1\Sigma_g^+$ ($^1\Sigma_g^+$)
RHF (18)	1.56	750		

Method				
MC-SCF/GVB-vdw (38)	3.06	110		$^1\Sigma_g^+$
MC-SCF (39)	3.14	92	0.35	($^1\Sigma_g^+$)
			$J = 90$ cm^{-1}	
LCGTO-Xα (40)	2.75	110	0.14	$^1\Sigma_g^+$
SC-LSD (41)	1.70		1.0	
GMO-CI (43; see also 22a)	1.73	396	1.80	
LSD + ECP (42)	1.7	470	2.8	($^1\Sigma_g^+$)
SCF + f functions (44)				$^1\Sigma_g^+$
CAS-SCF + 4f (22a)				

Mn$_2$

Experiment

Bulk metal	2.582			
Gas				
(46) mass spec.			2.98/at.	
Matrix				
(49) absorpt. at 650 nm, $\omega' = 111$ cm^{-1}			0.3(3)	
(50) no absorpt. assignable to Mn$_2$				
(47) ESR	3.4			
(48) opt. + MCD spec., bands at 347, 332 nm			$J = -9(3)$ cm^{-1}	$^1\Sigma$
(10) reson. Raman		124.6	$J = -10.3(6)$ cm^{-1}	

Theory

Semiempirical				
EH (12)	1.90		1.22	
Nonempirical				
SC-HF (45)	2.88		0.79	$^1\Sigma_g^+$
			($J = -4$ cm^{-1})	
LSD (16)	2.7			$^{11}\Pi_u$, $^{11}\Sigma_u^+$
RHF (18)	1.52	680	1.25	$^1\Sigma_g^+$

Table 1 (*continued*)

	r_e (Å)	ω_e (cm^{-1})	D_e (eV)	Ground electronic state
Fe$_2$				
Experiment				
Bulk metal α-Fe	2.478		4.29/at.	
Gas				
(51, 52) mass spec.			0.8(2)	$IP = 5.9(2)$ eV
(53) superson. beam				$IP = 5.6$-6.4 eV
Matrix				
(49) vis. absorption		$\omega' = 194, 218$		
(55, 56, 61) Mössbauer				
(54, 21) reson. Raman		300.3		
(59) argon, EXAFS	1.87(13)			
(60) neon, EXAFS	2.02(2)			
Theory (approx. chronological)				
Semiempirical				
EH (12)	1.9		3.10	$^3\Sigma_g^+$
EH (63)	3.99	calc. of hyperfine		
MEH (24)	2.4		0.66	
MINDO/SR (64)	2.0		1.31	$^1\Sigma$
Nonempirical				
LSD (16)	2.1	390	3.45	$^7\Delta_u$
RHF (18)	1.58	660		$^1\Sigma_g^+$
DVM-Xα (62)	(1.87)	calc. of hyperfine		
all-electron HF-CI (52)	2.40	204		$^7\Delta_u$
SCF-Xα-SW (58)	2.1	calc. of hyperfine		$S = 3$

TRANSITION METAL MOLECULES 299

Co_2			
Experiment			
Bulk metal	2.499		
Gas (65, 66) mass spec.		4.39/at.	
		1.0(3)	
Theory			
Semiempirical			
EH (12)	2.30	370	
Nonempirical			
LSD (16)	2.07	360	$^1\Sigma_g^+$
RHF (18)	2.64	200	$^1\Sigma_g^+$ $^5\Delta_g$
all electron HF-CI (66)	~2.4	~240	$^5\Sigma_g^+$
Ni_2			
Experiment			
Bulk metal (fcc)	2.487		
Surf. cluster (77)	2.24		
Gas			
(69, 68) mass spec.		4.44/at.	
(72a) superson. beam	2.200(7)	2.0(2)	
		2.07(1)	
Matrix			
(54, 70) UV-vis. absorpt.			$^1\Gamma_g$ or $^3\Gamma_u$
(72) fluoresc.		381	
Theory (approx. chronological)			
Semiempirical			
EH (12)	2.21	370	
EH (78)	2.5		π_g^2 config.
CNDO (78)	2.75		
MEH (24, 79)	2.21		
EH, Xα (70, 71)		395	π_g^2 config.
	compared with electronic trans.		

Table 1 (*continued*)

	r_e (Å)	ω_e (cm^{-1})	D_e (eV)	Ground electronic state
Nonempirical				
SCF-Xα-SW (80, see also 75)	2.49			$^1\Sigma_g^+$ $^3\Sigma_u^+$ (?)
MC-SCF/ECP (75)	2.04	344	2.9	
GVB/POLCI/ECP (76)	2.18	320	2.70	$^3\Sigma_g^-$ (π-π) $^1\Sigma_g^+$
LSD (16)	2.20	289	1.42	
all electron SCF-CI (74)	2.26	(190)	1.89	$^3\Sigma_u^+$
GVB-CI/ECP (68)	2.33	211	1.43	$^1\Sigma_g^+$ $^3\Sigma_u^+$
MC-SCF-CI/ECP (73)	2.28	240	3.42	
RHF (18)			(2.49) comparison of valence-level ioniz.	
SCF/ECP (81)				

Cu$_2$

Experimental

Bulk metal (fcc)	2.551		3.50/at.	
Gas				
(83a–d) mass spec.			2.1(1)	
(82a–c) opt. spec.	2.220	265		$^1\Sigma_g^+$
(88, 90) superson. beam	2.220	266.1		$IP = 7.894(15)$
(89, 91) superson. beam	(five new excited states)			
Matrix				
(180) UV-vis. absorpt.				
(90, 92, 92a) emission (see text)				

Theory (approx. chronological)			
Semiempirical			
EH (93)	2.19	200	2.19	IP = 7.53 eV
EH (94)	2.1	390	4.14	IP = 9.21 eV
CNDO (94)	3.0	510	2.17	IP = 7.21 eV
MEH (95)	2.26	325	2.05	
VB (96)	1.59		1.76	
2 val. electr./SECP (87a)	2.26	262	1.95	IP = 8.09 eV
Nonempirical				
all electron SCF (min. CI) (97)	2.24		1.7	
all electron SCF (98)	2.343	243	0.84	IP = 6 eV
valence elect. SCF-CI (99)				
LSD (16)	2.27	280	2.30	
SCF-Xα-SW (100)	2.17	272(5)	2.86(4)	
all electron SCF (101)	2.41	219	1.90	
RHF (18)	2.32	210		IP = 6.5 eV
ECP/SCF-CI (84)	2.25	265		
LCAO-Xα (102)	2.55		2.01	
DVM-Xα (103)	2.22			IP = 9.49 eV
ECP (min.-bas.-adapt.)/SCF-CI (86)				
all electron SCF (104)	2.41	235	0.681	
MC-SCF-CI (85)	2.32			
Bond order (105)				
2 val. electr./ECP + CVP (87)	2.14	269	1.54	
CAS-SCF/SDCI vs. POLCI (106)	2.35	227	1.99	IP = 6.23 eV
SCF-MRD-CI (109)	2.22	265	1.75	
all electron HF-CI (110)	2.40	200	2.07	
SCF-Xα (numer. basis) (111)	2.22	286	2.10	
Relativistic effects (106–108, but see also 108a)	−0.05	+15	+0.06	

Table 1 (continued)

	r_e (Å)	ω_e (cm^{-1})	D_e (eV)	Ground electronic state
Nb$_2$				
Experiment				
Bulk metal	2.853		7.47/at.	
matrix (26) UV-vis., absorption				
Theory				
Dissociation energy (112, 113)			5.2(1), 4.8(4)	
Semiempirical				
EH (26)	2.20		1.52	
EH (114)	2.08		6.44	
Nonempirical				
SCF-Xα-SW (115)				$^{1,3}\Sigma_g$
Mo$_2$				
Experiment				
Bulk metal	2.720		6.81/at.	
Gas				
(29) opt. spec.	1.929	477.1	4.1(6)	
(117, 113) mass spec.			4.2(2), 3.8(4)	
(121) superson. beam	1.938(9)			
Matrix				
(26, 36, 118, 119) UV-vis. spec.	(1.9)	476(5)		
(120) fluoresc. spec.				
Theory (approx. chronological)				
Semiempirical				
EH (26, 36)	2.1		3.4	

TRANSITION METAL MOLECULES

Method (Ref)	r_e	ω_e	D_e	State/Notes
Nonempirical				
SCF-Xα-SW (116)	2.3			
SCF-Xα-SW (36)	(2.1)			
SCF-Xα-SW (37)	(1.929)			
MC-SCF-CI (17)	2.1			
MC-SCF-CI (122)	1.97	475	(see also 39)	
MC-SCF-CI (123, 39)	2.01	388	0.86	
SC-CMS (124)	2.15			
GVB-vdw (38)	1.97	455	1.41	
LSD (41)	1.95(5)	520	4.35	
LSD (41)	1.95(5)	520	4.35	
LSD+ECP (42)	2.1	360	4.2	
SCF + f functions (44)				
Ru_2				
Experiment				
Bulk metal	2.645		6.615/at.	
Theory				
Dissoc. energy (113, 126, 127)	2.54–2.59		3.2, 3.39	
all electron SCF-CI (125)		1.16	0.64	$^7\Delta_u$
Pd_2				
Experiment				
Bulk metal (fcc)	2.745		3.936/at.	
Theory				
Dissoc. energy (23, 113, 127)			1.1(2)	
EH (131)				
RECP-MCSCF (128)	2.81	216	0.25–0.76	$^{1,3}\Gamma_{g,u}(\delta\delta)$
Pt_2				
Experiment				
Bulk metal	2.769		5.852/at.	
Gas				
(129) mass spec.			3.71(16)	

Table 1 *(continued)*

	r_e (Å)	ω_e (cm^{-1})	D_e (eV)	Ground electronic state
Matrix				
(130) absorpt. spec.		$\Delta G'_{1/2} = 216$		
Theory				
RECP-MCSCF (128)	2.58	267	0.93	$^{1,3}\Gamma_{g,u}\ (\delta\delta)$
Ag$_2$				
Experiment				
Bulk metal	2.884		2.96/at.	
Gas		192.4		$^1\Sigma_g^+$
(132a,b) opt. spec.		1.65(6)		
(83a–d) mass spec.	2.469(4) estimated			
(133) opt. spec.	2.482			
(133a, 144a) opt. spec.				
Matrix				
(134, 135) absorpt. spec.		194 ± 0.5		
(136) Raman				
(137) MCD				
(144) opt. spec., ice and wax matrices				
Theory (approx. chronological)				
Semiempirical				
EH (94, 138)	2.5, 2.1	410	1.82, 174	
CNDO (94, 138)	3.0	500	3.02	$IP = 7.23$ eV

Method			
VB-hard core model (96)	1.64		
EH (141)	2.6		
SH (140)	3.76		
van der Waals (139)		0.84 (1.74)	$IP = 7.68$ eV
Nonempirical			
SCF-Xα-SW (100)	2.84	1.56	187(2)
RECP/SCF-CI (142)	2.62	1.12	242
RECP/SCF-CI (143)	2.68	0.94	131
all electron HF-CI (110)	2.76	1.05	134
all electron MCSCF/CI (144a)	2.72	1.13	
NRMP (144b)	2.80		130.1
QRMP (144b)	2.73		145.4
relativistic effects (110, 142)	−0.15	+0.15	+20

Au$_2$

Experiment
Bulk metal	2.878			
Gas (149a,b) opt. spec. (83a–d, 145, 146) mass spec.	2.47		191	2.3(1) $^1\Sigma_g^+$

Theory
Semiempirical				
VB-hard core model (96)	1.30	2.02		$^1\Sigma_g^+$
EH (94)	2.1	2.0	464	$IP = 10.3$ eV
CNDO (94)	3.25	2.5	500	$IP = 7.7$ eV
Nonempirical				
AREP/MCSCF-CI (147a,b)	2.37	2.27	165	
RECP/SCF-CI (142)	2.60	1.34	166	

(1.685 Å) and a $^1\Sigma_g$ ground state. This has now been corroborated by other experiments and by theory. The effective bond order has been termed sextuple, and a relatively strong bond is supported by its dissociation energy, $D_e = 1.56 \pm 0.3$ eV (30, 33), and high vibrational frequency, 470 cm^{-1} (31, 32), compared to Cu_2, and by the dominance of the $s\sigma_g^2 d\sigma_g^2 d\pi_u^4 d\delta_g^4$ electronic configuration in the theoretical description. Recently Cr_2 spectra have been measured in the gas phase, produced by either pulsed laser vaporization of solid chromium using resonant two-photon ionization of very cold molecules ($T_{rot} \cong 4$ K) prepared in a supersonic molecular beam (33, 34), or by fluorescence excitation of thermally quenched molecules (~ 80 K) in flowing helium gas (32). The $A^1\Sigma_u^+ \leftrightarrow X^1\Sigma_g^+$ transition was rotationally analyzed in all cases to provide accurate values of the bond distance and vibrational frequencies. Two bands attributed to Cr_2 have also been observed in argon matrices at 455 nm (sharp) and 260 nm (broad) (35). Also in argon matrices, laser resonance Raman measurements yielded a vibrational frequency for Cr_2 (31) shifted from the gas-phase value [$\Delta G_{1/2} = 396$ versus 452.3 cm^{-1} (32)], which now makes the matrix value suspect. The bond lengths in the A and X states are nearly the same, only 0.003 Å longer in the A state. This is relevant to the bond order in the ground state, as discussed in more detail for Mo_2. In the most recent matrix optical studies, Pellin & Gruen (35a) have apparently detected transitions to the excited $B^1\Pi_u$ (known for Mo_2) and $^3\Sigma$ states.

There has been considerable controversy among ab initio theorists about the bonding in Cr_2. The early empirical studies (12, 24, 26, 36) and ab initio calculations (36, 37, 17, 18) yielded a short interatomic distance and strong bond, but Goodgame & Goddard (38) made a provocative MC-SCF/GVB calculation including correlations for van der Waals interactions, favoring a bond with $r_e = 3.06$ Å and a small D_e. The bonding would involve antiferromagnetic coupling of two high-spin ($S = 3$) Cr atoms [see Mn_2] with $J \cong 90$ cm^{-1}. This has spurred theorists on to further investigations with the general conclusion that correlation effects are very large in Cr_2 and an extended basis set is required to do an adequate job. The necessity for the inclusion of $4f$ functions has been recently emphasized (44, 22a). Support has also been given (44) to the experimental proposal of Smalley et al (see Mo_2) that the valence s electrons are nonbonding.

The application of LSD theory to Cr_2 (and Mo_2) has recently been defended vigorously and effectively by Delley et al (41) and by Bernholc & Holzwarth (42). In comparing their SC-LSD calculations with the GVB treatment (38), the former authors found the equilibrium bond distance to be 1.70 Å and $D_e = 1.80$ eV, in good agreement with experiment. Also important was the finding of no double-well potential function in the ground state as suggested by the GVB treatment. Bernholc & Holzwarth reported LD plus pseudopotential calculations with results in general agreement with Delley et al.

Although essential agreement between theory and experiment has now been achieved in several calculations, a need still remains for establishing the shape of the potential energy curve at bond distances longer than r_e (44). The consensus of theory does agree that a quintuple bond is an appropriate description for Cr_2 (and Mo_2).

Mn_2: Applying approximate ab initio all-electron HF theory, Nesbet (45) in 1964 predicted that Mn_2 would have a $^1\Sigma_g^+$ lowest state but with the two atoms weakly exchange-coupled antiferromagnetically ($J = -4$ cm^{-1}) at the bond distance of 2.88 Å. The weak bond was subsequently supported by the mass spectrometrically measured dissociation energy of 0.3(3) eV (46).

Nesbet's predictions were essentially confirmed when it was demonstrated by ESR measurements of Mn_2 trapped in rare-gas matrices (ideal media for this experiment) that Mn_2 could be observed at ~ 10 to 70 K in successive states with total spin $S = 0, 1, 2, 3$ separated by the Landé interval spacing $J, 2J, 3J$ with the isotropic constant $J = -9 \pm 3$ cm^{-1} (47). The observed anisotropic exchange, for this small J, can be assumed to be due solely to magnetic dipolar interaction, varying as r^{-3}. Thus r was derived to be 3.4 Å, the long distance expected of a van der Waals bond and considerably longer than the nearest neighbor distances in manganese metal.

More recent optical and MCD work by Vala and co-workers on Mn species in rare gas matrices has confirmed the ESR findings and places J at -10.3 ± 0.6 cm^{-1} (48). This value was obtained from the temperature variation of a transition at 347 nm assigned to excitation from the $S = 2$ level. Also in argon matrices, resonance Raman spectra (10) yielded three progressions, all with ω_e values near 124 cm^{-1}, attributed to excitation of different spin states.

Fe_2: The ground electronic state of Fe_2 is not known, but it is suspected to be $^7\Sigma_g$, $^7\Delta_u$, or possibly $^7\Pi_u$. The only gas-phase data on this molecule are mass spectrometric measurements of dissociation energy (51, 52) and the determination of its ionization potential (51), recently also in supersonic beam studies (53). In rare-gas matrices there have been reported measurements of electronic absorption bands (49), resonance Raman spectra (54, 21), Mössbauer spectra (55–58), extended-x-ray-absorption-fine-structure (EXAFS) (59, 60), and x-ray absorption near-edge structure (XANES) (60).

The most recent mass spectrometry (52), and revision of older measurements (51), yields the dissociation energy of Fe_2 as a 0.8(2) eV, indicating a rather weak bond. Resonance Raman spectra in solid argon at 11 K gives the ground state vibrational frequency as 300.3 cm^{-1} (54, 21). EXAFS of Fe in argon matrices gave $r_e = 1.87 \pm 0.13$ Å (59) and in neon 2.02 ± 0.02 Å (60); even the latter distance seems short considering that the bond is weak and the interatomic distance in bulk α-Fe is 2.48 Å.

Iron is, of course, the classic Mössbauer element, and a virtue of matrix isolation is that it allows Mössbauer spectroscopy to be applied to such species as Fe_2, as was first demonstrated by Barrett et al (55, 56). Signals due to Fe_2 are observed providing characteristic values of the isomer shift (IS), quadrupole splitting (QS), and hyperfine field (H_F), at the ^{57}Fe nucleus. Two detailed attempts have now been made to use these data to learn about the electronic structure of Fe_2 [and similarly for FeM diatomics, where M = Cr, Mn, Co, Ni, Cu (see below)] (58, 62). All electron DVM-Xα (62) and SCF-Xα-SW (58) methods have been applied to calculate the electric field gradients (EFG), derived from QS, the electronic charge density at the nucleus $\rho(0)$ (obtained from IS), and H_F. It is clear from those attempts that an accurate calculation of the Mössbauer parameters is difficult and not a straightforward means for determining the ground state properties of Fe_2.

An all-electron HF-CI calculation (52) identified 112 lowest electronic states of Fe_2, which all lie within an energy range of 0.54 eV. The lowest among these are $^7\Sigma_g^-$ and $^7\Delta_u$ (separated by 0.04 eV, spin-orbit interaction neglected) with the suggested ground state $^7\Delta_u$ having the configuration $(3d\sigma_g)^{1.57}(3d\pi_u)^{3.06}(3d\delta_g)^{2.53}(3d\delta_u)^{2.47}(3d\pi_g)^{2.89}(4s\sigma_g)^{2.00}$. The bond is essentially single due to the $4s\sigma_g$ MO with the $3d$ orbitals well localized on the atoms.

Co_2: The only experimental datum on this molecule is the dissociation energy determined mass spectrometrically (65), which has been corrected by reconsideration of the electronic partition function to 1.0(3) eV (66), indicating a bond of about the strength of Fe_2. An attempt to observe Co_2 in rare-gas matrices via X-band ESR at 4 K yielded no signals (67).

The most recent ab initio calculation is the all-electron HF-CI work of Shim & Gingerich (66) locating the 84 low-lying (within 0.42 eV) electronic states resulting from the interaction between Co atoms in their $^4F(3d)^8(4s)^1$ terms at 2.50 Å (the bulk metal distance). The lowest states among these were $^5\Sigma_g^+$, $^5\Sigma_u^-$, ^1I_g, $^3\Sigma_g^-$, $^3\Sigma_u^+$, ^3I_u, $^1\Sigma_g^+$, $^1\Sigma_u^-$, ^5I_g lying within a range of 0.036 eV. Among these the $^5\Sigma_g^+$ was lowest (spin-orbit coupling not included) but unbound with respect to the ground-state atoms $^4F(3d)^7(4s)^2$. After CI and bond distance optimization, and with estimates of correlation errors from similar calculations on Cu_2, the parameters given in Table 1 were obtained. From these calculations the chemical bond in

Co_2 appears to be due principally to the $4s\sigma_g$ MO as found by these authors for Ni_2, Fe_2, and Cu_2.

Ni_2: The meager experimental data for Ni_2 (see, however, below) have been reviewed and revised by Noell et al (68). From mass spectromentry the dissociation energy was found to be 2.36 eV (69), but this value has been recalculated to be 2.02 eV by a reevaluation of the electronic partition function based on theory and fluorescence data (68). Optical spectra have only been observed for matrix-isolated molecules (54, 70–72) and the transitions rationalized using EH and Xα theory (70, see also 71). A value of $\omega_e'' = 381$ cm^{-1} for $^{58}Ni_2$ in solid argon has been obtained from laser excited fluorescence of a 0-0 band at 22,250 cm^{-1} (72). Matrix-isolated Ni_2 was not detected by X-band ESR at 4 K (67).

The state of theoretical calculations on Ni_2 as of 1980 has been summarized by Basch et al (73). The approximate ground state configuration is $(3d\delta_g)^3(3d\delta_u)^3(3d\pi_u)^4(3d\pi_g)^4(3d\sigma_g)^4(3d\sigma_u)^2(4s\sigma_g)^2$. A single s-s bond, opposed somewhat by the antibonding σ_u MO, with two $d\delta$-$d\delta$ holes is the general way of referring to its electronic properties. In the most accurate calculations there is general agreement that there are six δ-hole states [$^3\Sigma_g^-$, $^3\Sigma_u^+$, $^1\Sigma_g^+$, $^1\Sigma_u^-$, $^1\Gamma_g$, $^3\Gamma_u$] which are so close together in energy (<0.2 kcal/mole) that prediction of the lowest is not presently feasible, particularly since spin-orbit effects have usually been neglected (68, 73–76). Upton & Goddard (76) do make an approximate calculation of spin-orbit effects on these lowest states but still conclude that the energy differences are too small for reliable ordering. Then the overall picture of Ni_2 is bonding by a single $4s\sigma_g$ orbital but complicated by exchange coupling between the d orbitals split by the axial symmetry such that there are six almost degenerate low-lying states.

[After the above was written there was communicated to the authors the unpublished work of Morse et al (72a) on the electronic spectrum of jet-cooled Ni_2. A complex spectrum between 600–900 nm was observed (not in agreement with the matrix spectra), indicating a large number of low-lying electronic states. Among the theoretically predicted ground states only the $^1\Gamma_g$ and $^3\Gamma_u$ are consistent with the observations. Rotational analysis yielded $r_e'' = 2.20$ Å; this long bond length is close to that predicted and indicates a single s-s bond. $D_e = 2.07(1)$ eV is proposed from observation of an onset of predissociation.]

Cu_2: Cu is an alkali-like metal atom ($^2S_{1/2}$, $3d^{10}4s^1$) and forms a closed-shell diatomic molecule with a single s-s bond and $^1\Sigma_g^+$ ground state. Its properties are well known from optical spectroscopy (82a–c) and mass spectrometry (83a–d), so that Cu_2 is used as a benchmark in attempting to

understand the electronic properties of transition-metal diatomics. Obviously the molecule is not as simple as the above description implies; although not directly involved in the bonding, the d electrons do not act as a "frozen core."

Following the early calculations using semiempirical methods (EH, CNDO), there has been a steady ab initio effort to understand the essential elements of the bonding. The best calculations are those of Pelissier (84), utilizing ECP/SCF-CI, and Bauschlicher et al (85), using full valence MC-SCF/34 electron CI + 4f polarization. Theory indicates that the d electron correlation (mainly intra-atomic) is the most significant factor preventing the high accuracy obtainable in the calculations in lighter molecules. Pseudopotential methods are recognized as a generally useful approach where in an effective core potential (ECP) replaces all the electrons except the 22 4s + 3d electrons, or preferably all except the two 4s electrons. Recent probes in the latter direction have been made by Jeung et al (86, 87) and Stoll et al (87a). The latter authors have used semiempirical pseudopotentials (SECP in Table 1) adjusted to experimental energies of singly ionized cores with the inclusion of core polarization, core-overlap effects, and valence correlation. The results are encouraging.

Relativistic effects must be included in the calculations to obtain the correct bond distance, etc, even for Cu_2 (86, 107, 108, 108a). The corrections of Martin (108) are given in Table 1, but they may be overestimated (108a).

We would be amiss here not to mention recent gratifying progress on the experimental side. A new visible system of Cu_2 (called B'-X or C-X) (88–90) and five ultraviolet systems (91) have been observed in laser excitation spectra of supersonically expanded copper vapor. Temperatures as low as $T_{transl} = 1$ K, $T_{rot} = 4$ K, and $T_{vib} = 20$ K were attained in these expansions! The expected low-lying $^3\Sigma_u^+$ state of Cu_2 has been observed by Bondybey (92) in emission in neon matrices at 4 K, with a lifetime of 27 ms, and another long-lived (65 μs) state at higher energy may be $D^3\Pi$ (92a).

From Second- and Third-Row Elements

Nb_2: There is no firm information about the ground state and bonding in this molecule. Its dissociation energy has been measured via mass spectrometry to be about 5 eV (112, 113; see Table 1). Its UV-visible absorption spectrum has been observed in matrices and EH theory applied to aid in assignment (26). Recently a comparison has been made between EH and SCF-Xα-SW theory as applied to niobium clusters, including Nb_2 (114, 115).

Mo_2: As for Cr_2, Efremov et al (29) also observed the gas phase spectrum of Mo_2 prepared analogously by flash photolysis of the $Mo(CO)_6$ molecule.

Analysis of the $B^1\Pi_u \leftrightarrow X^1\Sigma_g^+$ system at 390 nm yielded $r'_e = 1.912$ Å, $\omega'_e = 412$ cm^{-1}, $r''_e = 1.929$ Å and $\omega''_e = 477.1$ cm^{-1}, indicating a high bond order, as suggested earlier by SCF-Xα-SW calculations (116) and by the high dissociation energy of 4 eV (29, 117, 113). Other absorptions occur at 520 nm (A system) and 320 nm (C system). The A system has also been observed in absorption in matrices (26, 36, 118–120), and from laser induced fluorescence in solid argon and krypton ω''_e was found to be 476(5) cm^{-1} (120), confirming the gas-phase value. The (0-0) band of the $A^1\Sigma_u^+ \leftarrow X^1\Sigma_g^+$ spectrum of laser-vaporized, jet-cooled Mo$_2$ was rotationally analyzed to give $r'_e = 1.937$ Å, $\omega'_e = 449.0$ cm^{-1}, and $r''_e = 1.938(9)$ Å (121), definitely establishing the short bond length in the ground state. Thus, the bond lengths in the X, A, B states, 1.938, 1.937, 1.91 Å, are practically constant. Hopkins et al (121) interpret this as indicating that the excitation is of the nonbonding $5s\sigma_g$ electron to a σ_u^* (A state) or a π_u (B state) orbital, which are also nonbonding. Thus the ground state configuration $(4d\sigma_g)^2(4d\pi_u)^4(4d\delta_g)^4(5s\sigma_g)^2$ is quintuply bonded in agreement with theory; however, theoretical interpretations do differ as to whether the $s\sigma$ or the $d\delta$ orbitals contribute more to the bonding. The most recent SCF calculations including f-type orbitals (44, see below) supports the interpretation of Hopkins et al.

Even prior to its synthesis and characterization, the concept of a sextuple bond in Mo$_2$ was proposed by Norman et al (116) as a result of an SCF-Xα-SW calculation at a bond distance estimated to be 2.12 Å. As for Cr$_2$, recent theoretical contributions have provided a new controversy. A calculation of the full potential curve was carried out by the generalized valence bond (GVB)-van der Waals (vdw) method with the aims of providing a wavefunction that dissociates properly and contains the dominant correlation effects (38, see 44, 22a). In this treatment, due to competition between 5s-5s bonding and 4d-4d bonding, two minima are found in the potential energy surfaces of the ground $^1\Sigma_g^+$ state, with the deepest minimum at 1.97 Å (in agreement with experiment) and the second minimum at 3.09 Å. The inner minimum has $d\sigma$-$d\sigma$, $d\pi$-$d\pi$, and $d\delta$-$d\delta$ bonding (a quintuple bond) where the s-s interaction is repulsive, while at the outer minimum there is a weak s-s bond with the high spin d^5 configuration on each atom exchange-coupled antiferromagnetically. However, SC-LSD calculations (41) found a $^1\Sigma_g^+$ lowest state with $r''_e = 1.95$ Å. Possible errors in the GVB-vdw approach are suggested: use of too few valence orbitals and the neglect of overlap effects in 4s, 4p semicore states. These calculations, and those of Bernholc & Holzworth (42), correct an earlier application of LSD theory to Cr$_2$ (16) and show that it provides valid results for Mo$_2$.

Because of the larger number of bonding electron pairs to be correlated, the usual methods applicable to light molecules are inadequate for Mo$_2$ and even more so for Cr_2. An apparent point of disagreement between experiment (39) and some theoretical results (37, 39) concerns the nature and relative magnitude of the $s\sigma$ and $d\delta$ bonding in Mo_2, and probably also Cr_2. As mentioned above for Cr_2, the most recent SCF calculations (44, 22a) point to a need for the inclusion of f-type polarization functions in the orbital bases where d-bonding is involved. These authors (44) also emphasize the need for inclusion of the long distance antiferromagnetic coupling (38) if a valid potential function is to be obtained.

Ru_2: The only information on this molecule is obtained from an ab initio SCF-CI calculation (125) yielding a $^7\Delta_u$ ground state with $r_e = 2.71$ Å, $D_e = 0.64$ eV, and $\omega_e = 116$ cm^{-1}. The calculated dissociation energy is low since extensive correlation effects were not included; other estimates place it near 3 eV (113, 126, 127). Relativistic effects should shorten the bond distance to about 2.54 to 2.59 Å. The bond involves the $5s\sigma_g$ MO plus some $4d$ bonding.

Pd_2: There are no experimental data to compare with the relativistic effective-core-potential calculation of Basch et al (128). The results are surprisingly similar to Ni_2 in that the lowest electronic states can be considered as formed from two $4d^95s^1$ atoms bonding via a $5s\sigma$ bond, with the $4d$ contributions described in terms of hole state configurations. As for Ni_2, the $\delta\delta$ hole states are lowest in energy in spite of possible formation of a $\sigma\sigma$ double bonded structure, which one might have been expected to be stabilized by the increased d-d overlap. The dissociation energy (~ 1.1 eV) (23, 113) is essentially in accord with a single bond and places Pd_2 closer to Ni_2 than to Pt_2 in its bonding. It appears that extensive configuration interaction will be required in the theoretical treatment if accurate molecular parameters are to be obtained.

Pt_2: This molecule has also been treated by Basch et al (128), although in this case there are a few experimental data. D_e has been measured to be 3.71(16) (129), considerably stronger than its two lighter diatomic homologues. Multiple s-s and d-d bonding must be occurring to account for this large dissociation energy, but D_e values of that magnitude are generally expected for third-row elements (127).

Ag_2: Gas-phase optical spectra (132a, b, 133) have established the $^1\Sigma_g^+$ ground state and $\omega_e'' = 192.4$ cm^{-1}, and the use of pure ^{107}Ag isotope in high dispersion spectra has allowed a rotational analysis to be made resulting in $r_e'' = 2.482$ Å (133a). A number of mass spectrometry measurements have been made of silver vapor, yielding $D_e \cong 1.65$ eV (83a–d, 126). Extensive matrix optical studies, mainly by Ozin (134), Schulze (135) and their collaborators, have been useful to theoreticians in testing the relative energies of electronic states. Matrix Raman spectroscopy has confirmed ω_e'' (136). MCD studies (137) of matrix-isolated Ag_2 have been effective in corroborating the assignments of the electronic transitions.

Baetzold (94, 138) was the first to apply EH and CNDO theory, with reasonable results, to Ag_2 and larger clusters. This was followed by further semi-empirical calculations (96, 139–141) and by a SCF-Xα-SW calculation (100). Relativistic effective-core-potential theory has been used by Basch (142, 143), with the agreement with experiments shown in Table 1. The model potential method is similar to the use of an ECP, and for Ag_2 it was found to agree with all-electron results, but only if the $4p$ shell was released from the core and treated explicitly (144b). All-electron calculations, including CI (110, 144a), confirm the essentially single s-s bond and provide a nonrelativistic basis for evaluating relativistic effects, which are significant in Ag_2 (see Table 1). Cu_2 and Ag_2 are very similar, with little hybridization and largely localization of d orbitals on the metal atoms. However, as stated by Basch, "Even for the s-bonded group IB metals, the quantitatively accurate description of the metal-metal bond is a nontrivial problem."

Au_2: As in Cu_2 and Ag_2, the primary bonding in Au_2 is s-s, but the dissociation energy is much stronger ($D_e = 2.3$ eV) than that in the two lighter molecules (83a–d, 145, 146). This is due to the large relativistic effects in a molecule containing such heavy atoms, which in Au_2 leads to an increase of ~ 1 eV in the bond strength, a decrease in r_e of more than 0.3 Å and a 25% increase in ω_e! In a series of papers beginning in 1977, K. S. Pitzer and co-workers (147a, b, 148) have been developing pseudopotential methods that include relativistic effects on the core electrons as well as the valence electrons, including spin-orbit effects (termed AREP, averaged relativistic effective core potential). They have used Au_2 as a test case just as Cu_2 has been employed as a nonrelativistic benchmark. As can be seen from Table 1, the ab initio calculations are in good agreement with experiment. Comparable agreement is found in two excited states.

HETERONUCLEAR DIATOMICS

The results of the studies of some of these heteronuclear diatomics are beginning to support a generalized form of the Brewer-Engel theory of transition-metal alloys, in which it is noted that strong multiple bonds, and therefore very stable alloys, will be formed between elements at opposite ends of the Periodic Table (150a, b, 151a, b). Brewer (152) has noted that this should also apply to molecules, and this has been confirmed by the measurements by Gingerich et al of D_e for the TiRh (153) and YPt (129) molecules. The theory is presently in only qualitative form, and electronic structure data are needed for better understanding. Shim & Gingerich (154) have published an ab initio calculation on YPd attempting to examine the implications of the theory. Among first-row molecules, recent data on ScNi and TiCo appear to also agree with the theory.

From First-Row Elements

Figure 1 shows the present status of knowledge of the ground electronic states of the diatomics formed from first-row metal atoms. Those that are not in bold frames in that array are obtained from theoretical calculations, inferred from Mössbauer data, or arrived at by extrapolation from the known electronic properties of adjacent molecules. The molecules along the diagonal are, of course, homonuclear diatomics, discussed above. The numbers in the boxes are dissociation energies (in eV), which in the majority of cases have not been measured but have been predicted by empirical models (126, 127, 113).

ScNi: The ESR spectrum observed in an argon matrix at 4 K (155) occurs near $g = 2.0$ and exhibits hyperfine structure produced by ^{45}Sc ($I = 7/2$). The molecule has a $^2\Sigma$ ground state with about 85% (35% $s\sigma$ and 50% $d\sigma$) of

Sc	Ti	V	Cr	Mn	Fe	Co	Ni	Cu	
$^5\Sigma$ 1.1	1.3	2.2	2.2	1.4	2.4	$^1\Sigma$ 3.0	$^2\Sigma$ 3.3	2.3	Sc
	$^1\Sigma$ 1.3	$^4\Sigma$ 1.9	1.7	1.0	$^1\Sigma$ 1.8	$^2\Sigma$ 2.4	$^1\Sigma$ 2.7	1.7	Ti
		$^3\Sigma$ 1.9	2.0	1.4	1.9	$^1\Sigma$ 2.4	$^4\Sigma$ 2.8	2.0	V
			$^1\Sigma$ 1.6	1.0	$^7\Sigma$ 1.4	1.7	2.1	$^4\Sigma$ 1.6	Cr
				$^1\Sigma$ 0.3	$^4\Pi$ 0.8	1.3	1.7	$^5\Pi$ 1.0	Mn
					$^7\Delta$ 0.8	$^6\Sigma$ 1.3	$^5\Pi$ 1.7	$^2\Pi$ 1.3	Fe
						$^5\Sigma$ 1.0	2.1	1.7	Co
							$^1\Gamma$ 2.0	$^2\Delta$ 2.1	Ni
								$^1\Sigma$ 2.1	Cu

Figure 1 An array of all diatomic molecules formed from first-row transition metal atoms. The ground electronic state and dissociation energy (in eV) of each molecule is indicated. Ground states have been definitely established only for molecules in boxes with bold frames.

the unpaired spin on the Sc atom. This low spin suggests that ScNi is multiply bonded, which supports the high heat of dissociation (3.3 eV) predicted by Miedema (127). The difference in electronegativities should lead to charge transfer between the two atoms and contribute to the stabilization. The dominant electron configuration is presumably $4s\sigma^2 3d\sigma^2 3d\pi^4 3d\delta^4 4s\sigma^{*1}$, where the $s\sigma$ and $d\sigma$ orbitals are highly hybridized.

TiCo: Here, similar to ScNi, the ^{59}Co ($I = 7/2$) hyperfine structure could be observed in the ESR spectrum, and enrichment with ^{47}Ti gave proof of the molecule being observed and also complete accounting for one unpaired electron (155). It is not surprising, since it is isoelectronic with ScNi, that the ground state is $^2\Sigma$ and most of the spin (66%) is on the lighter atom. The dissociation energy is predicted to be lower (2.4 eV) (127) than ScNi, in accord with Brewer's proposal since the elements are closer together in the Periodic Table.

VFe is also a member of this 13-valence-electron series, which then extrapolates to a more weakly bonded $^2\Sigma$ molecule with a more uniform spin distribution. D_e is predicted to be 1.9 eV (127), but electronic data for VFe have not yet been obtained.

Molecules with one more or one less valence electron than ScNi and TiCo might be expected to have closed shell $^1\Sigma$ ground states, as indicated in Figure 1. ScCu is also in this category, but it is more uncertain because of the stabilizing effect of the d^{10} shell in the alkali-like metal Cu. The predicted lower dissociation energy (127), indicating less multiple bonding in ScCu, is also suggestive of this, and it would then be of interest to learn its ground state properties.

TiV: ESR spectra in matrices at 4 K analyze to yield a $^4\Sigma$ ground state with a zero-field splitting parameter, $|D| \geqslant 1 \text{ cm}^{-1}$ (156). ^{51}V ($I = 7/2$) hyperfine structure dominates the spectrum, and ^{47}Ti ($I = 5/2$) hyperfine splittings were also observed by using titanium metal enriched to 80% in ^{47}Ti. In the wave-function of the unpaired spins there is 8% $s\sigma$ character at the V atom and $\geqslant 9\%$ $s\sigma$ at the Ti atom, with the remainder presumably in $d\sigma, d\pi$, and $d\delta$ molecular orbitals. The major electronic configuration may then be $s\sigma^2 d\sigma^1 d\pi^4 d\delta^2$ or $s\sigma^2 d\sigma^1 d\pi^2 d\delta^4$ with some $s\sigma$ hybridization of the $d\sigma$ MO. Some multiple bonding might also be inferred from a semiempirical D_e value of 1.9 eV (127).

VNi: This 15 valence-electron molecule was also found to have a $^4\Sigma$ ground state and a large zero-field splitting (156). Only ^{51}V hyperfine structure was observable in the ESR, providing only the information that the odd electrons have about 10% $s\sigma$ character at that nucleus. The distribution among 3d orbitals on vanadium is then not established, but a negative anisotropic hyperfine constant indicates a dominance of $d\delta$ there. The major electron configuration might be $s\sigma^2 d\sigma^2 d\pi^4 d\delta^4 s^{*1} d\pi^{*2}$ or ... $d\delta^{*2}$. If correct, a predicted dissociation energy of 2.8 eV (127) implies a strong bond, about that of TiCo, in line with Brewer's proposal. VNi then appears to be essentially TiCo with two unpaired spins added.

FeM (M = Cr, Mn, Co, Ni, Cu): These diatomics have been studied via Mössbauer spectroscopy in rare-gas matrices, largely by Montano and co-workers (57, 58, 61, 157–159). Ground states were deduced based upon a crystal-field approach (56), except in the case of FeCr where an SCF-Xα-SW calculation was performed at $r_e = 2.0$ Å (58). Guenzburger & Saitovitch (62) have applied DVM-Xα theory to the series, except for FeCr. Because states of different configurations lie close together, a simple interpretation of the electronic configuration of Fe is very difficult, and the deduction of the ground state from the Mössbauer parameters is generally complex.

An ab initio HF-CI calculation has been carried out on FeNi (160), yielding a $^5\Pi$ lowest state with $^5\Phi$, $^3\Pi$, and $^3\Phi$ very close in energy. The bonding is almost entirely due to the $4s\sigma$ molecular orbital. Resonance Raman spectra of FeNi isolated in a matrix yielded a vibrational frequency of $\omega_e'' = 320$ cm^{-1} (54, 21).

The ground states of these molecules as given in the array of Figure 1 are taken from the above sources and must be considered as tentative. The possibility that VFe might have a $^2\Sigma$ ground state has already been alluded to above.

NiCu: As for FeNi, Shim (161) has carried out an ab initio HF-CI calculation between a Ni atom in the $3d^94s^1$ and Cu atom in the $3d^{10}4s^1$ configurations. Spin-orbit coupling has been approximately treated. The chemical bond is mainly due to a $4s\sigma$ molecular orbital in the lowest $^2\Delta_{5/2}$ state with $r_e = 2.41$ Å, $\omega_e = 347$ cm^{-1}, and $D_e = 1.54$ eV. $^2\Pi$ and $^2\Sigma^+$ levels, in that order, are low-lying states within about 3000 cm^{-1}. Empirical theory (127) gives $D_e = 2.12$ eV.

Heavier Diatomics

CrCu, CrAg, CrAu: In the simplest model the combination of a $3d^54s$ and a $3d^{10}4s$ atom would produce a diatomic molecule with an s-s bond and a $^6\Sigma$ ground state. This has been found to be the case for the two heavier molecules but not for CrCu.

These three molecules have been observed via ESR in solid rare-gas matrices at 4 K (162). Isotopically enriched ^{53}Cr ($I = 3/2$) was used in the preparation of CrCu and CrAg, and distinctively different Cr hyperfine splittings were observed in the two molecules. Analysis of the spectra, including hyperfine and quadrupole interaction with ^{63,65}Cu, ^{107,109}Ag, and ^{197}Au, indicated that CrCu has a $^4\Sigma$ ground state, while the heavier molecules have the expected $^6\Sigma$ ground state. CrCu is then proposed to be triply bonded and therefore intermediate in its bonding between Cr$_2$ and Cu$_2$. Among the three molecules the zero-field splitting parameters were found to increase from $-0.005(1)$ to $(+)0.44(1)$ to $\geqslant 2$ cm^{-1} with the increasing spin-orbit coupling.

Ozin & Klotzbücher (141) assigned absorption bands at 220 and 283 nm to CrAg trapped in an argon matrix. EH theory indicated that the molecule is bonded by an $s\sigma$ bond, with little interaction between the d orbitals, with a bond distance of 2.9 Å, and $D_e = 1.5$ eV. Gingerich (126) gives the dissociation energy of CrCu as 1.61(16) eV and CrAu as 2.17(18) eV. Values of D_e from an empirical model (127) are 1.50, 1.15, and 1.65 eV, respectively, for the three molecules.

MnAg(MnCu, MnAu): MnAg was first observed by Klotzbücher & Ozin (50) in an argon matrix at 10 K via its UV-visible absorption spectrum and bands at 263/266 and 362 nm tentatively assigned to it. EH calculations, yielding $r_e'' = 2.8$ Å, were used to aid in assignment of those transition. An experimental value of $D_e = 1.0(2)$ eV has been measured mass spectrometrically (163, 126).

The ESR spectrum of the ^{55}Mn^{107,109}Ag molecule, measured in argon and krypton matrices (67), established its ground state as $^7\Sigma$ with a zero-field splitting parameter $D = (+)0.20$ cm^{-1}. The hyperfine splittings due to the two nuclei are in accord with an approximate $(3d\delta)^2_{Mn}(3d\pi)^2_{Mn}3d\sigma^1 4s\sigma^1$ configuration.

The ESR spectra of the MnCu and MnAu molecules were not detected although, as for MnAg, both metal atoms were observed to be present in the matrix, and, in fact, MnAg is probably the least stable of the three diatomics (127). It seems likely that the reasons for not detecting MnCu and MnAu are quite distinct: the lighter MnCu probably has a $^5\Pi$ (or $^5\Delta$), i.e. orbitally degenerate, ground state, preventing its detection via ESR; and MnAu is probably $X^7\Sigma$ but has a very large zero-field splitting because of the large spin-orbit coupling constant of Au (162).

CuAg, AgAu: Absorptions in matrices at 271 and 382 nm have been tentatively associated with the AgCu molecule (50). EH theory yielded $D_e = 1.4$ eV compared to 1.76(11) eV experimentally (83a–d, 126). There are apparently no experimental data on AgAu except a value of $D_e = 2.08(11)$ eV (126). It seems safe to assume that both of these molecules have $^1\Sigma$ ground states.

Basch (142) reports a relativistic effective-core-potential calculation on AgAu, yielding $r_e'' = 2.65$ Å, $\omega_e'' = 266$ cm^{-1} and $D_e = 0.90$ eV. These calculated values of D_e and ω_e'' are larger than calculated values for Ag_2 and Au_2; this does not, however, appear to be true among the experimental values of D_e.

CrMo: Efremov et al (164) produced this molecule in the gas phase in the same way that Cr_2 and Mo_2 (see above) were prepared but using mixed carbonyls. An absorption band possibly attributable to CrMo occurred at 480 nm. A band in this region with possible vibrational structure ($v \cong 147$ cm^{-1}) was also observed by Klotzbücher et al (165), in argon matrices prepared by simultaneous cocondensation of Cr and Mo atoms. EH and SCF-Xα-SW theories were used by the authors in assignments of the transitions. All of the calculated properties of CrMo were intermediate between those of Cr_2 and Mo_2.

NbIr: A brief report of a CMS-$X_{\alpha\beta}$ calculation (166) on this molecule yields $r_e'' = 2.83$ Å, with very strong d bonding determining the deepest potential minimum.

YPd: An all-electron ab initio HF calculation (154) treats the three simple crystal-field states $^2\Sigma^+$, $^2\Pi$, and $^2\Delta$ resulting from the occupation by the single Y $4d$ electron of either a $d\sigma$, $d\pi$, or $d\delta$ orbital on Y, the filled Pd orbitals being at much lower energy. "The chemical bond in all these states can be attributed to donation of charge from the filled $d\sigma$ and $d\pi$ orbitals of the Pd atom into empty or partially filled $d\sigma$ and $d\pi$ of the Y atom and a back-donation of charge from the filled $5s$ orbital of Y into the empty $5s$ orbital of Pd." The lowest state is $^2\Delta$ with $r_e = 2.69$, $\omega_e = 144$ cm^{-1}, $D_e = 1.87$ eV, and dipole moment = 0.51 D, with a small negative charge on the Y atom. The experimental value of $D_e = 2.46(16)$ (129) is, as expected, higher than the calculated value. The shift of charge, but not its magnitude, supports Brewer's view (150a, b) of the bonding in such alloy molecules.

TRIATOMICS

Sc_3, Y_3, La_3: Knight et al (168) have observed the ESR spectra of Sc_3 in argon and Y_3 in neon, argon, and krypton matrices at 4 K and higher temperatures, depending upon the matrix. From the hyperfine structure due to ^{45}Sc ($I = 7/2$) it was established that Sc_3 has an equilateral triangle structure at 4 to 30 K with a $^2A_1'$ ground state. It is possible that even at these low temperatures it is a pseudorotating (fluxional) bent molecule, as is believed to be the case for the Li_3 molecule (169a, b). A theoretical calculation of the barrier height is needed. This structure or a near

equilateral triangle is also indicated by the recent resonance Raman studies of Moskovits et al (10). They assigned bands at 246, 151, 145 cm^{-1} to the symmetric stretch, asymmetric stretch, and bending frequencies, respectively, of equilateral Sc_3. A second progression is observed and attributed to a low-lying excited state at only 395 cm^{-1}.

In contrast, the ESR spectrum of Y_3 indicates that one of the Y atoms is different from the other two, so that it is probably also a bent molecule with a $^2B_2$ ground state (168). It does not pseudorotate up to about 60 K. La_3 was not observed, although it was probably formed. This could occur if the ground state is linear and $^2\Pi$ or $^2\Delta$, since the orbital angular momentum would lead to large g-tensor anisotropy and broaden its ESR spectrum beyond detection.

Cr_3: Progressions observed in the Raman spectra of chromium-containing argon matrices are tentatively attributed to Cr_3 (31); 308, 226, and 123 cm^{-1} are assigned as the symmetric stretching, asymmetric stretching and bending frequencies. This implies C_{2v} symmetry, and the angle (69°) derived from approximate vibrational analysis leads the authors to suggest that the molecule may be slightly Jahn-Teller distorted from D_{3h} symmetry. A recent highly speculative report of matrix far-IR spectra attributes more than 15 lines to Cr_3 molecules bent at various angles from 60° to >140° in matrix sites (7a). Attempts to observe this molecule via ESR in neon and argon matrices using both natural and ^{53}Cr-enriched chromium were not successful (167).

Fe_3: Two sets of authors have identified this molecule in solid argon from its Mössbauer spectrum (61, 170). From the Mössbauer parameters Shamai et al (170) deduce that it has an equilateral triangle structure.

Ni_3: This molecule has apparently only been observed via UV-visible absorption (70) and resonance Raman spectroscopy (171, 21), both in argon matrices. In the Raman spectrum the line shapes are attributed to isotopic fine structure, and assuming that the observed normal mode is the symmetric stretching frequency ($\omega_e = 232$ cm^{-1}), they are fit best for a C_{2v} structure with bond angle 90 to 100°. This highly bent structure is postulated to be the result of distortion from a true linear Δ_u ground state [suggested by theory (24, 79)] by the matrix site or by the Renner-Teller effect. Such a large distortion from either cause is unexpected, and the experimental interpretation must be viewed as uncertain.

Early CNDO (172) theory found Ni_3 to have an equilateral triangle structure, but later calculations have favored a linear molecule (79, 73, 142). Recent ECP-SCF theory applied by Basch et al (73) yields a bond distance of 2.38 Å and a multiplicity of five, with three of the unpaired electrons largely in $d\delta$ orbitals. The atomic configuration $3d^9 4s^1$ is preferred, leading to essentially only the one 4s bonding electron per Ni atom.

Cu_3, Cu_2Ag, Ag_3: All of these molecules have been observed by ESR spectroscopy in matrices (173–175). Recently a band system of jet-cooled Cu_3 at about 530 nm has been analyzed in an extraordinary piece of gas-phase research (176). Where overlap occurs, the ESR and gas-phase studies agree on the ground state properties of Cu_3. In the numerous measurements of the optical spectra of Ag (134, 177) and Cu (90, 178–180) clusters in matrices, bands due to the triatomics have been identified and transitions assigned on the basis of theoretical calculations, and verified in some cases by MCD measurements (137). Raman spectra have been reported for Ag_3 (136) and Cu_3 (181) in matrices. Atomization energies have been determined as 3.05 eV for Cu_3 and 2.62 eV for Ag_3 (46, 182). [The value for Au_3, 3.80 eV (46), again demonstrates exceptionally strong bonding, as in the diatomics.]

Ag_3 was identified by Howard et al (175) in the X-band ESR spectrum of a C_6D_6 matrix at 77 K. The hyperfine structure shows large isotropic interactions of the one unpaired electron with two equivalent ^{107}Ag nuclei in the molecule and small interaction with a third unique atom. The spectrum is strikingly similar to the alkali metal trimers (183a, b). The strong localization of the s electron on the terminal atoms places it in an antibonding molecular orbital with a node at the central atom, implying either a $^2\Sigma_u^+$ linear molecule or a $^2B_2$ slightly bent species, the latter being favored by an observed orthorhombic g tensor. Pseudorotation does not occur up to at least 77 K; such fluxional behavior begins at ~ 20 K for K_3 in solid argon (183a, b). Raman spectra of Ag_3 in argon, showing only one line, suggested a linear molecule with $\nu_{sym\,str} = 120.5$ cm^{-1} (136), but if only slightly bent $\nu_{asym\,str}$ may be too weak to observe (171, see also 184).

Similar ESR spectra were obtained at 77 K for Cu_3 and CuAgCu in adamantane (173) and C_6D_6 (174) matrices, respectively. All of the derived parameters, and therefore the odd-electron distribution, in CuAgCu are similar to those of Ag_3. The terminal atom s-character in Cu_3 is considerably less than in Ag_3 or Cu_2Ag, totaling only 58%. The authors suggest a variation in the obtuse angle among the bent molecules as the cause.

A resonance Raman spectrum of copper-containing matrices is interpreted to be that of Cu_3 as a Jahn-Teller fluxional molecule at 12 K (181). The assignment posed problems and may be due to other species, particularly in light of the ESR study and the recent jet-cooled gas-phase work.

Supersonic expansion of laser-vaporized copper leads to the formation of cold gas-phase Cu_3 (T_{rot} perhaps 5 K) with an absorption band system in the

523–543 nm region (176). Two photon ionization and other spectroscopic techniques indicate that this spectrum arises from a Jahn-Teller distorted D_{3h} molecule in both the excited and the ground states, the latter in agreement with the matrix ESR (173). The upper C_{2v} state has a much lower stabilization energy than the ground state. These results were derived from vibronic analysis; rotational fine structure was not yet resolved. Three "hot" bands were identified by varying the degree of cooling of the expansion.

The early EH (93, 185, 179, 131) and modified EH (95) calculations found Cu_3 and Ag_3 to be linear ($^2\Sigma_u^+$) with bond distance of 2.3 Å. Later theory has vacillated between predictions of their structures as an acute isosceles triangle ($^2A_1$) (184, 186, 187) and the linear or nearly-linear obtuse isosceles triangle (139, 140, 102, 104). The best ab initio calculations are presumably those of Basch (RECP/SCF-CI) on Ag_3 (143) where a $^2B_2$ molecule was favored, and two all-electron SCF calculations on Cu_3 (98, 189, 188) where $^2\Sigma_u$ was more stable than 120° $^2B_2$. The most recent SCF-CI calculation on Cu_3 (188) finds a linear molecule to lie at the true minimum, but because of the small basis set used, concludes that the relative stabilities of the three possible structures could not really be determined.

Note that no transition from a Jahn-Teller bent structure to a pseudorotating equilateral-triangle structure has been experimentally observed for any of the transition-metal trimer molecules, in contrast to the corresponding alkali metal molecules.

PENTATOMICS

Mn_5: This molecule appears to be formed very readily in rare gas matrices as indicated by an extensive (more than 35 lines) ESR spectrum, even though Mn_3 and Mn_4 were not observed (47). All of the lines could be assigned as fine-structure transitions of an axial molecule containing 25 unpaired electrons ($S = 25/2$) with a zero-field splitting parameter $D = -0.013(1)$ cm^{-1}. The analysis was facilitated by the fact that the trapped molecule was highly oriented, since the spectrum was very dependent upon the orientation of the substrate in the magnetic field. The ^{55}Mn ($I = 5/2$) hyperfine structure was not detectable and lies within the line widths. This is unfortunate since those splittings would allow discrimination between several possible molecular structures. A linear structure can be eliminated because of the small hyperfine, and the most likely model is proposed to be a plane pentagon. The bonding is then visualized as involving essentially only the $4s$ electrons on each atom, with the $3d$ electrons remaining unpaired.

Cu_5, Ag_5: Howard et al have observed clusters in cyclohexane and adamantane matrices at temperatures ranging from 77 to 240 K (190, 191). The X-band ESR spectrum in each case is interpreted as that of an $S = 1/2$ molecule containing two sets of equivalent atoms and one odd set, as indicated by the hyperfine structure of ^{63}Cu and ^{107}Ag. The proposed structure is a trigonal bipyramid in which the trigonal base has been Jahn-Teller distorted into an isosceles triangle (sort of a bent M_3 molecule with an atom above and below the triangle), resulting in a $^2B_2$ ground state. In fact, as in Cu_3 and Ag_3, most of the s character in the odd-electron wavefunction is at the two equivalent atoms in the distorted triangle base.

Although there are features of the ESR spectrum that are curious, the overall analysis is reasonable and similar for the two molecules. Raising the temperature above 77 K produced no basic alteration in the spectrum, so a "fluxional" state of these molecules must require temperatures higher than 240 K.

A CLUSTER!

$Sc_{13?}$: Besides Sc_2 and Sc_3, a third molecule appears when scandium atoms are trapped in neon matrices at 4 K (168). The more than 60 hyperfine structure lines centered near $g = 2$ in the ESR spectrum indicate that this molecule contains at least nine equivalent atoms, the exact number being unspecified because the intensity drops off in the wings of the broad Lorentzian line. The molecule contains one unpaired spin, which, from the small hyperfine splittings, has little s character. It is proposed that the observed molecule is Sc_{13} in an exceptionally stable icosahedral structure and a ^2A_g ground state; however, it is also possible that even at 4 K it is a dynamic Jahn-Teller molecule with only a small barrier between distorted conformations.

Although there have been many calculations on clusters of 13 transition-metal atoms [most notably, perhaps, the all-electron calculation of Cu_{13} (189)], Sc_{13} appears to have been specifically studied in only one SCF-Xα-SW treatment in the solid hcp structure (192). A cluster with one unpaired spin results, in agreement with the ESR molecule. The icosahedron may have preferred stability relative to the hcp or fcc clusters, but all three are probably of comparable energy (168, 189, 193).

CONCLUSION

The paucity of experimental data on transition-metal molecules, particularly beyond diatomics, is a notable feature of this review. Larger molecules with $n > 6$ have been observed in matrices and in gas-phase expansions, but they remain uncharacterized as of 1983. Such data are needed in order to challenge the interests of the theorists and to progress toward an understanding of metallic clusters and small particles.

ACKNOWLEDGMENTS

The preparation of this review was supported by the National Science Foundation. We are also grateful to many colleagues working in this area of research for communications and preprints of their recent results.

Literature Cited

1. Cotton, F. A., Walton, R. A. 1982. *Multiple Bonds Between Metal Atoms.* New York: Wiley
2. Diatomic Metals and Metallic Clusters, 1980. *Faraday Symp. Chem. Soc.*, Vol. 14
3. Bourdon, J., ed. 1980. *Growth and Properties of Metal Clusters.* Amsterdam: Elsevier
4. Perenboom, J. A. A. J., Wyder, P., Meier, F. 1981. *Phys. Rep.* 78:173–292
5. Davis, S. C., Klabunde, K. J. 1982. *Chem. Rev.* 82:153–208
6. Gole, J. L., Stwalley, W. C., eds. 1982. *Metal Bonding and Interactions in High Temperature Systems, Am. Chem. Soc. Symp. Ser.* 179
7. Baetzold, R. C., Hamilton, J. F. 1983. *Prog. Solid State Chem.* 15:1–53
7a. Ozin, G. A., Mitchell, S. A. 1983. *Angew. Chem. Int. Ed. Engl.* 22:674–94
7b. Sattler, K. 1983. *Festkörperprobleme, Adv. Solid State Phys.* 23:1–12
8. Verhaegen, G., Smoes, S., Drowart, J. 1964. *J. Chem. Phys.* 40:239–41
9. Busby, R., Klotzbücher, W., Ozin, G. A. 1976. *J. Am. Chem. Soc.* 98:4013–15
10. Moskovits, M., DiLella, D. P., Limm, W. 1984. *J. Chem. Phys.* 80:626–33
11. Knight, L. B. Jr., Van Zee, R. J., Weltner, W. Jr. 1983. *Chem. Phys. Lett.* 94:296–99
12. Cooper, W. F., Clarke, G. A., Hare, C. R. 1972. *J. Phys. Chem.* 76:2268–73
13. Das, G. 1982. *Chem. Phys. Lett.* 86:482–86
14. Walch, S. P., Bauschlicher, C. W. Jr. 1983. *Chem. Phys. Lett.* 94:290–95
15. Walch, S. P., Bauschlicher, C. W. Jr. 1983. *J. Chem. Phys.* 79:3590–91
16. Harris, J., Jones, R. O. 1979. *J. Chem. Phys.* 70:830–41
17. Wood, C., Doran, M., Hillier, I. H., Guest, M. F. 1980. *Faraday Symp. Chem. Soc.* 14:159–69
18. Wolf, A., Schmidtke, H.-H. 1980. *Int. J. Quant. Chem.* 18:1187–1205
19. Kant, K., Lin, S.-S. 1969. *J. Chem. Phys.* 51:1644–47
20. Cossé, C., Fouassier, M., Mejean, T., Tranquille, M., DiLella, D. P., Moskovits, M. 1980. *J. Chem. Phys.* 73:6076–85
21. Moskovits, M., DiLella, D. P. 1982. See Ref. 6, pp. 153–75
22. Walch, S. P., Bauschlicher, C. W. Jr. Private communication
22a. Walch, S. P., Bauschlicher, C. W. Jr., Roos, B. O., Nelin, C. J. 1983. *Chem. Phys. Lett.* 103:175–79
23. Gingerich, K. A. 1971. *J. Cryst. Growth* 9:31–45
24. Anderson, A. B. 1976. *J. Chem. Phys.* 64:4046–55
25. Ford, T. A., Huber, H., Klotzbücher, W., Kündig, E. P., Moskovits, M., Ozin, G. A. 1977. *J. Chem. Phys.* 66:524–30
26. Klotzbücher, W., Ozin, G. A. 1977. *Inorg. Chem.* 16:984–87
27. Langridge-Smith, P. R. R., Morse, M. D., Hansen, G. P., Smalley, R. E., Merer, A. J. 1964. *J. Chem. Phys.* 80:593–600
28. Efremov, Y. M., Samoilova, A. N., Gurvich, L. V. 1974. *Opt. Spectrosc.* 36:381–82
29. Efremov, Y. M., Samoilova, A. N., Kozhukhovsky, V. B., Gurvich, L. V. 1978. *J. Mol. Spectrosc.* 73:430–40
30. Kant, A., Strauss, B. 1966. *J. Chem. Phys.* 45:3161–62
31. DiLella, D. P., Limm, W., Lipson, R. H., Moskovits, M., Taylor, K. V. 1982. *J. Chem. Phys.* 77:5263–66
32. Bondybey, V. E., English, J. H. 1983. *Chem. Phys. Lett.* 94:443–47
33. Michalopoulos, D. L., Geusic, M. E., Hansen, S. G., Powers, D. E., Smalley, R. E. 1982. *J. Phys. Chem.* 86:3914–16
34. Riley, S. J., Parks, E. K., Pobo, L. G., Wexler, S. 1983. *J. Chem. Phys.* 79:2577–82

35. Kundig, E. P., Moskovits, M., Ozin, G. A. 1975. *Nature* 254:503–4
35a. Pellin, M. J., Gruen, D. M. 1983. *J. Chem. Phys.* 79:5887–93
36. Klotzbücher, W., Ozin, G. A., Norman, J. G. Jr., Kolari, H. J. 1977. *Inorg. Chem.* 16:2871–77
37. Bursten, B. E., Cotton, F. A. 1980. *Faraday Symp. Chem. Soc.* 14:180–93
38. Goodgame, M. M., Goddard, W. A. III. 1981. *J. Phys. Chem.* 85:215–17; 1982. *Phys. Rev. Lett.* 48:135–38
39. Atha, P., Hillier, I. H. 1982. *Mol. Phys.* 45:285–93
40. Dunlap, B. I. 1983. *Phys. Rev. A* 27:2217–19
41. Delley, B., Freeman, A. J., Ellis, D. E. 1983. *Phys. Rev. Lett.* 50:488–91
42. Bernholc, J., Holzwarth, N. A. W. 1983. *Phys. Rev. Lett.* 50:1451–54
43. Kok, R. A., Hall, M. B. 1983. *J. Phys. Chem.* 87:715–17
44. McLean, A. D., Liu, B. 1983. *Chem. Phys. Lett.* 101:144–48
45. Nesbet, R. K. 1964. *Phys. Rev. A* 135:460–65
46. Kant, A., Lin, S.-S., Strauss, B. 1968. *J. Chem. Phys.* 49:1983–85
47. Baumann, C. A., Van Zee, R. J., Bhat, S. V., Weltner, W. Jr. 1981. *J. Chem. Phys.* 74:6977–78; 1983. *J. Chem. Phys.* 78:190–99
48. Rivoal, J.-C., ShakhsEmampour, J., Zeringue, K. J., Vala, M. 1982. *Chem. Phys. Lett.* 92:313–16
49. DeVore, T. C., Ewing, A., Franzen, H. F., Calder, V. 1975. *Chem. Phys. Lett.* 35:78–81
50. Klotzbücher, W. E., Ozin, G. A. 1980. *Inorg. Chem.* 19:3776–82
51. Lin, S.-S., Kant, A. 1969. *J. Phys. Chem.* 73:2450–51
52. Shim, I., Gingerich, K. A. 1982. *J. Chem. Phys.* 77:2490–97
53. Rohlfing, E. A., Cox, D. M., Kaldor, A. 1983. *Chem. Phys. Lett.* 99:161–66
54. Moskovits, M., DiLella, D. P. 1980. *J. Chem. Phys.* 73:4917–24
55. McNab, T. K., Micklitz, H., Barrett, P. H. 1971. *Phys. Rev. B* 4:3787–97
56. Montano, P. A., Barrett, P. H., Shanfield, Z. 1976. *J. Chem. Phys.* 64:2896–2900
57. Montano, P. A. 1980. *Faraday Symp. Chem. Soc.* 14:79–86
58. Nagarathna, H. M., Montano, P. A., Naik, V. M. 1983. *J. Am. Chem. Soc.* 105:2938–43
59. Montano, P. A., Shenoy, G. K. 1980. *Solid State Commun.* 35:53–56
60. Purdum, H., Montano, P. A., Shenoy, G. K., Morrison, T. 1982. *Phys. Rev. B* 25:4412–17
61. Dyson, W., Montano, P. A. 1979. *Phys. Rev. B* 20:3619–25; 1980. *Solid State Commun.* 33:191–94
62. Guenzburger, D., Saitovich, E. M. B. 1981. *Phys. Rev. B* 24:2368–79
63. Trautwein, A., Harris, F. E. 1973. *Phys. Rev. B* 7:4755–57
64. Blyholder, G., Head, J., Ruette, F. 1982. *Theor. Chim. Acta* 60:429–44
65. Kant, A., Strauss, B. 1964. *J. Chem. Phys.* 41:3806–8
66. Shim, I., Gingerich, K. A. 1983. *J. Chem. Phys.* 78:5693–98
67. Baumann, C. A., Van Zee, R. J., Weltner, W. Jr. 1984. *J. Phys. Chem.* 88:1815–20
68. Noell, J. O., Newton, M. D., Hay, P. J., Martin, R. L., Bobrowicz, F. W. 1980. *J. Chem. Phys.* 73:2360–71
69. Kant, A. 1964. *J. Chem. Phys.* 41:1872–76
70. Moskovits, M., Hulse, J. E. 1977. *J. Chem. Phys.* 66:3988–94
71. Ozin, G. A. 1977. *Catal. Rev.* 16:191–286
72. Ahmed, F., Nixon, E. R. 1979. *J. Chem. Phys.* 71:3547–49
72a. Morse, M. D., Hansen, G. P., Langridge-Smith, P. R. R., Zheng, L.-S., Geusic, M. E., Michalopoulos, D. L., Smalley, R. E. 1984. *J. Chem. Phys.* Submitted
73. Basch, H., Newton, M. D., Moskowitz, J. W. 1980. *J. Chem. Phys.* 73:4492–4510
74. Shim, I., Dahl, J. P., Johansen, H. 1979. *Int. J. Quant. Chem.* 15:311–31
75. Melius, C. F., Moskowitz, J. W., Mortola, A. P., Baillie, M. B., Ratner, M. A. 1976. *Surf. Sci.* 59:279–92
76. Upton, T. H., Goddard, W. A. III. 1978. *J. Am. Chem. Soc.* 100:5659–68
77. Griffith, J. S. 1956. *J. Inorg. Nucl. Chem.* 3:15–23
78. Baetzold, R. C. 1973. *J. Catal.* 29:129–37; 1976. *Adv. Catal.* 25:1–55
79. Anderson, A. B. 1977. *J. Chem. Phys.* 66:5108–11
80. Rösch, N., Rhodin, T. N. 1974. *Phys. Rev. Lett.* 32:1189–92; *Discuss. Faraday Soc.* 58:28–34
81. Newton, M. D. 1982. *Chem. Phys. Lett.* 90:291–95
82a. Åslund, N., Barrow, R. F., Richards, W. G., Travis, D. N. 1965. *Ark. Fys.* 30:171–85
82b. Lochet, J. 1978. *J. Phys. B* 11:L55–L57
82c. Kleman, B., Lindqvist, S. 1954. *Ark. Fys.* 8:333–39
83a. Schissel, P. 1957. *J. Chem. Phys.* 26:1276–80
83b. Ackerman, M., Stafford, F. E.,

Drowart, J. 1960. *J. Chem. Phys.* 33: 1784–89
83c. Smoes, S., Mandy, F., Vander Auwera-Mahieu, A., Drowart, J. 1972. *Bull. Soc. Chim. Belges* 81:45–56
83d. Gingerich, K. A. 1980. *Faraday Symp. Chem. Soc.* 14:109–23
84. Pelissier, M. 1981. *J. Chem. Phys.* 75:775–80
85. Bauschlicher, C. W. Jr., Walch, S. P., Siegbahn, P. E. M. 1982. *J. Chem. Phys.* 76:6015–17; 1983. *J. Chem. Phys.* 78:3347–48
86. Jeung, G. H., Barthelat, J. C., Pelissier, M. 1982. *Chem. Phys. Lett.* 91:81–85
87. Jeung, G. H., Barthelat, J. C. 1983. *J. Chem. Phys.* 78:2097–99
87a. Stoll, H., Fuentealba, P., Dolg, M., Szentpály, L. V., Preuss, H. 1983. *J. Chem. Phys.* 79:5532–42
88. Preuss, D. R., Pace, S. A., Gole, J. L. 1979. *J. Chem. Phys.* 71:3553–60
89. Powers, D. E., Hansen, S. G., Geusic, M. E., Puiu, A. C., Hopkins, J. B., Dietz, T. G., Duncan, M. A., Langridge-Smith, P. R. R., Smalley, R. E. 1982. *J. Phys. Chem.* 86:2556–60
90. Gole, J. L., English, J. H., Bondybey, V. E. 1982. *J. Phys. Chem.* 86:2560–63
91. Powers, D. E., Hansen, S. G., Geusic, M. E., Michalopoulos, D. L., Smalley, R. E. 1983. *J. Chem. Phys.* 78:2866–81
92. Bondybey, V. E. 1982. *J. Chem. Phys.* 77:3771–72
92a. Bondybey, V. E., English, J. H. 1983. *J. Phys. Chem.* 87:4647–50
93. Hare, C. R., Sleight, T. P., Cooper, W., Clarke, G. A. 1968. *Inorg. Chem.* 7:669–73
94. Baetzold, R. C. 1971. *J. Chem. Phys.* 55:4355–63
95. Anderson, A. B. 1978. *J. Chem. Phys.* 68:1744–51
96. Boeyens, J. C. A., Lemmer, R. H. 1977. *J. Chem. Soc. Faraday Trans. II* 73:321–26
97. Joyes, P., Leleyter, M. 1973. *J. Phys. B* 6:150–54
98. Bachmann, C., Demuynck, J., Veillard, A. 1978. *Gazz. Chim. Ital.* 108:389–91; 1980. *Faraday Symp. Chem. Soc.* 14:170–79; 1980. *Studies in Surface Science and Catalysis*, Vol. 4, *Growth and Properties of Metal Clusters*, ed. J. Bourdon, p. 269. Amsterdam: Elsevier
99. Dixon, R. N., Robertson, I. L. 1978. *Mol. Phys.* 36:1099–1112.
100. Ozin, G. A., Huber, H., McIntosh, D. F., Mitchell, S. A., Norman, J. G., Noodleman, L. 1979. *J. Am. Chem. Soc.* 101:3504–11
101. Tatewaki, H., Huzinaga, S. 1980. *J. Chem. Phys.* 72:399–405
102. Post, D., Baerends, E. J. 1982. *Chem. Phys. Lett.* 86:176–80
103. Guenzburger, D. 1982. *Chem. Phys. Lett.* 86:316–19
104. Tatewaki, H., Miyoshi, E., Nakamura, T. 1982. *J. Chem. Phys.* 76:5073–86
105. Pauling, L. 1983. *J. Chem. Phys.* 78:3346
106. Bauschlicher, C. W. Jr. 1983. *Chem. Phys. Lett.* 97:204–8
107. Ziegler, T., Snijders, J. G., Baerends, E. J. 1981. *J. Chem. Phys.* 74:1271–84
108. Martin, R. L. 1983. *J. Chem. Phys.* 78:5840–42
108a. Pelissier, M. 1983. *J. Chem. Phys.* 79:2099–2100
109. Witko, M., Beckmann, H.-O. 1982. *Mol. Phys.* 47:945–57
110. Shim, I., Gingerich, K. A. 1983. *J. Chem. Phys.* 79:2903–12
111. Delley, B., Ellis, D. E., Freeman, A. J., Baerends, E. J., Post, D. 1983. *Phys. Rev. B* 27:2132–44
112. Gupta, S. K., Gingerich, K. A. 1979. *J. Chem. Phys.* 70:5350–53
113. Brewer, L., Winn, J. S. 1980. *Faraday Symp. Chem. Soc.* 14:126–35
114. Seifert, G., Mrosan, E., Müller, H., Ziesche, P. 1978. *Phys. Status Solidi* 89(6):K175–8
115. Müller, H., Optiz, Ch., Seifert, G. 1982. *Z. Phys. Chem. Leipzig* 263:1005–15
116. Norman, J. G. Jr., Kolari, H. J., Gray, H. B., Trogler, W. C. 1977. *Inorg. Chem.* 16:987–93
117. Gupta, S. K., Atkins, R. M., Gingerich, K. A. 1978. *Inorg. Chem.* 17:3211–13
118. Ozin, G. A., Klotzbücher, W. 1977/1978. *J. Mol. Catal.* 3:195–206
119. Bates, J. K., Gruen, D. M. 1979. *J. Mol. Spectrosc.* 78:284–97
120. Pellin, M. J., Foosnaes, T., Gruen, D. M. 1981. *J. Chem. Phys.* 74:5547–57
121. Hopkins, J. B., Langridge-Smith, P. R. R., Morse, M. D., Smalley, R. E. 1983. *J. Chem. Phys.* 78:1627–37
122. Bursten, B. E., Cotton, F. A., Hall, M. B. 1980. *J. Am. Chem. Soc.* 102:6348–49
123. Atha, P. M., Hillier, I. H., Guest, M. F. 1980. *Chem. Phys. Lett.* 75:84–86
124. Castro, M., Keller, J., Mareca, P. 1981. *Int. J. Quant. Chem. Symp.* 15:429–35
125. Cotton, F. A., Shim, I. 1982. *J. Am. Chem. Soc.* 104:7025–29
126. Gingerich, K. A. 1980. *Faraday Symp. Chem. Soc.* 14:109–25
127. Miedema, A. R. 1980. *Faraday Symp. Chem. Soc.* 14:136–48
128. Basch, H., Cohen, D., Topiol, S. 1980. *Isr. J. Chem.* 19:233–41
129. Gupta, S. K., Nappi, B. M., Gingerich, K. A. 1981. *Inorg. Chem.* 20:966–69

130. Jansson, K., Scullman, R. 1976. *J. Mol. Spectrosc.* 61:299–312
131. Baetzold, R. C., Mack, R. E. 1975. *J. Chem. Phys.* 62:1513–20
132a. Ruamps, 1954. *Compt. Rend. Acad. Sci. Paris* 238:1489–91; 1959. *Ann. Phys. Paris* 4:1111–57
132b. Kleman, B., Lindqvist, S. 1955. *Ark. Fys.* 9:385–90
133. Brown, C. M., Ginter, M. L. 1978. *J. Mol. Spectrosc.* 69:25–36
133a. Srdanov, V. I., Pesić, D. S. 1981. *J. Mol. Spectrosc.* 90:27–32. See footnote 12 in Ref. 144a
134. Ozin, G. A. 1980. *Faraday Symp. Chem. Soc.* 14:7–64
135. Schulze, W., Becker, H. U., Abe, H. 1978. *Chem. Phys.* 35:177–86
136. Schulze, W., Becker, H. U., Minkwitz, R., Manzel, K. 1978. *Chem. Phys. Lett.* 55:59–61
137. Grinter, R., Armstrong, S., Jayasoorinya, U. A., McCombie, J., Norris, D., Springall, J. P. 1980. *Faraday Symp. Chem. Soc.* 14:94–101
138. Baetzold, R. C. 1976. *J. Phys. Chem.* 80:1504–9; 1978. *J. Chem. Phys.* 68:555–61
139. Mitchell, J. W. 1978. *Photogr. Sci. Eng.* 22:1–6
140. Sahyun, M. R. V. 1978. *Photogr. Sci. Eng.* 22:317–21; 1980. *J. Chem. Educ.* 57:239–42
141. Ozin, G. A., Klotzbücher, W. 1979. *Inorg. Chem.* 18:2101–8
142. Basch, H. 1980. *Faraday Symp. Chem. Soc.* 14:149–58
143. Basch, H. 1981. *J. Am. Chem. Soc.* 103:4657–63
144. Huber, H., Mackenzie, P., Ozin, G. A. 1980. *J. Am. Chem. Soc.* 102:1548–52
144a. McLean, A. D. 1983. *J. Chem. Phys.* 79:3392–3403
144b. Klobukowski, M. 1983. *J. Comput. Chem.* 4:350–61
145. Kordis, J., Gingerich, K. A., Seyse, R. J. 1974. *J. Chem. Phys.* 61:5114–21
146. Gingerich, K. A. 1981. See Ref. 6, pp. 109–23
147a. Lee, Y. S., Ermler, W. C., Pitzer, K. S., McLean, A. D. 1979. *J. Chem. Phys.* 70:288–92
147b. Ermler, W. C., Lee, Y. S., Pitzer, K. S. 1979. *J. Chem. Phys.* 70:293–98
148. Pitzer, K. S. 1979. *Acc. Chem. Res.* 12:271–76
149a. Ames, L. L., Barrow, R. F. 1967. *Trans. Faraday Soc.* 63:39–44
149b. Kleman, B., Lindquist, S., Selin, L. E. 1954. *Ark. Fys.* 8:505–10
150a. Brewer, L. 1962. *Electronic Structure and Alloy Chemistry of the Transition Elements*, ed. P. A. Beck, pp. 221–35. New York: Interscience
150b. Brewer, L. 1967. *Acta Metall.* 15:533–36
150c. Brewer, L. 1968. *Science* 161:115–22
151a. Engel, N. 1949. *Kem. Maanedsbl.* 30:53, 75, 97, 105, 113
151b. Engel, N. 1964. *Am. Soc. Metals, Trans. Q.* 57:610–19
152. Brewer, L. 1968. As referenced in Ref. 126, footnote 4
153. Cocke, D. L., Gingerich, K. A. 1974. *J. Chem. Phys.* 60:1958–65
154. Shim, I., Gingerich, K. A. 1983. *Chem. Phys. Lett.* 101:528–34
155. Van Zee, R. J., Weltner, W. Jr. 1984. *High Temp. Sci.* In press
156. Van Zee, R. J., Weltner, W. Jr. 1984. *Chem. Phys. Lett.* In press
157. Dyson, W., Montano, P. A. 1978. *J. Am. Chem. Soc.* 100:7439–41
158. Montano, P. A. 1978. *J. Appl. Phys.* 49:1561–63
159. Montano, P. A., Talarico, M. A. 1979. *J. Appl. Phys.* 50:2405–7
160. Shim, I. 1981. *Theor. Chim. Acta* 59:413–21
161. Shim, I. 1980. *Theor. Chim. Acta* 54:113–22
162. Baumann, C. A., Van Zee, R. J., Weltner, W. Jr. 1983. *J. Chem. Phys.* 79:5272–79
163. Kant, A. 1968. *J. Chem. Phys.* 48:523–25
164. Efremov, Y. M., Samoilova, A. N., Gurvich, L. V. 1976. *Chem. Phys. Lett.* 44:108–9
165. Klotzbücher, W., Ozin, G. A., Norman, J. G. Jr., Kolari, H. J. 1977. *Inorg. Chem.* 16:2871–77
166. Keller, J. 1983. *J. Mol. Struct.* 93:93–110
167. Van Zee, R. J., Baumann, C. A., Weltner, W. Jr. 1984. To be published
168. Knight, L. B. Jr., Woodward, R. W., Van Zee, R. J., Weltner, W. Jr. 1983. *J. Chem. Phys.* 79:5820–77
169a. Schumacher, E., Gerber, W. H., Härri, H. P., Hofmann, M., Scholl, E. 1982. *Am. Chem. Soc. Symp. Ser.* 179:83–107
169b. Garland, D. A., Lindsay, D. M. 1983. *J. Chem. Phys.* 78:2813–16
170. Shamai, S., Pasternak, M., Micklitz, H. 1982. *Phys. Rev. B.* 26:3031–37
171. Moskovits, M., DiLella, D. P. 1980. *J. Chem. Phys.* 72:2267–71
172. Blyholder, G. 1974. *Surf. Sci.* 42:249–60
173. Howard, J. A., Preston, K. F., Sutcliffe, R., Mile, B. 1983. *J. Phys. Chem.* 87:536–37
174. Howard, J. A., Sutcliffe, R., Mile, B. 1983. *J. Am. Chem. Soc.* 105:1394
175. Howard, J. A., Preston, K. F., Mile, B. 1981. *J. Am. Chem. Soc.* 103:6226

176. Morse, M. D., Hopkins, J. B., Langridge-Smith, P. R. R., Smalley, R. E. 1983. *J. Chem. Phys.* 79:5316–28
177. Schulze, W., Becker, H. U., Abe, H. 1978. *Ber. Bunsenges. Phys. Chem.* 82:138–39
178. Brewer, L., King, B. 1970. *J. Chem. Phys.* 53:3981–87
179. Moskovits, M., Hulse, J. E. 1977. *J. Chem. Phys.* 67:4271–78
180. Ozin, G. A., Mitchell, S. A., McIntosh, D. F., Mattar, S. M., Garcia-Prieto, J. 1983. *J. Phys. Chem.* 87:4651–65
181. DiLella, D. P., Taylor, K. V., Moskovits, M. 1983. *J. Phys. Chem.* 87:524–27
182. Hilpert, K., Gingerich, K. A. 1980. *Ber. Bunsenges. Phys. Chem.* 84:739–45
183a. Thompson, G. A., Lindsay, D. M. 1981. *J. Chem. Phys.* 74:959–68
183b. Lindsay, D. M., Garland, D., Tischler, F., Thompson, G. A. 1982. See Ref. 6, pp. 69–82
184. Richtsmeier, S. C., Gole, J. L., Dixon, D. A. 1980. *Proc. Natl. Acad. Sci. USA* 77:5611–15
185. Baetzold, R. C. 1971. *J. Chem. Phys.* 55:4363–70
186. del Condë, G., Bagus, P. S., Novaro, O. 1982. *Phys. Rev. B* 25:7843–45
187. Novaro, O., Garcia-Prieto, J. 1982. *Kinam* 4:421–43
188. Jeung, G. H., Pelissier, M., Barthelat, J. C. 1983. *Chem. Phys. Lett.* 97:369–72
189. Demuynck, J., Rohmer, M., Strich, A., Veillard, A. 1981. *J. Chem. Phys.* 75:3443–53
190. Howard, J. A., Sutcliffe, R., Tse, J. S. 1983. *Chem. Phys. Lett.* 94:561–64
191. Howard, J. A., Sutcliffe, R., Mile, B. 1983. *J. Phys. Chem.* 87:2268–71
192. Salahub, D. R. 1983. *Impact of Cluster Physics in Material Science and Technology*, ed. J. Davenas. The Hague: Nijhoff
193. Stone, A. J. 1980. *Mol. Phys.* 41:1339–54; 1981. *Inorg. Chem.* 20:563–71

DIFFERENTIAL ABSORPTION AND DIFFERENTIAL SCATTERING OF CIRCULARLY POLARIZED LIGHT: Applications to Biological Macromolecules

Ignacio Tinoco, Jr. and Arthur L. Williams, Jr.

Chemistry Department and Laboratory of Chemical Biodynamics, University of California, Berkeley, California 94720

INTRODUCTION

A chiral object is defined as an object that is not superimposable on its mirror image. (The group theoretical definition of a chiral object is one that has no rotation-reflection symmetry axes, S_n). Chiral objects can interact differently with left and right circularly polarized light. Therefore, it is of interest to study this interaction and to learn how this chiral interaction differs from the interaction with unpolarized, or linearly polarized, light. Our emphasis is on large chiral molecules, such as proteins, nucleic acids, and viruses. "Large" here means relative to the wavelength of light, but as circularly polarized X-radiation becomes available in synchrotron sources (1, 2), "large" will include any molecule.

THEORY FOR LARGE MOLECULES

For an electronic transition from initial state $\langle i|$ to final state $|f\rangle$, the probability of absorption of light is:

$$\text{Absorption } (a) \propto \mathbf{e}_0 \cdot \langle i|e^{i\mathbf{k}_0 \cdot \mathbf{r}}\mathbf{p}|f\rangle \langle f|e^{-i\mathbf{k}_0 \cdot \mathbf{r}}\mathbf{p}^*|i\rangle \cdot \mathbf{e}_0^* \qquad 1.$$

where $\mathbf{e}_0$ is a vector specifying the polarization of the incident light of wave vector $\mathbf{k}_0$ with magnitude $2\pi/\lambda$. The vectors $\mathbf{r}$ and $\mathbf{p}$ are the position and linear momentum operators of each electron, and a sum of the matrix elements over all electrons is implied. The asterisk represents the complex conjugate. If Eq. 1 is written for incident left and right circularly polarized light and subtracted, the expression for differential absorption is obtained (3).

Circular differential absorption $(a_\mathrm{L} - a_\mathrm{R}) \propto$

$$\mathbf{k}_0 \cdot \langle i|e^{i\mathbf{k}_0 \cdot \mathbf{r}}\mathbf{p}|f\rangle \times \langle f|e^{-i\mathbf{k}_0 \cdot \mathbf{r}}\mathbf{p}^*|i\rangle \qquad 2.$$

The differential absorption of circularly polarized light depends on the vector cross-product of the matrix elements. This introduces the chiral nature of the effect. The mirror image of the object will have the opposite sign of circular differential absorption; an achiral object will have zero circular differential absorption. Comparison of Eqs. 1 and 2 shows why one needs more accurate wavefunctions to calculate circular differential absorption. For wavelengths large compared to molecular dimensions (more precisely, for wavelengths large compared to the distance of significant electron correlation), the exponentials in Eq. 1 can be replaced by unity. This gives the dipole approximation for absorption. A similar substitution in Eq. 2 gives zero. Higher order terms in the expansion of the exponential must be retained (or no expansion made at all), and better wavefunctions are required for calculation of circular differential absorption. In Eqs. 1 and 2 we have omitted the contribution from the spin of the electrons; we consider the $\mathbf{A} \cdot \mathbf{p}$ term in the interaction of the molecule with the vector potential $\mathbf{A}$ of the light, but drop the $\nabla \times \mathbf{A} \cdot \mathbf{S}$ term.

The scattering of light can be written (4) in terms of the same matrix elements of Eqs. 1 and 2, but a classical approach is more useful (5). The scattering of light is related to the polarizability tensor at each point in the object. The polarizability tensor is in general complex, and it can include the effects of interactions with other parts of the sample. The scattered light intensity in the direction of wave vector $\mathbf{k}$ is:

$$\text{Intensity scattered } (I) \propto [\mathbf{e}_0^* \cdot \boldsymbol{\alpha}_i^\dagger \cdot (\mathbf{1} - \hat{\mathbf{k}}\hat{\mathbf{k}}) \cdot \boldsymbol{\alpha}_j \cdot \mathbf{e}_0] e^{i\Delta\mathbf{k} \cdot (\mathbf{r}_j - \mathbf{r}_i)} \qquad 3.$$

Here $\boldsymbol{\alpha}_i$ and $\boldsymbol{\alpha}_j$ are the polarizability tensors at $\mathbf{r}_i$ and $\mathbf{r}_j$; $\Delta\mathbf{k} = \mathbf{k} - \mathbf{k}_0$, and $\hat{\mathbf{k}}$ is the unit wave vector for the scattered light. The complex conjugate of the transpose of $\boldsymbol{\alpha}_i$ is designated $\boldsymbol{\alpha}_i^\dagger$. This is the general equation for scattering of radiation; the sum over i and j is understood. In the X-ray region at wavelengths shorter than any absorption bands, the polarizabilities become spherical and directly proportional to the electron density; the standard equation for electron diffraction is obtained. The circular differential scattering is obtained from Eq. 3 by explicitly introducing incident left and right circularly polarized light. The notation is easier if Eq. 3 is written in terms of principal axes of the polarizability tensors.

$$\alpha_i = \sum_{l=1}^{3} \alpha_{il} \mathbf{e}_{il} \mathbf{e}_{il}$$

We shall write $\alpha_i = \alpha_i \mathbf{e}_i \mathbf{e}_i$ and imply the sum over the three principal axes.

Intensity scattered $(I) \propto$

$$\alpha_i^* \alpha_j (\mathbf{e}_0^* \cdot \mathbf{e}_i)(\mathbf{e}_0 \cdot \mathbf{e}_j)[\mathbf{e}_i \cdot \mathbf{e}_j - (\mathbf{e}_i \cdot \hat{\mathbf{k}})(\mathbf{e}_j \cdot \hat{\mathbf{k}})] e^{i\Delta \mathbf{k} \cdot (\mathbf{r}_j - \mathbf{r}_i)} \qquad 4.$$

Differential scattering of incident circularly polarized light $(I_L - I_R) \propto$

$$\alpha_i^* \alpha_j (\mathbf{e}_i \times \mathbf{e}_j) \cdot \mathbf{k}_0 [\mathbf{e}_i \cdot \mathbf{e}_j - (\mathbf{e}_i \cdot \hat{\mathbf{k}})(\mathbf{e}_j \cdot \hat{\mathbf{k}})] e^{i\Delta \mathbf{k} \cdot (\mathbf{r}_j - \mathbf{r}_i)} \qquad 5.$$

As in Eq. 3 for differential absorption, the chirality of the effect is apparent from the cross product.

If we consider interactions of light with matter in which there is no change in energy of the light and in which only linear interactions occur (the incident light intensity is not too high), then the maximum information obtainable about the system is present in the Mueller matrix (6). The Mueller matrix is the 4×4 matrix, which operates on the 4 components of the Stokes vector of the incident light to produce the 4 components of the Stokes vector of the outgoing light. The Mueller matrix thus characterizes the effect of the system on the light. The Stokes vectors represent the intensity and polarization (in general, the orientation and ellipticity of elliptically polarized light) of the light. In this review we emphasize only one of the four Stokes parameters of the outgoing light (the intensity) as a function of two of the Stokes parameters of the incident light (the total intensity and the intensity of left or right circularly polarized light). We thus need only to consider two of the 16 components of the Mueller matrix to understand either circular dichroism (CD) or circular intensity differential scattering (CIDS). Most of the applications of CD and CIDS have been to molecules in solution, therefore it is necessary to average Eqs. 1–5 over all orientations of the molecule.

The orientation average for the circular dichroism of large molecules (Eq. 2) was first given by Tobias et al (7). The corresponding equation for small molecules (remember that small molecules have significant electron correlation over distances small compared to the wavelength of light) is the Rosenfeld (8) expression relating the CD to the dot product of the electric dipole and magnetic dipole transition moments. Harris et al (9) have presented a very general treatment relating the complete Mueller matrix to the molecular properties of the sample interacting with the light. By expanding their equations in spherical harmonics and using properties of tensor invariants, they obtain expressions for the orientation-averaged components of the Mueller matrix; this provides the orientation average of Eqs. 4, 5. In our notation (4)

Orientation-averaged circular differential scattering $\langle I_L - I_R \rangle \propto$

$$\alpha_i^* \alpha_j e_j \times \mathbf{e}_i \cdot \hat{\mathbf{R}}_{ij} \left\{ \frac{(\mathbf{e}_i \cdot \mathbf{e}_j)}{5} [j_3(q) - 4j_1(q)] \right. \\ \left. - (\mathbf{e}_i \cdot \hat{\mathbf{R}}_{ij})(\mathbf{e}_j \cdot \hat{\mathbf{R}}_{ij}) [j_3(q)] \right\} (\sin \beta - \sin^3 \beta). \qquad 6.$$

Here β is half the scattering angle (half of the angle between $\mathbf{k}$ and $\mathbf{k}_0$); $q = (4\pi/\lambda)R_{ij} \sin \beta$; $j_1(q)$ and $j_3(q)$ are spherical Bessel functions; $\hat{\mathbf{R}}_{ij}$ and R_{ij} are the unit vector and magnitude of the vector $\mathbf{r}_j - \mathbf{r}_i$. There is a similar equation for the orientation-averaged intensity (Eq. 4) which is a function of $j_0(q), j_2(q)$ and $j_4(q)$. The original definition of CIDS, which is the quantity measured experimentally, was given by Barron & Buckingham (10). They obtained a molecular expression for CIDS for small molecules.

$$\text{CIDS} = \frac{I_L - I_R}{I_L + I_R} \qquad 7.$$

The orientation-averaged CIDS is the average of the numerator, Eq. 6, divided by the average of the denominator (5). Bustamante et al have made calculations of the scattering patterns for oriented (11) and randomly oriented helices (5, 12). McClain et al (13) have calculated the angular dependence of the components of the Mueller matrix for various randomly oriented objects.

Equation 5 was obtained using the first Born approximation for the interaction of the light with the system. Each point in the system interacts with the incident field of the light. Better approximations can be made by including the interactions between polarizable groups in Eq. 5. Higher order polarizabilities can also be included in the interaction of each group with the light (14–16). The second and higher Born approximations are particularly important (9, 16) because they are required to calculate circular differential scattering in the forward direction. Equation 5 requires that $\langle I_L - I_R \rangle$ be zero in the forward direction ($\beta = 0$); it is the interaction between the groups that directly gives rise to circular differential effects (including circular dichroism and optical rotation) in the forward direction. However, circular differential scattering (as obtained from Eq. 5) can contribute to circular differential extinction in the forward direction, as we describe next.

Circular dichroism is usually thought of as circular differential absorption (Eq. 2), but what is actually measured is circular differential extinction, which is the sum of absorption and scattering effects.

$$\varepsilon_L - \varepsilon_R = (a_L - a_R) + (s_L - s_R) \qquad 8.$$

The scattering contribution is obtained by integrating the circular differential scattering (Eq. 6) over all angles (17). Clearly the light scattered at any angle not included in the acceptance angle of the detector will not reach the detector, and the acceptance angle can be made vanishingly small. Therefore, the integral of the circular differential scattering gives the scattering contribution to the circular dichroism. This scattering contribution has long been recognized (18), but only recently has an explicit molecular interpretation been provided (17). Outside of absorption bands only $(s_L - s_R)$ contributes to circular dichroism; it adds monotonically decreasing positive or negative tails to the CD spectra. Inside of absorption bands the ratio of the contributions to CD is approximately

$$\frac{(s_L - s_R)}{(a_L - a_R)} \sim \frac{R_{ij}^3}{\lambda^3}. \qquad 9.$$

This equation is valid for R_{ij} equal to or less than λ. It shows that large chiral objects such as viruses can provide a significant scattering contribution to CD; smaller molecules, such as proteins with no dimension larger than 1/10 the wavelength of light, should have negligible scattering contribution to CD.

Detailed calculations for real macromolecules using Eq. 2 for CD or Eqs. 5 or 6 for CIDS have not been performed. Instead, calculations have been done for model systems.

CALCULATIONS FOR LARGE OBJECTS

Balazs et al (19) have calculated the rotation of linearly polarized light by a free electron constrained to a curve of arbitrary shape and size. This is equivalent to calculating the circular dichroism, because the wavelength dependence of optical rotation (optical rotatory dispersion) is the Kronig-Kramers transform of the circular dichroism. Moore & Tinoco (3) calculated the optical rotation of a free electron on an oriented helix. The calculated signs and magnitudes of the rotation agreed well with experimental measurements of the rotation of microwaves by wire helices (20).

Maestre et al (21) measured the circular intensity differential scattering (CIDS) as a function of angle for a solution of sperm heads from the octopus *Eledone cirrhosa*. These sperm are uniform left-handed screws with a pitch of 650 nm, an outer radius of about 310 nm and an inner radius of 120 nm. A scanning electron micrograph is shown in Figure 1. The CIDS was measured with light from a HeCd laser at 442 nm; the data are shown in Figure 2. CIDS = $(I_L - I_R)/(I_L + I_R)$ is unitless with a value between plus and minus 1. Experimentally, left circularly polarized light was found to be preferentially scattered in the forward direction and right circularly polarized light was preferentially scattered to the side. A maximum positive value of 1.4×10^{-2} was found at about 35°.

The sperm-head screw was modeled by a helix of noninteracting point polarizabilities with tangential polarizability only; radial and axial components of the polarizability were zero (M. F. Maestre, personal communication). One hundred polarizabilities per turn were used to model a continuous helix, and 7.7 turns were included. The pitch was fixed at 650 nm, and the radius was varied to give the best fit to the experimental data. A left-handed helix with a radius of 250 nm gave a maximum positive value of 2.1×10^{-2} at 37°. The calculated curve is shown in Figure 3. The agreement

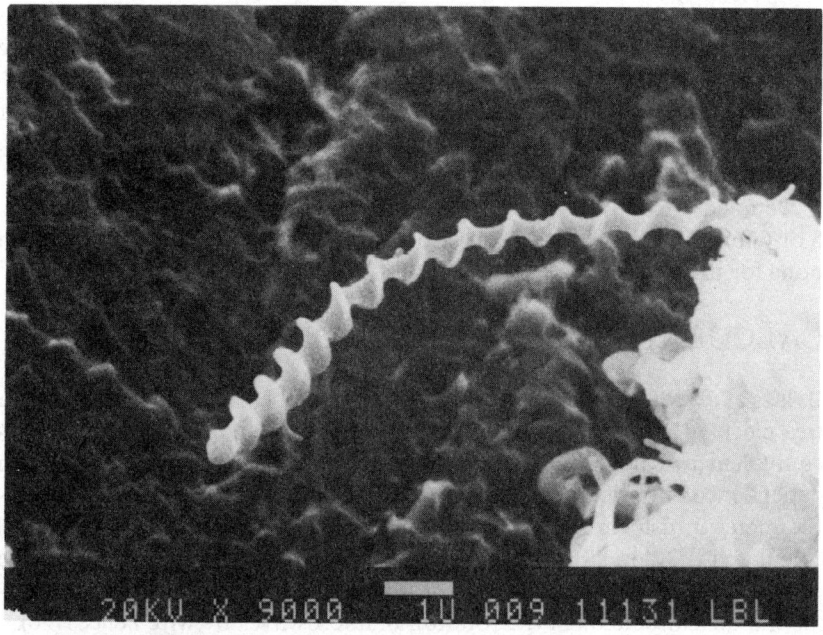

Figure 1 A scanning electron micrograph of the helical sperm head of the Mediterranean octopus *Eledone cirrhosa*. It is a left-handed screw with a pitch of 650 nm, an outer radius of 310 nm, and an inner radius of 120 nm. Reproduced with permission from Maestre et al (21), *Nature* 298: 773–74 (1982).

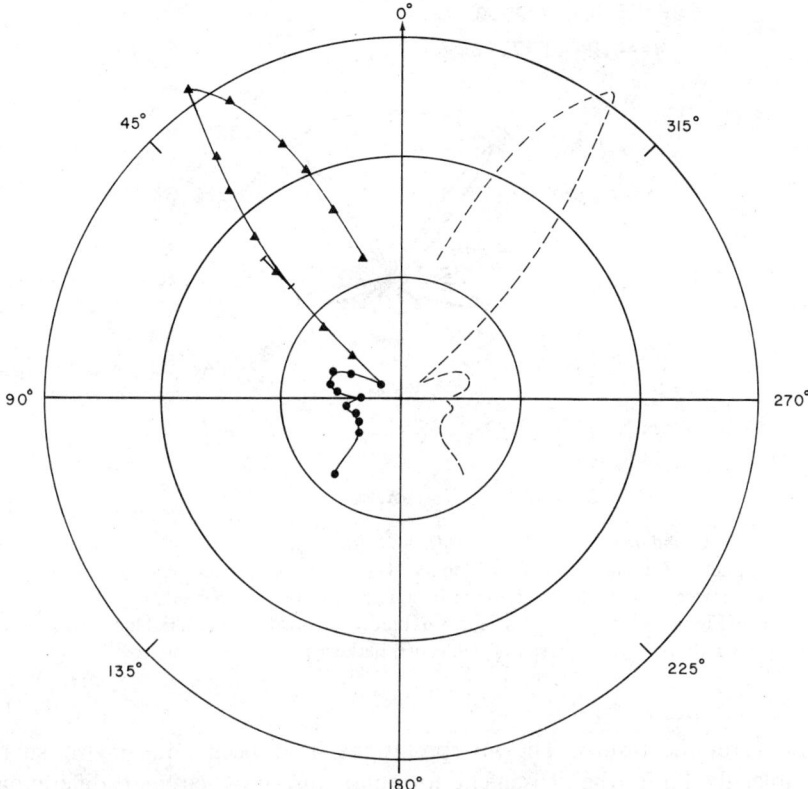

Figure 2 The measured circular intensity differential scattering pattern for a solution of the sperm head shown in Figure 1. The light (442 nm) is incident from the bottom of the picture. $(I_L - I_R)/(I_L + I_R)$ is plotted vs the scattering angle; (▲) labels positive values; (●) labels negative values. Reproduced with permission from Maestre et al (21) *Nature* 298:773–74 (1982).

with experiment in sign, magnitude, and general shape suggests that the interpretation of CIDS patterns in terms of simple theory is reasonable.

CIRCULAR DICHROISM OF PROTEINS AND NUCLEIC ACIDS

Most of the recent experimental work on chiral macromolecules has been measurements of the circular dichroism of proteins and nucleic acids. Research in the vacuum ultraviolet was reviewed in *Annual Reviews* in 1977 (22), and a more general review appeared in 1980 (18). Here we limit the discussion to measurements of circular dichroism and to attempts to

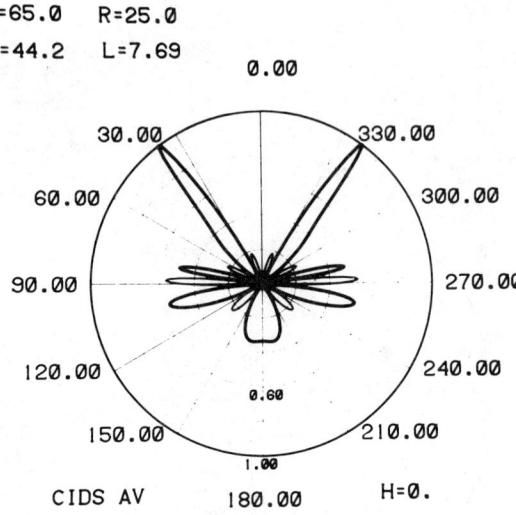

Figure 3 Calculated values of $(I_L - I_R)/(I_L + I_R)$ for a randomly oriented left-handed helix with a pitch of 650 nm and radius of 250 nm. The *heavy lines* are positive; the *light lines* are negative; note that the lobes alternate in sign. Satisfactory agreement with the measured values of Figure 2 is observed. This figure was kindly provided by Dr. M. F. Maestre, Lawrence Berkeley Laboratory, University of California, Berkeley.

interpret the results. The interpretations have been done in the small molecule limit (the Rosenfeld formula) and only circular differential absorption has been considered. Both of these assumptions are probably valid for the molecules considered.

Polypeptides and Proteins

SECONDARY STRUCTURE ANALYSIS   Circular dichroism (CD) has been widely used for studying the structure and conformation of proteins. Several methods have been developed to determine the secondary structure of a protein in solution from its measured CD. The general procedure for protein structural analysis by CD has been to find a correlation between measured CD spectra and basic secondary structures. If the CD spectra of these structures are linearly independent functions, then the CD of a protein at each wavelength, $\theta(\lambda)p$, can be calculated from the following:

$$\theta(\lambda)p = \sum_{i=1}^{N} f_i \theta_i(\lambda) \qquad 10.$$

where $\theta_i(\lambda)$ is the CD for the ith type of structure and f_i is the fraction of the protein in the ith type of structure.

The fundamental problem in the application of this analysis is the choice of $\theta_i(\lambda)$, the basis spectra. Two approaches (23–28) have been used to analyze protein structure by CD in terms of α-helix, β-sheet, and random coil fractions. One approach (23, 24) has been to evaluate $\theta_i(\lambda)$ from the CD data of synthetic model polypeptides of known structure. Another (25–28) has been to extract the α-helix, β-sheet, and random coil basis spectra from experimental CD spectra of proteins, whose secondary structure has been determined from x-ray diffraction of their crystals.

Analyses based on these two approaches have utilized a least-squares procedure to find the set of f_i which best fit the $\theta_i(\lambda)$ to the experimental CD spectrum. However, statistical analysis (29) of these approaches have shown that only the α-helix content can be determined with reasonable accuracy. Furthermore, most of these analyses have been restricted to a relatively narrow spectral range between 200 and 240 nm, where only a few electronic transitions of the polypeptide backbone occur.

It is well known that proteins in nature are not limited to the three basic secondary structures defined in these analyses. β-turns are widely recognized to play a significant role in the conformation of proteins (30). Chang et al (28) have obtained improved accuracy in determining secondary structure by including a β-turn reference spectrum into the basis spectra. Increasing the number of basis spectra further by splitting β-sheet into parallel and antiparallel classes, Bolotina et al (31, 32) have obtained greater accuracy.

When the CD spectral region was extended into the vacuum ultraviolet (VUV), a major improvement of the basis set occurred. Brahms & Brahms (33) have obtained CD data to 165 nm and have extended the basis set by including the contribution of three types of β-turns. The CD spectra of poly (Ala_2-Gly_2), L-Pro-D-Ala and N-acetyl-L-Pro-Gly-L-Leu-OH were used as reference spectra for β-turns. In addition, they have selected new reference spectra for α-helix, β-sheet, and unordered structures. For each type of secondary structure, the results of Brahms & Brahms (33) gave better agreement with experiment than Chang et al (28). However, the results for β-turns were poor when compared with available x-ray data (34).

A major assumption present in all these methods is that each type of secondary structure has a characteristic CD spectrum. In addition, the influence of the length of a segment of secondary structure, possible interactions among secondary structural elements, and contributions from nonpeptide chromophores have either been neglected or have been empirically corrected. Two differing approaches (35, 36) that avoid the above assumption and average out the other effects have been developed.

Provencher & Glockner (35) have developed a method of analyzing a protein CD spectrum directly as a linear combination of the CD spectra

(190 to 240 nm) of 16 proteins whose secondary structures are known. The fraction of each of the 16 protein spectra, γ, which sum to give the analyzed spectrum, is determined by a constrained regularization procedure. This procedure incorporates a least-squares term to minimize the difference between the calculated and measured spectrum, and a regularizor term which keeps γ small unless its corresponding CD spectrum happens to fit the measured spectrum well. Furthermore, the f_i values are constrained to be positive and sum to one. The f_i values are determined from a knowledge of the γ values and the fraction of residues of each protein in a particular conformational class. The method allows for a more flexible analysis of CD spectra, since freedom from defining basis spectra for each conformational class enables the determination of any number of classes that can be accurately assigned by x-ray data. Using their method, Provencher & Glockner (35) determined the f_i values of four conformational classes and found a very good correlation for both α-helix and β-sheet with x-ray data.

Hennessey & Johnson (36) have taken a different approach to the determination of a basis set of CD spectra. An orthogonal basis set of CD spectra was generated from the CD of 16 proteins (178 to 260 nm) by an eigenvector method of multicomponent matrix analysis. Each of the basis CD spectra have structural contributions of one or more secondary structural elements. The protein CD spectrum to be analyzed is reconstructed from the orthogonal set of CD spectra. Since each of the basis spectra corresponds to one or more structural elements, the coefficients from the analysis can be used to determine the secondary structure. The basis set was used by Hennessey & Johnson (36) to determine the f_i values of eight different types of secondary structure (α-helix, two types of β-sheet, four types of β-turn, and unordered structures). They have found that eliminating the 185 and 193 nm CD bands resulted in dramatic changes in the determination of the β-strand and unordered structures. Therefore, proper analysis must include information from the VUV. Overall, their secondary structural determinations gave a good correlation with x-ray determined results.

It should be noted that all the methods described are sensitive to the measured CD spectra of the reference proteins and polypeptides. Recently, Hennessey & Johnson (37) have investigated the types of errors that may interfere with the conformational analysis of protein by CD. The analyses were found to be extremely sensitive to the vertical shift of a CD spectrum, which occurs when the baseline is improperly aligned with the spectrum. Small errors in CD intensity at short wavelengths, due to low light intensity and high sample absorptions in this region, were found to contribute to poor analyses. Long-term baseline drift was also found to greatly affect the analyses. Thus, high stability and careful calibration of the instrument are necessary for successful and reproducible analyses to be carried out.

THEORETICAL STUDIES   Failure of CD structural analyses has been attributed to contributions of aromatic chromophores to the spectral region below 250 nm (33, 38). Many proteins have been found to exhibit positive CD bands in the 220–250 nm region [see Ref. (39) and references therein], which from their position are probably attributable to aromatic groups. Woody (39) has calculated the contribution of the L_a transition of phenylalanine and tyrosine side chains to the CD of peptides and proteins in the region 200–250 nm. His results demonstrated a strong bias favoring positive CD bands from the L_a transition in various backbone and side-chain conformations. The contribution was shown to be an order of 10% of the observed CD, even in proteins with rather strong amide contributions.

Goux & Hooker (40, 41) have calculated the CD of the enzyme ribonuclease (RNase) S. Their calculations have shown that the positive CD near 240 nm in the spectrum of RNase S includes major contributions from the L_a transitions of the tyrosine residues, especially from tyrosine-92. In addition, the $n - \sigma^*$ transitions of the disulfide chromophores coupling with the ^1B_b and ^1B_a transitions of the tyrosines and the peptide $\pi - \pi^*$ transitions make positive contributions to the CD in the 240–260 nm region (41). In another study, Goux & Hooker (42) have shown that the tyrosine L_a and the tryptophan ^1B_b transition give rise to positive contributions to the red of 224 nm and negative contributions to the blue of 224 nm in the calculated CD of lysozyme.

As well as not considering aromatic chromophores adequately, CD secondary structural analyses also fail to consider the dependence of the CD on the structure and conformation of these groups. The side chains of tyrosine and phenylalanine are able to rotate freely about the C^β–C^γ bond connecting the aromatic ring to the rest of the molecule. The L_b transition of tyrosine and phenylalanine is very sensitive to rotation about this bond (40, 42), which results from the fact that it is polarized perpendicularly to the rotation. The contribution of the tyrosine-76 L_b transition to the calculated CD of RNase S is two-fold larger in an energy-minimized conformer than the conformer found in the crystal (42). Similarly, the contribution of the tyrosine L_b transition to the near UV-calculated CD spectrum of adenylate kinase, based on the solid-state structure, was found to be greater than that observed in the measured spectrum (43).

Applequist (44, 45) has developed a model in which interactions between nonchromophoric groups and peptide $\pi - \pi^*$ transitions are taken into account. These studies (44–49) have demonstrated that there may be a very strong effect due to side-chain conformation on polypeptide CD. Variations of side-chain produced changes in band intensity by as much as two-fold, shifts in band positions as much as 10 nm, and sign reversal in others. Shifts in band intensity and position were also found for subtle changes in bond lengths of the backbone for spectra of helical polyproline I and II. While the calculations may have exaggerated the effects of these interactions, they nevertheless demonstrate that side-chain conformation does have some effect on the observed CD.

Woody (50) has recently studied the contribution to the CD of some regular polypeptide structures caused by deviations of the amide from planarity. Nonplanar amides are inherently chiral chromophores and are therefore expected to exhibit CD bands. Woody (50) has calculated the CD of chiral amides as a function of the angles describing the torsion about the peptide bond (ΔW) and the pyramidalization at the peptide nitrogen (θ_N). Considering the largest deviations from planarity expected for α-helices, he found the contributions to the rotational strength small when compared with the total rotational strength due to the $n-\pi^*$ and $\pi-\pi^*$ transitions. For positive values of ΔW, the calculated effects are of the same sign as those observed for β-pleated sheets and are sufficiently large to make a significant contribution to the observed CD. The calculations also indicated that chiral amide effects may be significant in understanding β-turn CD. The observed red-shift of the CD spectrum of the unordered form of polypeptides could be explained by negative values of ΔW. Unfortunately, there is no evidence that L-amino acids in polypeptides show such a bias.

Bowman et al (51) have recently measured the CD of three cyclic dimers of amino acids into the far-UV to discern what transitions can account for the observed CD. Their calculations have shown that amide $n-\pi^*$ and $\pi-\pi^*$ transitions could account for the CD at wavelengths longer than 180 nm. However, the $n-\sigma^*$, $\pi-\pi^*$, and $\pi-\sigma^*$ transitions of the amide were insufficient in accounting for the observed intensity of the CD bands at wavelengths shorter than 180 nm. Only the $\sigma-\sigma^*$ transitions of the backbone could account for the CD spectra observed below 180 nm. The lowest energy $\sigma-\sigma^*$ transition for the cyclic dimers, c(L-Ala-L-Gly) and c(L-Ala-L-Ala), only involved the transitions from the bonds in the backbone. However, the first two $\sigma-\sigma^*$ transitions in c(L-Pro-L-Pro) fell at 174 nm and involved the two prolyl rings. Thus, backbone and side-chain conformation will heavily influence the observed CD in the far-UV. Furthermore, since $\sigma-\sigma^*$ transition moments are directed along the bonds, these results indicate that the spectral range below 180 nm may provide a fruitful region to explore polypeptide secondary structure.

Polynucleotides

NEAREST-NEIGHBOR ANALYSIS The dependence of CD spectra of polynucleotides on sequence and conformational effects has been extensively studied for many years. In 1970, Gray & Tinoco (52) showed that if conformation is not a variable, any sequence dependent property of a polynucleotide chain can be expressed as a linear function of the properties of a "basis set" of polynucleotides of known sequence. The number of polymers in the basis set depends upon the furthest-neighboring base pairs that influence the sequence dependent property. If only nearest-neighbor interactions contribute to the observed CD of a polynucleotide, then a set of eight linearly independent polynucleotide CD spectra of known sequence are needed to (a) predict the CD of any polynucleotide of known sequence, or (b) determine the nearest-neighbor frequencies from a given polynucleotide CD spectrum.

According to the nearest-neighbor theory of Gray & Tinoco (52), the CD spectrum of any polynucleotide can be calculated from

$$\underset{n \times 1}{s} = \underset{n \times 10}{T} \underset{10 \times 1}{f} \qquad 11.$$

where s is a column vector of the CD spectrum at wavelength n, T is an $n \times 10$ matrix whose columns represent the basis set CD spectra of the ten nearest neighbors, and f is a column vector of the nearest-neighbor frequencies of the polynucleotide whose spectrum is s. For any set of m polynucleotide CD spectra, we have

$$\underset{n \times m}{S} = \underset{n \times 10}{T} \underset{10 \times m}{F} \qquad 12.$$

where the spectra, S, are recorded at the same wavelengths and under the same experimental conditions.

The fundamental problem with the application of Eq. 11 to find f for any polynucleotide is the determination of T, the basis set. Because of reentrant conditions on nearest-neighbor frequencies (53), there are only eight independent frequencies. Thus, the F matrix is of order 8 and Eq. 12 cannot be applied to find T from the CD spectra of polynucleotides of known sequence. As a result of the reentrant constraint, two formalisms, one based on single-strand units (54) and the other on double-stranded units (55), have been presented.

Allen et al (54) were the first to apply the theory of Gray & Tinoco (52) to estimate the nearest-neighbor frequencies of DNA from its CD spectrum. Allen et al (54) reasoned that because of the two reentrant conditions, there exists an $n \times 8$ subset matrix T^* of T. For m different measured polynucleotide CD spectra, where $m \geq 8$, T^* can be found from

$$T^* = SF^{*t}(F^*F^{*t})^{-1} \qquad 13.$$

where F^* is an $8 \times m$ matrix composed of the single-strand nearest-neighbor frequencies that include the reentrant conditions, F^{*t} is the transpose of F^*. Allen et al then obtain T from T^*, using the reentrant conditions. The values of f are found by Eq. 11 with the constraints that the components of f are all positive and sum to 1.0, and that Watson-Crick and reentrant conditions are obeyed. In addition, an optional base composition constraint could be incorporated.

Marck & Guschlbauer (55) have developed a formalism based on double-stranded units. They define a Z matrix that introduces the reentrant conditions and contracts F to an $8 \times m$ matrix, F'. Z is also used to define an $n \times 8$ matrix, T', which represents eight linear combinations of the ten base-pair nearest-neighbor CD spectra. T' is extracted from measured CD spectra in a manner analogous to T^*, Eq. 13. The eight independent nearest-neighbor frequencies, f', are determined directly from the CD spectrum of the polynucleotide by

$$f' = sT''(T'T'')^{-1}. \qquad 14.$$

The remaining two frequencies are determined from $f = Zf'$.

Using their procedure, Marck & Guschlbauer (56) obtained good estimates for the nearest-neighbor frequencies for five DNAs, which were not included in their basis set. However, although the determined frequencies obeyed the reentrant conditions, some frequencies were negative.

Marck & Guschlbauer (55–57) have questioned the validity of the method of Allen et al (54) for the determination of T. They have stated that the introduction of the reentrant conditions to determine T is both mathematically and physically unfounded. Recently, Allen et al (58) have reviewed both formalisms. They have shown that the T^* of Allen et al (54) and T' of Marck & Guschlbauer (55) are linear combinations of one another, related by a matrix of constants. Therefore, the two formalisms are essentially identical. However, the method of determining nearest-neighbor frequencies of Allen et al (54) uses a linear programming procedure that obeys a combination of constraints ignored by Marck & Guschlbauer (55).

Using the method of Allen et al (54), Gray et al (59, 60) combined the CD spectra of 12 different synthetic DNAs to give 21 basis sets of 8 spectra each. They tested these basis sets to find which combination of 8 spectra would lead to the best nearest-neighbor analysis of 14 different natural DNAs. The chosen basis set gave nearest-neighbor frequencies for the DNAs with an average absolute error between 3 and 13%. In a subsequent study, Gray et al (61) combined the CD spectra of 12 different synthetic RNAs to find an RNA basis set. However, by using the eigenvalue method (62, 63) they determined the number of independent spectral shapes in the 12 RNA spectra to be 7. Seven significant spectral components were also found in a reanalysis of the 12 DNA spectra. Thus, the data are not sufficient to determine all of the nearest-neighbor frequencies.

The mathematical formalism for nearest-neighbor analysis by CD is correct in principle (58). The inaccuracy of the method currently resides in the fact that the spectral components that comprise the T^* and T' matrices are not sufficiently independent. With only seven independent spectra, the current data do not fulfill the requirements of the nearest-neighbor theory of Gray & Tinoco (52). Extension of CD spectra into the VUV should increase the number of independent spectra of 8, and thus allow more accurate determinations.

Recently, Allen & Gray (64) refined a previous method (65) that described the use of CD measurements to analyze the nearest-neighbor units affected by ligand binding. The basic assumption of this method is that when a ligand binds to DNA, the electronic coupling between nearest-neighbors will be different in the presence and absence of the ligand. By following the manner in which the columns of the T^* matrix, Eq. 13, change as ligands are added to DNA, the nearest-neighbor unit or units that bind the ligand can be determined. Two of the nearest-neighbor units, ApA:TpT and GpG:CpC, may be unequivocally determined by this method. If fewer than six of the remaining eight nearest-neighbor units bind the ligand, then they too can be determined; otherwise, the binding can be characterized as nonspecific.

EXPERIMENTAL STUDIES CD has been widely used to distinguish between various types of double-stranded conformations of polynucleotides. A variety of conformations have been characterized by X-ray analysis of fibers in different relative humidities and counterions for DNA (66–68). RNA, however, has always been found in an A-type conformation (67). The CD of films of DNA at different relative humidities were studied by Tunis-Schneider & Maestre (69). Under controlled relative humidity conditions, they were able to correlate CD spectra with x-ray-determined A-, B-, and C-form conformations (Figure 4).

In 1972, Pohl & Jovin (70) reported that poly(dG-dC)·poly(dG-dC) underwent a salt-dependent, cooperative, reversible, conformational transition (Figure 5). The transition was found to be easily monitored by changes in the near-UV CD spectra. They labeled the transition R to L. Crystallographic (71–74) and Raman (75, 76) data have identified the high-salt L-form as a left-handed Z-form structure. The similarity in appearance of the low-salt R-form CD spectrum to that of the B-form DNA film CD spectrum has led to the suggestion that it is in the B-form secondary structure.

Sprecher et al (77) measured the CD spectra of the A-, B-, and C-forms of DNA into the VUV to 168 nm (Figure 6). A positive band at about 187 nm and a negative band at about 170 nm in the CD spectra were found to be sensitive to both the source of the DNA and the base-base interactions (indicative of DNA secondary structure). While B- and C-DNA have been observed to have very different CD spectra above 260 nm, their VUV CD spectra were found to be similar. This implies that their secondary structures are similar, confirming that C-DNA is a variant of B-DNA (68).

To determine whether the VUV CD was sensitive to the handedness of DNA, Sutherland et al (78) measured the CD of poly(dG-dC)·poly(dG-dC) in the B- and Z-forms to 180 nm (Figure 7). The B-form exhibited a positive

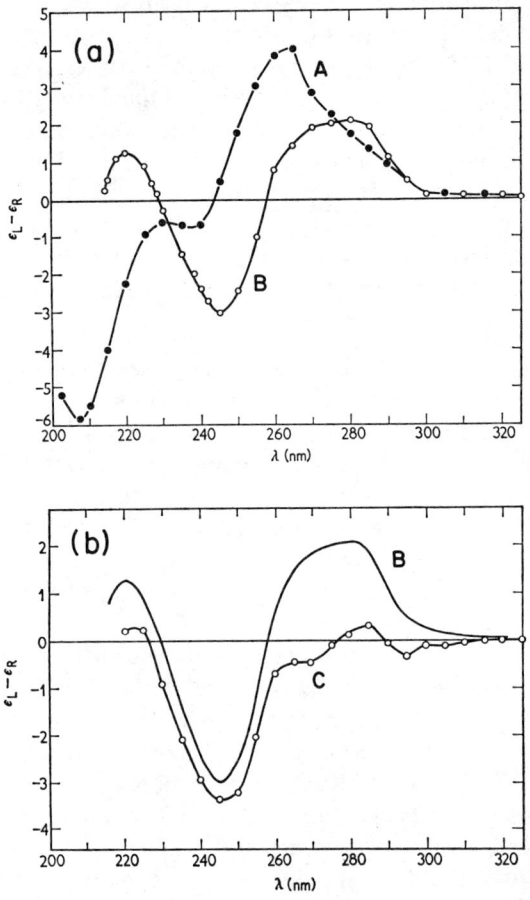

Figure 4 CD spectra of calf thymus DNA films under various conditions of humidity and salt content [Tunis-Schneider & Maestre (69)]. (*a*) Sodium salt of DNA in film as *A*-form at 75% relative humidity (r.h.) (—●—●—) and as *B*-form at 92% (r.h.) (—O—O—); (*b*) lithium salt of DNA as *C*-form at 75% (r.h.) (—O—O—) and as *B*-form at 92% (r.h.) (———). Reproduced with permission from *J. Mol. Biol.* 52: 521–41 (1970) (69).

band at 187 nm, characteristic of right-handed DNAs, whereas the Z-form spectra had a large negative peak at 194 nm followed by a positive band below 186 nm. The magnitude differences observed below 200 nm between the B- and Z-form spectra were ten times greater than those observed in the near UV, thus confirming the sensitivity of the VUV to DNA secondary structure. Brahms et al (79) have measured complementary single strand circles (form V DNA) to determine the handedness of this DNA. Based on VUV CD and Raman data, they were able to determine that form V DNA has left- and right-handed stretches. Recently, Sutherland & Griffin (80) measured the VUV CD spectrum of poly(dI-dC)·poly(dI-dC). Although the near UV CD of this polynucleotide resembles the near UV CD of the left-handed Z-form of poly(dG-dC)·poly(dG-dC), its VUV CD resembles that characteristic of right-handed structures. The VUV CD, as well as x-ray data (81, 82), provide no support for the belief that poly(dI-dC)·poly(dI-dC) is left-handed in solution.

Sprecher & Johnson (83) have monitored the intensities of the 275 nm and 190 nm CD bands as a function of temperature to test the sensitivity of the VUV band to melting. The 190 nm band underwent a 95%, 85%, and 55% decrease in intensity on melting for three DNAs with different $G+C$ contents. For the same DNAs, the 275 nm band underwent a 45%, 20%, and

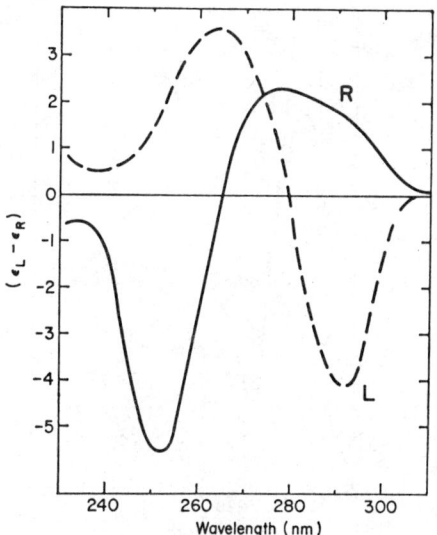

Figure 5 CD spectra of poly(dG-dC)·poly(dG-dC) in solution [Pohl & Jovin (70)]. As right-handed R-form (———) at 0.2 M NaCl, pH 7.2, 25°C; as left-handed L-form (– – –) after adding solid NaCl to 20% (W/W). Reproduced with permission from *J. Mol. Biol.* 67:375–96 (1972) (70).

5% decrease, respectively. However, the 190 nm band that underwent a 95% decrease on melting corresponded to a change in extinction coefficient of less than 0.5%, a small change when compared to hyperchromism monitored by normal absorption spectroscopy. Thus, despite the 190 nm band having ten times the intensity of the 275 band, they found its accuracy could not compare with normal absorption spectroscopy for monitoring DNA melting.

The CD of polynucleotides is dependent on solvent (77, 84–87), temperature (85–90), and the ionic strength as well as the nature of the salt (69, 88, 90–92). For natural DNA, increases in salt concentration lead to changes characteristic of a $B \rightarrow C$ conformational change (69). The positive 275 CD band decreases, even assuming negative values, with increasing ionic strength or decreasing temperature, while there is little change in the 245 nm band. Baase & Johnson (84) have correlated the salt dependence of DNA CD with changes in the average rotation of the bases about the helix axis. The intensity of the CD of DNA at 275 nm was directly related to the change in the helix winding angle, which was determined by a band

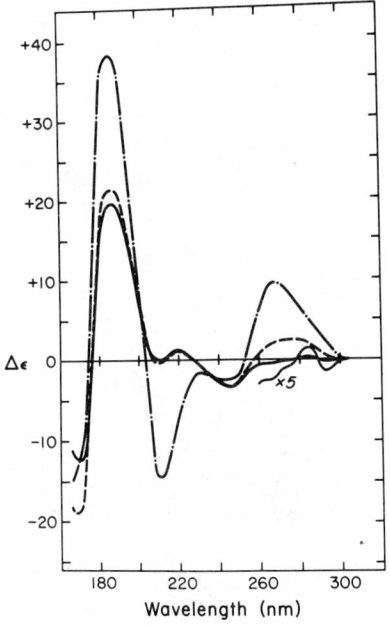

Figure 6 CD spectra of calf thymus DNA in solution [Sprecher et al (77)]. *A*-form in 80% trifluoroethanol, 0.667 mM $Na_{3/2}H_{3/2}PO_4$ (·····); *B*-form in 10 mM $Na_{3/2}H_{3/2}PO_4$, pH 7.0 (– – –); *C*-form in 6M NH_4F, 10 mM $Na_{3/2}H_{3/2}PO_4$, pH 7.0 (———). Reproduced with permission from *Biopolymers* 18:1009–19 (1979) (77).

counting method for superhelical DNA on agarose gels. They found that on going from low to high salt, the number of base pairs per turn changes by -0.22 ± 0.02, as compared with -0.7 expected for the $B \to C$ transition. Therefore, the assignment of the B- and C-form CD spectra should not be considered absolute.

With the recent identification of the L-form CD spectrum of poly(dG-dC)·poly(dG-dC) as a left-handed Z-form structure, environmental effects on the CD spectra of alternating purine-pyrimidine copolymers have been intensively studied (69, 78, 80, 93–104). Pohl (93) reported that the R to L transition could also occur in ethanol (0–56%). He further identified two other forms; one obtained in high ethanol concentration (80%) that resembles an A-form spectrum and another produced by 7.1 M CsCl or 3.8 M LiCl. Recently, Behe & Felsenfeld (94) have shown that poly(dG-m^{5}dC)·poly(dG-m^{5}dC) can undergo a $B \to Z$ transition at conditions close to physiological ionic strength.

Ivanov & Minyat (95) have done a thorough study of the CD of poly(dG-dC)·poly(dG-dC) as a function of trifluoroethanol (TFE) concentration. In solutions of 60% (v/v) or less, the polynucleotide is in the B-form, followed by a cooperative transition to Z-form at 60–66% (v/v). A noncooperative transition, which they attribute to an intrafamily transition within the Z-family, was observed at 66–78% TFE. At 80–84% TFE, a second cooperative transition, probably $Z \to A$ form, was found. Recently, Hall & Maestre (96) reported a $B \to Z$ form and a $Z \to Z'$ form transition for poly(dG-dC)·poly(dG-dC) as a function of temperature in moderate (10–20%) ethanolic solutions. The $B \to Z$ transition was found not to fit a two-state model and a number of intermediate forms were identified by CD. The noncooperative $Z \to Z'$ transition did show a reversible two-state transition. Furthermore, methanol was shown to induce the two transitions.

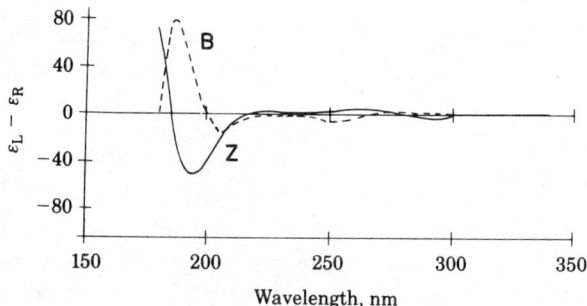

Figure 7 CD spectra of poly(dG-dC)·poly(dG-dC) in solution [Sutherland et al (78)]. As B-form (– – –) in 50% trifluoroethanol (TFE) and as Z-form (———) in 70% TFE. Reproduced with permission from *Proc. Natl. Acad. Sci. USA* 78:4801–9 (1981) (78).

Previously, a thermally driven interconversion of $B \to Z$ had been reported (97). The B-form was found to be stable at temperatures lower than 5°C and the Z-form above 20°C in dilute Mg^{2+} and Na^+ solutions.

Various divalent metal ions with organic solvents have been shown to induce the $B \to Z$ transition for poly(dG-dC)·poly(dG-dC). Zacharias et al (98) found that Co^{2+} (200 μM) or Mn^{2+} (250 μM) in the presence of ethylene glycol (25%) or ethanol (15%) were most effective in inducing the transition. The $B \to Z$ transition, as monitored by CD, was found to be different for the two metals. Mn^{2+} induced a monophasic transition, while Co^{2+} caused a biphasic conversion with at least one intermediate conformation, which was neither B- nor Z-like. Van de Sande & Jovin (99) reported a cooperative $B \to Z$ transition induced by 0.4 mM and 4 mM $MgCl_2$ in combination with 20% and 10% ethanol, respectively. However, this Mg^{2+}-ethanol Z-form had some properties usually attributed to B-form and showed signs of condensation and intermolecular aggregation. In both studies, the metal ions and the organic solvents acted synergistically.

Recently, Zacharias et al (100) have reported a $B \to Z$ cooperative transition of poly(dG-dC)·poly(dG-dC) induced by sodium acetate. The transition was monitored by CD and Raman spectroscopy. It was found to have a midpoint at 2.05 M. However, between 2.2 and 2.65 M sodium acetate, there was a many-fold increase of positive CD in the 250–340 nm range indicative of DNA aggregation to the highly condensed $\psi(+)$-type state. At concentrations greater than 2.7 M, the Z-form was found to be stable with no condensation. Hall & Maestre (96) in their study of poly(dG-dC)·poly(dG-dC) as a function of ethanol and methanol concentration found no evidence of scattering contributions to the CD. The scattering contributions were measured by the fluorescent detected circular dichroism (FDCD) or fluorescat techniques. The lack of such contributions indicated that little or no condensation or precipitation had occurred, in contrast to the Mg-ethanol and sodium acetate studies (99, 100).

This discussion of the $B \to Z$ transition of poly(dG-dC)·poly(dG-dC) should not lead the reader to assume that the Z-form CD spectra referred to in this review are identical to one another. It merely provides a nomenclature for the spectral change of a B-like spectrum to a Z-like spectrum, similar to that shown in Figure 5. In fact, there are differences in the shape of the various CD spectra of poly(dG-dC)·poly(dG-dC), which we refer to collectively as Z-form. Indeed, these spectral differences, along with the polymer's different chemical and physical properties under a variety of environmental conditions, strongly suggest the existence of a family of left-handed conformations and possible intermediates.

The intense interest in poly(dG-dC)·poly(dG-dC) has also spurred study of other alternating purine-pyrimidine copolymers. Oriented fibers of poly(dA-dC)·poly(dG-dT) have been observed to give A-, B-, and Z-form diffraction patterns (68). Upon increasing the ethanol concentration above 60%, the polynucleotide undergoes what has been characterized by CD as a $B \to A$ transition (101). Recent studies have demonstrated a $B \to Z$-like CD spectral transition for this polynucleotide as a function of increasing salt and ethanol concentration (102, 103), as well as temperature (102). Arnott et al (104) have reported that oriented fibers of poly(dA-ds^{4}T)·poly(dA-ds^{4}T) give a Z-form pattern. A noncooperative inversion of the long-wavelength region of the CD spectrum of poly(dA-dT)·poly(dA-dT), characteristic of $B \to Z$ transition of poly(dG-dC)·poly(dG-dC), has been observed at increasing concentration of CsF (105). Whether the Z-like CD spectra of any of these polynucleotides are due to left-handed conformations remains to be demonstrated.

THEORETICAL STUDIES The theoretical framework for understanding polynucleotide CD has been extensively developed over the past 20 years. The early work has been reviewed elsewhere (18, 106–108). The goal has been the development of a theoretical method capable of calculating the observed CD of polynucleotides, enabling the interpretation of the measured CD in terms of polynucleotide structure. In recent years, three theories [all-order polarizability theory (109–111), the matrix method (111–113), and linear response theory (114, 115)], have met with varying degrees of success toward that goal.

In the all-order, coupled oscillator theory developed by DeVoe (109, 110), polymer CD is obtained by assuming that the polymer is an aggregate of monomers, whose optical properties are due to its electric and magnetic oscillators. In the calculations discussed herein, the magnetic oscillators have been neglected. The theory incorporates the frequency dependence of the monomer absorption bands as complex polarizabilities. The electric oscillators are dynamically coupled to all order via electrostatic forces. Because the theory incorporates frequency-dependent monomer polarizabilities, it calculates circular dichroism directly at each frequency of interest.

Perturbation theory extended to all-order in V_{ij} has become known as the matrix method (111). In this method the polymer Hamiltonian $\mathcal{H} = \mathcal{H}_0 + V_{ij}$, where $\mathcal{H}_0$ is the sum of the unperturbed monomer Hamiltonians and V_{ij} is the interaction energy between transition i and j on different bases, is written in matrix form. The matrix is diagonalized to give the eigenvalues and eigenvectors of the polymer transition. Thus, only the polymer rotational strengths and transition frequencies are calculated, and a band shape must be assigned to each band.

Linear response theory (114, 115) is a more complete quantum mechanical approach to calculating CD. The theory assumes molecular separability (zero overlap between wavefunctions of neighboring unit cells) and applies the decorrelation approximation, which neglects transitions between excited states of the monomers. Thus, the polymer is treated as a set of linearly coupled oscillators. In addition to being all-order in V_{ij}, the polymer-response function arises in a natural fashion from the monomer-response function. Thus, the theory, like polarizability theory, is capable of predicting polymer CD bandshapes.

DeVoe's (109, 110) classical polarizability theory has been applied to the calculation of polynucleotide CD (111, 116–121). Calculations by Cech et al (111) for ApA and oligoadenylic acids of varying chain length gave good agreement with measured spectra. Cech et al (111) found that the matrix method and polarizability theory gave similar results for ApA, d(ApA), oligo(A) and poly(A). However, the ability to incorporate band shapes from the monomers gives polarizability theory an advantage in accuracy.

Calculated CD spectra, using polarizability theory, were found to be sensitive to both geometry and sequence, when applied to double-stranded polynucleotides of repeating sequence (116). The agreement was good with the measured CD of poly rA·poly rU and reasonably good for poly(rA-rU)·poly(rA-rU) and poly rG·poly rC. Calculations for other sequences and geometries were less satisfactory. They were particularly poor for poly(rG-rC)·poly(rG-rC), which was roughly the mirror image of the measured spectrum. The sensitivity of the calculations was demonstrated when the calculated CD of the four-stranded structures of poly rG and poly rI proposed by Arnott et al (122) gave better agreement with measured data than the structure proposed by Zimmerman et al (123). Subsequent calculations (118) for poly(dG-dC)·poly(dG-dC) in the B- and Z-form geometries suggested that the low-salt conformation is Z-DNA, in contradiction to experiment. Furthermore, the calculations were unable to account for the high-salt conformer.

Johnson et al (121), using the matrix method (111–113), have calculated the CD of an eleven base-pair helix in 68 different right-handed conformations. The dependence of the 275 nm CD band was correlated to five helical structural parameters (helix winding angle, base pair twist, distance between base-pairs, the distance of the base pairs from the helix axis, and tilt of the base pair relative to the helix axis). Good correlations were found for both the winding angle and the twist angle, while correlations for the other three were poor.

Linear response theory (114, 115), which is the quantum mechanical equivalent to DeVoe's (109, 110) polarizability theory, has also been applied to calculate the CD of poly(A) (115, 122–125). The calculated spectra were shown to be sensitive to both geometry and chain-length. The results are in reasonable agreement with measured spectra and comparable to those obtained by polarizability theory for poly(A) obtained by Cech et al (111).

Overall, these theoretical calculations of polynucleotide CD have been successful in accounting for qualitative features of polynucleotide CD. However, the results have not demonstrated the ability to correlate measured spectra with structure. These mixed results may be due to inadequate knowledge of monomer properties and geometry, or neglect of base-sugar interactions and n-π^* transitions. Recent theoretical studies (120, 126, 127) have begun to address some of these factors.

The magnitudes of the CD spectra of polynucleotides are generally of the same order as those for mononucleotides in the near UV. Thus, the sugar-phosphate groups may contribute significantly to the observed CD. Ito & I'Haya (126) have introduced an energy-dependent dielectric screening effect due to the sugar-phosphate groups into the formalism of linear response theory. They carried out CD calculations on three highly repeating sequence satellite DNAs. Their results showed that inclusion of the energy-dependent screening interactions did not change the qualitative features of the calculated CD compared with the results when an energy-independent screening interaction was considered. However, their method only qualitatively includes the effects due to the sugar-phosphate group, so a more definite result as to the effect of the sugar-phosphate on the near UV CD remains for future study.

The size of nucleic acid bases are large relative to the distance that separates them in polynucleotides. Because of this, the monopole representation of the nucleic acid base π-π^* transitions, rather than the point dipole representation, is more accurate in determining interaction potentials. A prerequisite for use of CD calculations for structural determination is the establishment of reliable transition monopoles. Recently, Williams & Moore (120) developed a semiempirical procedure to derive nucleic acid transition monopoles. The nucleic acid transition parameters are derived by adjusting them to optimize the fit between calculated and measured CD spectra of cyclic mononucleotides (120, 127, 129, 130).

The derived transition moment parameters were used in DeVoe (109, 110) theory CD calculations of 23 different synthetic oligomers, polymers, and gels of known sequence (120, 126, 129). Overall, the agreement (i.e. in terms of spectral signs, shapes, and magnitudes) that was found between the calculated (220–300 nm) and measured CD spectra was good. For poly rG · poly rC, there was good agreement between the sign of the calculated and measured CD bands; however, the agreement between shape and magnitude of the bands was poor (129). The results for other guanine and cytosine containing polymers were good; they were particularly accurate for poly dG · poly dC and poly rG-rC · poly rG-rC (120).

The derived guanine and cytosine monomer parameters of Williams & Moore (120) have been applied to calculate the CD spectra of poly(dG-dC) · poly(dG-dC) (127). The results gave good agreement between the calculated A-form CD spectrum and the measured spectrum for poly(dG-dC) · poly(dG-dC) in TFE concentrations greater than 80% (v/v) (95). Various right-handed geometries (131) produced results that are similar to an intermediate B-form CD spectrum identified by Hall & Maestre (96). However, none of the calculated Z-form CD spectra gave results similar to the CD spectrum that characterizes Z-form structures. Instead, the calculated CD spectra in Z-form geometries were found to be more in agreement with the measured CD spectra in 3.8 M CsCl or 7.1 M LiCl (93). Furthermore, none of the B-form geometries (131) that were tested gave a CD spectrum that has been attributed to the B-form geometry.

An alternative approach for adjusting transition monopoles has recently been reported by Rizzo & Schellman (128). They have introduced a constrained minimization of a starting set of monopoles and solve, with Lagrangian multipliers, for a set of transition monopoles that agree with experimentally determined values of the transition moments. Using the matrix method, favorable agreement was obtained between measured spectra of natural DNAs and RNAs with calculated CD spectra for random sequences of A-RNA, A-DNA, and B-DNA. The calculated CD spectrum of poly(dA-dT) · poly(dA-dT) in D-DNA geometry gave qualitative agreement in terms of shape with the low salt measured spectrum of that polynucleotide. However, the calculated spectrum for a Z-form structure compared almost quantitatively with low-salt (supposedly B-form) measured spectrum of poly(dG-dC) · poly(dG-dC).

In summary, the continual failure of theoretical CD calculations to identify the high salt form of poly(dG-dC) · poly(dG-dC) with Z-DNA may be due to the various factors mentioned above. All of the theoretical methods are sensitive to the nucleic acid monomer parameters or assumed CD bandshapes used in the calculations. Both of these factors can influence the sign, shape, and magnitude of calculated CD spectra. Also, the effect of static electric fields from the phosphates can make significant changes in the energies and transition moment directions of the bases. Furthermore, it is clear that poly(dG-dC) · poly(dG-dC) is the most polymorphic polynucleotide studied to date. Therefore, inadequate knowledge of the various conformations it may assume in solution could prove to be the deciding factor. Nevertheless, these and other factors must be explored more thoroughly if CD is to help explain the behavior of this or any other polynucleotide in solution.

Acknowledgments

We wish to thank Dr. Marcos F. Maestre for providing an unpublished figure and for helpful discussions. The research was supported in part by grant GM 10840 from the National Institutes of Health and by the US Department of Energy, Office of Energy Research Contract 03-82ER60090.000.

Literature Cited

1. Eyers, A., Heckenkamp, G., Schofers, F., Schonhense, G., Heinzmann, U. 1983. *Nucl. Instr. Meth.* 208:303–5
2. Kim, K.-J. 1984. *Nucl. Instr. Meth.* 219:425–29
3. Moore, D., Tinoco, I. Jr. 1980. *J. Chem. Phys.* 72:3396–3400
4. Tinoco, I. Jr., Keller, D. 1983. *J. Phys. Chem.* 87:2915–17
5. Bustamante, C., Tinoco, I. Jr., Maestre, M. F. 1982. *J. Chem. Phys.* 76:3440–46
6. van de Hulst, H. C. 1957. *Light Scattering by Small Particles*, pp. 40–58. New York: Wiley. 470 pp.
7. Tobias, I., Brocki, T. R., Balazs, N. L. 1975. *J. Chem. Phys.* 62:4181–83
8. Rosenfeld, L. 1928. *Z. Phys.* 52:161
9. Harris, R. A., McClain, W. M., Sloane, C. F. 1974. *Mol. Phys.* 28:381–98
10. Barron, L. D., Buckingham, A. D. 1971. *Mol. Phys.* 20:1111–19
11. Bustamante, C., Maestre, M. F., Tinoco, I. Jr. 1980. *J. Chem. Phys.* 73:6046–55
12. Keller, D. 1984. PhD thesis. Univ. Calif., Berkeley
13. McClain, W. M., Schaurte, J. A., Harris, R. A. 1984. *J. Chem. Phys.* 80:606–16
14. Harris, R. A., McClain, W. M. 1977. *J. Chem. Phys.* 67:269–70
15. Harris, R. A., McClain, W. M. 1977. *J. Chem. Phys.* 67:271–73
16. Bustamante, C., Maestre, M. F., Keller, D., Tinoco, I. Jr. 1984. *J. Chem. Phys.* 80:4817–23
17. Bustamante, C., Tinoco, I. Jr., Maestre, M. F. 1983. *Proc. Natl. Acad. Sci. USA* 80:3568–72
18. Tinoco, I. Jr., Bustamante, C., Maestre, M. F. 1980. *Ann. Rev. Biophys. Bioeng.* 9:107–41
19. Balazs, N. L., Brocki, T. R., Tobias, I. 1976. *Chem. Phys.* 13:141–51
20. Tinoco, I. Jr., Freeman, M. P. 1957. *J. Phys. Chem.* 61:1196–1200
21. Maestre, M. F., Bustamante, C., Hayes, T. L., Subirana, J. A., Tinoco, I. Jr. 1982. *Nature* 298:773–74
22. Johnson, W. C. Jr. 1978. *Ann. Rev. Phys. Chem.* 29:93–114
23. Greenfield, N., Davidson, B., Fasman, G. D. 1967. *Biochemistry* 6:1630–37
24. Greenfield, N., Fasman, G. D. 1969. *Biochemistry* 8:4108–16
25. Saxena, V. P., Wetlaufer, D. B. 1971. *Proc. Natl. Acad. Sci. USA* 68:969–72
26. Chen, Y. H., Yang, J. T., Martinez, H. M. 1972. *Biochemistry* 11:4120–31
27. Chen, Y. H., Yang, J. T., Chan, K. H. 1974. *Biochemistry* 13:3350–59
28. Chang, C. T., Wu, C. S. C., Yang, J. T. 1978. *Anal. Biochem.* 91:13–31
29. Siegel, J. B., Steinmetz, W. E., Lang, G. L. 1980. *Anal. Biochem.* 104:160–67
30. Chou, P., Fasman, G. D. 1977. *J. Mol. Biol.* 115:135–75
31. Bolotina, I. A., Chekhov, V. O., Lugauskas, V. Y. 1979. *Int. J. Quant. Chem.* 16:819–24
32. Bolotina, I. A., Chekhov, V. O., Lugauskas, V. Y., Ptitsyn, O. B. 1981. *Mol. Biol.* 15:167–75
33. Brahms, S., Brahms, J. 1980. *J. Mol. Biol.* 138:149–78
34. Levitt, M., Greer, J. 1977. *J. Mol. Biol.* 114:181–239
35. Provencher, S. W., Glockner, J. 1981. *Biochemistry* 20:33–37
36. Hennessey, J. P. Jr., Johnson, W. C. Jr. 1981. *Biochemistry* 20:1085–94
37. Hennessey, J. P. Jr., Johnson, W. C. Jr. 1982. *Anal. Biochem.* 125:177–88
38. Rozenkranz, H. 1974. *Z. Klin. Chem. Klin. Biochem.* 12:415–22
39. Woody, R. W. 1978. *Biopolymers* 17:1451–67
40. Goux, W. J., Hooker, T. M. Jr. 1975. *J. Am. Chem. Soc.* 97:1065–67

41. Goux, W. J., Hooker, T. M. Jr. 1980. *J. Am. Chem. Soc.* 102:7080–87
42. Goux, W. J., Hooker, T. M. Jr. 1980. *Biopolymers* 19:2191–2208
43. Snyder, R. W., Hooker, T. M. Jr. 1982. *Biopolymers* 21:547–63
44. Applequist, J. 1979. *J. Chem. Phys.* 71:4332–38
45. Applequist, J. 1980. *J. Chem. Phys.* 73:3521
46. Applequist, J. 1981. *Biopolymers* 20:387–97
47. Applequist, J. 1981. *Biopolymers* 20:2311–22
48. Applequist, J. 1982. *Biopolymers* 21:703–4
49. Applequist, J. 1982. *Biopolymers* 21:779–95
50. Woody, R. W. 1983. *Biopolymers* 22:189–203
51. Bowman, R. L., Kellerman, M., Johnson, W. C. Jr. 1983. *Biopolymers* 22:1045–70
52. Gray, D. M., Tinoco, I. Jr. 1970. *Biopolymers* 9:223–44
53. Josse, J., Kaiser, A. D., Kornberg, A. 1961. *J. Mol. Biol.* 236:864–75
54. Allen, F. S., Gray, D. M., Roberts, G. P., Tinoco, I. Jr. 1972. *Biopolymers* 11:853–79
55. Marck, C., Guschlbauer, W. 1978. *Nucleic Acid Res* 6:2013–31
56. Marck, C., Guschlbauer, W. 1978. *Biophys. J.* 24:575–78
57. Marck, C., Guschlbauer, W. 1978. *C. R. Acad. Sci. Ser. D* 286:713–16
58. Allen, F. S., Gray, D. M., Ratliff, R. L. 1984. *Biopolymers* 23: In press
59. Gray, D. M., Hamilton, F. D., Vaughn, M. R. 1978. *Biopolymers* 17:85–106
60. Gray, D. M., Lee, C. S., Skinner, D. M. 1978. *Biopolymers* 17:107–14
61. Gray, D. M., Liu, J.-J., Ratliff, R. L., Allen, F. S. 1981. *Biopolymers* 20:1337–82
62. Tinoco, I. Jr., Cantor, C. R. 1970. In *Methods in Biochemical Analysis*, ed. D. Glick, 18:81–203. New York: Wiley
63. Lloyd, D. A. 1969. PhD thesis. Univ. Calif., Berkeley
64. Allen, F. S., Gray, D. M. 1984. *Biopolymers* 23: In press
65. Allen, F. S., Jones, M. B., Hollstein, U. 1977. *Biophys. J.* 20:69–78
66. Arnott, S. 1970. *Prog. Biophys. Mol. Biol.* 21:265–319
67. Arnott, S. 1977. In *Proc. 1st Cleveland Symp. on Macromolecules and Properties of Biopolymers*, ed. A. G. Walton, pp. 87–104. Amsterdam: Elsevier
68. Leslie, A. G. W., Arnott, S., Chandrasekaran, R., Ratliff, R. L. 1980. *J. Mol. Biol.* 143:49–72
69. Tunis-Schneider, M. J. B., Maestre, M. F. 1970. *J. Mol. Biol.* 52:521–41
70. Pohl, F. M., Jovin, T. M. 1972. *J. Mol. Biol.* 67:375–96
71. Wang, A. H.-J., Quigley, G. J., Kolpak, F. J., van der Marel, G., van Boom, J. H., Rich, A. 1981. *Science* 211:171–76
72. Wang, A. H.-J., Quigley, G. J., Kolpak, F. J., Crawford, J. L., van Boom, J. H., van der Marel, G., Rich, A. 1979. *Nature* 282:680–86
73. Drew, H., Takano, T., Tanaka, S., Itakura, K., Dickerson, R. E. 1980. *Nature* 286:567–73
74. Crawford, J. L., Kolpak, F. J., Wang, A. H.-J., Quigley, G. J., van Boom, J. H., van der Marel, G., Rich, A. 1980. *Proc. Natl. Acad. Sci. USA* 77:4016–20
75. Pohl, F. M., Ranade, A., Stockburger, M. 1973. *Biochim. Biophys. Acta* 335:85–92
76. Thamann, T. J., Lord, R. C., Wang, A. H.-J., Rich, A. 1981. *Nucleic Acids Res.* 9:5443–57
77. Sprecher, C. A., Baase, W. A., Johnson, W. C. Jr. 1979. *Biopolymers* 18:1009–19
78. Sutherland, J. C., Griffin, K. P., Keck, P. C., Takacs, P. Z. 1981. *Proc. Natl. Acad. Sci. USA* 78:4801–4
79. Brahms, S., Vergne, J., Brahms, J. G., Di Capua, E., Koller, Th., Bucher, Ph. 1982. *J. Mol. Biol.* 162:473–93
80. Sutherland, J. C., Griffin, K. P. 1983. *Biopolymers* 22:1445–48
81. Davies, D. R., Baldwin, D. 1963. *J. Mol. Biol.* 6:251–55
82. Arnott, S., Chandrasekaran, R., Hukins, D. W. L., Smith, P. J. C., Watts, L. 1974. *J. Mol. Biol.* 88:523–33
83. Sprecher, C. A., Johnson, W. C. Jr. 1982. *Biopolymers* 21:321–29
84. Baase, W. A., Johnson, W. C. Jr. 1979. *Nucleic Acids Res.* 6:797–814
85. Brahms, J., Mommaerts, W. F. H. M. 1964. *J. Mol. Biol.* 10:73–88
86. Green, G., Mahler, H. R. 1971. *Biochemistry* 10:2200–16
87. Nelson, R. G., Johnson, W. C. Jr. 1970. *Biochem. Biophys. Res. Commun.* 41:211–16
88. Gray, D. M., Morgan, A. R., Ratliff, R. L. 1978. *Nucleic Acids Res.* 5:3679–95
89. Lewis, D. J., Johnson, W. C. Jr. 1974. *J. Mol. Biol.* 86:91–96
90. Tunis, M.-J., Hearst, J. E. 1968. *Biopolymers* 6:1218–23
91. Hanlon, S., Brudno, S., Wu, T. T., Wolf, B. 1975. *Biochemistry* 14:1648–60
92. Chan, A., Kilkuskie, R., Hanlon, S. 1979. *Biochemistry* 18:84–91
93. Pohl, F. M. 1976. *Nature* 260:365–66
94. Behe, M., Felsenfeld, G. 1981. *Proc. Natl. Acad. Sci. USA* 78:1619–23

95. Ivanov, V., Minyat, E. 1981. *Nucleic Acids Res.* 9:4783–98
96. Hall, K. B., Maestre, M. F. 1984. *Biopolymers* 23: In press
97. Roy, K. B., Miles, H. T. 1983. *Biochem. Biophys. Res. Commun.* 115:100–5
98. Zacharias, W., Larson, J. E., Klysik, J., Stirdivant, S. M., Wells, R. D. 1982. *J. Biol. Chem.* 257:2775–82
99. van de Sande, J. H., Jovin, T. M. 1982. *EMBO J.* 1:115–20
100. Zacharias, W., Martin, J. C., Wells, R. D. 1983. *Biochemistry* 22:2398–2405
101. Gray, D. M., Ratliff, R. L. 1975. *Biopolymers* 14:487–98
102. Zimmer, C., Tymen, S., Marck, C., Guschlbauer, W. 1982. *Nucleic Acids Res.* 10:1081–91
103. Vorlickova, M., Kypr, J., Stepanka, S., Sponar, J. 1982. *Nucleic Acids Res.* 10:1071–80
104. Arnott, S., Chandrasekaran, R., Birdsall, D. L., Leslie, A. G. W., Ratliff, R. L. 1980. *Nature* 283:743–45
105. Vorlickova, M., Kypr, J., Kleinwachter, V., Palecek, E. 1980. *Nucleic Acids Res.* 8:3965–73
106. Brahms, J., Brahms, S. 1970. In *Fine Structure of Proteins and Nucleic Acids*, ed. G. D. Fasman, S. N. Timasheff, pp. 191–270. New York: Decker
107. Bush, C. A. 1974. In *Basic Principles in Nucleic Acid Chemistry*, ed. P. O. P. T'so, 2:91–169. New York: Academic
108. Woody, R. W. 1977. *Macromolecular Rev.* 12:181–320
109. DeVoe, H. 1964. *J. Chem. Phys.* 41:393–400
110. DeVoe, H. 1965. *J. Chem. Phys.* 43:3199–3208
111. Cech, C. L., Hug, W., Tinoco, I. Jr. 1976. *Biopolymers* 15:131–52
112. Bayley, P. M., Nielsen, E. B., Schellman, J. A. 1969. *J. Chem. Phys.* 73:228–43
113. Madison, V., Schellman, J. 1972. *Biopolymers* 11:1041–76
114. Rhodes, W., Redmann, S. M. 1977. *Chem. Phys.* 22:215–20
115. Redmann, S. M., Rhodes, W. 1979. *Biopolymers* 18:393–409
116. Cech, C. L., Tinoco, I. Jr. 1976. *Biopolymers* 16:43–65
117. Cech, C. L., Tinoco, I. Jr. 1976. *Nucleic Acids Res.* 3:399–404
118. Vasmel, H., Greve, J. 1981. *Biopolymers* 20:1329–32
119. Levin, A. I., Tinoco, I. Jr. 1977. *J. Chem. Phys.* 66:3491–97
120. Williams, A. L., Jr., Moore, D. S. 1983. *Biopolymers* 22:755–86
121. Johnson, B. B., Dahl, K. S., Tinoco, I. Jr., Ivanov, V. I., Zhurkin, V. B. 1981. *Biochemistry* 20:73–78
122. Arnott, S., Chandrasekaran, R., Martilla, C. M. 1974. *Biochem. J.* 141:537–43
123. Zimmerman, S. B., Cohen, G. H., Davies, D. R. 1975. *J. Mol. Biol.* 92:181–92
124. Ito, H., I'Haya, Y. J., Eri, T. 1978. *Bull. Chem. Soc. Jpn.* 51:1341–49
125. Ito, H., Eri, T., I'Haya, Y. J. 1976. *Chem. Phys. Lett.* 39:150–56
126. Ito, H., I'Haya, Y. J. 1982. *J. Chem. Phys.* 77:6270–80
127. Williams, A. L. Jr. 1983. PhD thesis. Temple Univ., Philadelphia, Penna.
128. Rizzo, V., Schellman, J. A. 1984. *Biopolymers* 23:435–70
129. Moore, D. S., Williams, A. L. Jr. 1984. *Biopolymers* 23: In press
130. Moore, D. S. 1980. *Biopolymers* 19:1017–38
131. Zhurkin, V. B., Lysov, Yu. P., Ivanov, V. I. 1978. *Biopolymers* 17:377–412

EFFECTIVE POTENTIALS IN MOLECULAR QUANTUM CHEMISTRY[1]

Morris Krauss and Walter J. Stevens

Molecular Spectroscopy Division, National Bureau of Standards, Washington DC 20234

INTRODUCTION

The chemical similarities of the periodic elements arise because the valence electrons determine most of the chemical properties of molecules. The core electrons, being much more strongly bound to the atomic nuclei, are only slightly affected by the molecular environment and act predominantly to shield the nuclei and to provide an effective field in which the valence electrons move. These concepts, which really amount to a theoretical model, are pervasive in all of chemistry. They simplify our interpretation of chemical behavior and provide a framework for organizing and categorizing observations.

In quantum chemistry, the theoretical study of the electronic structure and properties of molecules and solids also can be made simpler if the atomic core electrons are considered invariant in the molecular environment. Since the complexity of ab initio calculations of electronic wavefunctions increases rapidly with the number of electrons, dramatic savings (especially for the heavier elements) can be realized if the core electrons can be removed and only the valence electrons considered explicitly. This requires quantification of the above concepts and the development of valence-only methods which, in order to be effective, must be competitive with all-electron treatments in terms of accuracy.

During the past several decades, many theories and approximations have been advanced that reformulate the Schrödinger equation in terms of

---

[1] The US Government has the right to retain a nonexclusive, royalty-free license in and to any copyright covering this paper.

valence electrons only. This is not an easy task, although formally one may write the essential equations rather trivially (1).

The wavefunction is written as a product of core and valence wavefunctions:

$$\Psi(1, 2, \ldots, n) = A\Phi_{core}(1, 2, \ldots, n_c)\Phi_{val}(n_c+1, n_c+2, \ldots, n) \qquad 1.$$

where A is the antisymmetrizer. The exact wavefunction cannot be factored this way in general, but certain approximate wavefunctions (e.g. the Hartree-Fock wavefunction) may be. The product wavefunction may be substituted into the Schrödinger equation, $\mathcal{H}\Psi = E\Psi$ where $\mathcal{H}$ is the all-electron Hamiltonian. If the core wavefunction, Φ_{core}, is known, a valence-only equation may be written:

$$\mathcal{H}_{val}\Phi_{val} = E_{val}\Phi_{val} \qquad 2.$$

where $E_{val} = E - E_{core}$ and $\mathcal{H}_{val}$ is given by:

$$\mathcal{H}_{val} = \sum_{i=i}^{n_v}\left(h_i + \sum_{j\neq i}^{n_v}\frac{1}{r_{ij}}\right) + \mathcal{V}_{core}. \qquad 3.$$

$\mathcal{V}_{core}$ is the operator arising from the core-valence interaction part of the Hamiltonian. It is a nonlocal integral operator because of the exchange terms introduced by the antisymmetrizer of Eq. 1. The exact form of $\mathcal{V}_{core}$ depends on the form of the core wavefunction.

A variational solution of Eq. 2 requires that the valence wavefunction be maintained orthogonal to the core wavefunction and that the core-valence interaction terms be evaluated. For a Hartree-Fock core, the orthogonality constraint is satisfied by requiring the valence wavefunction to be orthogonal to each core orbital, and the coulomb and exchange terms are evaluated directly for a given form of the valence wavefunction. This is the basis of the "frozen core" approximation in which an atomic core wavefunction is maintained invariant during the determination of a molecular valence wavefunction. The computational savings realized from the frozen core approximation are limited, however, since all of the core-valence two-electron integrals still must be evaluated and the orthogonality constraints must be explicitly included in the valence variational equations.

Much greater simplification and savings can be achieved if the core-valence orthogonality constraint and the nonlocal core-valence interaction term, $\mathcal{V}_{core}$, are replaced by a single operator in the valence Hamiltonian that prevents collapse of the valence wavefunction into the core region without explicitly imposing orthogonality, and also provides the correct core potential seen by the valence electrons. Such operators are known as *effective potentials* (EP) and are denoted $\mathcal{V}^{eff}$. There are many methods, both empirical and ab initio, for generating effective potentials. A detailed discussion of some of the methods is given in the second section. The most successful methods to date are based on Hartree-Fock calculations on atoms. Although the theoretical framework may be less than rigorous, the practical conclusion is that valence-only calculations employing effective potentials developed in the last ten years can produce quantitative results that compare favorably with all-electron calculations. A review of some comparative calculations is presented in the third section.

The true utility of effective potentials is that they provide a method for producing useful and quantitative theoretical results for systems that are otherwise untreatable at this time. Since most (if not all) molecular wavefunctions are represented by LCAO expansions, substantial reductions in calculational time for solving the Schrödinger equation are realized because the basis sets need not include functions that describe the atomic core orbitals. These reductions make calculations more tractable on larger molecules as well as on molecules containing heavier atoms.

Removal of the core from explicit consideration also suggests the extension of the EP method to relativistic calculations. It has been shown (2), via full all-electron Dirac-Fock calculations on atoms, that relativistic effects are predominantly core effects. This is true because only the core electrons spend most of their time near the nucleus, where effective velocities require relativistic correction. The valence electrons are affected primarily indirectly through their interaction with and orthogonality to the relativistic core. Methods have been developed for obtaining relativistic effective potentials (REP) that are analogous to those for nonrelativistic EP. A discussion of some of these methods is presented in the fourth section. While nonrelativistic EP are dependent on the orbital angular momentum (i.e. l-dependent), the REP are dependent on the total (spin plus orbital) angular momentum (i.e. j-dependent). The j-dependence of the REP implies that they contain information about spin-orbit coupling. Methods for extracting an effective spin-orbit potential have been proposed and are also discussed. It is possible to eliminate the j-dependence of the REP by averaging the $j = l + 1/2$ and $j = l - 1/2$ components to produce an l-dependent averaged relativistic effective potential (AREP), which can be used just like EP in molecular calculations. This provides a very simple method for introducing relativistic effects into molecular quantum chemistry.

EP, REP, and AREP are now being applied extensively in quantum mechanical calculations and compare well with all-electron methods. Although the potentials are usually derived from the ground-state wavefunctions of neutral atoms, they are transferable to a wide variety of molecular environments (covalent, ionic, weakly interacting). The known limitations and possible sources of error have not been serious enough to discourage their use. This review examines these potentials from the point of view of their application to molecular quantum chemistry. A brief review of the various forms of the potentials and methods for parameterizing them are given. Mention is made of the solid-state methods that closely parallel the molecular development. Although the solid-state approach is usually coupled with the density-functional theory of electronic structure, the forms of the resulting potentials and the conditions imposed on their determination are analogous to the molecular case.

The agreement with all-electron calculations and experimental data is the ultimate test for these potentials. We review several calculations of potential energy curves and spectroscopic constants that employ the most reliable forms of the nonrelativistic EP. For the relativistic case (REP and AREP), only a limited number of comparisons can be made with all-electron, Dirac-Hartree-Fock molecular calculations, but the results are encouraging. The calculation of fine structure splittings using effective spin-orbit potentials provides another successful test.

Finally, we discuss some of the areas that are just developing, such as spin-dependent potentials for open shell cores, valence-core polarization (correlation), valence properties, and fragment potentials.

THE DEVELOPMENT OF EFFECTIVE CORE POTENTIALS

Background

Hellmann (3, 4) and Gombas (5) were the first to suggest replacing the interaction of the valence and core electrons in an atom with an effective potential. Since that time, solid-state physics, particularly the study of metallic binding, has been a driving force behind the effective potential method. During the last ten years, however, there has been an increasing interest in the use of effective potentials in molecular structure calculations as well. New methods for obtaining the potentials have been developed, and our purpose here is to review those methods and their impact on molecular quantum chemistry. Only a brief review of earlier work is given and the reader is referred to several excellent reviews for more details (6–10).

In 1959, Phillips & Kleinman (11) provided a sound theoretical basis for replacing the explicit core-valence orthogonality constraints by a modification of the valence Hamiltonian. The eigenvalue equation for a single valence electron is written

$$\mathcal{H}_v \phi_v \equiv (h + \mathcal{V}_{\text{core}}) \phi_v = E_v \phi_v \qquad 4.$$

where ϕ_v satisfies the orthogonality constraint

$$\langle \phi_v | \phi_c \rangle = 0 \qquad 5.$$

for all core-like solutions of the same Hamiltonian

$$\mathcal{H}_v \phi_c = E_c \phi_c. \qquad 6.$$

If we write ϕ_v as

$$\phi_v = \chi_v - \sum_c \langle \chi_v | \phi_c \rangle \phi_c \qquad 7.$$

we see that for any χ_v, ϕ_v will be orthogonal to the core orbitals. Substituting Eq. 7 into Eq. 4, the valence eigenvalue equation may be rewritten

$$(\hbar + \mathcal{V}_{\text{core}} + \mathcal{V}^{\text{PK}})\chi_v = E_v \chi_v \qquad 8.$$

where

$$\mathcal{V}^{\text{PK}} = \sum_c [E_v - E_c] | \phi_c \rangle \langle \phi_c |. \qquad 9.$$

$\mathcal{V}^{\text{PK}}$ is called the *pseudopotential* and χ_v is the *pseudo-orbital*. It is important to note that $\mathcal{V}^{\text{PK}}$ is energy dependent because it depends on E_v, and it is nonlocal because it depends on χ_v. Also recall from the introduction that $\mathcal{V}_{\text{core}}$ is nonlocal due to exchange operators introduced by the antisymmetrizer. Equation 9 is satisfied by any function of the form

$$\chi_v = a_v \phi_v + \sum_c a_c \phi_c \qquad 10.$$

where the coefficients a_v and a_c are arbitrary except for the normalization constraint. Cohen & Heine (12) and Austin et al (13) have discussed the arbitrary nature of the pseudo-orbital and have suggested several ways to determine the coefficients. These include making χ_v as smooth as possible so that it may be represented by a minimal number of plane waves in solid state calculations.

The one-electron Phillips-Kleinman method was generalized for the case of many-valence electrons by Weeks & Rice (14). The relevant equations are

$$(\mathcal{H}_v + \mathcal{V}^{\text{GPK}})X_v = E_v X_v \qquad 11a.$$

$$\mathcal{V}^{\text{GPK}} = -\mathcal{H}_v \mathcal{P} - \mathcal{P}\mathcal{H}_v + \mathcal{P}\mathcal{H}_v\mathcal{P} + E_v \mathcal{P} \qquad 11b.$$

$$\mathcal{P} = \sum_c | \phi_c \rangle \langle \phi_c | \qquad 11c.$$

where X_v is a many-electron valence pseudo-wavefunction (that need not be orthogonal to the core), and $\mathcal{V}^{\text{GPK}}$ is the generalized Phillips-Kleinman pseudopotential. The valence Hamiltonian, $\mathcal{H}_v$, contains the interactions among the valence electrons as well as the core-valence interactions. Equation 11 is formidable, and solving it requires as much effort as the original all-electron problem (9, 15). The usefulness of the Phillips-

Kleinman method is in the theoretical basis it provides for the subsequent development of effective potentials. There have been many attempts to simplify the Phillips-Kleinman approach, and the reader is referred to reviews by Weeks et al (9) and Bardsley (10) for excellent discussions of the various methods and their applications to atomic and molecular structure calculations.

Model Potentials

Model potential methods have enjoyed widespread use as substitutes for the rigorous Phillips-Kleinman analysis (9, 10, 15–23). Model potentials are effective potentials that replace both the pseudopotential and the core-valence interaction terms in the valence Hamiltonian to give a one-electron valence equation of the form

$$(\hat{h} + \mathscr{V}_m)\phi_v = E_v \phi_v. \qquad 12.$$

The term *model potential* is used because $\mathscr{V}_m$ is given a specific form with adjustable parameters that are determined by some criterion such as matching the known eigenvalue spectrum of an atom. Such empirical schemes are possible only for one- or two-valence electron atoms and are not extendable to the entire periodic table.

Several model potential theories based on ab initio all-electron atomic calculations have been introduced (24–31). The method that has received the most widespread use in molecular calculations is due to Bonifacic & Huzinaga (24–27). In this theory, the valence Hamiltonian for a single valence electron is given by

$$\hat{h}_i = -\tfrac{1}{2}\nabla_i^2 - \frac{(Z-n_c)}{r_i} - \sum_j A_j \frac{e^{-\alpha_j r_i^2}}{r_i} + \sum_c B_c |\phi_c\rangle\langle\phi_c|. \qquad 13.$$

The exponential term is a local potential, which replaces the core-valence coulomb and exchange potentials. The last term has its origin in the orthogonality-constrained variation of the valence orbital (24) and ensures that the optimum valence orbital will be orthogonal to all core orbitals. Notice that the screening of the nuclear charge by the core electrons has been taken into account by absorbing n_c/r_i into the model potential. This is the usual practice in effective potential methods. The parameters A_j, α_j, and B_c are determined by any one of many methods. The B_c parameters can be related to the core orbital eigenvalues (24, 25, 28), but are often varied to produce the best fits to the valence orbitals and eigenvalues taken from atomic all-electron calculations (32). Systems with many valence electrons are handled by taking the total valence Hamiltonian to be a sum of one-electron Hamiltonians (cf Eq. 13) plus the r_{ij}^{-1} coulombic repulsion terms for all valence electrons. Since each valence orbital has overlap only with core orbitals of the same angular symmetry, the model potential is inherently l-dependent by virtue of the core projection term. The exchange contribution to the core-valence interaction is also l-dependent, but the exponential term in Eq. 13 is usually assumed to be independent of l. It is interesting to note that in the limit of a complete variational solution, a model potential of this type will produce valence orbitals that have all of the nodal structure needed to maintain orthogonality to the core orbitals. This could be a serious drawback in large molecular calculations where one wishes to restrict the size of the basis set as much as possible. On the other hand, the calculation of electronic properties that have significant contributions from the inner parts of the orbitals will be more reliable. Bonifacic & Huzinaga (25, 26) have suggested a simplification that restricts the sum of core projectors in Eq. 13 to include only the outer core orbitals in heavy atoms so that the basis set for the inner shells may be removed. Wahlgren and co-workers (33–37) have discussed the dangers associated with restricted basis sets in this model potential method. Pettersson & Wahlgren (37) have also proposed using a valence basis set that has been pre-orthogonalized to the core functions in order to reduce the dimensionality of the valence variational equations. This method requires an extra four-index transformation of the two-electron integrals.

Effective Core Potentials

In 1968 Goddard (38) introduced a new method for generating effective potentials. In an independent particle theory, the orbital equation may be written:

$$(\hat{h}+\mathcal{V})\phi_i \equiv \left(-\tfrac{1}{2}\nabla^2 - \frac{Z}{r} + \frac{l(l+1)}{2r^2} + \mathcal{V}\right)\phi_i = \varepsilon_i \phi_i \qquad 14.$$

where $\mathcal{V}$ is the nonlocal potential due to interactions with the other electrons in the system. If ϕ_i is nodeless, it is possible to define an equivalent equation

$$(\hat{h}+\mathcal{U}_i)\phi_i = \varepsilon_i \phi_i \qquad 15.$$

where $\mathcal{U}_i$ is a *local* potential given by

$$\mathcal{U}_i(r) = \varepsilon_i + \frac{z}{r} - \frac{l(l+1)}{2r^2} + \frac{\nabla^2 \phi_i(r)}{2\phi_i(r)}. \qquad 16.$$

The non-orthogonal, smooth, and nodeless atomic orbitals from Goddard's G1 (generalized SCF) method (39–41) can be used to provide a unique definition for $\mathcal{U}_i(r)$. G1 atomic effective potentials (GAEP) have been derived for a limited number of first row atoms and applied to some molecular calculations (42–45), but their use has not been widespread.

While the G1 method and GAEP have not been widely used, the form of Eq. 16 is very suggestive and has led to many advances in the determination of effective potentials for use in molecular calculations. Some advantages of effective potentials defined in this way:

1. They are local by definition and therefore easy to use.
2. They depend on the angular momentum of the valence orbital (l-dependent).
3. They can be defined for any nodeless orbital ϕ_i with energy ε_i.
4. Since ϕ_i is nodeless, it represents the lowest energy solution of Eq. 15.

In 1974 Melius & Goddard (46) suggested that Hartree-Fock (HF) orbitals could be used in place of G1 orbitals to define the local effective potentials for atoms. Recalling the Phillips-Kleinman pseudo-orbitals, they proposed that a linear combination of core and valence orbitals (as in Eq. 10) could be used to produce a nodeless pseudo-orbital that could be substituted, along with the valence HF eigenvalue, into Eq. 16. The problem of uniquely defining the linear coefficients in Eq. 10 was solved by requiring the kinetic energy of the pseudo-orbital to be minimized, as was proposed by Austin et al (13) some years earlier in connection with the non-uniqueness of the Phillips-Kleinman pseudo-orbital. The minimum kinetic energy orbitals were found to be nodeless and similar to G1 orbitals for a single valence electron.

For a single valence electron, the potential obtained by inverting the HF equation for a nodeless pseudo-orbital corresponds to replacing the nonlocal core potential and the pseudopotential of the Phillips-Kleinman equation with a local effective potential

$$(\hbar + \mathscr{V}_c + \mathscr{V}^{PK})\chi \equiv (\hbar + \mathscr{V}^{eff})\chi = \varepsilon\chi. \qquad 17.$$

This is similar to the model potential approach, but the local potential is determined uniquely by Eq. 16. The problem, therefore, is not to find an appropriate potential, but rather to find an appropriate way to specify the pseudo-orbital, χ. Since we are interested in molecular structure calculations, we can make a list of desirable characteristics of χ that have an impact on such calculations.

1. Because the difficulty and costs of molecular calculations rise rapidly with the number of primitive basis functions used to expand the orbitals, χ should be as smooth as possible in the core region so that it may be represented by as few basis functions as possible.
2. Because molecular interactions are often proportional to the overlaps of

the outer parts of the valence orbitals of the constituent atoms, χ should resemble the actual Hartree-Fock orbital as much as possible in that region.

3. Because localization of the potential "freezes in" exchange contributions that involve the valence orbital, the local potential is exact only for the pseudo-orbital that is used to generate it. To enhance the prospects for transferability of the potential to the molecular environment, the atomic state used to obtain χ must not differ dramatically from the normal range of valences and ionization level encountered in molecular structure.

If the third item above is to be enforced, then one must derive the effective potentials from many-valence electron atomic wavefunctions. A widely used method for accomplishing this was published in 1976 by Kahn et al (47). The Fock equation for a valence orbital with angular momentum l is

$$\left(-\tfrac{1}{2}\nabla^2 - \frac{Z}{r} + \frac{l(l+1)}{2r^2} + \mathscr{V}_{\text{val}} + \mathscr{V}_{\text{core}}\right)\phi_l = \varepsilon_l \phi_l \qquad 18.$$

where $\mathscr{V}_{\text{core}}$ and $\mathscr{V}_{\text{val}}$ represent the Coulomb and exchange potentials due to the core electrons and other valence electrons, respectively. This equation is replaced by

$$\left(-\tfrac{1}{2}\nabla^2 - \frac{Z_{\text{eff}}}{r} + \frac{l(l+1)}{2r^2} + \mathscr{V}'_{\text{val}} + \mathscr{V}_l^{\text{eff}}\right)\chi_l = \varepsilon_l \chi_l \qquad 19.$$

where χ_l is a pseudo-orbital constructed from the Hartree-Fock orbitals, and $\mathscr{V}_{\text{eff}}$ is the effective potential, which Kahn et al call an *effective core potential* (ECP). Z_{eff} is the effective nuclear charge equal to the nuclear charge minus the number of core electrons, so the shielding is included in $\mathscr{V}_{\text{eff}}$. The prime on $\mathscr{V}'_{\text{val}}$ indicates that the valence potential is evaluated over valence pseudo-orbitals rather than the original valence orbitals. This implies that $\mathscr{V}^{\text{eff}}$ must compensate for the difference between $\mathscr{V}_{\text{val}}$ and $\mathscr{V}'_{\text{val}}$, which, if large enough, can seriously affect the transferability of the ECP to the molecular environment where the valence-valence interactions are different (48). The ECP is obtained by inverting Eq. 19, which gives

$$\mathscr{V}_l^{\text{eff}} = \varepsilon_l + \frac{Z_{\text{eff}}}{r} - \frac{l(l+1)}{2r^2} + \frac{(\tfrac{1}{2}\nabla^2 - \mathscr{V}'_{\text{val}})\chi_l}{\chi_l}. \qquad 20.$$

Each valence orbital with a given angular momentum generates a unique potential. If core orbitals of the same angular momentum exist, then the potential is generally repulsive since it must prevent collapse of the pseudo-orbital into the core region. If there are no core orbitals with the same angular momentum, the potential is attractive and primarily represents unscreening of the nuclear charge as the valence electron penetrates the core region. The ECP also contains core-valence exchange interactions, but for high l-values, these are not large when compared to the Coulombic interactions. Thus, the potentials for all angular momenta larger than the highest value in the core are found to be very similar. The total ECP for an atom can be written as

$$\mathscr{V}^{\text{eff}} = \sum_{l=0}^{\infty} \mathscr{V}_l^{\text{eff}} \sum_m |lm\rangle\langle lm| \qquad 21.$$

where the projectors ensure that the l-dependent potentials operate only on the proper components of the valence wavefunction. Since the potentials with l-values larger than the maximum found in the core are nearly identical, one may approximate the ECP by

$$\mathscr{V}^{\text{eff}} = \mathscr{V}_{l_{\max}}^{\text{eff}} + \sum_{l=0}^{l_{\max}-1} (\mathscr{V}_l^{\text{eff}} - \mathscr{V}_{l_{\max}}^{\text{eff}}) \sum_m |l,m\rangle\langle l,m| \qquad 22.$$

where $l_{\max}$ is one more than the highest l found in the core. HF solutions from more than one atomic state may be needed to define potentials for all l-values in Eq. 22.

The atomic orbitals and eigenvalues needed in the construction of the ECP are easily obtained from exact numerical Hartree-Fock wavefunctions produced by codes such as the one developed by Froese-Fischer (49). Several prescriptions exist for producing nodeless pseudo-orbitals from the all-electron orbitals. Kahn et al (47) proposed taking a linear combination of core and valence orbitals (Eq. 10) and determining the linear coefficients by minizing the kinetic energy with a constraint to eliminate any residual cusp behavior at the origin. This procedure produces smooth and nodeless valence pseudo-orbitals, but the normalization constraint forces the coefficient a_v in Eq. 10 to be less than unity. Thus, pseudo-orbitals defined in this way will not match the all-electron orbitals even outside the core region (50), and the difference between $\mathscr{V}_{\text{val}}$ and $\mathscr{V}'_{\text{val}}$ mentioned above can cause $\mathscr{V}_l^{\text{eff}}$ to extend improperly beyond the radius where core-valence interactions are expected to be small. Such behavior is shown in Figure 1, where we compare $l = 1$ potentials for the oxygen atom, with the $2s$-$2p$ valence interactions evaluated with a $2s$ pseudo-orbital generated by the prescription of Kahn et al (47), and with the exact Hartree-Fock $2s$ orbital. The long tail on the first potential is due to the $\mathscr{V}_{\text{val}} - \mathscr{V}'_{\text{val}}$ difference when the nodeless $2s$ orbital is used. The inaccuracies introduced into molecular calculations by the long-range tails have been noted by Hay et al (51), who proposed smoothly truncating the tails in order to minimize the effects. Redondo et al (50) suggested avoiding the problem either by not normalizing the pseudo-orbital (i.e. $a_v = 1.0$ in Eq. 10), or by using a basis set expansion of the orbitals in which the coefficients of the outermost basis functions in the pseudo-orbital are restricted to be identical to those in the all-electron orbital.

Although the effective potential method is justified by the Phillips-Kleinman analysis, there is no reason that the construction of pseudo-orbitals must be restricted to linear combinations of core and valence orbitals as in Eq. 10. Several methods are available for constructing the pseudo-orbital so that it is smooth and nodeless while also minimizing the difference between $\mathscr{V}_{\text{val}}$ and $\mathscr{V}'_{\text{val}}$. Christiansen et al (48) have defined the pseudo-orbital as

$$\chi_l(r) = \sum_{i=0}^{4} c_i r^{i+N} \qquad r \leq r_m \qquad \text{23a.}$$

$$\chi_l(r) = \phi_l(r) \qquad r \geq r_m \qquad \text{23b.}$$

$$\langle \chi_l | \chi_l \rangle = 1 \qquad \text{23c.}$$

where r_m is a match radius and ϕ_l is the all-electron orbital. The coefficients, c_i are determined by matching the value and first three derivatives of χ_l and ϕ_l at r_m with the constraint that χ_l is normalized. The match radius is made

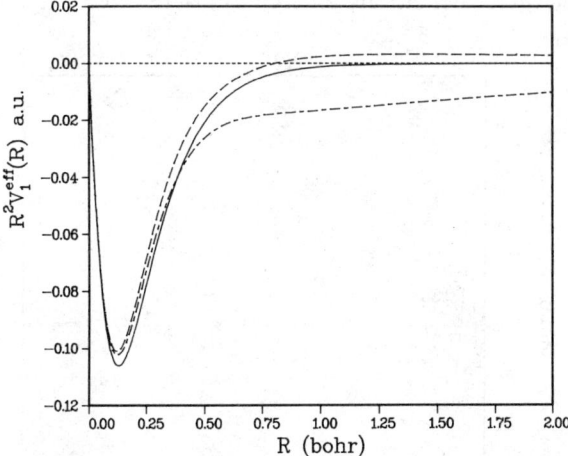

Figure 1 Comparison of $l = 1$ effective core potentials for oxygen generated from the $2p$ orbital using Eq. 19 while employing different $2s$ orbitals to construct the valence Coulomb and exchange term. *Solid line*: using the exact $2s$ HF orbital. *Double-dashed line*: using a minimum kinetic energy $2s$ pseudo-orbital of the form of Eq. 10. *Single-dashed line*: using a shape-consistent pseudo-orbital obtained from Eq. 23.

as small as possible, while satisfying the requirement that χ_l have only one maximum and two inflection points. In our own work we have found the optimum match radius to be consistently near the outermost maximum of the all-electron orbital, and we feel very little is lost by fixing it at the maximum. Orbitals constructed in this way are called *shape consistent*. Since the valence pseudo-orbitals are identical to the HF orbitals outside of the match radii, the difference between $\mathscr{V}_{val}$ and $\mathscr{V}'_{val}$ is minimized in that region. Figure 2 compares Xe 5s pseudo-orbitals generated by the minimum kinetic energy method of Kahn et al (47) and the shape-consistent method. Recently, Rappé et al (52) have proposed a procedure whereby the pseudo-orbital is constructed with a basis set expansion that is optimized to reproduce the long-range part of the HF orbital while directly minimizing the valence interaction difference. Although this could produce orbitals that are not necessarily smooth, they are guaranteed to be fit exactly by the chosen basis set. Potentials generated by this technique have been called shape and Hamiltonian consistent. Still another shape-consistent method has been developed by Durand & Barthelat and co-workers (53, 54), which also involves a basis set expansion of the pseudo-orbital, but minimizes the deviation from the HF orbital outside some match radius.

ECPs generated by Eq. 20 are usually tabulated on a numerical grid, which is coincident with the HF orbital grid if a numerical HF code was used to generate the all-electron wavefunction. However, to be generally useful in molecular calculations, the ECPs must be represented analytically.

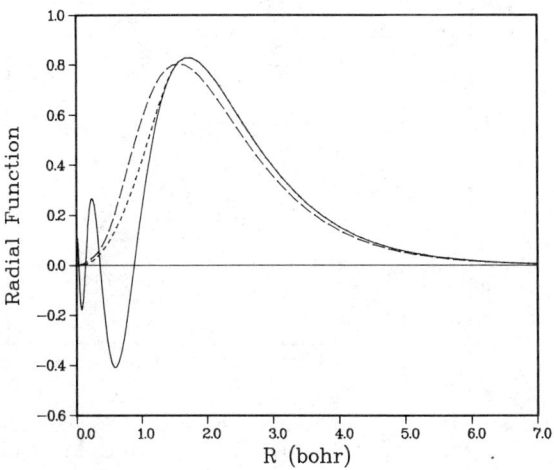

Figure 2 The Xe 5s orbital. *Solid line*: the exact 5s HF orbital. *Long dashed line*: the minimum kinetic energy pseudo-orbital of the form of Eq. 10. *Small dashed line*: a shape-consistent pseudo-orbital obtained from Eq. 23.

The most popular representation is a Gaussian expansion of the type

$$V_l(r) = \sum_i A_i \frac{e^{-\alpha_i r^2}}{r_i^{N_i}} \qquad 24.$$

where the parameters A_i, α_i, and N_i are optimized by fitting the numerical potential. Depending on how the pseudo-orbitals are constructed, the ECPs can have unusual functional forms, and the analytic expansion may require many terms for a good fit. In molecular calculations, the integrals over the ECPs and their corresponding projectors on each center can be very time-consuming when the number of terms for each potential is large.

There are alternatives to the use of Eq. 20 for generating the effective potentials. Melius et al (55) suggest determining a potential by minimizing

$$S = \sum_\mu [\langle \phi_\mu | \hat{h} + \mathscr{V}_l^{\text{eff}} + \mathscr{V}_{\text{val}}' - \varepsilon_l | \chi_l \rangle]^2 \qquad 25.$$

where $\{\phi_\mu\}$ is a suitably chosen set of basis functions. Reasonably compact expansions of $\mathscr{V}_l^{\text{eff}}$ have been achieved in this way (50, 56, 57). Barthelat et al (54) have proposed determining the effective potential by minimizing

$$\|\mathscr{O}\| = [\langle \chi_l | \mathscr{O}^2 | \chi_l \rangle]^{1/2} \qquad 26.$$

where the operator $\mathscr{O}$ is defined by

$$\mathscr{O} = \varepsilon_l' | \chi_l' \rangle \langle \chi_l' | - \varepsilon_l | \chi_l \rangle \langle \chi_l | \qquad 27.$$

where ε_l and χ_l are the "exact" eigenvalue and pseudo-orbital, and ε_l' and χ_l' are the eigenvalue and pseudo-orbital obtained by solving

$$\left(-\tfrac{1}{2}\nabla^2 - \frac{Z_{\text{eff}}}{r} + \frac{l(l+1)}{2r^2} + \mathscr{V}_{\text{val}}' + \mathscr{V}^{\text{eff}} \right) \chi_l' = \varepsilon_l' \chi_l'. \qquad 28.$$

The methods discussed above for generating effective core potentials are based on all-electron Hartree-Fock calculations and, as we shall see in a following section, they are easily extended to the Dirac-Fock method in order to incorporate relativistic effects. It should be noted that very similar progress has been made in solid state physics where "norm-conserving" ECP based on local density functional theory have been developed. The reader is referred to a paper by Bachelet, Hamann & Schlüter (58) that contains a good review of the methods and a tabulation of potentials for most of the periodic table.

COMPARISONS WITH ALL ELECTRON RESULTS

In molecular structure calculations, effective potentials can be useful tools for reducing costs and also for exploring new areas of chemistry that are untreatable by existing all-electron methods. In the latter category, the study of relativistic effects on molecular structure is one of the most exciting applications. We discuss some results in this area in the next section. By reducing the number of electrons that need to be considered and the number of basis functions needed to describe the orbitals, the cost savings to be realized by the use of effective potentials can be substantial for large molecules and molecules containing atoms beyond the first row. To be truly useful, however, the results must be reliable and equivalent to the corresponding all-electron calculations. If all-electron accuracy is not achievable, then the limitations must be well understood.

To this end many comparisons of effective potential and all-electron calculations have appeared in the literature. The studies fall into two groups: those involving main-group elements and those involving transition elements. For the main-group elements, the application of effective potential methods and the limitations in terms of accuracy are pretty well understood. For the transition elements, on the other hand, many questions arise about what constitutes the valence shell and whether the absence of nodal structure in the pseudo-orbitals affects the calculated molecular electronic structure.

The main-group elements are characterized by valence shells that are spatially well-separated from the underlying closed-shell cores. Consequently, two-center, core-valence, and core-core interactions should be minimal and a single valence shell representation in molecular calculations is probably adequate. For the most modern methods, in which the pseudo-orbitals are constructed to maintain the correct valence-valence interactions, effective potential results agree exceedingly well with comparable all electron calculations. Some representative examples follow, but this is by no means an exhaustive survey.

The model potential methods of Bonifacic & Huzinaga (24–27) and Wahlgrin and co-workers (33–37) have been compared to all-electron calculations for a variety of molecules containing first and second row elements (32–34, 59–62). When the basis sets are not truncated, and full orthogonality of the orbitals to the core is maintained, the results are impressive. Molecular SCF orbital energies and isomeric energy differences agree with all-electron values to within a few millihartrees. Optimized bond lengths generally agree to within 0.01 angstroms and bond angles to within a few degrees. Charge distributions are reliable, and calculated dipole moments are within a few hundredths of a Debye of all-electron values.

Comparisons of ECP and all-electron results is a little trickier. The pseudo-orbitals are nodeless and the comparability of the basis sets must be examined. Many promising results have been obtained (48, 50, 52, 54, 63–

67), and the agreement with all-electron calculations is similar to the model potential work described above. The heaviest systems for which direct comparisons between ECP and all-electron results have been made are Xe_2 and Xe_2^+ (63, 64). Our results for Xe_2^+ are shown in Figure 3. The calculational method used was first-order configuration interaction from a single reference configuration for each symmetry using a double-zeta plus polarization basis set. The agreement with the comparable all-electron calculations of Wadt (68) is seen to be very good even on the repulsive left-hand limbs of the potential curves. The ECP used in these calculations were generated from shape-consistent (48) pseudo-orbitals.

Recently, Stevens, Basch & Krauss (68a) have carried out careful comparisons of ECP and all-electron calculations for several first and second row diatomic and small polyatomic molecules. Very compact, analytic representations of the effective potentials were generated by a method similar to that of Durand & Barthelat (53, 54), using shape-consistent pseudo-orbitals. Small, four-Gaussian basis sets were found to give excellent descriptions of the s and p pseudo-orbitals when compared to the $9s5p$ and $11s7p$ all-electron basis sets of Huzinaga (69, 70). In fact, it was necessary to increase the all-electron basis sets to $10s5p$ and $12s8p$ to improve the description of the outer portions of the orbitals for a more valid comparison. Equilibrium geometries, atomization energies, and vibrational frequencies were calculated for several small hydrocarbons and

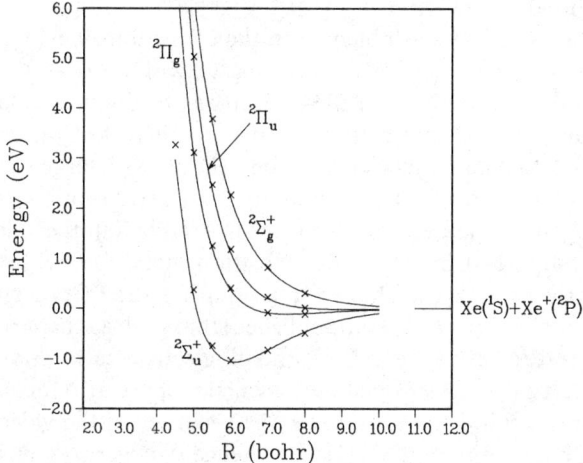

Figure 3 A comparison of all-electron (*solid line*) and shape-consistent ECP (X) calculations of the potential curves of Xe_2^+.

mixed first and second row diatomics. For the molecules tested, at both the SCF and correlated levels, the ECP results were comparable to the all-electron results in every way. Even for triply bonded systems, where one might expect the largest errors, the results are still quite acceptable. For example, for P_2, ECP and all-electron SCF calculations (with the basis sets augmented by a single polarization function on each center) yielded $R_e = 3.526\,(3.531)$ a.u., $D_e = 1.299\,(1.240)$ eV, and $\omega_e = 900\,(893)$ cm^{-1}, where the numbers in parentheses are taken from the ECP calculations. Similar results were obtained for other triply bonded systems. We conclude that for general application of the ECP method to molecules containing the main group elements, very little accuracy is lost versus all-electron calculations.

Studying compounds that contain one or more transition metal atoms is expensive if all-electron methods are used. Effective potentials can be useful for such studies, but care must be taken in their application. Certainly, for the transition metals on the left side of the third and fourth rows of the Periodic Table, an $(n-1)d$ subshell must be included in the valence space along with the ns shell. However, at least one attempt has been made (71) to include the d electrons in the core by defining a spin-dependent ECP for Ni to be used in a NiO calculation where the wavefunction was said to be almost purely ionic $Ni^{2+}O^{2-}$. At the end of the transition series (e.g. Cu, Zn, Ag, Cd) where the d-subshell is filled, it is possible to define a closed-shell core containing the d electrons (72), but this can be dangerous for Cu and Ag, since the correlation of the d electrons and the correct energetic positioning of the $d^{10}s$ and d^9s^2 atomic states can dramatically influence molecular bonding (73).

There is a more subtle problem with the transition elements that arises from the spatial overlap of the $(n-1)d$ subshell with the $(n-1)p$ and $(n-1)s$ subshells (74). The response of the d electrons to the molecular environment depends on the interactions with the other electrons in the penultimate shell. Such interactions between orbitals with large overlap is not properly accounted for by a local potential. This problem is circumvented by including all of the electrons from the $(n-1)$ shell in the valence space. This is accomplished by using the transition metal ion, with the outer s electrons stripped away, as the starting point for the ECP derivation. We have found in the case of FeO (unpublished) that such a scheme is necessary to achieve agreement between ECP and all-electron calculations.

Including the $(n-1)s$ subshell in the valence space also forces a node in the ns valence orbital. This improves the accuracy of the valence-valence interaction between ns and $(n-1)d$ and hence the energetic positioning of the d^ns^m, $d^{n+1}s^{m-1}$, and $d^{n-1}s^{m+1}$ states. All of the other problems that arise with transition metal compounds, such as valence shell correlation effects, remain in the ECP calculation as well as the all-electron calculation.

RELATIVISTIC EFFECTIVE POTENTIALS

Incorporation of Relativistic Effects

Pitzer (75) and Pyykko & Desclaux (76) have reviewed the chemical and electronic effects of relativity. However, molecular calculations of such effects using the all-electron Dirac-Fock (DF) method have proved to be very difficult. Very few many-center calculations have been done to date for molecules (77, 78) containing other than first row atoms (79, 80). The majority of molecular calculations have been for hydrides and have used a one-center expansion of the wavefunction (81). Effective potentials offer a way to incorporate relativistic effects into molecular calculations without the complexity of the DF method. As Z increases, the relativistic effects in the core region lead to a contraction of the core radial orbitals. Indirectly, because of orthogonality constraints and the change in the core potential, the relativistic effects are transmitted to the valence orbitals. However, even though direct relativistic effects in the valence orbitals are small, they have been shown to determine, by themselves, the decrease in equilibrium bond lengths from the calculated nonrelativistic values for Au_2, AuH, and AuCl (82, 83). Effective potentials must reflect both the indirect effects due to the core relativistic contraction and all direct relativistic effects, including the large spin-orbit interactions in the valence electrons of heavy atoms.

Relativistic four-component wavefunctions are obtained by solution of the Dirac-Fock equations (84). Large, P, and small, Q, component radial functions are obtained from the solution of coupled first-order equations. The small component, Q, is a small part of the valence orbital density, and a second order radial equation for only the large component can be obtained in the form (85)

$$\{\mathcal{H}_0(r) + \mathcal{H}_m(r) + \mathcal{H}_d(r) + \mathcal{H}_{so}(r) + \mathcal{H}(0(\alpha^4))\}P(r) = \varepsilon P(r) \qquad 29.$$

where $\mathcal{H}_0$ is the nonrelativistic Hamiltonian, $\mathcal{H}_m$ the mass-velocity correction to the kinetic energy, $\mathcal{H}_d$ the Darwin term, and $\mathcal{H}_{so}$ the spin-orbit coupling. Lee et al (86) have noted that an effective core potential that is correct within the Pauli approximation can replace the core part of $\mathcal{H}_0$ in Eq. 29, but that the relativistic terms in the valence Hamiltonian would then have to be included in subsequent molecular calculations. They opted instead to define a relativistic effective potential (REP) that includes all of the valence relativistic terms as well as the core potential by defining the REP by

$$[\mathcal{H}'_0 + \mathcal{V}^{REP}_{l,j}(r)]\chi_{l,j}(r) = \varepsilon \chi_{l,j}(r) \qquad 30.$$

where $\mathcal{H}'_0$ denotes the absence of core-valence interaction terms and $\chi_{l,j}(r)$ is a pseudo-orbital constructed from a renormalized large component solution of the atomic DF equation with the orbital energy ε. Deriving the REP is then entirely analogous to the nonrelativistic procedures for an ECP. The extent to which the small component is actually found to represent a small part ($\lesssim 1\%$) of the charge density of the valence orbitals obtained in DF calculations determines the validity of the REP. Since χ and ε are dependent on the total (spin plus orbital) angular momentum, $j = l \pm 1/2$, of the orbital, the REP is also j dependent. How well the local representation will mimic the explicit representation of the mass velocity, Darwin, and spin-orbit terms is not really known. Effective potentials for both large and small components (87) and other more accurate pseudopotential representations (88) have been proposed, but they have not been compared as yet with the more practical and extensively applied REP.

In order to insure that the relativistic orbitals and equations for an open shell atom go formally in the limit of $c \to \infty$ to the analogous nonrelativistic Schrödinger equation (89, 90), the DF calculation must be done for a degeneracy-weighted sum of all possible jj-coupled configurations arising from an LS configuration. If specific J states are used to determine the DF orbitals, "correlation" effects are included that complicate comparisons with nonrelativistic solutions and the determination of fine structure splittings between J states (91).

The norm-conserving potentials that have been applied to solid-state problems recently have also been extended to include relativistic effects in an analogous way (92). A number of other relativistic pseudopotential methods have been proposed in which an accurate representation of core functions are required for the determination of the core potential (93–95) and ultimately a model potential (93).

The spin-orbit contributions to the REP can be averaged away by defining an averaged relativistic effective potential (96–99), sometimes called AREP,

$$\mathscr{V}_l^{AREP}(r) = (2l+1)^{-1}[l\mathscr{V}_{l,l-1/2}^{REP}(r) + (l+1)V_{l,l+1/2}^{REP}(r)] \qquad 31.$$

$$\mathscr{V}^{AREP}(r) = \sum_l \sum_{m=-l}^{l} \mathscr{V}_l^{AREP}(r)|lm\rangle\langle lm|. \qquad 32.$$

Other averaged relativistic potentials that are determined from either j-averaged pseudo-orbitals (100) or from j-independent, one component relativistic atomic orbitals (101, 102) are analogous to the AREP in the sense that they all incorporate the core contraction effects and the direct valence relativistic effects with the exception of spin-orbit. The potentials based on j-independent, one component relativistic orbitals are not easier or harder to generate than those based on DF large component orbitals. The ease with which a spin-orbit potential can be extracted (see below) from the $\mathscr{V}_{l,j}^{REP}$ is a strong recommendation for this method.

The accuracy and transferability of all the relativistic potentials have been tested repeatedly at the atomic level. The comparison between all-electron and AREP excitation and ionization energies is invariably good as long as the ionization level of the atomic generating state does not differ too much from the states under consideration. The REP have been used for a number of single configuration j-dependent calculations (86, 103) that can be compared directly to DF results. For example, accurate values (error <0.1 eV) were obtained for the fine-structure splittings, excitation energies, and ionization potentials of Hg and Hg$^+$ (103).

Another test of the transferability of the potentials is the calculation of atomic dipole polarizabilities (103) which can be expressed as a sum over excited states

$$\alpha = 2 \sum_{i \neq 0} \frac{|\langle \psi_0 | \mu | \psi_i \rangle|^2}{E_i - E_0} \qquad 33.$$

where μ is the dipole moment operator. Calculation of the polarizability in effect then tests the accuracy of calculated transition moments and excitation energies. The calculated dipole polarizability for Hg using REP compares very favorably with the all-electron numerical DF calculation reported by Desclaux et al (104). Since the excitation energies for Hg are known to be calculated with good accuracy, an accurate polarizability implies accurate transition moments. This agrees with the Bates & Damgaard (105) analysis that an accurate description of the long-range part of the valence electronic distribution is sufficient for an accurate determination of transition moments.

Effective Spin-Orbit Potentials

Ermler et al (106) have shown that it is possible to extract an effective spin-orbit potential from the difference between the j-dependent REP and the j-independent AREP. The effective spin-orbit potential is given by

$$\mathcal{H}^{so} = \sum_{l=1}^{\infty} \Delta \mathcal{V}_l^{REP} \left[\frac{l}{2l+1} |l, l+\tfrac{1}{2}\rangle \langle l, l+\tfrac{1}{2}| + \frac{l+1}{2l+1} |l, l-\tfrac{1}{2}\rangle \langle l, l-\tfrac{1}{2}| \right] \qquad 34.$$

with

$$\Delta \mathcal{V}_l^{REP}(r) = \mathcal{V}_{l,l+1/2}^{REP}(r) - \mathcal{V}_{l,l-1/2}^{REP}(r). \qquad 35.$$

The operator has been tested on first and second row atoms (107), and applied to a variety of first and second row molecules (108) with good accuracy.

One difficulty that arises with effective spin-orbit potentials based on Dirac-Fock calculations is the absence of two-electron (Breit) contributions due to the approximate (non-covariant) handling of the two-electron part of the Hamiltonian in the many-electron Dirac-Fock method. Stevens & Krauss (107, 108) have devised a semiempirical scaling method that corrects for two-electron screening of the valence spin-orbit by the core (109), but the valence-valence screening, which depends on the state coupling, remains elusive. The overall Breit corrections to the spin-orbit splittings are sizeable (e.g. $\sim 15\%$ in first row, $\sim 5\%$ in the second row), but the bulk of the effect is due to core screening.

The effective spin-orbit potential provides a unique way for determining this property even though the pseudo-orbitals are nodeless and have arbitrary core behavior. Incorporation of the spin-orbit interaction through an empirical operator with a fixed r dependence (110) will work only if substantial charge transfer or orbital excitations are not present. A spin-orbit potential has been constructed from a perturbation calculation based on one-component relativistic orbitals (102) that has proved accurate for the manifold of ground and excited states of Ar_2 and InH (111).

Applications

The AREP have been implemented and tested in a variety of quantum chemistry programs that include the most modern MCSCF (including the complete active space variety) and CI approaches to the calculation of the correlation energy in the ground and excited states of molecules. Although the REP were quickly utilized in j-dependent SCF calculations (97, 98, 112), the extension to MCSCF or large CI methods has been more limited. An iterative method has been devised by Christiansen et al (113) in which the molecular orbitals are determined by MCSCF calculations using the j-independent AREP, and a CI calculation is done using the complete Hamiltonian including the effective spin-orbit potential. Schwarz and co-workers (114–116) have developed a relativistic generalization of the restricted SCF theory for two-component orbitals using j-dependent model potentials that are obtained empirically (99). Their calculations on the electronic structure of $HgCl_2$, $PbCl_2$, $BaCl_2$, and PbH_2 are the first polyatomic systems treated by j-dependent methods (117).

Much of the early application of the relativistic effective potentials in either the AREP or REP forms has been to test for the experimentally deduced relativistic effects outlined, for example, in the Pitzer review (75). Now being considered are the anomolous behavior of Au; the effect of large spin-orbit coupling in retarding s-p hybridization and causing more complicated effects of intermediate couplings in transition and other metals; relativistic effects on structure, orbital energies, and properties. Some of the properties of heavy elements and, particularly, metals are complicated by the appearance of f shells. This complicated chemical behavior is not easily ascribed to a single cause, especially since the calculations are complicated by intra- and interatomic valence correlation energies that vary substantially with bond formation. A bare beginning has been made on such problems, but it is a testament to the power of the effective potential method that such a beginning can seriously be made.

For molecules, few direct comparisons have been made between relativistic effective potential calculations and accurate DF calculations. The AgH DF calculations of Lee & McLean (77) provide reasonably accurate all-electron results against which the effective potential methods can be measured. Klobukowski (118) has studied the AgH ground state using a relativistic generalization of the Bonifacic & Huzinaga model potential (24–27), and the comparison with the all-electron results is good. We have also examined this molecule (unpublished) using an AREP to represent Ag with $4s$, $4p$, $4d$, and $5s$ in the valence space. The calculated R_e agrees to within 0.01 Å with the all-electron result, but the spectroscopic constants do show slight differences. Other properties, such as the dipole moment, are not available for comparison. Lee & McLean (77) also reported DF results for AuH, but the basis set was less complete than in the AgH case and they concluded the calculations are not as accurate. Nonetheless, the calculation exhibits the large shift in the R_e due to the relativistic effect (119).

The effective potential calculation of Hay et al (120) on AuH also shows a large shift in R_e, but their value is about 0.13 Å shorter than all-electron DF values obtained by either two-center or one-center DF calculations (77, 121). We have also calculated R_e for AuH using an AREP that excludes the $5s$ and $5p$ orbitals from the core. Our value (unpublished) of 1.60 Å is closer to the all-electron values (1.62–1.64 Å), but direct comparison is difficult due to the limitations of the basis sets and single-center expansions in the all-electron calculations. The origin of any differences is not clear, and comparisons of more properties than just R_e are needed.

Three reviews have appeared recently that describe the applications of the relativistic effective potentials. Two (122, 123) describe the research using one-component relativistic orbitals as the basis for constructing the pseudo-orbitals. In addition to tests of the potentials on molecules such as AuH, AuCl, Xe_2, and Xe_2^+ (124), these potentials have been applied to the conformations and orbital energies of a variety of uranium fluorides (125, 126); the conformational and energy surface behavior for a variety of Pt complexes (127–131); the spectroscopic properties of rare gas excimers (132) and mercury halides (120, 133, 134); the interaction of substrates with metal clusters (135); and the structural differences between ThO_2 and the uranyl ion, UO_2^{2+} (136).

In the third review, Pitzer (137) details the usefulness of j-dependent calculations for main group heavy metal molecules in critically analyzing experimental spectroscopic and thermodynamic data. The most extensive studies have been for TlH in a series of calculations that include more correlation energy at each step (112, 113, 138). There are slight discrepancies between the experimental and calculated spectroscopic constants for the TlH ground state, which can be attributed to d-correlation effects. As Pitzer notes, for internuclear distances larger than 4.5 bohr where the R-dependence of the intra-atomic correlation energy is negligible, the experimental and theoretical curves agree very well. The effect of d-correlation on transition metal compound energy curves is well documented in both all-electron (119, 139) and effective potential calculations (140, 73).

Tl_2 has also been examined in a succession of j-dependent calculations with increasing amounts of correlation taken into account (138, 141, 142). Christiansen (142) has used the theoretical results to analyze critically the conclusions of the high temperature mass spectrometric determination of the bond energy (143). The discrepancy between the calculated bond energy (0.16 eV) and the reported experimental value (0.63 eV) is shown to be due to the use of "inappropriate" spectroscopic constants in evaluating the thermal functions required in the mass spectrometric analysis. The need to reconsider high-temperature mass spectrometric determinations of bond energies is not unusual. Pitzer has warned of the pitfalls in the usual analysis (144). Ab initio calculations can now provide more accurate spectroscopic data for all of the thermally accessible states than is normally available from experiment, leading to a reconsideration for the bond energies in molecules such as Pb_2 (145, 146), Sn_2 (146), SnO (147), and, for an example from the lighter elements, Ni_2 (140).

Although Pb has a very large spin-orbit interaction, it has not been found necessary to consider it explicitly in the AREP-MCSCF calculation of the spectroscopic parameters and dipole moment for the ground state of PbO (148). Reasonable spectroscopic constants and comparisons with photoelectron spectra have been obtained for PbS and PbSe (112) from j-dependent SCF calculation. The MCSCF calculation shows that both σ and π bonding terms are included, but, as Lee et al note (112), these can both originate in the energetically favorable $p_{1/2}$ orbital. The AREP type of potential seems adequate to determine the interaction with the core. The empirical introduction of the spin-orbit in an "atoms-in-molecule" approximation then allows the calculation of the interaction potentials for various states of Pb with rare gases (149).

Assignments of transitions and states in the area of metal cluster structure, candidate laser molecules, and metal oxide spectra are beginning to appear. The electronic structure of a variety of metal-metal bonded systems such as Pt_2, Pd_2, Cu_2, Ag_2, Au_2, AgAu, $Ni_2C_2H_4$, and Ni_m ($m =$

1–6) was analyzed by Basch (150). He later extended these calculations to the Ag_3, Ag_3^-, and Ag_3^+ molecules (151) and showed that the variation in electron affinities and ionization potentials with cluster size found for larger clusters is also found in the diatomic and triatomic molecules. Quantitative differences in M–M and M–H bond strengths found for M=Pt, Pd, and Ni were related to photoemission energy distribution and difference spectra for the clean metal and hydrogen chemisorbed metal surfaces (152). A $^2\Delta$ ground state is found for NiH and PtH but $^2\Sigma^+$ is observed and calculated for PdH. Basch et al (152) note that this represents an example of an isoelectronic sequence of compounds down a column of the Periodic Table that do not all have the same ground state. Our calculations on FeH and RuH (unpublished) indicate $^4\Delta$ and $^4\phi$ ground states, respectively, which mirrors the variation in the transition metal atom ground states. More accurate calculations on ground and excited states of PtH and PtH$^+$ have recently appeared (153) that provide a model for photoemission properties of hydrogen chemisorbed on platinum.

The analysis of excimer laser molecule spectra would be enhanced by the development of j-dependent MCSCF and CI methods and codes. Although the calculations for HgCl (133, 154) provide reasonably accurate data for the known states and transitions, and provide the repulsive $^2\Pi$ energy curve and estimates for photodissociation of the ground state, analogous calculations on HgBr (133, 155) are not as accurate. The prediction of transition energies and strengths for XeF are also reasonably successful (132, 156), but the spin-orbit interaction was included by an atoms-in-molecule approximation that is not strictly valid when there is CI coupling between covalent and ionic states. The relative energies of the B and C states in rare gas halide excimers is of practical significance and requires a j-dependent CI calculation for correct analysis.

Metal oxide spectra are complex, and theoretical calculations are useful or necessary complements to an experimental analysis. Although the first row transition metal oxides can be studied by all-electron methods, the REP can still be useful because of the availability of an inexpensive spin-orbit analysis via the effective spin-orbit potential. The ability to compare isovalent systems down a column of the Periodic Table including heavy lanthanide or actinide molecules is also not possible today with all-electron calculations. With the use of the spin-orbit potentials and complete active space MCSCF calculations of $^5\Pi$ and $^5\Phi$ states in FeO, for example, we have been able to assign an electronic structure to these states (unpublished) in support of experimental assignments (157). This information is useful even though intra-atomic correlation energy errors in the d subshell have thus far prevented the determination of accurate excitation energies.

The energy curves and spectroscopic constants for the lowest states of

UH, UF, and UO⁺ have been estimated (158, 159). The ligand field relative energetic destabilizations of the s and d valence electrons proposed by Field (160) have been determined from AREP-SCF calculations and compared for Nd and U oxides (unpublished).

The availability of relativistic effective potentials permits the treatment of moderately large heavy metallo-organic molecules. For example, the interaction of Pt complexes with molecules such as imidazole and pyrimidine, as model organic bases, has been determined in our laboratory (unpublished); and calculations of metal complex interactions, and even energy surfaces, with biomolecule constituents are feasible with the codes in which the effective potentials are implemented.

OTHER TOPICS

Correlation Effects

For the alkali and alkaline earth elements, electron correlation between the valence and core electrons (core polarization) plays an important role in determining ionization potentials, electron affinities, and polarizabilities (161, 162). These effects, along with the contraction of the valence electronic distribution due to core-valence correlation, can significantly alter calculated binding energies, dipole moments, and other properties of molecules containing these elements (161, 163, 164).

For single valence electron systems it is possible to analyze the core-valence correlation, and the leading term in the expansion of the energy contribution may be shown to vary as r^{-4} (165–167). Hence, the effect is known as *core polarization*. Empirically determined model potentials, which are based on atomic spectroscopic data, by definition include core polarization effects (10, 168, 169); but transferability to molecular systems with a different number of valence electrons is complicated by two-electron screening terms that are difficult to incorporate properly (161).

Ab initio effective potentials derived from atomic Hartree-Fock wavefunctions do not contain core polarization contributions. This may be overcome by including the penultimate shell in the valence space in order to introduce the correlation terms explicitly in the wavefunction (170), or by adding empirically determined correction potentials with the correct functional form to the ab initio potentials (171, 172). A very promising empirical scheme has been introduced recently by Müller and co-workers (172a) that allows the incorporation of both one- and two-electron core polarization effects in ab initio calculations. Although their applications involved all-electron calculations, the method appears to be equally applicable to valence-electron-only calculations.

There are core-valence correlation effects that are significant in the transition metals, lanthanides, and actinides. However, as we pointed out in our discussion of the transition metals, the dominant question in these cases is how one defines the valence distribution. In the transition elements, the open d and f subshells have strong spatial overlap, not only with the outermost shell, but with the next inner shell as well. It is not clear that such interactions can be accounted for by a simple correction to the effective potential.

Properties

The calculation of properties within the effective potential method may require the construction of an effective form for the operator. For example, an effective spin-orbit operator naturally occurs in the REP schemes. The spin-orbit operator is limited by its derivation to one-center integrals and excludes Breit two-electron interactions between the valence electrons. However, because it is a one-electron operator, it may be easily used to evaluate higher order spectroscopic properties involving the spin-orbit interaction. Hafner & Schwarz (173) have shown that the dipole length operator can be used without modification in effective potential calculations to evaluate transition moments, since, for this representation, explicit reorthogonalization of the valence pseudo-orbitals to the frozen core orbitals contributes little to the calculated moment. For operators such as $\mathscr{L}/r^3$, which depend strongly on the charge density in the core region, reintroduction of the inner loops of the valence orbitals by reorthogonalization to the core would provide one approximation to the integral evaluation. Another possible method would be the construction of an effective operator (102).

Group Potentials

In order to simplify calculations on large polyatomic systems, various group potential methods are being proposed. Nicolas & Durand and co-workers (174–177) have constructed atomic carbon and hydrogen potentials from model molecular systems that have proved to be transferable to homologous series of saturated and unsaturated hydrocarbons. Applications have also been made to band structure calculations in polyethylene. Ohta et al (178) are developing potentials for groups of orbitals that are identified as participating primarily through electrostatic interactions in the molecular complex being investigated. In analogy with the effective potential method, which treats only the valence electrons explicitly, they propose an effective fragment potential that replaces the transferable, chemically inactive groups. Thus they treat explicitly only those electrons directly involved in the molecular interaction. The energy results for a pseudo-ammonia consisting of only the lone-pair electrons interacting with an all-electron ammonia or borazine appear good enough to warrant further tests and extensions of the method.

CONCLUSIONS

Model potentials and effective core potentials are having an impact on quantum chemistry by allowing accurate structure calculations to be performed for very large systems such as solid state systems, complex organic and biomolecules, and polymers. Relativistic effective potentials have allowed, for the first time, systematic theoretical exploration of heavy atom chemistry. The most appealing aspect of these effective potential methods is that they are ab initio, and, therefore, like the all-electron methods they simulate, extensible to all atomic and molecular systems.

The theoretical framework that underlies the determination of effective potentials is intuitively sound, but sometimes less than rigorous. This is especially true for the relativistic extensions that need to be examined much more carefully since direct comparisons with more rigorous all-electron results are difficult.

Acknowledgments

The authors would like to thank Professor Harold Basch of Bar Ilan University for many helpful discussions, and Drs. Frederick Mies and John Rumble for their comments and reading of this manuscript. We also thank Ms. Arvella Flynt for her careful and timely typing of this manuscript.

Literature Cited

1. Fock, V., Wesselov, W., Petraschen, M. 1940. *Zh. Eksp. Teor. Fiz.* 10:723–39
2. Desclaux, J. P., Kim, Y. K. 1975. *J. Phys. B* 8:1177–82
3. Hellmann, H. 1935. *J. Chem. Phys.* 3:61
4. Hellmann, H., Kassatotschkin, W. 1936. *J. Chem. Phys.* 4:325–26
5. Gombas, P. 1935. *Z. Phys.* 94:473–88
6. Ziman, J. M. 1964. *Adv. Phys.* 13:89–138
7. Harrison, W. A. 1966. *Pseudopotentials in the Theory of Metals.* New York: Benjamin. 336 pp.
8. Heine, V. 1966. *Optical Properties and Electronic Structure of Metals and Alloys,* ed. F. Abeles, pp. 16–21. New York: Wiley
9. Weeks, J. D., Hazi, A., Rice, S. A. 1969. *Adv. Chem. Phys.* 16:283–342
10. Bardsley, J. N. 1974. *Case Stud. At. Phys.* 4:299–368
11. Phillips, J. C., Kleinman, L. 1959. *Phys. Rev.* 116:287–94
12. Cohen, M. H., Heine, V. 1961. *Phys. Rev.* 122:1821–26
13. Austin, B. J., Heine, V., Sham, L. J. 1962. *Phys. Rev.* 127:276–82
14. Weeks, J. D., Rice, S. A. 1968. *J. Chem. Phys.* 49:2741–55
15. Abarenkov, I. V., Heine, V. 1965. *Philos. Mag.* 12:529–37
16. Animalu, A. O. E., Heine, V. 1965. *Philos. Mag.* 12:1249–70
17. Szasz, L., McGinn, G. 1965. *J. Chem. Phys.* 42:2363–70
18. Szasz, L., McGinn, G. 1966. *J. Chem. Phys.* 45:2898–2912
19. Simons, G., Mazzioti, A. 1970. *J. Chem. Phys.* 52:2449–55
20. Simons, G. 1971. *J. Chem. Phys.* 55:756–61
21. Chang, T. C., Habitz, P., Pittel, B.,

21. Schwarz, W. H. E. 1974. *Theor. Chim Acta* 34:263–75
22. Schwartz, M. E., Switalski, J. D. 1972. *J. Chem. Phys.* 57:4125–31
23. Schwartz, M. E., Switalski, J. D. 1972. *J. Chem. Phys.* 57:4132–36
24. Bonifacic, V., Huzinaga, S. 1974. *J. Chem. Phys.* 60:2779–86
25. Bonifacic, V., Huzinaga, S. 1975. *J. Chem. Phys.* 62:1507–8
26. Bonifacic, V., Huzinaga, S. 1975. *J. Chem. Phys.* 62:1509–12
27. Bonifacic, V., Huzinaga, S. 1976. *J. Chem. Phys.* 64:956–60
28. Höjer, G., Chung, J. 1978. *Int. J. Quant. Chem.* 14:623–34
29. Dixon, R. N., Hugo, J. M. V. 1975. *Mol. Phys.* 29:953–70
30. Dixon, R. N., Tasker, P. W., Balint-Kurti, G. G. 1977. *Mol. Phys.* 34:1455–71
31. Ewig, C. S., van Wazer, J. R. 1975. *J. Chem. Phys.* 63:4035–41
32. Sakai, Y. 1981. *J. Chem. Phys.* 75:1303–8
33. Wahlgren, U. 1978. *Chem. Phys.* 29:231–40
34. Wahlgren, U. 1978. *Chem. Phys.* 32:215–21
35. Gropen, O., Wahlgren, U., Pettersson, L. 1982. *Chem. Phys.* 66:453–58
36. Gropen, O., Wahlgren, U., Pettersson, L. 1982. *Chem. Phys.* 66:459–64
37. Pettersson, L., Wahlgren, U. 1982. *Chem. Phys.* 69:185–92
38. Goddard, W. A. III. 1968. *Phys. Rev.* 174:659–62
39. Goddard, W. A. III. 1967. *Phys. Rev.* 157:73–80
40. Goddard, W. A. III. 1967. *Phys. Rev.* 157:81–93
41. Goddard, W. A. III. 1968. *Phys. Rev.* 169:120–30
42. Kahn, L. R., Goddard, W. A. III. 1968. *Chem. Phys. Lett.* 2:667–70
43. Kahn, L. R., Goddard, W. A. III. 1972. *J. Chem. Phys.* 56:2685–2701
44. Melius, C. F., Goddard, W. A. III. 1972. *J. Chem. Phys.* 56:3342–48
45. Melius, C. F., Goddard, W. A. III. 1972. *J. Chem. Phys.* 56:3348–59
46. Melius, C. F., Goddard, W. A. III. 1974. *Phys. Rev. A* 10:1528–40
47. Kahn, L. R., Baybutt, P., Truhlar, D. G. 1976. *J. Chem. Phys.* 65:3826–53
48. Christiansen, P. A., Lee, Y. S., Pitzer, K. S. 1979. *J. Chem. Phys.* 71:4445–50
49. Fischer, C. F. 1969. *Comp. Phys. Commun.* 1:151–66
50. Redondo, A., Goddard, W. A. III, McGill, T. C. 1977. *Phys. Rev. B* 15:5038–48
51. Hay, P. J., Wadt, W. R., Kahn, L. R. 1978. *J. Chem. Phys.* 68:3059–66
52. Rappé, A. K., Smedley, T. A., Goddard, W. A. III. 1981. *J. Phys. Chem.* 85:1662–66
53. Durand, Ph., Barthelat, J. C. 1975. *Theor. Chim. Acta* 38:283–302
54. Barthelat, J. C., Durand, Ph., Serafini, A. 1977. *Mol. Phys.* 33:159–80
55. Melius, C. F., Olafson, B. D., Goddard, W. A. III. 1974. *Chem. Phys. Lett.* 28:457–62
56. Topiol, S., Moskowitz, J. W., Melius, C. F. 1976. Courant Inst. of Math. Sci., Rep. COO-3077-105
57. Topiol, S., Moskowitz, J. W., Melius, C. F. 1978. *J. Chem. Phys.* 68:2364–72
58. Bachelet, G. B., Hamann, D. R., Schlüter, M. 1982. *Phys. Rev. B* 26:4199–4228
59. Popkie, H. E., Kaufman, J. J. 1976. *Intern. J. Quant. Chem.* S10:47–57
60. Gropen, O., Huzinaga, S., McLean, A. D. 1980. *J. Chem. Phys.* 73:402–6
61. Sakai, Y., Huzinaga, S. 1982. *J. Chem. Phys.* 76:2537–51
62. Sakai, Y., Huzinaga, S. 1982. *J. Chem. Phys.* 76:2552–57
63. Christiansen, P. A., Pitzer, K. S., Lee, Y. S., Yates, J. H., Ermler, W. C., et al. 1981. *J. Chem. Phys.* 75:5410–15
64. Stevens, W. J., Krauss, M. 1982. *Appl. Phys. Lett.* 41:301–3
65. Pelissier, M., Durand, Ph. 1980. *Theor. Chim. Acta* 55:43–54
66. Teichteil, Ch., Malrieu, J. P., Barthelat, J. C. 1977. *Mol. Phys.* 33:181–97
67. Ortega-Blake, I., Barthelat, J. C., Costes-Puech, E., Oliveros, E., et al. 1982. *J. Chem. Phys.* 76:4130–35
68. Wadt, W. R. 1978. *J. Chem. Phys.* 68:402–14
68a. Stevens, W. J., Basch, H., Krauss, M. 1984. *J. Chem. Phys.* Submitted
69. Huzinaga, S. 1965. *J. Chem. Phys.* 42:1293–1302
70. Huzinaga, S., Sakai, Y. 1969. *J. Chem. Phys.* 50:1371–81
71. Sabelli, N. H., Kahn, L. R., Benedek, R. 1983. *J. Chem. Phys.* 73:6259–62
72. Jeung, G. H., Barthelat, J. C. 1983. *J. Chem. Phys.* 78:2097–2100
73. Pelissier, M. 1981. *J. Chem. Phys.* 75:775–80
74. Pettersson, L. G. M., Wahlgren, U., Gropen, O. 1983. *Chem. Phys.* 80:7–16
75. Pitzer, K. S. 1979. *Acc. Chem. Res.* 12:271–76
76. Pyykko, P., Desclaux, J. P. 1979. *Acc. Chem. Res.* 12:276–81
77. Lee, Y. S., McLean, A. D. 1982. *J. Chem. Phys.* 76:735–36

78. Wood, C. P., Pyper, N. C. 1981. *Chem. Phys. Lett.* 84:614–21
79. Mark, F., Lischka, H., Rosicky, F. 1980. *Chem. Phys. Lett.* 71:507–12
80. Malli, G., Oreg, J. 1980. *Chem. Phys. Lett.* 69:313–14
81. Desclaux, J. P. 1983. *Relativistic Effects in Atoms, Molecules and Solids*, ed. G. L. Malli, pp. 213–26. New York: Plenum. 543 pp.
82. Ziegler, T., Snijders, J. G., Baerends, E. J. 1980. *Chem. Phys. Lett.* 75:1–4
83. Snijders, J. G., Pyykko, P. 1980. *Chem. Phys. Lett.* 75:5–8
84. Desclaux, J. P. 1973. *At. Data Nucl. Data* 12:311–406
85. Bethe, H. A., Salpeter, E. E. 1957. *Quantum Mechanics of One- and Two-Electron Atoms*. New York: Academic. 368 pp.
86. Lee, Y. S., Ermler, W. C., Pitzer, K. S. 1977. *J. Chem. Phys.* 67:5861–76
87. Ishikawa, Y., Malli, G. 1981. *J. Chem. Phys.* 75:5423–31
88. Pyper, N. C. 1983. See Ref. 81, pp. 437–87
89. Wood, C. P., Pyper, N. C. 1980. *Mol. Phys.* 41:149–58
90. Detrich, J., Weiss, A. W. 1982. *Phys. Rev. A* 25:1203–5
91. Huang, K. N., Kim, Y. K., Cheng, K. T., Desclaux, J. P. 1982. *Phys. Rev. Lett.* 48:1245–48
92. Kleinman, L. 1980. *Phys. Rev. B* 21:2630–31
93. Datta, S. N., Ewig, C. S., van Wazer, J. R. 1978. *Chem. Phys. Lett.* 57:83–89
94. Das, G., Wahl, A. C. 1976. *J. Chem. Phys.* 64:4672–79
95. Das, G., Wahl, A. C. 1978. *J. Chem. Phys.* 69:53–62
96. Ermler, W. C., Lee, Y. S., Pitzer, K. S., Winter, N. W. 1978. *J. Chem. Phys.* 69:976–83
97. Lee, Y. S., Ermler, W. C., Pitzer, K. S., McLean, A. D. 1979. *J. Chem. Phys.* 70:288–92
98. Ermler, W. C., Lee, Y. S., Pitzer, K. S. 1979. *J. Chem. Phys.* 70:293–98
99. Hafner, P., Schwarz, W. H. E. 1978. *J. Phys. B* 11:217–33
100. Basch, H., Topiol, S. 1979. *J. Chem. Phys.* 71:802–14
101. Kahn, L. R., Hay, P. J., Cowan, R. D. 1978. *J. Chem. Phys.* 68:2368–72
102. Teichteil, C., Pelissier, M., Spiegelmann, F. 1983. *Chem. Phys.* 81:273–82
103. Stevens, W. J., Krauss, M. 1983. *J. Phys. B* 16:2921–30
104. Desclaux, J. P., Laaksonen, L., Pyykko, P. 1981. *J. Phys. B* 14:419–25
105. Bates, D. R., Damgaard, A. 1949. *Philos. Trans. R. Soc.* 242:101–22
106. Ermler, W. C., Lee, Y. S., Christiansen, P. A., Pitzer, K. S. 1981. *Chem. Phys. Lett.* 81:70–74
107. Stevens, W. J., Krauss, M. 1982. *Chem. Phys. Lett.* 86:320–24
108. Stevens, W. J., Krauss, M. 1982. *J. Chem. Phys.* 76:3834–36
109. Blume, M., Watson, R. E. 1963. *Proc. R. Soc. Ser. A* 271:565–78
110. Wadt, W. R. 1982. *Chem. Phys. Lett.* 89:245–48
111. Teichteil, C., Spiegelmann, F. 1983. *Chem. Phys.* 8:283–96
112. Lee, Y. S., Ermler, W. C., Pitzer, K. S. 1980. *J. Chem. Phys.* 73:360–66
113. Christiansen, P. A., Balasubramanian, K., Pitzer, K. S. 1982. *J. Chem. Phys.* 76:5087–92
114. Hafner, P., Schwarz, W. H. E. 1979. *Chem. Phys. Lett.* 65:537–41
115. Hafner, P. 1980. *J. Phys. B* 13:3297–3308
116. Esser, M., Butscher, W., Schwarz, W. H. E. 1981. *Chem. Phys. Lett.* 77:359–64
117. Hafner, P., Habitz, P., Ishikawa, Y., Wechsel-Trakowski, E., Schwarz, W. H. E. 1981. *Chem. Phys. Lett.* 80:311–15
118. Klobukowski, M. 1983. *J. Comp. Chem.* 4:350–61
119. McLean, A. D. 1983. *J. Chem. Phys.* 79:3392–3403
120. Hay, P. J., Wadt, W. R., Kahn, L. R., Bobrowicz, F. W. 1978. *J. Chem. Phys.* 69:984–97
121. Desclaux, J. P., Pyykko, P. 1976. *Chem. Phys. Lett.* 39:300–3
122. Hay, P. J. 1983. See Ref. 81, pp. 383–401
123. Kahn, L. R. 1984. *Int. J. Quant. Chem.* 25:149–83 Proc. Int. Symp. Relativistic Effects in Quantum Chemistry
124. Wadt, W. R., Hay, P. J., Kahn, L. R. 1978. *J. Chem. Phys.* 68:1752–59
125. Hay, P. J., Wadt, W. R., Kahn, L. R., Raffenetti, R. C., Phillips, D. H. 1979. *J. Chem. Phys.* 71:1767–79
126. Wadt, W. R., Hay, P. J. 1979. *J. Am. Chem. Soc.* 101:5198–5206
127. Kitaura, K., Obara, S., Morokuma, K. 1981. *J. Am. Chem. Soc.* 103:2891–92
128. Kitaura, K., Obara, S., Morokuma, K. 1981. *Chem. Phys. Lett.* 77:452–54
129. Noell, J. O., Hay, P. J. 1982. *Inorg. Chem.* 21:14–20
130. Noell, J. O., Hay, P. J. 1982. *J. Am. Chem. Soc.* 104:4578–84
131. Hay, P. J. 1981. *J. Am. Chem. Soc.* 103:1390–93
132. Hay, P. J., Dunning, T. H. 1978. *J. Chem. Phys.* 69:2209–20
133. Wadt, W. R. 1979. *Appl. Phys. Lett.* 34:658–60
134. Wadt, W. R. 1980. *J. Chem. Phys.* 72:2469–78

135. Martin, R. L., Hay, P. J. 1983. *Surf. Sci.* 130: L283-88
136. Wadt, W. R. 1981. *J. Am. Chem. Soc.* 103: 6053-57
137. Pitzer, K. S. 1983. See Ref. 81, pp. 403-20
138. Christiansen, P. A., Pitzer, K. S. 1980. *J. Chem. Phys.* 73: 5160-63
139. Walch, S. P., Bauschlicher, C. W. Jr. 1983. *J. Chem. Phys.* 78: 4597-4605
140. Noell, J. O., Newton, M. D., Hay, P. J., Martin, R. L., Bobrowicz, F. W., 1980. *J. Chem. Phys.* 73: 2360-71
141. Christiansen, P. A., Pitzer, K. S. 1981. *J. Chem. Phys.* 74: 1162-65
142. Christiansen, P. A. 1983. *J. Chem. Phys.* 79: 2928-31
143. Balducci, G., Piacente, V. 1980. *J. Chem. Soc. Chem. Commun.* 1980: 1287-90
144. Pitzer, K. S. 1981. *J. Chem. Phys.* 74: 3078-79
145. Pitzer, K. S., Balasubramanian, K. 1982. *J. Phys. Chem.* 86: 3068-70
146. Balasubramanian, K., Pitzer, K. S. 1983. *J. Chem. Phys.* 78: 321-27
147. Balasubramanian, K., Pitzer, K. S. 1983. *Chem. Phys. Lett.* 100: 273-76
148. Basch, H., Stevens, W. J., Krauss, M. 1981. *J. Chem. Phys.* 74: 2416-18
149. Basch, H., Julienne, P. S., Krauss, M., Rosenkrantz, M. E. 1980. *J. Chem. Phys.* 73: 6247-58
150. Basch, H. 1980. *Faraday Soc. Symp.* 14: 149-57
151. Basch, H. 1981. *J. Am. Chem. Soc.* 103: 4658-63
152. Basch, H., Cohen, D., Topiol, S. 1980. *Isr. J. Chem.* 19: 233-41
153. Wang, S. W., Pitzer, K. S. 1983. *J. Chem. Phys.* 79: 3851-58
154. Julienne, P. S., Konowalow, D. D., Krauss, M., Rosenkrantz, M. E., Stevens, W. J. 1980. *Appl. Phys. Lett.* 36: 132-34
155. Krauss, M., Stevens, W. J. 1981. *Appl. Phys. Lett.* 39: 686-88
156. Krauss, M., Stevens, W. J., Julienne, P. S. 1982. *J. Comp. Chem.* 3: 372-80
157. Cheung, A. S.-C., Lee, N., Lyyra, A. M., Merer, A. J., Taylor, A. W. 1982. *J. Mol. Spectrosc.* 95: 213-25
158. Krauss, M., Stevens, W. J. 1983. *J. Comp. Chem.* 4: 127-35
159. Krauss, M., Stevens, W. J. 1983. *Chem. Phys. Lett.* 99: 417-21
160. Field, R. W. 1982. *Ber. Bunsenges. Phys. Chem.* 86: 771-79
161. Stevens, W. J., Karo, A. M., Hiskes, J. R. 1981. *J. Chem. Phys.* 74: 3989-98
162. Partridge, H., Bauschlicher, C. W. Jr. 1983. *J. Chem. Phys.* 79: 1806-73
163. Partridge, H., Dixon, D. A., Walch, S. P., Bauschlicher, C. W. Jr., Gole, J. L. 1983. *J. Chem. Phys.* 79: 1859-65
164. Stevens, W. J., Konowalow, D. D., Ratcliff, L. B. 1984. *J. Chem. Phys.* 80: 1215-24
165. Bethe, H. A. 1933. *Hand. Phys.* 24: 273-560
166. Biermann, L. 1943. *Z. Astrophys.* 22: 157-61
167. Callaway, J. 1957. *Phys. Rev.* 106: 868-74
168. Bottcher, C., Dalgarno, A. 1974. *Proc. R. Soc. London Ser. A* 340: 187-98
169. Szasz, L., McGinn, G. 1968. *J. Chem. Phys.* 48: 2997-3008
170. Laskowski, B. C., Walch, S. P., Christiansen, P. A. 1983. *J. Chem. Phys.* 78: 6824-32
171. Jeung, G. H., Malrieu, J. P., Daudey, J. P. 1982. *J. Chem. Phys.* 77: 3571-77
172. Fuentealba, P., Preuss, H., Stoll, H., Von Szentpaly, L. 1982. *Chem. Phys. Lett.* 89: 418-22
172a. Müller, W., Flesch, J., Meyer, W. 1984. *J. Chem. Phys.* 80: 3297-3310
173. Hafner, P., Schwartz, W. H. E. 1978. *J. Phys. B* 11: 2975-99
174. Nicolas, G., Durand, Ph. 1979. *J. Chem. Phys.* 70: 2020-21
175. Andre, J. M., Burke, L. A., Dehalle, J., Nicolas, G., Durand, Ph. 1979. *Int. J. Quant. Chem. Symp.* 13: 283-91
176. Nicolas, G., Durand, Ph. 1980. *J. Chem. Phys.* 72: 453-63
177. Bredas, J. L., Chance, R. R., Silbey, R., Nicolas, G., Durand, Ph. 1982. *J. Chem. Phys.* 77: 371-78
178. Ohta, K., Yoshioka, Y., Morokuma, K., Kitaura, K. 1983. *Chem. Phys. Lett.* 101: 12-17

VELOCITY MODULATION INFRARED LASER SPECTROSCOPY OF MOLECULAR IONS[1]

Christopher S. Gudeman and Richard J. Saykally

Department of Chemistry, Materials and Molecular Research Division, Lawrence Berkeley Laboratory, University of California, Berkeley, California 94720

INTRODUCTION

The field of molecular ion spectroscopy is presently an area of remarkable vitality, in which important discoveries that affect many diverse subfields of science are being made at a truly impressive rate. Even the most current reviews of this subject (1–4), written as recently as mid-1983, are already obsolete. Exciting new technological progress, particularly, the advent of tunable narrowband lasers, the rapid evolution of quantum optics, and the microelectronics revolution, have provided the foundation for the development of new spectroscopic techniques with the inherent capability to study charged molecules—long regarded as the ultimate challenge to the spectroscopist.

Concurrent developments in other areas of science and technology, most notably, the astronomical investigation of interstellar dust clouds and the pervasive industrial utilization of reactive, low density plasmas, have spawned a widespread interest in the structures, properties, and dynamics of molecular ions. While even as late as 1974 only one ion (H_2^+) had been studied at a level of resolution transcending the optical Doppler limit, now over 25 charged molecules have been investigated by a wide variety of high resolution techniques, including microwave spectroscopy, laser magnetic

---
[1] The US Government has the right to retain a nonexclusive, royalty-free license in and to any copyright covering this paper.

resonance, laser spectroscopy of fast ion beams, laser induced fluorescence, Fourier transform infrared spectroscopy, and tunable infrared laser spectroscopy, employing difference frequency sources, diode lasers, and color center lasers. In this article we review the spectroscopic studies of gaseous molecular ions made with infrared techniques; it is this area in which we have witnessed the most dramatic evolution within the past year, principally because of the development of a general new approach to molecular ion spectroscopy—the velocity modulation laser absorption technique. We first present a general chronology of infrared spectroscopy of ions, then describe the development of velocity modulation spectroscopy and discuss the theoretical concepts upon which it is based.

INFRARED STUDIES OF $H_3O^+ \cdot nH_2O$ AND $NH_4^+ \cdot nNH_3$ GENERATED BY PULSED RADIOLYSIS

While low resolution infrared spectroscopy has been used extensively to study molecular ions in salts, liquids, and in cryogenic matrices (5), it was not until the important work of Schwartz (6) in 1977 that such experiments were carried out in the gas phase. (Ignoring the observation of vibration-rotation emission spectra of NO^+ observed during atmospheric nuclear weapons testing!) (7). In these pulsed radiolysis experiments, a beam of electrons from a Van de Graaf generator was used to ionize a sample of argon gas, kept near atmospheric pressure, and containing a variable amount of water vapor. The ions formed ($H_3O^+ \cdot nH_2O$ with $n = 0$–6) in this manner were measured at low resolution by infrared absorption spectroscopy, using a conventional glowbar source and signal averaging techniques.

The ionizing electron beam consisted of 2.1 MeV pulses, 2–10 μsec in duration and repeated 100–200 times per second, yielding an average beam current of 80 mA. The electron beam was directed into a steel 15 × 60 cm reaction cell in which a series of multipass mirrors were mounted in a White cell configuration. The infrared radiation from a glowbar operating near 1450°C made 40 traversals of the cell, corresponding to an optical path of 20 m, before entering a monochromator with a resolution near 40 cm^{-1}, and finally impinging on an InSb photoconductor detector. The output of the detector was amplified, analyzed by a waveform recorder, and signal averaged; approximately 10^4 individual pulses were averaged at each monochromator setting to produce a spectrum. Dried argon was flowed through the sample cell at atmospheric pressure from two different paths, one proceeding directly to the cell, and the other being saturated with water before merging with the first. Adjustment of the relative flow rates controlled the total amount of water in the cell. Infrared spectra of the

H_3O^+ ion and its various hydrates were obtained over the range from 2000 to 4000 cm^{-1} in this manner.

While the spectroscopic information derived from this experiment was not sufficiently precise to afford a definitive determination of the structures of the oxonium hydrates, it did serve to corroborate previous condensed phase data, and hence yielded measurements of vibrational transition frequencies for H_3O^+ ($v_3 = 3490$ cm^{-1}), H_2O^+ ($v_3 = 3290$ cm^{-1}), $H_5O_2^+$ (3170 cm^{-1}), $H_9O_4^+$ (2660 cm^{-1}, 3000 cm^{-1}), and for the slightly perturbed water molecules in the higher hydrates, $H_7O_3^+$, $H_9O_4^+$, etc (3620 cm^{-1} and 3710 cm^{-1}), all of which are free from the obfuscations resulting from solvent or matrix interactions associated with condensed phase spectra (although even Schwartz's assignments of some transitions are questionable). Furthermore, the spectroscopic evidence indicates that the structure of the hydrated proton should be considered to be an H_3O^+ entity with various numbers of water molecules hydrogen-bonded to the hydrogens. The fact that two fundamental bands are observed for $H_9O_4^+$ (excepting the H_2O stretches) implies that the symmetry of this species, the first solvation sphere of the proton, is C_{3V}, not D_{3h}.

In 1980 these measurements were extended to the series of solvated ammonium ions ($NH_4^+ \cdot nNH_3$, $n = 0$–4) (6), in which fundamental vibrational frequencies were again obtained for some of these important species (NH_4^+: $v_3 = 3335$ cm^{-1}, $NH_4^+ \cdot 3NH_3$: $v_3 = 2867$ cm^{-1}, $NH_4^+ \cdot 4NH_3$: 2682 cm^{-1}, 3365 cm^{-1}, 2790 cm^{-1}). These experiments, both on the oxonium and ammonium systems, represent an important achievement in ion spectroscopy, not only because they involved the first infrared measurements on gaseous molecular ions, but because they provided a degree of insight and guidance regarding the subjects of ion solvation and proton transfer—central questions in important areas of chemistry and biology.

INFRARED LASER SPECTROSCOPY OF FAST ION BEAMS

Although mass-selected ion beam techniques have been employed for various purposes in atomic and molecular physics for several decades, the first use of an ion beam for measurement of the vibration-rotation spectrum of a molecular ion was made in the pioneering work of Wing & co-workers (8) on HD^+ (1976).

In this experiment a fast (1–10 KeV) ion beam of HD^+ ions was crossed at a small angle with the beam from a line-tunable cw CO laser (5.3–6.0 μm), and the Doppler shift resulting from the fast ion motion in the lab frame was used to shift vibration-rotation transitions into coincidence with the laser frequency. The effect of laser excitation on the rate of charge transfer between the fast ions and a static buffer gas was used as an indirect means of detecting the transitions. (Typically the charge transfer cross section varies by a few percent with a unit change in vibrational quantum number of light ions, like HD^+, HeH^+, etc.) The spectral resolution in these measurements was sufficiently high ($\sim 10^{-7}$) to reveal the hyperfine splittings in HD^+. Subsequently, several other fundamental molecular ions [HeH^+ (9), D_3^+ (10), H_2D^+ (11)] have been investigated by Wing and co-workers (12) with this technique, and the extension to the vibration-rotation spectra of HeD^+ and HD_2^+ and to more v, J states in HD^+ has been made, although the analysis of these latter spectra have not yet been completed (W. H. Wing, private communication, 1984). The combination of mass selectivity and kinematic compression of the translational energy distributions inherent in this approach to high resolution spectroscopy have made fast ion beam spectroscopy one of the principal techniques for studying molecular ions—in the visible as well as in the infrared (1–4).

Carrington and his co-workers at Southampton have taken advantage of the unusually high degree of vibrational excitation observed for simple ions in low pressure ion sources to study vibration-rotation spectra of several important molecular ions in rovibrational states near their respective dissociation limits. In their experiments a mass selected and velocity-tuned ion beam, extracted from an electron impact or plasma ion source, was merged with the output of one or two cw CO_2 lasers, and transitions were induced from bound vibrational levels near the dissociation limit into either a repulsive electronic state, as in the case of HD^+, or into predissociating bound or quasibound levels above the dissociation limit. Again these spectra were detected indirectly; in this case, a dissociation fragment was monitored by a mass-selective detection system. By employing two CO_2 lasers in a double resonance configuration, high resolution vibration-rotation spectra could be obtained in the former case, whereas this was only necessary in the latter situation when the predissociating levels had a very short lifetime ($\leq 10^{-9}$ sec). Carrington and co-workers have utilized this fast ion beam photofragmentation apparatus to measure very extensive vibration-rotation spectra of HD^+ (14–16), HeH^+ (all 4 isotopes!) (17, 18), CH^+ (19), and H_3^+ (H_2D^+, HD_2^+, D_3^+) (20–23) near their dissociation limits. A thorough review of fast ion beam laser spectroscopy has been published recently by Carrington & Softley (24). In this paper we describe only the very latest developments in this field.

Carrington & Kennedy (23) have reported the results of an extensive study of the infrared predissociation spectrum of H_3^+. In this landmark work, fast ion beam spectroscopy was used in conjunction with photofragment detection to measure nearly 27,000 vibration-rotation transitions in

H_3^+ lying near or above the lowest dissociation limit [$H_3^+ \to H_2(v = 0, J = 0) + H^+$].

The H_3^+ ions were generated in an electron bombardment source at H_2 pressures near 10^{-3} torr, then accelerated, focused, and mass selected as in the earlier studies. The velocity-tuned ion beam intersected collinearly with the beam from a cw-CO_2 laser in the ion drift tube. Upon emerging from the drift tube, the ion beam entered an electrostatic energy analyzer, in which a given ion was mass selected and its kinetic energy determined. The vibration-rotation spectrum of H_3^+ was measured by monitoring the H^+ fragment produced when photodissociation was induced by a given CO_2 laser line. The spectral region from 872.120 to 1018.278 cm^{-1} was recorded nearly continuously in this fashion. Kinematic compression reduced the Doppler widths of H_3^+ transition to ~ 3 MHz, but in most cases linewidths were determined by the predissociation linewidth that varied from 3 to 60 MHz for the spectra recorded. States predissociating at a faster rate were not detectable by the ion beam velocity modulation scheme employed for subtraction of background ions produced from collision-induced ionization and nonresonant photoionization processes.

The predissociation lifetimes of both initial and final (laser excited) states of H_3^+ were constrained by the geometry of the apparatus and the ion beam velocity to the values: $\tau_{initial} \geq 3$ μsec, $\tau_{final} \leq 0.7$ μsec. Measurement of the kinetic energy of the H^+ photofragments allowed a determination of the energy difference between the predissociating state of H_3^+ and its appropriate $H_2 + H^+$ dissociation limit (no evidence of the more energetic $H_3^+ \to H_2^+ + H$ channel was observed in this work). The observation of large proton kinetic energy releases, over 3000 cm^{-1} in many cases, coupled with the fact that the exciting laser energy was between 870 and 1090 cm^{-1}, forced the authors to conclude that a very large portion of the lines observed in the infrared predissociation spectrum of H_3^+ must arise from transitions in which both the upper and lower (and necessarily metastable) levels of the ion lie well above the lowest dissociation limit.

Because this 27,000 line spectrum was far too complex to permit a detailed assignment, the authors synthesized a low resolution photodissociation spectrum by convolution of the 1934 strongest lines in the spectrum with a Gaussian linewidth function, in hope of revealing the large scale features of the spectrum. The striking result obtained from this procedure was that four relatively sharp maxima were produced in the convolution, and that the frequencies of these four maxima were in remarkable agreement with the $J = 3 \to 5$ rotational transitions of H_2 in its $v = 0, 1, 2,$ and 3 vibrational states. This suggests that the highly excited H_3^+ ions are best thought of as $H_2 \ldots H^+$ complexes, in which the vibrational and rotational quantum numbers of the H_2 entity are essentially conserved.

In order to obtain a qualitative understanding of the origin of these predissociation spectra, a model analogous to those used in studies of van der Waals complexes was adopted, in which H_3^+ is thought of as a weakly interacting H_2 and a proton. The H_3^+ potential surface of Giese & Gentry (25) was used as the basis of this model. This potential surface is a sum over diatomic potential functions in which the widths, depths, and minimum positions are allowed to vary from the values for H_2 as the perturbing H^+ approaches. The interaction energy of H^+ with specific v, J states of H_2 was computed and reexpressed as a Legendre expansion, as in recent treatment of van der Waals molecules. Vibration-rotation levels of $H_2...H^+$ were then calculated in an appropriate basis set according to the procedure of LeRoy & Liu (26). A histogram of the density of states from 500 cm^{-1} above the dissociation limit to 1000 cm^{-1} below it were calculated. This clearly indicated that transitions involving $J = 3 \to 5$ transitions of the H_2 entity do indeed show clustering about the unperturbed H_2 frequency. Moreover, the density of states in this region was found to be very high, with about 4000 rovibrational levels arising from $H^+...H_2(v = 2, J = 5)$ alone, and nearly 60 different $H^+...H_2(v, J)$ dissociation limits lying within the first 2 eV above the lowest $(v = 0, J = 0)$ dissociation limit. In spite of the semiquantitative nature of this model, resulting principally from the lack of detailed knowledge regarding the nature of the predissociation processes responsible for the H_3^+ spectrum, this formalism does reproduce the two most salient features of the H_3^+ predissociation spectrum: a very large level density (27,000 transitions in 222 cm^{-1}!) near the lowest dissociation limit and clustering of the H_3^+ transitions at frequencies near those of the $J = 3 \to 5$ transitions in unperturbed H_2 molecules.

Preliminary measurements of predissociation spectra were made for other isotopic forms of H_3^+, viz. H_2D^+, HD_2^+, and D_3^+. The results of these experiments further substantiated the general trends noted above. In addition, very interesting isotope effects on the predissociation process were noted; for both H_2D^+ and HD_2^+ only H^+ photofragments were observed; when the electrostatic analyzer was tuned to transit D^+ ions, no photofragment spectrum was detected. These observations again reinforce the need for theoretical developments of the photofragmentation process for H_3^+.

The exciting results obtained by Carrington & Kennedy (23) represent one of the most important achievements made in recent years with regard to understanding the structures and dynamics of molecules near dissociation. It is a beautiful example of how a high resolution spectroscopic technique can bridge the gap between "conventional" spectroscopy of molecules in quasi-equilibrium configurations and the powerful methods of modern reaction dynamics. The spectra observed for H_3^+ near its dissociation limit can be regarded as a detailed characterization of what molecular dynamicists term "shape resonances" and "Feshbach resonances." When answers to some of the important questions posed by this work—regarding processes leading to the formation of ions near or above their dissociation limits, the mechanisms of predissociation of such states, and theoretical description of the eigenstates of molecules in such states—are eventually obtained, our knowledge of molecules and the interactions between them will have been substantially advanced.

SPECTROSCOPY OF DC DISCHARGES WITH TUNABLE INFRARED LASERS

One of the most dramatic steps in molecular ion spectroscopy occurred in 1974, when Dixon & Woods (27) observed the microwave spectrum of CO^+ in a laboratory DC glow discharge; this work was soon followed by the measurement of microwave spectra of HCO^+ (28), HNN^+ (29), and their isotopic variants in similar discharges, thus indicating the promise of using low density plasmas as sources of molecular ions (and indeed, of excited species in general) for study by high resolution absorption spectroscopy.

In 1975, Oka initiated an ambitious project to observe the high resolution vibration-rotation spectrum of H_3^+ in a similar DC plasma by multipassing the output from a laser difference frequency infrared source through a long (2 meter) discharge cell. This important achievement in spectroscopy was indeed realized, but only after five years of grueling effort! In 1980, the infrared spectrum of H_3^+ was observed for the first time (30a,b). Subsequently, Oka's apparatus at the Herzberg Institute was used to study vibration-rotation spectra of HeH^+ [Bernath & Amano (31)] and NeH^+ [Wong, Bernath & Amano (32)], as described in earlier reviews (1–3, 24).

Davies and co-workers have employed a diode laser for the study of ions generated in a DC discharge. Davies & Hamilton (33) reported observing the $^2P_{1/2}$–$^2P_{3/2}$ infrared fine structure transition [$1431.644(1)$ cm^{-1}] in the ground ($3p^5$) state of $^{40}Ar^+$ by direct absorption in a multipassed 3 meter long discharge. The isotope shift for $^{36}Ar^+$ was measured to be 3×10^{-3} cm^{-1}. It is interesting to note that a weak magnetic dipole transition could be observed in this manner, although the much smaller partition function of an atomic ion in a plasma largely compensates for the reduced transition moment. Davies, Hamilton, Lewis-Bevan & Okamura (34) have also reported a measurement of the $R(8.5)$ transition of $HCl^+(X^2\Pi_{3/2})$ in a DC discharge, although no further details have been given.

The principal limitation in simply utilizing a discharge as a general source of molecular ions was demonstrated in a recent study by Amano &

co-workers with the difference frequency apparatus at the Herzberg Institute. In their study of the infrared spectrum of the NH_2 radical, Amano, Bernath & McKellar (35) present absorption spectra obtained in a NH_3 discharge near 3250 cm^{-1}. Strong lines from ammonia are observed, both with the discharge on and off, in addition to numerous overlapping transitions that are present only with the discharge turned on. The authors employ Zeeman modulation to discriminate against transitions arising from diamagnetic species, and thus obtain a beautiful infrared spectrum of NH_2, relatively free from interfering lines. However, the difficulty involved with studying diamagnetic molecules in such a chemically complicated discharge is clear—their spectra are often buried by transitions from the more abundant precursor, and the large number of lines arising from the many species present in a discharge through, for example, ammonia (NH_3, NH_2, NH, etc in electronically and vibrationally excited states, as well as in their ground states) makes assignments and identifications a perilous task. This problem is greatly exacerbated for the case of molecular ions, which are usually 10^4–10^6 times less abundant than the major neutral species in such plasmas.

FOURIER TRANSFORM INFRARED SPECTROSCOPY OF DC DISCHARGES

Fourier transform infrared spectroscopy is presently utilized for high resolution absorption measurements of stable gas phase molecules on an almost routine basis. Application of this technique to unstable molecules is considerably more difficult; nevertheless, several very reactive radical species have now been studied by FTIR methods (36–38), including three molecular ions. Brault & Davis (36) successfully obtained the infrared emission spectrum of ArH^+ by FTIR in 1982, employing a hollow cathode discharge with flowing argon at 3 torr to generate the ion. Impurity levels of hydrogen were sufficient to optimize the ArH^+ concentration. An extensive infrared spectrum was observed, consisting of five bands with rotational states as high as $J = 35$. Attempts to duplicate this experiment for ArD^+ were not successful. Details of this pioneering work have been reviewed by Carrington & Softley (24).

A preliminary report of Fourier transform infrared emission spectra of ArH^+, ArD^+, and KrH^+ has been given by Johns (37). His corresponding experiments on HeH^+ and NeH^+ were intriguingly unsuccessful. Hollow cathode discharge sources were also employed in this work. A detailed analysis of these interesting and important results is anxiously awaited. It appears that FTIR methods can now be generally applied to the study of molecular ions, although the level of difficulty still remains quite high.

VELOCITY MODULATION INFRARED LASER SPECTROSCOPY

The existence of a small, but detectable, Doppler shift ($\Delta v/v \sim 10^{-6}$) in the resonant absorption frequencies of molecular ions produced in DC glow discharges was first observed in 1975 by Woods et al (28), whose microwave measurement of the $J = 0 \rightarrow 1$ rotational transition frequency for HCO^+ was found to exhibit a small systematic deviation from astrophysical measurements (39) of this same transition. Based on simple gas phase mobility considerations (40, 41), the ion drift velocity in the axial electric field of the discharge was estimated to be ~ 1 Km/sec, which would result in a Doppler shift of 300 kHz for this transition at 89189 MHz. Indeed, after laboriously reconstructing their spectrometer so that the microwaves propagated anti-parallel to the positive ion drift direction (instead of parallel as in the previous measurements), the existence of a small but measurable Doppler shift (~ 100 kHz) was confirmed (42) (Figure 1). Over the next seven years this tedious and time-consuming rebuilding process was obviated by employing symmetric electrodes in the discharge cell and a DC power supply that could be reversed in polarity. These modifications, combined with refinements in microwave frequency control and data acquisition, have enabled reliable Doppler shift measurements to be made by microwave spectroscopy (Figure 2). A careful analysis of the Doppler

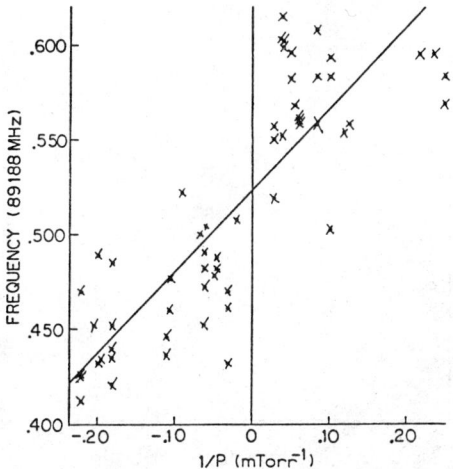

Figure 1 The first microwave Doppler shift measurements of an ion in DC plasma: the $J = 1 \leftarrow 0$ transition for HCO^+ by Woods et al (42). $P > 0$ corresponds to microwave propagation parallel to the ion drift velocity, whereas $P < 0$ corresponds to antiparallel propagation.

shifts is important from the standpoint of obtaining accurate rest frequencies for molecular ion transitions. Similar shifts have also been observed and measured for higher rotational transitions of HCO^+, NO^+, CO^+, and HNN^+ by DeLucia and co-workers (43–46).

Woods and co-workers have been able to exploit these Doppler shifts as a means of distinguishing between ionic and neutral spectra observed in DC glow discharges (47). For example, in their search for microwave vibrational satellite spectra of HCO^+, the Doppler shift in the transition assigned to the (001) $J = 0 \to 1$ satellite was observed to be similar to that for the ground state $J = 0 \to 1$ transition. This quite unambiguously proved that the carrier of the previously unobserved transition at 88599.5 MHz was indeed ionic.

These microwave results clearly demonstrated that non-random ionic drift velocities within the positive column of a hydrogen glow discharge are comparable in magnitude to the mean random thermal velocities, and that the resulting spectral shifts could not only provide a nonintrusive measure of the plasma electric field or ionic mobility, but could also be used to distinguish between ionic and neutral spectra. Furthermore, these results indicated, although not quite so evidently, that the extension of these techniques to the infrared spectroscopy of molecular ions would permit full advantage of the Doppler shift measurements to be taken, since collisional broadening—the dominant mechanism of microwave line-broadening—is usually much smaller than the Doppler broadening at infrared wavelengths. Moreover, the wide range of rovibrational quantum states

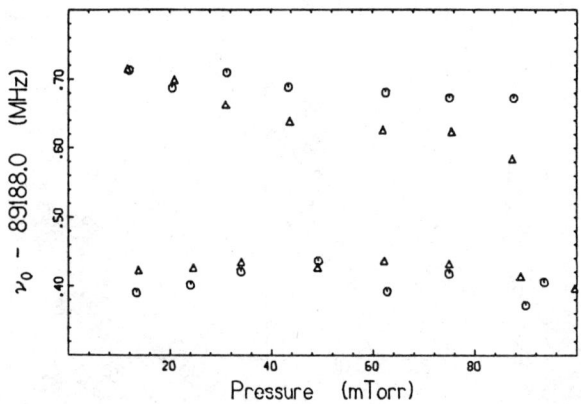

Figure 2 The second generation of Doppler shift measurements by Woods and co-workers (47). *Top curves*: copropagation; *bottom curves*: counter propagation. $\bigcirc = 300$ mA; $\triangle = 150$ mA discharge currents.

accessible by infrared studies within a relatively narrow bandwidth presents the possibility for the quantum state dependence of ionic transport properties (mobility and diffusion) to be examined quantitatively.

The extension of these concepts to infrared spectroscopy of molecular ions was pursued simultaneously by two research groups, both of which have been quite successful in this endeavor, even though their respective studies were directed toward different objectives. Haese, Pan & Oka (48), at the University of Chicago, have measured Doppler shifts in the fundamental and hot bands of ArH$^+$ at 2500 cm^{-1}. Their mobility spectrometer consisted of diode laser-generated counter-propagating single-pass infrared beams that probed the positive column of an air-cooled, predominantly helium DC discharge. The red- and blue-shifted absorption profiles of the ion were simultaneously observed by detecting each beam independently with HgCdTe detectors. High sensitivity was achieved by frequency modulating the diode laser source and demodulating the absorptions at $2f$ by phase-sensitive amplification. Electric field measurements made with platinum wire probes inserted into the plasma were combined with the spectroscopically derived drift velocities to obtain ion mobilities. Large Doppler shifts [Δv(red-to-blue) $\approx \Delta v$ (FWHM)] were observed (Figure 3), and the resulting mobilities were found to agree very nicely with classical measurements of ArH$^+$ mobilities in He based on drift tube data (49). In this first examination of quantum state dependent ionic mobilities, it was found that the rotational state dependence was less than 10%—the precision achieved in the shift measurements. On the other hand a comparison of the $(v, J) = (0, 3) \rightarrow (1, 2)$ and $(1, 2) \rightarrow (2, 3)$ spectral shifts revealed a marginally significant (10%) mobility dependence on vibrational state, suggesting that $v = 1$ excited state ArH$^+$ ions are more mobile than ground state ArH$^+$. This pioneering work of Haese et al has shown that state-specific mobilities can indeed be determined by high resolution laser

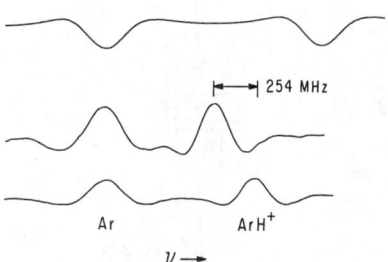

Figure 3 Doppler shift spectrum of the $P(5)$ fundamental transition of ArH$^+$ at 2479.4113 cm^{-1} etalon transmission curve (FSR = 1420 MHz) (*top*), red-shift (*middle*), and blue-shift (*bottom*) spectra. The unshifted line is neutral Ar. From Haese, Pan & Oka (48).

spectroscopy, although there exists a need for more precise frequency shift measurements before this technique can elucidate the weak vibrational and rotational state dependences, and achieve an overall level of precision competitive with that of classical drift tube studies.

Velocity modulation spectroscopy of molecular ions has been developed at the University of California at Berkeley by Saykally and co-workers as a means of exploiting the Doppler shifts of charged molecules to discriminate against the overwhelmingly more abundant neutral species that exist in a plasma, rather than as a method for measuring the Doppler shifts themselves. Early in 1983 Gudeman, Begemann, Pfaff & Saykally (50) reported the use of an audio frequency AC glow discharge and phase-synchronous demodulation of the absorbed infrared laser radiation for the detection and measurement of the v_1 fundamental (C–H stretch) of HCO^+. In this experiment (Figure 4) an infrared color center laser was used as a

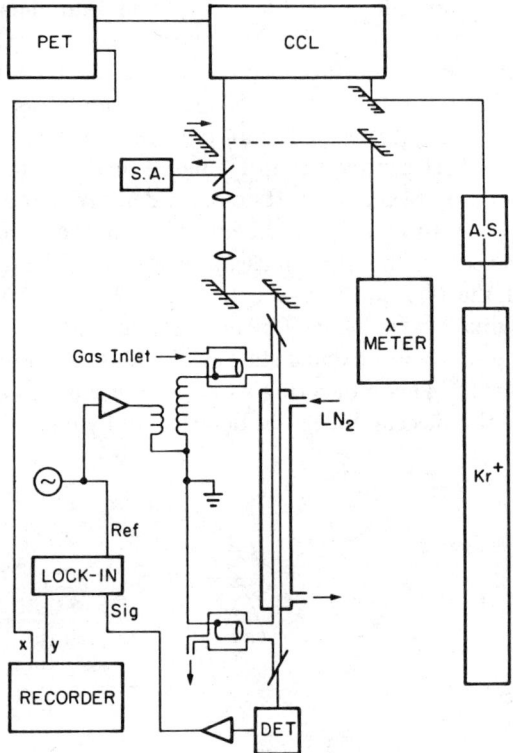

Figure 4 Schematic of the Color Center Laser—velocity modulation spectrometer of Saykally and co-workers (50, 51). A. S. = Amplitude Stabilizer, S. A. = Spectrum Analyzer, CCL = Color Center Laser.

source of tunable narrowband (1 MHz) radiation, which traversed the positive column of the AC plasma in a single pass. A 10/1 mixture of H_2 and CO at a total pressure of 800 mTorr was flowed through a 1 cm diameter × 100 cm long Pyrex discharge cell, the electrodes of which were placed off-axis in sidearms so that the infrared radiation probed only the positive column of the plasma. The absorption path was surrounded by a Pyrex jacket cooled with flowing liquid nitrogen. In these early experiments, the HCO^+ absorption spectrum could not be observed without this cooling. Discharge currents of 30–100 mA were achievable, with the maximum available current producing the strongest signals and the highest rotational temperatures. The total laser power traversing the plasma was typically ~50 μW (although far more power was available from the color center laser system), and was detected by a liquid nitrogen cooled InSb photovoltaic detector. When tuned off-resonance with an ionic rovibrational transition, the laser power at the detector would remain constant. When tuned through a molecular ion resonance, however, the power would be modulated at the audio frequency (af) of the discharge, since the symmetric oscillating electric field Doppler-shifts the absorption frequency of the ionic vibration-rotation transition in and out of resonance with the infrared radiation. Another viewpoint is that in the reference frame of the velocity-modulated ion, the laser radiation appears to be frequency modulated. Simple lock-in demodulation at $1f$ then provides substantial suppression (~20–40 dB) of the strong and numerous absorptions due to neutral species, which are essentially unaffected by the electric field, while absorptions due to molecular ions are nearly 100% amplitude modulated.

To date, the Berkeley color center laser velocity modulation effort has been successful in detecting and measuring spectra for six different molecular ions: H_3O^+, NH_4^+, H_2F^+, HCO^+, HNN^+, and H_3^+. Of these, the first five had never before been detected with high resolution infrared techniques, while the first three had not previously been observed by any form of high resolution spectroscopy. The two most important features of velocity modulation that have made these measurements possible are the inherent high sensitivity of narrow bandwidth phase-synchronous detection and the nearly complete suppression of neutral absorptions. The suppression has been found to be especially important in the detection of H_3O^+ and NH_4^+, where the O–H and N–H stretching modes of these species are intimately overlapped by the corresponding modes of their neutral precursors, simply because the bonds involved are very similar. Thus the suppression of H_2O and NH_3 transitions, which are extremely strong and numerous in this region, has been essential.

The original detection of the HNN^+ v_1 (N–H stretch) band at 3.1 μm (51) was made by Gudeman et al using the same approach employed for HCO^+.

These spectra revealed only simple P- and R-branches, and therefore provide further experimental confirmation that HNN$^+$, like HCO$^+$, is indeed linear. The observed HNN$^+$ spectra were quite strong (Figure 5); the transitions at the peaks of this band ($J'' = 7, 8$) absorbed $\sim 3\%$ of the infrared power and are the strongest infrared transitions observed for any molecular ion by velocity modulation.

The importance of H$_3$O$^+$ in chemistry, biology, astrophysics, and aeronomy has provided a great stimulus to high resolution spectroscopists for many years, but the strongly interfering H$_2$O vibrational spectrum has prevented a detailed study of this ionic species until very recently. The v_3 band (doubly degenerate, asymmetric stretch) of H$_3$O$^+$ was first observed and analyzed (52) by Begemann et al at Berkeley, using velocity modulation spectroscopy; H$_2$/O$_2$ and H$_2$/H$_2$O mixtures in air-cooled discharges and H$_2$/O$_2$ in liquid nitrogen cooled plasmas were found to yield comparable signal amplitudes. The rotation-vibration transitions of this band were, in general, much weaker (fractional absorption $\approx 0.1\%$) than those observed for the HCO$^+$ and HNN$^+$ because of the relatively higher density of states populated in a symmetric top species. The total integrated band strength suggested that the densities of HCO$^+$, HNN$^+$, and H$_3$O$^+$, in their respective optimum plasmas, were similar. Nearly 100 transitions in the liquid nitrogen cooled v_3 spectrum of H$_3$O$^+$ have now been measured and assigned (Figure 6). All of these have been observed as well in air-cooled plasmas, although in this latter case an additional ~ 240 transitions have also been detected. These cannot simply be assigned to higher-J v_3 transitions or to the v_1 band of H$_3$O$^+$. Additional work on these interesting spectra is currently in progress.

Protonated ammonia (NH$_4^+$) has almost the same general significance in chemistry as H$_3$O$^+$. The v_3 band of NH$_4^+$ (triply degenerate asymmetric stretch) near 3.0 μm was detected simultaneously by Crofton & Oka (53)

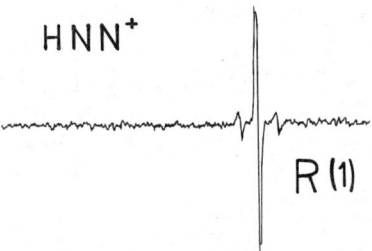

Figure 5 The velocity modulation spectrum of HNN$^+$. The strong $R(1)$ v_1 fundamental transition shown here has been clipped ($\div 10$) in order to reveal clearly the $v_2 = 1^1$ hot band-doublet and typical noise amplitude. From Gudeman & Saykally (to be published).

and by Schäfer et al (54) at room temperatures in mixtures of H_2 and NH_3. In more recent work over 200 transitions of NH_4^+ in this band have been observed and analyzed by Schäfer, Saykally & Robiette (55). The tetrahedral spherical top structure of NH_4^+ was clearly evident in the observed v_3 spectrum, which displayed remarkable similarities to that of the isoelectronic methane molecule.

The fluoronium ion H_2F^+, the ionic analog of water, has also been studied at Berkeley by Schäfer & Saykally (56). Over 300 transitions have been assigned to both the v_3 (asymmetric F–H stretch) and the v_1 (symmetric F–H stretch) modes. Air-cooled glow discharges through H_2/HF mixtures produced these bands, which were free from Fermi resonance because of symmetry restrictions, even though their vibrational frequencies are very nearly degenerate.

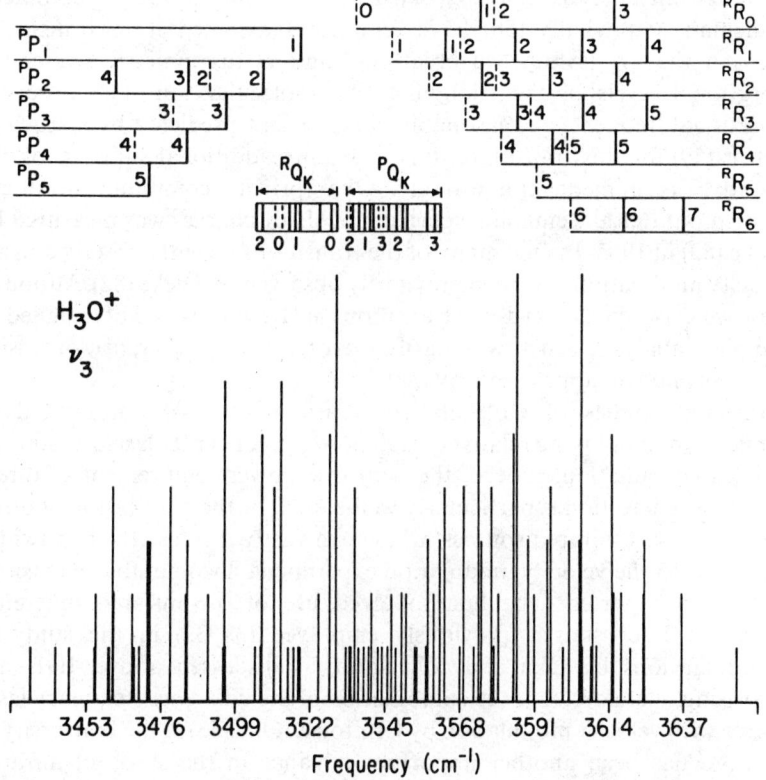

Figure 6 Reconstructed spectrum of the v_3 band of H_3O^+ measured in LN_2 cooled H_2/O_2 plasmas. From Begemann et al (52). In the assignment the *dashed lines* correspond to the asymmetric inversion band, the *solid* to the symmetric.

The chemical, astrophysical, and theoretical importance of H_3^+ demands that an entire review be dedicated to its discussion, and such a review has indeed been presented by Oka (57). In the course of the recent extensive H_3O^+ spectral search (52) by Begemann et al, however, two new broad ionic transitions were detected at a lock-in phase 180° different from all H_3O^+ absorption lines. These were first believed to be due to a negative ionic carrier, but a more extensive study of these species revealed that their signal strengths were maximized in liquid nitrogen cooled plasmas of pure H_2, wherein the H_3O^+ spectrum was not observable at all. A careful scan over 2/3 of the entire color center laser bandwidth yielded 41 transitions, of which approximately 1/2 were in-phase with the discharge current and 1/2 were 180° out of phase. Furthermore, $2f$ population modulation of these transitions indicated that those which were in-phase absorbed laser power, while the out-of-phase lines were observed in stimulated emission. Although an analysis of this spectrum at 2.76 μm is not yet complete, Gudeman & Saykally (58) have tentatively assigned it to transitions between the $2v_1$ (symmetric breathing) and v_2 (asymmetric stretching) vibrational levels of the triangular H_3^+ molecular ion. The observed vibrational frequency is reasonably close to that predicted by Carney & Porter (59) for this transition, thus providing additional support for the tentative assignment. Interestingly, Oka (private communication) has pointed out that the emission spectrum of H_2 discharges was measured by Dieke (82) in 1965. In fact, many of the transitions recently observed using velocity modulation had been previously observed by Dieke (82). Although these were ascribed to Rydberg transitions of H_2, most were not assigned. A complete analysis, which will clarify the origin of these transitions, is in progress and will appear shortly (58).

Infrared studies of molecular ions are, of course, not limited to applications employing relatively high power color center lasers as sources of infrared radiation. Indeed, the very low power requirement of direct absorption detection is particularly well-suited for the application of other state-of-the-art infrared sources, which can vastly increase the bandwidth accessible to the velocity modulation experiment. Two notable extensions have been reported. The first involved use of the infrared difference frequency laser source, previously employed (30–32) in the study of molecular ions by Oka and colleagues, as discussed above. It is not surprising, therefore, that the combination of the difference frequency laser source with velocity modulation by Crofton & Oka (53) at the University of Chicago has been another important advance in the study of infrared spectra of molecular ions, first demonstrated in their work (53) in the v_3 band of NH_4^+.

In the most recent application of the velocity modulation technique,

Altman, Crofton & Oka (60) have reported the detection and analysis of both the C–H stretching (v_2) mode of $HCNH^+$ and the N–H stretching (v_1) mode (T. Oka, private communication) which were observed in air-cooled H_2/HCN plasmas, using their difference frequency laser source spectrometer near 3.1 μm. This represents the first detection of this linear species by any spectroscopic technique, yielding a very accurate prediction of its $J = 1 \leftarrow 0$ pure rotational transition (74112.5±6.0 MHz), which will greatly facilitate laboratory and astrophysical microwave efforts in the search for this extremely important charged molecule.

A similar system, combining difference frequency infrared generation and velocity modulation has been employed by Nesbitt et al in a collaborative effort at the University of California, Berkeley. In this latter work, laser source noise subtraction has been found to be a very effective means of increasing detection sensitivity, even though power levels on the order of only 5 μW were available. Detailed studies of the deuterated ions DNN^+ (61) and D_3O^+ (62) have been made thus far in this joint effort. The combination of infrared diode laser sources with velocity modulation has also been pioneered in Chicago. Haese & Oka have recently reported (63) the detection of the H_3O^+ v_2 band (the umbrella mode) near 10 μm by use of a diode laser and HgCdTe photoconductive detector. This study serves not only as a key step in measuring the barrier to inversion for H_3O^+, but also provides infrared transition frequencies that are not obscured by atmospheric H_2O and can therefore be applied to an astrophysical search for this important ionic species.

Within one week of the first observation of HCO^+, the v_1 band of the isoelectronic molecular ion HNN^+ (protonated nitrogen) was also observed at Berkeley. Within one year of this first observation, a total of seven chemically distinct molecular ions, including 17 different vibrational bands of 12 unique isotopic variants, have been reported by the two groups employing velocity modulated infrared laser spectroscopy. This work yielded a total of ~1400 lines, 1100 of which have now been assigned and analyzed. Results of these numerous measurements include spectroscopic constants, molecular structures, and a more complete understanding of the molecular ion chemistry in partially ionized plasmas. In Table 1 we have compiled the spectroscopic and structural information for the various bands and isotopic variants that have been analyzed so far.

From a chemical point of view, the species observed have been produced in predominantly (~70%), $H_2(D_2)$ plasmas, with the remainder comprising the appropriate neutral to yield the proton adduct. The individual experiments provide very interesting chemical information, however, and these merit some additional discussion. As mentioned above, HCO^+ was not observed in an ambient temperature plasma, although HNN^+ was, albeit rather weakly. It has been shown by Haese in mass spectrometric studies of air-cooled H_2/CO discharges (64) that the H_3O^+ ion actually dominates HCO^+, thus indicating that water is rapidly formed from CO and H_2 and that liquid nitrogen cooling reduces the H_2O level sufficiently to allow CO to compete effectively with H_2O for the proton. The strong H_3O^+ signals observed at Berkeley in room temperature H_2/O_2, H_2/CO,

Table 1 Molecular ions studied by high resolution infrared spectroscopy

HD^+

(8)*   Fast ion beam, $v = 0, 1, 2, 3$, sampled with 7 transitions; no constants determined

(14, 16)   Two photon photodissociation, 9 rotational components of $v = 18–16$ measured, hyperfine splittings observed, but no hfs constants determined:
$\Delta G = 916.2081(4)$ $B_{16} = 7.33607(38)$ $D_{16} = 7.201(52) \times 10^{-3}$
$H_{16} = 3.1(3.5) \times 10^{-6}$ $B_{18} = 5.15628(38)$ $D_{18} = 7.772(52) \times 10^{-3}$
$H_{18} = 7.6(3.5) \times 10^{-6}$

(15)   Two photon photodissociation, 7 rotational components of $v = 17–14$ band and one component in $v = 15–17$ measured, hfs observed but no hfs constants determined:
$\Delta G = 1801.315(2)^a$ $B_{14} = 9.2859(8)$ $D_{14} = 7.26(12) \times 10^{-3}$
$H_{14} = 3.65 \times 10^{-6a}$ $B_{17} = 6.2837(6)$ $D_{17} = 7.40(8) \times 10^{-3}$
$H_{17} = 2.10 \times 10^{-6a}$

HeH^+

(9)   Fast ion beam, five transitions involving $v = 0, 1, 2$ measured, no constants determined

(17)   Fast ion beam photodissociation, two bound to quasibound transitions involving $v = 5, 6, 7$ observed, no constants determined

(18)   Fast ion beam photodissociation, bound to quasibound transitions measured for three isotopes of HeH^+: 5 transitions of $^4HeD^+$ involving $v = 4, 5, 6, 7, 9, 13$; one transition of $^3HeD^+$ involving $v = 5, 6, 7$; no constants determined

(31)   Nine vibration-rotations measured in the fundamental (0–1) band of HeH^+ by difference frequency laser spectroscopy of a discharge:
$v_0 = 2910.95681(64)$ $B_0 = 33.55841(21)$ $D_0 = 1.6210(12) \times 10^{-2}$
$H_0 = 5.81(15) \times 10^{-6}$ $B_1 = 30.83991(18)$ $D_1 = 1.5869(10) \times 10^{-2}$
$H_1 = 5.76(20) \times 10^{-6}$

NeH^+

(32)   Eleven transitions of $^{20}NeH^+$ and 8 transitions of $^{22}NeH^+$ have been measured for the $v = 1–0$ band by difference frequency laser spectroscopy of a discharge:

	$^{20}NeH^+$	$^{22}NeH^+$
ω_e	2900(6)	2894(6)
$\omega_e X_e$	111(6)	111(6)
B_e	17.88(2)	17.80(2)
α_e	1.08(4)	1.08(4)
$D_e \times 10^3$	2.704(17)	2.676(19)
$\beta_e \times 10^3$	$-0.0235(72)$	$-0.0232(72)$
r_e (Å)	0.9913(6)	0.9913(6)

Table 1 *(continued)*

KrH$^+$

(37) Fourier transform infrared emission spectrum observed in a hollow cathode discharge; unpublished

ArH$^+$

(36) Fourier transform infrared emission spectrum observed in a glow discharge for vibrational states up to $v = 4$ for J values up to 35. The leading Dunham coefficients have been determined:

$Y_{10} = 2710.9191$ $Y_{20} = -61.63075$ $Y_{30} = 0.502654$
$Y_{40} = -0.0010917$ $Y_{01} = 10.461276$ $Y_{11} = -0.378782$
$Y_{21} = 0.0030173$ $Y_{02} = -6.230 \times 10^{-4}$ $Y_{12} = 0.08297 \times 10^{-4}$
$Y_{03} = 1.584 \times 10^{-8}$ $Y_{13} = -0.0459 \times 10^{-8}$ $Y_{04} = -0.64 \times 10^{-12}$

(37) Fourier transform infrared emission spectrum observed for ArH$^+$, ArD$^+$; unpublished analysis

(48) Mobility of ArH$^+$ measured in a glow discharge for $v = 0$ and $v = 1$ from Doppler shifts of $P(3) - P(6)$ lines in fundamental band

HCl$^+$

(34) $R(8.5)$ transition of fundamental band observed for $X\,^2\Pi_{3/2}$ state of HCl$^+$ in a glow discharge with a diode laser; no analysis or constants given

CH$^+$

(19) Fast ion beam photodissociation spectra measured. Over 78 transitions observed near dissociation limit; no assignments

HCO$^+$

(50) Nineteen transitions in v_1 fundamental [$R(0)$ to $R(18)$] measured by velocity modulation laser absorption. $v_1 = 3088.727(3)$

(69) Ten additional transitions measured in v_1 P-branch. Improved constants determined:

$v_1 = 3088.73951(31)$ $B_1 = 1.475699(11)$ $D_1 = 2.68(11) \times 10^{-6}$
$B_0 = 1.48750974(7)^b$ $D_0 = 2.748(2) \times 10^{-6b}$

HNN$^+$

(51) Velocity-modulated laser absorption, 43 transitions measured in v_1 fundamental [$R(0)$–$R(19)$, $P(1)$–$P(23)$]:

$v_1 = 3233.9538(32)$ $B_1 = 1.541429(81)$ $D_1 = 3.07(14) \times 10^{-6}$
$B_0 = 1.541429(81)$ $D_0 = 3.11(14) \times 10^{-6}$

(61) Velocity modulation laser absorption, 48 transitions observed in v_1 fundamental of DNN$^+$ isotope, 70 transitions observed in $v_2 = 1$ hot band
$\underline{(000) \rightarrow (100)}$ $v_0 = 2636.983(50)$, $B_{000} = 1.286098(84)$,
$B_{100} = 1.274535(84)$, $D_{000} = 2.28(12) \times 10^{-6}$, $D_{100} = 2.17(12) \times 10^{-6}$,
$(B_{100} - B_{000}) = -0.011563(17)$
$\underline{(01'0) \rightarrow (11'0)}$ $v_0 = 2618.785(50)$, $B_{010} = 1.29068(11)$,
$B_{110} = 1.27999(11)$, $D_{010} = 1.98(24) \times 10^{-6}$, $D_{110} = 1.82(24) \times 10^{-6}$,
$(B_{110} - B_{010}) = -0.010692(25)$, $(B_{010} - B_{000}) = 0.00458(42)$,
$q_{010} = 0.2101(11)$ GHz, $q_{110} = 0.1892(11)$ GHz, $X_{12} = -18.90(10)$

H$_3^+$

(30a) Fundamental of the v_2 (E-type asymmetric stretch) band measured by difference frequency laser spectroscopy of a glow discharge; 15 transitions measured and fit to

Table 1 (continued)

	a Hamiltonian with 14 molecular parameters: $v_2 = 2521.564(135)$, $B_1 = 44.051(45)$, $C_1 = 19.689(38)$, $(\xi C)_1 = -18.527(50)$, $q = -5.380(74)$, $B_0 = 43.568(48)$, $C_0 = 20.708(48)$, $D_J(1) = D_J(0) = 0.047(7)$, $D_{JK}(1) = D_{JK}(0) = -0.099(18)$, $D_K(1) = D_K(0) = 0.040(3)$, $q_J = 0.018(5)$
(10)	Fast ion beam, 4 transitions measured for the v_2 fundamental measured, 4 others unassigned; no molecular constants given
(30b)	Fifteen additional transitions reported for H_3^+ in v_2 fundamental, no molecular constants given
(11)	Fast ion beam, 9 transitions measured for H_2D^+ isotope, no assignment given
(57)	Review on H_3^+, no new molecular constants given
(20)	Fast ion beam photodissociation, over 300 lines reported for H_3^+ from bound and metastable levels near the dissociation limit. No assignments
(23)	Fast ion beam photodissociation, $\sim 27{,}000$ lines measured for H_3^+ from 872 to 1094 cm^{-1}. No detailed assignments made; synthesized low resolution spectrum implies correspondence with $J = 2$ transitions in $H^+ \ldots H_2(V, J)$ with $J = 3$ and V varying from 0 to 3. Similar spectra were observed for D_3^+, HD_2^+, and H_2D^+.
(58)	Velocity modulation laser spectroscopy, 41 transitions measured in the $2v_1 - v_2$ band, over half of these occur in stimulated emission; no analysis available.

H_2F^+

(56)	Velocity modulation laser absorption, 150 transitions measured in v_3 fundamental and analyzed with S-reduced Hamiltonian, $v_3 = 3334.6718(32)$
(83)	Velocity modulation laser absorption, over 300 transitions measured in v_1 and v_3 fundamentals and analyzed with S-reduced Hamiltonian:

	(0, 0, 0)	(0, 0, 1)	(1, 0, 0)
v_0		3334.6895(26)	3348.7078(36)^c
A	34.5110(14)	33.1985(15)	33.9131(28)
B	12.88518(24)	12.66161(24)	12.59785(33)
C	9.08015(22)	8.90609(16)	8.89164(21)
$D_J \times 10^3$	0.9533(33)	0.9661(13)	0.9197(22)
$D_{JK} \times 10^3$	$-6.759(43)$	$-6.854(44)$	$-6.04(11)$
$D_K \times 10^3$	77.65(15)	71.55(14)	92.59(73)
$d_1 \times 10^3$	$-0.38992(80)$	$-0.39665(89)$	$-0.3917(13)$
$d_2 \times 10^3$	$-0.05003(50)$	$-0.04987(58)$	$-0.0567(11)$
$H_J \times 10^6$	0.059(17)	d	d
$H_{KJ} \times 10^6$	$-105.3(4.5)$	$-66.3(3.5)$	$-111(20)$
$H_K \times 10^6$	46.3(2.1)	d	1788(87)
$L_{KJ} \times 10^6$	$-3.24(17)$	$-0.652(69)$	$-4.64(96)$
$L_K \times 10^6$	d	d	44.0(3.6)
$S_{KJ} \times 10^9$	43.9(2.4)	d	d

H_3O^+

(6)	Low resolution gas phase study of $H_3O^+ \cdot nH_2O$
(52)	Velocity modulation laser absorption, 60 transitions measured in v_3 (E asymmetric stretch) fundamental:
Symmetric	$^*v_3' = 3530.165(55)$, $B' = 11.00(11)$, $B'' = 11.23(11)$, $C'(1-\zeta) = 5.60(11)$, $(C'-B')-(C''-B'') = 0.199(8)$, $C'' = 6.14(11)$, $C' = 6.11(11)$

Table 1 (*continued*)

Asymmetric	$*v'_3 = 3513.840(47)$, $B' = 10.788(36)$, $B'' = 10.957(36)$, $C'(1-\xi) = 5.693(36)$, $(C'-B')-(C''-B'') = 0.132(7)$, $C'' = 6.177(36)$, $C' = 6.140(36)$ $r_{OH} = 0.979(6)$ Å $<_{HOH} = 114.91(45)°$ $*v' = v_0 + C'(1-2\xi) - B'$
(63)	Velocity modulation laser absorption, 32 transitions measured in v_2 fundamental (umbrella mode): $v_0 = 954.417(14)$, $B_1 = 10.690(4)$, $B_0 = 11.253(4)$, $(C_1 - B_1) - (C_0 - B_0) = 0.694(15)$, $D_J(1) = 1.8(2) \times 10^{-4}$, $D_J(0) = 13(1) \times 10^{-4}$, $D_{JK}(1) = -1.9(1) \times 10^{-4}$, $D_{JK}(0) = -30(4) \times 10^{-4}$, $D_K(1) - D_K(0) = -20(10) \times 10^{-4}$
(62)	Velocity modulation laser absorption, over 380 transitions measured for v_3 fundamental of D_3O^+ isotope; no analysis available
NH_4^+	
(54)	Velocity modulation laser absorption, 73 transitions in v_3 (F-type asymmetric stretch) fundamental measured. Preliminary analysis
(53)	Velocity modulation laser absorption, 21 transitions measured in v_3 fundamental, preliminary analysis
(55)	Velocity modulation laser absorption, over 200 transitions in v_3 fundamental measured and analyzed with an effective 6th order Hamiltonian. 26 parameters determined. $v_3 = 3343.1399(21)$, $B_3 = 5.79904(18)$, $B_0 = 5.852$, $D_0 = 1.20 \times 10^{-4}$, $D_T = -t_{044} = 5.53 \times 10^{-6}$, $2B\xi_3 = 0.55495(40)$, $\alpha_{220} = -2.83(13) \times 10^{-3}$, $\alpha_{224} = -4.164(15) \times 10^{-3}$, $D_3 = 33.4(4.4) \times 10^{-6}$. See reference for remaining parameters.
$HCNH^+$	
(60)	Velocity modulation laser absorption, 27 transitions measured in v_2 (CH-stretch), v_1 (N–H stretch) observed but not yet analyzed: $v_2 = 3187.86382(39)$, $B'' = 1.236067(32)$, $D'' = 1.640(115) \times 10^{-6}$, $B' = 1.228493(30)$, $D' = 1.632(92) \times 10^{-6}$

* References.
[a] Theoretical results for H.
[b] Microwave values referenced above.
[c] Numbers in parenthesis are one standard deviation.
[d] These parameters were constrained to zero.

and H_2/CO_2 plasmas further suggest that any source of atomic oxygen in H_2 plasmas quickly reacts to form H_2O. Moreover, HNN^+ is only weakly observable at ambient temperatures, implying that ammonia is a stable product of H_2/N_2 plasmas. The production rate of NH_3, and thus NH_4^+, does, however, appear to be slower than that of water and H_3O^+ in similar H_2/O_2 mixtures, since NH_4^+ is only weakly observed in H_2/N_2 mixtures. This suggests that electron impact cleavage of N_2 is inefficient, and that proton transfer to atomic nitrogen is slow. This is in accord with gas phase ion-molecule kinetic studies (65) employing ion-cyclotron resonance and flowing afterglow techniques. Generally speaking, it is observed that the ion chemistry of these discharges occurs on a sufficiently rapid timescale to yield the most chemically stable ionic species as the dominant ionic constituent, which has been termed the "terminal ion" or, colloquially, as the "end-of-the-food-chain."

In the remainder of this section, several recent extensions and additional applications of velocity modulation laser spectroscopy are discussed. These include electronic spectroscopy of molecular ions, using tunable dye lasers; spectroscopy of neutral radicals, using discharge population modulation; and dynamical studies of plasmas based on observed rotational, vibrational, and translational distribution of the ions and neutrals within the plasma.

The extension of velocity modulation to visible wavelengths follows quite naturally from a simple consideration of Doppler shifts and Doppler linewidths—both are proportional to the transition frequency, and the ratio of shift to width will therefore be independent of frequency, provided the linewidth of the transition remains Doppler-limited. Moreover, the highly sophisticated technology available at visible wavelengths, including tunable lasers, fast detectors, and electro-optic devices, permits more elaborate development of these spectroscopic methods to be made with considerably more ease, compared with the technology presently available in the infrared. Gudeman, Martner & Saykally (66) have recently reported the extension of velocity modulation to electronic spectroscopy in their detection of the $A^2\Pi \leftarrow X^2\Sigma$ systems for both CO^+ and N_2^+ in direct absorption. The $B \leftarrow X$ system of N_2^+ has long been the prototypical species for the development of techniques for molecular ion electronic spectroscopy; the $A \leftarrow X$ system, on the other hand, is inherently much weaker due to the long radiative lifetime of the A state and the additional intensity dilution resulting from the Π state partition function. Nevertheless, several vibronic bands have been observed by velocity modulation for N_2^+ and CO^+. The CO^+ (Figure 7) and N_2^+ spectra were easily observed in AC plasmas composed mainly of He (1–8 Torr) with relatively small amounts of CO or N_2 (50–150 mTorr), and the absorption signal strengths were surprisingly insensitive to chemical composition within this range. In accord with the large Doppler shifts observed (48) for ArH^+ in He by Haese et al in DC plasmas, He correspondingly serves well as a buffer gas for velocity modulation in AC plasmas.

The methodology of these visible studies is similar to that described above in connection with infrared techniques, except that the tunable infrared source has been replaced with a single mode dye laser. Also, laser output amplitide noise, which typically limits the detection sensitivity, has been greatly reduced by employing a spatial double beam configuration.

The wide range of vibrational and rotational states accessible by electronic spectroscopy provides a convenient quantum state-selective probe of the plasma environment. Rotational and vibrational distributions have been determined (67) for both CO^+ and N_2^+ from relative intensity measurements, and the translational energy has been estimated from the Doppler-limited linewidths. Furthermore, these nonintrusive investigations can potentially provide spatial distributions of these dynamical parameters and can be combined with radiative lifetimes and Franck-Condon factors to determine absolute densities of the ionic species. The extension of the velocity modulation technique to ultraviolet wavelengths is anticipated in the near future, where techniques for nonlinear mixing sum frequency generation will be applicable. This will permit the first observation of allowed singlet-singlet electronic transitions of the chemically important closed shell proton adducts, which are presently accessible only by vibrational or rotational spectroscopy.

The application of AC discharge modulation actually goes well beyond that of velocity modulation spectroscopy of molecular ions. Population modulation of short-lived radicals in 50% duty-cycle pulsed discharges can be combined with phase-synchronous detection of the absorbed radiation to yield the difference in populations in these species between the on and off portions. For very short-lived species ($\tau \ll 1/f$, where τ is the effective lifetime of the radical and f is the discharge pulse rate), the population will closely follow the discharge current, and spectral transitions will essentially be fully modulated. For longer lived species, however, the modulation depth will decrease, and a phase lag in the population with respect to the discharge current will become evident. Amplitude and phase of absorption signals therefore provide important discriminatory information, which can

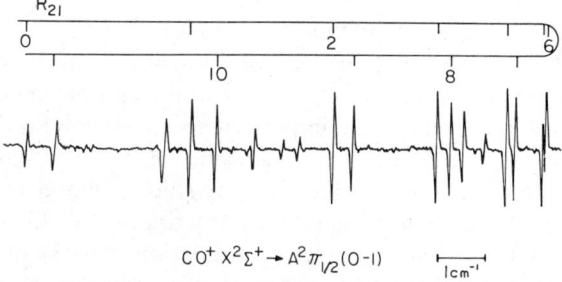

Figure 7 Bandhead of the R_{21} branch in the CO^+ $A^2\Pi_{1/2}(v=1) \leftarrow X^2\Sigma^+(v=0)$ system. Numbers in the assignment are the N (pure rotation) quantum number. The approximately half of the unassigned transitions are due to N_2 impurities (high $vB \leftarrow XN_2^+$); the remainder are likely due to X(high v) $\leftarrow X(v=0)$ CO^+ transitions that borrow intensity from the A state. From Martner et al (67).

be crucial in the assignment of congested and multicomponent spectra observed in these plasmas.

Population modulation in pulsed discharges was first noted in 1981 by Hirota (68) and co-workers in Okazaki, where the infrared spectra of neutral SF radical within OCS/CF_4 plasmas have been studied. This method is also applicable to molecular ions and has been used very recently by Amano (69) in a follow-up study of the Berkeley work on the v_1 band in HCO^+.

Symmetric bipolar AC discharges have also been found to lend themselves very conveniently to population modulation. For an ideal symmetric discharge current waveform, the population of all species in the plasma is equal during each half cycle and the suppression of neutral absorption signals is therefore complete for $1f$ detection. For $2f$ demodulation, however, the plasma density passes through a minimum once each ($2f$) cycle, and the absorption signal is equivalent (except for the amplitudes of the Fourier components making up the absorption signal waveform) to that observed in the pulsed discharge modulation discussed above. Conventional lock-in amplifiers are easily modified for $2f$, as well as $1f$, demodulation. Hence, the experimenter can virtually turn the neutral absorption signals off and on by simply switching from $1f$ to $2f$ detection. Many neutral species, both stable and short-lived ones, have been observed and assigned based on $2f$ detection and phase discrimination. In the Berkeley color center laser effort, these include OH ($v = 1 \leftarrow 0$), CO ($\Delta v = 2, v'' = 0, 19, 20,$ and 21), HCN (v_1), H_2O (v_1 and v_3), NH_3 (v_1 and v_3), and atomic Ar, He, and H Rydberg transitions.

THE THEORY OF VELOCITY MODULATION LASER SPECTROSCOPY

The success of the velocity modulation technique depends on the fact that drift velocities comparable to random velocities can be realized for molecular ions in discharge plasmas. This certainly is not true for discharge plasmas in general; hence, it is important to consider quantitatively the parameters that determine the magnitude of the drift and random velocities, or correspondingly, the Doppler shifts (δv) and widths (Δv) of molecular ion absorption profiles. This description can then be used to assess the effectiveness of ion velocity modulation in terms of modulation depth ($\delta v/\Delta v$). Beyond the importance of this technique as a means for discrimination against neutral absorption signals, we are also interested in its applicability for quantitative measurements of the spectral linewidths, lineshifts, and intensities, which, in turn, permit dynamical aspects of the plasma environment to be examined on a quantum state resolved basis.

Extraction of such dynamical information from observed velocity and population modulated absorption profiles depends on having a model that accounts for the dependence of the velocity modulation lineshapes on the discharge plasma parameters. In this section we discuss such a model in a semiquantitative fashion.

The average drift velocity, v_d (cm/sec), of an ion moving through a neutral gas under the accelerating force of an external electric field E (volts/cm), can be conveniently expressed (40, 41) as the product of the field and the ionic mobility K (cm^2–volt^{-1}–sec^{-1}),

$$v_d = KE. \qquad 1.$$

We must note that K is not necessarily independent of the field strength, although at low and intermediate fields ($E/P \leq 10$ volts-cm^{-1}–torr^{-1}), this is typically the case. A quantitative definition of high and low fields can be found in the treatise of McDaniel & Mason (40), but for the plasmas presently employed for velocity modulation in which the drift energy of the ion is comparable to its random thermal energy, the intermediate field case applies. The magnitude of K depends upon the nature of the electrodynamic interaction between the ion and the neutral species, and upon the efficiency of momentum transfer between the two masses. For the simplest case, in which the neutral is a polarizable sphere and the ion is a charged sphere, the interaction is governed by the monopole-induced dipole attractive potential, which varies as r^{-4}, and is proportional to the dipolar polarizability (Å)^3 of the neutral. A "reduced mobility," K_0, can be expressed as

$$K_0 = 13.876/(\alpha\mu)^{1/2}, \qquad 2.$$

the derivation of which is generally credited to Langevin (70, 71) in 1905; it is therefore known as the "Langevin" mobility. Here the reduced mass μ(amu) of the ion-neutral collision pair reflects the efficiency of momentum transfer. The effective mobility can then be expressed in terms of K_0 as

$$K = K_0[760/p(\text{torr})][T(\text{Kelvin})/273.16]. \qquad 3.$$

We can estimate the magnitude of Doppler shifts (δv) expected for ionic transitions with the standard result

$$\delta v = v_0(v_d/c) \qquad 4.$$

and estimates of the field and neutral molecule translational temperature within the plasma. For HCO$^+$ in 1 torr H$_2$, the axial field of the positive column has been estimated to be 20 volts-cm^{-1} using high impedance probes inserted into the plasma, while the rotational distribution of HCO$^+$ in this environment is consistent with a neutral temperature of 200 K (assuming that the ionic rotational populations are equilibrated with the neutral gas translation temperature). Equations 1–4 then predict

$$\delta v = 390 \text{ MHz } (\alpha_{H_2} = 0.808 \text{ Å}^3, v_0 = 3089 \text{ cm}^{-1}).$$

This shift must be comparable to or larger than the Doppler broadened linewidth if the ionic absorptions are to be fully modulated and maximum sensitivity is to be achieved. The Doppler width (Δv(HWHM)) is given by (72)

$$\Delta v = 3.581 \times 10^{-7} v_0 \, [T(\text{Kelvin})/m(\text{amu})]^{1/2}, \qquad 5.$$

and extending the HCO^+ example, we obtain $\Delta v = 90$ MHz, which indeed indicates that ionic absorptions will be significantly overmodulated! In this limit, where $\delta v > \Delta v$, the demodulated line profile would appear as two isolated Gaussians separated by $2\delta v$, and with equal but opposite intensities and equal widths. As the drift velocity is decreased, either by reducing the electric field or by decreasing the ion mobility, the two Gaussians move closer together, with overlap of the red- and blue-shifted components ($\delta v < \Delta v$), thus leading to cancellation of intensity and the loss of sensitivity. It is convenient to define the modulation depth, $\delta v/\Delta v$, which is completely analogous to that commonly used in derivative spectroscopy (since in the limit where $\delta v/\Delta v \lesssim 1$, velocity modulation is, in fact, simply derivative spectroscopy). The ratio of the fully modulated absorption amplitude to the amplitude of an undermodulated absorption is roughly equal to $\delta v/\Delta v$; positions of the positive and negative maxima of the derivative profile occur at $v_0 \pm (\Delta v/\sqrt{2 \ln 2})$ (where v_0 is the rest frequency of the ionic transition), and are independent of δv for $\delta v/\Delta v < 1$. Lineshift information is virtually lost in this case, and is very strongly correlated to the loss of amplitude due to the overlap.

This apparently desperate situation is easily rectified, however, since the intensity information is recoverable from the $2f$ population modulated signal when a symmetric bipolar double square wave signal is used to drive the AC discharge. The demodulated molecular ion absorption profile for this $2f$ case is composed of two overlapping Gaussians of equal intensity, sign, and width, separated by $2\delta v$. This is similar to the demodulated $1f$ signal, except that in the $1f$ case the two profiles are opposite in sign. For $\delta v/\Delta v < 1$, the $2f$ absorption profile consists of an unresolved doublet, and thus appears as a single line. A simultaneous fit of both the $1f$ and $2f$ demodulated line profiles to two Gaussians, which are described by a single width, intensity, separation, and average rest frequency, then provides sufficient information to determine the three relative lineshape parameters with very small correlations. In the absence of the requisite nonlinear least squares fitting routines, fairly good estimates of the shift, width, and intensity can be made in the limits where $\delta v/\Delta v > 1$ or $\delta v/\Delta v < 1$. For a modulation depth greater than one the estimates are trivial; for $\delta v/\Delta v < 1$, the 1f derivative maxima are separated by $2\Delta v/\sqrt{2 \ln 2}$, and the ratio of the peak-to-peak 1f signal amplitude to the 2f amplitude is approximately

$$\frac{2(2 \ln 2)^{1/2}(\delta v/\Delta v)e^{-0.5}}{\exp[(\delta v/\Delta v)^2 \ln 2]}.$$ 6.

For the intermediate case in which $\delta v/\Delta v \approx 1$ (which, in fact, usually prevails), nonlinear least squares analysis is necessary for accurate estimates. While making the simultaneous 1f and 2f absorption measurements, it is critical that the double square wave discharge current waveform have a duty cycle $\lesssim 50\%$, i.e. the discharge must shut off for at least a full 1/4 cycle twice each cycle. Otherwise the 2f population changes will be undermodulated. A more complicated 2f lineshape has been observed in Ar plasmas and in high frequency ($\gtrsim 15$ kHz) H_2 plasmas (67), where a large fraction of the molecular ion population persists throughout the "off" $\frac{1}{4}$ cycles of the discharge. A complete analysis of population modulation phenomena, which can reveal the effective ionic lifetime and yield "field free" ion spectra, is in progress and will appear in a separate publication (67).

It was predicted above that 1f absorptions will be overmodulated by a factor of $\delta v/\Delta v \approx 4$, which is in marked contrast to those actually observed for HCO^+ and similar molecular ions in the H_2 discharges. The widths and shifts are approximately the same for these systems. As noted in the previous section, however, Haese et al have detected large shifts for ArH^+ in DC He plasmas (48) and have found that a model identical to that given above describes the observed shifts quite sufficiently.

In general, it appears that this simple model cannot adequately explain our observations, and that the limitations that are imposed upon the velocity modulation technique cannot yet be fully assessed. Several qualitative observations and considerations are helpful at this stage. First, it must be noted that the positive column of H_2 plasmas, and of molecular plasmas in general, are visibly striated (73). The effects of striations upon the drift motion of charged molecules had been considered previously (47) in connection with microwave Doppler shift observations by Gudeman & Woods, who concluded that the non-uniform vacillating axial electric field will produce bunches of ions concentrated in the low field wells because total current flux must be conserved. A two-fold reduction in the modulation depth $\delta v/\Delta v$ results in this case, due first to a decreased Doppler shift, since the field in the neighborhood of the ion bunch is lower than the average axial field, and second to broadening and distortion of the idealized Gaussian profile, because marked spatial drift velocity dispersion exists along the striation.

It must also be realized that estimates of neutral gas translational temperatures in the plasma are very difficult to make. Conventional thermometry can measure only a weighted average of the neutral, ionic, and electron temperatures. Although the ions and electrons are far lower in abundance, average electron temperatures of several eV (10^4–10^5 Kelvin) are typical, resulting in an overestimation of the neutral temperature. Doppler-broadened linewidths of probe molecules within the plasma are, in principle, a more reliable measure of their translation temperature, but such reliable linewidth measurements are notoriously difficult to make. Rotational temperatures of neutral species tend to be well equilibrated with their translational energy, and therefore serve as a more convenient means of estimating the neutral temperature, since they depend simply on the relative intensities of resolved rotational transitions. Thus, it is anticipated that rotational intensity distributions will be a very important diagnostic in future studies of mobility in active plasmas.

The limitations of the simple Langevin ion-neutral interaction potential assumed above must also be considered; only the simplest collisional interaction has been incorporated in the present treatment. Bowers and co-workers (74) have shown that the monopole-dipole interaction can greatly increase chemical reaction rates, which suggests that ionic mobilities in polar gases will be substantially reduced. For high fields the simple Langevin picture has also been shown to be inadequate (40) and must be modified to reflect a $E^{1/2}$ dependence of the ionic drift velocity. For systems in which resonant charge transfer or proton transfer can occur, mobilities well below those predicted from monopole-induced dipole potentials have been observed in conventional time-of-flight measurements (40). These limitations occur often in chemically important plasmas, and the need for a more complete model that can accurately treat these effects in conjunction with the spatial variations of the plasma (striations) is urgently needed.

A simple extension of the monopole-induced dipole model permits the mobility of an ion in a known mixture of gases to be treated (40). This extension, known as Blanc's Law, can be expressed as

$$1/[K_{mix}] = \chi_1/[K_1] + \chi_2/[K_2] + \cdots, \qquad 7.$$

where K_i's are the mobilities of the ion in the corresponding pure gases and the χ_i's are the molar fractions. The importance of this law becomes evident when one considers the change of mobility of HCO^+ in going from pure H_2

to a 10/1 mixture of H_2 and CO. Assuming Langevin mobilities for K_{H_2} and K_{CO}, one finds that $K_{mix} = 0.8\ K_{H_2}$. However, resonant proton transfer, monopole-dipole, and monopole-quadrupole interactions of HCO^+ with CO are likely to lower K_{CO} significantly from the Langevin prediction and ultimately yield $K_{mix} \sim 0.5\ K_{H_2}$, even for this 10/1 mixture.

CONCLUDING REMARKS

The first velocity modulation spectra of a molecular ion (HCO^+) were measured on Christmas Eve of 1982. Only one year after its inception, it is clear that velocity modulation laser spectroscopy has already had a substantial impact on the study of molecular ions, having produced high quality spectroscopic data for HCO^+, HNN^+, H_3O^+, NH_4^+, H_2F^+, H_3^+, $HCNH^+$, D_3O^+, and DNN^+ in the infrared and for CO^+ and N_2^+ in the visible ranges. We emphasize that the experimental realization of velocity modulation of molecular ions in low density plasmas represents the latest stage of evolution of the original ideas put forth by Woods et al (28) in 1974.

The measurements, assignments, and analyses of vibrational spectra of molecular ions discussed in this article were greatly facilitated by the existence of high quality ab initio calculations of vibrational frequencies and intensities, made by several theory groups. The calculations of Hennig, Kraemer & Dierksen (75) have been extremely valuable for studies of HCO^+, HNN^+, and their isotopes; Botschwina's results (76) for H_2F^+ have been of enormous help in our assignment and analysis of the v_1 and v_3 band measurements, and Carney & Porter's work on H_3^+ (59) has been important to the success of both the Wing and Oka experiments on that ion. Bunker, Kraemer & Spirko (77), Botschwina (78), and Colvin, Raine, Schaefer & Dupuis (79) have all published important results for H_3O^+. The calculations of the Schaefer group at Berkeley have been of particular significance to this work. Both the Berkeley and Chicago spectroscopy groups have relied heavily on their results for H_3O^+ (79), NH_4^+ (80), and H_2CN^+ (81). It is most encouraging to note the excellent agreement that has been found between the ab initio results of these groups and the experimental measurements. This indicates that similarly reliable theoretical calculations will be available for guiding future experiments on increasingly larger and more complex molecular ion systems, where such guidance is likely to be an absolute necessity. We encourage molecular theoreticians to press forward with detailed calculations on complexed ions, such as $H_3O^+ \cdot nH_2O$, and on negative ions, like CCH^-, as these will probably become experimentally accessible in the relatively near future. The former case presents two major obstacles to theoreticians—

computation of an accurate potential surface, including accurate barriers to tunneling and isomerization, and a rigorous treatment of the large amplitude vibrations that will clearly occur in complexed ions. The resolution of these difficult problems will substantially increase our comprehension of the structures and dynamics of these important prototypes of solvation and proton transfer.

ACKNOWLEDGMENTS

The Berkeley Infrared Spectroscopy effort is supported by the National Science Foundation, Grant # CHE 82-07307. The visible laser velocity modulation experiments were supported by the Director, Office of Energy Research, Office of Basic Energy Sciences, Chemical Sciences Division of the US Department of Energy under Contract No. DE-AC03-76SF00098. C.S.G. was supported by the Petroleum Research Fund, Grant # 12096-G6, and by a fellowship from IBM. The dye laser equipment was rented from the San Francisco Laser Center, operating under a grant from the National Science Foundation Grant #CHE 79-16250.

We thank Ms. Cordelle Yoder for typing this manuscript under the usual emergency conditions.

Literature Cited

1. Saykally, R. J., Woods, R. C. 1981. *Ann. Rev. Phys. Chem.* 32:403
2. Amano, T. 1983. *Bull. Soc. Chim. Belg.* 92:565
3. Miller, T., Bondybey, V. E. 1982. *Appl. Spectros. Rev.* 18:105
4. Miller, T. A., Bondybey, V. E., eds. 1983. *Molecular Ions: Structure and Chemistry.* Amsterdam: North-Holland
5. Andrews, L. 1980. In *Molecular Ions: Geometric and Electronic Structure,* NATO Adv. Study Inst., ed. J. Berkowitz, K. Groeneveld. New York: Plenum
6. Schwarz, H. A. 1977. *J. Chem. Phys.* 67:5525; Schwarz, H. A. 1980. *J. Chem. Phys.* 72:284
7. Huber, K. P., Herzberg, G. H. 1979. *Molecular Spectra and Molecular Structure IV: Constants of Diatomic Molecules.* New York: Van Nostrand-Reinhold
8. Wing, W. H., Ruff, G. A., Lamb, W. E., Spezeski, J. J. 1976. *Phys. Rev. Lett.* 36:1488
9. Tolliver, D. E., Kyrala, G. A., Wing, W. H. 1979. *Phys. Rev. Lett.* 43:1719
10. Shy, J.-T., Farley, J. W., Lamb, W. E., Wing, W. H. 1980. *Phys. Rev. Lett.* 45:535
11. Shy, J.-T., Farley, J. W., Wing, W. H. 1981. *Phys. Rev. A* 24:1146
12. Wing, W. H. 1977. *Springer Ser. in Optical Sci.,* Vol. 7: *Laser Spectroscopy III,* ed. J. L. Hall, J. L. Carlsten. Berlin: Springer-Verlag
13. Deleted in proof
14. Carrington, A., Buttenshaw, J. 1981. *Mol. Phys.* 44:267
15. Carrington, A., Buttenshaw, J., Kennedy, R. 1983. *Mol. Phys.* 48:775
16. Carrington, A., Buttenshaw, J., Kennedy, R. A. 1982. *J. Mol. Struct.* 80:47
17. Carrington, A., Buttenshaw, J., Kennedy, R. A., Softley, T. P. 1981. *Mol. Phys.* 44:1233
18. Carrington, A., Kennedy, R. A., Softley, T. P., Fournier, P. G., Richard, E. G. 1983. *Chem. Phys.* 81:251
19. Carrington, A., Buttenshaw, J., Kennedy, R. A., Softley, T. P. 1982. *Mol. Phys.* 45:747
20. Carrington, A., Buttenshaw, J., Kennedy, R. 1982. *Mol. Phys.* 45:753
21. Carrington, A., Kennedy, R. A. 1982. *Ions and Light,* ed. M. Bowers. New York: Academic
22. Carrington, A. 1979. *Proc. R. Soc. London Ser. A* 367:433

23. Carrington, A., Kennedy, R. A. 1984. Preprint
24. Carrington, A., Softley, T. P. 1983. See Ref. 4, pp. 49–72
25. Giese, C. F., Gentry, W. R. 1974. *Phys. Rev. A* 10:2156
26. LeRoy, R. J., Liu, W.-K. 1978. *J. Chem. Phys.* 69:3622
27. Dixon, T. A., Woods, R. C. 1975. *Phys. Rev. Lett.* 34:61
28. Woods, R. C., Dixon, T. A., Saykally, R. J., Szanto, P. G. 1975. *Phys. Rev. Lett.* 35:1269
29. Saykally, R. J., Dixon, T. A., Anderson, T. G., Szanto, P. G., Woods, R. C. 1976. *Astrophys. J. Lett.* 205:L101
30a. Oka, T. 1980. *Phys. Rev. Lett.* 45:531
30b. Oka, T. 1981. *Philos. Trans. R. Soc. London A* 303:543
31. Bernath, P., Amano, T. 1982. *Phys. Rev. Lett.* 48:20
32. Wong, M., Bernath, P., Amano, T. 1982. *J. Chem. Phys.* 77:693
33. Davies, P. B., Hamilton, P. A. 1983. *Chem. Phys. Lett.* 94:565
34. Davies, P. B., Hamilton, P. A., Lewis-Bevan, W., Okamura, M. 1983. *J. Phys. E Sci. Instrum.* 16:289
35. Amano, T., Bernath, P. F., McKellar, A. R. W. 1982. *J. Mol. Spectrosc.* 94:100
36. Brault, J. W., Davis, S. P. 1982. *Phys. Script.* 25:268
37. Johns, J. W. C. 1982. *37th Symp. Mol. Spectrosc., Columbus, OH, June,* p. 41
38. Knights, J. C., Schmitt, J. P. M., Perrin, J., Guelachvili, G. 1982. *J. Chem. Phys.* 76:3414
39. Snyder, L. E., Hollis, J. M., 1976. *Astrophys. J. Lett.* 204:L139
40. McDaniel, E. W., Mason, E. A. 1973. *The Mobility and Diffusion of Ions in Gases.* New York: Wiley
41. von Engel, A. 1965. *Ionized Gases.* London: Oxford Univ. Press. 2nd ed.
42. Woods, R. C., Saykally, R. J., Anderson, T. G., Dixon, T. A., Szanto, P. G. 1981. *J. Chem. Phys.* 75:4256
43. Sastry, K. V. L. N., Herbst, E., DeLucia, F. C. 1981. *J. Chem. Phys.* 75:4169
44. Bowman, W. C., Herbst, E., DeLucia, F. C. 1982. *J. Chem. Phys.* 77:4261
45. Sastry, K. V. L. N., Helminger, P., Herbst, E., DeLucia, F. C. 1981. *Astrophys. J. Lett.* 250:L91
46. Sastry, K. V. L. N., Helminger, P., Herbst, E., DeLucia, F. C. 1981. *Chem. Phys. Lett.* 84:286
47. Gudeman, C. S. 1982. PhD thesis. Univ. Wisc., Madison. Unpublished
48. Haese, N. N., Pan, F.-S., Oka, T. 1983. *Phys. Rev. Lett.* 50:1575
49. Lindinger, W., Albritton, D. L. 1975. *J. Chem. Phys.* 62:3517
50. Gudeman, C. S., Begemann, M. H., Pfaff, J., Saykally, R. J. 1983. *Phys. Rev. Lett.* 50:727
51. Gudeman, C. S., Begemann, M. H., Pfaff, J., Saykally, R. J. 1983. *J. Chem. Phys.* 78:5837
52. Begemann, M. H., Gudeman, C. S., Pfaff, J., Saykally, R. J. 1983. *Phys. Rev. Lett.* 51:554
53. Crofton, M. W., Oka, T. 1983. *J. Chem. Phys.* 79:3157
54. Schäfer, E., Begemann, M. H., Gudeman, C. S., Saykally, R. J. 1983. *J. Chem. Phys.* 79:3159
55. Schäfer, E., Saykally, R. J., Robiette, A. G. 1984. *J. Chem. Phys.* 80:3969
56. Schäfer, E., Saykally, R. J. 1984. *J. Chem. Phys.* 80:2973
57. Oka, T. 1983. See Ref. 4, pp. 73–90
58. Gudeman, C. S., Saykally, R. J. 1984. In preparation
59. Carney, G. D., Porter, R. N. 1976. *J. Chem. Phys.* 65:3547
60. Altman, R. S., Crofton, M. W., Oka, T. 1984. Preprint
61. Nesbitt, D. J., Petek, H., Gudeman, C. S., Moore, C. B., Saykally, R. J. 1984. *J. Chem. Phys.* Submitted
62. Petek, H., Nesbitt, D. J., Begemann, M. H., Gudeman, C. S., Moore, C. B., Saykally, R. J. 1984. In preparation
63. Haese, N. N., Oka, T. 1984. *J. Chem. Phys.* 80:572
64. Haese, N. N.. 1981. PhD thesis. Univ. Wisc., Madison. Unpublished
65. Huntress, W. T. 1977. *Astrophys. J. Suppl.* 33:495
66. Gudeman, C. S., Martner, C. C., Saykally, R. J. 1984. In preparation
67. Martner, C. C., Gudeman, C. S., Saykally, R. J. 1984. In preparation
68. Endo, Y., Nagai, K., Yamada, C., Hirota, E. 1983. *J. Mol. Spectrosc.* 97:213
69. Amano, T. 1983. *J. Chem. Phys.* 79:3595
70. Langevin, P. 1905. *Ann. Chim. Phys.* 5:245
71. McDaniel, E. W. 1964. *Collision Phenomena in Ionized Gases.* New York: Wiley
72. Townes, C. H., Schawlow, A. L. 1975. *Microwave Spectroscopy.* New York: Dover
73. Garscadden, A. 1978. In *Gaseous Electronics,* Vol. 1, ed. M. N. Hirsh, H. J. Oskam. New York: Academic
74. Su, T., Su, E. C. F., Bowers, M. T. 1978. *J. Chem. Phys.* 69:2243
75. Hennig, P., Kraemer, W. P., Diercksen, G. H. F. 1978. Max Planck Institut für

Physik und Astrophysik. Unpublished report
76. Botschwina, P. 1980. In *Molecular Ions: Geometric and Electronic Structures*, NATO ASI, ed. J. Berkowitz, K. Groeneveld. New York: Plenum
77. Bunker, P. R., Kraemer, W. P., Spirko, V. 1984. In preparation
78. Botschwina, P. 1983. In preparation
79. Colvin, M. E., Raine, G. P., Schaefer, H. F. III, Dupuis, M. 1983. *J. Chem. Phys.* 79:1551
80. Yamaguchi, Y., Schaefer, H. F. III, 1980. *J. Chem. Phys.* 73:2310
81. Lee, T. J., Schaefer, H. F. III. 1984. *J. Chem. Phys.* 80:2977
82. Crosswhite, H. M., ed. 1972. *The Hydrogen Molecule Wavelength Tables of Gerhard Heinrich Dieke*. New York: Wiley-Interscience
83. Schäfer, E., Saykally, R. J. 1984. *J. Am. Chem. Soc.* Submitted

SIMULATION OF POLYMER MOTION

A. Baumgärtner

Institut für Festkörperforschung der Kernforschungsanlage Jülich, D-5170 Jülich, West Germany,* and IBM Research Laboratory, San Jose, California 95193

INTRODUCTION

Presented here is a review of computer simulations for the dynamics of polymeric systems. Even restricting the review to the more recent results, the subject is so large that some omissions are necessary. The review is concerned with the present theoretical understanding of irreversible processes in polymeric systems. Accordingly, I do not discuss the chain dynamics associated with conservative vibrational motions in either crystalline or liquid states.

More detailed reviews on the static and dynamic properties of polymers are given in (1–3). Details concerning Monte Carlo simulations of polymers are given in (4).

One of the most interesting and unusual features of linear macromolecules is their dynamical behavior. Due to its connectivity, the motion of a flexible chain is quite different from the motion of small molecules in ordinary liquids. Two fascinating phenomena result from this property: the internal Brownian motion of an isolated chain related to its elasticity, and the collective motion of many chains governed by the topological restriction on a given chain imposed by many other chains. I discuss the former aspect in the first section of the review, the latter in the following.

BROWNIAN MOTION OF THE SINGLE CHAIN

Two aspects of the dynamical behavior of an isolated polymer are discussed in this section: local rapid relaxation processes at a molecular level, and the

* Permanent address.

diffusional behavior of isolated polymers which is dominant at times large compared to microscopic relaxation times.

Macroscopic dynamical properties of polymers in dilute solutions are successfully described by models representing extensions of the classical theory of Brownian motion (5). The Rouse model (6) successfully incorporates the Brownian diffusion forces in a dynamical theory of the bead-spring polymer model without destroying the essential topological connectivity of the chain. Pre-averaged hydrodynamic interactions have been included in the Zimm model (7). Later embellishments have been many and several reviews are now available (1, 2, 8–10).

Monte Carlo simulations including bare hydrodynamic forces have not yet been performed. A possible starting point in this direction may be the "Brownian-Dynamic" technique (11, 12) proposed recently. Molecular dynamics simulations investigating the hydrodynamical aspect of polymer motion are reported in (13, 14).

Local Modes in Flexible Chains

The fastest processes, even in dilute solution, are related to conformational transitions of the chain backbone from one rotational isomeric state to another, and clearly correspond nearly to free-draining conditions in which hydrodynamic effects are deliberately neglected. Initially the motivation for the studies of internal motions was the interest in the dynamics of very special polymers, e.g. proteins (15). But more universal aspects have been investigated recently, too (16–20). As a particular bond rotates, the attached tails cannot rigidly follow without experiencing a huge frictional resistance. Consequently, local motions are correlated only over a short range, which is determined by transition probabilities between the rotational isomeric states of consecutive bonds. One of the most important tasks is to derive from torsional potentials dynamic quantities such as transition probabilities, relaxation times, etc.

Standard Monte Carlo simulations are less suitable for the study of short-time motions of polymers and have not been performed; but molecular dynamics simulations have proven to be very helpful (17, 18). The Brownian motion is simulated by solving a coupled set of Langevin equations for the Cartesian degrees of freedom. These equations take the form

$$m\frac{d^2\mathbf{r}_i}{dt^2} = -\zeta\frac{d\mathbf{r}_i}{dt} - \nabla_i V + \mathbf{A}_i(t), \qquad 1.$$

$$\langle \mathbf{A}_i(t)\mathbf{A}_j(t')\rangle = 2\zeta k_B T \delta(t-t')\delta_{ij}\mathbf{1}. \qquad 2.$$

The acceleration of each carbon center i equals the sum of three forces: a frictional force, a potential force, and a stochastic force. The latter is taken to be Gaussian white noise; this is the only way temperature enters the problem. Using Runge-Kutta methods, the coupled Eqs. 1 and 2 have been solved numerically at zero acceleration (17), which corresponds to the high friction limit.

From the hazard analysis (17) of the times between the transitions of one bond from the *trans* state to one of the gauche [*gauche*(+) and *gauche*(−)] states, one can determine the corresponding transition rates. The results (17) indicate a cooperativity of transitions of pairs of second neighbor bonds. These pair transitions can be classified by configurational types and also by the state of the central bond. Two types of pair transitions have been found (17) to account for most of the observed cooperativity:

$$g^{\pm}tt \rightleftarrows ttg, \qquad 3.$$

$$ttt \rightleftarrows g^{\pm}tg^{\mp}. \qquad 4.$$

That means, for example, that if the initial state consists of a pair of second neighbor trans bonds separated by a trans *bond* (*ttt*), the resulting state after a short time span is, for example, *gauche*(+)-*trans*-*gauche*(−) (Eq. 4).

Although many of the transitions are the cooperative pair type, individual transitions are also frequently observed (17). The problem of local conformational transitions of a polymer chain has also been considered using analytical methods, and results are in reasonable accord with molecular dynamics simulations (19, 20).

Long Time Relaxation Effects

The free-draining polymer chain model, in which different polymer chains can be considered as dynamically independent, should be a reasonable approximation for the dynamics in concentrated solutions and undiluted polymer liquids, provided entanglement effects can be neglected. For chains larger than a characteristic chain length ("entanglement length") the influence of entanglements dominate.

THE VERDIER-STOCKMAYER MODEL   The influence of excluded volume on the dynamics of an isolated polymer chain was first studied using Monte Carlo simulations by Verdier & Stockmayer (21). One of the simplest models of a polymer chain in which the excluded volume effect can be investigated is the representation of a chain of $N+1$ units as a random path of N steps on a simple cubic lattice. The $N+1$ occupied lattice sites on the path are the polymer "beads," and the excluded volume interaction imposes the condition that no two beads of the chain can occupy the same lattice site. Verdier and co-workers (21–24) examined a stochastic generalization of this model in which the Brownian motion of the polymer beads is simulated by stochastic jumps of single beads ("kink-jumps"). A bead is picked at random and its position r_n on the lattice is determined. This bead is then "flipped" from its position to the new position $\mathbf{r}'_n = \mathbf{r}_{n+1} + \mathbf{r}_{n-1} - \mathbf{r}_n$. Clearly, a flip preserves the connectivity of the chain. The computer simulations show that the Verdier-Stockmayer model without excluded volume behaves remarkably similar to the predictions of the Rouse (6) model, which is the simplest analytically solved model for polymer motion in the absence of excluded volume and entanglement effects. This agreement has been explained in subsequent papers (25, 26).

However, the Monte Carlo calculations for the Verdier-Stockmayer model with excluded volume show a very different behavior. The surprising result of this "lattice-kink-jump" dynamics was that the introduction of the excluded volume condition caused a strong slowing down of the relaxation of the squared end-to-end distance. In their original work, the relaxation time τ was defined as the required time for a smooth graph of R_N^2 against the number of Monte Carlo cycles to approach its final average value to within $1/e$ of the difference between the values of the initial stretched-out configuration and the final equilibrium configuration. Subsequent work (22–24) considered the equilibrium autocorrelation function of the squared end-to-end distance

$$\phi_{RR}(t) = [\langle R_N^2(t_0) R_N^2(t_0+t) \rangle - \langle R_N^2 \rangle^2]/[\langle R_N^4 \rangle - \langle R_N^2 \rangle^2]. \qquad 5.$$

The simulations led to $\tau \sim N^3$, which is in contrast to the conjecture (27)

$$\tau \propto \langle R_N^2 \rangle / D \propto N^{2\nu+1} \qquad 6.$$

where $D \sim 1/N$ is the center of mass diffusion constant and $\nu = 3/5$ is the excluded volume exponent in three space dimensions. It has been shown recently by Hilhorst, Deutch & Boots (28, 29) that the discrepancy is a consequence of the inability of the "lattice-kink-jump" model to permit local extrema in the chain conformation to pass each other or to disappear without passage of these extrema all the way to one of the ends of the chain. This type of "defect-motion" is closely related to the motion of a flexible chain in a dense, randomly distributed system of fixed obstacles, introduced some years ago by de Gennes (30). Subsequent Monte Carlo calculations (31) on the cubic-lattice model, including "crankshaft" kinetics in order to avoid pure reptation-like motion, had supported this interpretation. The power law Eq. 6 has been confirmed by calculating the half-time $\tau_2 \sim \tau$ of the autocorrelation function, Eq. 6, $\phi_{RR}(\tau_2) = 1/2$ assuming an exponential decay of ϕ_{RR} in that time range. It should be noted that the Verdier-Stockmayer model is not equivalent to the reptation model (30), because the models differ in their time-dependence of the monomer displacements

$$g(t) = \langle [r_n(0) - r_n(t)]^2 \rangle \qquad 7.$$

of the bead n at intermediate times $t \ll \tau$, where one expects for the Verdier-Stockmayer and the Rouse models $g(t) \sim t^{1/2}$, whereas in the reptation model, $g(t) \sim t^{1/4}$. (The reptation model is discussed below.)

THE ROUSE MODEL The general "kink-jump" procedure for modeling the Rouse dynamics of lattice polymer chain models is as follows:

1. To preserve the underlying lattice structure, in general more than two links have to be moved simultaneously. For the diamond lattice, three or four successive links can be moved, keeping the end points of this part of the chain fixed (32, 33).
2. If there are k choices to perform an n-link motion [i.e. two choices of four-link motion on the diamond lattice (33)], all possible choices and motions can be combined. But in order to fulfill the "detailed balance" condition (4), the types of motion during the Monte Carlo procedure have to be selected at random.

A generalization of the "kink-jump" dynamics for lattice polymer models to off-lattice polymer models has been proposed recently (34, 35). The stochastic motion of the bead-rod polymer chain results from successive local jumps randomly distributed along the chain; each of them affects a very small part of the chain. The simplest dynamics of local character for the bead-rod model consist in trying rotations of two neighboring links around the axis joining their end points through an angle ϕ chosen randomly from the interval $(-\Delta\phi, +\Delta\phi)$. The parameter $\Delta\phi$ is arbitrary, but it is convenient to choose it so that about one-half of the attempted rotations are successful. If an end point of the chain is chosen, the link is rotated to a new position by specifying two randomly chosen angles (ϕ, θ) in three space dimensions, with $\cos\theta$ being equally distributed in the interval $-1 < \cos\theta \leq +1$. For the two-dimensional case it is only the angle ϕ that can then be randomly chosen.

Thus one starts a suitable initial configuration of the chain, and then chooses randomly one of the beads for rotation. The transition probability $W(\mathbf{r}_i \to \mathbf{r}'_i)$ for this rotation is calculated and compared to a random number $0 < a < 1$. If $W > a$, the rotation is actually performed and the new configuration is accepted; otherwise this rotation is rejected, and the old configuration is counted once more in averaging. In this manner, a sequence of configurations is generated. Since the system tends to equilibrium by construction (4), there is a correspondence between the time lapse and the number of configurations. In order that the time unit does not depend on the chain length N, this unit is defined as a sequence in which, on the average, any bead has the possibility to move once. This Monte Carlo step per particle contains a sequence of N chain configurations. Thus, for describing the evolution of the system we may use a parameter t, called time, which takes on the sequential values $t_k = k/N, k = 1, 2, \ldots$. The dynamics of the "kink-jump" procedure applied to the freely jointed bead-rod model is that of the Rouse model. This has been shown by simulations of the latter model self-interacting via a Lennard-Jones potential (35). At high temperatures the relaxation time exhibits $\tau \sim N^{2.2}$ in agreement with Eq. 6. The correlation function $g(t)$ (Eq. 7) varies as

$$g(t) \sim \begin{cases} t^{1/2}, & t < \tau \\ t/N, & t > \tau. \end{cases} \qquad \begin{matrix}\text{8a.}\\\text{8b.}\end{matrix}$$

The Rouse behavior has also been found in simulations of the "bead-spring" model (11, 12, 36–39). Neglecting inertial terms (high viscosity limit) and hydrodynamic forces, it is assumed that the time evolution of the polymer probability density is given by a Smoluchowski equation,

$$\partial f(\mathbf{R}, t)/\partial t = D \sum_{j=1}^{N} \nabla_j [\nabla_j f(\mathbf{R}, t) + \beta f(\mathbf{R}, t) \nabla_j U(\mathbf{R})] \qquad 9.$$

where the coordinates of the beads are denoted by $\mathbf{R} = \{\mathbf{r}_1, 1 \leq i \leq N\}$. Here $U(\mathbf{R})$ is the total potential energy of the chain, consisting of a Lennard-Jones type potential and a modified harmonic potential for the springs, and D is the diffusion constant. The solution of the diffusion equation is simulated by a special Monte Carlo scheme (12). The two most essential steps are the following:

1. The new position $\mathbf{r}_i$ for the randomly selected particle i is sampled from the equation

$$\mathbf{r}_i(t+\tau) = \mathbf{r}_i(t) - \tau\beta D\nabla_i U + \chi_i \qquad 10.$$

where $\langle \chi_i \chi_j \rangle = 6D\tau$.

2. The move is accepted with probability $q(\mathbf{R}, \mathbf{R}')$ where

$$q(\mathbf{R}, \mathbf{R}') = \min\{1, [\exp(-\beta U(\mathbf{R}'))G(\mathbf{R}', \mathbf{R}, \tau)]/ $$
$$[\exp(-\beta U(\mathbf{R}))G(\mathbf{R}, \mathbf{R}', \tau)]\}, \qquad 11.$$

which is constructed so that $q(\mathbf{R}, \mathbf{R}')G(\mathbf{R}, \mathbf{R}')$ satisfies the detailed balance condition. The probability density $G(\mathbf{R}, \mathbf{R}')$ for the polymer to diffuse from point $\mathbf{R}$ to $\mathbf{R}'$ in time τ is given by

$$G(\mathbf{R}, \mathbf{R}') = (2\pi\tau D)^{-3N/2} \exp\{-[\mathbf{R} - \mathbf{R}' + \tau D\nabla U(\mathbf{R}')]^2/4D\tau\}.$$

It is possible to generalize the above method to include a hydrodynamical interaction matrix (Oseen tensor), but this has not yet been done.

THE KRON MODEL The dynamics of the "slithering-snake" model originally proposed by Kron (40, 41) and by Wall & Mandel (42) in order to generate equilibrium configurations of lattice polymers [but also applicable to off-lattice polymers (33, 43)], has been suggested to be in a different dynamic universality class (44, 45) than the Rouse or the Verdier-Stockmayer model.

Technically, the dynamics are as follows. Starting from an arbitrary configuration, one first selects one of the ends of the chain at random and then removes this end link of the chain and adds it to the other end, randomly specifying the orientation of the link. The resulting state is accepted as a new configuration, if the excluded volume condition is not violated. [For the thermal case the Metropolis criterion has to be considered (4).] This mechanism corresponds to a "slithering snake-like" motion of the chain along itself.

Using a dynamic Monte Carlo renormalization group method, Muthukumar & Banavar (44, 45) found $\tau \sim N^{2.04}$, which is rather close to the Rouse exponent, 2.2. The behavior of the monomer correlation function $g(t)$ has not yet been studied.

THE REPTATION MODEL The fact that the chain cannot intersect itself or other chains is not explicitly included in any of the dynamic models discussed above. This topological interaction is important for dilute polymer solutions under theta or poor solvent conditions [self-intersections (46, 47)], for ring-polymers, in concentrated solutions, and in polymer melts.

The reptation model was constructed originally to describe the motion of a single chain inside a network of fixed obstacles (30). Although this model is important for a few experimental realizations (e.g. gel chromatography), the reptation concept has become even more important in recent discussions on the dynamics of polymer melts. There are indications that the reptation process remains dominant for a polymer melt in which the network of fixed obstacles is released and replaced by mobile chains of comparable length (48–52).

Several computer simulations have been performed in order to clarify the applicability of the reptation concept to polymer melts (discussed below). However, the reliability of Monte Carlo simulations of polymer melts depends strongly on the verifiability of the predictions of the original reptation model for fixed obstacles.

One can summarize the characteristic features of reptation (2, 3, 30, 50): At any moment of time, the chain is trapped inside a certain "tube." If N_e is the number of monomers between entanglement points, and the chain has an ideal configurational statistic (i.e. end-to-end distance $R_0 \sim N^{1/2}$), the tube diameter is of the order of $d \sim N_e^{1/2}$ and the length of the tube is $L \sim N N_e^{-1/2}$. The fundamental relaxation time τ_d ("disengagement time") corresponds to diffusion along the tube over a length of the order L:

$$\tau_d \sim N^3/N_e \quad (N \gg N_e). \qquad 12.$$

For $N \simeq N_e$, τ_d goes back to the Rouse relaxation time $\sim N^2$. The translational diffusion coefficient of the chain is

$$D \cong \langle \mathbf{R}_0^2 \rangle / \tau_d \sim N_e N^{-2} \quad (N \gg N_e).\qquad 13.$$

At small scales ($\mathbf{r} < \mathbf{R}_0$) the dynamics become completely different from the Rouse model. An appropriate quantity to describe this behavior is the monomer correlation function $g(t)$ (Eq. 7). This is summarized in Figure 1. At short times ($t < N_e^2$) the entanglements are ineffective and the conventional Rouse behavior is dominant, $g(t) \sim t^{1/2}$ (Eq. 8). At times $N_e^2 < t < N^2$ the chain feels the dynamical restriction due to the fixed obstacles and has to adjust accordingly. This leads to a very slow motion $g(t) \sim t^{1/4}$. In the time regime $N^2 < t < N^3/N_e$ we can still think of the chain moving as a whole in a fixed tube, (since $t < \tau_d$) with $g(t) \sim (t/N)^{1/2}$. Finally, at times $t > \tau_d$ the diffusive behavior is dominant according to $g(t) \sim Dt \sim t/N^2$.

The reptation model has been studied by several authors using Monte Carlo simulations applied to various polymer models (53–56). In one of the first studies (53), the motion of one chain through a network of a "frozen" polymer melt was considered. Both the test chain and the "frozen" chains consist of $N = 15$ rigid links of length l freely jointed together (bead-rod model). As an interaction between any pairs of beads at distance r a Lennard-Jones potential $U(r) = 4\varepsilon[(h/r)^{12} - (h/r)^6]$ was assumed. The width of the "effective hard core" was chosen as $h/l = 0.4$ because then the end-to-end distance remains ideal, $\mathbf{R}_0 \sim N^{1/2}$, down to very small values of N. The local "kink-jump" dynamics were introduced as discussed above for the Rouse model (34, 35, 53). A trial move was accepted only if the transition probability W exceeded a random number $0 < a < 1$, otherwise it was rejected and the old configuration was counted once more as a new one. The transition probability $W = W_H W_T$ was constructed so that it satis-

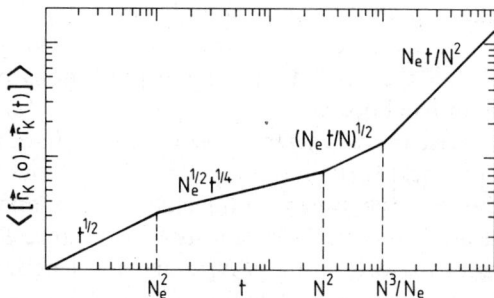

Figure 1 Log-log diagram of the monomer displacement versus time for the reptation model as propounded by de Gennes (30, 50).

fied a detailed balance with the equilibrium probability distribution $P_0 \sim \exp(-H/kT)$, where H is the total energy of the system, and satisfied the "topological" requirement of prohibition of chain intersections during one attempted rotation: if $\Delta H = H_{old} - H_{new} \geq 0$, $W_H = 1$, if $\Delta H < 0$ then $W_H = \exp(\Delta H/kT)$; and $W_T = 0$, if the rotation required a link intersection, and $W_T = 1$ otherwise.

The network was equilibrated at a concentration of beads $c = 2.5$ and at a reduced temperature $kT/\varepsilon = 3$. The predicted behavior $g(t) \sim t^{1/4}$ was seen over more than three decades of time. The correlation function of the center-of-mass $\mathbf{R}_{CM}(t) = \sum_{i=1}^{N} \mathbf{r}_i(t)/N$

$$g_{CM}(t) = \langle [\mathbf{R}_{CM}(0) - \mathbf{R}_{CM}(t)]^2 \rangle \qquad 14.$$

exhibits $g_{CM}(t) \sim t^{1/2}$, in contrast to the Rouse model in which $g_{CM}(t) \sim t$, but in agreement with the reptation model [although it was not explicitly predicted in the original work of de Gennes (30)]. Of course, at large times the conventional diffusive behavior $g_{CM}(t) \sim t$ was also observed. Two objections could be made against the simulation of one chain confined to a random "frozen" network (53, 54, 56), which have not yet been tested:

1. Since the test chain is a part of the equilibrated "frozen" melt at time $t = 0$, the chain moves at later times through a tube, leaving "holes" behind it, which remain available, while an opening at the other end may or may not exist. There is some evidence that this effect slows the diffusion by an amount that depends on the ratio of bead diameter to chain length.
2. In de Gennes' reptation model (30) the chain is assumed to obey Gaussian statistics, which might be essential to obtain $g(t) \sim t^{1/4}$. The configurational statistics of the simulated reptating chains (53, 54, 56) had not yet been examined in detail. Moreover, the influence of the "hole" effect (see above) on the chain configurational statistics for times $t < \tau_d$ could likewise be important.

Similar findings for $g(t)$ and $g_{CM}(t)$ in agreement with reptational behavior have been reported from simulations of the bead-spring model (55). With this model, chain crossing is not strictly excluded. Chain lengths of between 20 and 32 beads have been considered. At the lower density of the network of 0.1, the exponents are very similar to the Rouse model. However, the tendency to reptational behavior was seen with increasing density of the network: at density 0.5, a chain of 32 beads exhibited $g(t) \sim t^{0.37}$ and $g_{CM}(t) \sim t^{0.62}$.

A different approach, namely simulations of a polymer as a random walk on a cubic lattice and confined by a regular obstacle lattice, has been performed recently (54). The stochastic motion of the chain was achieved by a kink-jump procedure similar to the original method of Verdier & Stockmayer (21). The self-avoiding walk condition for the reptating chain was deliberately neglected. Of course, the statistics of this chain were Gaussian, in agreement with the reptation model (30). Chain lengths of between 5 and 80 beads were used. The chains diffused through cubic lattice cages that varied in spacing between 1 and 10. Good agreement was found for the diffusion constant $D \sim N^{-2}$, defined as

$$D = \lim_{t \to \infty} g_{\text{CM}}(t)/6tl^2 \qquad 15.$$

and disengagement time $\tau_d \sim N^{3.1}$, defined according to

$$\langle \mathbf{R}(t)\mathbf{R}(0) \rangle \sim \exp(-t/\tau_d) \qquad 16.$$

where $\mathbf{R}(t)$ is the end-to-end vector at time t. The correlation function $g(t)$ was almost consistent with reptation predictions (Figure 1); but no indication of the existence of the initial $t^{1/2}$ behavior was found. For obstacle spacings >2 it was not possible to discern $t^{1/4}$ behavior either. Whether the crossover from Rouse to reptation behavior, which appeared in all quantities D, τ_d and $g(t)$ during the reduction of obstacle spacings, occurs discontinuously or not, could not be decided. More refined finite size considerations are necessary to clarify this last point.

More recently, simulations of a chain moving on the tetrahedral lattice (56) through an equilibrated "frozen" melt of comparable chains have been performed. At a concentration of 34% occupied lattice sites, the chain of length $N = 200$ exhibited a behavior compatible with Figure 1. In contrast to simulations of Ref. (54), here the self-avoiding condition was fully included. The local dynamic used for this simulation was the modified "kink-jump" technique for lattice polymers (33), discussed above in connection with the Rouse model.

STAR MOLECULES Recently, computer simulations of star polymers diffusing in a fixed-obstacle matrix have been reported (57, 58). The three-arm star polymer was modeled as a set of $(3N + 1)$ points on a cubic lattice. Self-avoiding walk conditions were neglected. Points along the chain were chosen at random and moved such that connectivity was preserved and no portion of the chain had moved through any obstacles present. The obstacles were infinitely long rods located on a cubic lattice shifted from the polymer lattice in each of the Cartesian directions by one half of a lattice spacing. Evans found (57) that the variations of the arm relaxation time τ_s and the diffusion coefficient D_s could be fitted by power laws $\tau_s \sim N^4$ and $D_s \sim N^{-3}$. Needs & Edwards (58), however, extending the previous simulation to obtain better statistics, found variation of τ_s and D_s with N significantly more rapid than a power relation, and better fitted by an exponential law $\tau_s \sim N^{1.9} \exp(0.19\,N)$ and $D_s \sim N^{-0.59} \exp(-0.37\,N)$. These results are in qualitative agreement with de Gennes' predictions (59), $\tau_s \sim N^3 \exp(AN)$ and $D_s \sim f(N) \exp(-BN)$, and with a more recent conjecture by Graessley (3), $D_s \sim N^{-2} \exp(-CN)$. However, the exponential suppression of D has been found recently in experiments made by Klein & Fletcher (60).

DYNAMICS OF ENTANGLED POLYMER CHAINS

The viscoelastic properties of polymeric liquids have been measured for many years, starting from the pioneering work of Ferry (61). The viscosity exhibits a crossover from $\eta \sim N$ to $\eta \sim N^{3.4}$ with increasing chain length at a critical value N_e. If a polymer melt is subjected to rapidly varying perturbations, it behaves like an elastic rubber, whereas at somewhat lower frequencies, it flows like conventional liquids. The characteristic crossover time τ_d (or characteristic frequency) exhibits the same behavior as the viscosity: $\tau_d \sim N$ for $N < N_e$ and $\tau_d \sim N^{3.4}$ for $N > N_e$. The meaning of τ_d was qualitatively understood many years ago (51, 61). At times $t < \tau_d$ the entanglements force the system to behave as a permanently cross-linked network. At times $t > \tau_d$ the "knots" open up by Brownian motion; the chains can slip with respect to each other and flow is observed. However, the diffusive behavior is quite different from conventional flow with respect to the diffusion coefficient D. Measurements (48, 49, 52) gave some indication for $D \sim N^{-2}$.

To date, the reptation concept is the most successful attempt to explain the dynamics of polymer liquids over more than six orders of magnitude of time. The "tube" model, proposed by Edwards (62), was the first attempt to include explicitly the topological effect of entanglement constraint, where the tube is the topological skeleton of the chain under consideration, diffusing through slip links or entanglement points with other chains. This concept was quantified in the "reptation model" by de Gennes (30), where the surrounding medium remains essentiallly fixed. This reptation concept led to the construction of a very elegant theory for the viscoelastic behavior of melts by Doi & Edwards (63).

However, since the reptation concept leads to a somewhat weaker exponent in viscosity and relaxation time, $\eta \sim \tau_d \sim N^3$, the situation is unsatisfactory and controversial. Various explanations have been proposed for this discrepancy: polydispersity effects, cross-over effect from nonentangled to entangled regimes (64, 65), or tight knots with lifetimes much larger than τ_d.

Computer experiments could play a key role in either supporting the present theory or improving it conceptually. In fact, the intention of computer experiments are twofold: (a) as in real experiments, we can test the predictions of theory, e.g. η, τ_d, $g(t)$, etc; (b) because the de Gennes-Doi-Edwards theory is essentially of the mean-field type, we can ask for the limitations reflected in some qualitative and quantitative (e.g. power laws) deviations based on subtle microscopic phenomena undetectable by real experiments. Questions such as the precise microscopic meaning of entanglements and the importance of tube fluctuations (or tube renewal) are examples.

However, after the first simulation on the dynamics of polymer melts were performed (53), it turned out that the simulation itself became a controversial subject:

1. Were the chains under consideration too short, i.e. $N \ll N_e$, to demonstrate reptation (53, 55, 56)?
2. How can we preclude reptation effects related unequivocally to inherent features of the local model dynamics (66a–c)?

In the first simulation (53), $n = 10$ freely jointed bead-rod model chains, consisting of $N = 16$ beads each and interacting by a Lennard-Jones potential with effective hard core $h = 0.4l$, were put into a box of size $L = 4l$. Periodic boundary conditions were applied to simulate a macroscopic system. Dynamics were simulated by local kink-jumps. The effect of entanglements were introduced by the prohibition of chain intersections during the attempted rotations, as discussed above for the reptation model. The results were in agreement with the Rouse model (Eq. 8a,b). No intermediate reptation regime for $g(t)$ (cf Figure 1) was observed, in contrast to the corresponding simulation of a single chain moving through a "frozen" melt (53), discussed above. Also, the incoherent and coherent dynamic structure factors of one of the chains in the melt

$$S_{\text{inc}}(\mathbf{q}, t) = \left\langle \sum_{i=1}^{N} \exp\{i\mathbf{q} \cdot [\mathbf{r}_i(0) - \mathbf{r}_i(t)]\} \right\rangle \bigg/ N \qquad 17.$$

$$S_{\text{coh}}(\mathbf{q}, t) = \left\langle \sum_{i=1}^{N} \sum_{j=1}^{N} \exp\{i\mathbf{q} \cdot [\mathbf{r}_i(0) - \mathbf{r}_j(t)]\} \right\rangle \bigg/ N \qquad 18.$$

were in quantitative accord with the Rouse model (67)

$$S_{\text{coh}}(\mathbf{q}, t)/S_{\text{coh}}(\mathbf{q}, 0) = f[\mathbf{q}^2 l^2 (Wt)^{1/2}/6] \qquad 19.$$

where W is the transition rate of a bead and $f(u)$ is a function with $f(u) \sim \exp(-u)$ for $u \gg 1$. Experimental findings from inelastic neutron scattering and computer simulations were in nice agreement (68). This is in contrast to a recent conjecture of de Gennes (69) for the dynamic structure factor of a single reptating chain in polymer melts

$$-[S_{\text{coh}}(\mathbf{q},t)/S_{\text{coh}}(\mathbf{q},0)] \propto \begin{cases} \mathbf{q}^2 l^2 u; & u \ll 1 \\ -\ln[1-(\mathbf{q}d/6)^2] - \{[(\mathbf{q}d/6)^2 - 1]u\}^{-1}; & u \gg 1 \end{cases}$$

20.

where u is the Rouse variable, and d is the tube diameter.

These simulations (53) may have failed to detect reptational behavior for the monomer correlation function $g(t)$ for several reasons: (a) The chains were too short, (b) the density of $c = 2.5$ beads per unit volume was too small, (c) the temperature $kT/\varepsilon = 3$ (scaled by the Lennard-Jones parameter ε) was too high, or (d) the reptation concept is inadequate. Indeed, simulations at a higher density ($c = 10$) and at a lower temperature ($kT/\varepsilon = 0.4$) exhibit reptational features [e.g. $g(t) \sim t^{1/4}$], but only in a nonliquid state, where the displacements in that regime are extremely small [$g(t) \sim 10^{-2} l^2$], and hence may be interpreted as small adjustments of the bead positions in a sort of glassy state.

Likewise, in simulations of the bead-spring model (55) consisting of up to 50 beads at concentration $c = 0.5$, there was no evidence for a "reptation" regime in the melt.

Recently, first evidence for reptation from a computer simulation was reported (66a) in which all the molecules were free to move. In that model, the polymers were confined to move on a cubic lattice. The essential condition was that beads on any one polymer chain were not allowed to occupy the same lattice site as a different polymer chain, but any one polymer chain was deliberately allowed to intersect itself. As a local dynamic, the Verdier-Stockmayer (21) kink-jump process was introduced. Since self-avoiding walk conditions were neglected, the concentration itself was a fluctuating quantity. Indeed, the chains exhibited a reptational behavior $g(t) \sim t^{1/4}$ over more than two orders of magnitude of time. However, this result has been criticized (66b) as having been obtained by a dynamic model with inherent features intimately related to the Verdier-Stockmayer model (21) and de Gennes' original reptation model (30) for an isolated chain, where local conformational transitions are unable to pass each other and so are forced to move like "defects" along the chain's contour all the way up to the ends (28, 29). This artificial mobility of each chain, it is argued (66b), is responsible for the reported reptational motion. However, this argument has been rejected recently (66c).

Indeed, later simulations (56), considering a system of seven self-avoiding chains of chain length $N = 200$ on the diamond lattice at concentration $c = 0.34$, exhibit simple Rouse behavior over more than four orders of magnitude of time. From corresponding simulations of the reptation model, i.e. a single chain moving in a "frozen" melt (56) as discussed above, the entanglement length was estimated as $N_e \approx 20$. (Actually this has been done by fitting the Monte Carlo data to Figure 1.) Since $N_e < N$, one would expect to see reptation for the melt, too. However, it is not clear whether the entanglement length is the same for the melt and its frozen-in counterpart.

More recently, extensive Monte Carlo simulations on a system of "pearl-necklace" polymer chains have been performed (70). Ensembles of up to 36 chains, each consisting of up to $N = 162$ hard spheres of diameter $h = 0.9l$ freely jointed together by $N-1$ rigid links of length l, have been simulated with periodic boundary conditions at bead concentration $c = 0.7$. Analogous to previous works (53), the local dynamics were modeled by stochastic kink-jumps. Entanglement effects were taken into account by choosing the diameter of the hard spheres as large as $h = 0.9l$: It can be shown that for $h/l > (3/4)^{1/2}$ chain intersections are impossible; thus the entanglement restriction is related to an excluded volume condition. This is in close analogy to the entanglement mechanism supposed in real polymers.

In Figure 2, the resulting correlation function $g(t)$ is plotted versus time. The dynamics are not inconsistent with reptation: rough estimates of diffusion coefficient D and disengagement time are consistent with $D \sim N^{-2}$ and $\tau_d \sim N^3$, respectively (cf the table inserted in Figure 2). However, the error for the exponents is (at least) 10%, which precludes a

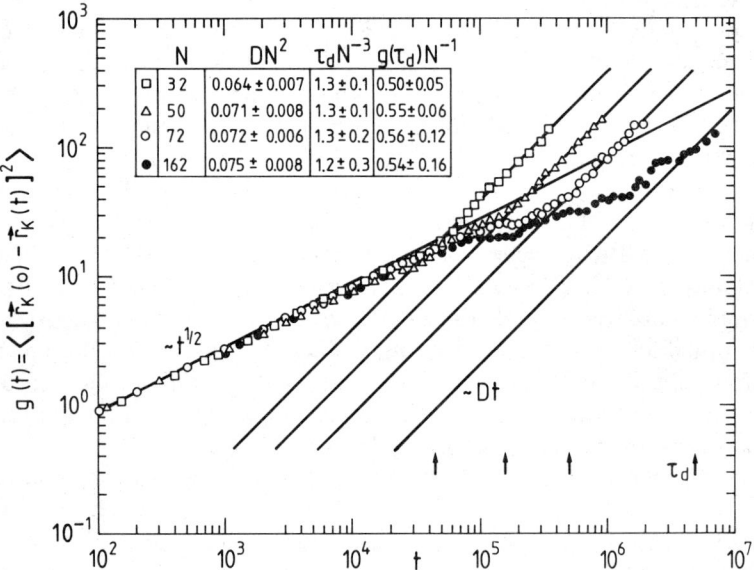

Figure 2 Log-log plot of Monte Carlo data for the monomer displacement $g(t)$ versus time, obtained from simulations of the "pearl-necklace" model (70).

conclusive answer as to whether $\tau_d \sim N^{3.4}$. Likewise, assuming a power law $g(t) \sim t^x$ within the intermediate regime between Rouse behavior $\sim t^{1/2}$ and diffusive behavior $\sim Dt$, the exponent is estimated to $x = 0.27 \pm 0.04$. The results presented in Figure 2 do not support the interpretation of preliminary data given in (71, 72) concerning the intermediate behavior of $g(t)$. Assuming adequacy of the reptation model (30, 50) to polymer melts, one can estimate the entanglement length $N_e = 30 \pm 6$ by comparing Figure 1 and Figure 2.

Recently the self-consistency of the Doi-Edwards theory has been tested by Monte Carlo calculations (73). The dynamics of the "primitive chain," which is the basic concept in the Doi-Edwards theory and which is defined as the center line of the tube, were extracted from the simulations of a single reptating chain on the cubic lattice with fixed obstacles (54). Evans & Edwards (73) found values for steady state viscosity, rigidity modulus, and disengagement time of $\eta \sim N^3 c^3$, $G \sim N^0 c^2$, and $\tau_d \sim N^3 c$, respectively. These are to be compared with experimental results (61): $\eta \sim N^{3.4} c^{5.1}$, $G \sim N^0 c^{2.2}$, and $\tau_d \sim N^{3.4} c^{1.5}$.

Recently Doi (65) conjectured that the discrepancy between the predicted and measured viscosity exponents can be reconciled by considering fluctuations, sometimes called breathing motions, in the length of the primitive chain of the polymer. This was tested recently by Needs (74), using a special Metropolis algorithm that allowed fluctuations in the number of primitive chain segments N about a mean $\bar{N}$ obeying a distribution in which the probability P of a primitive path of N segments was given by $P(N) \sim \exp(-3(N-\bar{N})^2/2\bar{N})$. Although the viscosity is lowered by the inclusion of these fluctuations, the Monte Carlo data do not provide an explanation of the experimental $\eta \sim N^{3.4}$.

CONCLUDING REMARKS

Of course, this will not be the last review on simulation of polymer motion. Several problems, more complicated than the reptation problem for linear polymers for example, have still to be considered by means of Monte Carlo simulations. Two important areas are (a) whether the kinetic effects involved in polymer crystallization from the melt are influenced by reptation-like motions (75, 76); (b) the dynamics of stiff polymers (rigid long rods) at high densities in their isotropic as well as in their anisotropic phase (77, 78).

ACKNOWLEDGMENTS

I would like to thank D. Y. Yoon for his hospitality at IBM San Jose Research Laboratory, where parts of this review were written, and IBM

World Trade of Germany for granting a postdoctoral fellowship in 1982–1983.

Literature Cited

1. Stockmayer, W. 1976. In *Fluides Moléculaires*, ed. R. Balian, G. Weill, pp. 107–49. New York: Gordon & Breach
2. De Gennes, P. B. 1979. *Scaling Concepts in Polymer Physics*. Ithaca: Cornell Univ. Press
3. Graessley, W. W. 1982. *Adv. Poly. Sci.* 47:68–117
4. Baumgärtner, A. 1984. In *Monte Carlo Methods in Statistical Physics 2*, ed. K. Binder. Berlin/Heidelberg/New York: Springer
5. Chandraskhar, S. 1943. *Rev. Mod. Phys.* 5:1–89
6. Rouse, P. E. 1953. *J. Chem. Phys.* 21:1272–80
7. Zimm, B. H. 1956. *J. Chem. Phys.* 24:269–78
8. Zwanzig, R. 1974. *J. Chem. Phys.* 60:2717–20
9. Bixon, M. 1976. *Ann. Rev. Phys. Chem.* 27:65–84
10. Edwards, S. F. 1976. See Ref. 1, pp. 151–208
11. Ceperley, D., Kalos, M. H., Lebowitz, J. L. 1978. *Phys. Rev. Lett.* 41:313–16
12. Ceperley, D., Kalos, M. H., Lebowitz, J. L. 1981. *Macromolecules* 14:1472–79
13. Fixman, M. 1978. *J. Chem. Phys.* 69:1527–37, 1538–45
14. Zimm, B. H. 1980. *Macromolecules* 13:592–602
15. McCammon, J. A., Karplus, M. 1980. *Ann. Rev. Phys. Chem.* 31:29–45
16. Helfand, E. 1979. *J. Chem. Phys.* 71:5000–7
17. Helfand, E., Wasserman, Z. R., Weber, T. A. 1980. *Macromolecules* 13:526–33
18. Gotlib, Yu. Ya., Balabaev, N. K., Darinskii, A. A., Neelov, I. M. 1980. *Macromolecules* 13:602–8
19. Helfand, E. 1971. *J. Chem. Phys.* 54:4651–61
20. Helfand, E., Skolnick, J. 1982. *J. Chem. Phys.* 77:5714–24
21. Verdier, P. H., Stockmayer, W. H. 1962. *J. Chem. Phys.* 36:227–35
22. Verdier, P. H. 1966. *J. Chem. Phys.* 54:2118–28
23. Kranbuehl, D. E., Verdier, P. H. 1972. *J. Chem. Phys.* 56:3145–49
24. Verdier, P. H. 1973. *J. Chem. Phys.* 59:6119–27
25. Iwata, K., Kurata, M. 1969. *J. Chem. Phys.* 50:4008–13
26. Orwoll, R. A., Stockmayer, W. H. 1969. *Adv. Chem. Phys.* 15:305–24
27. De Gennes, P. G. 1976. *Macromolecules* 9:587–98
28. Hilhorst, H. J., Deutch, J. M. 1975. *J. Chem. Phys.* 63:5153–61
29. Boots, H., Deutch, J. M. 1977. *J. Chem. Phys.* 67:4608–10
30. De Gennes, P. G. 1971. *J. Chem. Phys.* 55:572–79
31. Lax, M., Brender, C. 1977. *J. Chem. Phys.* 67:1785–87
32. Geny, F., Monnerie, L. 1979. *J. Poly. Sci. Phys. Ed.* 17:131–63
33. Kremer, K., Baumgärtner, A., Binder, K. 1982. *J. Phys. A* 15:2879–97
34. Baumgärtner, A., Binder, K. 1979. *J. Chem. Phys.* 71:2541–45
35. Baumgärtner, A. 1980. *J. Chem. Phys.* 72:871–79
36. Bruns, W., Bansal, R. 1981. *J. Chem. Phys.* 74:2064–71
37. Bruns, W., Bansal, R. 1981. *J. Chem. Phys.* 75:5149–52
38. Bishop, M., Kalos, M. H., Frisch, H. L. 1979. *J. Chem. Phys.* 70:1299–1304
39. Bishop, M., Kalos, M. H., Frisch, H. L. 1983. *J. Chem. Phys.* 79:3500–4
40. Kron, A. K. 1965. *Polymer Sci. USSR* 7:1361–67
41. Kron, A. K., Ptitsyn, O. B. 1967. *Polymer Sci. USSR* 9:847–53
42. Wall, F. T., Mandel, F. 1975. *J. Chem. Phys.* 63:4592–95
43. Baumgärtner, A. 1982. *Polymer* 23:334–35
44. Muthukumar, M. 1983. *J. Stat. Phys.* 30:457–65
45. Banavar, J. R., Muthukumar, M. 1982. *Chem. Phys. Lett.* 93:35–37
46. Brochard, F., De Gennes, P. G. 1977. *Macromolecules* 10:1157–61
47. Baumgärtner, A. 1981. *Polymer* 22:1308–9
48. Klein, J. 1978. *Macromolecules* 11:852–58
49. Klein, J., Briscoe, B. J. 1979. *Proc. R. Soc. London Ser. A* 365:53–73
50. De Gennes, P. G. 1980. *J. Chem. Phys.* 72:4756–63
51. Graessley, W. W. 1974. *Adv. Polym. Sci.* 16:1–179
52. Tanner, J. E. 1971. *Macromolecules* 4:748–50
53. Baumgärtner, A., Binder, K. 1981. *J. Chem. Phys.* 75:2994–3005
54. Evans, K. E., Edwards, S. F. 1981. *J.*

Chem. Soc. Faraday Trans. 2 77:1891–1912
55. Bishop, M., Ceperley, D., Frisch, H. L., Kalos, M. H. 1982. J. Chem. Phys. 76:1557–63
56. Kremer, K. 1983. Macromolecules 16:1632–38
57. Evans, K. E. 1981. J. Chem. Soc. Faraday Trans. 2 77:2385–99
58. Needs, R., Edwards, S. F. 1983. Macromolecules 16:1492–95
59. De Gennes, P. G. 1975. J. Phys. 36:1199–1203
60. Klein, J., Fletcher, D. 1983. Nature 304:526–27
61. Ferry, J. D. 1980. Viscoelastic Properties of Polymers. New York: Wiley
62. Edwards, S. F. 1967. Proc. Phys. Soc. 92:9–16
63. Doi, M., Edwards, S. F. 1978. J. Chem. Soc. Faraday Trans 2 74:1789–1832
64. Graessley, W. W. 1980. J. Polym. Sci. Polym. Phys. 18:27–34
65. Doi, M. 1981. J. Polym. Sci. Lett. 19:265–73
66a. Deutsch, J. M. 1982. Phys. Rev. Lett. 49:926–29
66b. Kremer, K. 1983. Phys. Rev. Lett. 51:1923
66c. Deutsch, J. M. 1983. Phys. Rev. Lett. 51:1924
67. De Gennes, P. G. 1967. Physics 3:37–45
68. Richter, D., Baumgärtner, A., Binder, K., Ewen, B., Hayter, J. 1981. Phys. Rev. Lett. 47:109–12; 48:1695
69. De Gennes, P. G. 1981. J. Phys. 42:735–40
70. Baumgärtner, A. To be published
71. Baumgärtner, A. 1982. Proc. Workshop on Dynamics of Macromolecules, Santa Barbara
72. Baumgärtner A., Kremer, K., Binder, K. 1983. Faraday Discuss. Chem. Soc. 18:In press
73. Evans, K. E., Edwards, S. F. 1981. J. Chem. Soc. Faraday Trans. 2 77:1929–38
74. Needs, R. J. 1984. Macromolecules 17:437–41
75. Klein, J., Ball, R. C. 1979. Faraday Discuss. 68:198–209
76. Hoffman, J. D. 1982. Polymer 23:656–70
77. Doi, M. 1983. Faraday Discuss. Chem. Soc. 18: In press
78. Frenkel, D., Maguire, J. F. 1983. Mol. Phys. 49:503–41

ELECTRON TRANSFER REACTIONS IN CONDENSED PHASES

Marshall D. Newton and Norman Sutin

Chemistry Department, Brookhaven National Laboratory, Upton, New York 11973

INTRODUCTION

In the last decade the study of electron transfer reactions has been very actively pursued both theoretically and experimentally. This interest is not surprising considering the ubiquity of electron transfer processes in chemistry, physics, and biology. The theoretical developments include semiclassical extensions of the classical formalisms as well as quantum-mechanical treatments deriving from applications of Fermi's golden rule. Despite the multitude of formalisms, there is general agreement that the crux of the electron transfer problem is the change in equilibrium nuclear configuration that occurs when a molecule or ion gains or loses an electron. In recent developments attention has been focused on the dynamics of these nuclear configuration changes, and, in addition, on the electronic factors determining the electron transfer rate. In parallel with these theoretical developments there have been a large number of experimental studies. These studies have not only elucidated the factors determining electron transfer rates, but have in many cases directly tested the theories and suggested modifications to the theories where appropriate.

It is impossible in an article such as this to cover the entire area of electron transfer reactions. Instead we shall concentrate on those aspects of the problem in which we are particularly interested. Steady-state schemes for the diffusion, activation, and electron transfer steps in bimolecular reactions are discussed first. This is followed by a description in terms of Born-Oppenheimer states and surfaces and a discussion of the classical, semiclassical, and quantum mechanical formalisms. Recent experimental studies bearing on the questions raised are presented in the final section. Throughout we restrict the discussion to localized or trapped systems, specifically to systems in which the electronic interaction of the initial and final states is sufficiently small so that the electron transfer can be described in terms of the electronic properties of the unperturbed reactants and products. More detailed treatments of some of these topics can be found in (1–6), while (7–10) contain recent applications of the (classical) electron transfer formalism to a variety of processes, including electrochemical systems (7), proton and atom transfer reactions (8, 9), and gas phase S_N2 reactions (10). A recent issue of *Progress in Inorganic Chemistry: An Appreciation of Henry Taube* (11), and a *Faraday Society Discussion on Electron and Proton Transfer* (12) contain a great deal of relevant material.

While considerable recent activity has focused on the dynamical and electronic aspects of the problem, by and large the recent findings substantiate the classical formalism (13–15) for reactions at room temperature for which the free energy change (driving force) is in the small-to-moderate range and the redox centers are in close contact. The classical theory is also valid for reactions with large driving forces provided the relevant frequencies of the motion (vibrational and rotational) are not high and the temperature is not too low. Nuclear tunneling effects are important when these conditions are not satisfied, and tunneling corrections to the classical model are considered in some detail in this article.

STEADY-STATE DESCRIPTION

The probability of electron transfer increases with increasing electronic interaction between the redox centers. Consequently, electron transfer is favored by small separations of the two reactants. In addition the equilibrium nuclear configurations of the two reactants will generally be different before and after electron transfer: reorganization of the reactants prior to the electron transfer is necessary so that the transfer can occur with minimal change in nuclear configuration and momentum. These dual requirements of close approach of the two reactants and nuclear reorganization are embodied in the following scheme for bimolecular electron transfer (13–17).

Scheme 1:

$$A + B \underset{k_{-12}}{\overset{k_{12}}{\rightleftharpoons}} A|B \underset{k_{-n}}{\overset{k_n}{\rightleftharpoons}} [A|B]^* \underset{v_{\text{eff}}}{\overset{v_{\text{eff}}}{\rightleftharpoons}} [A^+|B^-]^* \overset{k'_{-n}}{\longrightarrow}.$$

The first step is the diffusion together of the separated reactants A and B to form the precursor complex A|B. The second step is the reorganization of the precursor complex to a configuration appropriate to electron transfer.

Electron transfer within the reorganized precursor complex [A|B]* yields the reorganized successor complex [A$^+$|B$^-$]*; the frequency of interconversion of [A|B]* and [A$^+$|B$^-$]* is the effective electron-hopping frequency v_{eff}.[1] Vibrational relaxation of the reorganized successor complex and its dissociation yields the separated products A$^+$ and B$^-$.

The rate expressions, Eqs. 1–3, can be derived for the above scheme by making a steady-state approximation for the concentrations of the various intermediates (13–17)

$$\frac{1}{k_{\text{obsd}}} = \frac{1}{k_{\text{diff}}} + \frac{1}{k_{\text{act}}} \qquad 1.$$

where

$$k_{\text{diff}} = k_{12} \qquad 2.$$

$$k_{\text{act}} = \frac{K_A k_n k'_{-n} v_{\text{eff}}}{(k_{-n} + k'_{-n}) v_{\text{eff}} + k_{-n} k'_{-n}}. \qquad 3a.$$

In these expressions k_{diff} is the diffusion-controlled rate constant for the formation of the precursor complex, $K_A = k_{12}/k_{-12}$ is the stability constant of the precursor complex, and k_{act} is the activation-controlled rate constant for the electron transfer reaction (that is, the rate constant that would be observed if the precursor complex remained in equilibrium with the separated reactants). Equation 3a simplifies to

$$k_{\text{act}} = \frac{K_A k_n v_{\text{eff}}}{2 v_{\text{eff}} + k_{-n}} \qquad 3b.$$

when $k'_{-n} = k_{-n}$. Similar expressions are given in (16, 17, 25). Two limiting forms of Eq. 3b are of interest:

1. When the effective electron-hopping frequency is relatively high, the second term in the denominator of Eq. 3b may be neglected and the activation-controlled rate constant is given by

$$k_{\text{act}} = K_A K_n k_{-n}/2 \qquad 4a.$$

$$= (K_A k_{-n}/2) \exp(-\Delta G_n^*/RT) \qquad 4b.$$

where $K_n = k_n/k_{-n} = \exp(-\Delta G_n^*/RT)$ and ΔG_n^* is the reorganization free energy. In this case the rate constant depends upon the dynamics of the vibrational relaxation and the magnitude of the nuclear reorganization barrier but not upon the electron-hopping frequency.

---

[1] As defined here, this frequency does not in general correspond to an electron tunneling frequency. Cases where it does correspond to an electron tunneling frequency can be found in (2).

2. When the effective electron-hopping frequency is relatively low, the activation-controlled rate constant is given by

$$k_{act} = K_A K_n v_{eff} \qquad \qquad 5a.$$

$$= K_A v_{eff} \exp(-\Delta G_n^*/RT) \qquad \qquad 5b.$$

and now depends upon the effective electron-hopping frequency rather than upon the dynamics of the vibrational relaxation.

Inner- and Outer-Shell Reorganization

The nuclear reorganization that occurs prior to the electron transfer involves all the coordinates that change as a result of the electron transfer. In the case of electron transfer between metal complexes it is convenient to distinguish two classes of configuration change: changes in bond lengths and angles in the inner coordination shells of the reactants, and changes in the orientation of the solvent molecules surrounding the reactants (13–15, 18–20). These are generally called inner-shell and outer-shell configuration changes, respectively; through an elaboration of the treatment presented above (17a), these configuration changes can be incorporated into Scheme 1 as follows:

Scheme 2:

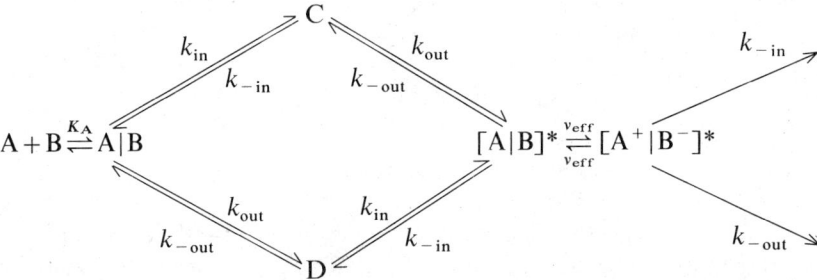

The steady-state approximation for the concentration of the various intermediates again gives Eq. 1, but with the activation-controlled rate constant now given by Eq. 6

$$k_{act} = \frac{K_A k_{out} k_{in} v_{eff}}{v_{eff} Q + k_{-out} k_{-in}} \qquad \qquad 6.$$

with

$$Q = \frac{2k_{-out}k_{-in}S + k_{out}k_{-out}(k_{-out}+k_{in}) + k_{in}k_{-in}(k_{out}+k_{-in})}{(k_{-out}+k_{-in})S}$$

$$S = k_{out} + k_{-out} + k_{in} + k_{-in} = 1/\tau_{out} + 1/\tau_{in}$$

where

$$1/\tau_{out} = (k_{out} + k_{-out}), \quad 1/\tau_{in} = (k_{in} + k_{-in}),$$

and, as before, $K_A = k_{12}/k_{-12}$. Certain limiting forms of Eq. 6 are of interest.

1. When the effective electron-hopping frequency is relatively high and k_{-in} and k_{-out} are each larger than both k_{in} and k_{out}, then

$$k_{act} = K_A K_{out} K_{in}(k_{-out} + k_{-in})/2 \qquad 7a.$$

$$= \frac{K_A(k_{-out} + k_{-in})}{2} \exp\left[-\frac{(\Delta G_{out}^* + \Delta G_{in}^*)}{2}\right] \qquad 7b.$$

where $K_{in} = k_{in}/k_{-in} = \exp(-\Delta G_{in}^*/RT)$ and

$$K_{out} = k_{out}/k_{-out} = \exp(-\Delta G_{out}^*/RT).$$

Two interesting cases can be distinguished depending on the relative magnitudes of k_{-in} and k_{-out}. If $k_{-out} \gg k_{-in}$ (i.e. if $\tau_{out} \ll \tau_{in}$), then the upper pathway in Scheme 2 predominates and some inner-shell reorganization precedes the solvent reorganization. On the other hand, if $k_{-out} \ll k_{-in}$ (i.e. if $\tau_{out} \gg \tau_{in}$), then the lower pathway in Scheme 2 predominates and some solvent reorganization occurs first. Note that in each case the second (rate-determining) step is associated with the process that has the shorter relaxation time (17a).

2. When the effective electron-hopping frequency is relatively low, the activation-controlled rate constant is given by Eq. 8.

$$k_{act} = K_A K_{out} K_{in} v_{eff} \qquad 8a.$$

$$= K_A v_{eff} \exp[-(\Delta G_{out}^* + \Delta G_{in}^*)/RT]. \qquad 8b.$$

Evidently, various limiting rate-constant expressions can be distinguished, depending on the relative rates of diffusion, outer-shell reorganization, inner-shell reorganization, and electron hopping. The rates of these processes are discussed below; however, it is convenient at this stage to anticipate one of the results, namely, that $k_{-in} > k_{-out} \gg k_{in}, k_{out}$ when $k_{in}/k_{-in} < 5 \times 10^{-2}$ (or $\Delta G_{in}^* \gtrsim 2$ kcal mol^{-1}). Under these conditions the major pathway for the reaction is provided by the lower sequence in Scheme 2, with the solvent reorganization present as a preequilibrium (17a). This pathway is shown in Scheme 3.

Scheme 3:

$$A + B \underset{}{\overset{K_A}{\rightleftharpoons}} A|B \underset{}{\overset{K_{out}}{\rightleftharpoons}} D \underset{k_{-in}}{\overset{k_{in}}{\rightleftharpoons}} [A|B]^* \underset{v_{eff}}{\overset{v_{eff}}{\rightleftharpoons}} [A^+|B^-]^* \xrightarrow{k_{-in}}$$

The steady-state approximation for this scheme yields

$$k_{act} = \frac{K_A K_{out} k_{in} v_{eff}}{2 v_{eff} + k_{-in}}. \qquad 9.$$

This is an important and convenient result since it means that provided the simplifying assumptions are justified, the dynamics of the reaction can be discussed in terms of Scheme 3 and Eq. 9 rather than the more cumbersome Scheme 2 and Eq. 6.

Formation of the Precursor Complex

In any particular system, bimolecular electron transfer will occur over a range of separation distances. If any angular dependence is neglected, or removed by appropriate averaging,[2] then k_{act} is obtained by integration over the equilibrium distribution of separation distances, each weighted by a characteristic first-order electron-transfer rate constant (14, 17a, 17b, 21–26).

$$k_{act} = \int_0^\infty \frac{4\pi N r^2}{1000} \exp\left[-\frac{w(r)}{RT}\right] k_{el}(r) \, dr. \qquad 10.$$

In the above equation $w(r)$ is the work required to bring the reactants reversibly to the separation distance r, and $k_{el}(r)$ is the first-order rate constant for electron transfer at this separation distance. If the integrand peaks over a small range of r values, then Eq. 10 can be factored to give

$$k(r_m) = K_A(r_m) k_{el}(r_m) \qquad 11.$$

with

$$K_A = \frac{4\pi N r_m^2 \delta r}{1000} \exp\left[-\frac{w(r_m)}{RT}\right] \qquad 12.$$

where r_m is the value of the separation distance corresponding to the maximum value of the integrand and δr is the range of distances over which the rate is appreciable; a value of 0.8 Å for δr has been proposed (24, 25), and similar values may be inferred from detailed calculations for the $Fe(H_2O)_6^{3+}-Fe(H_2O)_6^{2+}$ exchange (17a, 23). In terms of Scheme 3, K_A is, as before, also equal to k_{12}/k_{-12}, and k_{el} is given by

$$k_{el} = \frac{K_{out} k_{in} v_{eff}}{2 v_{eff} + k_{-in}}. \qquad 13.$$

This expression for the rate constant for electron transfer between the

---

[2] Angular dependence can be allowed for by introducing a steric factor into the rate constant expression (17a).

precursor and successor complexes is, of course, also applicable to (weak-interaction) intramolecular electron-transfer processes.

Time-Dependent Rate Constants

Implicit in the derivation of the expressions for the rate constants for Schemes 1–3 is the assumption that the steady state is established relatively rapidly and that the kinetic measurements are made after the steady state has been established. As a consequence the derived rate constants are independent of time, since they refer to the system under steady-state conditions. By contrast, the observed rate constants will be time dependent if kinetic data taken prior to the establishment of the steady state are included. Thus when reactions that are fast compared with diffusion are initiated by a "sudden" perturbation (for example, by flash photolysis or pulse radiolysis), the initial rate constants will be larger [$k_{obsd} \to k_{act}$ as $t \to 0$] than the later steady-state values [$k_{obsd} \to k_{diff}$ when $t > 0.1$ ns for $k_{act}/k_{diff} \sim 10^2$ and the sum of the diffusion coefficients $\sim 10^{-5}$ cm^2 s^{-1}, based on equations in (27)]; the time dependence of the rate constants arises from the fact that the distribution of distances separating the reactants is changing with time (the closest reactants react first). Indeed, when $k_{diff} \ll k_{act}$, as can be the case, for example, for reactions in viscous media or in glasses at low temperature, the steady-state distribution of distances may not be established and the observed rate constant may remain time dependent throughout the course of the reaction, which may be quite long. Examples of this behavior are given later when the distance dependence of the electronic factor is considered.

STATES AND ENERGIES

The formalisms describing electron transfer processes can be analyzed in terms of a suitable Hamiltonian operator and its eigenvalues and matrix elements, as defined by an appropriate basis of vibronic wavefunctions (1, 2). The Hamiltonian must in some sense reflect the influence of all the electrons and nuclei of the reactive system (i.e. the supermolecule comprising the interacting redox centers together with tightly held ligands, as well as surrounding solvent or other medium).

It is convenient to decompose the full Schrödinger vibronic Hamiltonian $\mathcal{H}(r_{el}, q_n)$, into a Born-Oppenheimer (i.e. rigid nuclei) electronic Hamiltonian, $H(r_{el}, q_n)$, and the nuclear kinetic energy operator, $T(q_n)$

$$\mathcal{H}(r_{el}, q_n) = H(r_{el}, q_n) + T(q_n) \qquad 14.$$

where r_{el} and q_n are the electronic and nuclear coordinates, respectively. In the spirit of the previous discussion, the relevant q's are those coordinates for which the equilibrium values or associated frequencies differ in the precursor and successor complexes. In general, q also includes the separation of the redox centers (r, in Eq. 10); this separation is treated as a classical parameter in the present context. In other words, the $T(q_n)$ terms do not include contributions from the relative motion of the reactants along the separation coordinate.

The vibronic wavefunctions $\Psi(r_{el}, q_n)$ are of the form

$$\Psi(r_{el}, q_n) = \sum_i \phi_i(r_{el}, q_n) \chi_i(q_n) \qquad 15.$$

where $\{\phi_i(r_{el}, q_n)\}$ are the electronic functions and $\{\chi_i(q)\}$ are vibrational functions. Equations 14 and 15, in conjunction with the vibronic Schrödinger equation

$$\mathcal{H}(r_{el}, q_n)\Psi(r_{el}, q_n) = E\Psi(r_{el}, q_n) \qquad 16.$$

lead (28) to the following set of equations for the nuclear functions $\chi_i(q_n)$

$$[T + T''_{ii} + H_{ii} - E]\chi_i(q_n) = -\sum_{j \neq i} (H_{ij} + T'_{ij} + T''_{ij})\chi_j(q_n). \qquad 17.$$

The expectation value of $H(r_{el}, q_n)$,

$$H_{ii}(q_n) = \int \phi_i(r_{el}, q_n) H(r_{el}, q_n) \phi_i(r_{el}, q_n) \, d\tau_{el} \qquad 18.$$

defines the Born-Oppenheimer potential energy surface (PES) associated with state ϕ_i. Similarly, the diagonal matrix element of $T(q_n)$,

$$T''_{ii}(q_n) = \int \phi_i(r_{el}, q_n) T(q_n) \phi_i(r_{el}, q_n) \, d\tau_{el} \qquad 19.$$

can be considered as an additional contribution to the potential energy.

The coupling of the simple vibronic products in Eq. 15 arises from the combined effect of the three matrix elements on the r.h.s. of Eq. 17

$$H_{ij}(q_n) = \int \phi_i^*(r_{el}, q_n) H(r_{el}, q_n) \phi_j(r_{el}, q_n) \, d\tau_{el} \qquad 20.$$

$$T'_{ij}(q_n) = -\sum_q (\hbar^2/M_q) \left[\int \phi_i^*(r_{el}, q_n) \nabla_q \phi_j(r_{el}, q_n) \, d\tau_{el} \right] \cdot \nabla_q \qquad 21.$$

where M_q is the effective mass associated with the qth coordinate, and

$$T''_{ij}(q_n) = \int \phi_i^*(r_{el}, q_n) T(q_n) \phi_j(r_{el}, q_n) \, d\tau_{el}. \qquad 22.$$

Note that $T(q_n)$ operates only on ϕ in Eqs. 19 and 22.

The above results assume that $\{\phi_i\}$ is an orthonormal set; this restriction is relaxed in subsequent sections as the need arises. The T'_{ij} and T''_{ij} terms are frequently called the nuclear coupling terms or the nuclear coupling matrix elements (also the Born-Oppenheimer breakdown or the nonadiabaticity operators). The H_{ij} term is generally called the electronic coupling term or the electronic coupling matrix element. The sum of H_{ij}, T'_{ij}, and T''_{ij} is denoted by $\mathcal{H}_{ij}$; in order to distinguish it from the H_{ij} term, we refer to $\mathcal{H}_{ij}$ as the transition operator or simply as the transition matrix element.

Choice of Electronic Basis

The relative magnitudes of the three terms contributing to $\mathcal{H}_{ij}$ depend on the particular choice adopted for the electronic basis, $\{\phi_i(r_{el}, q_n)\}$ (28). In studying electron transfer reactions, one frequently employs a nonstationary [i.e. one that does not diagonalize $H(r_{el}, q_n)$], two-state basis consisting of wavefunctions $\phi_p(r_{el}, q_n)$ and $\phi_s(r_{el}, q_n)$, which can be directly identified in a valence-bond-structure sense with the precursor and successor complexes, respectively. These functions, which are not, of course, uniquely defined, may be thought of as eigenfunctions of suitable zeroth-order electronic Hamiltonians, sometimes referred to as channel Hamiltonians in which, for example, the interaction between the transferring electron and the accepting center has been removed (1). Once $\phi_p(r_{el}, q_n)$ and $\phi_s(r_{el}, q_n)$ have been specified, the corresponding PES, $H_{pp}(q_n)$ and $H_{ss}(q_n)$, are obtained as expectation values of the full Born-Oppenheimer electronic Hamiltonian, $H(r_{el}, q_n)$, according to Eq. 18. If necessary, the ϕ_p, ϕ_s basis may be supplemented with appropriate excited state functions ϕ_p^* and ϕ_s^*, associated with the precursor and successor complexes, respectively, or with other intermediate states, ϕ_I (e.g. charge-transfer intermediates involving ligands or solvent intervening between the redox centers) (1, 21).

While the $\{\phi_i\}$ in general exhibit some degree of q-dependence, this dependence is expected to be small for ϕ_p and ϕ_s since by construction ϕ_p and ϕ_s are constrained to correspond to precursor and successor complexes, respectively, at all q. For such a "diabatic" basis we may confidently neglect the "nonadiabatic" terms, T'_{ps} and T''_{ps} (Eqs. 21 and 22) relative to the nonzero H_{ps} terms in most situations (see below), obtaining the following "Born-Oppenheimer approximation" for Eq. 17,

$$[T + H_{pp} - E]\chi_{pv}(q_n) = H_{ps}\chi_{sw}(q_n). \qquad 23.$$

For completeness, we also specialize Eq. 17 for the case of a stationary basis, i.e. the "adiabatic" basis, which (by definition) diagonalizes $H(r_{el}, q_n)$ (28). Such a basis is commonly employed in the analysis of static or spectroscopic properties, and may in some cases provide a useful alternative to ϕ_p and ϕ_s in treating electron transfer processes. In the two-state model, we designate the "adiabatic" counterparts of ϕ_p and ϕ_s as ϕ_1 and ϕ_2, and obtain the following expression for Eq. 17

$$[T + H_{11} - E]\chi_{1v}(q_n) = T'_{12}\chi_{2w}(q_n), \qquad 24.$$

where $H_{11}(q_n)$ is now an eigenvalue of $H(r_{el}, q_n)$ (i.e. $H_{12} = 0$) and T''_{12} has been neglected relative to T'_{12}.

Figure 1 displays schematically the diabatic potential energy profiles, H_{pp} and H_{ss}, and the adiabatic curves, H_{11} and H_{22}. Near the crossing or avoided crossing (denoted by q^*), ϕ_p and ϕ_s maintain their chemical identity (reactants and products, respectively), whereas the adiabatic states change from mostly ϕ_p to mostly ϕ_s, or vice versa. The splitting of the two adiabatic surfaces at q^* is twice the electronic matrix element, H_{ps}, in the ϕ_p, ϕ_s basis. Figure 1 provides examples of both the normal and abnormal (or inverted) energy regions, corresponding to situations in which q^* is situated, respectively, between, and to the side of, the equilibrium configuration of the precursor and successor complexes (q_p^0 and q_s^0) (13–15).[3] Quantitatively, in terms of the energies defined in Figure 1 we have the normal region when $-\Delta E_0 < E_\lambda$ and the abnormal region when $-\Delta E_0 > E_\lambda$ where ΔE_0 is defined to be negative for an exergonic reaction.

Thermal Reactions

In either the normal or abnormal regions, a thermal electron transfer reaction involves a nonradiative transition from the vicinity of q_p^0 to the vicinity of q_s^0. Such a reaction can be analyzed in terms of either the diabatic (ϕ_p, ϕ_s) or adiabatic (ϕ_1, ϕ_2) basis, since the two representations and their associated PES essentially coincide near q_p^0 and q_s^0, as shown in Figure 1. Thus in the *diabatic* representation the thermal reaction schematically corresponds to the transition

$$\phi_p(q \sim q_p^0) \rightarrow \phi_s(q \sim q_s^0). \qquad 25a.$$

In the *adiabatic* representation, by contrast, it is necessary to distinguish between the normal

$$\phi_1(q \sim q_p^0) \rightarrow \phi_1(q \sim q_s^0) \qquad 25b.$$

and abnormal cases

$$\phi_1(q \sim q_p^0) \rightarrow \phi_2(q \sim q_s^0). \qquad 25c.$$

In the normal region the adiabatic basis (ϕ_1, ϕ_2) is perhaps the most convenient for visualizing the process since, as revealed in Eq. 25b, a

---

[3] The rate increases with increasing driving force in the normal region but is predicted to decrease with increasing driving force in the abnormal region.

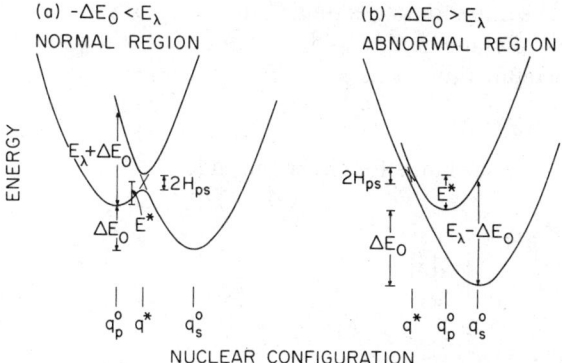

Figure 1 Plot of potential energy as a function of nuclear configuration: *continuous solid curves* are the adiabatic states of the system, the splitting at the intersection is equal to $2H_{ps}$ where H_{ps} is the electronic coupling matrix element, the difference in the energies of the final and initial states is ΔE_0 (negative for an exergonic reaction), and the difference in the energies of the intersection region and initial state is E^*. Case 1(a): normal region, $-\Delta E_0 < E_\lambda$; case 1(b): abnormal region, $-\Delta E_0 > E_\lambda$.

successful reaction may be accounted for in terms of a single adiabatic PES (i.e. H_{11}). However, for the purposes of implementing kinetic formalisms, it is simpler in general to employ ϕ_p and ϕ_s; in this representation the transition operator is dominated by the electronic matrix element (H_{ps}) in both the normal and abnormal regions. The diabatic representation is emphasized in the remainder of this review.

Adiabatic vs Nonadiabatic Behavior

Electron transfer reactions are generally categorized as being "adiabatic" or "nonadiabatic" (1, 2). In an "adiabatic" process, the "nonadiabatic" coupling terms (T'_{ij} and T''_{ij}) can be safely neglected, irrespective of basis ($\phi_{p,s}$, or $\phi_{1,2}$). If the stationary basis, ϕ_1, ϕ_2 is adopted, then the entire transition matrix element, $\mathscr{H}_{12}$, is zero, and we need only consider a single PES, H_{11}. On the other hand, if no representation exists that permits the neglect of the transition matrix elements (i.e. the r.h.s. of Eq. 23), then the reaction is intrinsically a two-state or multi-state process and is said to be "non-adiabatic." The term "nonadiabatic electron transfer" tends to be confusing: e.g. in the case of the normal region (Figure 1a) the nonadiabatic events are actually nonreactive (i.e. nonadiabatic hopping from H_{11} to H_{22}, if viewed in an adiabatic basis; otherwise, simply remaining on H_{pp}), and only occasionally does the system "react" as defined above. Conversely, in the abnormal region (Figure 1b) only the nonadiabatic events are capable of yielding vibrationally relaxed products, as in Eqs. 25a and 25c.

Before proceeding with the analysis, we return to Eq. 23 and examine in detail the justification for neglecting T'_{ps} and T''_{ps}. For simplicity we consider only a single stretching mode q and we characterize the q dependence of the electronic wavefunction by

$$|\partial\phi/\partial q| \lesssim |\phi|/\alpha \qquad 26.$$

The following upper limits for the magnitude of $T'\chi$ and $T''\chi$ can then be derived

$$|T'\chi| \lesssim 2\left[\frac{h^2 E'}{2M\alpha^2}\right]^{1/2}|\chi| \qquad 27a.$$

$$|T''\chi| \lesssim \frac{h^2}{2M\alpha^2}|\chi| \qquad 27b.$$

where E' is the local nuclear kinetic energy (28). In evaluating these expressions as well as some that follow, we shall employ as parameters those characteristic of hexaaquo (or hexaammine) electron exchange reactions at room temperature. Thus, with $M = 18$ a.m.u. (the mass of the water ligand), and with a mean value of ~ 200 cm^{-1} for E', Eq. 27 yields $|T'\chi| \lesssim (25/\alpha)|\chi|$ and $|T''\chi| \lesssim (1/\alpha^2)|\chi|$, where the energy coefficients of $|\chi|$ are in cm^{-1} and the distance scale, α, is in Å. Thus if a conservative estimate of α is chosen (> 10 Å) for ϕ_p and ϕ_s, we see that the BO approximation (as in Eq. 23) is valid provided that $H_{ps} \gtrsim 5$ cm^{-1}, which is the case for many electron transfer reactions of interest. The above estimate of α is reasonable since in many systems ϕ_p and ϕ_s tend to be concentrated on centers that are not strongly involved in the relevant q coordinates: e.g. in electron transfer reactions of transition metal complexes the important q's typically involve solvent and ligand motion whereas the metal atoms, which are the primary redox centers, tend to remain essentially stationary in the q space of interest. If ligand-to-metal and metal-to-ligand charge transfer is entirely absent in such systems, then $\alpha \to \infty$ and $|T'\chi|$ and $|T''\chi| \to 0$. It should also be noted that the preceding analysis of nuclear motion in terms of Eqs. 26 and 27 in effect uses unity for the electronic overlap element, $S_{ps} = \int \phi_p \phi_s \, d\tau$. Since S_{ps} may be considerably less than unity (22), values of H_{ps} well below 5 cm^{-1} may still satisfy the BO requirements.

Having justified Eq. 23 for use with the two-state ϕ_p, ϕ_s basis, we now assess the adiabatic/nonadiabatic character of electron transfer in terms of the semiclassical Landau-Zener (LZ) model (1, 29); specifically, we require an estimate of the probability of the systems actually undergoing the reactive transitions depicted in Eq. 25: i.e. what fraction of the flux proceeding from the reactant region q_p^0 through the crossing region q^* actually ends up on the product surface near q_s^0. The basic element in the analysis is the quantity,

$$P = 1 - \exp(-2\pi\gamma) \qquad 28a.$$

with

$$\gamma = 2\pi H_{ps}^2/[h\dot{q}|(\partial(H_{ss}-H_{pp})/\partial q|] \qquad 28b.$$

where $\dot{q}$ is the velocity (assumed constant) with which the system passes through the intersection region. The probability for hopping between surfaces each time the crossing region is traversed is given by P in the diabatic representation ($\phi_p \leftrightarrow \phi_s$) and by the complementary expression $(1-P)$ in the adiabatic representation ($\phi_1 \leftrightarrow \phi_2$). In order to obtain the overall reaction probability it is necessary to consider multiple passages through the intersection region: an infinite sum of events (oscillations inside the two inner turning points) for the normal region, and a two-passage process for the abnormal region, analogous to the classic case of Na+Cl scattering (29). The various states (species) relevant to the crossing process are illustrated in Figure 2, and their interconversions can be considered in terms of Schemes 4 and 5. In each case, passages below the barrier (nuclear tunneling) are ignored (extension to include nuclear tunneling is presented below).

Scheme 4:

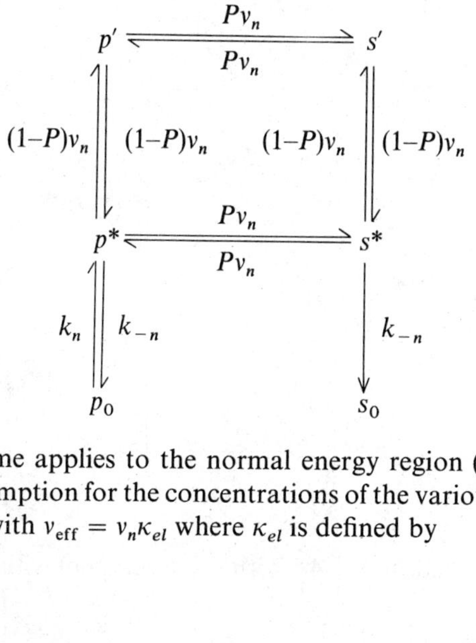

$$\begin{array}{ccc}
p' & \underset{Pv_n}{\overset{Pv_n}{\rightleftarrows}} & s' \\
(1-P)v_n \updownarrow \quad (1-P)v_n & (1-P)v_n & (1-P)v_n \updownarrow \quad (1-P)v_n \\
p^* & \underset{Pv_n}{\overset{Pv_n}{\rightleftarrows}} & s^* \\
k_n \updownarrow k_{-n} & & \downarrow k_{-n} \\
p_0 & & s_0
\end{array}$$

The above scheme applies to the normal energy region (Figure 2a). The steady state assumption for the concentrations of the various intermediates leads to Eq. 4a with $v_{\text{eff}} = v_n \kappa_{el}$ where κ_{el} is defined by

$$\kappa_{el} = \frac{2P}{1+P} \qquad 29a.$$

and v_n is the frequency of passage (nuclear motion) through the intersection region; κ_{el} has been termed the "transmission coefficient" or "electronic factor" and is the probability of electron transfer once the nuclear configuration appropriate to the intersection region has been achieved.

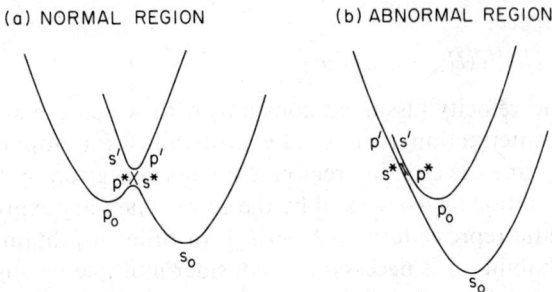

Figure 2 States of the precursor (p) and successor (s) complexes implicated in Schemes 4 (Figure 2a) and 5 (Figure 2b): p_0 and s_0 are the initial (equilibrium) states, p^* and s^* are the reorganized states with configurations appropriate to electron transfer in the intersection region, and p' and s' are the states resulting from the nuclear configuration fluctuation if no electron transfer had occurred in the intersection region. Note that in Schemes 4 and 5 the forward and reverse rate constants for the interconversion of p^*, p', s^*, and s' are equated: this derives from the fact that all four states have the same total energies (and entropies).

Scheme 5 describes the multiple crossings in the intersection region in the abnormal energy region (Figure 2b).

Scheme 5:

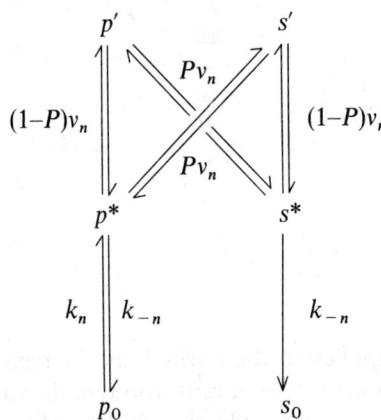

This scheme also leads to Eq. 4a, with $v_{\rm eff} = v_n \kappa_{el}$ but with κ_{el} now given by

$$\kappa_{el} = 2P(1-P). \qquad 29b.$$

In each case $\kappa_{el} \to 2P$ when $P \ll 1$ (i.e. when $2\pi\gamma \ll 1$). Under these conditions the exponential in Eq. 28a may be expanded and $\kappa_{el} = 4\pi\gamma$. Note that Eqs. 29a and 29b can also be derived using simple summation procedures (29). The utility of the present approach, however, is that it incorporates the multiple crossings into the general steady-state framework.

An electronic factor of close to unity ($\kappa_{el} \geqslant 0.6$) is achieved for typical transition metal redox systems when $H_{ps} \geqslant 200$ cm^{-1} (based on a value of 3×10^4 cm s^{-1} for $\dot{q}$ and 2×10^8 eV cm^{-1} for $[\partial(H_{ss}-H_{pp})/\partial q]_{q^*}$). In the case of the abnormal region, the reaction is of necessity nonadiabatic (as implied by Eq. 25b).[4]

As emphasized above, the distinction between an adiabatic and nonadiabatic process is independent of basis. The crucial quantity γ can be expressed in terms of ϕ_1 and ϕ_2 (31), yielding

$$\gamma = \pi(H_{11}-H_{22})_{q^*}/(4h\dot{q}\langle\phi_1|\partial/\partial q|\phi_2\rangle) \qquad 30.$$

where $\langle\phi_1|\partial/\partial q|\phi_2\rangle$ is the electronic factor in T'_{12} (Eq. 21). The standard parameters defined above yield a value of ~ 20 Å^{-1} for $\langle\phi_1|\partial/\partial q|\phi_2\rangle$. This value may be compared with the analogous electronic factor in the p,s basis, which, according to Eq. 26 and our conservative estimate of $1/\alpha$, is <0.1 Å^{-1}. Such a difference in magnitudes is not surprising since, as noted elsewhere (32), T'_{12} is expected to be considerably larger than T'_{ps} in view of the fact that ϕ_1 and ϕ_2 change character rapidly near q^*. This large T'_{12} element is, of course, still compatible with $\kappa_{el} \sim 1$, since this value was derived above for the standard parameters.

KINETIC FORMALISMS

A variety of formalisms—classical, semiclassical, and quantum mechanical—are available for treating the activation and electron hopping events represented in Eq. 5 by the product

$$v_{\text{eff}}K_n = v_{\text{eff}} \exp(-\Delta G_n^*/RT) \qquad 31.$$

where v_{eff} is the effective electron-hopping frequency (equal to $v_n\kappa_{el}$) and ΔG_n^* is the free energy of activation for the reaction. Further justification for employing a free energy instead of an internal energy is given below.

Transition State Theory

In the classical formalism (13–16, 18–20, 33) the electronic interaction energy is assumed to be large enough so that κ_{el} has its maximum value of unity (the factor $1/2$ for the abnormal region is generally neglected) and v_{eff} is simply put equal to v_n. Transition state theory (TST) is used, and the transition state (or activated complex) is identified with the crossing point (q^* in Figures 1 and 2) where the precursor and successor states are degenerate (or equivalently, where the minimum occurs in the separ-

---

[4] Equation 29b carries the interesting implication that there is an optimum magnitude of H_{ps} for electron transfer in the abnormal region (corresponding to $\kappa_{el} = P = 1/2$) and that the electron transfer rate will *decrease* when H_{ps} is increased beyond this optimum (30).

ation between adiabatic surfaces). The barrier height, E^*, is simply $H_{pp}(q^*) - H_{pp}(q_p^0)$. In the remainder of this review the q-dependence of H_{pp} and H_{ss} is assumed to be harmonic: detailed calculations indicate that the effects of anharmonicity are generally quite small, since contributions from the individual reactants tend to cancel (34, 35). The barrier height E^* can be written in terms of the "vertical" reorganization parameters E_λ and the net energy change for the reaction, ΔE_0 (negative for an exothermic reaction), as follows:

$$E^* = (E_\lambda + \Delta E_0)^2 / 4E_\lambda \qquad 32.$$

where

$$E_\lambda = \sum_i E_i = \frac{1}{2} \sum_i \bar{f}_i [(q_i^0)_p - (q_i^0)_s]^2. \qquad 33a.$$

The equilibrium values of the various q-coordinates for the precursor and successor states are denoted by $(q_i^0)_p$ and $(q_i^0)_s$, respectively, and a single reduced force constant $\bar{f}_i$ is employed (14) for each q_i:

$$\bar{f}_i = 2f_i^p f_i^s / (f_i^p + f_i^s). \qquad 33b.$$

This approximation is generally quite satisfactory (14, 36–38). Clearly only those modes associated with nonzero $\{(q_i^0)_p - (q_i^0)_s\}$ values need be considered. A two-mode model is commonly employed (36, 40–42), consisting of a low-frequency solvent or medium mode, and a high-frequency inner-shell mode (specifically, the antisymmetric combination of ligand-metal breathing modes when the reactants are solvated ions).

Recognizing that the location q_i^* of the transition state is given by (14),

$$\frac{[q_i^* - (q_i^0)_p]}{[(q_i^0)_p - (q_i^0)_s]} = \frac{1}{2}\left[1 + \frac{\Delta E_0}{E_\lambda}\right] \qquad 34.$$

we can assign the q_ith contribution to the barrier height E^* as

$$E_i^* = \frac{E_i}{4}\left[1 + \frac{\Delta E_0}{E_\lambda}\right]^2 \qquad 35a.$$

since some compensation occurs between the two species (14); however, this Eq. 33b would not in general be appropriate for the reorganization of only a single species (39). Equation 35a can be rearranged (25) to display the effective fraction of ΔE_0 associated with E_i:

$$E_i^* = \frac{\left[E_i + \left(\dfrac{E_i}{E_\lambda}\right)\Delta E_0\right]^2}{4E_i}. \qquad 35b.$$

If the frequencies are assumed to be the same for the precursor and successor complexes and also temperature independent, then E^*, E_λ, and ΔE_0 do not differ from the corresponding free-energy quantities, ΔG^*, λ, and ΔG^0, respectively, and, anticipating the conclusions reached below, it is convenient to reexpress Eqs. 32 and 35a as

$$\Delta G^* = (\lambda + \Delta G^0)^2/4\lambda \qquad \text{36a.}$$

$$\Delta G_i^* = \frac{\lambda_i}{4}\left[1 + \frac{\Delta G^0}{\lambda}\right]^2 \qquad \text{36b.}$$

where $\Delta G^* = \sum \Delta G_i^*$ and $\lambda = \sum \lambda_i$.[5a,b] In terms of the two-mode model, $\Delta G^* = (\Delta G_{in}^* + \Delta G_{out}^*)$ and $\lambda = (\lambda_{in} + \lambda_{out})$. Since frequency changes are not explicitly considered, λ_{in} is given by (cf Eq. 33a)

$$\lambda_{in} = \frac{1}{2}\sum_i \bar{f}_i[(q_i^0)_p - (q_i^0)_s]^2. \qquad \text{36c.}$$

Although the solvent reorganization is often formally treated in terms of the harmonic oscillator model [using Fourier components of the polarization vector (40, 45)], ultimate appeal to dielectric continuum theory leads to the following free-energy expression (40, 45–47) for the "vertical" solvent reorganization energy:

$$\lambda_{out} = \frac{(\Delta e)^2}{8\pi}\left(\frac{1}{D_{op}} - \frac{1}{D_s}\right)\int [(\mathbf{D}_0)_p - (\mathbf{D}_0)_s]^2 \, dV \qquad \text{37.}$$

where D_{op} and D_s are the high frequency (optical) and static dielectric constants of the medium (dispersion in D_s is neglected), and $(\mathbf{D}_0)_p$ and $(\mathbf{D}_0)_s$ are the Maxwell displacement fields associated with the charge distributions of the precursor and successor complexes, respectively (the subscript "zero" denotes the usual approximation of neglecting image effects, i.e. the influence of the polarized dielectric on the $\mathbf{D}$ field). The integral in Eq. 37 is over the dielectric volume characterized by D_s and excludes the cavities that contain the charged species and that are dielectrically saturated with respect to orientational polarization ($D_s = D_{op}$). Due to limitations of simple dielectric continuum theory, a common set of cavities must be assigned to the precursor and successor complexes. The most commonly employed model for the two reactants is that of two

---

[5a] The free energy of activation will have an additional entropic contribution (36, 43) $T\Delta S' = -RT \ln(kT/hv_n)$, but this contribution is usually small (typically ~ 0.4 kcal mol^{-1} at room temperature).

[5b] It has been pointed out by Marcus (44) that the quadratic expression for the free energy of activation (Eq. 36a) does not require harmonic oscillator potential-energy surfaces. Indeed Marcus notes that the solvational potential-energy surfaces are highly anharmonic.

spherical cavities (13), which leads to the familiar Marcus expression

$$\lambda_{\text{out}} = (\Delta e)^2 \left[\frac{1}{2a_1} + \frac{1}{2a_2} - \frac{1}{r} \right] \left[\frac{1}{D_{\text{op}}} - \frac{1}{D_s} \right] \qquad 38.$$

where the cavity radii a_1 and a_2 are equal to the radii of the two reactants (for solvated ions, a_1 and a_2 typically correspond to the radii of the bare ions plus the thickness of the first solvent shell). As the inner shells of the two reactants begin to interpenetrate (17a, 17b, 22, 23), a single ellipsoidal cavity becomes a viable alternative to the two-sphere model, and relevant formalisms and applications are available (17a, 46, 47). Attempts have also been made to extend Eq. 37 to include image effects and frequency and spatial dispersion in D_s (1, 45–47).

At the TST level, the vibration frequency v_n is given by (36, 43, 48)

$$v_n = \left[\left(\sum_i v_i^2 E_i \right) \Big/ \sum_i E_i \right]^{1/2} \qquad 39.$$

which clearly tends to be dominated by the highest frequencies if the various E_i are comparable in magnitude (36).

The derivation of the expression for v_n depends upon the concept of the reaction coordinate, q_r, which follows a steepest descent path (in mass-adapted Cartesian coordinates) from q^* to q_p^0 (or q_s^0). This is illustrated in Figure 3. For harmonic oscillators the direction cosines of the tangents to the reaction path at q^* are given by

$$(2\pi v_i)(E_i/2)^{1/2} / \sum [(2\pi v_i)^2 (E_i/2)]$$

and the above expression for v_n then follows straightforwardly from the following TST identity:

$$v_n = \frac{kT}{h} \Pi'(Q_i^*) / \Pi(Q_i^p)$$

$$= \frac{kT}{h} \exp(\Delta S'/R) \qquad 40.$$

where the Q_i^* and Q_i^p are the "classical" harmonic oscillator partition functions (kT/hv_i) and $\Delta S'$ is the entropy referred to in the footnote following Eq. 36. The Q_i^p pertain to the precursor complex in its initial (equilibrium) configuration, whereas the Q_i^* refer to the normal coordinates at q^* (but excluding q_r, as indicated by the "prime"). As shown in Figure 3 (and as discussed above in the steady-state treatment), the initial motion along q_r, starting at q_p^0, follows the solvent coordinate, while the motion in the final ascent is primarily along the inner-shell coordinate, i.e. some solvent reorganization precedes the inner-shell reorganization (36).

As an alternative to the harmonic oscillator formulation, the solvent (outer) component of the reaction coordinate can be expressed as a functional of fluctuations in the solvent polarization vector (17b, 43). The dynamics of solvent fluctuations play a central role in a variety of recent kinetic models that transcend the limitations of the usual TST models (17b, 17c, 49–58). For the most part the dynamical models deal only with solvent reorganization; however, in one case intramolecular modes have been included (55). The continuum model given in (49) is formulated within the adiabatic framework for both large and small values of H_{ps} relative to kT (in practice, the case in which $H_{ps} \ll kT$ may not correspond to the adiabatic regime).

In many cases of interest, observed ΔG^* and ΔG^0 values contain appreciable entropy contributions arising primarily from solvation changes outside the inner coordination shells of the reactants (59, 60). In principle, one could attempt to account for these within the harmonic oscillator

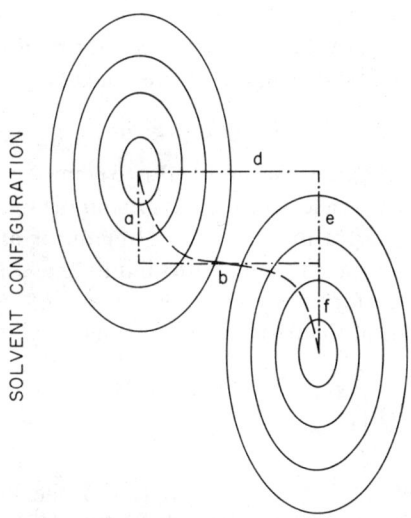

Figure 3 Equipotential sections through the elliptical basins describing the potential energy of the precursor complex (*top left-hand set*) and of the successor complex (*bottom right-hand set*) as a function of the configuration of the inner coordination shells of the reactants and products and the state of polarization of the surrounding solvent: *dashed line*, reaction coordinate; *dash-dot path abf* describes passage through or over the inner-shell barrier once the solvent configuration appropriate to the intersection region has been attained; *dash-dot path def* describes passage through or over the inner-shell barrier without any prior change of solvent configuration.

model in terms of $\Delta S'$, as defined by Eq. 40, and the additional terms arising from the relaxing of the constraints (temperature independent mean frequencies) that were employed in obtaining Eq. 36. However, oscillator (anharmonic as well as harmonic) models do not appear to be capable of yielding realistic estimates of the observed entropy changes (37). In practice this difficulty is circumvented (37) by retaining the harmonic oscillator formalism for v_n and λ_{in}, but using the classical expression, Eq. 38 (which is derived rigorously in terms of free energies), or some other expression derived from Eq. 37 [the temperature dependence of λ_{out}, via D_s, is quite small (14, 36, 37)] to calculate λ_{out}; this λ_{out} is then combined with the experimental ΔG^0 (which, of course, includes the solvation entropy changes) to yield the activation free energy associated with the solvent reorganization (Eq. 36b). The entropy changes associated with λ_{in} are generally small (37) and the inner-shell mode is satisfactorily treated within the (reduced-frequency) harmonic oscillator approximation, which, again in conjunction with the experimental ΔG^0, yields the free energy for the inner-shell reorganization (Eqs. 36d and 36b). These free energy quantities are then substituted into Eq. 31, with $\Delta G_n^* = (\Delta G_{in}^* + \Delta G_{out}^*)$. The above considerations apply not only to the TST formalism but also to the quantum mechanical formalisms considered next (see also 50 and 61).

Quantum Mechanical and Semiclassical Approaches

The simple TST implementation of Eq. 31 must be modified when nonadiabatic effects or nuclear tunneling are important. The latter effect is generally not large except in the inverted region or at low temperature. Nuclear tunneling in the nonadiabatic limit ($2\pi\gamma \ll 1$; Eq. 28) can be accounted for straightforwardly (1, 2, 36, 40–42, 45, 62–67) using time-dependent perturbation theory (the golden rule)

$$v_{eff}\kappa_n = \frac{4\pi^2}{hZ_p} \sum_{p,v,w} \exp\left[-\frac{E_{pv}}{RT}\right] \sum_w |H_{pv,sw}|^2 \delta(E_{pv} - E_{sw} - \Delta E_0) \qquad 41.$$

where K_n has been replaced by κ_n to reflect the fact that quantum-mechanical tunneling is being included. In Eq. 41 E_{pv} and E_{sw} are the vibronic energies of the precursor and successor complexes, respectively, $H_{pv,sw}$ is a vibronic matrix element

$$H_{pv,sw} = \int \chi_{pv}^0 H_{ps} \chi_{sw}^0 \, dq \qquad 42.$$

based on the electronic coupling term (Eq. 20) and zeroth-order nuclear functions, χ_0, obtainable from Eq. 23 (or the analogous equation for the successor complex) by neglecting the r.h.s., and Z_p is the nuclear partition function for the precursor complex. In most applications the vibronic matrix element is factored (the Condon approximation) into the product of a nuclear overlap integral and the electronic coupling element, $H_{ps}(q)$. For consistency with requirements of detailed balance, H_{ps} should be evaluated at the crossing point (q^* in Figure 1) (23, 61, 68).

Very compact expressions can be obtained by expressing the Dirac delta functions in Eq. 41 in terms of their Fourier time integral representations (Eq. 43) and evaluating the resulting time integral for $v_{eff}\kappa_n$ by means of a saddle-point approximation[6] in conjunction with the so-called generating function method (40, 69). [Equivalent results are obtained (40, 64) by casting Eq. 41 in terms of Bessel functions, which are approximated by asymptotic expansions.]

$$\delta(E_{pv} - E_{sw} - \Delta E_0) = \frac{2\pi}{h} \int_{-\infty}^{+\infty} dt \, \exp(-it(E_{pv} - E_{sw} - \Delta E_0)). \qquad 43.$$

Through the use of the above approximations the evaluation of Eq. 41 is straightforward, and various special cases have been discussed extensively in the recent literature, including very-low-temperature situations (40, 41, 64–67). If fixed frequencies are used (Eq. 33b), exact (and simple) analytic solutions for the saddle-point time, t_s, are obtained for thermoneutral ($\Delta G^0 = 0$) and barrierless ($\Delta G^0 = \pm E_\lambda$) harmonic multi-mode cases, and for all harmonic one-mode cases (37, 69, 70). Numerical solutions for t_s can easily be obtained for any harmonic case (37, 69).

For the case of a single harmonic mode, a simple semiclassical theory (SCT) based on an energy saddle-point approximation gives results identical with the above quantum mechanical theory (QMT) over a wide range of temperatures and exothermicities (63, 71): specifically, the product of a surface-hopping frequency (which coincides with the nonadiabatic Landau-Zener expression) and a barrier tunneling probability is integrated over a Boltzmann distribution of velocities, yielding

$$v_{eff}\kappa_n = \frac{2H_{ps}^2}{h} \left\{ \frac{2\pi^3}{E_\lambda h\nu [\Delta E_0^2/E_\lambda^2 + \text{csch}^2(h\nu/2kT)]^{1/2}} \right\}^{1/2}$$

$$\times \exp\left(-\frac{\Delta E_0}{2RT} - \frac{E_\lambda}{h\nu}\left\{\coth\left(\frac{h\nu}{2kT}\right) - \left[\frac{\Delta E_0^2}{E_\lambda^2} + \text{csch}^2\left(\frac{h\nu}{2kT}\right)\right]^{1/2}\right.\right.$$

$$\left.\left. + \left(\frac{\Delta E_0}{E_\lambda}\right) \sinh^{-1}\left[\left(\frac{\Delta E_0}{E_\lambda}\right)\sinh\left(\frac{h\nu}{2kT}\right)\right]\right\}\right). \qquad 44.$$

[6] The saddle-point time, t_s, is the value of t that maximizes the integrand of the integral representation of Eq. 41 (40, 69).

Equation 44 is valid over a very broad range of parameters, including very low temperatures, provided that

$$\left(\frac{\Delta E_0}{E_\lambda}\right) \sinh\left(\frac{hv}{2kT}\right) < 1. \qquad 45.$$

It should be noted that a discrete one-mode model (as in Eq. 44) implies some sort of coarse graining (36, 64, 72); e.g. replacing the Dirac delta function in Eq. 41 by appropriate Kronecker delta functions divided by hv.

For electron exchange reactions ($\Delta E_0 = \Delta G^0 = 0$), the single-mode SCT expression (Eq. 44) becomes

$$v_{\text{eff}}\kappa_n = \frac{2H_{ps}^2}{h}\left(\frac{2\pi^3}{(E_\lambda hv_\lambda)\,\text{csch}\,(hv/2kT)}\right)^{1/2} \exp\left[-\left(\frac{E_\lambda}{hv}\right)\tanh\left(\frac{hv}{4kT}\right)\right]. \qquad 46.$$

The multi-mode generalization, obtained by replacing E_λ, v by E_i, v_i and summing the exponential argument and the denominator of the prefactor over all modes with nonzero E_i's, yields a result identical with the exact QMT saddle-point result. For the familiar two-mode case (a classical solvent mode and a high-frequency inner-shell mode) we have (25, 42)

$$v_{\text{eff}}\kappa_n = \frac{2H_{ps}^2}{h}\left(\frac{2\pi^3}{2E_{\text{out}}RT + (E_{\text{in}}hv_{\text{in}})\,\text{csch}\,(hv_{\text{in}}/2kT)}\right)^{1/2}$$

$$\times \exp\left[-\frac{E_{\text{out}}}{4RT} - \left(\frac{E_{\text{in}}}{hv_{\text{in}}}\right)\tanh\left(\frac{hv_{\text{in}}}{4kT}\right)\right]. \qquad 47.$$

Equations 44, 46, and 47 all yield Eq. 31 in the high-temperature limit $(hv \ll kT)^7$ with the prefactor

$$v_{\text{eff}} = \kappa_{el}v_n = 4\pi\gamma v_n \qquad 48.$$

where $2\pi\gamma$ (Eq. 28) for the harmonic oscillator is given by (36)

$$2\pi\gamma = \frac{H_{ps}^2}{hv_n}\left(\frac{\pi^3}{E_\lambda RT}\right)^{1/2} \qquad 49.$$

with v_n defined by Eq. 39. Thus the nonadiabatic prefactor, $\kappa_{el}v_n$, is independent of v_n and in some applications $\kappa_{el}v_n$ is denoted by v_{el} to emphasize its v_n independence (6, 24, 25). A semiclassical device (36, 62) for extending the above equations so as to accommodate arbitrary values of $2\pi\gamma$ is to replace $4\pi\gamma$ in Eq. 48 by the general Landau-Zener expression for

---

[7] These equations yield the correct high-temperature limit in spite of the fact that the saddle-point approximation upon which they are based is nominally restricted to the energy region below the crossing point (73).

κ_{el} (Eq. 29). A related but still more general approach, based on the methodology of (29), has been implemented in conjunction with the extended Hückel model (31).

Returning now to the lower temperature regime ($kT \lesssim h\nu$) and considering first the nonadiabatic limit ($2\pi\gamma \ll 1$), we see that nuclear tunneling leads to a reduced, temperature-dependent, free energy of activation that can be expressed in terms of a nuclear tunneling factor, $\Gamma_\lambda \geqslant 1$,

$$RT \ln \Gamma_\lambda = \Delta G_n^* - \Delta G_n^*(T) \qquad 50.$$

where $\Delta G_n^*(T)/RT$ is equated to the entire argument of the exponential in Eqs. 44, 46, or 47, and at the high-temperature limit, ΔG_n^*, is defined in Eq. 31. For example, the expression for $\Delta G^*(T)$ implicated by Eq. 44 is

$$\Delta G^*(T) = \frac{\Delta E_0}{2} + \frac{E_\lambda kT}{h\nu} \left\{ \coth\left(\frac{h\nu}{2kT}\right) - \left[\frac{\Delta E_0^2}{E_\lambda^2} + \operatorname{csch}\left(\frac{h\nu}{2kT}\right)\right]^{1/2} \right.$$
$$\left. + \left(\frac{\Delta E_0}{E_\lambda}\right) \sinh^{-1}\left[\left(\frac{\Delta E_0}{E_\lambda}\right) \sinh\left(\frac{h\nu}{2kT}\right)\right] \right\}. \qquad 51.$$

Nuclear tunneling also leads to a more complex prefactor. Analogously to the high-temperature case (Eq. 48), in the nonadiabatic limit one may equate the prefactor to $\kappa_{el}\nu_n = 2\pi\gamma\nu_n$, thereby defining[8] a generalized γ once ν_n has been specified. This in turn allows more generalized prefactors to be generated in terms of κ_{el} (Eqs. 28–29). When nuclear tunneling is important, the best choice of ν_n and its associated coordinate, q_n, is somewhat uncertain. In line with the earlier steady-state analysis of activation within the two-mode model [solvent and inner-shell, Eqs. 7–9; see also (17a, 74)], one is led to postulate a preequilibrium of species activated in the low-frequency classical mode, thus leaving the high-frequency mode as a plausible candidate for ν_n (this choice will in many cases differ little from that given by Eq. 39). Comparison of this approach (classical solvent combined with Eq. 44 for the inner-shell) with the exact treatment of the two-mode case yields good agreement down to 100 K and $\Delta E_0 = -4$ eV ($\nu_{in} = 432$ cm^{-1}, $E_{in} = 1.5$ eV, $E_{out} = 1.1$ eV) (75). Alternatively, the single-mode expression (Eq. 44) may be used for the two-mode case by substituting ($E_{in} + E_{out}$) for E_λ and defining ν_n in terms of the effective T-dependent barriers; for the example cited above, calculation of ν_n from

$$\nu_n^2 = \frac{\nu_{out}^2 \Delta G_{out}^* + \nu_{in}^2 \Delta G_{in}^*(T)}{\Delta G_{out}^* + \Delta G_{in}^*(T)} \qquad 52.$$

---

[8] The factor 2π is used here instead of 4π (as in Eq. 48) since multiple passage effects in the crossing regime are absent for passage below the barrier (29, 71, 73).

with $v_{out} = 30$ cm^{-1} (which corresponds to 10^{12} s^{-1}, see below), $v_{in} = 432$ cm^{-1}, $\Delta G^*_{out} = E_{out}/4$ and $\Delta G^*_{in}(T)$ given by

$$\Delta G^*_{in}(T) = E_{in}(kT/hv_{in}) \tanh(hv_{in}/4kT) \qquad 53.$$

improves the agreement with the two-mode model even further (75).

Constraints on the Validity of Nonadiabatic Rate Constants

In order for the thermally averaged nonadiabatic rate constants given in the previous section to serve as valid models for k_{act} (Eq. 3b), several requirements must be met (40). The conditions for perturbation theory and the golden rule must, of course, be satisfied, i.e. the duration of reactive events interconverting the precursor and successor complexes, t_r, must be $< \hbar/H_{ps}$ and the density of final states must be high and roughly constant over an adequate energy range ΔE (defined by $h/\Delta E < t_r$). In addition, the reactive widths of the states χ_{pv} (see Eq. 23) should be small relative to the interstate spacing (hv) and kT. Rapid vibrational relaxation (represented by k_{-n} in Scheme 1) guarantees that a Boltzmann distribution of precursor states is maintained and also ensures irreversibility as soon as the equilibrium region of the successor complex (i.e. $q \sim q_s^0$) is reached (cf Eqs. 3b and 5). The vibrational relaxation time t_v (defined as $1/k_{-n}$) serves to provide an upper limit for t_r (51, 76, 77).

Various generalized nonadiabatic formalisms have been developed by the use of exponential or Gaussian damping terms, but still within the context of a Boltzmann distribution (51, 69, 73, 77, 78a,b). For example, incorporation of the factor $\exp[-(t/t_r)^2]$ into the integral of Eq. 41, where t_r is found to correspond to a special type of vibrational relaxation time [t_r is denoted by t_c and t_a in (73) and (77), respectively], leads to the conclusion that the discrete one-mode model that serves as the basis for the rate constant displayed in Eq. 44 is valid provided that $(2\pi t_r v)^2 \ll 1$ (i.e. due to lifetime broadening, the discrete manifold becomes effectively a continuum) (73, 77).

A Gaussian damping model has also been used to define an effective reaction exothermicity, $\Delta E'_0 = \Delta E_0 + \Delta E \exp[-(t_s/\tau)^2]$, which decreases in magnitude as some rotational relaxation time, τ, of the medium increases; t_s is the saddle-point time and ΔE could, for example, be the reorganization energy, E_i, for the relevant rotational mode, q_i (69). In a sufficiently viscous medium, the mode would be completely frozen out (i.e. $\Delta E'_0 = \Delta E_0 + \Delta E$). For consistency, it appears that one should allow the reorganization energy, as well as the exothermicity, to vary with τ so that $\Delta E_0 \to \Delta E_0 + E_i$ and $E_\lambda \to E_\lambda - E_i$ as $\tau \to \infty$; (e.g. in the two-mode model, freezing out of the solvent mode would correspond to the horizontal reaction path d in Figure 3).

In the low-damping regime, the assumption of a Boltzmannn distribution should be relaxed. For thermal reactions this has been achieved by chemical kinetics or steady-state models [see (17a, 17c) and Eqs. 3 and 6], by a mean-passage time analysis, based on a fluctuating cavity field model (17b), and by other dynamical analyses (e.g. 49). These analyses allow the competition between the frequency of entering the crossing region and the rate of relaxation to be displayed explicitly (e.g. v_{eff} vs k_{-n} in Eq. 3b). Similar competition has been analyzed for excited state (photo-induced) electron transfer (24, 72).

ELECTRONIC FACTORS

The electronic interaction of the redox centers, as represented by the electronic coupling element, H_{ps}, is an important factor determining electron transfer rates (e.g. as in Eq. 44). The magnitude of this interaction can be calculated if suitable expressions for ϕ_p, ϕ_s, and H are available. It may be necessary to augment the p, s basis by including contributions from electronically excited states in addition to the customary ground-state interactions (38, 62, 79–84). If the mixing is small, this can be accomplished by using standard perturbation theory. Such mixing is required if the exchange is a so-called "spin forbidden" process (38, 83). For example, $Co(NH_3)_6^{3+}$ is low spin while $Co(NH_3)_6^{2+}$ is high spin; as a consequence the $Co(NH_3)_6^{3+/2+}$ exchange involves not only an electron transfer, but also the redistribution of the remaining d electrons. Specifically, the exchange in this case is a three-electron process (i.e. ϕ_p and ϕ_s differ in the occupation of three orbitals) and is called spin-forbidden since H_{ps} is zero except for small contributions arising from nonzero orbital overlap integrals. The exchange becomes allowed through mixing with electronically excited (ligand field) states via the spin-orbit Hamiltonian (38). Mixing with excited-state configurations can also be important in spin-allowed processes: for example, it has been proposed that the electronic coupling in the spin-allowed $Ru(bpy)_3^{3+/2+}$ exchange is enhanced by contributions from the metal-to-ligand charge-transfer excited state of $Ru(bpy)_3^{2+}$ (85). Similarly, the electron transfer between positively charged complexes can be catalyzed by added anions, and it has been proposed (86) that this catalysis arises, in part, from the mixing with charge-transfer excited states, e.g. mixing of $M^{2+}XM^{2+}$ (and/or $M^{3+}X^{2-}M^{3+}$) with the ground-states $M^{3+}X^-M^{2+}$ and $M^{2+}X^-M^{3+}$ (the latter two species being described by ϕ_p and ϕ_s, respectively). Similar effects can also occur when X is a neutral bridging ligand, and in some cases X may even be the solvent.

The effect on the electronic coupling element of charge interactions of the type described above can be encompassed within a superexchange formalism (1, 62, 79–82). For example, in the general case of n sequential bridging units (capable of producing charge-transfer states), one can employ (80b, 82b, 87, 88) for the superexchange contribution to H_{ps}[9]

$$H_{ps}^{se} = \left(\prod_{i=0}^{n} H_{i,i+1}\right) \Big/ \left(\prod_{i=1}^{n} \Delta E_i\right) \qquad 54.$$

where the bridging species are coupled by the nearest-neighbor energy matrix elements $H_{i,j}$ either to each other ($H_{i,i+1}$) or, in the case of terminal species, to the redox centers ($H_{0,1}, H_{n,n+1}$), and ΔE_i is the vertical difference in energy between the ground state (the mean for ϕ_p and ϕ_s) and the charge transfer state associated with species i [generalizations of Eq. 54 are discussed in (88a)]. The mixing of interest occurs at the nuclear configuration appropriate to the intersection region, i.e. at $q = q^*$, and in a harmonic approximation the energy difference ΔE_i at q^*, denoted $\Delta E^\ddagger$, is given by (89)

$$\Delta E^\ddagger = E_{CT} - 2[E^*(E_{CT} - \Delta E_{CT}^0)]^{1/2} \qquad 55.$$

where E_{CT} is the vertical difference between the energy of the excited state and the energy of the (reactants') ground state at its equilibrium configuration (i.e. at q_p^0), ΔE_{CT}^0 is the difference between the energies of the excited state and the (reactants') ground state at their equilibrium configurations, and the force constants for the excited- and ground-state surfaces are assumed to be the same. Evidence for a dependence of $\log k_{obsd}$ on $\Delta E^\ddagger$ (calculated assuming that the minimum in the excited state surface lies directly above the minimum in the products' potential energy surface) has recently been found (86) for the anion-catalyzed reaction of $Co(PP)^{3+}$ with $Co(sep)^{2+}$ (PP = polypyridine derivatives, sep = sepulchrate), thus suggesting a nonadiabatic process dominated by H_{ps}^{se} (as in Eq. 54 with $n = 1$). In the study cited, $\Delta E^\ddagger$ is the energy of the vertical $Co(PP)_3^{3+}$, $X^- \to Co(PP)_3^{2+}$, X transition at $q = q^*$.

The electronic interaction of the two redox centers decreases with increasing separation. It is frequently assumed that this distance depen-

---

[9] This generalization of H_{ps} may be understood in terms of perturbation theory. The direct and double exchange contributions associated with the zeroth order ϕ_p and ϕ_s functions are augmented by the superexchange terms arising from first and higher order admixture of charge transfer components. Provided that the excited states being mixed do not lie too close to the ground state, little additional q-dependence will be introduced by this admixture (62). Henceforth, the term H_{ps} is understood to include all relevant contributions.

dence is exponential and that H_{ps} and κ_{el} are given by

$$H_{ps} = H_{ps}^0 \exp[-\beta(r-r_0)/2] \qquad \qquad 56a.$$

$$\kappa_{el} = \kappa_{el}^0 \exp[-\beta(r-r_0)] \qquad \qquad 56b.$$

where H_{ps}^0 (or κ_{el}^0) is the value of H_{ps} (or κ_{el}) when $r = r_0$, with r_0 generally being the most probable separation (or the effective contact distance) of the reactants (cf Eq. 11). Numerous models support the exponential behavior depicted in Eq. 56a [(82b, 87, 88); see also below]. For example, when the superexchange mechanism (Eq. 54) is dominant with all $H_{i,i+1} = \bar{H}$ and all $\Delta E_i = \overline{\Delta E}$, this exponential dependence can be displayed (87) by identifying β and r with $(2/d) \ln (\overline{\Delta E}/\bar{H})$ and nd, respectively, where d is the length of the bridging group. The prefactor H_{ps}^0 is given by $\bar{H}(\bar{H}/\overline{\Delta E})^{n-n_0}$, where n_0 is equal to r_0/d.

The competition between a large number of bridging units and a smaller number of units interacting over larger (through-space) distances can be compactly analyzed in terms of the above quantities if we assume in addition that the exchange integrals in Eq. 54 obey the relation $\bar{H}(d) = \bar{H}(d_0) \exp[-\beta'(d-d_0)/2]$ (compare Eq. 56a) where β' governs the through-space interaction of the bridging groups, and d_0 is a reference distance analogous to r_0. For example, the ratio of the coupling elements for two coordinated ligands (or metal centers) interacting either via a bridging group or directly is $\bar{H}(d_0)[\exp(\beta'd_0/2)]/\overline{\Delta E}$. Clearly, interaction via a bridging group (rather than through space) is favored when this ratio is larger than unity, i.e. when β' is less than β, where the latter governs the through-bridge interactions [see related discussion in (82b, 88b)].

If the distance dependence of the reorganization energies can be neglected compared with the distance dependence of the coupling element then (90)

$$k_{el} = k_0 \exp[-\beta(r-r_0)] \qquad \qquad 56c.$$

where k_0 is the value of the first-order rate constant k_{el} at $r = r_0$. Note that on the basis of Eq. 56c, the reaction thickness δr (Eq. 12) may be identified with $1/\beta$ (24).

A variety of procedures are available for evaluating the electronic coupling element and its distance dependence.

1. Ab initio molecular orbital calculations: Ab initio calculations of H_{ps} have been carried out for $Fe(H_2O)_n^{3+} - Fe(H_2O)_n^{2+}$ supermolecules ($n = 1$ to 6) designed to serve as models for the transition state in the aqueous $Fe(H_2O)_6^{3+} - Fe(H_2O)_6^{2+}$ exchange (17b, 22, 23, 68). Charge-localized self-

consistent field (SCF) wavefunctions were obtained for ϕ_p and ϕ_s,[10] and the matrix element then evaluated as

$$H'_{ps} = (H_{ps} - S_{ps}H_{pp})/(1 - S_{ps}^2) \qquad 57.$$

(in the following, the prime is suppressed). As an alternative to this direct construction of H_{ps} from the diabatic functions, one may define (23, 68)

$$H_{ps} = (H_{11} - H_{22})/2 \qquad 58.$$

where H_{11} and H_{22} are based on ϕ_1 and ϕ_2, the adiabatic counterparts of ϕ_p and ϕ_s (cf Eq. 24). In an ab initio SCF approach, ϕ_1 and ϕ_2 can be calculated by imposing appropriate symmetry constraints if the reacting species are related by a symmetry element (23, 68). It was estimated that inclusion of the influence of the solvent would have a rather small effect on the calculated H_{ps} values (17b) (see also 61). In order to maintain the desired relation, $H_{ps} = H_{sp}$, Eq. 57 should be evaluated at the crossing q^*, where $H_{pp} = H_{ss}$ (68).

The principal conclusions of the ab initio study are as follows: 1. H_{ps}^0 depends strongly on the relative orientation of the quasi-octahedral inner-shell reactants (17b, 22, 23, 68). The interpenetration[11] allowed by a staggered face-to-face approach yields an estimated r_m of ~ 5.3 Å, with $H_{ps}^0 \sim 115$ cm^{-1} and $\kappa_{el}^0 \sim 0.2$ (17a, 17b, 68). This nonadiabatic encounter has been shown capable of accounting for the observed kinetic data, using a pair distribution function for the reactants that simultaneously allows good account of the kinetics of a comparable bimolecular process, namely, spin-relaxation involving aqueous ions of the same charge type $(2+, 3+)$. By contrast, the apex-to-apex approach yields $r_m \sim 7.3$ Å and $H_{ps}^0 \sim 25$ cm^{-1}. 2. Over the range of 5.5–7.0 Å, the face-to-face orientation corresponds to $\beta \sim 2.4$ Å^{-1} (Eq. 56a) (68, 94). 3. Eqs. 57 and 58 yield similar values for H'_{ps} (68). Moreover, the calculated H'_{ps} values are rather insensitive to several variations in the orbital basis set and are not expected to be greatly affected by electron correlation (68). 4. The Condon approximation was tested for the apex-to-apex approach and found accurate to within a few percent (68). 5. While the above SCF calculations provide a suitable energy-optimized pair of diabatic (or adiabatic) states, a price to be paid is that direct, double, and superexchange contributions are combined in a rather nontransparent manner. Comparison with results based on a simple crystal-field model for the water ligands (22), and

---

[10] The high-spin (ferromagnetic) supermolecule state was considered; other spin-couplings would yield somewhat smaller H'_{ps} values (91a,b).

[11] Additional support for interpenetration of aqueous transition initial ions is provided by NMR (92) and neutron scattering data (93).

examination of electronic orbital populations of the reactants (23), suggest that direct exchange is important for the aqueous $Fe(H_2O)_6^{3+/2+}$ system, whereas ligand involvement is appreciably more important for the $Ru(H_2O)_6^{3+/2+}$ and $Ru(NH_3)_6^{3+/2+}$ couples.

2. Other theoretical models: Numerous other theoretical estimates of H_{ps} based on a variety of approximations and (in many cases) empirical data are available for both inorganic and organic systems (20, 82, 88, 95–104). Many bridged first-row transition metal ion systems have been studied (88, 99, 100) using extended Hückel theory (EHT) (105) and a molecular orbital (MO) analogue of Eq. 58 in which H_{ps} is equated to half the difference in orbital energies of symmetric and antisymmetric pairs of MO's involving primarily the 3d orbitals [extension to the case in which the reacting centers are not related by a symmetry element has also been proposed and applied (100)]. Calculated H_{ps} values for the $Fe(H_2O)_6^{3+/2+}$ couple (88b) agree quite well with comparable ab initio calculations (68). The EHT results for bridging groups suggest that π-electron transfer along unsaturated bridges is adiabatic even for very long bridges, and that σ-electron transfer is often competitive with π-electron transfer for both saturated and unsaturated bridges (here π and σ correspond to t_{2g} and e_g metal 3d-orbitals, respectively). In spite of the approximations involved, the EHT approach appears to be quite useful as an exploratory tool in assessing the coupling in a large number of systems.

Facile coupling of metal centers by σ bonding pathways in saturated hydrocarbons [specifically, dithiaspiro-bridged ruthenium(II)–(III) systems] has also been proposed (82a,b) on the basis of CNDO (complete neglect of differential overlap) and EHT calculations. However, the manner in which the relevant metal orbitals (t_{2g}) become coupled to the σ system of the bridging group has not been adequately analyzed. The local site symmetry (C_{2v}) for the Ru centers in the model calculations does not lead to mixing of the (t_{2g} or πd) orbitals and the σ orbitals of the bridge.

We now turn to interactions between unsaturated organic groups. On the basis of π-electron molecular orbital calculations similar to those performed on crystalline solids (104), it has been proposed (72) that the electronic coupling between two parallel aromatic molecules A^-A, where A is naphthalene or anthracene, can be expressed by Eq. 56a with $H_{ps}^0 = 1.0 \times 10^5$ cm^{-1} and $\beta = 2.0$ Å^{-1}, and with r equal to the distance between the centers of the aromatic molecules (and r_0 equal to 0). The apparent lack of dependence of H_{ps} on the nature of A can be understood as follows. If the relevant molecular orbital is symmetrically distributed among the N π-atomic orbitals of A, then the MO coefficient for each orbital is $N^{-1/2}$. In a parallel stacked configuration there are N equivalent nearest-neighbor contributions to H_{ps}, each given by $N^{-1/2}N^{-1/2}\bar{H}_{CC}$, where $\bar{H}_{CC}$ is the carbon-carbon exchange integral, and hence the magnitude of H_{ps}^0 is independent of N.

The calculations described so far were carried out within the Born-Oppenheimer framework. An alternative approach is to employ a one-dimensional electron tunneling model (101) in which the electron tunnels from one square well to another through a rectangular barrier of height $\overline{\Delta E}$. In this model, β is given by $4\pi\sqrt{2m\overline{\Delta E}}/h$ where m is the electron mass and $\overline{\Delta E}$ is typically identified with the height of the conduction band of the medium (101). In a study of the interaction of two heme groups, a value of 1.44 Å^{-1} has been obtained for β (101), and, in addition, H_{ps}^0 has been evaluated as $N^{-1/2}N^{-1/2}\overline{H}_{CC}$ in analogy with the above analysis. Here an edge-to-edge approach involving a single nearest-neighbor contact [equal to $(r-r_0)$ in Eq. 56] was assumed and N was taken as 20, derived on the assumption that the transferring electron is symmetrically delocalized over the 20 conjugated carbon atoms of the heme porphyrin rings in each redox center; $\overline{H}_{CC}$ was evaluated using the value of the exchange integral (8.0 × 10^3 cm^{-1}) for two carbon atoms separated by the C–C aromatic bond distance ($d = 1.4, d_0 = 0$ Å), yielding $H_{ps}^0 = 1.1 \times 10^2$ cm^{-1}. This approach involves a number of assumptions and approximations and its shortcomings have recently been discussed (5). This edge-to-edge value of H_{ps}^0 is considerably smaller than the value noted above for electron transfer between two parallel aromatic groups—a result that seems reasonable in view of the larger number of interacting centers when the interacting rings are parallel.

The electron tunneling model, like the Born-Oppenheimer models, obeys Eq. 56a and, in common with the superexchange model, it includes the electronic influence of the medium through the $\overline{\Delta E}$ dependence. However, this dependence is different for the two models, with β varying as $(\overline{\Delta E})^{1/2}$ and ln $(\overline{\Delta E}/\overline{H})$ for the tunneling and superexchange models, respectively, and it has been argued (87) that the superexchange model is more flexible in reconciling the available binding energy $(\overline{\Delta E})$ and distance dependence (β) data for electron transfer between cations and neutrals.

3. Intensities of intervalence bands in mixed-valence systems: Many mixed-valence systems feature an intervalence absorption band in the near-infrared region of the spectrum and H_{ps} may be estimated from the intensity of this band using the relation (19, 106)

$$H_{ps} (\text{cm}^{-1}) = \frac{2.06 \times 10^{-2}}{r} \{\varepsilon_{max}\bar{\nu}_{max}\Delta\bar{\nu}_{1/2}\}^{1/2} \qquad 59.$$

where r is in Å and $\bar{\nu}$ is in wavenumbers. As expected, the H_{ps} values depend upon the nature of the bridging group; typical values for diruthenium systems are in the range 10–200 cm^{-1} (107). Note that although a bridging group is useful, it is not required for the observation of an intervalence absorption band; such bands are also seen in outer-sphere ion pairs (108). The value of H_{ps} estimated from Eq. 59 is, of course, for the optical electron transfer; this value is not necessarily the same as H_{ps} for the thermal electron transfer, particularly if ϕ_p and ϕ_s vary with nuclear configuration (q). Thus $(H_{ps})_{op}$ depends upon ϕ_p and ϕ_s at $q = q_p^0$, while $(H_{ps})_{th}$ depends upon ϕ_p and ϕ_s at $q = q^*$. Depending upon the magnitude of the q dependence of ϕ, $(H_{ps})_{op}$ and $(H_{ps})_{th}$ can be quite different (109, 110). Although the electronic coupling in nonlinear bridged systems may be provided by circuitous through-bond routes, the formalism relating the oscillator strength to the matrix element still involves the direct (through-space) metal–metal separation (82b).

4. Magnetic exchange coupling. The magnetic coupling element J can be obtained from magnetic or other spectroscopic measurements on suitable systems. This element can be related to H_{ps} through superexchange models (111, 112) and has been used to estimate H_{ps} in a number of systems (113, 114).

5. Measurements of electron transfer rates as a function of separation distance. Estimates of β for the reactions of anion radicals or trapped electrons with suitable electron acceptors in frozen media have been obtained from pulse radiolysis studies (87, 115–118). A distance dependence of the rate constant consistent with Eq. 56c is assumed and a time-dependent reaction volume centered on the electron acceptor is defined by $V(t) = 4\pi(R^3 - r_0^3)/3$ where R is given by

$$R = r_0 + [\ln (gk_0 t)]/\beta \qquad \text{60a.}$$

with g equal to 1.9. The probability of survival of a radical (or trapped electron) at time t is then

$$P(t) = \exp[-4\pi c(R^3 - r_0^3)/3] \qquad \text{60b.}$$

where c is the number of acceptors per unit volume at time t. This procedure yields $\beta = 0.9$ Å^{-1} for the reaction of trapped electrons with $Cu(en)_3^{2+}$ (115) and $\beta = 1.2$ Å^{-1} for the reaction of tetramethyl-p-phenylenediamine (TMPD) with the pyrene radical cation (87). For a broad class of reactions studied by this technique $\beta = 1.0$ to 2.0 Å^{-1}.

Values of β have also been obtained from studies of electron transfer quenching of excited states produced by continuous irradiation or flash photolysis in glassy media (119, 120). For example, the kinetic data for the steady-state quenching of the emission of the singlet state of TMPD by phthalic anhydride (119) can be fit to Eq. 60b with $P(t) = I/I_0$, where I and I_0 are the emission intensities in the presence and absence of the phthalic anhydride, respectively. This procedure yields $R \approx 17$ Å. Assuming that R is given by Eq. 60 with $r_0 = 6$ Å, $k_0 = 10^{13}$ sec^{-1}, and $t = \tau_0$, the lifetime of the excited state (7 ns), gives $\beta = 1.0$ Å^{-1}. In another study, the time-resolved quenching of the emission of excited Ru[(CH$_3$)$_4$phen]$_3^{2+}$ by MV^{2+} in a glycerol medium gave $\beta = 1.5 \pm 0.1$ Å^{-1} (121, 122).

Measurements of *intramolecular* electron transfer rates have also yielded values of β (123–125). For example, pulse radiolysis and flash photolysis techniques have been used to initiate reduction of the heme center of cytochrome c by a Ru(NH$_3$)$_5^{2+}$ moiety bound at histidine-33 of the metalloprotein (124, 125). Analysis of the kinetic data yields $\beta = 1.4 \pm 0.1$ Å^{-1} if $(r - r_0) = 12$ Å and $k_0 = 10^{13}$ sec^{-1}. In another study, pulse radiolysis was used to generate a biphenyl radical anion from a biphenyl group rigidly bound through a steroid molecule to various acceptors (117). This study gave $\beta \leqslant 0.9$ Å^{-1} for the most exothermic reaction ($k > 1 \times 10^9$ sec^{-1}), again assuming $k_0 = 10^{13}$ sec^{-1}. Pulse radiolysis techniques have also been used to study electron transfer between donors and acceptors bound to a steroid molecule (118). A flash photolysis study of electron transfer from an excited zinc prophyrin to a quinone ring rigidly held ~ 10 Å above the porphyrin plane yielded $k \sim 10^9$ sec^{-1}, consistent with $\beta = 1.1$ Å^{-1} if $k_0 = 10^{13}$ sec^{-1} (126).

To summarize, the assumption that $k_0 = 10^{13}$ sec^{-1} implicates β's ranging from ~ 0.9 to 2.0 Å^{-1}. Although some dependence of β on the relative orientation of the donor and acceptor and on the nature of the intervening medium is expected, no definite trends can be discerned at this stage. The value 0.8 Å proposed for δr (Eq. 12) corresponds to a β value of 1.2 Å^{-1}. For $\beta = 1.2$ Å^{-1} and $k_0 = 10^{13}$ sec^{-1}, electron transfer will occur in 10^{-8} sec if $(r - r_0) = 10$ Å and in 10^{-5} sec if $(r - r_0) = 15$ Å. This is the range of timescales and edge-to-edge distances appropriate to electron transfer in many biological systems (127).

6. Temperature dependence of the rate constant: κ_{el} can be estimated from the entropy of activation for the electron transfer reaction: small κ_{el} values will tend to make ΔS^* more negative. This procedure must be used with caution since nuclear tunneling contributions and the temperature dependence of the electrostatic work terms will also tend to decrease ΔS^*. Note that the conventional Debye-Hückel expressions, while satisfactory for calculating work terms, yield very poor estimates of the electrostatic contributions to the entropy of activation (17a, 128).

7. The limiting rate constant at high driving force: Normally $k_{obsd} \rightarrow k_{diff}$ as the driving force increases (i.e. as $\Delta G^* \rightarrow 0$). However, rate saturation below the diffusion-controlled limit may be observed if $\kappa_{el} \ll 1$, since under these conditions $k_{obsd} \rightarrow K_A \kappa_{el} v_n$ as $\Delta G^* \rightarrow 0$.[12] Thus κ_{el} can be calculated

---

[12] However, see (130) for a comment on the preexponential factor in barrierless reactions.

from the maximum rate constant at high driving force (6, 129). A recent application of this procedure yielded $\kappa_{el} \sim 10^{-3}$–10^{-4} for the reactions of Ru(bpy)$_3^{3+}$ with cobalt(II) cage compounds (89).

RELAXATION TIMES AND PREEXPONENTIAL FREQUENCIES

The reorganization rate constants appearing in the steady-state expressions are measures of the rate of formation and destruction of the configurations appropriate to electron transfer. The rate of decay of the reorganized medium is primarily determined by the solvent dynamics, specifically by the rate of reorientation of the solvent molecules. The latter, in turn, can be related to the frequency dependence of the dielectric constant of the medium and is governed by a relaxation time. A large part of the dielectric relaxation of water and similar polar solvents is accounted for by the Debye equation

$$D(\omega) = D_\infty + \frac{D_s - D_\infty}{1 - i\omega\tau_D} \qquad 61.$$

where D_∞ and D_s are the high- and low-frequency (static) dielectric constants, respectively, ω is the angular frequency of the applied field and τ_D is the Debye relaxation time. (Note that for H$_2$O and small alcohols, D_∞ is larger than the optical dielectric constant D_{op}, usually taken equal to n^2, where n is the refractive index of the medium. This arises because the high frequency defined in Eq. 61 is the applied frequency at which the solvent dipoles are no longer able to respond to the changing field, i.e. the reorientation of the dipoles no longer contributes to the measured susceptibility. This frequency is lower than optical frequencies because of absorptions in the visible or near infrared.)

Even though the frequency dependence of the dielectric function is characterized by a single relaxation time τ_D, this relaxation time and a second relaxation time, called the longitudinal relaxation time τ_L, are required to describe the time dependence of the polarization of the medium (131–133). The two relaxation times are related in a Debye model by the Fröhlich equation

$$\tau_L = \frac{D_\infty}{D_s} \tau_D. \qquad 62.$$

The longitudinal relaxation time is evidently much shorter than the Debye (transverse) relaxation time. For H$_2$O at 25°C, $\tau_D = 8.5 \times 10^{-12}$ sec, $D_s = 78.5$ and $D_\infty = 5.5$ so that $\tau_L = 0.60 \times 10^{-12}$ sec.

An interesting illustration of the difference between τ_D and τ_L has been given by Fröhlich (131) and recently elaborated by Friedman (134). Theoretical analysis shows that $1/\tau_D$ is the rate constant governing the change of the charge density on the plates of a capacitor (filled with dielectric) after a single voltage step while $1/\tau_L$ is the rate constant governing the change of the voltage across the capacitor after a single step of the charge on the plates: the different relaxation times arise essentially from the differing amounts of charge transported in the two cases.

Clearly, it is important to know whether τ_D or τ_L (or some other relaxation time) is the relevant parameter in any given instance. Since reorganization of the solvent involves a change in polarization at constant charge, it seems reasonable to identify the relaxation time for this reorganization with τ_L. In terms of the steady-state schemes considered, the rate constants for the solvent reorganization and τ_L are related by

$$k_{out} + k_{-out} = 1/\tau_L \qquad 63.$$

and since $k_{out} \ll k_{-out}$ ($K_{out} \ll 1$) it follows that

$$k_{-out} \sim 1/\tau_L. \qquad 64.$$

Most electron transfer measurements are performed in the presence of an added electrolyte. Because of dielectric saturation effects arising from the added electrolyte and the charges on the reactants and products (133, 135a), the dielectric constant of such a medium may differ from that of the pure solvent. The relaxation time for the field gradient in Ni^{2+} solutions (~ 1 M ionic strength) is 2.2×10^{-12} sec (135b) and it has been proposed that this may provide an estimate of the relaxation time of the cavity field (17b). Although more analysis is needed, this time may be tentatively identified with τ_L and is close to the value for the pure liquid.

The value of k_{-in} is determined by the relaxation time of the inner-shell modes. For a single symmetrical breathing mode, the relaxation time can be estimated from the width (δv) of the Raman absorption line. This width gives the transverse relaxation time $T_2 (= 1/\pi \delta v)$, while the relevant time for the relaxation is the longitudinal (or so-called spin-lattice) relaxation time T_1. Since $T_1 \geqslant T_2$ and since $k_{in} \ll k_{-in}$, we have

$$k_{-in} < \pi \delta v. \qquad 65.$$

The measured linewidth for $Mg(H_2O)_6^{2+}$ is 39 cm^{-1} (136) so that k_{-in} for the symmetrical breathing motion of this aquo ion is $\leqslant 5 \times 10^{12}$ sec^{-1}. The linewidths are larger for higher frequency modes and k_{in} and k_{-in} will be proportionately increased. Allowing for contributions from intraligand modes, we may conclude that $k_{-in} \sim 1 \times 10^{13}$ sec^{-1}.

From the above discussion of the rate constants governing the solvent and inner-shell reorganization we conclude that the assumption made in simplifying Scheme 2, namely that $k_{-in} > k_{-out} \gg k_{in}, k_{out}$, is justified provided $K_{in} < 5 \times 10^{-2}$ (so that $k_{-out} \gg k_{in}$). Under these conditions use of Scheme 3 and Eq. 9 is appropriate. On the other hand, when ΔG_{in}^* is very small the above condition may no longer be satisfied and it is then necessary to revert to Scheme 2. In the limit when ΔG_{in}^* is small enough to be neglected $(K_{in} > 5 \times 10^{-2})$, Eq. 4 with $k_n = k_{out}$ and $k_{-n} = k_{-out}$ can be used. Regardless of which scheme is used, the activated complex expression, Eq. 8b, will be valid, provided $v_{eff} < 5 \times 10^{12}$ sec^{-1}. When this condition is not satisfied,[13] the reaction will exhibit some degree of relaxation control, and it is then necessary to use the more general steady-state expressions. More rigorous treatments of this problem for the single-mode (solvent only) case have recently been published (17c, 17d, 49, 50, 58).

Probably the best example to date of a system where the electron transfer rate is controlled by medium relaxation processes is provided by studies of the effect of solvent on the light-induced intramolecular electron transfer rates in (phenylamino)naphthalenesulfonate derivatives (137–141). Excitation produces a nonplanar $S_{1,np}$ excited state, largely localized on the naphthalene, which undergoes intramolecular electron transfer to form a charge-transfer state $S_{1,ct}$. The important observation was made (141) that the fluorescence lifetime τ_{fl} of the $S_{1,np}$ state is equal to τ_L (calculated assuming that D_∞ is equal to the square of the refractive index) for a variety of solvents. This result can be understood if the intramolecular electron transfer is in the barrierless regime so that from Eq. 4b, $\tau_{fl} \approx k_{-n} \sim \tau_L$. However, reservations about this simple interpretation have been expressed (130).

When the reaction is not relaxation-controlled, then the preexponential frequency is v_{eff} (Eqs. 5, 8). As discussed above, v_{eff} is equal to $\kappa_{el} v_n$ where v_n is given by Eq. 39. When ΔG_{in}^* is small, $v_n \sim v_{out}$ (Eq. 39) and consequently $v_{eff} \sim v_{out}$ provided $\kappa_{el} \sim 1$. This situation generally obtains for outer-sphere ruthenium(II)-ruthenium(III) [but not cobalt(II)-cobalt(III)] electron transfers. It has also been proposed (142) that $v_{eff} \sim v_{out}$ for intramolecular electron transfer in a pyrazine-bridged, asymmetric diruthenium system, and more detailed calculations based on this model have been presented (49). It should be noted that when H_{ps} is large it is no longer appropriate to expand the exponential in Eq. 28a. If this is not recognized, unreasonable, high values of v_{eff} may be calculated (143).

NUCLEAR FACTORS

Exchange Reactions

Most applications of the classical and semiclassical formalisms have been in the calculation of electron exchange rates of metal complexes in solution.

---

[13] For all practical purposes, this condition is satisfied by the conventional activated-complex preexponential factor kT/h.

The reactants and products of exchange reactions are identical (differing merely by the interchange of an electron), so there are no net thermodynamic changes associated with the electron transfer, and the molecular properties of only one redox couple (rather than of two different redox couples) need be known. As a consequence, the theoretical expressions for the rates of exchange reactions have a special simplicity.

Good agreement between the observed and calculated exchange rates of metal complexes is generally found using the classical model (with $\kappa_{el} = 1$) and assuming that r is equal to $2a$, the sum of the radii of the two reactants (25). Support for the latter assumption is provided by the fact that the logarithm of the (precursor corrected) rate constants for the $Ru(NH_3)_6^{3+/2+}$, $Ru(NH_3)_5(py)^{3+/2+}$, $Ru(NH_3)_4(bpy)^{3+/2+}$, $Ru(NH_3)_2(bpy)_2^{3+/2+}$, and $Ru(bpy)_3^{3+/2+}$ exchange reactions, in which the sizes of the reactants are systematically increased, show a linear dependence upon $1/2a$ (144). For these ruthenium systems ΔG_{in}^* is small, so $v_n \to v_{out}$ and the rates are primarily determined by the solvent reorganization barriers. A linear dependence of $\log k_{obsd}$ on $1/r$ has also been observed in intramolecular electron transfer in which the separation of the metal centers is systematically varied by changing the bridging group (145). The dependence of the exchange rate on $(1/D_{op} - 1/D_s)$ predicted by the model is found for some (146–148) but not all (149) of the systems studied. Specific ion-pairing and solvation effects, as well as other shortcomings of the two-sphere model (47), could be responsible for the different solvent dependences found.

The major factor responsible for the large range of exchange rates observed in practice (more than 15 orders of magnitude) is variations in the inner-shell reorganization barriers. Values of the difference in the metal-ligand bond lengths Δd_0 in the two oxidation states and the exchange rates for various couples are presented in Table 1 (128, 150–166). It will be seen that Δd_0 ranges from -0.02 to 0.22 Å and that Δd_0 is large when the two oxidation states differ in the population of the antibonding σd orbitals (loosely speaking, when transfer of a σd electron is involved) (128). The Δd_0 changes are smaller, but not absent, when a πd electron is transferred. Detailed ab initio calculations (22, 23, 167) for $Fe(H_2O)_6^{3+/2+}$, $Ru(H_2O)_6^{3+/2+}$, and $Ru(NH_3)_6^{3+/2+}$ systems indicate that the dominant factor determining Δd_0 in these systems is actually a σd effect: specifically, the reduction in charge caused by addition of a πd electron to the higher oxidation state inhibits the ligand-to-metal charge transfer interaction, resulting in a σ-bond weakening that surpasses the direct weakening caused by the addition of the electron to the (weakly antibonding) πd orbitals.

In Figure 4 (128) the rate constants for a variety of exchange reactions have been corrected for ΔG_{out}^*, for nuclear tunneling effects,[14] and for

---

[14] The nuclear tunneling corrections are not large, increasing the rates by less than an order of magnitude.

differences in the stabilities of the precursor complexes, and plotted as $[\ln(k_{obsd}/K_A v_n \Gamma_\lambda) + \Delta G^*_{out}/RT]$ vs $(\Delta d_0)^2$. A good linear correlation with a small intercept (corresponding to an average κ_{el} of 10^{-1}–10^{-2}) is obtained; apart from one glaring exception, the comparatively minor deviations could arise, in part, from differences in the normal-mode force constants, and from the neglect of changes in the *intraligand* bond lengths and angles (128). Overall, the correlation does not leave room for any especially large nonadiabaticity in the cobalt(III)–cobalt(II) exchanges (25, 128). The one exception is the $Co(H_2O)_6^{3+/2+}$ exchange reaction, which proceeds about 10^7 times more rapidly than predicted. Clearly there is a more favorable pathway for this reaction than the simple outer-sphere exchange considered. Various possibilities have been proposed, including a preequilibrium spin change (85) and an inner-sphere water-bridged mechanism (172).

The effect of intraligand C–C bond distance changes has been considered in the $Co(bpy)_3^{2+/+}$ exchange (163), a system in which the ligands acquire appreciable bpy^- character upon reduction of the cobalt(II) complex. Including the intraligand changes does not alter the rate constant appreciably: the decrease in rate resulting from the increase in ΔG^*_{in} is largely offset by the accompanying increase in v_n. More recently (168), a molecular mechanics approach, in which a comprehensive set of intraligand distances and angles was included, was used to calculate the reorganization energy in

Table 1 Differences in the metal-ligand bond lengths in the two oxidation states and second-order rate constants for electron exchange in aqueous solution at 25°C[a]

Couple	Electronic configuration	Δd_0, Å	k, $M^{-1} s^{-1}$	Refs.
$Cr(H_2O)_6^{2+/3+}$	$(\pi d)^3 (\sigma d)^1 - (\pi d)^3$	(0.20)[b]	$\leqslant 2 \times 10^{-5}$	(128, 158)
$Fe(H_2O)_6^{2+/3+}$	$(\pi d)^4 (\sigma d)^2 - (\pi d)^3 (\sigma d)^2$	0.13	4.2	(128, 151)
$Co(H_2O)_6^{2+/3+}$	$(\pi d)^5 (\sigma d)^2 - (\pi d)^6$	0.21	3.3	(152, 153)
$Ru(H_2O)_6^{2+/3+}$	$(\pi d)^6 - (\pi d)^5$	0.09	(10^2)[c]	(154, 155)
$Ru(NH_3)_6^{2+/3+}$	$(\pi d)^6 - (\pi d)^5$	0.04	3.2×10^3	(156, 157)
$Fe(phen)_3^{2+/3+}$	$(\pi d)^6 - (\pi d)^5$	0.00	3×10^8	(128, 158)
$Ru(bpy)_3^{2+/3+}$	$(\pi d)^6 - (\pi d)^5$	0.00	4.2×10^8	(128, 159)
$Co(NH_3)_6^{2+/3+}$	$(\pi d)^5 (\sigma d)^2 - (\pi d)^6$	0.22	$\geqslant 10^{-7}$	(128, 160)
$Co(en)_3^{2+/3+}$	$(\pi d)^5 (\sigma d)^2 - (\pi d)^6$	0.21	7.7×10^{-5}	(128, 161)
$Co(bpy)_3^{2+/3+}$	$(\pi d)^5 (\sigma d)^2 - (\pi d)^6$	0.19	18	(128, 162, 163)
$Co(sep)^{2+/3+}$	$(\pi d)^5 (\sigma d)^2 - (\pi d)^6$	0.17	5.1	(164)
$Co(bpy)_3^{+/2+}$	$(\pi d)^6 (\sigma d)^2 - (\pi d)^5 (\sigma d)^2$	-0.02	10^9	(128, 163)
$Ni(bpy)_3^{2+/3+}$	$(\pi d)^6 (\sigma d)^2 - (\pi d)^6 (\sigma d)^1$	(0.13)[b]	2×10^3	(165, 166)

[a] Abbreviations used: en, ethylenediamine; bpy, 2,2′-bipyridine; phen, 1,10-phenanthroline; sep, 1,3,6,8,10,13,16,19-octaazabicyclo[6.6.6]eicosane ("sepulchrate").
[b] Average value over the six metal-ligand bonds.
[c] Estimated from cross-reaction data.

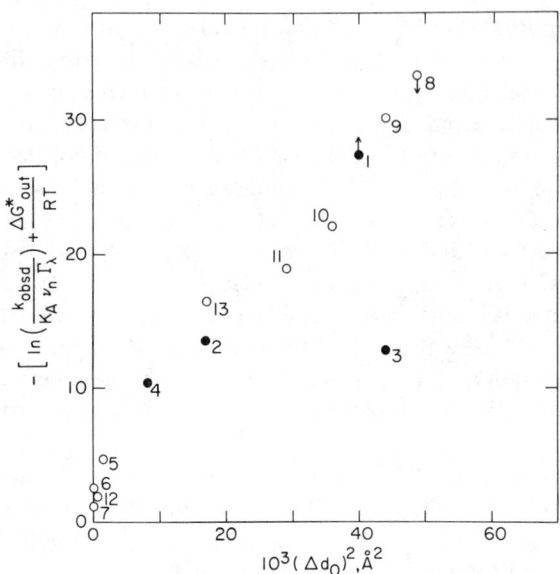

Figure 4 Natural logarithm of the observed exchange rate constant, corrected for the stability of the precursor complex, the effective nuclear frequency and the inner-sphere nuclear tunneling factor, plus the outer-sphere barrier divided by RT plotted against the square of the difference of the metal-ligand bond distances in the two oxidation states: *closed circles*: 1, $Cr(H_2O)_6^{3+/2+}$; 2, $Fe(H_2O)_6^{3+/2+}$; 3, $Co(H_2O)_6^{3+/2+}$; 4, $Ru(H_2O)_6^{3+/2+}$; *open circles*: 5, $Ru(NH_3)_6^{3+/2+}$; 6, $Fe(phen)_3^{3+/2+}$; 7, $Ru(bpy)_3^{3+/2+}$; 8, $Co(NH_3)_6^{3+/2+}$; 9, $Co(en)_3^{3+/2+}$; 10, $Co(bpy)_3^{3+/2+}$; 11, $Co(sep)^{3+/2+}$; 12, $Co(bpy)_3^{3+/+}$; 13, $Ni(bpy)_3^{3+/2+}$. After (128).

the $Co(sep)^{3+}$–$Co(sep)^{2+}$ exchange. No major difference with the results of the simpler calculation was found: the calculated reorganization energy was only about 10% (0.8 kcal mol^{-1}) lower than the value obtained in the simpler calculation in which only Co–N bond distance changes were considered (25). Finally it should be noted that while intraligand O–H, C–H, and N–H modes may not be of major importance in determining reorganization barriers, these high-frequency modes can be important in determining isotope effects (78b, 94, 169–171).

Free-Energy and Cross-Relations

When the electron transfer is accompanied by a net chemical change, the classical formalism (13–15) predicts a linear relationship between log k and log K provided that $-\Delta G^0 \ll \lambda$. More generally, plots of log k vs log K are predicted to be curved, with a local slope of $0.50[1 + \Delta G^0/\lambda]$. In the normal region ($-\Delta G^0 < \lambda$) the predicted dependence of log k upon log K is found for a number of systems (173–179). On the other hand, there is very little evidence for the decrease of log k with increasing log K predicted for the abnormal region ($-\Delta G^0 > \lambda$), at least in the case of bimolecular reactions (6) (but see 190). Reasons for the absence of the predicted rate decrease include the formation of electronically excited products (27), nuclear tunneling (179), a change in mechanism from electron to atom transfer (180), underestimation of the diffusion-controlled region (181), and rapid relaxation through adiabatic channels in the barrierless region (182). The formation of electronically excited products (chemiluminescence) has been observed in certain electron-transfer reactions (183a,b) and could account for the absence of inverted behavior in some systems. Although not small, the nuclear tunneling corrections are, in general, not large enough to eliminate the rate decrease predicted for the highly exothermic region (37). Nuclear tunneling effects do, however, become important with increasing exothermicity (24) and result in the logarithm of the rate constant depending on the first power of the energy gap, rather than on the second power, as predicted by the classical theory. Evidence for the energy gap law comes from the radiationless decay rates of excited states (184, 185): such decays can be viewed as intramolecular electron transfer processes. Moreover, certain intramolecular electron transfer reactions, initiated by pulse radiolysis (189), do show abnormal behavior. Interestingly, evidence for the abnormal region is also provided by studies of "bimolecular" electron transfer in frozen media where diffusion is absent (110); however, the solvent barrier is time dependent and the λ and ΔG^0 values used in this work still need to be justified.

Probably the most frequently used free-energy expression is the cross-relation, Eq. 66 (13–15), which relates the rate constant for a cross-reaction, k_{12}, to the rate constants for the component exchange reactions, k_{11} and k_{22}, the equilibrium constant for the cross-reaction, K_{12}, and a factor f_{12}, which is defined in terms of k_{11}, k_{22}, K_{12} and the preexponential factors and which is important only at large K_{12}.

$$k_{12} = (k_{11}k_{22}K_{12}f_{12})^{1/2} \qquad 66.$$

There have been numerous tests of the cross-relation (3, 4, 60, 175–178, 187, 188). The agreement between theory and experiment is frequently very good: the observed and calculated rate constants generally agree within an order of magnitude. The disagreement does, however, show a trend: with few exceptions, the observed rates are slower than the calculated rates, and the deviations become larger when the driving force is very large. In the latter case the agreement is improved when corrections for nonadiabaticity are incorporated into the cross-relation (24). In other cases corrections for electrostatic and nonelectrostatic work terms are necessary (188). The latter appear to be important, for example, in the cross-reactions of complexes with aromatic ligands (bipyridine or phenanthroline) with aquo ions: in these systems the calculated rates are two to three orders of magnitude faster than the observed rates (175, 181, 186). In other systems a change in mechanism may be involved (178). The cross-relation has proved useful in correlating and interpreting electron transfer rates.

CONCLUSIONS

The rates of electron transfer in solution are primarily determined by the configuration changes accompanying the electron transfer, that is, by the nuclear factors for the reaction. These are adequately described by the classical model, except at low temperature and high exothermicities, where nuclear tunnelling corrections can be important. The electronic factors can become dominant when the reorganization barrier is reduced by the exothermicity of the reaction or when the reactants are kept far apart, for example, in biological systems or in frozen media; for such systems $-d(\ln \kappa_{el})/dr$ values generally lie in the range 0.9–2.0 Å^{-1}, with 1.2 Å^{-1} being a typical value. Although electron exchange between low-spin cobalt(III) and high-spin cobalt(II) complexes are "spin forbidden," the experimental data do not leave room for any especially large nonadiabaticity: the measured cobalt(III)–cobalt(II) exchange rates can be accounted for in terms of the relatively large inner-shell reorganization barriers associated with the bond distance changes in these systems.

When the nuclear and electronic factors are both very favorable, the dynamics of the solvent reorganization may become rate determining: a few systems displaying this type of behavior have been reported and it is expected that more examples will be found in the future.

Acknowledgments

We wish to thank Dr. Bruce Brunschwig for helpful discussions and for assistance with the steady-state derivations and Drs. J. Miller and E. Kosower for providing preprints of their work (110). Helpful discussions with Drs. C. Creutz, H. Friedman, and R. A. Marcus are also gratefully acknowledged. This work was performed at Brookhaven National Laboratory under contract with the US Department of Energy and supported by its Office of Basic Energy Sciences.

Literature Cited

1. Ulstrup, J. 1979. *Lecture Notes in Chemistry*. Berlin: Springer-Verlag. 419 pp.
2. Chance, B., DeVault, D. C., Frauenfelder, H., Marcus, R. A., Schrieffer, J. R., Sutin, N. eds. 1979. *Tunneling in Biological Systems*. New York: Academic. 758 pp.
3. Cannon, R. D. 1980. *Electron Transfer Reactions*. London: Butterworths. 351 pp.
4. Pennington, D. E. 1978. In *Coordi-*

nation Chemistry, ed. A. E. Martell, 2:476–590. Am. Chem. Soc. Monogr. No. 174. 718 pp.
5. DeVault, D. 1980. *Q. Rev. Biophys.* 13:387–564
6. Sutin, N. 1982. *Acc. Chem. Res.* 15:275–82
7. Weaver, M. J., Hupp, J. T. 1982. *Am. Chem. Soc. Symp. Ser.* 198:181–208
8. Albery, W. J. 1980. *Ann. Rev. Phys. Chem.* 31:263–77
9. Albery, W. J., Kreevoy, M. M. 1978. *Adv. Phys. Org. Chem.* 6:87–157
10. Pellerite, M. J., Brauman, J. I. 1980. *J. Am. Chem. Soc.* 102:5993–99
11. *Prog. Inorg. Chem.* 1983. Vol. 30. 528 pp.
12. *Faraday Discuss. Chem. Soc.* 1982. No. 74. 413 pp.
13. Marcus, R. A. 1956. *J. Chem. Phys.* 24:966–78
14. Marcus, R. A. 1965. *J. Chem. Phys.* 43:679–701
15. Marcus, R. A. 1964. *Ann. Rev. Phys. Chem.* 15:155–96
16. Sutin, N. 1973. In *Inorganic Biochemistry*, ed. G. L. Eichhorn, 1:611–53. New York: American Elsevier. 652 pp.
17a. Tembe, B. L., Friedman, H. L., Newton, M. D. 1982. *J. Chem. Phys.* 76:1490–1507
17b. Friedman, H. L., Newton, M. D. 1982. *Faraday Discuss. Chem. Soc.* 74:73–81
17c. Northrup, S. H., Hynes, J. T. 1980. *J. Chem. Phys.* 73:2700–14
17d. Grote, R. F., Hynes, J. T. 1980. *J. Chem. Phys.* 73:2715–32
17e. Marcus, R. A. 1960. *Faraday Discuss. Chem. Soc.* 29:129–30
18. Hush, N. S. 1961. *Trans. Faraday Soc.* 57:557–80
19. Hush, N. S. 1967. *Prog. Inorg. Chem.* 8:391–441
20. Hush, N. S. 1968. *Electrochim. Acta* 13:1105–23
21. Reynolds, W. L., Lumry, R. W. 1966. *Mechanisms of Electron Transfer*, p. 112. New York: Ronald Press. 175 pp.
22. Newton, M. D. 1980. *Int. J. Quant. Chem. Symp.* 14:363–91
23. Newton, M. D. 1982. *Am. Chem. Soc. Symp. Ser.* 198:255–79
24. Sutin, N., Brunschwig, B. S. 1982. *Am. Chem. Soc. Symp. Ser.* 198:105–125
25. Sutin, N. 1983. *Prog. Inorg. Chem.* 30:441–98
26. Rosseinsky, D. 1984. *Comments Inorg. Chem.* 3:153–70
27. Marcus, R. A., Siders, P. 1982. *J. Phys. Chem.* 86:622–30
28. O'Malley, T. F. 1971. *Adv. At. Mol. Phys.* 7:223–49
29. Nikitin, E. E. 1974. *Theory of Elementary Atomic and Molecular Processes in Gases*, transl. M. J. Kearsley, pp. 99–178. Oxford: Clarendon. 472 pp.
30. Child, M. S. 1972. *Faraday Discuss. Chem. Soc.* 53:18–26
31. Larsson, S. 1981. *Theor. Chim. Acta* 60:111–25
32. Wong, K. Y., Schatz, P. N. 1981. *Prog. Inorg. Chem.* 28:369–449
33. Truhlar, D. G., Hase, W. L., Hynes, J. T. 1983. *J. Phys. Chem.* 87:2664–82
34. Søndergaard, N. C., Ulstrup, J., Jortner, J. 1976. *Chem. Phys.* 17:417–22
35. Newton, M. D. 1982. *Faraday Discuss. Chem. Soc.* 74:101
36. Brunschwig, B., Logan, J., Newton, M. D., Sutin, N. 1980. *J. Am. Chem. Soc.* 102:5798–5809
37. Siders, P., Marcus, R. A. 1981. *J. Am. Chem. Soc.* 103:741–47
38. Buhks, E., Bixon, M., Jortner, J., Navon, G. 1979. *Inorg. Chem.* 18:2014–18
39. Delahay, P. 1983. *Chem. Phys. Lett.* 96:613–18
40. Kestner, N. R., Logan, J., Jortner, J. 1974. *J. Phys. Chem.* 78:2148–66
41. Webman, I., Kestner, N. R. 1982. *J. Chem. Phys.* 77:2387–98
42. Buhks, E., Bixon, M., Jortner, J., Navon, G. 1981. *J. Phys. Chem.* 85:3759–62
43. Dakhnovskii, Yu. I., Ovchinnikov, A. A. 1983. *Chem. Phys.* 80:17–27
44. Marcus, R. A. 1982. *Faraday Discuss. Chem. Soc.* 74:307–8
45. Levich, V. G. 1966. *Adv. Electrochem. Electrochem. Eng.* 4:249–371
46a. Marcus, R. A. 1965. *J. Chem. Phys.* 43:1261–74
46b. Cannon, R. D. 1977. *Chem. Phys. Lett.* 49:299–304
47. German, E. D., Kuznetsov, A. M. 1981. *Electrochim. Acta* 26:1595–1608
48. Dogonadze, R. R., Urushadze, Z. D. 1971. *J. Electroanal. Chem.* 32:235–45
49. Calef, D. F., Wolynes, P. G. 1983. *J. Phys. Chem.* 87:3387–3400
50. Calef, D. F., Wolynes, P. G. 1983. *J. Chem. Phys.* 78:470–82
51a. Efrima, S., Bixon, M. 1979. *J. Chem. Phys.* 70:3531–35
51b. Bixon, M. 1982. *Faraday Discuss. Chem. Soc.* 74:103–4
52. Alexandrov, I. V. 1980. *Chem. Phys.* 51:449–57
53. Zusman, L. D. 1980. *Chem. Phys.* 49:295–304
54. Helman, A. B. 1983. *Chem. Phys.* 79:235–44

55. Helman, A. B. 1982. *Chem. Phys.* 65: 271–79
56. Shushin, A. I. 1981. *Chem. Phys.* 60: 149–60
57. Yakobson, B. I., Burshtein, A. I. 1980. *Chem. Phys.* 49: 385–95
58. Van der Zwan, G., Hynes, J. T. 1982. *J. Chem. Phys.* 76: 2993–3001
59. Marcus, R. A., Sutin, N. 1975. *Inorg. Chem.* 14: 213–16
60. Weaver, M. J., Yee, E. L. 1980. *Inorg. Chem.* 19: 1936–46
61. Efrima, S., Bixon, M. 1976. *J. Chem. Phys.* 64: 3639–47
62. Kuznetsov, A. M., Ulstrup, J. 1981. *J. Chem. Phys.* 75: 2047–55
63. Holstein, T. 1959. *Ann. Phys.* 8: 343–89
64. Jortner, J. 1976. *J. Chem. Phys.* 64: 4860–67
65. Ulstrup, J., Jortner, J. 1975. *J. Chem. Phys.* 63: 4358–68
66. Webman, I., Kestner, N. R. 1979. *J. Phys. Chem.* 83: 451–56
67. Dogonadze, R. R., Kuznetsov, A. M., Zakaraya, M. G., Ulstrup, J. 1979. See Ref. 2, pp. 144–71
68. Logan, J., Newton, M. D. 1983. *J. Chem. Phys.* 78: 4086–91
69. Van Duyne, R. P., Fischer, S. F. 1974. *Chem. Phys.* 5: 183–97
70. Buhks, E., Bixon, M., Jortner, J. 1981. *J. Phys. Chem.* 85: 3763–66
71. Scher, H., Holstein, T. 1981. *Philos. Mag. B* 44: 343–56
72. Jortner, J. 1980. *J. Am. Chem. Soc.* 102: 6676–86
73. Holstein, T. 1978. *Philos. Mag. B* 37: 49–62
74. Dogonadze, R. R., Kuznetsov, A. M., Marsagishvili, T. A. 1980. *Electrochim. Acta* 25: 1–28
75. Brunschwig, B. 1984. To be published
76. Efrima, S., Bixon, M. 1979. *Chem. Phys.* 36: 161–69
77. Holstein, T. 1978. *Philos. Mag. B* 37: 499–526
78a. Schmidt, P. P. 1976. *J. Chem. Soc. Faraday Trans. 2* 72: 1736–40
78b. Bixon, M., Jortner, J. 1982. *Faraday Discuss. Chem. Soc.* 74: 17–29
79. George, P., Griffith, J. S. 1959. In *The Enzymes*, ed. P. D. Boyer, H. Lardy, K. Myrbäck, 1: 347–89. New York: Academic. 785 pp.
80a. Halpern, J., Orgel, L. E. 1960. *Discuss. Faraday Soc.* 29: 32–41
80b. McConnell, H. M. 1961. *J. Chem. Phys.* 35: 508–15
81a. Ratner, M. A. 1978. *Int. J. Quantum Chem.* 14: 675–94
81b. Ratner, M. A., Ondrechen, M. J. 1976. *Mol. Phys.* 32: 1233–45
82a. Stein, C. A., Lewis, N. A., Seitz, G., Baker, A. D. 1983. *Inorg. Chem.* 22: 1124–28
82b. Beratan, D. N., Hopfield, J. J. 1984. *J. Am. Chem. Soc.* 106: 1584–94
83. Jortner, J., Ulstrup, J. 1979. *J. Am. Chem. Soc.* 101: 3744–54
84. Buhks, E., Navon, G., Bixon, M., Jortner, J. 1980. *J. Am. Chem. Soc.* 102: 2918–23
85. Creutz, C., Sutin, N. 1984. In *Inorganic Reactions and Methods*, ed. J. J. Zuckerman. Weinheim: Verlag-Chemie. In press
86. Endicott, J. F., Ramasami, T., Gaswick, D. C., Tamilarasan, R., Heeg, M. J., Brubaker, G. R., Pyke, S. C. 1983. *J. Am. Chem. Soc.* 105: 5301–10
87. Miller, J. R., Beitz, J. V. 1981. *J. Chem. Phys.* 74: 6746–56
88a. Larsson, S. 1981. *J. Am. Chem. Soc.* 103: 4034–40
88b. Larsson, S. 1984. *J. Phys. Chem.* 88: 1321–23
89. Mok, C. Y., Zanella, A. W., Creutz, C., Sutin, N. 1984. *Inorg. Chem.* In press
90. Marcus, R. A. 1981. *Int. J. Chem. Kin.* 13: 865–72
91a. Cox, P. A. 1980. *Chem. Phys. Lett.* 69: 340–43
91b. Girerd, J.-J. 1983. *J. Chem. Phys.* 79: 1766–75
92. Hirata, F., Friedman, H. L., Holz, M., Hertz, H. G. 1980. *J. Chem. Phys.* 73: 6031–38
93. Neilson, G. W., Enderby, J. E. 1984. To be published
94. Newton, M. D. 1982. *Faraday Discuss. Chem. Soc.* 74: 108–11
95. Richardson, D. E., Taube, H. 1983. *J. Am. Chem. Soc.* 105: 40–51
96. Dolin, S. P., Dongonadze, R. R., German, E. D. 1977. *J. Chem. Soc. Faraday Trans. 1* 73: 648–54
97. Hush, N. S. 1982. *Am. Chem. Soc. Symp. Ser.* 198: 301–29
98. Khan, S. U. M., Wright, P., Bockris, J. O'M. 1977. *Sov. Electrochem.* 13: 774–83 (Engl. Transl.)
99. Larsson, S. 1982. *Chem. Phys. Lett.* 90: 136–39
100. Larsson, S. 1983. *J. Chem. Soc. Faraday Trans. 2* 79: 1375–88
101. Hopfield, J. J. 1974. *Proc. Natl. Acad. Sci. USA* 71: 3640–44
102. Linderberg, J., Ratner, M. A. 1981. *J. Am. Chem. Soc.* 103: 3265–71
103. Lee, C.-Y., DePristo, A. E. 1983. *J. Am. Chem. Soc.* 105: 6775–81
104a. Jortner, J., Rice, S. A. 1965. In *Physics of Solids at High Pressures*, ed. C. T. Tomizuka, R. M. Emrick, pp. 63–168. New York: Academic. 595 pp.
104b. Silbey, R., Jortner, J., Rice, S. A., Vala,

M. T. 1965. *J. Chem. Phys.* 42:733–37; 43:2925–26
105. Summerville, R. H., Hoffman, R. 1976. *J. Am. Chem. Soc.* 98:7240–54
106. Mulliken, R. S. 1942. *J. Am. Chem. Soc.* 74:811–24
107. Creutz, C. 1983. *Prog. Inorg. Chem.* 30:1–73
108. Curtis, J. C., Meyer, T. J. 1982. *Inorg. Chem.* 21:1562–71
109. Redi, M., Hopfield, J. J. 1980. *J. Chem. Phys.* 72:6651–60
110. Miller, J. R., Beitz, J. V., Huddleston, R. K. 1984. *J. Am. Chem. Soc.* In press
111. Goodenough, J. B. 1963. *Magnetism and the Chemical Bond*, New York: Wiley. 393 pp.
112. Anderson, P. W. 1959. *Phys. Rev.* 115:2–13
113. Haberkorn, R., Michael-Beyerle, M. E., Marcus, R. A. 1979. *Proc. Natl. Acad. Sci. USA* 76:4185–88
114. Okamura, M. Y., Fredkin, D. R., Isaacson, R. A., Feher, G. 1979. See Ref. 2, pp. 729–43
115. Zamaraev, K. I., Khairutdinov, R. F., Miller, J. R. 1978. *Chem. Phys. Lett.* 57:311–15
116. Huddleston, R. K., Miller, J. R. 1982. *J. Phys. Chem.* 86:200–3
117. Calcaterra, L. T., Closs, G. L., Miller, J. R. 1983. *J. Am. Chem. Soc.* 105:670–71
118. Huddleston, R. K., Miller, J. R. 1983. *J. Chem. Phys.* 79:5337–44
119. Miller, J. R., Hartman, K. W., Abrash, S. 1982. *J. Am. Chem. Soc.* 104:4296–98
120. Miller, J. R., Peeples, J. A., Schmitt, M. J., Closs, G. L. 1982. *J. Am. Chem. Soc.* 104:6488–93
121. Guarr, T., McGuire, M., Straugh, S., McLendon, G. 1983. *J. Am. Chem. Soc.* 105:616–18
122. Straugh, S., McLendon, G., McGuire, M., Guarr, T. 1983. *J. Phys. Chem.* 87:3579–81
123. McGourty, J. L., Blough, N. V., Hoffman, B. M. 1983. *J. Am. Chem. Soc.* 105:4470–72
124. Winkler, J. R., Nocera, D. G., Yokom, K. M., Bordignon, E., Gray, H. B. 1982. *J. Am. Chem. Soc.* 104:5798–5800
125. Isied, S. S., Worosila, G., Atherton, S. J. 1982. *J. Am. Chem. Soc.* 104:7659–61
126. Lindsey, J. S., Mauzerall, D. C., Linschitz, H. 1983. *J. Am. Chem. Soc.* 105:6528–29
127. Marcus, R. A., Sutin, N. 1984. *Biochim. Biophys. Acta*. In preparation
128. Brunschwig, B. S., Creutz, C., Macartney, D. H., Sham, T.-K., Sutin, N. 1982. *Faraday Discuss. Chem. Soc.* 74:113–27
129. Balzani, V., Scandola, F., Orlandi, G., Sabbatini, N., Indelli, M. T. 1981. *J. Am. Chem. Soc.* 103:3370–78
130. Friedman, H. L. 1982. *Faraday Discuss. Chem. Soc.* 74:198–99
131. Fröhlich, H. 1958. *Theory of Dielectrics*, pp. 72–73. London: Oxford Univ. Press. 180 pp. 2nd ed.
132. Hubbard, J., Onsager, L. 1977. *J. Chem. Phys.* 67:4850–57
133. Hubbard, J. B. 1978. *J. Chem. Phys.* 68:1649–64
134. Friedman, H. L. 1983. *J. Chem. Soc. Faraday Trans.* 79:1465–67
135a. Friedman, H. L. 1981. *J. Chem. Phys.* 76:1092–1105
135b. Friedman, H. L. 1978. In *Protons and Ions Involved in Fast Dynamic Motion*, ed. P. Laszlo, pp. 27–42. Amsterdam: Elsevier. 452 pp.
136. Bulmer, J. T., Irish, D. E., Ödberg, L. 1975. *Can. J. Chem.* 53:3806–11
137. Kosower, E. M., Dodiuk, H., Kanety, H. 1978. *J. Am. Chem. Soc.* 100:4179–89
138. Huppert, D., Kanety, H., Kosower, E. M. 1981. *Chem. Phys. Lett.* 84:48–53
139. Huppert, D., Kanety, H., Kosower, E. M. 1982. *Faraday Discuss. Chem. Soc.* 74:161–75
140. Kosower, E. M., Kanety, H., Dudluk, H., Striker, G., Jovin, T., Boni, H., Huppert, D. 1983. *J. Phys. Chem.* 87:2479–84
141a. Kosower, E. M. 1982. *Faraday Discuss. Chem. Soc.* 74:199–200
141b. Kosower, E. M., Huppert, D. 1983. *Chem. Phys. Lett.* 96:433–35
142. Creutz, C., Kroger, P., Matsubara, T., Netzel, T. L., Sutin, N. 1979. *J. Am. Chem. Soc.* 101:5442–44
143. Meyer, T. J. 1979. *Chem. Phys. Lett.* 64:417–20
144. Brown, G. M., Sutin, N. 1979. *J. Am. Chem. Soc.* 101:883–92
145. Haim, A. 1983. *Pure Appl. Chem.* 55:89–98
146. Li, T. T.-T., Brubaker, C. H. Jr. 1981. *J. Organomet. Chem.* 216:223–34
147. Chan, M.-S., Wahl, A. C. 1980. *J. Phys. Chem.* 86:126–30
148. Brandon, J. R., Dorfman, L. M. 1970. *J. Chem. Phys.* 53:3849–56
149. Yang, E. S., Chan, M.-S., Wahl, A. C. 1980. *J. Phys. Chem.* 84:3094–99
150. Anderson, A., Bonner, N. A. 1954. *J. Am. Chem. Soc.* 76:3826–30
151. Silverman, J., Dodson, R. W. 1952. *J. Phys. Chem.* 56:846–51
152. Beattie, J. K., Best, S. P., Skelton, B. W., White, A. H. 1981. *J. Chem. Soc. Dalton Trans.*, pp. 2105–11
153. Habib, H. S., Hunt, J. P. 1966. *J. Am. Chem. Soc.* 88:1668–71

154. Bernhard, P., Bürgi, H. B., Hauser, J., Lehmann, H., Ludi, A. 1982. *Inorg. Chem.* 21:3936–41
155. Böttcher, W., Brown, G. M., Sutin, N. 1979. *Inorg. Chem.* 18:1447–51
156. Meyer, T. J., Taube, H. 1968. *Inorg. Chem.* 7:2369–79
157. Stynes, H. C., Ibers, J. A., 1971. *Inorg. Chem.* 10:2304–8
158. Ruff, I., Zimonyi, M. 1973. *Electrochim. Acta* 18:515–16
159. Young, R. C., Keene, F. R., Meyer, T. J. 1977. *J. Am. Chem. Soc.* 99:2468–73
160a. Geselowitz, D., Taube, H. 1982. *Adv. Inorg. Bioinorg. Mechanisms* 1:391–407
161. Dwyer, F. P., Sargeson, A. M. 1961. *J. Phys. Chem.* 65:1892–94
162. Neumann, H. M., quoted in Farina, R., Wilkins, R. G. 1968. *Inorg. Chem.* 7:514–18
163. Szalda, D. J., Creutz, C., Mahajan, D., Sutin, N. 1983. *Inorg. Chem.* 22:2372–79
164. Sargeson, A. M. 1979. *Chem. Brit.* 15:23–27
165. Szalda, D. J., Macartney, D. H., Sutin, N. 1984. *Inorg. Chem.* 23: In press
166. Macartney, D. H., Sutin, N. 1983. *Inorg. Chem.* 22:3530–34
167. Logan, J., Newton, M. D., Noell, J. O. 1984. *Int. J. Quant. Chem. Symp.* 18: In press
168. Endicott, J. F., Brubaker, G. R., Ramasami, T., Kumar, K., Dwarakanath, K., Cassel, J., Johnson, D. 1983. *Inorg. Chem.* 22:3754–62
169. Guarr, T., Buhks, E., McLendon, G. 1983. *J. Am. Chem. Soc.* 105:3763–67
170. Weaver, M. J., Nettles, S. M. 1980. *Inorg. Chem.* 19:1641–46
171. Jortner, J. 1982. *Faraday Discuss. Chem. Soc.* 74:111
172. Endicott, J. F., Durham, B., Kumar, K. 1982. *Inorg. Chem.* 21:2437–44
173. Ford-Smith, M. H., Sutin, N. 1961. *J. Am. Chem. Soc.* 63:1830–34
174. Dulz, G., Sutin, N. 1963. *Inorg. Chem.* 2:917–21
175. Lin, C.-T., Böttcher, W., Chou, M., Creutz, C., Sutin, N. 1976. *J. Am. Chem. Soc.* 98:6536–44
176. Brunschwig, B. S., Sutin, N. 1979. *Inorg. Chem.* 18:1731–36
177. Macartney, D. H., Sutin, N. 1983. *Inorg. Chim. Acta.* 74:211–28
178. Hupp, J. T., Weaver, M. J. 1983. *Inorg. Chem.* 22:2557–64
179. Efrima, S., Bixon, M. 1974. *Chem. Phys. Lett.* 25:34–37
180. Marcus, R. A. 1968. *J. Phys. Chem.* 72:891–99
181. Brunschwig, B., Sutin, N. 1978. *J. Am. Chem. Soc.* 100:7568–77
182. Hush, N. S. 1980. In *Mixed-Valence Compounds*, ed. D. B. Brown, pp. 151–88. Dordrecht: Reidel. 519 pp.
183a. Wallace, W. L., Bard, A. J. 1979. *J. Phys. Chem.* 83:1350–57
183b. Balzani, V., Boletta, F., Ciano, M., Maestri, M. 1983. *J. Chem. Ed.* 60:447–50
184. Caspar, J. V., Sullivan, B. P., Kober, E. M., Meyer, T. J. 1982. *Chem. Phys. Lett.* 91:91–95
185. Meyer, T. J. 1983. *Prog. Inorg. Chem.* 30:389–440
186. Chou, M., Creutz, C., Sutin, N. 1977. *J. Am. Chem. Soc.* 99:5615–23
187. Frese, K. W. Jr. 1981. *J. Phys. Chem.* 85:3911–16
188. Haim, A., Sutin, N. 1976. *Inorg. Chem.* 15:476–78

References added in proof:

160b. Hammershøi, A., Geselowitz, D., Taube, H. 1984. *Inorg. Chem.* 23:979–82
189. Miller, J. R., Calcaterra, L. T., Closs, G. L. 1984. *J. Am. Chem. Soc.* 106:3047–49
190. Allen, A. O., Gangwer, T. E., Holroyd, R. A. 1975. *J. Phys. Chem.* 79:25–31

HUMAN EFFECTS ON THE GLOBAL ATMOSPHERE[1]

Harold S. Johnston

Department of Chemistry, University of California, and Materials and Molecular Research Division, Lawrence Berkeley Laboratory, Berkeley, California 94720

INTRODUCTION

This review considers whether human activities can significantly change important functions of the global atmosphere by altering the amount or distribution of certain trace species. It deals with three specific topics: stratospheric ozone, the role of species other than carbon dioxide on the "greenhouse effect," and certain recently recognized atmospheric consequences of a large scale nuclear war (1).

This subject is interdisciplinary and currently under active investigation. The present review is not especially intended for experts in this interdisciplinary field, but rather it attempts to point out recent developments to interested onlookers, especially to chemists.

In view of the great mass of the atmosphere relative to that of all human objects and in view of the large amount of energy delivered by the sun to the earth each day compared to that of all human activities, one is inclined to assume that the global atmosphere is impervious to human intervention. It is interesting to display some of these numbers. Assuming four billion people on earth, the mass of the entire atmosphere is about 1.3 million tons per person, the mass of carbon dioxide in the atmosphere is 670 tons per person, the mass of atmospheric ozone is 1.2 tons per person, and the mass of stratospheric nitrogen oxides (calculated as NO_2) is 2.4 kg per person.

---

[1] The submitted manuscript has been authored by a contractor of the US Government under contract No. DE-AC03-76SF00098. Accordingly, the US Government retains a nonexclusive, royalty-free license to publish or reproduce the published form of this contribution, or allow others to do so, for US Government purposes.

From these numbers one feels certain that human activity will not change the mass of the atmosphere as a whole; it is surprising that human activity has apparently increased atmospheric carbon dioxide, but it is not surprising that stratospheric nitrogen oxides or other stratospheric species in similar amounts could be changed by human enterprises.

Nitrogen oxides play a major role in catalytically destroying stratospheric ozone; stratospheric ozone has a strong influence on the temperature structure of the atmosphere and provides a vital shield for the surface of the earth against ultraviolet radiation; and through this linkage it is plausible that a relatively small quantity of nitrogen oxides (one within the scope of human activities) could have an impact on the global atmosphere (2–4). In the stratosphere, atomic chlorine and chlorine oxide, ClO, catalytically destroy ozone in a manner similar to that by nitrogen oxides (5–7). During the past decade the release of the chlorofluorocarbons, CF_2Cl_2 and $CFCl_3$, to the atmosphere has been at a rate of about a billion kilograms per year (8), and the atmospheric lifetime of these species is several decades (9). On these general grounds it is plausible that humanly possible releases of chlorofluorocarbons (CFC) could reduce stratospheric ozone, and large research programs are underway examining this situation.

In the greenhouse effect, added infrared active gases in the atmosphere decrease the temperature of the stratosphere and increase the temperature of the earth's surface. These temperature changes are parallel effects in a complex mechanism, not a matter of one causing the other. The major natural greenhouse gases are water vapor, carbon dioxide, and ozone. Between the absorption bands of these major infrared absorbers there are several "windows" through which thermal radiation from the surface is readily radiated to space (10). Any substance with a very strong infrared absorption coefficient in one of these windows can make a significant contribution to the greenhouse effect if it is present even at a few parts per billion in the atmosphere. The chlorofluorocarbons are just such a case (10–13). Methane and nitrous oxide, N_2O, are examples of naturally occurring trace species occupying spots in some of the windows between the major greenhouse gases. These "greenhouse windows" provide situations wherein the thermal structure of the global atmosphere is sensitive to special substances in amounts readily provided by human activities.

The energy from the sun that falls on the earth every hour is larger than the energy that would be released to the atmosphere by a major nuclear war involving 10,000 MT of bombs, and on this basis one tends to dismiss the climatic impact of a nuclear war. However, recently Crutzen & Birks (1) have pointed out that such a war would cause forest, city, and oil-well fires and that these would form a persistent smoke cloud over the northern hemisphere. Incoming solar radiation would be absorbed high in the atmosphere by such smoke and by bomb-raised dust, instead of at the surface of the earth. It has been found that this displaced absorption of sunlight would probably have a profound effect on the climate for several months (14–16). This topic is discussed in the last section of this review.

OBSERVED ATMOSPHERIC TRENDS

Trace Species

During this review occasional reference is given to chemical reactions and to certain groups of such reactions. These reactions are assembled in Table 1 and assigned Equation numbers 1–13(a–mm).

Recent atmospheric measurements have established that several trace species are now increasing in amount. The entirely synthetic substances, CF_2Cl_2 and $CFCl_3$, were detected in the atmosphere by Lovelock et al (17) in the early 1970s, and these and related compounds have been found to be increasing in the atmosphere (9). These observed increases are generally consistent with photochemical models, which calculate the atmospheric concentrations from known manufacturing rates and photochemical destruction in the stratosphere. An example of the manufacturing history of CF_2Cl_2, for example, is given by Figure 1, curve A; the other curves are defined in a later section.

There is agreement between several workers that methane has increased at a rate of about 1 to 2% per year during the past four years (18–21), and there is weak evidence for an increase at a rate of about 0.5% per year between 1965 and 1975 (18). Air trapped in ice cores indicate a long-term increasing trend for methane (22). The primary source of methane is the natural biological carbon cycle (23, 19), which may be perturbed by such human activities as agriculture, forestry, and cattle raising. The main sink for atmospheric methane is oxidation by hydroxyl radicals in the troposphere and stratosphere. It is conceivable that human activity has decreased atmospheric hydroxyl radicals by combustion-generated nitrogen oxides, including aircraft, by way of reactions (y) and (z). The methane trend may be affected by human modification of methane source or sink, or both.

An increase of atmospheric carbon dioxide since 1958 has been well established (24, 25). Air samples from ice cores (26) indicate that atmospheric carbon dioxide is about 1.25-fold greater than it was a century ago, and it may be twice as great as it was 20,000 years ago.

Weiss (27, and quoted in 28) discovered that nitrous oxide has been increasing in the atmosphere since 1962 at a rate of about 0.2% per year. Others have confirmed this trend (9). There are two quite different sources of nitrous oxide. It is a by-product of the natural nitrogen cycle, involving considerations of soils, forests, fresh waters, and oceans (29–31). In 1976 two articles were published (32, 33) that found significant amounts of N_2O in smoke stacks of power plants. There has been a curious lack of further work on this source. Since the manufacture of nitrogen fertilizers fixes nitrogen (29) at more or less the global rate of all natural sources (34), the increase of nitrous oxide may be related to such manufacture, but the magnitude and time scale of such a source is quite uncertain (35). The observed increasing trend could be caused by combustion, by manufactured fertilizers, by both, or by natural processes.

Table 1 Photochemical reactions in the atmosphere classified in various ways

Ozone production from solar radiation	1.
(a) $O_2 + hv$ (below 242 nm) $\to O + O$	
(b) $O + O_2 + M \to O_3 + M$ (twice)	
net: $3O_2 \to 2O_3$	
Electronic states of atomic oxygen	2.
(c) $O_3 + hv$ (below 310 nm) $\to O_2 + O(^1D)$	
(d) $O_3 + hv$ (above 310 nm) $\to O_2 + O(^3P)$	
(e) $O_3 + hv$ (any) $\to O_2 + O$ (any)	
Ozone destruction by atomic oxygen	3.
(f) $O_3 + O \to 2O_2$	
NO_x catalyzed ozone destruction	4.
(g) $NO + O_3 \to NO_2 + O_2$	
(h) $NO_2 + O \to NO + O_2$	
net: $O_3 + O \to 2O_2$	
ClX catalyzed ozone destruction after CFC plus ultraviolet yields atomic chlorine	5.
(i) $Cl + O_3 \to ClO + O_2$	
(j) $ClO + O \to Cl + O_2$	
net: $O_3 + O \to 2O_2$	
NO_x null cycle	6.
(g) $NO + O_3 \to NO_2 + O_2$	
(l) $NO_2 + hv \to NO + O$	
(b) $O + O_2 + M \to O_3 + M$	
net: null	
HCl as ClX reservoir	7.
(m) $Cl + CH_4 \to HCl + CH_3$	
(n) $HO + HCl \to H_2O + Cl$	
HNO_3 as NO_x reservoir	8.
(p) $HO + NO_2 + M \to HNO_3 + M$	
(q) $HNO_3 + hv \to HO + NO_2$	
ClX and NO_x mutual reservoir	9.
(r) $ClO + NO_2 + M \to ClONO_2 + M$	
Production of HO_x and of NO_x	10.
(s) $O(^1D) + H_2O \to 2HO$	
(t) $O(^1D) + N_2O \to 2NO$	
(u) $O(^1D) + M \to O(^3P) + M$	

Loss of HO_x
- (v) $HO + HOO \rightarrow H_2O + O_2$
- (w) $HOO + HOO \rightarrow H_2O_2 + O_2$
- (x) $HO + H_2O_2 \rightarrow H_2O + HOO$
- (y) $HO + HNO_3 \rightarrow H_2O + NO_3$
- (z) $HO + HOONO_2 \rightarrow H_2O + O_2 + NO_2$

HO_x catalyzed ozone destruction
- A.
  - (aa) $HO + O_3 \rightarrow HOO + O_2$
  - (bb) $HOO + O_3 \rightarrow HO + 2O_2$
  - net: $2O_3 \rightarrow 3O_2$
- B.
  - (aa) $HO + O_3 \rightarrow HOO + O_2$
  - (cc) $HOO + O \rightarrow HO + O_2$
  - net: $O_3 + O \rightarrow O_2 + O_2$
- C.
  - (dd) $HO + O \rightarrow H + O_2$
  - (ee) $H + O_3 \rightarrow HO + O_2$
  - net: $O_3 + O \rightarrow O_2 + O_2$

Methane-NO_x-smog reactions
A. Methane to carbon monoxide
- (ff) $CH_4 + HO \rightarrow CH_3 + H_2O$
- (gg) $CH_3 + O_2 + M \rightarrow CH_3OO + M$
- (hh) $CH_3OO + NO \rightarrow CH_3O + NO_2$
- (ii) $CH_3O + O_2 \rightarrow H_2CO + HOO$
- (jj) $H_2CO + h\nu \rightarrow H_2 + CO$
- (kk) $HOO + NO \rightarrow HO + NO_2$
- (l) $NO_2 + h\nu \rightarrow NO + O$ (twice)
- (b) $O + O_2 + M \rightarrow O_3 + M$ (twice)
- net: $CH_4 + 4O_2 \rightarrow CO + H_2 + H_2O + 2O_3$

B. Carbon monoxide to carbon dioxide
- (ll) $CO + HO \rightarrow CO_2 + H$
- (mm) $H + O_2 + M \rightarrow HOO + M$
- (kk) $HOO + NO \rightarrow HO + NO_2$
- (l) $NO_2 + h\nu \rightarrow NO + O$
- (b) $O + O_2 + M \rightarrow O_3 + M$
- net: $CO + 2O_2 \rightarrow CO_2 + O_3$

C. Hydrogen to water
- (pp) $H_2 + HO \rightarrow H_2O + H$
- (mm) $H + O_2 + M \rightarrow HOO + M$
- (kk) $HOO + NO \rightarrow HO + NO_2$
- (l) $NO_2 + h\nu \rightarrow NO + O$
- (b) $O + O_2 + M \rightarrow O_3 + M$
- net: $H_2 + 2O_2 \rightarrow H_2O + O_3$

D. Combined CH_4 to CO_2 and H_2O
- net: $CH_4 + 8O_2 \rightarrow CO_2 + 2H_2O + 4O_3$

E. Competition in all cases
- (kk) $HOO + NO \rightarrow HO + NO_2$
- (bb) $HOO + O_3 \rightarrow HO + 2O_2$

F. CH_4 to $CO_2 + H_2O$ with (kk), replaced by (bb)
- net: $CH_4 + 2O_3 \rightarrow CO_2 + 2H_2O + O_2$

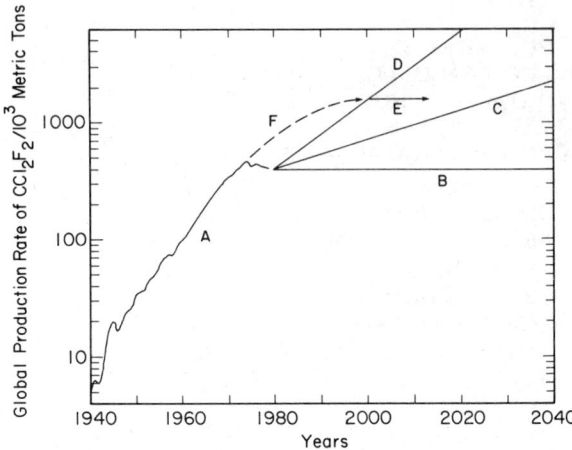

Figure 1 A. History of annual rate of manufacturing of CF_2Cl_2. B. Assumed constant rate of CFC production after 1980. C. Assumed increase in CFC production after 1980 at 3% per year. D. Assumed 7% per year increase after 1980. E. Assumed 7% per year increase between 1980 and 2000 with constant rate thereafter. F. Hypothetical "uninterrupted" growth curve from 1974 to 2000 with saturation at the level of E, which is a factor 3.3 greater than 1974 production rate.

Wuebbles (36) examined the currently measured trends of various atmospheric species, and he estimated the trace-gas composition of the atmosphere back to 1850. For four CFC species, nitrous oxide, methane, and carbon dioxide, the smoothed trends derived by Wuebbles are given by Figure 2 where the logarithm of the mole fraction is plotted against time for 1960 to 1982. This figure demonstrates the rapid recent fractional growth of the chlorinated species compared to the fractional rate of growth of nitrous oxide, methane, and carbon dioxide. The three panels of Figure 2 have widely different scales of mixing ratio by volume (mole-fraction): parts per trillion (10^{-12}), parts per billion (10^{-9}), and parts per million (10^{-6}), respectively.

Ozone Trends

Stratospheric ozone is strongly variable in altitude, latitude, and season. Intense efforts have been made to detect temporal trends in the ozone vertical column, especially during the period 1970–1980. Since 1958 there have been 30 or more ground-based ozone-observing stations, with the large majority being in the northern hemisphere. In analyzing these data Komhyr et al (37) detected an increase in ozone over the period, 1960–1970, in many of these stations. London & Oltmans (38) constructed global maps of ozone over the period 1957–1975, and they found a significant increase of ozone in the northern hemisphere (NH) and no significant trend in the southern hemisphere (SH). Other empirical studies of the total ozone record (39, 40) are in essential agreement with these results. Advanced statistical techniques have been applied to the total ozone data in order to separate long-term trends from short-term cycles and from noise. On the basis of 36 ozone observing stations, estimates were made of the ozone trends over the ten year period 1970–1979 (28). In terms of percentage change of total ozone for the decade, the results were:

Bloomfield et al (41)	$+1.7 \pm 2.0$	14.
St. John et al (42)	$+1.1 \pm 1.2$	
Reinsel (43)	$+0.5 \pm 1.4$	

All three studies indicate a small ozone increase for the decade, even though the trend is not significant to the 95% confidence level. In a later section these indicated trends are compared with the predictions of theoretical models.

The vertical distribution of ozone is measured from ground-level stations by spectroscopic observation of scattered sunlight at various solar zenith

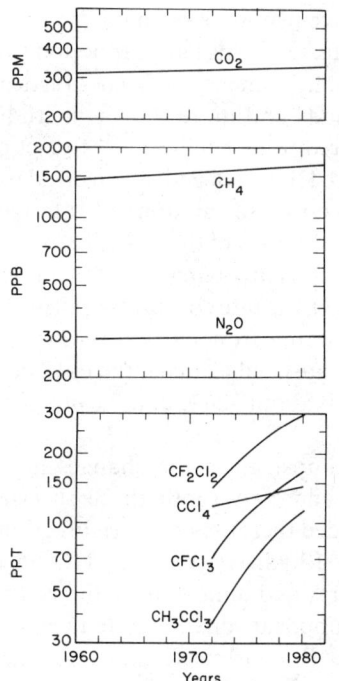

Figure 2 Trends in observed atmospheric trace species. Smoothed observed mole fractions of four chlorinated carbon compounds, nitrous oxide, methane, and carbon dioxide as a function of time (36).

angles, in situ with balloons, and from satellites by spectroscopic observation of backscattered solar radiation. The data have gaps with respect to time and are noisy. These data are being examined for trends, especially for the period 1970–1980 (44, 46), with the following preliminary results:

	Ozone change	15.
Middle troposphere, NH	+7%	
Middle troposphere, SH	0%	
Upper stratosphere	−4%	

LABORATORY PHOTOCHEMISTRY

Critical Data Tabulations

Approximately once a year the National Aeronautics and Space Administration issues an updated edition of *Chemical Kinetics and Photochemical Data for Use in Stratospheric Modeling*, through the Jet Propulsion Laboratory, Pasadena, California (47). Evaluation number 6 was published in September 1983. The last five of these volumes spaced over the years 1979–1983, similar tables in a 1977 report by the National Bureau of Standards (48), and the final report in 1975 of the Climatic Impact Assessment Program (49) give a history of unparalleled growth of this branch of science. These volumes give convenient tables of data and critical discussions of the reactions and of new data. Over this ten-year period some rate coefficients have been confirmed; but some have been drastically changed as the result of improved methods and more thorough studies. Several new reactions have been discovered. Rate coefficients of reactions by free radicals derived from water, H, HO, HOO, collectively referred to as HO_x, have undergone many revisions, and these revisions of rate parameters have had major impacts on the predictions of theoretical models of atmospheric photochemistry. For 11 reactions involving HO_x free radicals from Table 1 (*s, v, w, y, z, aa, bb, cc, dd, ee, kk*), the rate constants appropriate to 30 kilometer altitude are plotted against time from 1975 to 1983 (Figure 3). These are the values derived from the rate coefficients in the data tables (47–49). At 30 kilometers the temperature is about 226 K and the pressure is about 0.012 atmospheres.

By a surprising double coincidence almost all of the changes in rate coefficients that occured between 1975 and 1979 (both those shown in Figure 3 and the other reactions not included there) acted in the direction of increasing the calculated concentration of hydroxyl radicals, HO, in the troposphere and in the lower stratosphere; and almost all of the changes between 1979 and 1983 acted in the opposite direction, reducing the calculated amount of HO in these regions. For various technical reasons it has not been possible to measure the concentration of hydroxyl radicals in the lower stratosphere; and this quantity, which is critical to many key atmospheric processes, is known largely through model calculations. The HO_x family is produced primarily by reaction (s) and destroyed by reactions (v, w, x, y, z). The calculated rate of HO_x production has remained relatively constant over the ten-year period of Figure 3, but the rate constants for HO_x destruction, k_y and especially k_v, decreased between 1975 and 1977. Between 1977 and 1983, the apparent rate constant for reaction (v) has steadily crept upward; and by 1981 k_y was found to be much larger, especially at low temperatures. Late in the 1970s the newly recognized species, peroxy-nitric acid, was included in the tabulations, and in 1981 there was a dramatic increase in its tabulated rate constant.

Hydroxyl radicals are important in scavenging many species in the troposphere, such as hydrocarbons, CO, CH_3Cl, CH_3CCl_3, NO_2. They play a key role in determining the efficiency of catalytic ozone destruction, both by the oxides of nitrogen and by chlorine radicals; and they initiate the methane-NO_x-smog reactions, which are an important source of ozone in the upper troposphere.

Catalytic Cycles

The NO_x catalytic cycle that reduces ozone, Eq. 4, involves NO and NO_2. The partitioning of nitrogen oxides between the catalytically active form

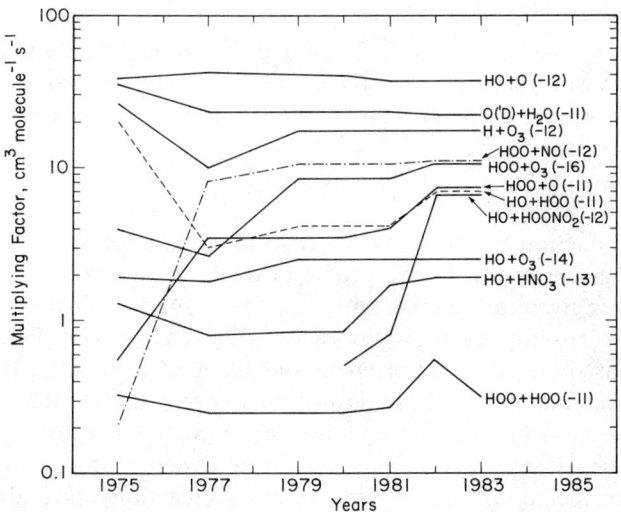

Figure 3 Trends in values of tabulated rate constants from 1975 to 1983. The rate constant appropriate to 30 km altitude (226 K, 0.012 atm) is given by the numerical factor times 10 to the power given in parentheses.

NO_2 and the inert reservoir form HNO_3 is controlled by the reactions, $HO + NO_2 = HNO_3$, $HNO_3 + h\nu = HO + NO_2$, Eq. 8. The ClX (Cl + ClO) catalytic cycle that reduces ozone, Eq. 5, it isomorphic with the NO_x catalytic cycle, Eq. 4. The partitioning of chlorine radicals between the catalytically active form Cl and the inert reservoir form HCl is largely controlled by the reactions $Cl + CH_4 = HCl + CH_3$, $HO + HCl = H_2O + Cl$, Eq. 7. Thus the hydroxyl radical plays an opposite role in the NO_x and ClX systems: it binds a catalytically active NO_x radical into the inactive acid form HNO_3, but it releases a catalytically active ClX radical from its inert acid form HCl (50).

Methane-Smog Reactions

Hydroxyl radicals, and to a lesser extent chlorine atoms and singlet oxygen atoms, convert methane to the methyl radical, whose irreversible oxidation constitutes the CH_4–NO_x–smog reactions (51). These reactions are presented in stages as Eq. 13. There are minor complications and alternate paths to this mechanism, but these reactions give the essence of the process whereby methane is consumed and ozone is formed. This process is catalytic in NO_x and in HO_x, but it is consumptive, not catalytic, so far as methane is concerned. If one makes the steady-state assumption for the catalysts, the expression for ozone formation is

$$[O_3]/dt = k_{ff}[HO][CH_4]Y, \qquad 16.$$

$$Y = (4A - 2B)/(A + B), \qquad 17.$$

where $A = k_{kk}[NO]$, $B = k_{bb}[O_3]$, and Y is the ozone yield per molecule of methane consumed. If $2k_{kk}[NO] > k_{bb}[O_3]$, the methane-smog reactions produce ozone, but if $2k_{kk}[NO] < k_{bb}[O_3]$, then the methane-smog reactions destroy ozone.

Ozone Production Profile

Ozone production in the atmosphere from the photolysis of molecular oxygen is given by Eq. 1. The profile of ozone production from oxygen photolysis is given in Figure 4, and the rate of ozone production from the methane-NO_x-smog reactions, Eq. 16, is included on the same figure. [This figure is based on a detailed print-out of 1983 model result (D. Wuebbles, personal communication).] The *dashed curve* corresponds to the maximum possible ozone yield (Eq. 17) from this mechanism, that is $2A \gg B$. The *solid curve* on the left side corresponds to the rate of ozone production by this mechanism, including the competition between reactions (*kk*) and (*bb*). The natural upper and middle troposphere is low in nitric oxide (52), so the competition between (*kk*) and (*bb*) is unfavorable to ozone production. An

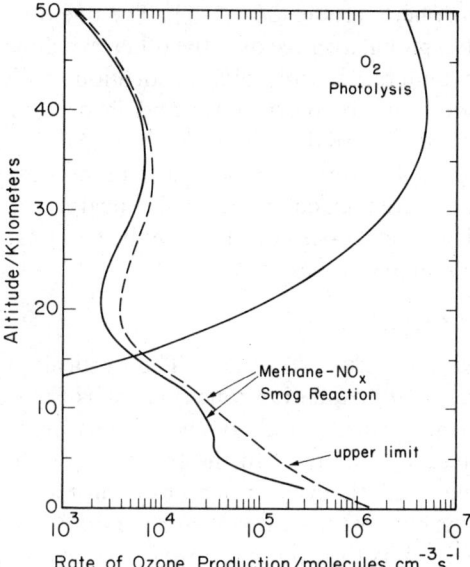

Figure 4 Vertical profile of two rates of ozone production: The rate from oxygen photolysis and the rate from the methane-NO_x-smog reactions. The *dashed curve* is the maximum rate, where $k[HOO][NO] \gg k'[HOO][O_3]$, and the *solid curve* includes the competition between NO and O_3 for HOO.

injection or increase of nitric oxide in the middle and upper troposphere has the potential of tilting the yield, $(4A - 2B)/(A + B)$, from a negative value to a positive value, or it can substantially increase the ozone yield. A comparison of the rate of ozone production by the methane-smog reaction and vertical ozone transport rates in the upper troposphere indicates that the methane-smog reactions are an important but not dominant source of ozone there.

THEORETICAL MODELS

This section reports the results of calculations by theoretical models. It does not in every paragraph include the warning that the facts stated are model results, not necessarily proven, real world events. A large literature and several recent reviews (28, 36, 53–55) describe the interpretations, verifications, and predictions by theoretical models of atmospheric transport, photochemistry, and radiation transfer. For internal consistency, this section uses results from a single theoretical model, the Lawrence Livermore National Laboratory (LLNL) one-dimensional radiative-convective atmospheric model (36, 55–57). The one dimension is vertical, and it represents a global average over the other two dimensions. Incoming solar radiation undergoes molecular absorption, molecular scattering, scattering by clouds and by particulates, and absorption and reflection by the earth's surface. Infrared radiation is emitted by the surface and absorbed and emitted by infrared active gases in the atmosphere. Above 12 km the air temperature is calculated, and chemical reactions occur at these temperatures. The model does not calculate the temperature below 12 km, but uses fixed boundary values there.

Stratospheric Ozone

NATURAL OZONE SOURCES AND SINKS The vertical profile of the two principal sources of atmospheric ozone is given as Figure 4. According to Wuebbles (36), the vertical profile of the various processes that destroy ozone are as given by Figure 5. In the troposphere, the most important ozone-destroying catalytic cycle is one based on the HO and HOO free radicals. From the base of the stratosphere to about 40 kilometers, the NO_x catalytic cycle (Eq. 4) is the dominant ozone-destroying process, and it is over this range that most of the ozone column resides. The ClX catalytic cycle, Eq. 6, is close behind the NO_x cycle, between 35 and 45 kilometers. At about 45 kilometers NO_x, ClX, HO_x, and direct reaction by oxygen atoms are about equally important in destroying ozone. Figures 4 and 5 are useful in qualitatively interpretating the results of complex model calculations.

EFFECTS OF SINGLE PERTURBATIONS As can be seen from Figure 2, many trace species are increasing at once in the atmosphere, and any attempts

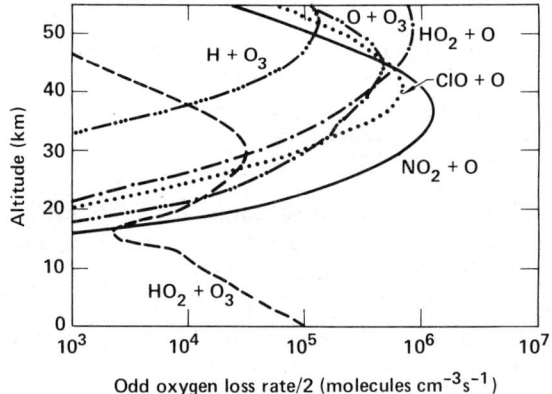

Figure 5 Calculated ozone destruction from major catalytic cycles as a function of altitude (36).

realistically to calculate the effect of future postulated or proposed injections into the atmosphere must include the role of these gases that are also increasing. Even so, there is a long record of diagnostic calculations that consider the increase of only one species at a time.

NO_x injections At Lawrence Livermore National Laboratory, calculations of the long-term ozone changes have been made for a given injection of nitric oxide at 20 km, using then current chemistry (compare Figure 3), for each year since 1974 (28). Originally this injection was related to some proposed large fleet of supersonic transports (4), but the calculation is continued as a demonstration of the effect of changing chemistry on model predictions. The history of these changing predictions for fixed assumed perturbation is given as the upper curve in Figure 6, which was derived from (28). Ozone decreases of about 7% were predicted in 1976, ozone increases of about 2% were predicted by the chemistry of 1979, and since 1981 ozone decreases of about 7% are again obtained.

CFC increase At the same time, LLNL made model predictions for the effect of the single perturbation of chlorofluorocarbons being added at the

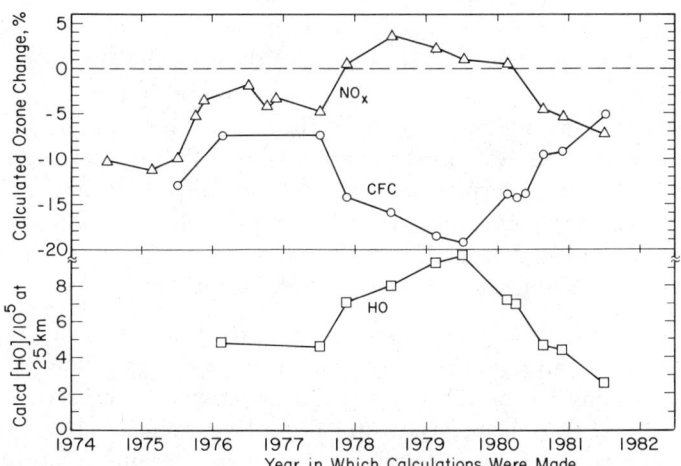

Figure 6 History of results of model calculations that differ only in the differences in chemistry, such as that illustrated by Figure 3. The *upper curve* is the calculated steady-state percentage change of total ozone for an injection of 2000 molecules cm^{-3} s^{-1} of NO between 19.5 and 20.5 km as the only perturbation (an artificial representation of a large fleet of stratospheric aircraft). The CFC curve is the calculated future steady-state ozone change for the sole perturbation of CFC manufacture and release at the 1974 level. The *lower curve* is the calculated concentration of hydroxyl radicals at 25 km, showing correlation and anti-correlation with the NO_x and CFC curves.

steady rate obtained in 1974. In all calculations an ozone reduction was predicted: about 9% in 1976, about 18% in 1979, and about 5% in 1981 (28). These model calculations are given as the CFC curve in Figure 6.

Interactions A notable feature of Figure 6 is the almost mirror image relation between the curves for chlorine injections and for nitrogen oxides injections. The primary explanation of this negative correlation between the two curves is given by the opposite effect of hydroxyl radicals on the NO_x and Cl_x systems as given by Eqs 7(n) and 8(p). A major effect of the changes in rate constants shown in Figure 3 was to change the calculated hydroxyl radical concentrations in the lower stratosphere and troposphere. The calculated hydroxyl radical concentrations at 26 km (personal communication from D. Wuebbles) for the same time period and same laboratory is also presented in Figure 6. It can be seen that low calculated hydroxyl radical concentrations are associated with small ozone reductions by chlorine and that large concentrations of HO are associated with large calculated ozone reductions. The opposite relation exists between hydroxyl radicals and the calculated ozone perturbations by nitrogen oxides. Hydroxyl radicals bind active NO_2 into inert HNO_3 and free active chlorine atoms from inert HCl.

There is another strong interaction between added NO_x and added chlorine to the stratosphere, the formation of chlorine nitrate (Eq. 9), $ClO + NO_2 + M = ClONO_2 + M$. In regions of the stratosphere where ozone destruction by NO_x is dominant (compare Figure 5), the formation of chlorine nitrate by added ClX reduces the ozone reducer, NO_x, to give a local ozone increase by action of the double negative. In such regions, an increase of NO_x also forms additional chlorine nitrate so that the ozone reduction by NO_x is less than it would have been if there were no chlorine. Cicerone et al (58) gave a detailed analysis of such effects. Chlorine nitrate is important in the 20 to 35 kilometer altitude range.

Predictions with 1982 rate constants The growth of the most abundant CFC, CF_2Cl_2, is given in Figure 1A, and its history is typical of that of the CFC industry as a whole (8). The global production rate of this compound decreased after 1974, when its potential hazard was pointed out (6, 7), but its production rate increased slightly in 1981 (36). Various future growth scenarios have been proposed (59) for CFCs as a whole, and four of them are indicated on Figure 1: *B*, constant production at the 1980 level with no increase or decrease; *C*, increase after 1980 at the compounded rate of 3% per year; *D*, increase after 1980 at 7% per year; *E*, increase at 7% per year between 1980 and 2000 with constant production rate after that time. For a commercial product such as CFCs, exponential growth cannot go on indefinitely, because it would eventually saturate the market or deplete the resources. Thus growth curves C and D eventually become untenable. The actual growth curve A was still rapidly increasing between 1970 and 1973, and it increased a factor of three between 1962 and 1974. In terms of potential global market for these substances, increase by another factor of three or four might be expected before the growth curve saturated. Such a situation is indicated by F on Figure 1, and it joins smoothly onto curve E. On these considerations of the CFC market momentum and eventual saturation, curve E (7% growth of the entire CFC production from 1980 to 2000 and constant production thereafter) as an approximation to A-F-E is perhaps a realistic estimate of what the uninterrupted market would have done.

Figure 7 gives the calculated change (56) of the total ozone column out to the year 2050 for these four scenarios of CFC manufacture, but in the model the amount of atmospheric CO_2, CH_4, and N_2O was held constant. Curve B in Figure 7 indicates about a 3% ozone reduction by 2030 and a steady-state reduction of about 4% (compare Figure 6). Curve C indicates an 8% ozone reduction by 2030, a 20% reduction by 2050, and with an indefinitely large reduction after 2050. For Curve D the calculated ozone reductions are 10% by 2015, 30% by 2020, and off scale for later times. The calculated curve E, which approximates a supposed uninterrupted economic growth, A-F-E in Figure 2, gives the following predictions for ozone reductions in Figure 7: 5% by 2008, 10% by 2022, 15% by 2040, and leveling off after 2100 at about 30%. These particular calculations predict a modest ozone reduction by CFC at present rates of production, but very large ozone reduction if the production rate should increase appreciably.

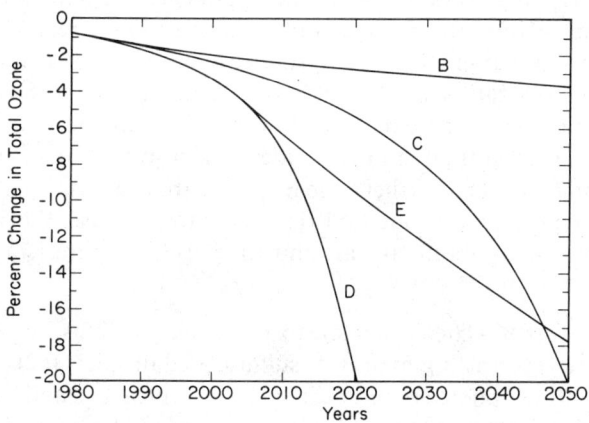

Figure 7 Calculated development with time of percentage ozone reduction with each CFC growth curve B, C, D, or E of Figure 1 as the sole assumed perturbation.

Species in Figure 2 Single perturbations will be considered for the other species in Figure 2 before full realistic scenarios are discussed. An increase in atmospheric carbon dioxide leads to a temperature increase at the earth's surface but to a temperature decrease in the stratosphere. Because of the sensitivity of certain rate constants to temperature, an increase in carbon dioxide causes an increase of the ozone column (13, 36, 57, 60, 61). A doubling of atmospheric carbon dioxide is estimated to increase the ozone column by about 3% (36).

An increase of methane affects ozone in at least two ways. It increases ozone by the methane-NO_x-smog reactions (Eq. 16). Also, it increases ozone by the double negative effect on the ClX system; it reduces the ozone reducer, atomic chlorine, by converting it to hydrogen chloride (Eq. 7). A doubling of atmospheric methane is expected to increase ozone by about 3% (57).

An increase of atmospheric nitrous oxide increases stratospheric NO_x by way of Eq. 10(t). In the natural atmosphere this source of NO_x is the primary agent for ozone loss in the region between 20 and 40 km (Figure 5). Over the years, many model calculations have been made for an arbitrarily assumed doubling of atmospheric nitrous oxide, and with 1983 chemistry about a 12% ozone reduction is calculated for this single perturbation.

The calculated effect of separately doubling the non-CFC species is given by the chart:

	Doubling CO_2	Doubling CH_4	Doubling N_2O
Ozone change	+3%	+3%	−12%

Tropospheric aircraft Another single perturbation not yet considered in this review is nitric oxide in the exhaust gases from commercial aircraft operating mostly in the middle and upper troposphere. Whether the methane smog reactions destroy or produce ozone depends strongly on the nitric oxide-ozone ratios (Eq. 17 and Figure 4). In the clean troposphere (52) the concentration of nitric oxide is so low that the methane-smog reactions destroy ozone. Even though nitric oxide is short lived in the troposphere, the air traffic in the northern hemisphere (but not in the southern hemisphere) produces enough nitric oxide to increase the calculated amount of ozone by about 10% around 10 km [(36), Figure 5.7a; compare Eq. 15].

Particulate matter The lower stratosphere has a diffuse layer of particulate matter, consisting mainly of sulfuric acid droplets (62). The sulfur reached the stratosphere as sulfur dioxide from volcanoes and as carbonyl sulfide (OCS) from forest fires and some other forms of combustion (63). So far, none of these particles has been found to have an effect on atmospheric ozone.

Summary The expected qualitative effect on ozone of these various single agencies is summarized in the following outline.

Single agency:	Present status:	Ozone effect:	18.
CO_2	Increasing	Increase	
CH_4	Increasing	Increase	
N_2O	Increasing	Decrease	
Chlorofluorocarbons	Increasing	Decrease	
Tropospheric aircraft	Increasing	Increase	
Stratospheric aircraft	Few	Decrease	
Particulate matter	Increasing	?	

Fortunately there are compensating features between these various single causes of ozone change. Interesting questions are: What is the past net effect of all of these perturbations acting together? What will be the net effect of these combined perturbations in the future?

EFFECTS OF MULTIPLE PERTURBATIONS On the basis of the trends shown in Figure 2, the history of ozone has been calculated back to the year 1850 (36), and, on the basis of various assumptions about the future rate of manufacture of CFC, the changes of ozone through the next century have been calculated (56, 57). Atmospheric observations can be compared with model predictions for the period 1970 to 1980 (36); such comparisons are outlined below in terms of percentage change of ozone at various altitudes and for total ozone.

	At 10 km:	At 40 km:	Total:	19.
CFC only (calc)	0%	−5.6%	−0.1%	
CFC + CO_2 + N_2O + CH_4 + aircraft (calc)	+6%	−4.4%	+0.4%	
Observations	+7%	−4%	+0.5 ± 1.4%	

From this example it appears that one must include the concurrent trends in other trace species to obtain observed results for tropospheric ozone and for total ozone when considering the effect of increasing CFC.

For scenario *B* of Figure 1, i.e. long-term continuation of CFC release at the 1980 level of production, the calculated ozone changes for the next 70 years are given in Figure 7 for the hypothetical case of changing CFC only (56) and in Figure 8 for concurrent increase of CO_2, N_2O, and commercial aircraft (57). For the assumed single perturbation, CFC, there is a slow 3% ozone decrease (Figure 7), but for the combined perturbations there is an ozone increase of a few tenths percent. Figure 8 also shows predicted ozone changes (56) for pattern *E* of Figure 1. This growth curve is similar to what has been called the expected economic growth curve for global CFC production, *A-F-E*. The calculated ozone changes are for ozone increases between 1980 and 1995, an ozone reduction of 6.5% by 2030, and substantial further decreases thereafter. The comparison of curves B and E of Figure 8 illustrates the finding of Cicerone et al (58) that the response of total ozone to CFC is strongly nonlinear. The compensating factors to CFC destruction of ozone are formation of chlorine nitrate (Eq. 9), CO_2, CH_4, and tropospheric NO_x. After these effects are saturated by CFC, major control of stratospheric ozone transfers from nitrogen oxides to chlorine, and then further additions of CFC strongly reduce total ozone (Figure 8).

The compensating effects of the other trace gases on ozone reduction by CFC are illustrated by vertical profiles of local ozone change, Figure 9. This figure gives the calculated changes between 1950 and 1983 (36). The curve marked "CFC ONLY" is based on N_2O, CO_2, and CH_4 at their 1983 levels, and the input of CFC in the calculation is based on the known history of CFC production. The local ozone reduction between 30 and 55 kilometers is caused by chlorine catalysis (Eq. 5), and the ozone increase between 15 and 30 kilometers is caused largely by chlorine reduction of ozone-destroying nitrogen oxides via formation of chlorine nitrate (Eq. 9). The other curve in Figure 9, which includes increases of N_2O, CH_4, and CO_2 between 1950 and 1983, shows a small decrease of the absolute magnitude of the ozone reduction between 30 and 55 km, which is caused by the reduced stratospheric temperature and by the increase of methane, which removes atomic chlorine, Eq. 7(m). The large increase in ozone in the troposphere is primarily caused by the methane-NO_x-smog reactions, Eq.

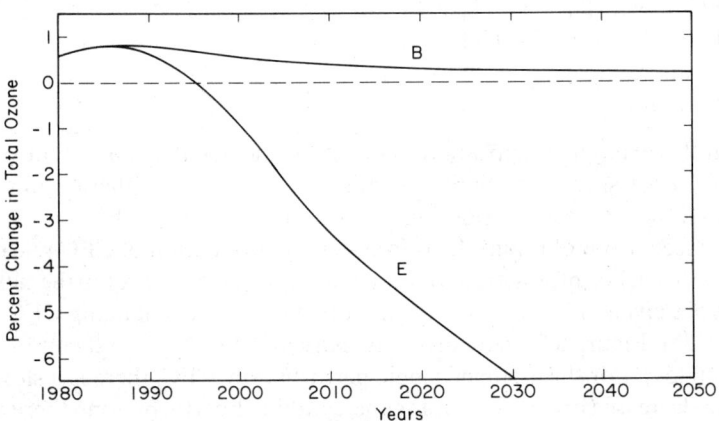

Figure 8 Calculated time development of future ozone changes when CO_2 and N_2O increase at the rates shown in Figure 2, when tropospheric aircraft injection of NO_x is included, and for two CFC future production rates: B, CFC production at the 1980 level; E, CFC increases 7% per year from 1980 to 2000 and continues at the 2000 rate.

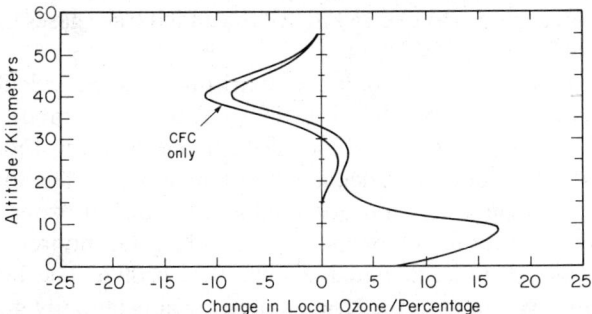

Figure 9 Calculated vertical profile of percentage differences in local ozone between 1950 and 1983: one curve includes only the known history of CFC production; the other curve includes this CFC production, the growth curves of CO_2, N_2O, and CH_4 in Figure 2, and increases of commercial aviation.

16, which are promoted both by increase in methane and increase of nitric oxide from aircraft. The extra increase of ozone between 20 and 30 km results from the chemical effect of methane on atomic chlorine and on the lowered local temperature from the greenhouse effect.

These interpretations, of course, concern the status of this subject in 1983. A glance at Figures 3 and 6 leads one to suspect that this picture will again be strongly modified within another five years. Quantitative modeling of the troposphere is especially uncertain. Other modelers find that general combustion produced NO_x is more important than that from aircraft (64) in the troposphere.

Greenhouse Effect by Species Other than Carbon Dioxide

Figure 10 shows two calculated profiles of changes in the local temperature between 1950 and 1983; this case is the same as that for Figure 9. One profile is for the input of CFC into the atmosphere with no other perturbation. The other involves CFC input, the increases of CO_2, CH_4, and N_2O as shown in Figure 2, and the increase of all forms of aviation over the period. The surface temperature was not calculated in this model, and it was held constant at its 1950 value. For the case of added CFC only, there is a large temperature decrease with maximum value at 42 km. There are two major causes of this temperature decrease: reduced local ozone (Figure 9), resulting in less heating from absorption of solar radiation, and the greenhouse effect of CFC. There is an increase in temperature between 15 and 30 km, which is caused largely by increased absorption of solar radiation by the increased local ozone there (Figure 9). For the case of added CO_2, CH_4, N_2O, and tropospheric NO_x, there is a temperature decrease at all altitudes above 15 km. All the added trace gases increase the greenhouse effect.

For the period 1970–1980, a comparison between calculated and observed ozone and temperature is given by Table 2. The model uses 1983 chemistry, includes the effects of all trace species shown in Figure 2, and through radiative coupling calculates the temperature. The model gives fairly good agreement between observed and calculated ozone, including the vertical distribution of ozone. The model gives poorer agreement between observed and calculated temperature profile. The temperature decreases observed high in the stratosphere are substantially greater than those calculated. Between 26 and 35 km, the observed temperature changes are over five times greater than those calculated; between 38 and 45 km the discrepancy is about a factor of two; and between 48 and 55 km the factor is about three. If the radiation-transfer model should be adjusted to give the observed temperature changes, the calculated ozone changes, which now agree with observations, would be altered. It is never surprising to see some "loose ends" associated with atmospheric model calculations. The subject is still evolving.

Other authors have calculated increased surface temperature as a function of increased trace gases (10), for example:

Species:	Assumed change:	Temperature rise, K:	20.
CO_2	300 to 600 ppm	2.9	
CH_4	1600 to 3200 ppb	0.26	
N_2O	280 to 560 ppb	0.65	
$CFCl_3$	0 to 2 ppb	0.29	
CF_2Cl_2	0 to 2 ppb	0.36	

These calculations illustrate the very large specific activity of CFC as

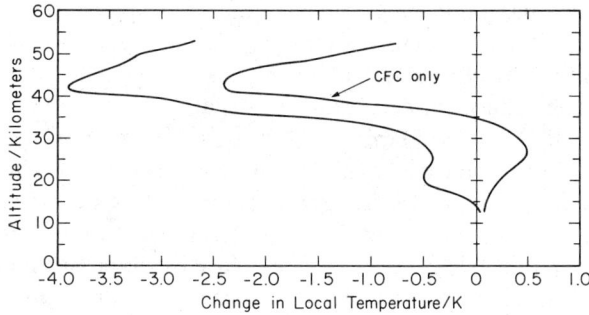

Figure 10 Calculated vertical profile of local temperature changes for the same conditions as the two curves in Figure 9.

Table 2 Calculated and observed changes between 1970 and 1980: Total ozone, ozone vertical distribution, and temperature vertical distribution

Changes in ...	Calculated[a]	Observed	Ref.
Total ozone (%)	+0.45	+1.1 ± 1.2	(41)
		+1.7 ± 2.0	(42)
		+0.5 ± 1.4	(43)
Ozone profile (%)			
2–8 km	+6	+7	(44)
8–16 km	+3	Some +	
16–31 km	+1	Little change	
38–43	−3.0 to −4.4	−4.6	(45)
		−3 to −4	(46)
Temperature profile (K)			
2–16 km	∼0	Increase	(44)
16–26 km	−0.1		
26–35 km	−0.2 to −0.6	−1.5 to −3	(44)
38–45 km	−1.2 to −1.7	−2.5 to −3.5	
48–56 km	−1.4 to −1.5	−3.5 to −5	

[a] Wuebbles 1983, Ref. (36), Table 5.6.

greenhouse gases. They have large infrared absorption cross sections in the atmospheric infrared "window" between 8 to 12 μm. The increase of these species was measured between 1970 and 1980, and it is interesting to compare the calculated increase in surface temperature for CO_2 and for the other trace species:

Species:	1970–1980 change:	Temperature rise, K:	21.
CH_4	1500 to 1650 ppb	0.032	
N_2O	295 to 301 ppb	0.016	
$CFCl_3$	0.045 to 0.180 ppb	0.20	
CF_2Cl_2	0.125 to 0.315 ppb	0.034	
	Total:	0.102	
CO_2	325 to 337 ppm	0.14	

The calculated increase in the earth's surface temperature due to the four trace gases is about 70% that due to CO_2 over this decade.

The optical properties of particulate matter in the atmosphere are complicated, and their role with respect to heat transfer depends on these optical properties. For a large class of particulates the net effect is to absorb visible and ultraviolet solar radiation and to transmit surface thermal radiation. In this case the atmosphere containing the particulates is heated, and the surface of the earth is cooled. The effect of such particulate matter is opposite to that of the greenhouse gases.

Summary

The section on THEORETICAL MODELS can be concisely summarized by the following chart:

Effect of increase in ...	Total O_3 change:	Surface temp.:	22.
CO_2	+	+	
CH_4	+	+	
N_2O	−	+	
CFC	−	+	
Aircraft below 13 km	+	+	
Aircraft above 16 km	−	?	
Particulate matter	?	−	

This chart brings out a significant recent realization: Whereas the trends in atmospheric species display compensating features as far as total ozone is concerned, all of them, except particulate matter, move in one direction so far as the greenhouse effect is concerned. As illustrated by Figures 3 and 6, this field is rapidly evolving with major surprises every few years, and there is no reason to assume that current quantitative features are final.

CLIMATIC EFFECTS OF NUCLEAR WAR

The energy that would be released by exploding all nuclear arsenals is very small compared to the energy of a hurricane, or a winter storm, or sunlight for a day. It has been common knowledge that however devastating a nuclear war would be in terms of blast, heat, radiation, and radio-activity, it would have negligible effect on the global weather and climate. Recently it has been found that this common knowledge is incorrect (1, 14). Crutzen & Birks (1) assessed the quantity and distribution of the smoke, soot, and dust that would be expected from a large nuclear war. The burning of cities, forests, agricultural fields, oil fields, and oil storage tanks would produce multiple plumes of smoke that soon would overlap to blanket most of the northern hemisphere. There would also be large dust clouds thrown up by some of the explosions and by the local windstorms produced by blasts and fires. The smoke and dust clouds would absorb sunlight, heat the local air, and create mid-air turbulence that would further lift these clouds. These effects have now been studied by many atmospheric scientists and additional insights have been provided (16).

Although there are many possibilities in terms of the number of bombs exploded, the size distribution of the bombs, and the nature of the area hit, there is consensus that major climatic effects would occur (16). The smoke and dust particles transmit much infrared radiation but, especially the smoke particles, absorb visible solar radiation. For a large but not the maximum nuclear war (14), the surface of the earth would receive less than 1% of the normal amount of solar radiation, and it would lose much infrared radiation to space. Within a few days after such a war, the surface temperature of most land masses in the northern hemisphere would drop to 20°C or more below the freezing point of water, subfreezing temperatures would persist for about three months, and subnormal temperatures would last for about a year. The ocean temperatures would be largely unaffected, so that there would be large horizontal temperature gradients between the oceans and the continents, which would probably produce violent weather.

The site of absorption of the large continuous input of solar energy would be displaced from the surface of the earth to the smoke clouds 10 to 20 km above the earth. The relatively small energy of the nuclear bombs would very significantly displace the altitude of absorption of sunlight to produce atmospheric temperature distributions and weather patterns strongly different from anything experienced by the human race. These smoke and dust clouds would be far heavier than that produced by the largest volcano during geological history, but it would be far less dense than the dust cloud from the asteroid impact that has been suggested to be responsible for the extinction of the dinosaurs (16). Survivors of a major nuclear war and their crops, even for those far from the site of hostilities, would be exposed to a "nuclear winter."

ACKNOWLEDGMENT

This work was supported by the Director, Office of Energy Research, Office of Basic Energy Sciences, Chemical Sciences Division of the US Department of Energy under Contract Number DE-AC03-76SF00098.

Literature Cited

1. Crutzen, P. J., Birks, J. W. 1982. *Ambio* 11:114–25
2. Crutzen, P. J. 1970. *Q. J. R. Meteorol. Soc.* 96:320–35
3. Crutzen, P. J. 1971. *J. Geophys. Res.* 76:7311–27
4. Johnston, H. S. 1971. *Science* 173:517–22
5. Stolarski, R. S., Cicerone, R. J. 1974. *Can. J. Chem.* 52:1610–15
6. Molina, M. J., Rowland, F. S. 1974. *Nature* 249:810–12
7. Rowland, F. S., Molina, M. J. 1975. *Rev. Geophys. Space Phys.* 13:1–35
8. Chemical Manufacturers Association. 1982. Report FPP 83-F, Washington DC
9. Prinn, R. G., Simmonds, P. G., Rasmussen, R. A., et al. 1983. *J. Geophys. Res.* 88:8353–69
10. Lacis, A., Hanson, J., Lee, P., et al. 1981. *Geophys. Res. Lett.* 8:1035–38
11. Ramanathan, V. 1976. *J. Atmos. Sci.* 33:1330–46

12. Ramanathan, V. 1980. *Interactions of Energy and Climate*, pp. 269–80. Dordrecht: Reidel
13. Callis, L. B., Natarajan, M., Boughner, R. E. 1983. *J. Geophys. Res.* 88:1401–26
14. Turco, R. P., Toon, O. B., Ackerman, T. P., et al. 1983. *Science* 222:1283–92
15. Ehrlich, P. R., Harte, M., Harwell, M. A., et al. 1983. *Science* 222:1293–1300
16. Symposium at National Meeting of the American Geophysical Union. San Francisco. December 1983. *EOS* 64:658–59, 666–67 (Abstr.)
17. Lovelock, J. E., Maggs, R. J., Wade, R. J. 1973. *Nature* 241:194–96
18. Ehhalt, D. H., Zander, R. J., Lamontagne, R. A. 1983. *J. Geophys. Res.* 88:8442–46
19. Blake, D. R., Mayer, E. W., Tyler, S. C., et al. 1982. *Geophys. Res. Lett.* 9:477–80
20. Fraser, P. J., Khalil, M. A. K., Rasmussen, R. A., et al. 1981. *Geophys. Res. Lett.* 9:461–64
21. Rasmussen, R. A., Khalil, M. A. K. 1981. *J. Geophys. Res.* 86:9826–32
22. Craig, H., Chou, C. C. 1982. *Geophys. Res. Lett.* 9:1221–24
23. Sheppard, J. C., Westberg, H., Hopper, J. F., et al. 1982. *J. Geophys. Res.* 87:1305–12
24. Keeling, C. D., Bacastow, R. B., Whorf, T. P. 1982. *Carbon Dioxide Review*, pp. 377–85. New York: Oxford Univ. Press
25. Bern CO_2 Symposium. 1983. *J. Geophys. Res.* 88:1257–1359
26. Neftel, A., Oeschger, H., Schworder, J., et al. 1982. *Nature* 295:220–23
27. Weiss, R. F. 1981. *J. Geophys. Res.* 86:7185–95
28. World Meteorological Organization. *WMO Global Ozone Research and Monitoring Project Report* No. 11, 1982, Geneva, Switzerland
29. Delwiche, C. C. 1982. Nitrogen fertilizers. In *Stratospheric Ozone and Man*, 2:65–77. Boca Raton: CRC Press
30. Keller, M., Goreau, T. J., Wofsy, S. C., et al. 1983. *Geophys. Res. Lett.* 10:1156–59
31. McElroy, M. B., Wofsy, S. C., Yung, Y. L. 1977. *Philos. Trans. R. Soc. London Ser. B* 277:159–81
32. Weiss, R. F., Craig, H. 1976. *Geophys. Res. Lett.* 3:751–53
33. Pierotti, D., Rasmussen, R. A. 1976. *Geophys. Res. Lett.* 3:265–67
34. Bauer, E. 1982. Dept. Transportation, Federal Aviation Report No. FAA-EE-82-7
35. Hahn, J., Junge, C. 1977. *Z. Naturforsch. Teil A* 32:190–220
36. Wuebbles, D. J. 1983. *A theoretical analysis of past variations in global atmospheric composition and temperature structure*. PhD thesis. Univ. Calif., Davis, Calif. 162 pp.
37. Komhyr, W. D., Barrett, E. W., Slocum, G., et al. 1971. *Nature* 232:390–91
38. London, J., Oltmans, S. J. 1978. *Pure Appl. Geophys.* 117:345–54
39. Angell, J. K., Korshover, J. 1978. *Mon. Weather Rev.* 101:63–75
40. Johnston, H. S., Whitten, G., Birks, J. W. 1973. *J. Geophys. Res.* 78:6107–35
41. Bloomfield, P., Thompson, M. L., Watson, G. S., et al. 1981. Techn. Rep. 182, Dept. Statistics, Princeton Univ., Princeton, NJ
42. St. John, D. S., Bailey, S. P., Fellner, W. H., et al. 1981. *J. Geophys. Res.* 86:7299–7311
43. Reinsel, G. C. 1981. *Geophys. Res. Lett.* 8:1227–30
44. Angell, J. K. 1982. *Interpretation of Climate and Photochemical Models, Ozone, and Temperature Measurements*, pp. 197–200, 241–45. New York: Am. Int. Phys.
45. Heath, D. F., Schlesinger, B. M. 1982. *Secular and Periodic Variations in Stratospheric Ozone from Satellite Observations 1970–1979*. Greenbelt, MD: NASA Goddard Space Flight Center
46. Reinsel, G. C., Tiao, G. C., Lewis, R. 1982. Univ. Wisc. Tech. Rep. No. 681, Madison, WI
47. *Chemical Kinetics and Photochemical Data for Use in Stratospherical Modeling.* NASA Jet Propulsion Lab., Pasadena, Calif. No. 6, Sept. 1983; No. 5, July 1982; No. 4, Jan. 1981; No. 3, NASA Ref. Publ. 1049, Dec. 1979; No. 2, April 1979
48. Hampson, R. F., Garvin, D. 1977. Natl. Bureau of Standards Special Publ. 513, Washington DC
49. CIAP Monograph 1. 1975. *The Natural Stratosphere of 1974* DOT-TST-75-51, Natl. Technical Information Serv., Springfield, Va.
50. Johnston, H. S. 1975. *Ann. Rev. Phys. Chem.* 26:315–38
51. Crutzen, P. J. 1973. *Pure Appl. Geophys.* 106:1385
52. Liu, S. C., McFarland, M., Kley, D., et al. 1983. *J. Geophys. Res.* 88:1360–68
53. Natl. Research Council. 1982. *Causes and Effects of Stratospheric Ozone Reduction: An Update.* Washington DC: Natl. Acad. Sci.
54. *Stratospheric Ozone and Man*, Vols. 1, 2. 1982. Boca Raton, FL: CRC Press
55. Chang, J. S., Duewer, W. H. 1979. *Ann. Rev. Phys. Chem.* 443–69
56. Wuebbles, D. J. 1983. *J. Geophys. Res.* 88:1433–43

57. Wuebbles, D. J., Luther, F. M., Penner, J. E. 1983. *J. Geophys. Res.* 88:3647–61
58. Cicerone, R. J., Walters, S., Liu, S. C. 1983. *J. Geophys. Res.* 88:3647–61
59. OECD. *Scenarios for Chlorocarbons.* 1981. ENV/CHEM/PJC/81.93. Paris, France: Organization for Economic Cooperation and Development
60. Luther, F. M., Wuebbles, D. J., Chang, J. S. 1977. *J. Geophys. Res.* 82:4935–42
61. Groves, K. S., Mattingly, S. R., Tuck, A. F. 1978. *Nature* 273:711–15
62. Junge, C. E. 1963. *Air Chemistry and Radioactivity.* New York: Academic
63. Crutzen, P. J., Heidt, L. E., Krasnec, J. P., et al. 1979. *Nature* 282:253–56
64. Crutzen, P. J., Gidel, L. T. 1983. *J. Geophys. Res.* 88:6641–61

THE ELECTROMAGNETIC THEORY OF SURFACE ENHANCED SPECTROSCOPY

Horia Metiu and Purna Das

Department of Chemistry, University of California, Santa Barbara, California 93106

Introduction

Surface enhanced spectroscopy (SES) was born in 1974 with the measurement of Raman spectra of molecules adsorbed on a roughened electrode surface (1a–e). Given the smallness of the Raman cross section, the detection of a Raman signal should have generated some excitement. This did not happen, but was delayed until 1977 when Jeanmaire & van Duyne (2) and Albrecht & Creighton (3) showed that the rough silver surface enhances the Raman cross section by a factor ranging between 10^4 and 10^6. This started a great outpouring (4a–i) of theoretical and experimental work, whose central theme is to understand how the presence of a solid surface modifies the spectroscopic and photochemical properties of a molecule located nearby.

We divide the effects of the surface in two categories: electromagnetic, which can be described by solving Maxwell's equations, and chemical, which belong to quantum chemistry. The enhancement of the local laser field due to the polarization of the surface is an example of an electromagnetic effect. The appearance of a new excited state caused by chemisorption, leading to an enhancement of the Raman cross section through resonance Raman scattering (which would not be expected on the basis of the gas phase properties of the molecule), is an example of a chemical effect.

While there is no doubt that chemisorption modifies the optical response of adsorbed molecules, the magnitude of the modification is still a subject of controversy. We don't know whether we should expect large modifications for all molecules, or for a small class (e.g. those with π orbitals); or whether there are cases in which chemisorption-induced effects are a necessary condition for SES or just another factor that must be included, along with the electromagnetic ones. The present article is concerned mainly with electromagnetic effects, since they are better understood theoretically and documented experimentally. Comments concerning chemical effects are made occasionally, only when the data seem to indicate strongly that some nonelectromagnetic factors might be at work; the purpose is mainly to illustrate why it is so difficult to gather convincing evidence for the existence of large chemical effects. A thorough discussion of various chemical effects can be found in the work of Otto (4e,i).

An Outline of the Electromagnetic Model

The presence of a solid surface electromagnetically modifies the spectroscopic properties of a molecule in three ways: the incident electromagnetic field acting on the molecule is increased through the addition of a field caused by the polarization of the surface; the intensity of the emitted light is increased because the polarized molecule is electromagnetically driving the surface, making it radiate; the lifetimes of the excited states of the adsorbed molecules are modified by energy transfer from the molecule to the surface.

To understand better the origin of these effects it is useful to examine the outline of the electromagnetic theory in the form developed by Efrima & Metiu (5a,b). We assume that a molecule-surface system is irradiated by light of frequency ω, having the electric field $\mathbf{E}_0(\mathbf{r}, t) = \mathbf{E}_0(\mathbf{r}, \omega)e^{i\omega t}$. A molecule located at $\mathbf{r}_0$ is exposed to the electric field

$$\mathbf{E}(\mathbf{r}_0; \omega) = (\mathscr{I} + \mathscr{R}(\mathbf{r}_0; \omega)) \cdot \mathbf{E}_0(\mathbf{r}_0; \omega), \qquad 1.$$

where the reflected field $\mathscr{R} \cdot \mathbf{E}_0$ is caused by the polarization of the surface by the laser. The reflection tensor $\mathscr{R}$ is a shorthand notation indicating that by solving Maxwell's equations we can express the field caused by surface polarization in terms of the incident field $\mathbf{E}_0$ and the properties of the surface (contained in $\mathscr{R}$). In the case of a perfectly conducting planar surface the effect of $\mathscr{R}$ amounts to putting a mirror behind the molecule, to increase the amount of light incident on it.

Since in almost all spectroscopic measurements the light emission process can be reduced to dipolar radiation, we can compute the electric field due to this emission from

$$\mathbf{E}(\mathbf{R}_d; \omega) = \{\mathscr{G}_0(\mathbf{R}_d, \mathbf{r}_0; \omega) + \mathscr{G}_s(\mathbf{R}_d, \mathbf{r}_0; \omega)\} \cdot \boldsymbol{\mu}(\omega), \qquad 2.$$

where $\boldsymbol{\mu}(\omega)$ is the emitting dipole (monochromatic, with frequency ω) located at $\mathbf{r}_0$, $\mathbf{R}_d$ is the detector position, $\mathscr{G}_0 \cdot \boldsymbol{\mu}$ is the electric field at $\mathbf{R}_d$ in the absence of the solid surface, and $\mathscr{G}_s \cdot \boldsymbol{\mu}$ is the additional field due to emission by the polarized surface. In this paper we call the latter the *surface emission* field. For the case of a perfectly conducting flat surface (5), the surface effect contained in $\mathscr{G}_s \cdot \mu$ is, essentially, the additional intensity that would be obtained by placing a mirror behind the emitter.

The change in the emitter lifetime, caused by the presence of the surface, is slightly more subtle (6a,b). The molecular dipole $\mu(\omega)$ induced by light polarizes the surface and creates at the dipole location $\mathbf{r}_0$ an electric field given by

$$\mathbf{E}(\mathbf{r}_0;\omega) = \mathscr{G}_s(\mathbf{r}_0;\omega) \cdot \mu(\omega). \qquad 3.$$

We call this the *image field*, even in those situations in which Eq. 3 does not give the simple image formula.

To determine the lifetime change induced by the surface, we use for simplicity the empirical observation, made by Drude (7a,b), that the molecular dipole $\mu_0(\omega)$ induced by the external field $\mathbf{E}_0$ is given by

$$\mu_0(\mathbf{r}_0;\omega) = \{\omega_0^2 - \omega^2 - i\omega\Gamma_0\}^{-1}(e^2/m)f_0 \cdot \mathbf{E}_0 \equiv \vec{\alpha}_0(\mathbf{r}_0;\omega) \cdot \mathbf{E}_0 \qquad 4.$$

where f_0, ω_0, Γ_0, and $\vec{\alpha}_0$ are the oscillator strength tensor, the absorption frequency, the linewidth, and the polarizability of the molecule, respectively, when the surface is absent. The electron mass and charge are m and e. When the surface is present, the field acting on the molecule is

$$\mathbf{E}(\mathbf{r}_0;\omega) = [\mathscr{I} + \mathscr{R}(\mathbf{r}_0;\omega)] \cdot \mathbf{E}_0(\mathbf{r}_0;\omega) + \mathscr{G}_s(\mathbf{r}_0;\omega) \cdot \mu(\mathbf{r}_0;\omega). \qquad 5.$$

Replacing $\mathbf{E}_0$ with $\mathbf{E}$ in Drude's equation leads to

$$\mu(\mathbf{r}_0,\omega) = \{\tilde{\omega}_0^2 - \omega^2 \mathscr{I} - i\omega\vec{\Gamma}(\omega)\}^{-1} \cdot \frac{e^2}{m} f \cdot \mathbf{E}_0(\mathbf{r}_0;\omega) = \vec{\alpha}(\mathbf{r}_0;\omega) \cdot \mathbf{E}(\mathbf{r}_0,\omega). \qquad 6.$$

with

$$\tilde{\omega}_0^2 = \omega_0^2 \mathscr{I} - (e^2/m)f_0 \cdot Re\mathscr{G}_s(\mathbf{r}_0;\omega), \qquad 7.$$

$$\vec{\Gamma} = \Gamma_0 \mathscr{I} + (e^2/m\omega)f_0 \cdot Im\mathscr{G}_s(\mathbf{r}_0;\omega), \qquad 8.$$

and

$$f \equiv f_0 \cdot \{\mathscr{I} + \mathscr{R}(\omega)\}. \qquad 9.$$

Although Drude's model has certain limitations, it correctly shows that the surface changes the transition frequency from ω_0 to $\tilde{\omega}_0$ [this is a small effect (8) that is neglected here], increases the line-width from Γ_0 to $\vec{\Gamma}(\omega)$, and the oscillator strength from f_0 to $f(\omega)$. Note that the lifetime is modified by the image tensor and the oscillator strength by the reflection tensor.

As is well known, the theory of most linear spectroscopic processes can be expressed in terms of the polarizability tensor $\vec{\alpha}(\mathbf{r}_0;\omega)$ or its derivatives with the normal coordinates. Assuming that the proper polarizability is used, the quantity measured in spectroscopy is the intensity of the dipole emission, which is proportional to $|E(\mathbf{R}_d;\omega)|^2$. The field reaching the detector is given by Eqs. 2 and 6:

$$\mathbf{E}(\mathbf{R}_d;\omega') = \{\mathscr{G}_0(\mathbf{R}_d;\omega')+\mathscr{G}_s(\mathbf{R}_d;\omega')\}\cdot\vec{\alpha}(\omega';\mathbf{r}_0)\cdot\mathbf{E}_0(\omega). \qquad 10.$$

The emission frequency ω' could differ from that of the incident light, as in the case of Raman scattering or fluorescence.

Formula 10 contains, with the proper adjustments of $\vec{\alpha}$ to correspond to the desired spectroscopic situation, the whole of the electromagnetic theory of surface enhanced spectroscopy. The effects of the surface are (a) an increase of the oscillator strength due to the reflection tensor $\mathscr{R}$ (Eq. 9); (b) an increase in the emitted field due to the tensor $\mathscr{G}_s(\mathbf{R}_d,\mathbf{r}_0;\omega')$ (Eq. 10); and (c) an increase of the absorption linewidth (i.e. a decrease of the excited state lifetime) caused by the image tensor (Eq. 8).

The chemical effects, when present, can modify the "bare" parameters ω_0, f_0, Γ_0 appearing in the polarizability; chemisorption can create a new state, so that a value of ω_0, which does not appear when the molecule is in a vacuum, is present in the polarizability of the adsorbed molecule; it can shift the states, thus changing ω_0; it broadens them by coupling the localized state to the continuum of states in the solid, thus increasing Γ_0; it can increase f_0 by, for example, delocalizing the "molecular" electrons.

The effects of these chemical changes are entangled with the electromagnetic ones and must be examined carefully. For example, since the largest term in $|E(R_d)|^2$ (see Eq. 10) contains $|G_s|^2|f_0|^2|R|^2$, an enhancement of f_0 multiplies the electromagnetic enhancement due to $|G_s|^2|R|^2$. Assume now that $G_s \sim 4$, and $R \sim 4$, which leads to an electromagnetic enhancement factor of 256, which is not, for many experimental groups, enough to permit signal detection. Now assume that chemical effects increase f_0 by a factor of 2 for one Raman mode and 3 for another. This leads to enhancements of 1024 and 2304 for these modes. If only the latter signal is above the detection limit the chemical factor controls the signal even though it is the smallest factor involved. This will give the misleading impression that the process is dominated by chemical effects that enhance only one mode, while in reality this effect takes importance only because we are seeing only the "tip of the iceberg."

Another confusing situation arises when chemisorption causes the formation of a new excited state. One would then observe resonant Raman scattering at frequencies at which one expects nonresonant Raman, and would interpret the resonant enhancement as an unusual "chemical" enhancement of the regular Raman signal.

Given the uncertainties in surface morphology and the orientation of the molecules at the surface, which affect the magnitude of the electromagnetic effects, it is difficult to calculate accurately the electromagnetic enhancement. Since the observed signal is (in almost all spectroscopic techniques) a product of electromagnetic and chemical effects, it is very difficult to document the existence of the latter when the former are only approximately known. A large body of experimental and theoretical work supports the electromagnetic model. In several systems the deviations from the electromagnetic predictions suggest the existence of large chemical effects.

A General Discussion of Electromagnetic Effects

It is useful, in discussing electromagnetic effects, to analyze the processes in which the frequency-dependent polarizability of the molecule has a resonance, separately from the nonresonant ones. In a nonresonant process $\omega_0^2 - \omega^2$ appearing in Eq. 6 is usually much larger than $\vec{\Gamma}(\omega)$, and the latter can be neglected. This eliminates $\vec{\Gamma}$ from the discussion and leaves only two sources of electromagnetic enhancement: the enhanced oscillator strength f (through the reflection tensor $\mathscr{R}$) and the enhanced emission (through the emission factor $\mathscr{G}_s$). In a resonant process $\omega_0 \sim \omega$ and the molecular polarizability is proportional to $\vec{\Gamma}^{-1}$. The two enhancement factors discussed above are divided by $\vec{\Gamma}$, which is also enhanced by the presence of the surface. Thus the overall surface enhancement of resonant molecular processes (6b, 9) is smaller than that of the nonresonant ones. This explains, for example, why resonant Raman scattering is enhanced (2, 10) by a factor of approximately 1000 on surfaces on which the Raman signal goes up by a factor of 10^6.

For nonresonant processes the only relevant factors are the enhancement of the local field [through the reflection tensor $\mathscr{R}(\mathbf{r}_0; \omega)$] and of the molecular emission [through the emission tensor $\mathscr{G}_s(\mathbf{R}_d, \mathbf{r}_0; \omega)$]. They are related through a reciprocity theorem (11) and, as a result, one can crudely state that the two effects must have the same order of magnitude (H. Metiu, unpublished).

This magnitude depends on several factors: the shape of the surface; the dielectric properties of the solid material; the dielectric properties of the material with which the surface is in contact (i.e. the electrolyte, the vacuum, etc.); the molecule surface distance; the orientation of the molecule with respect to the surface. Since these conditions can be broadly varied, the electromagnetic theory makes a large number of specific, qualitative predictions that can be tested experimentally.

By far the most important aspect of the behavior of $\mathscr{R}(\mathbf{r}_0; \omega)$ and $\mathscr{G}_s(\mathbf{R}_d, \mathbf{r}_0; \omega)$ is the excitation of electromagnetic resonances. Suggestions that surface resonances play a role in enhancement spectroscopy appeared in several experimental papers, but the first to make this the central point of a theoretical development was Moskovits (12a,b). While the method used by him has not survived, his ideas have been incorporated into later developments (13–16) that emphasized the role of resonances. The current form of the theory is the Efrima-Metiu scheme, outlined above, in which resonances appear in $\mathscr{G}_s(\mathbf{R}_d, \mathbf{r}_0; \omega)$, $\mathscr{R}(\mathbf{r}_0; \omega)$, and $\vec{\Gamma}(\mathbf{r}_0; \omega)$. Physically, these resonances appear when the external fields drive and excite collective excitations of the surface. The most common excitations appearing in surface enhanced spectroscopy are various surface plasmons; however, surface polaritons or excitons can also be used. The excitation of surface resonances by the incident light leads to large surface polarization, creating large local fields near the surface. When the driving field is caused by an oscillating molecular dipole, the excitation of surface resonances leads to intense radiation by the surface.

Although the properties of these resonances depend strongly on the shape and size of the surface and the dielectric constant of the material, they all have certain properties in common. Mathematically they appear because $\mathscr{G}_s(\mathbf{R}_d, \mathbf{r}_0; \omega)$ and $\mathscr{R}(\mathbf{r}_0; \omega)$ contain terms of the form $\{(f_i + g_i \varepsilon \omega)\}^{-1}$, which are maximized [and take the value $(Im\varepsilon(\omega))^{-1}$] when $Re\varepsilon(\omega) = -f_i/g_i$. The quantities f_i and g_i depend on the surface shape and size. Generally they have the same sign and the resonance condition $Re\varepsilon(\omega_i) = -(f_i/g_i)$ is fulfilled only in a frequency range in which the real part of the dielectric constant takes negative values. A given surface could have many resonances corresponding to various terms of the form $\{f_i + g_i \varepsilon(\omega)\}^{-1}$. These resonances lead to a large increase in $\mathscr{G}_s$ or $\mathscr{R}$ only if the value $Im\varepsilon(\omega_i)$ is small. Since f_i/g_i tends to vary (in a known way) between 1 and 3, depending on surface shape and size, we can guess for each material the frequency at which the enhancement takes place and predict its rough magnitude: the frequency is in the region in which $-1 < Re\varepsilon(\omega) < -3$ and the enhancements of $|\mathscr{R}|^2$ and $|\mathscr{G}_s|^2$, which appear in spectroscopic studies, are each proportional to $[Im\varepsilon(\omega)]^{-2}$. For example, nonresonant Raman scattering, which consists in photon absorption and emission, should have (assuming no "chemical" changes in the derivative of the polarizability) an enhancement factor proportional to $[Im\varepsilon(\omega_i)]^{-4}$. This observation is very useful as a qualitative rule for finding materials that enhance in a desired frequency region, and for estimating the magnitude of the enhancement. It explains, for example, why Ag enhances the Raman spectrum at higher frequency than Cu and Au and why Ag causes larger enhancement than the other two metals.

A second prediction made by the general theory concerns the excitation spectrum of a given spectroscopic measurement. For nonresonant Raman scattering the excitation spectrum $I(\omega)$ (i.e. the dependence of the Raman intensity of a given mode on the frequency of the incident light) is proportional to ω^4. The electromagnetic model predicts that in the presence of the surface $I(\omega)/\omega^4 \propto |G_s(\mathbf{R}_d, \mathbf{r}_0; \omega - \omega_v)|^2 |R(\mathbf{r}_0; \omega)|^2$. We have simplified the exact formula by ignoring the tensorial character of G_s and R. If the incident frequency is varied in a frequency range in which the surface has an electromagnetic resonance, the reflection tensor components $R(\omega)$ peak at the resonance frequency ω_r, while those of the emission tensor G_s peak when $\omega - \omega_v = \omega_r$. Thus, instead of having a constant value, as in a gas phase measurement, the function $I(\omega)/\omega^4$ varies with the incident frequency in a way that depends on the surface geometry and its dielectric properties. This leads to three powerful qualitative predictions.

1. The excitation spectrum of various modes tends to be very similar; exceptions can appear if the mode frequencies differ by an amount comparable with the width of the electromagnetic resonance, or if the Raman polarizabilities (i.e. the derivative of the polarizability tensor with the amplitude of the normal mode) of the modes being compared have different orientations with respect to the surface. No clear interpretation has been provided for the cases when exceptions from this rule have been observed (17).
2. The excitation spectrum of the nonresonant Raman signals from two different molecules adsorbed on the same surface ought to be identical (except perhaps when the two molecules bind to different surface sites having different local roughness). This has been observed experimentally (18a,b).
3. If the same surface-molecule system is placed in contact with different materials (e.g. vacuum, or an electrolyte or a film covering the molecules) the electromagnetic resonance of the surface, and with it the Raman excitation spectrum of the molecule, is red shifted as the dielectric constant of the medium is increased.

Another striking outcome (4h, 19) of the electromagnetic model is that it permits us to use a study of one kind of surface enhanced spectroscopy to predict roughly the behavior of another spectroscopic measurement, for the same molecules adsorbed on the same surface. For example (4h), the frequency dependence of the intensity of the light emitted by a rough surface bombarded with electrons should be similar to the Raman excitation spectrum of molecules deposited on that surface. For surfaces with small random roughness the intensity of the light emitted as a result of electron bombardment is proportional (20) to the mean square height δ^2 of the surface, while the Raman intensity (20) varies as δ^4. This has been verified in an elegant experiment (21) in which the two intensities were measured for samples with different values of δ^2.

For all surface geometries and materials, the local field created near the surface by the excitation of an electromagnetic resonance decays slowly, on the scale of a molecular diameter, with the distance from the surface. Contrary to chemical effects the electromagnetic enhancement extends (except for the case of very small roughness) beyond the first molecular layer. This prediction has been tested for various geometries such as rough, granular surfaces (22–25) and gratings (26a,b). The cases in which only short range enhancement has been observed (27–29) could be explained by the assumption that the surface roughness is of small size (~ 50 Å); however, other explanations (27) cannot be ruled out.

Finally, we should mention that signal enhancement is caused not only by electromagnetic resonances but also by high surface curvature (13). This effect appears at all frequencies at which the conductivity of the surface is high.

Before concluding this section we return briefly to those enhanced spectroscopic processes which involve a molecular resonance, such as absorption (9), fluorescence (8), and resonance Raman (6b, 9), as well as a host of possible resonant nonlinear processes. The major difference between resonant and nonresonant processes stems from the following: When the resonance condition $\omega \sim \omega_0$ is fulfilled the molecular polarizability is proportional to $\tilde{\Gamma}(\omega)^{-1}$. The emitted intensity I_d depends on the specific process but in all cases the term $I_d \sim |R|^2 |G_s(\mathbf{R}_d, \mathbf{r}_0; \omega)|^2$ (again, we ignore the tensorial character of R, G_s and Γ) appearing in the enhancement factor for nonresonant processes is replaced either by $\Gamma^{-1}|R|^2|G_s|^2$ or by $\Gamma^{-2}|R|^2|G_s|^2$. The linewidth Γ, given by Eq. 8, is the sum of the molecular width Γ_0, plus an additional term due to the presence of the surface. If other chemisorption effects are ignored, the magnitude of the latter is proportional to the imaginary part of the image field. When the molecule is close to the surface, the image field, and therefore Γ, are very large; this depresses the surface enhanced signal. Thus, the enhancement factor (but not necessarily the absolute signal) for a resonant process is smaller than that for a nonresonant one. For example, in the case of Raman scattering the enhancement factor $\mathscr{E}_{NR}$ for the nonresonant process is roughly proportional to $|R|^2|G_s|^2$, while the enhancement factor $\mathscr{E}_R$ for the resonant Raman is $(\Gamma_0|R|^2|G_s|^2/\Gamma^2)$; since $\Gamma_0/\Gamma \ll 1$ we have $\mathscr{E}_R/\mathscr{E}_{NR} = \Gamma_0/\Gamma \ll 1$. Note that a measurement of this ratio can lead to an independent measurement of fluorescence lifetime. The predictions made above have been tested experimentally: Stacy & van Duyne (10) found that the enhancement of the resonant Raman intensity is about 1000, while on the same surface the nonresonant Raman signal is enhanced by 10^6.

The distance dependence of the enhancement factor is interesting because R, G_s, and Γ become small when the molecule-surface distance is increased, but Γ is diminished more rapidly. Therefore, for resonant processes, the enhancement factor can have a maximum at a distance of 10–20 Å from the surface. Detailed calculations show that this is not a general feature and it depends on the magnitude of the molecular parameters (P. Das, J. Redding, H. Metiu, unpublished).

The manner in which the surface modifies the excitation spectrum, etc for resonant processes can be discussed by following the presentation made for the nonresonant ones.

The Predictions of Electromagnetic Theory for Various Systems

As we have emphasized, the electromagnetic resonances play a fundamental role in enhancing the spectroscopic signals of adsorbed molecules. All the properties of these resonances depend strongly on the shape of the surface. It is therefore necessary to discuss the predictions of the electromagnetic theory for various surfaces and see how they fare when compared to experiment.

It seems that in most surface enhanced experiments it is not possible to control surface morphology perfectly, or to document convincingly that such control has been achieved: flat surfaces have kinks, steps, and sometimes small (50 Å) boulders on them; well-made gratings can have small roughness or boulders on them; colloids aggregate and the shape and size of the particles may vary, etc. It is therefore important to examine the behavior of the surface enhanced signals on as many surface morphologies as possible. Whereas the unknown surface features (i.e. boulders, small roughness etc) will give rise to puzzling "secondary" signals, the main signal should behave according to the predictions of the electromagnetic theory. Although experimental limitations prevented the performance of a detailed, quantitative verification of the theoretical predictions, the mass of "circumstantial" evidence generated by the study of a wide diversity of surfaces gives credence to the theory.

Surface Enhanced Spectroscopy on Flat Surfaces

The electromagnetic resonance of a flat surface is the surface plasmon. This is a surface wave (the variable is the electric field) whose amplitude decays exponentially with the distance from the surface, and oscillates along the surface. The wave has only two kinematic parameters, the parallel momentum $\mathbf{K}_\parallel$ and the frequency $\omega(\mathbf{K}_\parallel)$, which are related through the dispersion relation

$$\mathbf{K}_\parallel^2 = (\omega/c)^2 Re[\varepsilon_0 \varepsilon(\omega)[\varepsilon_0 + \varepsilon(\omega)]^{-1}]. \qquad 11.$$

Here $\varepsilon(\omega)$ is the dielectric constant of the material used as enhancer and ε_0 is the dielectric constant of the adjacent medium (i.e. vacuum, electrolyte, etc).

If $\varepsilon_0 = 1$ kinematic constraints forbid plasmon excitation by the incident field, since the dispersion relation of the photon is such that the equations $\omega_\parallel^{ph} = \omega_\parallel^{pl}$ and $\mathbf{K}_\parallel^{ph} = \mathbf{K}_\parallel^{pl}$ (here the superscript ph and pl denote photon and plasmon) cannot be simultaneously satisfied. Therefore, the reflection tensor will not have a resonant behavior. Its main effect is essentially (5a,b) that of a lossy mirror placed behind the molecule, thus increasing the total laser intensity on it.

The interaction between the emitting dipole and the plasmon affects both the emission intensity and the dipole lifetime. The emitting dipole is not kinematically forbidden to transfer its energy to the plasmon since no momentum conservation is required in transferring energy from a point source to an extended, two-dimensional wave. The excited plasmon, however, is prevented by momentum and energy conservation from emitting this energy as a photon. Therefore the energy transfer from the dipole to the plasmon is a radiationless transition: It diminishes the fluorescence lifetime (8) and broadens the absorption linewidth (6b, 9), but it does not affect strongly the emission intensity (in a steady state experiment). Therefore, there is no large, resonant emission enhancement; the effect of the surface on the emission [through $\mathscr{G}_s(\mathbf{R}_d; \mathbf{r}_0; \omega)$] is that of a lossy mirror.

A nice illustration of the role of kinematic restriction in this dipole-plasmon interplay is provided by the following example (30, 31). Consider first a very thin frozen layer of N_2 molecules on a flat Ag surface. Bombarding the N_2 layer with electrons, we can form excited N atoms, which emit green light. On the flat surface no emission is observed: The excited atoms prefer to transfer their energy to the plasmon, which cannot radiate it. We can test whether this is the case by making on the surface a grating of wave length L, chosen to permit the plasmon to radiate; this is possible because the momentum conservation condition becomes

$$\mathbf{K}_\parallel^{ph} = \mathbf{K}_\parallel^{pl} + 2\pi n/L \qquad n = 0, 1, 2 \ldots \qquad 12.$$

and since $\mathbf{K}_\parallel^{ph} = (\omega/c)\sin\theta$, one can find a length L and an integer n for which Eq. 12 is satisfied. This equation fixes the emission angle θ (measured with respect to the normal to the surface) and the photons are emitted at fixed angles giving rise to sharp "angular resonances" (30), which have been observed experimentally (31). The grating opens up a new channel that allows a part of the molecular energy, which has been transferred to the plasmon and would have been dissipated if the surface was flat, to be recovered as radiation.

To summarize, the flat surface provides a very minor enhancement factor [through $\mathscr{R}$ and $\mathscr{G}_s(\mathbf{R}_d, \mathbf{r}_0)$] that can range from 4 to 30, depending on the experimental conditions; the fluorescence lifetime is substantially decreased (the absorption line width is increased), especially if the resonance conditions $Re\varepsilon(\omega) = -\varepsilon_0$ with the plasmon is fulfilled (6a,b) and the value of $Im\varepsilon(\omega)$ at the resonance frequency is very small. Therefore, the electromagnetic theory predicts a minor enhancement for the nonresonant Raman, and substantial quenching of the resonance Raman, absorption, and fluorescence signals.

Although there have been several reports of surface enhanced Raman measurements on flat surfaces (32–38) it has been suspected that much of the Raman signal might come from small Ag "boulders" lying on the surface. In some cases this suspicion is based on the fact that the measured enhancement factors are very large; if the surface was indeed flat the enhancement must be attributed to nonelectrodynamic effects. This implies an enormous molecular Raman cross section for the adsorbed molecules; using such a large molecular cross section for rough surfaces on which large electromagnetic effects are expected leads to predictions of signals that are much larger than those observed experimentally. Such inconsistencies cause doubts regarding the perfect flatness of the surfaces used experimentally. Recently Campion's group in Texas (36–38) carried out measurements on single crystal Ag surfaces that seem to be free of roughness; the enhancement factor, the angular dependence of the emission, the depolarization ratio, and the selection rules found experimentally are those expected from the electromagnetic calculations of Efrima & Metiu (5b).

The study of the fluorescence lifetime of molecules located near flat surfaces agrees very well with the electromagnetic theory (6a, 8). The recent work (39–50) has focused on the question of whether local electromagnetic theory, and a point dipole model, can be used when the molecule is very close to the surface (45–53a,b). There is experimental evidence (39–44) that supports the claim (48) (based on a nonlocal electrodynamic calculation) that the proportionality of the fluorescence lifetime with d^3 predicted by the local electrodynamic model (6a,b) (d is the molecule surface distance) holds even when d is only several Å.

We should emphasize that energy transfer is not the only mechanism by which a surface quenches an excited state. If the ionization potential of the excited molecule is smaller than the work function of the surface, the excited molecule can be ionized (54) by charge transfer to the metal. This lowers the excited state lifetime and broadens the absorption lineshape. Molecular beam experiments with metastable noble gases (55a,b), as well as X-ray absorption experiments for adsorbed noble gases (56a–c), have shown that this mechanism can operate very competitively with quenching by energy transfer.

Surface Enhanced Spectroscopy on Gratings

One method for exciting surface plasmons optically is to use surfaces on which a periodic structure is etched (57). This alters the parallel momentum conservation condition and removes the kinematic constraints that—in the case of the flat surface—prevent plasmon excitation by photons. When the plasmon is excited by the incident light the adsorbed molecules interact with the evanescent plasmon field, which is much larger than the field of the incident laser; this leads to enhancement of the spectroscopic signals from adsorbed molecules.

One can prepare a film with a periodic surface by making a grating on photoresist and then condensing metal vapors on it. An alternative (26a) is to deposit a polymer stencil (PMMA) on a single crystal face and use X-ray lithography and chemical etching to create the grating on the surface. This can be done in an ultrahigh vacuum system, where the grating can be sputtered and annealed to create a surface that (according to LEED) has 90% single crystal terraces. The purpose of this careful preparation is to insure that the grating surface is as smooth as possible, so that the electromagnetic effects due to small scale roughness are eliminated. Unfortunately, there are no measurements, other than high resolution electron microscopy or tunneling electron microscopy, that can guarantee the absence of small roughness or of small (~ 50 Å) metal particles on the surface of the grating. Such measurements have not yet been done, so some uncertainty regarding the small length scale morphology of the surface still exists.

Since the electromagnetic theory for periodic surfaces makes very specific predictions, and since those surfaces are, in the context of SES among the best characterized, they received extensive attention from both theorists (58–73) and experimentalists (26a, 74–79).

The early theory of SES on gratings (58, 60–62a–d) is based on a perturbation method introduced by Rayleigh (59) and developed by others (60–73). The local field enhancement (i.e. the reflection tensor) was computed by Jha et al (58), and the dipolar emission (i.e. $G_s(\mathbf{R}_d, r_0; \omega)$) by Aravind et al (30).

In the simplest terms, the periodic surface affects the electromagnetic properties of the radiation because it diffracts light. This diffraction can be incorporated into the solution of Maxwell's equations by using the procedures developed for flat surfaces: we take reasonable forms for the electric fields inside and outside the material and force them to satisfy the boundary conditions and Maxwell's equations in both media. This differs from the flat surface case in only two respects: to the usual incident, reflected and transmitted waves we must add the diffracted ones; and the equations generated by the boundary conditions contain a function describing the shape of the periodic surface. The latter circumstance introduces complications that are greatly reduced by the use of a perturbation theory in which it is assumed that the height and the curvature of the grating are smaller than the wavelength of the light.

Methods that avoid the use of perturbation theory are available (80) and they are currently being applied to SES (60–73). The dipole emission problem is more difficult since the near field of the dipolar radiation, which drives the grating, is not a planar wave. One can reduce the problem to that of many planar waves by using Fourier transform methods (30).

The main qualitative predictions made by such calculations, for a sinusoidal grating, are summarized below.

1. If the resonance condition

$$k^2 \sin^2 \theta \pm 2kK_n \sin \theta \sin \phi + 2K_n^2 = k^2 \, \text{Re} \left\{ \frac{\varepsilon_0 \varepsilon(\omega)}{\varepsilon_0 + \varepsilon(\omega)} \right\} \qquad 13.$$

is satisfied, an evanescent wave (the surface plasmon) is excited at the surface. Here $\varepsilon(\omega)$ and ε_0 are the dielectric constants of the two media forming the interface; θ and ϕ are the angles of incidence of light (θ is the angle with the normal to the surface and ϕ is the azimuthal angle; this is zero when the plane of incidence is parallel to the grooves of the sinusoidal grating and $\pi/2$ when it is perpendicular to them); $K_n = 2\pi n/L$, ($n = +1, 2, \ldots$) and $k = \omega/c$ are the wave vectors of the grating and the light. If K_n and ω are fixed we can vary the angles of incidence θ and ϕ and attempt to satisfy Eq. 13: depending on conditions [i.e. $\varepsilon(\omega)$, ε_0, ω, L, etc) we may find one, several (for different values of n), or no solution.

2. At angles for which Eq. 13 is fulfilled, the evanescent wave is excited and the electric field at the grating surface is increased. Since the evanescent wave is coupled to electron-hole pairs, its energy is dissipated with a rate controlled by the magnitude of $Im\varepsilon(\omega)$. This gives the plasmon resonance a width proportional to $Im\varepsilon(\omega)$ and as a consequence the enhanced local field is proportional to $[Im\varepsilon(\omega)]^{-1}$. Again $Im\varepsilon(\omega)$ is the factor controlling the enhancing power of a given material. Detailed calculations (58) provide an explicit and simple formula for the relative enhancements corresponding to different resonances on the same grating, or to the same grating etched on different materials, or to surfaces made of the same material but having different grating periods or heights. No detailed experimental investigation of the validity of these relations exists.

3. In all cases in which an electromagnetic resonance is excited, the theory predicts a relationship between light absorption by the surface and the enhancement; in the case of a grating, a dip in reflectance corresponds to enhanced Raman intensity, and vice versa. This has been observed experimentally (57, 78c).

4. Within perturbation theory the local field enhancement is proportional to $\xi^2(a+b\xi^2)^{-1}$, where ξ is the height of the grating and a and b are constants. The denominator appears only if the effect of plasmon collision

with the grating is taken into account. This dependence has not been investigated experimentally. Since this relation breaks down when perturbation theory fails, it can be used to study the domain of validity of perturbation theory.

5. If the plane of incidence of the incoming light is perpendicular to the grooves of the grating, the s-polarized light (the electric field vector is parallel to the grooves) cannot excite the plasmon, regardless of the value of the polar angle θ or the incident frequency. Therefore, if the grating surface is perfectly smooth and there is no chemical enhancement, the enhancement factor should be close to that of a flat surface (i.e. between 4 and 30). In surface enhanced Raman experiments carried out under such incidence conditions (78c) it was found that the enhancement factor is independent of θ, but it was much larger than that expected for a smooth grating surface. Depending on the personal bias of the interpreter, this observation is perceived as either evidence for large "chemical" effects or as an indication that small scale roughness is present on the surface. Since it is notoriously difficult to make smooth Ag films by vapor condensation the latter interpretation is credible, though not compelling.

6. If the plane of the incidence is parallel to the grooves, both s- and p-polarized light can excite the plasmon and enhance the spectrum.

7. The theory also predicts that the electric field dependence on the molecule-surface distance is a long range (many hundreds of angstroms) exponential. This prediction has been tested (26a) by measuring the Raman signal as a function of the number of pyridine layers frozen on the surface of the grating. The expected distance dependence has been confirmed by these measurements. By extrapolating the exponential function back to the first layer, it was found that in that layer the measured enhancement is larger than the extrapolated one by a factor of about 100. Given the fact that these particular measurements were carried out on one of the most carefully characterized and thoroughly studied systems in SES research (electron energy loss, photoelectron spectra, LEED, and thermal desorption measurements are available), one should examine this discrepancy seriously even though small roughness has not been conclusively ruled out. A plausible explanation for this enhancement has been suggested by Avouris et al (81–83), who showed, by electron energy loss spectroscopy, that the absorption of pyridine (81) and pyrazine (82) on Ag(111) leads to the formation of a broad excited state, which has no counterpart in the gas phase spectrum. It is suspected that this corresponds to what can be roughly described as a transition of an electron from the Fermi level into an empty orbital of the adsorbed molecule to form a transient "negative ion." A crude estimate of the frequency at which such a transition might be located is given by the difference between the work function of the surface minus the electron

affinity of the molecule. The latter depends on the distance of the molecule from the surface because of the "image" correction needed when the temporary negative ion is formed. This places this state at frequencies at which surface enhanced Raman is performed; therefore, we expect it to produce a weak (because the state is broad) resonant Raman signal that can be the source of the additional enhancement. Other mechanisms (58), as well as local roughness, can also be invoked.

8. Finally we mention that specific phenomena have been predicted (30) for the emission process: in particular, enhanced emission can appear at sharply defined detection angles, whose properties depend in an interesting way on the frequency of the emitter and the azimuthal angle with respect to the grating. The polarization of the resonantly emitted light depends on the direction of detection: if the collection plane is perpendicular to the grooves the detected light is p-polarized; the emission at other azimuthal angles has both components. The existence of these angular resonances in emission has been verified by both fluorescence (31) and Raman (78c) experiments. The polarization was not studied.

In comparing the theoretical predictions to experiment one should not expect detailed quantitative agreement, since perturbation theory has been used. Even when the grating height and curvature are smaller than the wave length of light the theory exaggerates the magnitude of the fields because it does not take into account the effect of the grating on the plasmon. This interaction tends to broaden the plasmon, depressing the electric field intensity. Furthermore, the theory is weakest at computing the fields at positions located deep in the grooves, which are the ones of most interest for SES. Exact calculations are possible (80) and we hope that they will make a detailed, quantitative reexamination of SES on gratings worthwhile.

Attenuated Total Reflection

Another method for gaining access to the surface plasmon for spectroscopic purposes places the surface in contact with a medium whose refractive index n is chosen so that the photon parallel momentum $(\omega n/c) \sin \theta$ (here θ is the angle of incidence with respect to the normal) can match that of the plasmon. This method was invented by Otto (84) and modified by others (85a,b, 86a,b). Several reviews of its theory and applications are available (86b). The first application to SES was made by Burstein et al (87a,b), who used attenuated total reflection (ATR) to enhance the Raman and the CARS signal of molecules sandwiched between the two media. Amusingly, this work preceded SES and it was connected to it only at a later time when the electromagnetic model took hold in the field.

Since the basic physical processes involved in ATR are very similar to those used in the work with gratings, we summarize them very briefly. The enhancement of the surface plasmon takes place when the parallel momentum conservation

$$(\omega n/c) \sin \theta = (\omega/c)^2 \, Re\left[\varepsilon_0 \varepsilon(\omega)(\varepsilon_0 + \varepsilon(\omega))^{-1}\right] \qquad 14.$$

is satisfied. If ω and n are carefully chosen, this equation gives one angle θ at which plasmon excitation by p-polarized radiation takes place. An s-polarized beam does not couple to the plasmon.

The same condition also controls photon emission by the excited plasmon, and the emitted light is p-polarized. Raman measurements (87–92a–c) in the ATR configuration qualitatively confirm these predictions. There is, however, substantial additional enhancement because the metal film surface is rough. Furthermore, the enhancement due to the plasmon is smaller than that predicted by the theory; the likely reason is that surface roughness damps the plasmon, depressing both the local and the emitted fields.

The ATR studies (42, 93–98) of the influence of the metal film on molecular fluorescence are very ingenious. By using Langmuir-Blodgett layers (99) as spacers one can vary the distance between the fluorescing dye and the metal surface. The measured emission intensity has the distance dependence predicted by the electromagnetic model. A detailed summary of this work can be found in Ref. (4h).

Small Roughness

The optical properties of films grown by condensation can be, in many cases, understood in terms of a model (62a, 100–102) in which the height ζ of the surface, with respect to a mean surface plane, is a random Gaussian function with a two point correlation function proportional to $\exp(-a^2 \cdot \mathbf{k}_\parallel^2)$; here $\mathbf{k}_\parallel$ is the parallel momentum of the Fourier component of $\zeta(x,y)$ and a is the correlation length of the surface. It is believed (103a,b) that one can produce such surfaces by depositing a thin metal film on a CaF_2 surface. One can usefully think of this surface as a random superposition of gratings whose wavelength is distributed around the value $2\pi/a$. If the height and the curvature of each grating are smaller than the wavelength of light, one can solve Maxwell's equations for each grating by perturbation theory. The sum of the electric fields corresponding to each grating can be used to compute observable quantities; these must be averaged with a statistical weight describing how much each grating contributes to the true shape of the surface. Such averaging destroys the perfect coherence present when the surface is sinusoidal. Therefore, the predictions made for a surface with random roughness are a "smeared out" version of those made for a grating. The enhancement factors must be somewhat smaller than those predicted for a grating whose wavelength equals $2\pi/a$, where a is the coherence length of the Gaussian distribution.

The manner in which a small roughness (SR) surface modifies fluorescence intensity (20) and lifetime (104) have been computed by Aravind & Metiu. Light-emitting tunneling junctions, in which the tunneling current drives the rough surface electromagnetically and makes it emit light, have been examined by Laks & Mills (62c, 62d), who found reasonable agreement with the experiment (105, 106). The only disagreement (107) concerns the magnitude of driving source and its physical origin [i.e. tunneling current fluctuations or hot electrons (107)] but not the frequency dependence of the orientation of the driving current, which determines most of the characteristics of the emitted light. Thus the two phenomena (i.e. fluorescence and light emission driven by the tunneling current) have the same mechanism: The rough surface is driven by an oscillating electric field and made to radiate. In one case the oscillating field is caused by the fluctuations in the tunneling current [or by hot electrons (107)], in the other it is caused by the near field of the emitting molecular dipole. The similarity between the Raman signal of molecules sandwiched in a tunneling junction and the intensity of light emission from the same junction as functions of mean roughness height has been demonstrated (58, 108) experimentally. Such measurements convincingly illustrate the statement that when $G_s(\mathbf{R}_d, \mathbf{r}_0; \omega)$ is large the light is emitted by the surface. The properties of observed emission are given by a convolution of the properties of G_s and those of the driving source (i.e. molecular dipole or fluctuating tunneling current). The above experiment (58, 108) looks at the intensity of surface emission caused by two different sources and compares how they change when the surface is systematically modified. Eesley's work (21), discussed above, is based on the same principle: in his case the two sources are Raman emission and electron bombardment and the surface parameter varied is the mean height of the roughness.

Surface Enhanced Spectroscopy on Isolated Particles

A large body of work has been dedicated to the study of SES on isolated metal particles. For the purpose of testing the electromagnetic theory one would like to work with an ensemble of noninteracting particles, having either a random or a known periodic spatial distribution, atomically smooth surfaces, identical sizes and shapes, with the latter selected to permit accurate electromagnetic calculations. Clearly it is very difficult to satisfy all these conditions, but several systems provide a crude approximation to this ideal.

Carefully prepared colloidal solutions contain spherical particles that will enhance the Raman spectrum of the molecules located nearby. These systems, first introduced by Creighton et al (109), are very popular (89, 110–134). Unfortunately the colloidal particles tend to aggregate, creating conglomerates whose electromagnetic properties are radically different from those of the isolated spheres. Similar systems can be created by trapping in a solid matrix metal particles prepared by evaporation (135–138). While it seems possible (139) to obtain uniform, spherical particles in this manner, essentially by trial and error, the task is not easy and requires careful electron microscopy studies to test the outcome. In principle, the matrix-isolated systems are cleaner and more flexible than the colloidal ones, since one can make a matrix from the molecules whose spectroscopic properties are to be studied, thus eliminating solvent, electrolytes, unknown surface impurities, and uncertainties regarding whether the molecules are in contact with the sphere.

Another interesting system can be prepared (140) by condensing metal vapor on the tops of a periodic array of vertical pillars made on a photoresist surface. The advantage is that by spacing the pillars properly, the magnitude of the electromagnetic interactions between the metal particles can be controlled.

In experiments whose purpose is to enhance various spectral signals from surface molecules, one can use as enhancers the metal islands (29, 141–164) that tend to form by the break up of very thin films obtained by vapor deposition.

There are numerous calculations (15a–c, 16, 165–199) of the enhancing properties of isolated metal particles. They are confined to surface geometries that permit a separation of variables in Maxwell's equations (200), such as spheres, (14–15b, 165–185), ellipsoids (15c, 16, 185–196), and spherical and ellipsoidal protuberances on flat surfaces (13, 197–199). The study of other shapes would require complicated numerical work. In what follows we present, as an example that is simplest and most illuminating, the case of a sphere whose diameter is smaller than the wavelength of light. This restriction eliminates the need for considering retardation, and it reduces electrodynamics to electrostatics.

The content of the formula for rhe reflection tensor, which gives the local field enhancement, is described by the following sequence of processes: The incident field $\mathbf{E}_0(\omega)$ induces a dipole moment $\mathbf{p}(\omega)$ at the center of the sphere; this creates a field $\mathbf{E}(\mathbf{r}_0; \omega)$ at the location $\mathbf{r}_0$ of the molecule. The enhancement occurs because $\mathbf{E} \gg \mathbf{E}_0$. The dipole $\mathbf{p}(\omega)$ is given by (201)

$$\mathbf{p}(\omega) = \beta(\omega)\mathbf{E}_0(\omega) = [\varepsilon(\omega)-\varepsilon_0][\varepsilon(\omega)+2\varepsilon_0]^{-1}a^3\mathbf{E}_0(\omega), \qquad 15.$$

where $\varepsilon(\omega)$ and a are the dielectric constant and the radius of the sphere and ε_0 is the dielectric constant of the surrounding medium. The field created by this dipole at the molecular location $\mathbf{r}_0$ (the origin of the coordinate system is in the center of the sphere) is (201)

$$\mathbf{E}(\mathbf{r}_0;\omega) = \mathscr{T}(\mathbf{r}_0)\cdot\mathbf{p}(\omega) = (3\hat{r}_0\hat{r}_0-\mathscr{I})r_0^{-3}\cdot\mathbf{p}(\omega), \qquad 16.$$

where $\hat{r}_0 = \mathbf{r}_0/r_0$ and $\hat{r}_0\hat{r}_0$ is a dyadic [i.e. $\hat{r}_0\hat{r}_0\cdot\mathbf{p} = \hat{r}_0(\hat{r}_0\cdot\mathbf{p})$]. Combining Eqs. 15 and 16 with the definition of the reflection tensor given by Eq. 1, we obtain

$$\mathscr{R}(\mathbf{r}_0;\omega) = \beta(\omega)\mathscr{T}(\mathbf{r}_0). \qquad 17.$$

This reflection tensor has an electromagnetic resonance [which minimizes the denominator in Eq. (15)] at the frequency ω_r given by

$$Re\varepsilon(\omega_r) = -2\varepsilon_0. \qquad 18.$$

We note that a small sphere is capable of sustaining an infinite number of resonances (202) located at the frequencies ω_n given by

$$Re\varepsilon(\omega_n) = -[(n+1)/n]\varepsilon_0, \qquad n = 1, 2, \ldots \qquad 19.$$

The resonance displayed by the reflection tensor corresponds to $n = 1$. The other resonances do not appear in $\mathscr{R}$ because the incident field $\mathbf{E}_0(\omega)$ is spatially smooth over the region occupied by the sphere and therefore can excite only the dipolar resonance $n = 1$. Conversely, this is the only resonance that can emit light. An electric field that varies rapidly over a distance scale of order a, such as the field generated by an induced molecular dipole located near the surface, can excite all these resonances by transferring energy to them; a part of the energy transferred to the $n = 1$ resonance can be radiated by the sphere; that transferred to the others is entirely lost to heat. In the language of molecular spectroscopy, the excitations of the resonances $n > 1$ are radiationless transitions.

The intensity of light radiated by the molecular dipole through energy transfer to the resonance $n = 1$, which represents the enhanced emission, can be easily computed. The molecular dipole exerts the field $\mathscr{T}(\mathbf{r}_0)\cdot\boldsymbol{\mu}(\omega')$ on the sphere inducing the dipole $\mathbf{p}(\omega') = \beta(\omega')\mathscr{T}\cdot\boldsymbol{\mu}(\omega')$, which radiates according to the customary (201) dipole emission formula. Note that the quantity $\beta(\omega')\mathscr{T}$, which was identified with the reflection tensor above, appears also in the emission formula; if the excitation frequency ω and the emission frequency ω' are close to each other (i.e. $|\omega'-\omega| <$ the width of the electromagnetic resonance) the two processes are equally enhanced. In Raman scattering $\omega-\omega'$ is the vibrational frequency, and for high frequency modes it can happen that the emission and the excitation process cannot be resonant simultaneously with the sphere; such modes would be much less enhanced than the low frequency ones, which benefit from simultaneous resonant enhancement of both excitation and emission.

One can easily verify that the small sphere system has all the general properties mentioned above. Because this specific example can illuminate the general statements, we consider it in some detail. The enhancement factor depends on the position of the molecule and that of the detector, and we simplify the discussion by placing both so as to maximize the enhancement. For comparison with the experiment one should sum up the emission from all molecules and average over molecular orientations. For the most favorable situation the maximum enhancement factor (at resonance) can be obtained by using Eqs. 15–17; the result is $[2(a/r_0)^3(3\varepsilon_0/Im\varepsilon(\omega_r))]^2$ for both emission and local field enhancement. Since Raman scattering benefits, for low frequency modes, from both, the square bracket appears in the expression for Raman intensity to the fourth power. This equation shows that the enhancement factor is diminished slowly when the molecule-surface distance is increased; that a change in ε_0 red-shifts the resonance and modifies the enhancement factor through $\varepsilon_0/Im\varepsilon(\omega_r)$ (ω_r depends on ε_0 through Eq. 18); and that materials with small values of $Im\varepsilon(\omega)$ in the frequency region where $Re\varepsilon(\omega) = -2$ produce a large enhancement.

Although there is no detailed quantitative verification of the predictions of the theory for small particle systems, there are numerous studies of Raman scattering (89, 109–134) or fluorescence (140, 143, 145, 146, 150, 153–155) in which some of the qualitative predictions, made by no other model, have been verified. However, some of the experimental observations are puzzling. Given the uncertainties in the experimental systems, one cannot draw firm conclusions from the data. To exemplify the nature of these difficulties we discuss two cases in which apparent deviations from the electromagnetic theory have been observed.

1. The first case is a measurement of Raman scattering in a colloidal solution. One of the serious problems with colloidal systems is their tendency to coagulate. This must be avoided, since the electromagnetic properties of the coagulated particles are very different from those exhibited in isolation (203). There is no universal prescription for avoiding coagulation, but optical and electron microscopy studies can be used to determine whether the absence of coagulation has occurred by accident. In one such case (116) it was found that the Raman excitation spectrum of molecules located at the surface of the colloidal particles has a peak at the "wrong" frequency and no signal at frequency where one is expected on the basis of the one particle theory. Many plausible "explanations" can be offered for this: The "wrong" peak appears because of a molecular resonant Raman effect induced by chemisorption; the dielectric constant of a small sphere differs (204–206) from the bulk value (used in the electromagnetic calculation), and this might explain either the shift in frequency or the

inexistence of the expected signal, or both; some aggregates have escaped detection and they are responsible for the observed signal; the predicted signal is large but just below the detection limit; etc.

2. A similar situation occurs in experiments (135) that measure the Raman spectrum of CO molecules, enhanced by imbedding Ag spheres in a CO matrix. The excitation spectrum of the CO stretch seems to agree with the predictions made by the theory; however, the excitation spectrum of a low frequency mode, assigned to the C–Ag stretch, has the "wrong" frequency dependence. This is in qualitative disagreement with the electromagnetic theory, which requires that both excitation spectra have similar shape. Unfortunately, the degree of control in this experiment is not sufficient to prevent a defender of the "purely electromagnetic point of view" from raising the following objections: Perhaps CO chemisorbs at special sites that sample a peculiar local field; perhaps the local electrodynamic theory used for calculations breaks down (40–50) in the surface region where the C–Ag bond lies; perhaps the C–Ag stretch interacts with the electrons in the metal, using them to enhance resonantly its Raman cross section. These examples illustrate the complexity of these systems: too poor in details and too rich in ambiguity to permit a compelling choice among various interpretations.

We conclude this section with the discussion of surface enhanced fluorescence and of the possibility of enhancing the rate of photochemical reactions by placing molecules near a small metal particle (207, 208). We consider a three level system in which the $1 \to 2$ transition is pumped by the incident laser and $2 \to 3$ is caused by a first order molecular process. If the state 3 is a continuum, the model simulates photochemistry; if a radiative $3 \to 1$ transition is allowed, the model can be used to discuss the fluorescence corresponding to the $3 \to 1$ emission. Naively one might think that if the incident frequency excites an electromagnetic resonance of the sphere, the rate of the $1 \to 2$ transition, and hence that of the photochemical process, is enhanced. This does not take into account that the photochemical and the fluorescence yields are both diminished by energy transfer from the molecule to the sphere. Another way of stating this is that the sphere and the two level system (for understanding absorption we can focus on the levels 1 and 2 only) are strongly coupled: as a result they absorb jointly more energy than the isolated molecule, but this is distributed between the particle and the sphere in a way that does not necessarily place a large amount of energy in the molecule. If it happens that most of the energy goes into the sphere, its presence can quench the photochemical processes or the $3 \to 1$ fluorescence. The interplay between two effects (i.e. enhanced absorption and energy redistribution) is expressed mathematically through the reflection tensor $\mathscr{R}$ (characterizing the local field) and the image tensor

$\mathcal{G}_s$ (characterizing the width of level 2). Since $\mathcal{R}$ increases the rate and $\mathcal{G}_s$ decreases it, the two tend to work against each other: $\mathcal{R}$ pumps energy in the molecule and $\mathcal{G}_s$ controls how much of it is shared with the sphere. The role of the theory is to provide a quantitative measure of the outcome of this competition. The strategy used in earlier work (207–209) was to replace (for simplicity we neglect in what follows the vectorial and the tensorial character of all quantities) E_0 with $(1+R)E_0$, and Γ_0 with Γ (given by Eq. 8) in the equation describing the dynamic behavior of the three level system in the absence of the sphere. Thus, the isolated molecule theory is "renormalized" to include the effects of the sphere. The resulting equations predict large enhancements (207–209) of the phtochemical rate. Recent work (P. C. Das, J. Redding, H. Metiu, unpublished) examining a quantum model of the joint molecule-sphere system concluded that the enhancement is much smaller. Since a detailed discussion and comparison of the two models would be rather tedious, we illustrate here how they reach different results by examining the Drude model. The power W absorbed by the $1 \to 2$ transition is given by the formula

$$W = -\frac{\omega}{2} Im\mu^*(\omega)E(\omega). \qquad 20.$$

For the isolated molecule we replace $\mu(\omega)$ with $\mu_0(\omega)$ given by Eq. 5 and $E(\omega)$ with the incident field E_0. We obtain, at resonance when $\omega = \omega_0$, the equation

$$W = (e^2/2m)(f_0/\Gamma_0)|E_0|^2. \qquad 21.$$

Earlier work (209) replaces Γ_0 with Γ and E_0 with $(1+R)E_0$; since $\Gamma \sim (e^2/m\omega)f_0 Im G_s$ (we use Eq. 8) and $(1+R)E_0 \sim RE_0$, we obtain $W = (\omega_0|R|^2 E_0^2)/(2Im G_s)$. Since $|R|^2 \gg Im G_s$ at all but perhaps the shortest molecule-surface distances, this leads to enhanced power absorption, hence photochemistry and fluorescence. However, if instead of using the above "renormalization" we compute the absorbed power by using Eq. 20 with the total local field $E = (1+R)E_0 + G_s \cdot \mu(\omega)$ and the induced dipole $\mu(\omega)$ given by Eq. 6, we obtain (on resonance, and using the same simplifications as above)

$$W = (e^2/m)\frac{|R|^2|E_0|^2\Gamma_0\omega_0}{2Im G_s[(e^2/m\omega_0)f_0 Im G_s]}. \qquad 22.$$

We see that this is equal to the result of the early work multiplied with $\Gamma_0/[e^2/m\omega_0)f_0 Im G_s]$. This is the ratio (see Eq. 8) $\Gamma_0/(\Gamma-\Gamma_0)$ between the total linewidth of the isolated molecule and the change in linewidth caused by the presence of the sphere. In many cases $\Gamma_0 \ll \Gamma-\Gamma_0$, which means essentially that the molecule would rather transfer its energy to the sphere than undergo the molecular process described by Γ_0 (i.e. photochemical break up, a radiationless transition, or a radiative decay by spontaneous emission). In such cases the recent theory of Das et al (unpublished work mentioned above) predicts that the sphere quenches photochemistry.

We emphasize that both earlier (207–209) and more recent works are more complex than the simple example discussed above might suggest; however, the example illustrates the kind of difficulties one runs into when the addition of the effect of the sphere is included as an "afterthought" following an analysis of the behavior of a single molecule.

To conclude this discussion we emphasize, as a practical matter, that the best enhancer (Ag) has resonances in a frequency region around 3.2 eV, where many molecules do not undergo photochemistry. The I_2 molecule, which is often used as an example, reacts with Ag when irradiated, complicating matters. Other metals that have high frequency plasmons have a larger value for $Im\varepsilon(\omega)$, and therefore cannot lead to large enhancements. It seems that two-photon photochemistry (P. C. Das, J. Redding, H. Metiu, unpublished), enhanced by Ag particles, has a better chance of yielding interesting results. Recent experimental work (210) seems to have detected such effects.

Finally we note that compared to photochemical processes the $3 \rightarrow 1$ fluorescent emission benefits from an additional enhancement through the enhanced emission process discussed at several places in this article. In the small sphere this comes about through energy transfer from the dipole corresponding to the $3 \rightarrow 1$ transition to the dipole of the sphere, which then radiates the energy very effectively. If the $3 \rightarrow 1$ frequency corresponds to the frequency of the $n = 1$ resonance, this enhancement can be sizeable. Such effects have been observed experimentally (153–155, 209).

The Interaction Between Electromagnetic Resonances

While detailed calculations have been carried out for isolated particles, in most real systems the particles are fairly close to each other so that their resonances couple and interact. Because the evanescent fields excited at resonance reach far (several hundred angstroms) outside the particles, the interaction between resonances is long ranged. This results in frequency shifts, in the weak coupling regime (i.e. at large interparticle distances), or in the creation of completely new resonances, in the strong coupling limit. Since the interaction takes place because the strong local field of one particle polarizes the other, there is a propensity to build enormous fields in the region between particles, which can lead to very large enhancements of the spectroscopic signals from molecules located there.

Examples of such interactions were analysed by Aravind, Nitzan & Metiu for two spheres (203), for a perfect metal sphere (which cannot sustain resonances) interacting with a flat surface (211, 212), and for the interaction between the resonances of a sphere and that of a plane surface (213). We illustrate the physical effects of such interactions with the example of the sphere-plane system (213). The resonance of the isolated plane surface is the surface plasmon, which is a planar wave whose frequency varies from zero (for zero parallel momentum) to a maximum value given by $Re\varepsilon(\omega) = -\varepsilon_0$. The sphere has an infinite number of resonances whose properties are described above. The resonances corresponding to large n have the same frequency as the flat surface plasmon; the one for lowest n has the frequency $Re\varepsilon(\omega) = -2\varepsilon_0$. For a Drude dielectric constant these limits correspond to $\omega_p/\sqrt{2}$ (flat) and $\omega_p/\sqrt{3}$.

If the sphere-plane distance is large, the only resonance that can be excited by light is the $n = 1$ dipolar resonance of the sphere. Although the light cannot excite the surface plasmon of the flat surface, because of momentum conservation, the dipole induced in the sphere can do it rather effectively. So the outcome is that because of the presence of the sphere, the light excites a resonance that is a mixture of the $n = 1$ resonance of the sphere and a localized wave packet made up of the high parallel momentum plasmon states. When the sphere is brought closer to the surface, it interacts with the image of its own dipole and this leads to the excitation of the $n > 1$ resonances of the sphere; these interact with the flat surface plasmon, exciting it more. At very small surface-sphere distances the result of all these effects is the build up of a strong local field between the sphere and the plane. The interaction between two spheres can be described similarly.

Such calculations serve two purposes:

1. They address the "engineering" problem (213) of finding the surface configurations leading to largest enhancements. It seems that the use of small Ag spheres as local laser enhancers is enormously efficient when the properties of the materials are used to couple different resonances (213). Such coupling also broadens considerably the frequency range at which the enhancement takes place, providing much needed flexibility.
2. They dramatize that for the rough surfaces used in electrochemical cell experiments, the interaction between the small "structures" composing the roughness of the surface is very important; single particle estimates of surface behavior might be irrelevant for such systems.

The problem of interaction between various small, resonance-sustaining structures is intellectually very intriguing. Our experience has been that large enhancements are produced when the structure absorbs the photon and localizes it. Gratings and flat surfaces (in the ATR configuration) absorb the photon and "store" the electromagnetic energy in the surface plasmon; this is delocalized in the direction parallel to the surface but localized in the perpendicular one. This increases the electromagnetic energy density near the surface. A sphere localizes the photon, by plasmon excitation, in all directions, and the resulting concentration of electromagnetic energy is larger than that produced by a grating. A sphere-plane structure concentrates the energy in the small gap between the surfaces leading to even higher enhancements. The interesting questions connected to localization can be illustrated by considering an ensemble of small cubes (of order 300 Å). If we put them next to each other to construct a flat surface, we obtain practically no enhancement; if we can space them with a specified periodicity, to make a sort of "grating," we can generate moderate enhancements; if we can use pairs of cubes we can obtain enormous enhancements in the gap between them. Clearly the problem of maximizing enhancement is related to that of obtaining maximum photon localization in the structure created by the solid surfaces. As a general problem this is of interest in other areas of physics, and little serious work (214–222) has been done on it in the context of surface enhanced spectroscopy.

Concluding Remarks

There is no doubt that electromagnetic effects play a fundamental role in surface enhanced spectroscopy; nor could there be any doubt that the optical response of a chemisorbed molecule differs from that of a free one. The only question is whether such "chemical" effects are small (e.g. leading to a factor of less than 10 in intensity) or large; or whether large chemical enhancements appear rarely, frequently, or always. Throughout this article we have tried to indicate why it is not easy to obtain an unambiguous answer to such questions. This is an area that needs a lot of experimental work, since reliable theoretical predictions concerning the optical properties of chemisorbed molecules are not available.

The electromagnetic theory is rather mature, but several fundamental problems still remain: We need better "engineering" to find the best enhancing surface configurations and materials; we need a more thorough exploration of the possibility of enhancing nonlinear and photochemical phenomena; we need to understand better the electrodynamics of ensembles of particles and especially the localization of the electromagnetic fields by them; and finally we need to know whether the use of molecular point dipoles (51–53a,b) and local electrodynamics (45–50, 223–230) when computing optical properties of a molecule near the surface is good within an order of magnitude, or is dubious, or foolish.

Acknowledgments

We are grateful to the Sloan, Dreyfus, and National Science Foundations and the Office of Naval Research for partial support of this work, and to our colleagues W. Kohn, R. Schrieffer, D. Scalapino, and P. Hansma for providing a stimulating environment for this work. Various members of our group, S. Efrima, T. Maniv, P. K. Aravind, E. Hood, and J. Redding, have contributed to our understanding of these phenomena.

Literature Cited

1a. Fleischmann, M., Hendra, P. J., McQuillan, A. J. 1973. *J. Chem. Soc. Commun.*, p. 80
1b. Fleischmann, M., Hendra, P. J., McQuillan, A. J. 1974. *Chem. Phys. Lett.* 26:163–66
1c. McQuillan, A. J., Hendra, P. J., Fleischmann, M. 1975. *J. Electroanal. Chem.* 65:933
1d. Fleischmann, M., Hendra, P. J., McQuillan, A. J., Paul, R. L., Reid, E. S. 1976. *J. Raman Spectrosc.* 4:269
1e. Paul, R. L., McQuillan, A. J., Hendra, P. J., Fleischmann, M. 1975. *J. Electroanal. Chem.* 66:248
2. Jeanmaire, D., Van Duyne, R. P. 1977. *J. Electroanal. Chem.* 84:1
3. Albrecht, M. G., Creighton, J. A. 1977. *J. Am. Chem. Soc.* 99:5215
4a. Van Duyne, R. P. 1978. In *Chemical and Biochemical Applications of Lasers*, ed. C. B. Moore, 4:101. New York: Academic
4b. Burstein, E., Chen, C. Y., Lundquist, S. 1979. In *Proc. Joint US-USSR Symp. on Theory of Light Scattering in Condensed Matter.* New York: Plenum
4c. Efrima, S., Metiu, H. 1979. *Isr. J. Chem.* 17:18
4d. Burstein, E., Chen, C. Y., Lundquist, S. 1979. In *Light Scattering in Solids*, eds. J. L. Birman, H. Z. Cummins, K. K. Rebane, p. 479. New York: Plenum
4e. Otto, A. 1980. *Appl. Surf. Sci.* 6:309
4f. Furtak, T. E., Reyes, J. 1980. *Surf. Sci.* 93:351
4g. Chang, R. K., Furtak, T. E., eds. 1982. *Surface Enhanced Raman Scattering.* New York: Plenum
4h. Metiu, H. 1984. *Prog. Surf. Sci.* In press
4i. Otto, A. 1983. In *Light Scattering in Solids*, Vol. 4, ed. M. Cardona, G. Guntherodt. New York: Springer
5a. Efrima, S., Metiu, H. 1978. *Chem. Phys. Lett.* 60:54
5b. Efrima, S., Metiu, H. 1979. *J. Chem. Phys.* 70:1602, 2297
6a. Chance, R. R., Prock, A., Silbey, R. 1978. *Adv. Chem. Phys.* 37:1
6b. Efrima, S., Metiu, H. 1979. *J. Chem. Phys.* 70:1939
7a. Drude, P. 1900. *Ann. Phys.* 4:437
7b. Born, M., Wolf, E. 1970. *Principles of Optics*, p. 91. New York: Pergamon. 4th ed.
8. Chance, R. R., Prock, A., Silbey, R. 1975. *Phys. Rev. A* 12:1448
9. Efrima, S., Metiu, H. 1980. *Surf. Sci.* 92:417
10. Stacy, A. M., Van Duyne, R. P. 1983. *Chem. Phys. Lett.* 102:365
11. Jones, D. S. 1964. *The Theory of Electromagnetism.* New York: Pergamon
12a. Moskovits, M. 1978. *J. Chem. Phys.* 6:4159
12b. Moskovits, M. 1979. *Solid State Commun.* 32:59
13. Gersten, J. I., Nitzan, A. 1980. *J. Chem. Phys.* 73:3023
14. McCall, S. L., Platzman, P. M., Wolff, P. A. 1980. *Phys. Lett. A* 77:381
15a. Wang, D. S., Kerker, M., Chew, H. 1980. *Appl. Opt.* 19:2135, 4159
15b. Wang, D. S., Kerker, M., Chew, H. 1980. *Appl. Opt.* 19:2256
15c. Wang, D. S., Kerker, M. 1981. *Phys. Rev. B* 24:1777
16. Adrian, F. J. 1981. *Chem. Phys. Lett.* 78:45
17. DiLella, D. P., Gohin, A., Lipson, R., McBreen, P., Moscovits, M. 1980. *J. Chem. Phys.* 73:4282
18a. Blatchford, C. G., Campbell, J. R., Creighton, J. A. 1982. *Surf. Sci.* 120:435
18b. Bachackashvilli, A., Efrima, S., Katz, B., Priel, Z. 1983. *Chem. Phys. Lett.* 94:571
19. Metiu, H. 1982. See Ref. 49, pp. 1–34
20. Aravind, P. K., Metiu, H. 1980. *Chem. Phys. Lett.* 74:301
21. Eesley, G. L. 1981. *Phys. Rev. B* 24:5477
22. Rowe, J. E., Shank, C. V., Zwemer, D. A., Murray, C. A. 1980. *Phys. Rev. Lett.* 44:1770
23. Zwemer, D. A., Shank, C. V., Rowe, J. E. 1980. *Chem. Phys. Lett.* 73:201
24. Murray, C. A., Allara, D. L., Rheinwine, M. 1981. *Phys. Rev. Lett.* 46:57
25. Murray, C. A. 1982. See Ref. 4g, pp. 203–21
26a. Sanda, P. N., Warlaumont, J. M., Demuth, J. E., Tsang, J. C., Christman, K., Bradley, J. A. 1980. *Phys. Rev. Lett.* 45:1519
26b. Sanda, P. N., Demuth, J. E., Tsang, J. C., Warlaumont, J. M. 1982. See Ref. 4g, pp. 189–202
27. Pockrand, I., Otto, A. 1980. *Surf. Sci.* 35:861
28. Wood, T. H., Zwemer, D. A., Shank, C. V., Rowe, J. E. 1981. *Chem. Phys. Lett.* 82:5
29. Seki, H., Philpott, M. R. 1980. *J. Chem. Phys.* 73:5376

30. Aravind, P. K., Hood, E., Metiu, H. 1981. *Surf. Sci.* 109:95
31. Adams, A., Moreland, J., Hansma, P. K. 1981. *Surf. Sci.* 111:351
32. Pettinger, B., Wenning, U. 1978. *Chem. Phys. Lett.* 56:253
33. Pettinger, B., Wenning, U., Kolb, D. M. 1978. *Ber. Bunsenges. Phys. Chem.* 82:1326
34. Schultz, S. G., Janik-Czachor, M., Van Duyne, R. P. 1981. *Surf. Sci.* 104:419
35. Udagawa, M., Chou, C.-C., Hemminger, J., Ushioda, S. 1981. *Phys. Rev. B* 23:6843
36. Campion, A., Brown, J. K., Grizzle, V. M. 1982. *Surf. Sci.* 115:L153
37. Campion, A., Mullins, D. R. 1983. *Chem. Phys. Lett.* 94:576
38. Mullins, D. R., Campion, A. 1984. *J. Phys. Chem.* 88:8
39. Adams, A., Rendell, R. W., Garnett, R. W., Hansma, P. K., Metiu, H. 1980. *Opt. Commun.* 34:417
40. Adams, A., Rendell, R. W., West, W. P., Broida, H. P., Hansma, P. K., Metiu, H. 1980. *Phys. Rev.* 21:5565
41a. Rosetti, R., Brus, L. E. 1980. *J. Chem. Phys.* 73:572
41b. Rosetti, R., Brus, L. E. 1982. *J. Chem. Phys.* 76:1146
42. Eagen, C. F., Weber, W. H., McCarthy, S. L., Terhune, R. W. 1980. *Chem. Phys. Lett.* 75:274
43a. Campion, A., Gallo, A. R., Harris, C. B., Robota, H. J., Whitmore, P. M. 1980. *Chem. Phys. Lett.* 73:447
43b. Whitmore, P. M., Robota, H. J., Harris, C. B. 1982. *J. Chem. Phys.* 76:740
43c. Whitmore, P. M., Robota, H. J., Harris, C. B. 1982. *J. Chem. Phys.* 77:1560
44. Pockrand, I., Brillante, A., Mobius, D. 1980. *Chem. Phys. Lett.* 69:499
45. Feibelman, P. J. 1981. *Phys. Rev. B* 23:2629
46. Fuchs, R., Barrera, R. G. 1981. *Phys. Rev. B* 24:2940
47. Weber, W. H., Ford, G. W. 1980. *Phys. Rev. Lett.* 44:1774
48. Korzeniewski, G., Maniv, T., Metiu, H. 1980. *Chem. Phys. Lett.* 73:212
49. Korzeniewski, G., Maniv, T., Metiu, H. 1982. *J. Chem. Phys.* 76:2697
50. Persson, B. N. J., Lang, N. D. 1982. *Phys. Rev. B* 26:5409
51. Hilton, P. R., Oxtoby, D. W. 1980. *J. Chem. Phys.* 72:6346
52. Palke, W. 1980. *Surf. Sci.* 97:L331
53a. Efrima, S., Metiu, H. 1981. *Surf. Sci.* 108:329
53b. Efrima, S., Metiu, H. 1981. *Surf. Sci.* 109:109
54. Hagstrum, H. D. 1954. *Phys. Rev.* 96:336
55a. Bozso, F., Yates, J. T. Jr., Arias, J., Metiu, H., Martin, R. M. 1983. *J. Chem. Phys.* 78:4256
55b. Bozso, F., Arias, J., Hanrahan, C., Martin, R. M., Yates, J. T. Jr., Metiu, H. 1984. *Surf. Sci.* 136:257
56a. Cunningham, J. A., Greenlaw, D. K., Flynn, C. P. 1980. *Phys. Rev. B* 22:717
56b. Flynn, C. P., Chen, Y. C. 1981. *Phys. Rev. Lett.* 46:447
56c. Lang, N. D., Williams, A. R., Himpsel, F. J., Reihl, B., Eastman, D. E. 1982. *Phys. Rev. B* 26:2810
57. Tsang, J. C., Kirtley, J. R., Bradley, J. A. 1979. *Phys. Rev. Lett.* 43:772
58. Jha, S. S., Kirtley, J. R., Tsang, J. C. 1980. *Phys. Rev. B* 22:3973
59. Rayleigh, J. W. S. 1907. *Proc. R. Soc. London Ser. A* 79:399; 1907. *Philos. Mag.* 14:70
60. Celli, V., Marvin, A., Toigo, F. 1975. *Phys. Rev. B* 11:1779
61. Marvin, A., Toigo, F., Celli, V. 1975. *Phys. Rev. B* 11:2777
62a. Maradudin, A. A., Mills, D. L. 1975. *Phys. Rev. B* 11:1392
62b. An error in Ref. 62a has been corrected by Agarwal, G. S. 1976. *Phys. Rev. B* 14:846
62c. Laks, B., Mills, D. L. 1979. *Phys. Rev. B* 20:4962
62d. Laks, B., Mills, D. L. 1980. *Phys. Rev. B* 21:5175
63a. Vincent, P. 1980. See Ref. 80, p. 101
63b. Neviere, M. 1980. See Ref. 80, p. 123
64. Glass, N. E., Maradudin, A. A. 1981. *Phys. Rev. B* 24:595
65. Reinisch, R., Neviere, M. 1982. *J. Opt. Paris* 13:81
66. Arya, K., Zeyher, R., Maradudin, A. A. 1982. *Sol. State Commun.* 42:461
67. Sheng, P., Stepleman, R. S., Sanda, P. 1982. *Phys. Rev. B* 26:2907
68. Mills, D. L., Weber, M. 1982. *Phys. Rev. B* 26:1075
69. Glass, N. E., Maradudin, A. A., Celli, V. 1982. *Phys. Rev. B* 26:5357, and references therein
70. Neviere, M., Reinisch, R. 1982. *Phys. Rev. B* 26:5403
71a. Glass, N. E., Maradudin, A. A., Celli, V. 1983. *Phys. Rev. B* 27:5150
71b. Glass, N. E., Maradudin, A. A. 1982. *Surf. Sci.* 114:240
72. Weber, M., Mills, D. L. 1983. *Phys. Rev. B* 27:2698
73. Arya, K., Zeyher, R. 1983. *Phys. Rev. B* 28:4090
74. Metcalfe, K., Hester, R. E. 1983. *Chem. Phys. Lett.* 94:411
75. Adams, A., Moreland, J., Hansma, P. K., Schlesinger, L. 1982. *Phys. Rev. B* 25:3457
76. Inagaki, T., Motosuga, M., Yamamori,

K., Asakawa, E. T. 1983. *Phys. Rev. B* 28:1740
77. Tsang, J. C., Kirtley, J. R., Theis, T. N. 1980. *Solid State Commun.* 35:667
78a. Girlando, A., Gordon, J. G. II, Heitmann, D., Philpott, M. R., Seki, H., Swalen, J. D. 1980. *Surf. Sci.* 101:417
78b. Girlando, A., Knoll, W., Philpott, M. R. 1981. *Solid State Commun.* 38:895
78c. Girlando, A., Philpott, M. R., Heitmann, D., Swalen, J. D., Santo, R. 1980. *J. Chem. Phys.* 72:5187
78d. Knoll, W., Swalen, J. D., Girlando, A. 1981. *J. Chem. Phys.* 75:4795
79. Numata, H. 1982. *J. Phys. Soc. Jpn.* 51:2575
80. Petit, R., ed. 1980. *Electromagnetic Theory of Gratings*. Berlin: Springer
81. Demuth, J., Sanda, P. N. 1981. *Phys. Rev. Lett.* 47:57
82a. Avouris, Ph., Demuth, J. E. 1981. *J. Chem. Phys.* 75:4783
82b. Demuth, J. E., Avouris, Ph. 1981. *Phys. Rev. Lett.* 47:61
83. Schmeisser, D., Demuth, J. E., Avouris, Ph. 1982. *Chem. Phys. Lett.* 87:324
84. Otto, A. 1968. *Z. Phys.* 216:398
85a. Kretschmann, E. 1971. *Z. Phys.* 241:313; 1978. *Opt. Commun.* 26:41
85b. Brillante, A., Pockrand, I., Philpott, M. R., Swalen, J. D. 1978. *Chem. Phys. Lett.* 57:395
86a. Philpott, M. R. 1979. In *Topics in Surface Chemistry*, ed. E. Kay, P. S. Bagus, p. 329. New York: Plenum
86b. Otto, A. 1976. In *Optical Properties of Solids—New Developments*, ed. B. O. Seraphin, p. 677. Amsterdam: North Holland
87a. Chen, Y. J., Chen, W. P., Burstein, E. 1976. *Phys. Rev. Lett.* 36:1207
87b. Chen, W. P., Ritchie, G., Burstein, E. 1976. *Phys. Rev. Lett.* 37:993
88. Pettinger, B., Tadjeddine, A., Kolb, D. M. 1979. *Chem. Phys. Lett.* 66:544
89. Dornhaus, R., Benner, R. E., Chang, R. K., Chabay, I. 1980. *Surf. Sci.* 101:367
90. Tadjeddine, A., Kolb, D. M., Kotz, R. 1980. *Surf. Sci.* 101:277
91. Otto, A. 1980. *Surf. Sci.* 101:99
92a. Sakoda, K., Ohtaka, K., Hanamura, E. 1982. *Solid State Commun.* 41:393
92b. Ushioda, S., Sasaki, Y. 1983. *Phys. Rev. B* 27:1401
92c. Tom, H. W. K., Chen, C. K., de Castro, A. R. B., Shen, Y. R. 1982. *Solid State Commun.* 41:259
93. Gerbshtein, Y. M., Merkulov, I. A., Mirlin, D. A. 1974. *JETP Lett.* 22:35
94. Weber, W. H., Eagen, C. F. 1979. *Opt. Lett.* 4:236
95. Benner, R. E., Dornhaus, R., Chang, R. K. 1979. *Opt. Commun.* 30:145
96. Pockrand, I., Brillante, A., Mobius, D. 1981. *Nuovo Cimento* 63B:350
97a. Lukosz, W., Meier, M. 1981. *Opt. Lett.* 6:251
97b. Lukosz, W., Kunz, R. E. 1979. *Opt. Commun.* 31:42
97c. Lukosz, W. 1980. *Phys. Rev. B* 22:3030
98. Kunz, R. E., Lukosz, W. 1980. *Phys. Rev. B* 21:4814
99. Kuhn, H., Mobius, D., Bucher, H. 1972. *Techniques of Chemistry*, Vol. 1, *Physical Methods of Chemistry*, ed. A. Weissberger, B. W. Rossiter, Part 2b, Ch. 7. New York: Wiley-Interscience
100. Crowell, J., Ritchie, R. H. 1970. *J. Opt. Soc. Am.* 60:795
101. Elson, J. M., Ritchie, R. H. 1974. *Phys. Status Solidi B* 62:461
102. Kretschmann, E. 1969. *Z. Phys.* 222:412
103a. Endriz, J. G., Spicer, W. E. 1970. *Phys. Rev. Lett.* 24:64
103b. Endriz, J. G., Spicer, W. E. 1971. *Phys. Rev. B* 4:4144
104. Arias, J., Aravind, P. K., Metiu, H. 1982. *Chem. Phys. Lett.* 85:404
105. Adams, A., Wyss, J. C., Hansma, P. K. 1979. *Phys. Rev. Lett.* 42:912
106. Mills, D. L., Weber, M., Laks, B. 1982. In *Tunneling Spectroscopy: Capabilities, Applications, and New Techniques*, ed. P. K. Hansma, Ch. 5. New York: Plenum
107. Kirtley, J. R., Theis, T. N., Tsang, J. C., DiMaria, D. J. 1983. *Phys. Rev. B* 27:4601
108. Tsang, J. C., Kirtley, J. R. 1979. In *Light Scattering in Solids*, ed. J. L. Birman, H. Z. Cummins, K. K. Rebane, p. 499. New York: Plenum
109. Creighton, J. A., Blatchford, C. G., Albrecht, M. G. 1979. *J. Chem. Soc. Faraday Trans.* 2:75,790
110. Kerker, M., Siiman, O., Bumm, L. A., Wang, D. S. 1980. *Appl. Opt.* 19:3253, 4137
111. Regis, A., Corset, J. 1980. *Chem. Phys. Lett.* 70:305
112a. Lippitsch, M. E. 1980. *Chem. Phys. Lett.* 74:125
112b. Lippitsch, M. E. 1981. *Chem. Phys. Lett.* 79:224
113. Wetzel, H., Gerischner, H. 1980. *Chem. Phys. Lett.* 76:460
114. McQuillan, A. J., Pope, C. G. 1980. *Chem. Phys. Lett.* 71:349
115. Lombardi, J. R., Knight, E. A. S., Birke, R. L. 1981. *Chem. Phys. Lett.* 79:214
116. Von Raben, K. U., Chang, R. K., Laube, B. L. 1981. *Chem. Phys. Lett.* 79:465
117. Akins, D. E. 1982. *J. Colloid Interface Sci.* 90:373
118. Mabuchi, M., Takenaka, T., Fujiyoshi,

Y., Uyeda, N. 1982. *Surf. Sci.* 119:150
119. Creighton, J. A., Albrecht, M. G., Hester, R. E., Matthew, J. A. D. 1978. *Chem. Phys. Lett.* 55:55
120. Siiman, O., Bumm, L. A., Callaghan, R., Blatchford, C. G., Kerker, M. 1983. *J. Phys. Chem.* 87:1014
121. Suh, J. S., DiLella, D. P., Moskovits, M. 1983. *J. Phys. Chem.* 87:1540
122a. Garrell, R. L., Shaw, K. D., Krimm, S. 1981. *J. Chem. Phys.* 74:4155
122b. Garrell, R. L., Shaw, K. D., Krimm, S. 1983. *Surf. Sci.* 124:613
123. Blatchford, C. G., Siiman, O., Kerker, M. 1983. *J. Phys. Chem.* 87:2503
124a. Siiman, O., Lepp, A., Kerker, M. 1983. *Chem. Phys. Lett.* 100:163
124b. Siiman, O., Lepp, A., Kerker, M. 1983. *J. Phys. Chem.* 87:5319
125. Creighton, J. A. 1983. *Abstr. Pap. Am. Chem. Soc.* 185:23
126. Heard, S. M., Grieser, F., Barraclough, C. G. 1983. *Chem. Phys. Lett.* 95:154
127. Benner, R. E., Von Raben, K. U., Lee, K. C., Owen, J. F., Chang, R. K. 1983. *Chem. Phys. Lett.* 96:65
128. Kneipp, K., Hinzmann, G., Fassler, D. 1983. *Chem. Phys. Lett.* 99:503
129. Lee, P. C., Meisel, D. 1983. *Chem. Phys. Lett.* 99:262
130. Lee, E. H., Benner, R. E., Fenn, J. B., Chang, R. K. 1978. *Appl. Opt.* 17:1980
131. Benner, R. E., Dornhaus, R., Long, M. B., Chang, R. K. 1979. In *Microbeam Analysis*, ed. D. E. Newbury, p. 191. San Francisco: San Francisco Press
132. Lee, E. H., Benner, R. E., Fenn, J. B., Chang, R. K. 1979. *Appl. Opt.* 18:862
133. Benner, R. E., Barber, P. W., Owen, J. F., Chang, R. K. 1980. *Phys. Rev. Lett.* 44:475
134. Benner, R. E., Owen, J. F., Chang, R. K. 1980. *J. Chem. Phys.* 84:1602
135. Abe, H., Manzel, K., Schulze, W., Moskovits, M., DiLella, D. P. 1981. *J. Chem. Phys.* 74:792
136. Krasser, W., Kettler, U., Brechthold, P. S. 1982. *Chem. Phys. Lett.* 86:223
137. Manzel, K., Schulze, W., Moskovits, M. 1982. *Chem. Phys. Lett.* 85:183
138. Froben, F. W., Schulze, W., Kloss, U. 1983. *Chem. Phys. Lett.* 99:500
139. Abe, H., Schulze, W., Tesche, B. 1980. *Chem. Phys.* 47:95, 2200
140. Liao, P. F., Bergman, J. G., Chemla, D. S., Wokaun, A., Melngailis, J., Hawryluk, A. M., Economou, N. P. 1981. *Chem. Phys. Lett.* 82:355
141. Chen, C. Y., Burstein, E., Lundquist, S. 1979. *Solid State Commun.* 32:63
142. Chen, C. Y., Davoli, I., Ritchie, G., Burstein, E. 1980. *Surf. Sci.* 101:363
143. Ritchie, G., Burstein, E. 1981. *Phys. Rev. B* 24:4843
144. Eagen, C. F. 1981. *Appl. Opt.* 20:3035
145. Glass, A. M., Liao, P. F., Bergman, J. G., Olson, D. H. 1980. *Opt. Lett.* 5:368
146. Glass, A. M., Wokaun, A., Heritage, J. P., Bergman, J. G., Liao, P. F., Olson, D. H. 1981. *Phys. Rev. B* 24:4906
147. Seki, H. 1981. *J. Vac. Sci. Technol.* 18(2):633
148. Seki, H. 1982. *J. Vac. Sci. Technol.* 20:584
149. Yamaguchi, S. 1960. *J. Phys. Soc. Jpn.* 15:1577
150. Garoff, S., Weitz, D. A., Gramila, T. J., Hanson, C. D. 1981. *Opt. Lett.* 6:245
151. Bergman, J. G., Chemla, D. S., Liao, P. F., Glass, A. M., Hart, R. M., Olson, D. H. 1981. *Opt. Lett.* 6:33
152. Seki, H., Chung, T. J. 1982. *Solid State Commun.* 44:473
153. Weitz, D. A., Garoff, S., Hanson, C. D., Gramila, T. J., Gerste, J. I. 1982. *Opt. Lett.* 7:89
154. Weitz, D. A., Garoff, S., Hanson, C. D., Gramila, T. J., Gersten, J. I. 1981. *J. Lumin.* 24/25:83
155. Garoff, S., Weitz, D. A., Alverez, M. S. 1982. *Chem. Phys. Lett.* 93:283
156. Seki, H. 1982. *Solid State Commun.* 42:695
157. Weitz, D. A., Garoff, S., Gramila, T. J. 1982. *Opt. Lett.* 7:168
158. Wokaun, A., Lutz, H. P., King, A. P., Wild, U. P., Ernst, R. R. 1983. *J. Chem. Phys.* 79:509
159. Seki, H. 1982. *J. Chem. Phys.* 76:4412
160. Seki, H. 1983. *J. Electroanal. Chem.* 150:425
161. Seki, H. 1983. *J. Electron. Spectrosc.* 29:413
162. Lyon, S. A., Worlock, J. M. 1983. *Phys. Rev. Lett.* 51:593
163. Seki, H., Chuang, T. J. 1983. *Chem. Phys. Lett.* 100:393
164. Chen, C. Y., Davoli, I., Ritchie, G., Burstein, E. 1980. In *Proc. Int. Conf. on Nontraditional Approaches to the Study of the Solid-Electrolyte Interface*, ed. T. E. Furtak, K. L. Kliewer, D. W. Lynch. Amsterdam: North-Holland
165. Kerker, M., McNulty, P. J., Sculley, M., Chew, H., Cooke, D. D. 1978. *J. Opt. Soc. Am.* 68:1676
166. Chew, H., Sculley, M., Kerker, M., McNulty, P. J., Cooke, D. D. 1978. *J. Opt. Soc. Am.* 68:1686
167. Chew, H., McNulty, P. J., Kerker, M. 1976. *Phys. Rev. A* 13:396
168. Kerker, M., Druger, S. D. 1979. *Appl. Opt.* 18:1172
169. Messinger, B. J., Von Raben, K. U.,

Chang, R. K., Barber, P. W. 1981. *Phys. Rev. B* 24:649
170. Ruppin, R. 1982. *J. Chem. Phys.* 76:1681
171. Ruppin, R. 1982. *Phys. Rev. B* 26:3440
172. Kerker, M. 1982. *Aerosol. Sci. Technol.* 1:275
173. Ohtaka, K., Inoue, M. 1982. *J. Phys. C* 15:6463
174. Ohtaka, K., Inoue, M. 1982. *Phys. Rev. B* 25:677
175. Ohtaka, K., Inoue, M., Yanagawa, S. 1982. *Phys. Rev. B* 25:689
176. Inoue, M., Ohtaka, K. 1982. *Phys. Rev. B* 26:3487
177. Chew, H., Wang, D. S. 1982. *Phys. Rev. Lett.* 49:490
178. Chew, H., Wang, D. S., Kerker, M. 1983. *Phys. Rev. B* 28:4169
179. Greenler, R. G., Snider, D. R., Witt, D., Sorbello, R. S. 1982. *Surf. Sci.* 118:415
180. Creighton, J. A. 1983. *Surf. Sci.* 124:209
181. Wang, D. S., Kerker, M. 1982. *Phys. Rev. B* 25:2433
182. Kerker, M., Blatchford, C. G. 1982. *Phys. Rev. B* 26:4052
183. Kerker, M., Wang, D. S. 1984. In press
184. Chew, H., Kerker, M., McNulty, P. J. 1976. *J. Opt. Soc. Am.* 66:440
185. Kerker, M. 1969. *The Scattering of Light and Other Electromagnetic Radiations.* New York: Academic
186. Wang, D. S., Kerker, M., Chew, H. 1980. *Appl. Opt.* 19:1573
187. Wang, D. S., Kerker, M., Chew, H. 1980. *Appl. Opt.* 19:2315
188. Gersten, J. I. 1980. *J. Chem. Phys.* 72:5779, 5780
189. Gersten, J. I., Nitzan, A. 1981. *J. Chem. Phys.* 75:1139
190. Little, J. W., Ferrell, T. L., Callcott, T. A., Arakawa, E. T. 1982. *Phys. Rev. B* 26:5953
191. Barber, P. W., Chang, R. K., Massoudi, H. 1983. *Phys. Rev. B* 27:7251
192. Barber, P. W., Chang, R. K., Massoudi, H. 1983. *Phys. Rev. Lett.* 50:997
193. Krauss, W. A., Schatz, G. C. 1983. *Chem. Phys. Lett.* 99:353
194. Gersten, J. I., Weitz, D. A., Gramila, T. J., Genack, A. Z. 1980. *Phys. Rev. B* 22:4562
195. Chen, C. Y., Burstein, E. 1980. *Phys. Rev. Lett.* 45:1287
196. Wokaun, A., Gordon, J. P., Liao, P. F. 1982. *Phys. Rev. Lett.* 48:957
197. Berreman, D. W. 1970. *Phys. Rev. B* 1:381
198. Ruppin, R. 1981. *Solid State Commun.* 39:903
199. Das, P. C., Gersten, J. I. 1982. *Phys. Rev. B* 25:6281
200. Morse, P. M., Feshbach, H. 1953. *Methods of Theoretical Physics*, Vols. 1, 2. New York: McGraw-Hill
201. Jackson, J. D. 1975. *Classical Electrodynamics.* New York: Wiley. 2nd ed.
202. Stratton, J. A. 1941. *Electromagnetic Theory.* New York: McGraw-Hill
203. Aravind, P. K., Nitzan, A., Metiu, H. 1981. *Surf. Sci.* 110:189
204. Dasgupta, B. B., Fuchs, R. 1981. *Phys. Rev. B* 24:554
205. Apell, P., Penn, D. R. 1983. *Phys. Rev. Lett.* 50:1316
206. Penn, D. R., Rendell, R. W. 1982. *Phys. Rev. B* 26:3047
207. Nitzan, A., Brus, L. E. 1981. *J. Chem. Phys.* 74:5321
208. Nitzan, A., Brus, L. E. 1981. *J. Chem. Phys.* 75:2205
209. Weitz, D., Garoff, S., Gersten, J. I., Nitzan, A. 1983. *J. Chem. Phys.* 78:5324
210. Goncher, G. M., Harris, C. B. 1982. *J. Chem. Phys.* 77:3767
211. Aravind, P. K., Rendell, R. W., Metiu, H. 1982. *Chem. Phys. Lett.* 85:396
212. Aravind, P. K., Metiu, H. 1982. *J. Phys. Chem.* 86:5076
213. Aravind, P. K., Metiu, H. 1983. *Surf. Sci.* 124:506
214. Laor, U., Schatz, G. C. 1981. *Chem. Phys. Lett.* 82:566
215. Laor, U., Schatz, G. C. 1982. *J. Chem. Phys.* 76:2888
216. Kotler, Z., Nitzan, A. 1983. *Surf. Sci.* 130:124
217. Gerardy, J. M., Ausloos, M. 1983. *Phys. Rev. B* 27:6446
218. Gerardy, J. M., Ausloos, M. 1982. *Phys. Rev. B* 26:4703
219. Kotler, Z., Nitzan, A. 1982. *J. Phys. Chem.* 86:2011
220. Bergman, D. J., Nitzan, A. 1982. *Chem. Phys. Lett.* 88:409
221. Aspnes, D. E. 1982. *Phys. Rev. Lett.* 48:1629
222. Persson, B. N. J., Liebsch, A. 1983. *Phys. Rev. B* 28:4247
223. Metiu, H. 1983. *Isr. J. Chem.* 22:329
224. Maniv, T., Metiu, H. 1980. *Phys. Rev. B* 22:4731
225. Maniv, T., Metiu, H. 1982. *J. Chem. Phys.* 76:2697
226. Maniv, T., Metiu, H. 1982. *J. Chem. Phys.* 76:696
227. Maniv, T., Metiu, H. 1980. *J. Chem. Phys.* 72:1996
228. Feibelman, P. J. 1975. *Phys. Rev. B* 12:1319
229. Feibelman, P. J. 1975. *Phys. Rev. B* 12:4282
230. Feibelman, P. J. 1982. *Prog. Surf. Sci.* 12:287

ELECTRONIC PROPERTIES OF SURFACES

Marvin L. Cohen

Department of Physics, University of California, and Materials and Molecular Research Division, Lawrence Berkeley Laboratory, Berkeley, California 94720

Steven G. Louie

Department of Physics, University of California, and Lawrence Berkeley Laboratory, Berkeley, California 94720

INTRODUCTION

Surface physics and chemistry is one of the oldest branches of material science. Experimental studies have been broad and varied. Despite the age of the field and the large body of knowledge that has been accumulated, many researchers consider this area to be a young and vigorous one. One of the main reasons for this impression is that surface science had a renaissance of sorts in the 1970s. It was around this time that experimental studies of clean surfaces were judged to be adequately reproducible. In addition, new experimental and theoretical techniques evolved that became widely available and a concerted effort was undertaken to attempt to understand the properties of at least a few simple systems. This effort has grown and activity has increased, making this subfield one of the largest branches of condensed matter science.

This review focuses on the theoretical accomplishments in surface physics. Experimental results are discussed only insofar as they bear on a theoretical result. In addition, because of space considerations, emphasis is placed on only a few theoretical techniques—primarily the pseudopotential approach. Some prototype systems are considered to illustrate the accomplishments of the theory for semiconductors, simple metals (i.e. *sp* metals), and transition metals. Some details of calculations and results are given, but the reader should go to the original papers for most specifics.

The electronic properties of surfaces and interfaces are dominated by electronic states that are localized near the surface or interface. Early work on these states include studies by Tamm (1), Shockley (2), Davydov (3), Mott (4), Schottky (5), and Bardeen (6). Specific calculations for model systems were done in some early studies by Goodwin (7), but it wasn't until the 1970s that realistic models for real systems were analyzed in detail. These studies, together with experimental probes like low energy electron diffraction to study surface geometry and photoemission to determine electronic structure, ushered in a new era of surface science. The results have been impressive in some areas, but not as broad as many would desire. In particular, a lot of effort has been focused on a few representative materials—the rationalization being that understanding a specific system in depth will yield useful information about surface physics in general. This review adopts this posture. However, the alternative view of examining a wide class of materials and phenomena with emphasis on materials having more technical than scientific relevance is easy to justify and useful particularly in surface chemistry.

Hence a major motivation of the theoretical work in this area is to search for surface states and characterize their properties. Once these are understood for clean surfaces, one can examine their role in determining the properties of interfaces, adsorbate-covered surfaces, and nonperfect surfaces with steps and defects. The most studied surface states are associated with semiconductor and transition metal surfaces because in these systems electrons can be more localized near the surfaces or interfaces. It is the opportunity to pile up charge in a semiconductor bond or a transition metal d-state that localizes surface states compared to extended states found in simple free-electron-like metals. Some discussion of the latter case is presented using Al as a prototype, but it was not too long ago that there were debates among researchers as to whether it was possible to have surface states on Al. It is possible, but the states are not as localized near the surface as they are in semiconductors and transition metals. Because of this, most researchers in this area have focused on these latter systems, and we also concentrate on these subareas in this review.

We use silicon surfaces as the prototype for the semiconductor case, while surfaces of niobium and palladium serve as representatives of the transition metal cases. Interfaces receive little attention, but reviews (8) are available illustrating the features of metal-semiconductor (Schottky barriers) and semiconductor-semiconductor (heterojunctions) interfaces.

SURFACES OF SEMICONDUCTORS

Theoretical research in this area has focused on Si and particularly on the Si(111) surface. Although this surface probably has been studied more than any other, the details of the atomic and electronic structure are still considered open subjects. Experimental interest remains high because it is possible to cleave Si in vacuum and produce clean surfaces that can be studied with a variety of techniques. Proportionally, theoretical activity is probably even larger than experiment because many theorists feel it imprudent to tackle more complicated systems until the "simple" cases like Si(111) are understood.

A ball and stick model of the Si(111) surface appears in Figure 1. The sticks at the top of the figure represent the cut bonds resulting from the cleaving of the crystal. Some major questions that arise are: How do the electrons in the bonds react to the formation of the surface? Do they stay localized in half-bonds or do they spread out? Is it possible to describe and locate the energy of the surface states? Will the surface geometry remain ideal, i.e. (1 × 1), or will reconstruction of the surface occur? Many of these questions have been answered through collaborative experimental and theoretical studies. Briefly, the electronic charge adjusts to the surface and smooths out to heal the cut. Surface states do exist and the most prominent one has an energy corresponding to a state in the semiconductor gap. Its charge is localized near the surface dangling bond. The surface does reconstruct into (2 × 1) and (7 × 7) patterns. These are the answers in general, but as mentioned above, many details are still unsettled.

To examine the questions and answers discussed above using theory, we require a solution of the quantum mechanical problem of a system of electrons interacting with a terminated lattice of atomic cores. Two major constraints are the lack of translational symmetry because of the surface and the necessity to do this calculation self-consistently. The self-consistent requirement means that the electrostatic potential resulting from the rearrangement of the electrons at the surface must be fed back and added to the core potential to produce a total potential for the electrons.

Two major methods were used to account for the lack of translational invariance. Appelbaum & Hamann (9) matched extended electronic

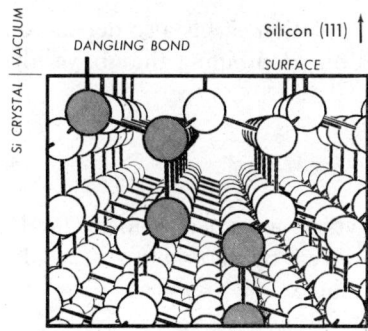

Figure 1 Schematic "ball and stick" model of a Si surface. This side view has the surface at the top of the figure. *Shaded discs* represent atomic cores in the total electronic charge density plots of Figure 3.

wavefunctions from the bulk crystal to decaying states representing the surface. Schlüter et al (10) used a supercell technique that modeled the crystal with a thin slab having two surfaces. The slab was mathematically repeated infinitely in a manner similar to a unit cell of a crystal. This allowed the use of standard band structure techniques employed for bulk crystals. It is this approach that we describe here. The self-consistency requirement is dealt with in a similar fashion for both approaches, and it has become a standard tool.

To compute the eigenstate energies and wavefunctions for the problem posed, it is necessary to construct a potential for the electrons. Here we adopt a pseudopotential model in which the $3s^2 3p^2$ valence electrons of Si are considered to flow freely through the material, interacting among themselves and with the atomic cores. The positive atomic core is composed of a nucleus and the $1s^2 2s^2 2p^6$ core electrons, which are assumed to be unchanged from their atomic states. A crystalline pseudopotential (11), V_p, is used to represent the array of core potentials and it can be constructed from ionic pseudopotentials, V_{ion},

$$V_p(\mathbf{r}) = \sum_{\mathbf{R}_n, \tau} V_{ion}(\mathbf{r} - \mathbf{R}_n - \tau) \qquad 1.$$

where $\mathbf{R}_n$ is the lattice vector to a crystal cell and τ is the basis vector to atoms in the cell.

The core potentials are screened by electron-electron interactions that can be evaluated using a Hartree potential, V_H, and an exchange-correlation potential, V_X. The Hartree potential can be determined using Poisson's equation

$$\nabla^2 V_H(\mathbf{r}) = -4\pi e^2 \rho(\mathbf{r}) \qquad 2.$$

where $\rho(\mathbf{r})$ is the electronic charge density. An example of an exchange potential is the Slater form

$$V_X(\mathbf{r}) = -3\left(\frac{3}{8\pi}\right)^{1/3} \alpha e^2 [\rho(\mathbf{r})]^{1/3}, \qquad 3.$$

where α is a predetermined parameter. Correlation effects also depend on $[\rho(\mathbf{r})]^{1/3}$ in this local density approximation. Combining the above, the total Hamiltonian is

$$\mathcal{H} = \frac{p^2}{2m} + V_p + V_H + V_X. \qquad 4.$$

Using a basis set of plane waves and analytical or numerical forms for V_p, the resulting Schrödinger equation is solved by setting up a secular equation and diagonalizing the resulting matrix.

Self-consistency is achieved by fixing V_p but allowing V_H and V_X to adjust to the changes in the electronic density, $\rho(\mathbf{r})$. This is illustrated in Figure 2, which contains a flow chart describing the method for achieving self-consistency. A cycle is started with an empirical pseudopotential, V_{EMP}, to give an initial estimate of $\rho(\mathbf{r})$ and hence V_H and V_X. The sum of these potentials is the screening potential, V_{SCR}, which is then added to the ionic pseudopotential to produce a total potential, V_T, to initiate the next cycle. When input and output potentials or charge densities are the same, self-consistency has been achieved. Typically this requires around six iterations.

The above method was applied (10) to Si(111)-(1 × 1). A plot of $\rho(\mathbf{r})$ is given in Figure 3. As illustrated, the charge smooths out and "heals" the cut bonds. Channels exist that represent openings and paths along which foreign impurities can travel into the crystal. As shown in Figure 3, the charge density near the surface does not differ significantly from that of the bulk after the first layer, thus indicating that the surface perturbation does not greatly affect the nature of the bonding beyond the first few layers of the crystal.

Since the total charge density shown in Figure 3 contains contributions from all electronic states, the surface state contributions are small and hence they are not discernable in a total $\rho(\mathbf{r})$ plot. These states can be located in energy by examining the local density of states (10), which illustrates the number of states within an energy interval as a function of

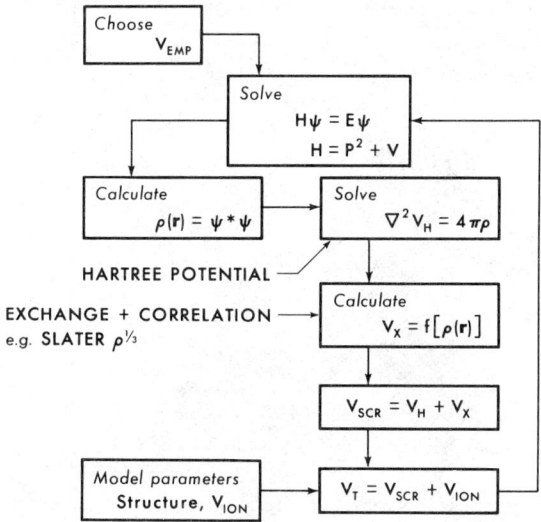

Figure 2 Block diagram indicating the self-consistent loop used for calculating the electronic structure of surfaces.

distance into the crystal. This function clearly demonstrates the existence of a surface state having energy in the semiconductor energy gap region. A charge density plot (10) places this state spatially near the cut surface bonds, and this "dangling bond surface state" is the most prominent surface state for Si.

Despite the usefulness of the ideal (1 × 1) calculation in giving an overall picture of the Si(111) surface, it is not an adequate model of the real surface. The ideal surface is left with one electron per bond and is therefore metallic contrary to experiment. If the (1 × 1) atomic geometry is retained, then one possible origin of semiconducting behavior is a gap coming from electron-electron interactions. A possible antiferromagnetic state has been predicted (12–15) but not observed. For an ideal geometry, the antiferromagnetic ordering gives a lower energy (Figure 4), but with relaxation it is difficult to determine whether the occurrence of this state is feasible. As shown in Figure 4, theoretical calculations (16) place the paramagnetic minimum energy slightly below that of the antiferromagnetic state. However, the energy difference is sensitive to the approximations used in the calculation. Perhaps a magnetic state can exist on the (111) surface of C, Si, or Ge, but it is considered unlikely at this point.

A more conventional method of attaining a semiconducting surface is to reconstruct the lattice. Low energy electron diffraction studies find a (7 × 7)

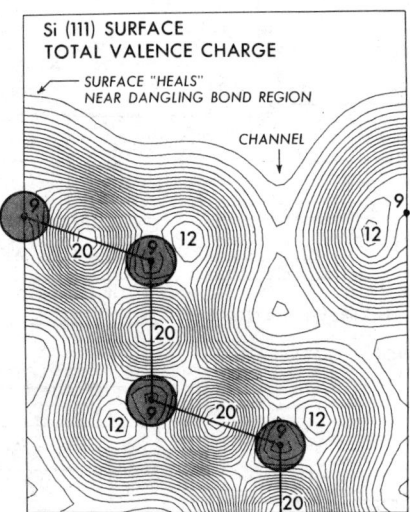

Figure 3 Total valence electron charge density for the Si(111)-(1 × 1) surface. A (110) plane is shown with the top of the diagram representing the semiconductor surface. The cores are shown as *shaded discs* corresponding to the schematic in Figure 1 and *heavy lines* represent bonds. Charge density contours are normalized to e/Ω_c where Ω_c is the bulk unit cell.

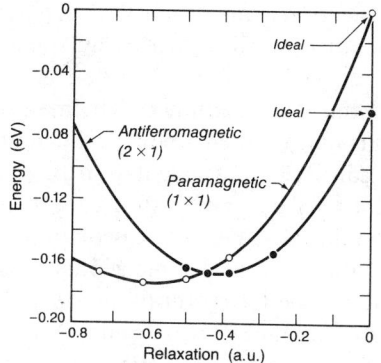

Figure 4 Energy as a function of relaxation of the surface layer the antiferromagnetic 2 × 1) and the paramagnetic (1 × 1) states of Si(111). The zero of energy is the ideal paramagnetic surface.

and (2 × 1) pattern for this surface. Because of the complexity of the (7 × 7) geometry, only recently have theorists focused on this case. Most theoretical research has been directed to the (2 × 1) modification. The geometry of this reconstruction has been a major area of research in this field. A buckling model (Figure 5) in which alternate rows of surface atoms are raised and lowered was the most popular choice. This model, based on low energy electron diffraction data, has two inequivalent atoms per unit cell and hence a semiconductor band structure. Theoretical calculations like those of Ref. (10) used this structural model as input and appeared to derive satisfactory electronic structure for the reconstructed geometry.

More recently the buckling model has been challenged and it is not likely to be correct. Angular-resolved-photoemission measurements give an energy dispersion curve, $E(\mathbf{k})$, for the surface state that is in disagreement with theory constrained to a buckled (2 × 1) reconstruction. In addition, some significant advances occurred on the theoretical side that allowed structural determination so that in principle experimental structural information was no longer required input. The theoretical scheme used pseudopotentials obtained from atomic wavefunctions (17–22), an accurate method (23) for calculating the total energy or forces for a system of cores

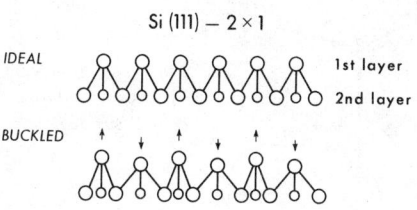

Figure 5 The (2 × 1) buckling model for Si (111).

and valence electrons, and a density functional approach (24, 25) for computing the electron-electron interactions. Applications to bulk materials (26, 27) were very successful.

For surfaces the approach involved the computation of Hellmann-Feynman forces (28–32) on surface atoms and total energies for different surface configurations. The procedure consisted of a calculation of the residual forces on the surface atoms for the ideal geometry. Atoms are then moved in the direction of these forces and the self-consistent calculation is done again. This procedure is repeated until a near zero force geometry is arrived at, and this is considered to be the best structural choice. For Si(111), as discussed above, a meta-stable minimum was found (15). If displacements were assumed to promote a buckled geometry, restoring forces developed as illustrated in Figure 6. Hence the Si(111) surface is predicted to be stable with respect to a buckling distortion, suggesting that the observed (2 × 1) geometry was caused by a distortion of another kind. The lowest energy geometry found thus far (31) is a π-bonded chain geometry suggested by Pandey (33). A schematic ball and stick model of the ideal and π-bonded chain configurations appears in Figure 7. The characteristic six-fold ring geometry and its decomposition into five-fold and seven-fold rings for the π-bonded chain case are illustrated.

At first the π-bonded chain model was greeted with some scepticism because it appeared that too many bonds needed to be broken to produce this geometry. However, it was shown (31) that at least one path existed going from the ideal geometry to the π-bonded chain with a barrier of only ~ 0.01 eV. The cleaving process could easily supply enough energy for this transition, and the resulting chain geometry is lower in energy by ~ 0.2 eV than the ideal geometry. Once the geometry was determined using the minimum energy constraint, it was possible to compute $E(\mathbf{k})$ for the surface state. These results (31, 33) were found to be in good agreement with existing and subsequent measurements (34–36) using angular resolved photoemission (Figure 8). The agreement between theory and experiment is

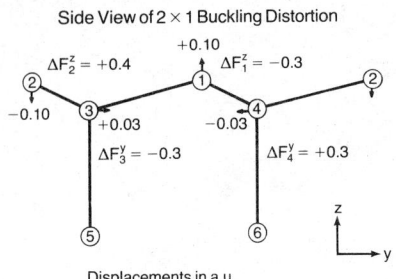

Figure 6 Side view of a (2 × 1) buckling distortion used to test the stability of the ideal (1 × 1) surface.

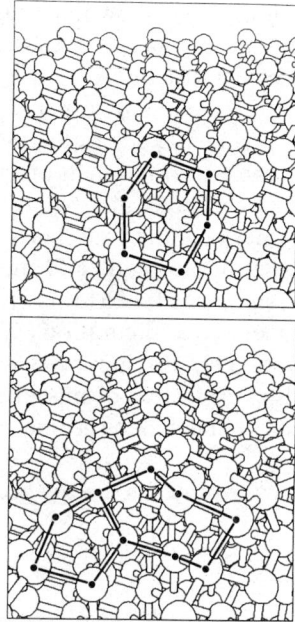

Figure 7 (*Upper*) Ideal geometry of Si(111) with six-fold rings. (*Lower*) The π-bonded chain geometry with five-fold and seven-fold rings.

impressive considering that only the atomic number and some geometrical restrictions were used to generate the theoretical results.

Agreement between theory and experiment for $E(\mathbf{k})$ and the proof of a low energy structure are necessary conditions for determining surface reconstructions but not sufficient conditions. Other geometries may give

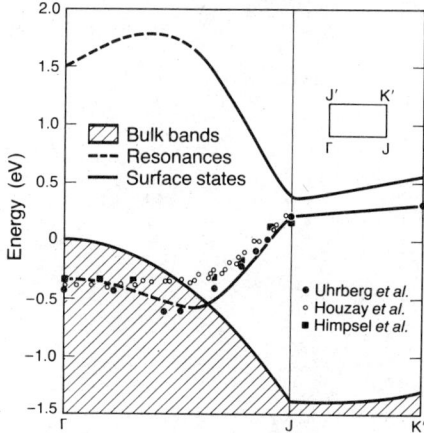

Figure 8 Calculated electron energy dispersion curve for Si(111) in the energy optimized π-bonded chain structure. References for the experimental points are given in the text.

similar $E(\mathbf{k})$ curves, and at present there are no tests to show that a given structure corresponds to the absolute minimum and not a meta-stable minimum in energy. Final decisions for specific structures require collaborative experimental-theoretical studies and definitive experiments. New techniques (37) may give more direct experimental determinations.

Another surface of Si that has attracted a great deal of attention is the (001) surface. When the ideal surface is formed, two surface atoms (labeled 1 and 1' in Figure 9) are believed to move together and dimerize. There was early evidence for a reconstruction (38) of this surface and early models relied on symmetric dimer models. Chadi (39) suggested that a lower energy geometry could be achieved if an asymmetric dimer configuration (Figure 10) were operative. Total energy calculations (29, 30) support Chadi's suggestion and many measurements are also supportive of this picture. Once again the details of the model and specifics related to the geometry are still not completely settled.

Because the asymmetric dimer can have two equivalent energy geome-

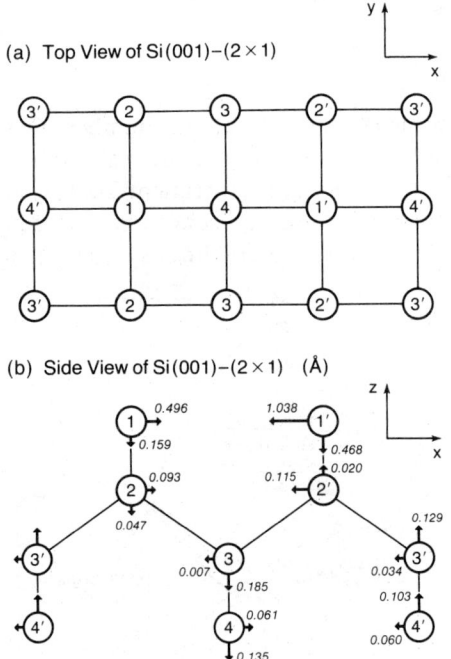

Figure 9 (a) Top view of Si(001) and (b) side view illustrating the atomic geometry. Atoms 1, 2, 3, 4 and 1', 2', 3', 4' are in the first, second, third, and fourth layers. The *arrows* and *numbers* refer to the directions and displacements (Å) from the ideal positions for a (2 × 1) reconstruction. Fourth layer displacements are only approximate because of supercell restrictions.

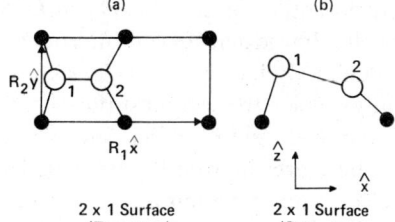

Figure 10 (a) Top view and (b) side view of an asymmetric dimer geometry for the Si(001)-(2 × 1) surface.

tries by interchanging atoms 1 and 2 in Figure 10, this system can have domains and interesting statistical temperature-dependent properties. An analysis of the structural properties and calculation of the related phase diagram has recently been done (40). Future calculations of this type could prove valuable. Pseudopotentials can be used to compute microscopic parameters, and renormalization group techniques can then be applied to study statistical behavior.

GaAs(110) is one of the most studied and best understood semiconductor surfaces; total energy pseudopotential calculations (41) have been done for the clean and adsorbate covered surface, i.e. Al-GaAs(110). These results indicate that adsorbate geometries can be determined using techniques developed for clean surfaces. Energy surfaces for Al on GaAs(110) were determined and comparisons made between various possible adsorbate sites.

Earlier work (42, 43) on Si used comparisons with photoemission data to determine chemisorption geometries. Similar approaches were used for the study of defects, like steps, on Si surfaces (44, 45). It is likely that total energy studies will be applied in this area to determine the lowest energy reconstructions near defects such as steps.

SURFACES OF METALS

The theoretical study of metal surfaces has lagged somewhat behind that of semiconductor surfaces. Most work has been done only on the electronic properties. Ab initio determination of surface structures via total energy calculations similar to those discussed in the previous section are yet to be done. This situation exists because of computational difficulties in obtaining accurate total energies for metals (especially for transition metals). Nevertheless, as we see from this section, electronic structure calculations have given much insight into the nature of these surfaces and, in many cases, provide detailed explanations of experimental observations.

The plane wave pseudopotential approach can be applied rather straightforwardly to simple metal surfaces. Figure 11 shows the calculated charge density and the planar averaged potential for the Al(111) surface (46). As in the case of the semiconductors, the charge density smooths out as it spreads out into the vacuum. The calculated charge density is significantly perturbed from the bulk charge only outside the second surface layer of the aluminum ions. The averaged charge distribution has an excess oscillatory amplitude near the surface. These are the well-known Friedel oscillations (47a,b), which result from electronic screening at the surface. Surface states do exist for the s-p metal surfaces. They, however, tend to be much more delocalized (along the direction perpendicular to the surface) when compared to semiconductor or d-band metal surface states. For the

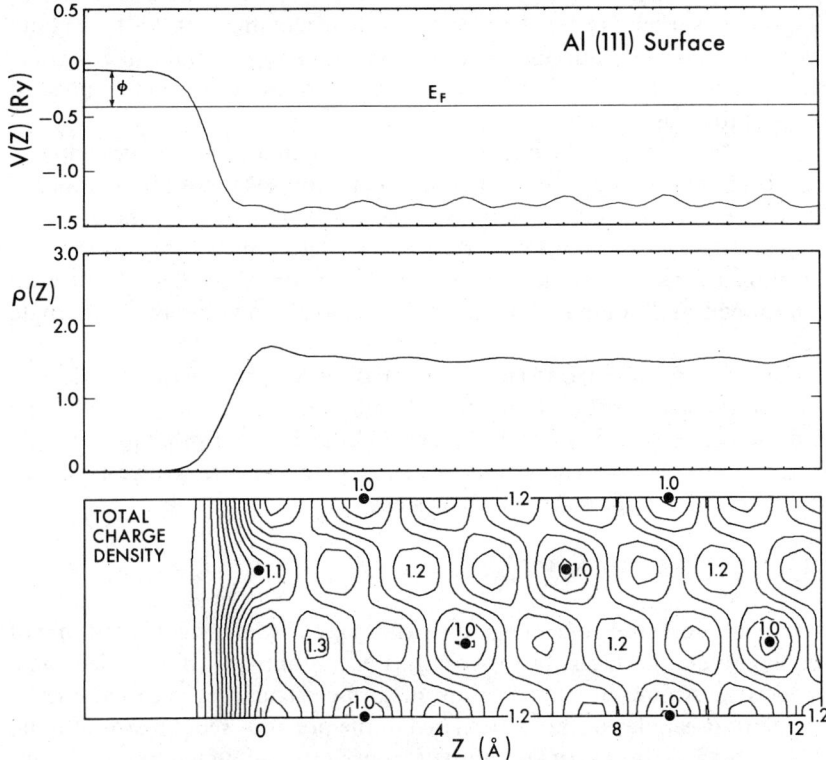

Figure 11 The top figure indicates the self-consistent potential averaged parallel to the surface and plotted as a function of distance into the bulk. The middle figure shows a similarly averaged total charge density (normalized to one electron per unit supercell, $\Omega_{cell} = 300$ Å). The bottom figure shows the total charge density in the (110) plane with the same normalization; the contour spacing is in units of 0.15. Only the minima of the charge density are labeled. The ionic positions are indicated by the *black dots*.

Al(111) surface, even the most localized surface state extends four to five layers into the bulk (46).

In recent years research in metal surfaces has been focused primarily on the transition metals and in particular on the fcc group-VIII members. Transition metal surfaces are, however, much more difficult to treat theoretically because of the concurrence of very highly localized d-electrons and delocalized s-p electrons. Several methods (48–54) have been developed to calculate the surface electronic structure self-consistently. All of these involve modeling the surfaces by thin slabs (or by repeated slabs in the case of the supercell approach) and expanding the electron wavefunctions in some basis sets.

In the pseudopotential approach, the most judicious set of basis functions is a combined set of plane waves and localized functions. In this mixed basis scheme (49), the electron wavefunctions are expanded as

$$\psi_{\mathbf{k}}(r) = \sum_{\mathbf{G}} \alpha_{\mathbf{G}}(\mathbf{k}) \exp\left[i(\mathbf{k}+\mathbf{G})\cdot\mathbf{r}\right] + \sum_{i} \beta_i(\mathbf{k})\phi_i(\mathbf{k},\mathbf{r}) \qquad 5.$$

where the $\mathbf{G}$'s are reciprocal lattice vectors and the ϕ_i's are Bloch sums of d-like Gaussian orbitals (55). Figure 12 depicts the calculated charge density for a Nb(001) surface, which serves to illustrate the efficiency of a mixed basis representation. The plane waves are useful to describe the delocalized s-p electrons in the interstitial regions and those leaking out to the vacuum; the localized orbitals are useful to describe the atomic-like character of the d-electrons near the ion cores. Note that the d-charge of Nb is quite directional and the charge on the surface atoms is altered from that of the bulk. Some of the noticeable changes on the surface are (a) the atoms on the second layer have a slightly higher charge density, which can be understood in part in terms of Friedel oscillations; (b) the density at the surface atoms becomes less directional and more s-like.

In the following discussion we use the Nb(001) and Mo(001) surfaces as examples of bcc transition metal surfaces and use the Pd(111) surface as an example of an fcc transition metal surface. Chemisorbed hydrogen on the Pd(111) surface is used as a prototypical chemisorption system. With the exception of the Nb surface, which was calculated with a plane wave basis, the calculations were performed using the mixed-basis method in the repeated slab geometry, with slab sizes typically 7 to 11 atomic layers thick. Self-consistency was achieved through using the procedure described in the previous section.

Figure 13 shows the calculated $E(\mathbf{k})$ for electrons at the Nb(001) surface along high symmetry lines in the two-dimensional Brillouin zone. The vertical and horizontal cross-hatching is used to show the allowed bulk states (the projected band structure) of various symmetries. The dash curves are the surface bands (bona fide or strong resonance). This work demonstrated, for the first time in a fully self-consistent calculation, that transition metal surfaces support a large number of surface states. They are of different angular momentum character, existing over a wide range of energies and over different regions of the two-dimensional Brillouin zone.

The surface states shown in Figure 13 are mainly atomic d-like in character and are highly localized in the surface region. For example, one of the most prominent surface states on this surface is associated with the band labeled $T1$ that extends over a large portion of k-space and has very flat dispersion. As seen in Figure 14, states in the $T1$ band are $d_{3z^2-r^2}$-like and are almost completely localized on the surface layer. They have practically no overlap with the charge from nearby surface atoms.

A surface electronic structure very similar to that of the Nb surface was

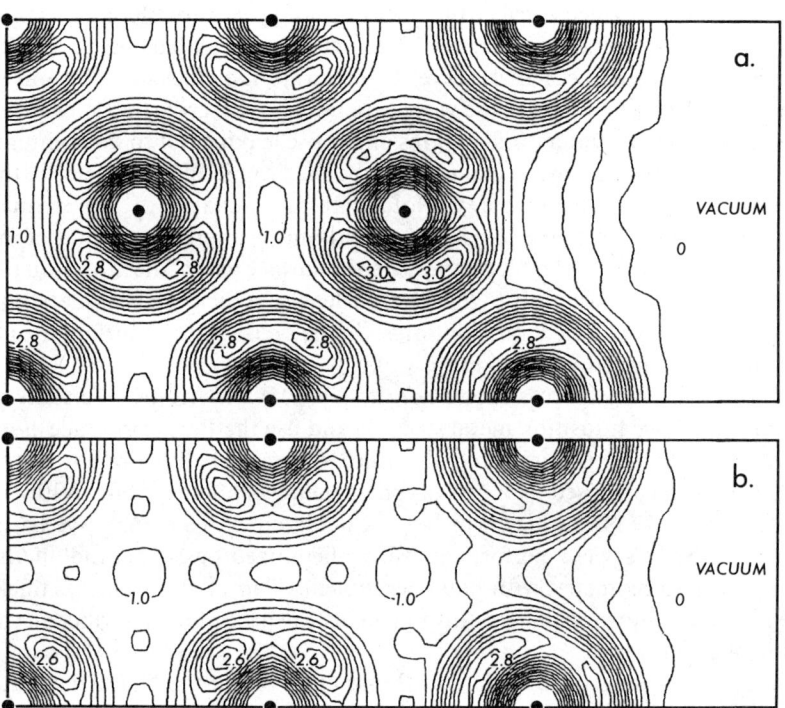

Figure 12 Total valence charge density of the Nb(001) surface plotted on (*a*) the (110) plane and (*b*) the (100) plane. The charge density is normalized to one electron per unit supercell.

obtained for the Mo(001) surface (56). Surface states of similar nature are found to exist in approximately the same energies and regions in k-space. It is found that the existence of some of these states are sensitive to the details of the self-consistent potential. The wavefunction symmetries and energy positions of the calculated surface states have been used to interpret many of the surface features found in angle-resolved photoemission measurements (57). The general agreement is very good. Hence, contrary to proposals suggested before the self-consistent calculations, relativistic, many-body, and surface-contraction effects are not needed to explain the observed spectra for the Mo(001) surface.

The surface band structure for the Pd(111) surface (49) is given in Figure 15. Also presented on the same figure is the surface band structure for a monolayer of H adsorbed on the surface three-fold sites. For the clean surface, as in the case of the bcc metals, many surface states are predicted. They are again mostly d-like and highly localized. An exception is the surface state ~ 2 eV above the Fermi level at near $\bar{\Gamma}$. It is an sp-like surface state that penetrates some distance into the bulk.

Because of the highly localized nature of their wavefunctions, these surface states strongly influence the spectroscopic properties of the surface. This can be seen from the local density of states (LDOS), $N_i(E)$, which is defined for a given volume, Ω_i, in space by

$$N_i(E) = \sum_{\mathbf{k}_{\parallel},n} \int_{\Omega_i} |\psi_{\mathbf{k}_{\parallel},n}(\mathbf{r})|^2 d^3r \delta[E - E_n(\mathbf{k}_{\parallel})] \qquad 6.$$

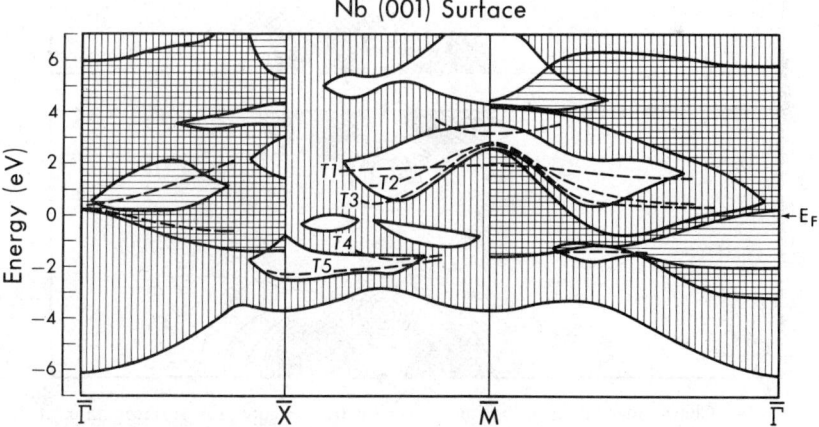

Figure 13 Surface bands (*dashed curves*) and the projected band structure for the Nb(001) surface.

where $\mathbf{k}_\parallel$ is the wavevector parallel to the surface, n is the band index, and ψ is the electron wavefunction. The LDOS gives the energy spectrum for electrons in the region Ω_i. Figure 16 shows the calculated LDOS for the fourth layer from the surface, the second layer from the surface, the surface layer, and a region one layer thick beyond the surface layer for the Pd(111) surface. The fourth layer LDOS is essentially identical to the bulk DOS of Pd, indicating that significant perturbations arising from the surface do not penetrate deeper than three layers. The surface layer LDOS, however, is very different from that of the bulk. There is a large enhancement in the density of states in the region just below the Fermi level E_F at the expense of the state density near the bottom of the d-bands. This is a direct consequence of the existence of the many surface states near E_F. (See Figure 15.) The second moment of the surface layer LDOS is also 16% smaller than that of the bulk layers.

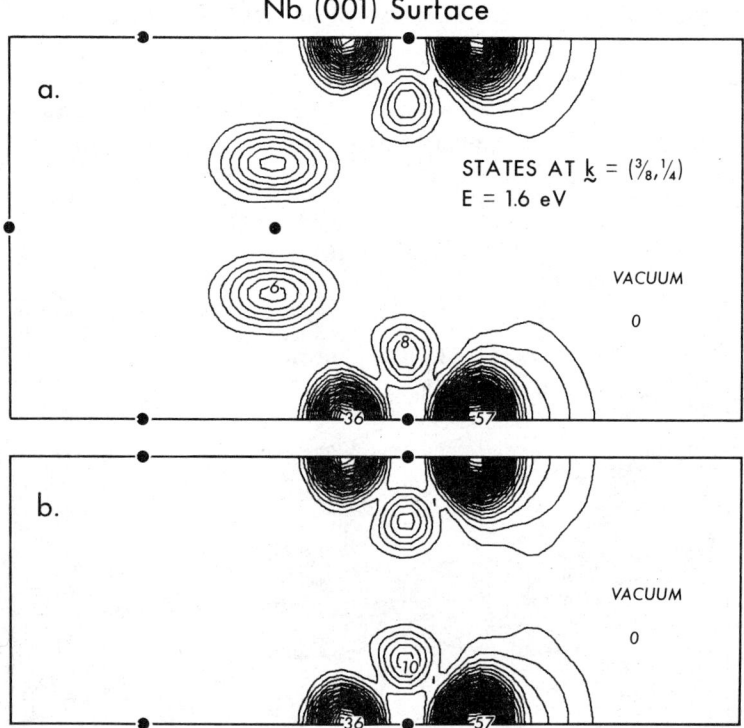

Figure 14 Charge density distribution of a *T*1 (see Figure 13) surface state at **k** = $(3/8, 1/4)2\pi/a_c$ (with a_c the bulk crystal lattice constant) plotted on (*a*) the (110) plane and (*b*) the (100) plane. Charge density is given in relative units.

These results explain a characteristic adsorbate independent reduction of angle-integrated photoemission signal from 0–2 eV below E_F observed when different adsorbates are chemisorbed on the Pd(111) surface (58, 59, J. E. Demuth, unpublished). (Similar effects are also observed in Ni and Pt.) Since these experiments measure the LDOS near the surface, it is interpreted that this reduction is largely due to the removal (shifting to lower energies) of surface states and resonances by the adsorbates. Figure 17(a) shows the difference between the LDOS of layer 4 and that of the surface layer. The negative parts of the curve correspond to energy regions in which there are excess electronic states at the surface. This is to compare with the curves in Figure 17(b), which are measured differences in photoemission intensity between clean and adsorbate-covered Pd(111) (59, J. E. Demuth, unpublished). The highly localized character of the wavefunctions of these surface states is consistent with the interpretation of their high reactivity with adsorbates. The effect is adsorbate-independent because it involves removal of states that are characteristic of the substrate. As we see below, this picture has since been confirmed in detail by a direct calculation of H chemisorption on the Pd(111) surface.

The calculations also yield very valuable information on the origin and character of the individual surface states. For example, the four intrinsic

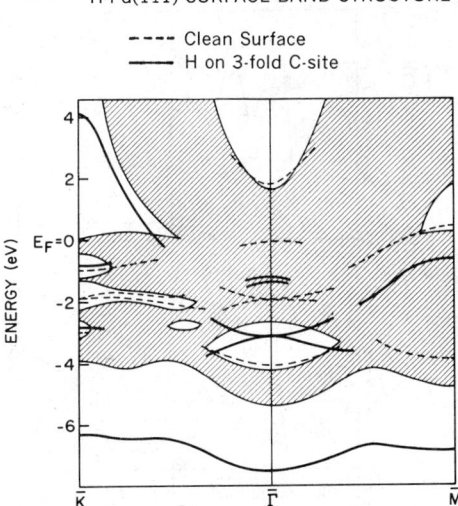

Figure 15 Localized states at the Pd(111) surface. *Dash curves* are for states of the clean surface, and *solid curves* are for states of the H-covered surface with H in the three-fold centered C-site.

surface states at $\bar{\Gamma}$ shown in Figure 15 are derived from the four sets of bulk bands along Γ to L in the three-dimensional Brillouin zone. The surface states at ~ 2 and ~ -4 eV are of Λ_1 symmetry and can only be detected by p-polarized light in a photoemission experiment. The states at ~ 0 and -2 eV are of Λ_3 symmetry and can be excited by s-polarized light. Physically these states are split off from the bulk bands by the slightly less attractive potential felt by a surface electron. Recent angle-resolved photoemission measurements (60, 61) have been performed to search for the surface states predicted in Figure 15. The agreement between experiment and theory is very good in general for both energy positions and the symmetry of the states (see Table 1). Moreover, it is shown that using a rigid-band interpretation one can understand the surface states found on the Cu(111) surface in terms of the Pd(111) results (62).

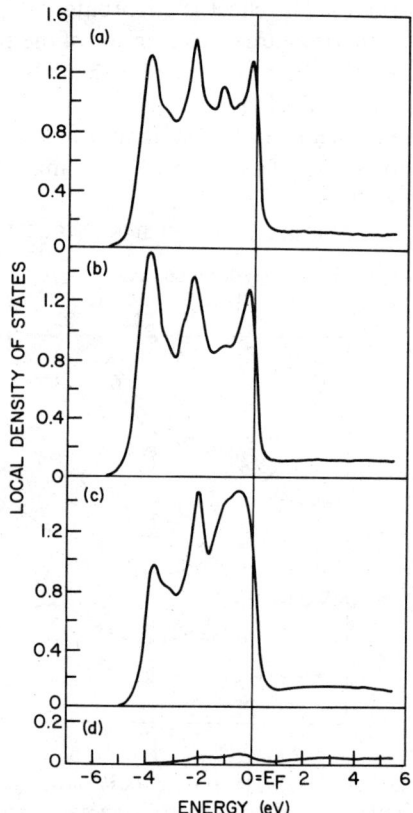

Figure 16 Calculated local density of states for (*a*) layer 4, (*b*) layer 2, (*c*) layer 1 (the surface layer), and (*d*) a region one layer thick beyond the surface layer for the Pd(111) surface.

There have been several self-consistent calculations of hydrogen chemisorption on transition metal surfaces (63–65). The major objectives of these studies are to obtain a microscopic understanding of the chemisorption bonds and to determine the adsorption geometries. Experimentally, it is found that hydrogen adsorbs on the Pd(111) surface dissociatively and forms a (1 × 1) monolayer at low temperature (58). For the H–Pd(111) calculations (63), three structural models for H monolayer coverage have been studied in an attempt to achieve the above objectives. The results are compared with photoemission data for a determination of the preferred adsorption site. The calculations were carried out by fixing the H–Pd bond length at the sum of the Pd metallic radius and the H covalent radius. The three geometries considered are (A) H on top of every surface Pd atom; (B) H on a three-fold site over a hole in the second layer of Pd atoms; and (C) H on a three-fold site over an atom in the second Pd layer.

The calculated results show that sites B and C produce very similar spectra that agree well with photoemission data. Site A, on the other hand, is ruled out as the preferred site. Figure 15 shows the surface band structure for H on site C. As evident from the figure, the H adatoms induce extensive changes in the surface band structure of clean Pd(111), indicating a strong

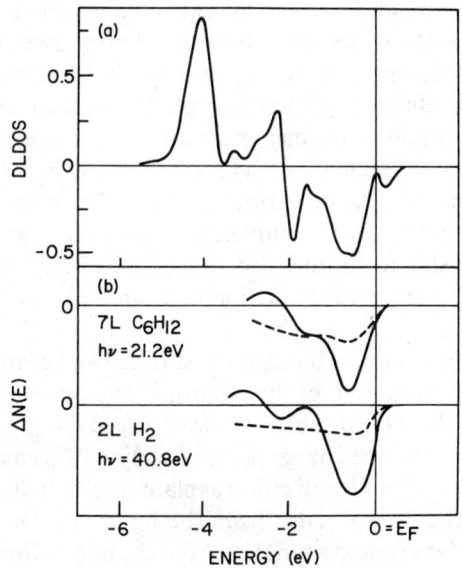

Figure 17 (a) Calculated difference in LDOS between layer 4 and the surface layer for the Pd(111) surface. (b) Adsorbate-induced differences in photoemission intensities from (58). The *dash curves* indicate the estimated attenuation if uniform attenuation of the d-band were to occur.

Table 1 Localized states on the Pd(111) surface. (Energies are in eV measured relative to E_F)

	Clean		H-adsorbed	
	Theory	Expt.	Theory	Expt.
$\bar{\Gamma}$	1.9	1.7[a]	$\left.\begin{array}{r}-1.3\\-1.4\end{array}\right\}$	-1.2[b]
	-0.2	>-0.3[b]	-3.2	-3.1[b]
	-2.0	-2.2[b]	-7.5	-7.9[b]
	-4.1	—		
$\bar{K}$	-1.0	>-0.3[b]	-0.8	-1.0[b]
	-1.9	-2.1[b]	-2.8	-2.8[b]
	-3.0	—	-6.3	-5.9[b]
$\bar{M}$	—	-1.0[b]	-0.6	—
	-3.8	—	—	-4.1[b]
			-6.8	-6.4[b]

[a] Ref. (61).
[b] Ref. (60).

surface chemical bond. The prominent H-induced features are an H–Pd bonding adsorbate band below the Pd bulk d bands and a dispersive band of antibonding H–Pd states just above the Fermi level in a gap in the projected band structure (near $\bar{K}$). The effects of hydrogen on the intrinsic surface states are different depending on the orbital character of the individual states. Some states disappear and contribute to the adsorbate states, while others have their character intact but move to lower energies because of the lowering of the surface potential by the adsorbates. Some states change their character completely. Note that near E_F all intrinsic surface states are shifted to lower energies, confirming the interpretation given above on the universal reduction of photoemission intensity in this energy region.

The nature of the chemical bond at the surface can be studied directly by examining the wavefunctions of the electrons. The bonding and antibonding character of the H–Pd states is clearly seen in the charge density contours in Figure 18. The charge is completely localized on the H atoms and the first Pd layer. This localization explains the virtually identical band structures for sites B and C. Also from the figure, we see that the strong binding at the surface is predominantly from the interaction between the H $1s$ and the Pd $4d$ orbitals. In fact, the results suggest that the interaction is stronger than the d-band width, leading to the formation of well-defined surface molecular bonding and antibonding states.

Figure 19 depicts the theoretical surface density of states together with the bulk density of states. The bonding Pd–H band appears as a distinct peak at -6.5 eV. There is also a drastic reduction of state density near E_F, which is primarily due to removal of intrinsic surface states from this energy region. Another important result from the calculation is the absence of any peak corresponding to antibonding H–Pd states as would be predicted by simple chemisorption models (66, 67). This is a band structure effect. The antibonding states are not visible because of their dispersion and because they can exist as true localized states only in a rather limited region of the two-dimensional Brillouin zone.

Figure 20 presents the theoretical photoemission difference spectra ΔN for the three chemisorption geometries together with the experimental curve (59). The theoretical ΔNs are calculated from the difference in the LDOS, weighted by an escape depth and convoluted in energy by an experimental lifetime and instrument broadening factor (63). Final-state

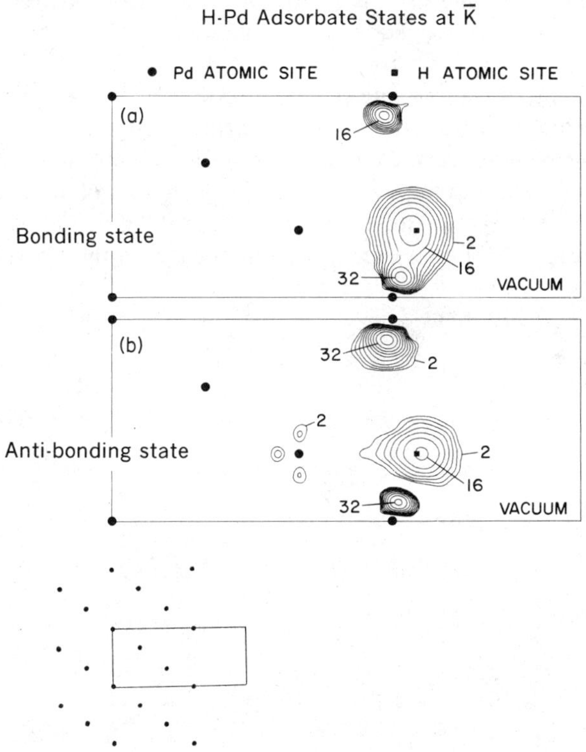

Figure 18 Charge density contour plots for adsorbate states at $\bar{K}$ at (a) -6.3 eV and (b) 4.1 eV [H/Pd(111) at C sites]. The charge densities are given in relative units and plotted for a $(1\bar{1}0)$ plane.

and matrix element effects are not included. Of the four main features in the experimental data, the structure (*d*) is reproduced in all three theoretical curves. As discussed above, this structure is primarily brought about by the removal of surface states or resonances from this particular energy region; it should, therefore, be insensitive to the nature of the adsorbate or to its position. The case with H on top of each Pd atom (*A*-site) is clearly ruled out. The spectra for sites *B* and *C*, on the other hand, reproduce the experimental spectrum very well. We may, therefore, conclude that hydrogen prefers a three-fold site, but we cannot discriminate between the two three-fold sites *B* and *C*.

Since the surface band structures of the two three-fold sites are virtually identical, the interpretation of the experimental features is the same for both geometries. The peak (*a*) corresponds to the H–Pd bonding band at this energy (Figures 15 and 19). Peak (*b*) arises mainly from the appearance of the two surface bands in the gap in the project band structure near $\bar{\Gamma}$. (See Figure 15.) Correspondingly, the valley (*C*) arises from the removal of the above-mentioned surfaces states and of the surface state about $\bar{K}$ from this energy region.

In addition to the above angle-integrated data at low temperature (59, J. E. Demuth, unpublished), the surface electronic structure for H on Pd(111) has been measured recently by high resolution angle-resolved photoemission measurements at temperatures ranging from that of liquid nitrogen to above room temperature (60). It is shown that the predicted surface band structure (Figure 15) is in excellent agreement with the low temperature

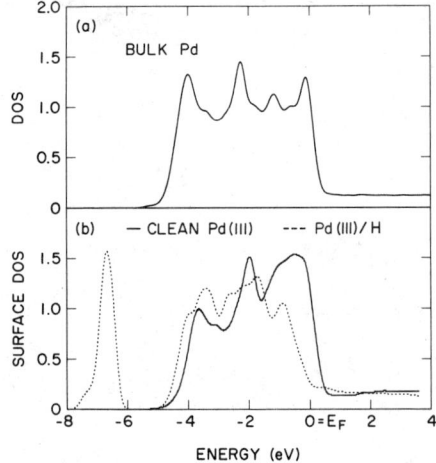

Figure 19 Calculated LDOS for (*a*) bulk Pd and (*b*) Pd(111) surface with and without a monolayer of H in the *C* site.

data. A comparison between theory and experiment is shown in Figure 21. (Also see Table 1.) The only possible discrepancy is for states near $\bar{M}$. The theory, therefore, gives a very accurate description of the chemisorption process at low temperature.

At higher temperatures, the situation is more complex. There appear to be two chemisorption phases with the more stable high temperature phase

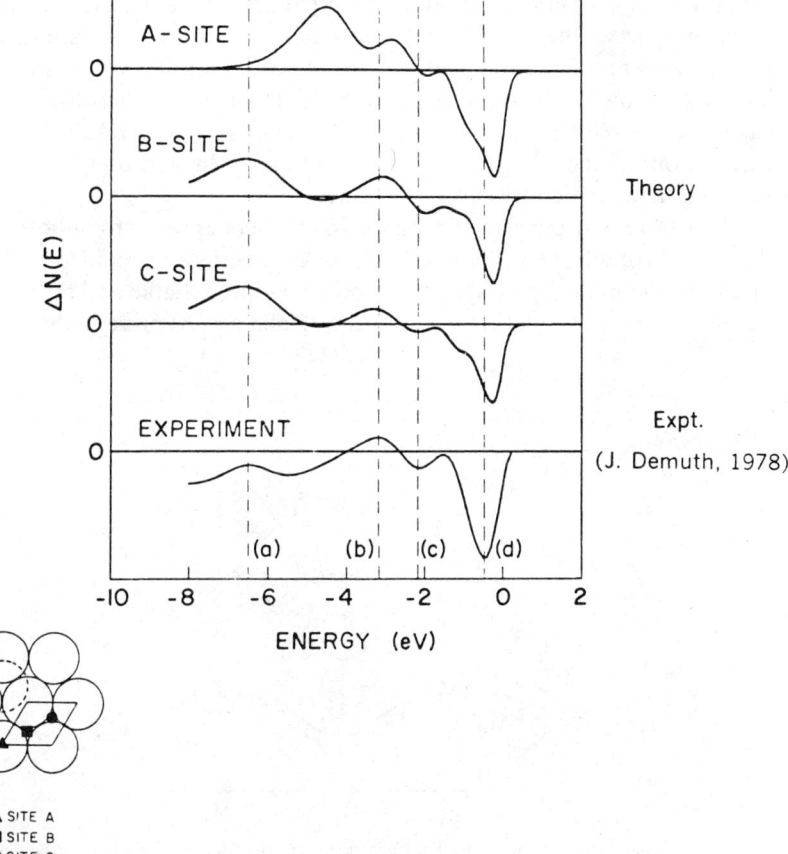

Figure 20 Comparison of the calculated H-induced photoemission difference spectra with the experimental difference curve.

showing no sign of any H-induced features, at least in near normal emission spectra (60). These surprising measurements suggest a structural transformation with the H moving away from the three-fold site into some subsurface positions as the temperature rises. The same phenomenon is observed also for H on the (111) surfaces of Ni and Pt. At present there is no satisfactory description of the nature of the high temperature phase. Total energy calculations on these systems would help in clarifying the situation.

CONCLUSIONS

One objective of this review is to report on the past successes of theoretical investigations in surface physics. Another is to demonstrate to the reader that this area is a very active one and a great deal of research is in progress now. In a sense, the total energy approaches described here are in their infancy with regard to applications. For the cases considered, the methods have been shown to be successful and useful. However, this approach is not sufficiently developed to a point at which one would believe results based on calculations if they disagreed with experiment. In addition, few if any startling new predictions have been made.

We believe that the rate of progress in this field appears to indicate that the above conditions will come to pass. As experience is gained with the methods, theoretical predictions will be even more reliable and the present usefulness of the calculational methods will increase. At present, theoretical

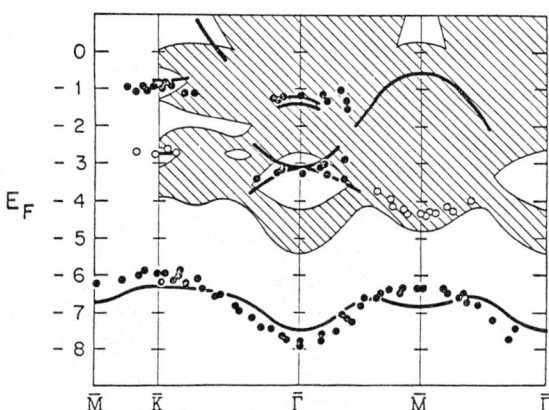

Figure 21 Calculated and measured surface states for the H(1 × 1) Pd(111) surface at low temperature phase. The *shaded regions* are the calculated projection of the bulk bands. The *heavy lines* are calculated surface states or resonances; the *circles* are data. The *open circles* indicate experimental peaks with some uncertainty.

results have helped considerably in interpreting experimental data. In the future, it is likely that complex materials under conditions difficult to simulate experimentally will be analyzed on computers, and useful predictions will be made.

For semiconductor surface and solid-solid interfaces based on semiconductors, interfacial geometries will be determined theoretically. Adsorbate and defect geometries are also possible to discern if sufficient computer memory is available or simpler computational methods, like the localized orbital approach, are developed further.

For transition metal surfaces, past calculations have concentrated solely on studying the electronic band structures, without considering total energies. As illustrated by the examples in this review, a great deal can be and has been learned using this approach. However, for further progress in understanding metal surfaces, especially chemisorption systems, total energy calculations are a necessity. At present many groups are developing such total energy methods. It is certain that many total energy calculations at the level of present day semiconductor studies will soon appear for transition metal surfaces.

ACKNOWLEDGMENTS

This work was supported by National Science Foundation Grant No. DMR8319024. M. L. C. acknowledges support by the Director, Office of Energy Research, Office of Basic Energy Sciences, Materials Sciences Division of the US Department of Energy under Contract No. DE-AC03-76SF00098. S. G. L. acknowledges support from the Director's Program Development Fund of Lawrence Berkeley Laboratory.

Literature Cited

1. Tamm, I. 1932. *Phys. Z. Sowetunion* 1:733
2. Shockley, W. 1939. *Phys. Rev.* 56:317
3. Davydov, B. 1939. *J. Phys. Moscow* 1:167
4. Mott, N. F. 1939. *Proc. R. Soc. London Ser. A* 171:27
5. Schottky, W. 1939. *Z. Phys.* 113:367
6. Bardeen, J. 1947. *Phys. Rev.* 71:717
7. Goodwin, E. T. 1939. *Proc. Cambridge Philos. Soc.* 35:205
8. Cohen, M. L. 1980. *Advances in Electronics and Electron Physics*, p. 1. New York: Academic
9. Appelbaum, J. A., Hamann, D. R. 1974. *Rev. Mod. Phys.* 48:3
10. Schlüter, M., Chelikowsky, J. R., Louie, S. G., Cohen, M. L. 1975. *Phys. Rev. B* 12:4200
11. Cohen, M. L., Heine, V. 1970. *Solid State Physics*, Vol. 24, p. 37. New York: Academic
12. Allan, G., Lannoo, M. 1977. *Surf. Sci.* 63:11
13. Duke, C. B., Ford, W. K. 1981. *Surf. Sci.* 111:L685
14. Del Sole, R., Chadi, D. J. 1981. *Phys. Rev. B* 24:7430
15. Northrup, J. E., Ihm, J., Cohen, M. L. 1981. *Phys. Rev. Lett.* 47:1910
16. Northrup, J. E., Cohen, M. L. 1984. *Phys. Rev. B* 29:5944
17. Starkloff, T., Joannopoulos, J. D. 1977. *Phys. Rev. B* 16:5212
18. Zunger, A., Cohen, M. L. 1978. *Phys. Rev. B* 18:5449
19. Hamann, D. R., Schlüter, M., Chiang, C. 1979. *Phys. Rev. Lett.* 43:1494

20. Kerker, G. 1980. *J. Phys. C* 13:L189
21. Yin, M. T., Cohen, M. L. 1982. *Phys. Rev. B* 25:7403
22. Louie, S. G., Froyen, S., Cohen, M. L. 1980. *Phys. Rev. B* 26:1738
23. Ihm, J., Zunger, A., Cohen, M. L. 1979. *J. Phys. C* 12:4401
24. Hohenberg, P., Kohn, W. 1964. *Phys. Rev. B* 136:864
25. Kohn, W., Sham, L. J. 1965. *Phys. Rev. A* 140:1133
26. Cohen, M. L. 1982. *Phys. Script.* T1:5
27. Yin, M. T., Cohen, M. L. 1982. *Phys. Rev. B* 26:3259, 5668
28. Ihm, J., Cohen, M. L. 1980. *Phys. Rev. B* 21:1527
29. Ihm, J., Cohen, M. L., Chadi, D. J. 1980. *Phys. Rev. B* 21:4592
30. Yin, M. T., Cohen, M. L. 1981. *Phys. Rev. B* 24:2303
31. Northrup, J. E., Cohen, M. L. 1982. *Phys. Rev. Lett.* 49:1349
32. Northrup, J. E., Cohen, M. L. 1983. *Phys. Rev. B* 27:6553
33. Pandey, K. C. 1981. *Phys. Rev. Lett.* 47:1913
34. Zehner, C., White, C. W., Heimann, P., Reihl, B., Himpsel, F. J., Eastman, D. E. 1981. *Phys. Rev. B* 24:4875
35. Uhrberg, R. I. G., Hansson, G. V., Nicholls, J. M., Flodstrom, S. A. 1982. *Phys. Rev. Lett.* 48:1032
36. Houzay, F., Guichar, G., Pinchaux, R., Jezequel, G., Solal, F., Barsky, A., Steiner, P., Petroff, Y. 1983. *Surf. Sci.* 132:40
37. Binnig, G., Rohrer, H., Gerber, Ch., Weibel, E. 1983. *Phys. Rev. Lett.* 50:120
38. Schlier, R. E., Farnsworth, H. E. 1959. *J. Chem. Phys.* 30:917
39. Chadi, D. J. 1979. *Phys. Rev. Lett.* 43:43
40. Ihm, J., Lee, D. H., Joannopoulos, J. D. 1983. *Phys. Rev. Lett.* 51:1872
41. Ihm, J., Joannopoulos, J. D. 1982. *Phys. Rev. B* 26:4429
42. Larsen, P. K., Smith, N. V., Schlüter, M., Farrell, H. H., Ho, K. M., Cohen, M. L. 1978. *Phys. Rev. B* 17:2612
43. Ho, K. M., Cohen, M. L., Schlüter, M. 1977. *Phys. Rev. B* 15:3888
44. Schlüter, M., Ho, K. M., Cohen, M. L. 1976. *Phys. Rev. B* 14:550
45. Chadi, D. J., Chelikowsky, J. R. 1981. *Phys. Rev. B* 24:4892
46. Chelikowsky, J. R., Schlüter, M., Louie, S. G., Cohen, M. L. 1975. *Solid State Commun.* 17:1103
47a. Lang, N. D., Kohn, W. 1970. *Phys. B* 1:4555
47b. Lang, N. D., Kohn, W. 1971. *Phys. Rev. B* 3:1215
48a. Louie, S. G., Ho, K. M., Chelikowsky, J. R., Cohen, M. L. 1976. *Phys. Rev. Lett.* 37:1289
48b. Louie, S. G., Ho, K. M., Chelikowsky, J. R., Cohen, M. L. 1977. *Phys. Rev. B* 15:5627
49. Louie, S. G. 1978. *Phys. Rev. Lett.* 40:1525
50. Appelbaum, J. A., Hamann, D. R. 1978. *Solid State Commun.* 27:881
51. Wang, C. S., Freeman, A. J. 1979. *Phys. Rev. B* 11:793
52. Smith, J. R., Gay, J. G., Arlinghaus, F. J. 1980. *Phys. Rev. B* 21:2201
53. Krakauer, H., Posternak, M., Freeman, A. J. 1979. *Phys. Rev. B* 19:1706
54. Jepsen, O., Madsen, J., Andersen, O. K. 1978. *Phys. Rev. B* 18:605
55. Louie, S. G., Ho, K. M., Cohen, M. L. 1979. *Phys. Rev. B* 19:1774
56. Kerker, G. P., Ho, K. M., Cohen, M. L. 1978. *Phys. Rev. Lett.* 40:1593
57. Wang, S. L., Gustafsson, T., Plummer, E. W. 1977. *Phys. Rev. Lett.* 39:822
58. Conrad, H., Ertl, G., Kuppers, J., Latta, E. E. 1976. *Surf. Sci.* 58:578
59. Demuth, J. E. 1977. *Surf. Sci.* 65:369
60. Eberhardt, W., Louie, S. G., Plummer, E. W. 1983. *Phys. Rev. B* 28:465
61. Johnson, P. D., Smith, N. V. 1982. *Phys. Rev. Lett.* 49:290
62. Louie, S. G., Thiry, P., Pinchaux, R., Petroff, Y., Chandesris, D., LeCante, J. 1980. *Phys. Rev. Lett.* 44:549
63. Louie, S. G. 1979. *Phys. Rev. Lett.* 41:476
64. Kerker, G. P., Yin, M. T., Cohen, M. L. 1979. *Solid State Commun.* 32:433
65. Feibelman, P. J., Hamann, D. R. 1980. *Phys. Rev. B* 21:1385
66. Newns, D. M. 1959. *Phys. Rev.* 178:1123
67. Schonhammer, K. 1977. *Solid State Commun.* 22:51

QUANTUM ERGODICITY AND SPECTRAL CHAOS[1]

E. B. Stechel[2]

Sandia National Laboratories, Albuquerque, New Mexico 87185

E. J. Heller

Los Alamos National Laboratories, Los Alamos, New Mexico 87554

INTRODUCTION

In this review, we highlight certain aspects of the field of quantum chaos from the literature of the past few years. Due to space limitations, we could not make this review encyclopedic lest we render it bland. Instead, we adopt the point of view that the classical concept of ergodicity is the most relevant and meaningful aspect of chaos for quantum mechanics, and we steer our way through the literature (and our own work) with this in mind.

Why is chaos relevant to chemistry? What are the definitions and measures of quantum chaos? What role does the correspondence principle play in linking classical and quantum aspects of chaos? It is fair to say that, amongst those who have worked in the field, there is no consensus on these three questions. Perhaps if we could get a consensus on the first question, all else would fall into place. For example, the impossibility for a quantum system to exhibit some important features of classical chaos (e.g. mixing, a finite Kolmogorov entropy) makes quantum chaos a nonsubject for some researchers. Others (including ourselves) retreat to weaker properties of chaos, e.g. ergodicity, and ask the question: How much of the molecule's available phase space will be covered, given some initial nonstationary state

---

[1] The US Government has the right to retain a nonexclusive, royalty-free license in and to any copyright covering this paper.
[2] This work performed at Sandia National Laboratories supported by US Department of Energy under contract DE-AC04-76-DP00789.

of the system? This question corresponds to the situation in collisionless gas-phase spectroscopy, where spectral features (such as local mode overtone bands) correspond to a localized excitation that decays into other modes of the molecule.

Our opinion is that the basic hallmark of ergodicity under this kind of preparation is what we have come to call "spectral chaos": the spectral intensities of individual eigenstates underneath the band are totally random, within the constraint that together, the intensities must combine to reproduce the correct spectral envelope. Totally random here means a chi-squared distribution. We elaborate on this point below and in the literature cited.

In Part I, we briefly review aspects of classical ergodicity. In Part II, some semiclassical correspondence principle issues are raised. In Part III, fully quantum mechanical theories of ergodicity are discussed. In Part IV, the numerical evidence for quantum ergodicity is reviewed; and in Part V, a short discussion of the emerging experimental evidence for quantum chaos and quantum ergodicity is presented.

CLASSICAL ERGODIC THEORY

Classical ergodic theory (1) was developed in the 1930s as a branch of measure theory (2). This provided the mathematical definitions of ergodicicity, weakly mixing, and mixing and provided the first important theorems. Primarily it related to asymptotic properties of measure-preserving transformations. However, despite the remarkable and intuitive appeal of the development, the ideas laid dormant for several decades as concepts whose time was yet to come; this mostly was due to its deficiency in being unable to provide a means for determining necessary or sufficient conditions that a certain class of transformations be ergodic, weakly mixing or mixing. This situation was not to be remedied until the late 1950s and early 1960s (3–8).

The basic concept (1, 2) is that of a measure space or of a measure algebra[3] and transformations of the measure algebra onto itself. A measure algebra is a set M together with a specified sigma-algebra X of subsets of M and a measure μ defined on that algebra. An abstract dynamical system (M, X, μ, ϕ_t) is a measure algebra (M, X, μ) with a one parameter group of automorphisms, ϕ_t, of (M, X, μ) depending measurably on t, which is to say that for any measurable subsets A and B of M in X, $\mu(\phi_t A \cap B)$ is a measurable function of t and that $\mu(\phi_t A) = \mu(A)$; the transformation is measure-preserving. An algebra has two set operations, multiplication and

---

[3] The difference between a measure space and a measure algebra is only in the treatment of sets of measure zero.

addition. In a sigma-algebra the operations are commutative, associative, and idempotent; the last meaning $A * A = A$ and $A + A = 0$. The multiplication operator is typically that of intersection and the addition operator is typically that of symmetric difference.[4]

The intuitive notion of ergodicity is credited to Boltzmann (9a–c, 10a,b) who assumed that the phase space trajectories for any isolated physical system having a constant total energy, would pass through each and every point on the constant energy hypersurface. This would then justify equating the microcanonical phase space average of an observable to the infinite time average of the same observable. The mathematician's formal and rigorous framework for this intuitive picture states that a system is ergodic if and only if it is metrically transitive, which is to say that trajectories on the constant energy hypersurface cannot be decomposed into two invariant sets, each having measure greater than zero. If we denote the energy hypersurface as M and let the transformation be the flow resulting from Newton's equations of motion, then this can also be defined in the measure algebra as:

$$\lim_{\Gamma \to \infty} 1/\Gamma \int_0^\infty \exp(-t/\Gamma) \mu(\phi_t A \cap B) \, dt = \mu(A)\mu(B)/\mu(M),$$

for any subregions A and B of M.

Concurrent with Boltzmann's intuitive development of the concept of ergodicity, Gibbs (9a–c, 11) formulated an intuitive notion of statistical mechanical mixing: quite like the stirring of a martini leads eventually to a uniform mixture of gin and vermouth, one can visualize a nonequilibrium initial state evolving through "mixing" to a uniform equilibrium distribution (microcanonical) on the constant energy surface (at least in a coarse-grained sense). The formal and rigorous mathematical definition due to Hopf (9c, 12) is

$$\lim_{t \to \infty} \mu(\phi_t A \cap B) = \mu(A)\mu(B)/\mu(M),$$

which in words says: after a sufficiently long time the fractional amount of the time-evolving set $\phi_t A$ in any set B of positive measure is the same as the original fractional amount of A in M.

A third concept, which is slightly weaker than mixing but retains most of the physical appeal, is the notion of weakly mixing, defined as:

$$\lim_{\Gamma \to \infty} 1/\Gamma \int_0^\infty \exp(-t/\Gamma) |\mu(\phi_t A \cap B) - \mu(A)\mu(B)/\mu(M)| \, dt = 0.$$

This allows the system occasionally to deviate from equilibrium.

---

[4] For those more familiar with the set operations of union and simple differences, the symmetric difference is defined as $A \triangle B = (A - B) \cup (B - A)$ or $A \triangle B = (A \cup B) - (A \cap B)$.

We turn now to how this does (and does not unambiguously) relate to the classical mechanics of nonlinear coupled oscillator systems, which leads us immediately to the general theory of invariant tori due to Kolmogorov, Arnol'd & Moser (KAM) (3–6). This theory refers to Hamiltonians of the form, $\mathcal{H}(\mathbf{q}, \mathbf{p}) = \mathcal{H}_0(\mathbf{q}, \mathbf{p}) + \lambda \mathcal{H}'(\mathbf{q}, \mathbf{p})$ where $\mathcal{H}_0$ is an integrable Hamiltonian and $(\mathbf{q}, \mathbf{p})$ forms a $2d$-dimensional phase space (for d degrees of freedom). The KAM theorem states that if the perturbation $\lambda \mathcal{H}'$ is small enough, that is if $\mathcal{H}$ is sufficiently close to $\mathcal{H}_0$, then the phase space of $\mathcal{H}$ is almost always dominated by invariant tori. Trajectories confined to invariant tori are called *quasiperiodic*. When the transition from quasiperiodic [which is typically ergodic on a torus of dimension d embedded in a $(2d-1)$-dimensional energy hypersurface in the $2d$-dimensional phase space] to "chaotic" occurs, the chaotic trajectory wanders in a $(2d-1)$-dimensional manifold, which is not necessarily the entire energy surface and certainly is not if there are coexisting quasiperiodic trajectories at the same energy. Then, if the quasiperiodic regions fill an area on the energy hypersurface that has measure greater than zero, the transformation is nonergodic on that energy hypersurface. Yet at the same time, if the energy surface (set M) is decomposable into one or more invariant quasiperiodic sets, Q_i and one or more invariant chaotic sets C_i, then within Q_i (considered now as a measure algebra) flow has no stronger stochastic properties than ergodicity, while within C_i the flow does possess stronger stochastic properties, including exponential separation of trajectories and consequently a positive K-entropy (5a,b). Probably more significant is that the flow is mixing within each C_i. From the chemist's viewpoint the relevant issues are (a) how large a measure is each C_i relative to a unit cell (h^d) and (b) how fast do subsets of C_i relax to "equilibrium." Seen from this point of view, the transition from quasiperiodic to chaotic in classical mechanics is not at all abrupt. Furthermore, this allows for the introduction of "vague tori" (13a,b) or approximate constants of the motion. We return to this below.

SEMICLASSICAL ERGODIC THEORY

Introduction

Much of the work on chaos in quantum mechanics is properly described as semiclassical. In this section we review the qualitative and quantitative predictions made possible by appealing to the correspondence limit of quantum mechanics. We shall reserve until the next section ("quantum ergodic theory") most of our own work on spectral intensities and quantum ergodicity. Originally, much of the motivation for this theory was both qualitative and semiclassical, but more recently, one of us (14a) has been able to put the spectral intensity theory on a firm mathematical footing and to show that it is indeed a quantum theory of ergodicity.

We divide our semiclassical considerations into several topics. First, we examine energy level spacings, and second, wavefunctions and other matrix elements involving wavefunctions. We then consider Kay's (14b) semiclassical theory, and finally quantization in the irregular regions of phase space.

Energy Level Spacings

Percival (15a,b) introduced into chemistry the concepts of regular and irregular spectra. (The word "spectra" is here taken to mean eigenvalues only, not spectral intensities.) His was a semiclassical construct, based upon the regular (integrable) and irregular (nonintegrable, possibly ergodic) regions of classical phase space. More recently, Berry & Tabor (16) have considered the energy level spacings expected from integrable systems. Semiclassically, $E_J = \mathcal{H}(\mathbf{J})$, where $\mathcal{H}$ is the classical Hamiltonian expressed in terms of the classical actions, and $J_i = (n_i + \alpha_i/4)\hbar$, where J_i is the ith action, n_i the quantum number (d of these for d degrees of freedom), α_i the Maslov index (17), and E_J is the semiclassical eigenvalue. This form for E_J shows that the spectrum consists of independent progressions obtained by incrementing the n_i's. The spacings are then governed by the classical frequencies $\omega_i = \partial \mathcal{H}/\partial J_i$. If the system is not harmonic, the result is a Poisson distribution of energy levels (16). This is the distribution that would arise from a completely random choice of eigenvalues. If $\langle s \rangle$ is the local, average spacing, then $P(s) = 1/\langle s \rangle \exp(-s/\langle s \rangle)$.

The study of the eigenvalues of random matrices (18–21) suggests that for strongly coupled systems, the level spacing distribution should be the Wigner surmise (22, 23); i.e. $P(s) \, ds = \pi s/2\langle s \rangle^2 \exp(-\pi s/4\langle s \rangle^2) \, ds$. This distribution is nearly devoid of very large or very small spacings relative to the average spacing (contrary to the Poisson distribution). Loosely speaking, a mutual repulsion of eigenvalues has occurred and they are trying to avoid each other as much as possible, with some fluctuations remaining. The Wigner surmise can also be obtained by assuming a linear repulsion between adjacent levels. That argument goes as follows (23, 24): Given a level at E, let $\chi(s) \, ds$ be the conditional probability differential that the next energy level falls in the range $[E+s, E+s+ds]$. Then

$$P(s) \, ds = \chi(s) \exp\left(-\int_0^s \chi(s') \, ds'\right) ds.$$

The Poisson distribution follows from assuming that the sequence of energy levels is random; i.e. $\chi(s)$ is independent of whether there is a level at E or not, hence $\chi(s) = 1/\langle s \rangle$ where $\langle s \rangle$ is the mean local spacing. The Wigner surmise follows from assuming linear repulsion, then $\chi(s) = \pi s/2\langle s\rangle^2$. The coefficient in both cases was chosen so

$$\int_0^\infty sP(s)\,ds = \langle s\rangle.$$

Other distributions, namely Brody distributions, come from assuming that $\chi(s)$ is proportional to s^q. (The Poisson distribution and the Wigner surmise are Brody distributions with $q = 0$ and $q = 1$, respectively.) For q greater than zero, Brody distributions exhibit "level repulsion," and for q less than or equal to zero they exhibit "level clustering." Furthermore, the larger the value of q the smaller the dispersion in the observed spacings.

Pechukas (25) provided an interesting and novel derivation of the level spacing distribution for small $\hbar$, for a system with energy eigenstates from the "irregular spectrum." His is a dynamical derivation of the random-matrix Gaussian orthogonal ensemble (GOE). The nearest-neighbor spacing distribution for the GOE has been shown to fit well a Wigner surmise (21a,b).

Recently, Bohigas et al (26) have considered the connections between the "old" literature, which has discussed the Poisson, Wigner and (more recently) Brody distributions motivated by nuclear physics, and the "new" literature growing up around chemical dynamics. Nuclei have unknown, presumably strongly coupled Hamiltonians. Molecules have more or less known Hamiltonians (for Born-Oppenheimer motion on potential energy surfaces). This has encouraged many workers to relate spacing distributions directly to the underlying classical dynamics (e.g. Percival's (15a,b) regular and irregular spectra). This is the modern aspect of the energy level distribution problem.

In this regard, Marcus and co-workers (27–31) have made some fascinating semiclassical arguments relating classical resonance zones to quantum avoided crossings of energy levels as a function of a parameter, and overlapping classical resonance zones [associated by Chirikov and co-workers (32–34) and by Ford and co-workers (35, 36) with the onset of chaos] to overlapping avoided crossings. While it seems difficult to make their arguments quantitative, the qualitative features of level repulsion in the chaotic domain and strong state mixing due to repeated avoidances is appealing and intuitively correct.

Regular and Irregular Semiclassical Eigenfunctions

In a beautiful analysis of the two classes of wavefunctions (regular and irregular), Berry (37) showed that the nature of the wavefunction as $\hbar$ approaches zero is distinctly different, depending on the class to which the wavefunction belongs. The main differences are in the behavior near the boundaries of the classically allowed regions and in the nature of the quantum oscillations. His conclusions are that states of Percival's (15a,b) regular spectrum, which correspond to tori in the classical phase space, exhibit "vivid patterns of regular interference fringes and violent fluctuations in intensity associated with caustics of the classical motion." On the other hand, states of Percival's (15a,b) irregular spectrum, which correspond to ergodic motion in the classical phase space, exhibit random patterns of interference extrema with more "temperate intensity fluctuations" and "anticaustics" on the boundaries of the classically allowed motions. By "anticaustics" he means that for $d > 2$ the amplitude of the wavefunction vanishes on the boundaries and for $d = 2$ the amplitude approaches a constant value on the boundaries. The origin for these properties lies in the nature of the underlying trajectories. Regular or quasiperiodic trajectories are distributed smoothly on the scale of a wavelength (or $\hbar$), while irregular or chaotic trajectories show detailed structure on all scales. This enlightening analysis of the wavefunctions leads Berry (37) to conjecture that irregular (ergodic) wavefunctions should be "Gaussian random" functions of the coordinates. The semiclassical reason is that, at a given value for the coordinate vector $\mathbf{q}$, there is a well-defined value for the kinetic energy $T = E - V(\mathbf{q})$. Now $\langle \mathbf{q} | n \rangle$ can be thought of as being a superposition of many plane waves with random phases; $\langle \mathbf{q} | n \rangle \propto Re\{\sum_k \exp(i/\hbar(\mathbf{p}_k \cdot \mathbf{q}) + i\phi_k)\}$, where $(\mathbf{p}_k \cdot \mathbf{p}_k)/2m = T$. This then leads to a function that is Gaussian random in $\mathbf{q}$.

Heller and co-workers (38–42) have examined aspects of wavefunctions other than their coordinate space representations. They were motivated by the pioneering work of Nordholm & Rice (43–45) and were interested in spectral intensities such as Franck-Condon factors (FCFs), and in the information that they might contain about dynamics. A spectral intensity (p_n^a) is simply the squared projection of the initial state (multiplied by the transition moment) (call it $|a\rangle$) onto an eigenstate $|n\rangle$; $(p_n^a = |\langle n|a\rangle|^2)$. In an experimental spectrum, $|a\rangle$ is fixed and $|n\rangle$ varies, so that the overlaps probe the propensity of the eigenstates to be found "near" $|a\rangle$. If the system is ergodic, then every $|n\rangle$ under the spectral envelope ought to exist in the vicinity of $|a\rangle$, hence the spectrum should be as dense as possible; i.e. every state should have nonzero intensity. However, if the system is quasiperiodic, then many (and at higher energies or with many degrees of freedom, by far most) $|n\rangle$ would "miss" $|a\rangle$ altogether, while other states $|n'\rangle$ would overemphasize the region "near" $|a\rangle$.

These qualitative conclusions are borne out by the formulae for flow between two states $|a\rangle$ and $|b\rangle$. If the system is ergodic, then the spectral intensities p_n^a and p_n^b should be qualitatively similar for (nonarbitrary) $|a\rangle$ and $|b\rangle$, leading to full flow from $|a\rangle$ to $|b\rangle$. This will be reviewed below.

Berry's (37) conjecture about eigenfunctions has immediate implications for spectral intensities. Since $\langle a|n\rangle = \int d\mathbf{q}\langle a|\mathbf{q}\rangle \langle \mathbf{q}|n\rangle$, a Gaussian random $\langle \mathbf{q}|n\rangle$ will lead to a Gaussian random $\langle a|n\rangle$. Hence the spectral intensities (after accounting for their position underneath the envelope, since a line in the wings is not expected to have comparable intensity to a line near the absorption maximum) should follow a chi-squared distribution of one degree of freedom. It is fascinating that this is the famous Porter-Thomas distribution (46, 47) of channel widths from nuclear physics. There are deep reasons for this, but we simply note here that this is another example of the "modern" literature making contact with the older ideas generated in nuclear scattering theory.

Kay's Theory

Kay (14b) has developed a semiclassical ergodic theory by pursuing the analog of analytic ergodic theory for semiclassical mechanics. Analytic ergodic theory studies linear operators that are induced by a transformation in Hilbert space, as opposed to geometric or algebraic ergodic theories, which study automorphisms on a measure space or measure algebra, respectively. His analysis of the properties of wavefunctions and of matrix elements of a class of operators in the limit that $\hbar$ goes to zero is consistent with Berry's (37) conjecture and the earlier conjectures of Percival (15a,b). Kay (14b) deals only with what he calls "acceptable" operators. Crudely put, an "acceptable" operator is any operator whose classical analog [$\hbar$ goes to zero limit of the Wigner equivalent (48a,b)] is square integrable on the energy surface. This eliminates pure-state density operators, which become delta functions as $\hbar$ goes to zero. Kay (14b) finds that for an ergodic system, the diagonal matrix elements (in the energy representation), A_{nn}, for any "acceptable" operator $\hat{A}$, vary smoothly with energy in the classical limit. Furthermore, the value of A_{nn} tends to the classical microcanonical average of the corresponding classical function on the energy surface. He also concludes that for systems that have a finite fraction of degenerate levels as $\hbar$ goes to zero, any matrix element of an "acceptable" operator connecting degenerate states must be identically zero. The ergodicity condition as applied to an operator that projects onto a connected region of finite measure in coordinate space implies that the magnitude squared of a wavefunction, coarse-grained and averaged over the coordinate, is the same for all states having similar energies. Imposing stronger ergodic properties such as weak mixing, Kay (14b) further concludes that an "acceptable" operator (with some further restrictions) $\hat{A}$ couples almost every eigenstate $|n\rangle$ to an infinite number of states $|n'\rangle$ of similar energy. Kay's (14b) results further imply that for a weakly mixing system (but not necessarily for an ergodic system), the nodal patterns of wavefunctions have complexity on arbitrarily small, but measurable scale.

This, of course, is consistent with Berry's (37) conjecture that the wavefunctions are Gaussian random functions of their arguments.

Since Kay (14b) deals only with mixed-state density operators, his conclusions cannot be applied to many experiments that strive to produce pure quantum states [see for example Refs. (49–51)].

Quantization in the Irregular Region

As Michael J. Berry once stated, quantization in the irregular region is the "holy grail" of semiclassical theory. Much of the work of Berry & Tabor (52a,b, 53) and of Gutzwiller (54–57a,b) on this problem stems from the "trace" formula:

$$d(E) = \int d\mathbf{q} \int dt \, \exp(iEt/\hbar) \langle \mathbf{q} | \exp(-iHt/\hbar) | \mathbf{q} \rangle.$$

The function $d(E)$ has simple poles at all the eigenvalues of H. Semiclassically, since a diagonal coordinate-space matrix element of the propagator is involved, only periodic orbits contribute to $d(E)$. Gutzwiller (57a,b) has been able to quantize a classically ergodic system (the anisotropic Kepler problem) by enumerating all the periodic orbits. Berry and co-workers (58) have made much progress along these lines for billiard problems.

Swimm & Delos (59–61) and Jaffe & Reinhardt (62) have successfully used smoothed versions of the dynamics [via introduction of high-order "regular" approximations in finite perturbation theory (63, 64)] and have quantized the smoothed dynamics via the usual Einstein-Brillouin-Keller conditions.

Very recently, E. J. Heller and N. DeLeon (in preparation) have directly quantized an actual chaotic trajectory (actually a finite-time segment thereof) using their spectral quantization method (42). Both the perturbation method and the spectral (or "color") quantization method are useful only if the chaos is not extensive; e.g. if it is confined to narrow bands in phase space. It may also be useful when the chaos takes considerable time to develop classically, even if the system is eventually ergodic. The reason is that bound quantum systems have finite level spacings and therefore quantum mechanics has only finite time to evolve into new regions. Once the time $2\pi/\delta$ is reached (where δ is the smallest spacing between energy levels), essentially nothing new can happen quantum mechanically. This is referred to as the "finite time effect" by Heller (38), and is related to the concept of the "vague torus" introduced by Shirts & Reinhardt (13a). It may also be related to the "break-time" phenomenon in the periodically-driven quantum pendulum (65–71), around which a growing, and somewhat controversial, literature is developing.

Zaslavsky (71, 72) has considered trajectories that are periodic in a coarse-grained sense, when all invariant tori have been destroyed. He arrived at a quantization condition, from which he concluded that the level spacing distribution $P(s)$ should go as a power of s for $s \ll \langle s \rangle$, a power that is dependent on the Kolmogorov entropy (5a,b, 7a–c, 73) of the system. The physical interpretation is that with increasing instability (larger Kolmogorov entropy) the exponent decreases. This, in turn, implies that the repulsion between levels is weakened. The larger the Kolmogorov entropy, the shorter the time for two trajectories with nearly the same initial conditions to become stochastically independent. Similarly [by Zaslavsky's (71) analysis] the correlation between eigenvalues should also weaken.

It is not unappealing that the level interactions should depend on the Kolmogorov entropy. This seems to imply that as the Kolmogorov entropy vanishes, the level repulsion becomes stronger. This leaves open the possibility that an ergodic system that is not mixing would have a level distribution very different from that of a Wigner surmise, and possibly even have a distribution very strongly peaked about the mean spacing. After all, this is the distribution for a class of systems that are ergodic and are not mixing, namely one-dimensional systems.

QUANTUM ERGODIC THEORY

Introduction

It would be desirable to have a consistent theory of quantum ergodicity that is formulated without reference to classical or semiclassical mechanics. At the same time, one would hope that such a theory would possess a correspondence limit that related to classical mechanics. Quantum mechanics itself has these properties (it was formulated independently of classical mechanics and it has a correspondence limit).

Quite often in the past, however, purely quantum mechanical theories of ergodicity have had no correspondence to classical mechanics. Von Neumann (9a–c, 74), for example, held that any quantum system with a discrete, non-degenerate spectrum is ergodic. Since it is possible to construct low-symmetry, classically quasiperiodic systems without any quantum degeneracies, Von Neumann's ergodicity is seemingly independent of the underlying classical dynamics. (There is a caveat, however. See below.)

More recently, Kosloff & Rice (75, 76) and Pechukas (77) have considered the quantum analog of the classical Kolmogorov entropy (5a,b, 7a–c, 73). These workers proceed by analogy with the classical Kolmogorov entropy, but their definitions are fully quantum mechanical. The major conclusion of Kosloff & Rice (75, 76) was that a quantum system with a discrete spectrum could not execute chaotic motion, independent of the complexity of the motion of the underlying classical system. On the other hand, by allowing the measurement process to interfere with the quantum evolution, Pechukas (77) defines a Kolmogorov entropy that is always finite, but which is not linear in t (where t is the time between measurements). For small t, he finds that the value for the Kolmogorov entropy depends only on the level spacing distribution and, furthermore, the larger the variance of the distribution the larger his Kolmogorov entropy. Oddly enough, the expected level distribution for an integrable system has a significantly larger variance than that for the expected level distribution for a chaotic system. The Kolmogorov entropy, as defined by Pechukas (77), is dependent only on the eigenvalue spectrum and is not in any way dependent on the eigenfunctions. Furthermore, the fine features of the energy distribution can only manifest themselves after very long times, and thereby cannot be the entire explanation. Thus, these more recent contributions are also independent of the underlying classical dynamics.

Spectral Criterion—A Quantum Criterion with Classical Correspondence

The spectral criterion for quantum chaos was born from considering "flow" in phase space and is largely independent of the eigenvalue distribution. For the most part, it rests only on properties of the eigenfunctions and in this sense is completely complementary to the Kolmogorov approach. The motivation arose from the concept that a classically ergodic system will uniformly sample all of its phase space (phase space is an example of a measure space), subject to known constants of the motion (e.g. energy conservation, angular momentum conservation, etc). In the formulation of the spectral criterion, phase space in a quantum system can be envisioned via the Wigner transcription (48a,b), which represents quantum density operators in classical phase space $(\mathbf{q}, \mathbf{p})$, or via the coherent state representation (78–81). In the latter representation the density operator is traced with phase space localized ($\Delta q \Delta p = h/2$) density operators. We review here the fundamental principles leading to the spectral criterion.

The first question is: If two nonstationary localized states $\hat{\rho}^a$ and $\hat{\rho}^b$ have the same mean energy and the same spread in energy, then what do we expect from the dynamics of these two states, if the system is ergodic? Drawing from classical analogy, we expect that the long time average $P(a, a)$ should equal $P(a, b)$ should equal $P(b, b)$ where

$$P(a, a') \equiv \lim_{T \to \infty} 1/T \int_0^T Tr\{\hat{\rho}^{a'} \hat{\rho}^a(t)\} \, dt.$$

$P(a, a')$ is the time averaged probability of starting in $\hat{\rho}^a$ and being found in $\hat{\rho}^{a'}$ at some later time. This dynamic quantity can be determined from the high resolution spectrum of $\hat{\rho}^a$ and $\hat{\rho}^{a'}$ (if the energy spectrum is nondegenerate) since

$$\varepsilon_a(\omega) = \omega\kappa \int \exp(i\omega t) Tr\left\{\exp(-i\mathcal{H}t/\hbar)\hat{\rho}^a\right\} dt = 2\pi\omega\kappa \sum_n \rho^a_{nn}(\omega - E_n/\hbar)$$

where κ is a constant, $\rho^a_{nn} = p^a_n$ (diagonal elements of the density operator in the energy representation), with $\mathcal{H}|n\rangle = E_n|n\rangle$ and $Tr\{\hat{\rho}^a\} = 1$. Then $P(a, a') = \sum_n p^a_n p^{a'}_n$ which holds for $a' = a$ as well.

The second question is: How do we determine what the "available" phase space of a state $\hat{\rho}^a$ is, without executing the complete dynamics. To answer this question we note that the concept of "accessible" phase space (classically or quantum mechanically) must include all of phase space (all states $\hat{\rho}^a$) consistent with the known constants of the motion ($\langle E^k \rangle = Tr\{\hat{\rho}^a \mathcal{H}^k\}$), known conserved symmetries, total angular momentum (if present and conserved), etc. The spectral moments ($\langle E^k \rangle$) are conveniently summarized by the envelope or strength function $S(\omega)$. $S(\omega)$ is the low resolution version of the spectrum. The "accessible" space is then all states having the same envelope, i.e. the same $S(\omega)$. However, quantum systems of more than one degree of freedom can never be as rigorously ergodic as classical ones. The reason can be traced to quantum interference effects and is reviewed briefly below. This has been discussed somewhat by Heller & Sundberg (40) and will be discussed in greater detail by E. B. Stechel (in preparation) and by E. B. Stechel and R. L. Sundberg (work in progress).

We can actually make contact with the Von Neumann (9a–c, 74) definition of quantum ergodicity by taking $\hat{\rho}^a$ to be the density of an eigenstate. Thus the envelope $S(\omega)$ is a delta function and if the state is nondegenerate, it trivially evolves into itself and into all states underneath the envelope, which is also the accessible phase space. Hence it is trivially ergodic. This brings us to the caveat mentioned above. In classical mechanics, trajectories typically cover whatever subdomain that they sample in phase space, ergodically. For example, a two dimensional harmonic oscillator with irrational frequency ratio (ω_1/ω_2) covers a 2-torus ergodically. The 2-torus is the classical analog of the eigenstate. As an entity in phase space, the 2-torus is time invariant under the classical dynamical propagator. However, we don't usually think of such motion as ergodic, since the 2-torus fails to sample the whole energy shell, which is three dimensional. Clearly, when we speak of the property of ergodicity, we must specify the prior constraints; i.e. we must specify the measure space. If the only constraint is the total energy, the measure space is the constant energy hypersurface, and the two dimensional harmonic oscillator is nonergodic. However, if the measure space is the actual torus on which the trajectory moves (both actions are constrained), then the system is ergodic.

It should now be evident that Von Neumann's (9a–c, 74) theory, dealing as it did with individual eigenstates (which are the stationary states and analogous to the 2-torus above), was really the definition of ergodicity with respect to the torus. In this sense, Von Neumann's definition does have a classical correspondence limit. However, this sense is not especially enlightening. Often we know just the energy and we want to find out if the trajectory covers the energy shell. This we discover by time-evolving the trajectory. Whatever region the trajectory does actually cover, it covers it ergodically. Indeed, quantum mechanics has the exact same property. Suppose $\hat{\rho}^{a'} = \exp(-i\mathcal{H}\tau/\hbar)\hat{\rho}^a \exp(i\mathcal{H}\tau/\hbar)$ then $P(a,a) = P(a,a') = P(a',a')$ for all τ. This is the case because $\hat{\rho}^a$ and $\hat{\rho}^{a'}$ have exactly the same spectral characteristics. Simply put, wherever a state $\hat{\rho}^a$ goes, it goes ergodically.

These concepts were based on the classically rigorous ideas of phase space flow and of the available phase space, carried over, using physical arguments, to quantum mechanics. The missing element was the rigorous concept of a measure space or measure algebra, which is at the heart of ergodic theory. The "accessible" space has all the intuitive appeal of a measure space and is (with hindsight) closely related to a measure space. We elaborate on this in the next section.

A Quantum Measure Algebra

A measure algebra is a time invariant set M under an evolution operator ϕ_t with a sigma-algebra X of subsets of M and a measure defined on that algebra (1, 2). This was touched upon in the first section. A quantum measure algebra (QMA) (14a) is a set M and an algebra X of subsets of M. Associated with M is a time invariant state $\hat{\rho}^M$. Time invariance implies that in the energy representation, $\hat{\rho}^M_{nm} = x_n \delta_{nm}$. The quantum measure of a set A with an associated state $\hat{\rho}^A$ was defined to be $\mu(A) \equiv 1/Tr\{(\hat{\rho}^A)^2\}$ implying that $\mu(M) = 1/(\sum_n x_n^2) = 1/(\mathbf{x} \cdot \mathbf{x})$. Basically the measure counts the number of pure states comprising a given state. Thus $\mu(M)$ is essentially the number of eigenstates underneath the envelope $\mathbf{x}$. A necessary condition for A to be a set in X is that A is a subset of M [$\mu(A \cap M) = \mu(A)$]. The definition of the measure of the intersection of two sets A and B with associated states $\hat{\rho}^A$ and $\hat{\rho}^B$, respectively, was defined to be

$$\mu(A \cap B) \equiv Tr\{\hat{\rho}^A \hat{\rho}^B\}/(Tr\{(\hat{\rho}^A)^2\} Tr\{(\hat{\rho}^B)^2\}) = \mu(A)\mu(B) Tr\{\hat{\rho}^A \hat{\rho}^B\}.$$

Now, without loss of generality and in the energy representation $\rho^A_{nn} = x_n + \delta^A_n$ ($\mathbf{x}$ is the envelope and δ^A is the fluctuation of A), implying that $\sum_n \delta^A_n = 0$ (normalization, all density operators are taken to have unit trace) and $\sum_n x_n \delta^A_n = 0$ ($\mathbf{x} \cdot \delta^A = 0$, which is the condition that A be a subset of M). This formulation sets up the mathematical framework in which ergodicity, weakly mixing, and strongly mixing can be studied as parallel concepts to those in abstract (or classical) dynamical systems. For example, in an abstract dynamical system, $[M, X, \mu, \phi_t]$, ergodicity can be defined as

$$f(A, B) \equiv \lim_{\Gamma \to \infty} 1/\Gamma \int_0^\infty \exp(-t/\Gamma)\mu(\phi_t A \cap B)\, dt = \mu(A)\mu(B)/\mu(M),$$

for all A and B in X. Furthermore, this holds for B equal to A. In a quantum dynamical system, $[M, X, \mu, \phi_t]$,

$$f(A,B) = \mu(A)\mu(B)[\mathbf{x}\cdot\mathbf{x} + \delta^A \cdot \delta^B]$$
$$= [1 + (\delta^A \cdot \delta^B)/(\mathbf{x}\cdot\mathbf{x})]\mu(A)\mu(B)/\mu(M).$$

Whatever the state $\hat{\rho}^A$, the average of $f(A, B)$ over all B is $\mu(A)\mu(B)/\mu(M)$. Nonetheless, if there is a distribution of fluctuation vectors in the measure algebra, then there will be dispersion about the mean value. Now, considering the self-correlation, it can be seen that

$$f(A, A) = [1 + (\delta^A \cdot \delta^A)/(\mathbf{x}\cdot\mathbf{x})]\mu(A)\mu(A)/\mu(M).$$

This is strictly greater than the ergodic value unless $\delta^A = 0$ (mod 0). We review below the motivation for why, even in the most chaotic of quantum systems, we expect there always to be a distribution of fluctuation vectors. We do this by appealing to the plausible conjecture that the eigenfunctions of a quantum chaotic system are Gaussian random. Quantum ergodic theory then becomes a study of the distribution of fluctuation vectors. Nonetheless, when the fluctuation vectors are distributed randomly (we elaborate on this below) the expectation for $f(A, A)$ can be as much as three times that of the ergodic value.

Connection to Earlier Theories

There are numerous connections and similarities between the measure theory approach to quantum ergodic theory and earlier theories; most directly that of our earlier work (38–42), but also that of Nordholm & Rice (43–45) as well as that of Kay (82). We discuss our work first because the measure theoretic framework was simply a recasting of our earlier intuitive development (38), which was based on the concept that a time-evolving state of an ergodic system should spend equal time in any other equivalent state. We defined equivalence in terms of the low resolution envelope (38). A smooth envelope (low resolution envelope) has been shown to create a measure algebra (83a). This then lends additional meaning to both the low resolution envelope and to the idea of equivalence. It also provides a mathematically direct (as opposed to physically reasonable) connection to the set-theoretic definition of ergodicity. Nordholm & Rice (43–45) were also operating on the physically appealing concept that nonstationary localized states (basis functions) in a quantum ergodic system should become uniformly distributed over the energy surface. Kay (82), in introducing this concept again, but in a coarse-grained sense, used mixed-state density operators to create the analog of a connected set of finite measure in phase space, which should then sample all the available phase space (the energy surface) if the flow is ergodic. This again is the set-theoretic definition of ergodicity. The only thing missing in terms of a rigorous definition of ergodicity was the concept of a measure algebra in quantum mechanics. However, rigorous ergodicity is a classical concept and as such is not directly applicable to quantum dynamics of pure states, where interference effects persist even when $\hbar$ is allowed to become as small as we like. E. B. Stechel (in preparation) and E. B. Stechel and R. L. Sundberg (work in progress) have considered a measure algebra defined by the low resolution envelope together with random fluctuations about that envelope. Some interesting results emerge.

Statistical Fluctuations

Here we discuss more quantitatively the fluctuations mentioned above. Suppose we take Berry's (37) conjecture literally. That is, suppose that all the eigenstates are Gaussian random. Then most states in the measure algebra have fluctuations that are statistical. By statistical we mean that δ_n^A and δ_n^B are uncorrelated and that δ_n^A and $\delta_{n'}^A$ are likewise uncorrelated for $n \neq n'$. Since the statistical ensemble average for p_n^A must be x_n (to reproduce the envelope), then the distribution function for δ_n^A has mean zero and some unknown variance. The variance determines the deviation from ideal (or classical) ergodicity. Appealing to the plausible conjecture that each state $|n\rangle$ is Gaussian random, we expect $\langle n|a \rangle$ [for typical (real) states $|a\rangle$] to be Gaussian random, as well, with the expected value of $p_n^A = |\langle n|a\rangle|^2$ being x_n. Further arguments (E. B. Stechel, in preparation) suggest that p_n^A/x_n is distributed with a normalized[5] chi-squared distribution of $2z_A$ degrees of freedom, where $\mu(A) \leqslant 2z_A \leqslant 2\mu(A)$. In other words, the effective number of degrees of freedom is between $\mu(A)$ and twice $\mu(A)$. The range comes from allowing for the states $|a\rangle$ to be complex. This, however, assumes that the state is observed with infinite spectral resolution; finite resolution also has the effect of increasing the number of effective degrees of freedom. Evidently the effective number of degrees of freedom for the observed fluctuation

---

[5] A distribution, related to the chi-squared distribution, also parameterized by a single parameter, z, arises frequently. That distribution is $P(y;z) = z^z y^{z-1} \exp(-zy)/\Gamma(z)$. The expected value of y is $\int_0^\infty y P(y;z) \, dy = 1$, and the variance is $\int_0^\infty (y-1)^2 P(y;z) \, dy = 1/z$. We call $P(y;z)$ a normalized chi-squared distribution of $2z$ degrees of freedom.

depends on spectral resolution (determined by the number of states under the unresolved spectral line), on purity resolution [determined by $\mu(A) = 1/Tr\{(\hat{\rho}^A)^2\}$] and to a lesser extent on whether the state is real or complex. Furthermore, as the number of degrees of freedom increases, the variance (fluctuations about the mean) decreases. Hence, the observed fluctuations disappear quickly with decreasing resolution and/or purity.

Implications for Pure States and Mixed States

For a normalized ergodicity function $\bar{f}(A,B) \equiv f(A,B)\mu(M)/[\mu(A)\mu(B)]$ a classically ergodic system would have $\bar{f}(A,B) = 1$. In a quantum measure algebra $\bar{f}(A,B) = 1 + (\delta^A \cdot \delta^B)/(\mathbf{x} \cdot \mathbf{x})$ for A, B in X. If the $p_n^A/x_n = (1 + \delta_n^A/x_n)$ are independently distributed with normalized chi-squared distributions of $2z_A$ degrees of freedom, then the mean and the variance of $\bar{f}(A,B)$ are $\langle \bar{f}(A,B) \rangle = 1$ and $\text{Var}[\bar{f}(A,B)] = 1/(\mu(M)z_A z_B)$, respectively, and the mean of the self-correlation $\langle \bar{f}(A,A) \rangle$ is $1 + 1/z_A$. The startling result is, for a pure real state this implies that $\langle \bar{f}(A,A) \rangle = 3$.

To make connections to phase space flow, we further note that $1/\bar{f}(A,A)$ is the fraction of the measure space that the time evolving set A explores. This means that a pure real state is expected to sample only one third of the available space, independent of the value of $\hbar$, even in a system that is expected to be highly chaotic, i.e. having only Gaussian random eigenfunctions. This is a quantum interference effect. For a complex pure state this would be somewhere between one third and one half, still missing a sizeable fraction. Since this fraction is $z_A/(z_A+1)$ it deviates less from one as the value of z_A increases, which among other entities depends on the purity of the state. So with increasing coarse-graining [$\mu(A)$ increases], the state samples a larger fraction of the available space.

This result is reminiscent of Pechukas' (83b) finding that the infinite time average of

$$\langle P(t) \rangle = \lim_{T \to \infty} 1/T \int_0^T |\langle \Psi | \Psi(t) \rangle|^2 \, dt$$

is almost twice statistical for the "typical" state. He chose the envelope to be a square wave and examined random fluctuations about this envelope. Of course, he included all states, real and complex, and, furthermore, they were pure states. Hence the factor of two is consistent with the results stated here. We return to his results in the next part of this review.

Although the expected values for $\bar{f}(A,A)$ and $\bar{f}(A,B)$ are independent of $\hbar$, their variances are not. In particular their variances are inversely proportional to $\mu(M)$ (the number of states underneath the envelope). Hence, even in the quantum regime, the variance for $\bar{f}(A,B)$ tends to be small, if the density of states is reasonably high. One additional relevant point is that there is no reason to expect that pure states must have the proper classical limit, since the stated definition of ergodicity is only applicable to sets of finite measure. In the classical limit a pure state has measure zero [since $\mu(A)/\mu(M)$ goes to zero]. One proper way to take the classical limit is to hold $\mu(A)/\mu(M)$ finite, in which case $\mu(A)$ must become arbitrarily large in the classical limit and consequently z_A also becomes arbitrarily large. Accordingly, $\langle \bar{f}(A, A) \rangle$ approaches one (as expected).

A Quantum Analog to the KAM Theorem

The KAM (3–6) theorem established the existence of quasiperiodic motion for nonintegrable Hamiltonian systems that are "close enough" to an integrable system. Another way of viewing this is, if $\mathcal{H}$ is a nonintegrable Hamiltonian (such that $\mathcal{H} = \mathcal{H}_0 + \lambda V$, where $\mathcal{H}_0$ is an integrable Hamiltonian), then for λ small enough or energy low enough, most initial conditions lead to quasiperiodic motion and consequently to invariant tori. The proof rests on convergence of a perturbation series in λ, beginning from an invariant torus of $\mathcal{H}_0$. Hose & Taylor (84, 85) have put forth what appears to be the quantum generalization of the KAM theorem based on quantum perturbation theory. The upshot is that, if $\mathcal{H}_0$ exists such that a perturbation-iteration expansion in λ (beginning from an eigenstate of $\mathcal{H}_0$) converges to an eigenstate of $\mathcal{H}$, then that eigenstate is quantum quasiperiodic (unambiguously assignable by d quantum numbers). The criterion for convergence is $|\langle \Phi | \Psi \rangle|^2 > 0.5$, where Φ is an eigenstate of $\mathcal{H}_0$ and Ψ is an eigenstate of $\mathcal{H}$.

NUMERICAL STUDIES

Introduction

The quasi-ergodic hypothesis was thought to be valid only for a large number of degrees of freedom, but it is now known that systems with as few as two degrees of freedom can exhibit chaotic behavior. Much insight into classical chaos was gained from numerical experiments on Hamiltonian systems of few degrees of freedom (86–90). The now classic paper by Henon & Heiles (87) demonstrated the KAM theorem and showed that quasi-periodic motion exists for fairly large perturbations. The success of those numerical experiments has led theorists to conduct analogous numerical experiments on the Schrödinger equation to search for a quantum analog to classical chaos and to elucidate the features of this so-called "quantum chaos." How intramolecular energy transfer occurs is perhaps obvious from a classical standpoint, but it is a more subtle and possibly problematic question from a quantum standpoint.

Here we review some of the numerical studies that are relevant to the issue of quantum chaos in molecules. Space limitations force us to be selective rather than comprehensive. First we examine the sensitivity of unimolecular rates to the existence of quasiperiodic versus chaotic motion in the quasibound state. Next we examine the "survival" probability, $P(t)$, and the related phase space flow measures $P(a, a)$, $P(a, b)$, etc. These first two topics are dynamical in viewpoint. Next we examine properties of stationary states in chaotic domains, including the intriguing localization that occurs around temporary classical motion. Finally we examine eigenvalue spacing distributions.

Unimolecular Rates

The influence of quasiperiodic and chaotic classical motion on quantum mechanical unimolecular rates due to dissociation by tunneling or by predissociation has been the topic of a series of papers. The possibility of using lasers to induce mode-specific chemistry is the motivation for these studies. The question is: Will unimolecular chemistry occur before intramolecular relaxation has been able to destroy the specificity of the excitation? In other words, do the unimolecular rate constants depend on the type of excitation or only on the energy of excitation? At first glance it might seem that mode-specific chemistry should be expected at low energies (where quasiperiodic classical dynamics dominates), while at higher energies (where chaotic dynamics begin to be prevalent) statistical behavior might be expected. However, recent studies have shown that the dynamics are more subtle than this. Waite & Miller (91) determined the energies and the lifetimes of the resonant states for a system of two nonlinearly coupled oscillators, for which one of the oscillators is able to dissociate. They found that the system displayed mode-specificity when the coupling was weak. However, they also found that mode-specificity was destroyed by both increased coupling and frequency degeneracy. Furthermore, they found no relation between the statistical versus mode-specific behavior and the stochastic versus regular classical mechanics for the same system. In a similar study (92) for the Henon-Heiles potential energy surface, they found that the unimolecular decay rates could be adequately described by a simple statistical model [basically RRKM (93–96) with tunneling]. They found this to be true for the entire energy range. In other words, quasiperiodic motion was irrelevant to the decay rates. In actuality they found that within any given symmetry class the rate constants followed a smooth function of the total energy. Nonetheless, they found that symmetry plays an important role in the question of mode-specificity. (The Henon-Heiles Hamiltonian has C_{3v} symmetry and thus three equivalent exit channels.) Examining a similar Hamiltonian (but with only one exit channel) they found that the unimolecular decay constants were highly mode specific for the entire energy range. They also found that that system was classically quasiperiodic for energies up to dissociation. More recently, work by Bai et al (97) questioned some of the conclusions, especially the lack of mode-specificity, claimed by Waite & Miller (92). By considering a Henon-Heiles potential with 99 bound states [instead of the 35 in Ref. (92)], Bai et al (97) found considerable (and intuitively reasonable) mode-specificity for the decay rates. Apparently, a potential with 35 bound states, which is very few for two degrees of freedom (large $\hbar$), was too far from the correspondence limit to see the intuitively expected behavior.

Shapiro & Child (98) examined intramolecular and dissociation rates in a modified Henon-Heiles oscillator (modified so that it is completely bound) coupled to a dissociative channel via a doorway state. Their definition of the "continuum-doorway" coupling guarantees that the rate of decay is determined by the golden rule. This, in turn, allows the authors to use the decay rates as a probe of properties of the undistorted bound states of the system. Now since only the doorway component of the wavefunction determines the rate of dissociation, they expected [from their classical intuition, as did Waite & Miller (91, 92)] in the regions where the classical mechanics is primarily quasiperiodic that the decay rates should fluctuate considerably. At energies where the classical mechanics are primarily chaotic they expected a decreasing sensitivity. Nonetheless, they found to the contrary that the decay rates strongly fluctuated with the eigenstates for the entire energy range. However, this can be attributed to the doorway state being located on a stable periodic orbit embedded in a mostly chaotic surrounding phase space.

In work related to these prior studies, Sundberg & Heller (99) demonstrated that predissociation rates can be sensitive to the change from localized eigenstates to delocalized eigenstates. They associate the localized eigenstates, which have d good quantum numbers, to classically quasiperiodic motion, which is confined to a d-dimensional torus (in the $2d$-dimensional phase space). They further associate the delocalized eigenstates that cover most of their available phase space with chaotic motion. They found that at low energies where the classical mechanics is quasiperiodic, the dispersion in the predissociation rates can indeed be large. Furthermore, at higher energies, where the classical mechanics is predominantly chaotic, the dispersion substantially decreases. They also demonstrated that the dissociative surface (predissociation occurs by a dissociative surface crossing a bound surface) can be chosen so that the predissociative rates are insensitive to the underlying classical mechanics of the bound surface. Sensitivity occurred when the crossing surface was perpendicular to the bound state normal-mode motion, and insensitivity occurred when the crossing surface was oblique to that motion. The bottom line is that the rates are functions of both the nature of the state and the "probe" (in this case, the crossing surface).

$P(t)$, $P(a,a)$, $P(a,b)$, and Franck-Condon Factors

Another series of papers has examined the survival probability ($P(t)$) and related long time averages, $P(a,a)$, $P(a,b)$, and Franck-Condon factors (FCFs). The motivation for these studies is the question: If we specifically place energy initially in bond A, will it flow to bond B and/or will it preferentially stay in bond A (memory)? In other words, where in the molecule does the initial energy go? Perhaps this has been the most controversial of all work in this area.

The description and classification of preparable states is of critical interest. A given molecular property, such as excitation along one bond (essential to many chemical reactions) or a dipole transition (essential to IR absorption), does not necessarily correspond to an approximate eigenstate, rather it corresponds to some nonstationary state (a coherent superposition of eigenstates). The results of any analysis of $P(a,a)$, $P(a,b)$, or FCFs will depend intimately on the choice of the initial state. This has led to much debate and controversy. Furthermore, it has not gone unnoticed that features of $P(a,a)$, $P(a,b)$, and FCFs depend on infinite time averages and hence, unlike the rate constants, they say little to nothing about how fast intramolecular redistribution occurs. Furthermore, they are independent of the details of the eigenvalue spectrum. Nonetheless, these quantities do address more directly the question of ergodicity. That is, does the initial energy find its way to all available (within energy constraints) modes of the molecule? We have taken the standpoint that meaningful questions can be addressed if the initial state $|a\rangle$ and the probe states $|b\rangle$ are somehow localized and hence initially nonstatistical states. Examples of such states are minimum-uncertainty Gaussian wavepackets, which are "maximally" localized at a phase space point, and $|n_x, \mathbf{0}\rangle$, which is assumed to be an eigenstate of a zeroth order Hamiltonian with n_x quanta in the x-mode and zero quanta in all remaining modes. This latter type of state initially places the energy in one molecular mode. Through the mode-coupling, the subsequent dynamics determine whether the energy finds itself to all other modes. Studies of these types of states and their dynamics seem to address the question of where in the molecule a mode-specific excitation goes—but they do not address the question of how fast.

The exact distribution of states formed by some excitation process is a crucial question. Along this vein Taylor & Brumer (100) have asked and investigated the question: What states can be prepared by laser excitation?

Pechukas (83b) examined the effect of the details of the eigenvalue spectrum on the survival probability, ($P(t)$). His results and his approach are intriguing, but they address only the question: Can the details of the eigenvalue spectrum distinguish regular from irregular quantum dynamics? He chose his initial state "at random" from an N-dimensional subspace of eigenstates. To model a regular Hamiltonian he also chose the eigenvalue spectrum "at random." To model an irregular Hamiltonian he took the eigenvalues to be equally spaced (with spacing ΔE). After averaging over initial states he found first that the long-time average of the ensemble averaged $\langle\langle P(t)\rangle\rangle$ is $2/(N+1)$, independent of the eigenvalue spectrum (The classically statistical expectation would have been $1/N$.) He also found that $\langle\langle P(t)\rangle\rangle$ falls monotonically and quickly to its long-time average for the regular spectrum, whereas it falls and oscillates about $1/N$ (for times less than $h/\Delta E$) for his model irregular spectrum. Of course, the expected distribution for the irregular spectrum is probably the Wigner surmise and unequally spaced. It would be of interest to know whether the Wigner surmise $\langle\langle P(t)\rangle\rangle$ behaved more like the "regular" $\langle\langle P(t)\rangle\rangle$ or the model "irregular" $\langle\langle P(t)\rangle\rangle$.

Heller & Sundberg (40) and R. L. Sundberg & E. J. Heller (in preparation) studied the phase-space flow of minimum uncertainty Gaussian wavepackets in a new two-dimensional, highly anharmonic, completely bound, nondegenerate Hamiltonian system

$$\mathcal{H}(u,s) = 1/2 p_u^2 + 1/2 p_s^2 + 1/2\omega_u^2 u^2 + 1/2\omega_s^2 s^2 + \lambda u^2 s + \beta(u^4 + s^4),$$

with $\omega_u = 1.1$, $\omega_s = 1.0$, $\lambda = -3.24$ and $\beta = 0.324$. This system (termed "demonic") possesses extensive classically chaotic regions without the possibility of dissociation. They found, in regions showing no sizeable quasiperiodic classical dynamics, that the spectra of the wavepackets are highly congested and are statistically similar. However, the fluctuations of the FCFs did show more "missing states" than expected of a "quantum ergodic" (QE) system (Gaussian random eigenfunctions). The $P(a,b)$s also showed greater dispersion than that expected of a QE system. Nonetheless, this is probably the most nearly quantum ergodic system with a smooth potential studied to date. Heller & Sundberg further verified the prediction that pure real states sample approximately one third of the available space (in a QE system), whereas complex states (Gaussians displaced in both coordinate and in momentum space) sample approximately one half of the available space. Davis & Heller (101a) examined the time evolution of the phase space flow in quasiperiodic regions of the simple Barbanis (90) potentials, and found generally excellent classical-quantum correspondence. (The factor of two or three is only expected from chaotic dynamics.)

Moiseyev & Certain (101b) took as an indicator of quantum chaos the time-dependent energy in a zeroth-order vibrational mode of the Henon-Heiles Hamiltonian. In measuring the deviation of the averaged energy in the "x-mode," they found an abrupt transition. But the transition was energy dependent on the type of excitation, that is on whether it was pure overtone or combination overtone. It would be interesting to measure the classical counterpart to the averaged deviation of the energy to see whether the classical and quantum agreed.

Moiseyev & Peres (102) showed that there is a marked difference in the "rate of spreading of wavepackets" moving in the regular versus the chaotic regions of phase space. They pointed out that a quantum (Gaussian) wavepacket should spread with the same rate as a classical Gibbs ensemble of particles in the same potential. They attributed the different rates of spreading of the wavepackets to qualitatively different structures of the energy spectra. A system that is classically integrable has families of nearly equidistant energy levels, whereas no such regularity exists for the quantum energy spectrum of a classically chaotic system. Centering their wavepackets on periodic orbits, they found that a regular wavepacket (on a stable periodic orbit) follows the classical orbit (with almost the same period) and slowly spreads around it, while a chaotic wavepacket (on an unstable periodic orbit) diffuses through the entire available phase space in a very short time. They further observe that the energy levels involved in the regular wavepacket are nearly equidistant so motion is nearly periodic. On the other hand, there is no regularity in the energy levels involved in the chaotic wavepacket. This regularity (for wavepackets centered on stable periodic orbits) has been observed in several other studies as well, in particular those by Noid et al (103) and Wyatt et al (104). It can also be observed in the work of Shapiro & Child (98) discussed above, whose doorway state was centered on the $x = 0$, $P_x = 0$ periodic orbit of the Henon-Heiles Hamiltonian. The strong components (long-lived eigenstates) form a sequence that is very nearly equidistant. In Davis et al (39a), the large components in the spectral decomposition of the Gaussian wavepacket (which was centered on another periodic orbit of the Henon-Heiles Hamiltonian) can also be seen to be nearly equally spaced. Moiseyev & Peres (102) also expected that as $\hbar$ becomes very small, "different energy levels" will be involved in the two wavepackets. This has already been seen to be the case in Davis et al (39a), where the wavepacket placed in the chaotic region (of the Henon-Heiles Hamiltonian at 75% dissociation) does not yield a fully stochastic set of FCFs due to the coexisting quasiperiodic regions at the same energy. Detailed examination of the FCFs of the two wavepackets (the second placed in one of the quasiperiodic islands) shows an anticorrelation. That is, large FCFs in one case correlate to small FCFs in the other. This again was seen in Heller & Sundberg's (40) study of the "demonic" potential at an energy at which both classically chaotic and classically quasiperiodic motion exist. The spectrum of the wavepacket placed in a quasiperiodic island is very sparse and the large FCFs can be seen to be nearly evenly spaced, whereas the spectrum of the wavepacket placed in the chaotic sea is highly congested. Nonetheless, the strong lines in the former spectrum are conspicuously missing in the latter spectrum, again due to anticorrelation.

Eigenfunctions in the Irregular Region

Heller & Sundberg (40) found that quite a few eigenstates in the almost totally chaotic regime of their "demonic" potential localized around classical trajectories that resemble quasiperiodic motion for short times (at least a few vibrational periods). This localization is clearly related to the concept of a "vague torus" introduced by Shirts & Reinhardt (13a). Shirts & Reinhardt used the concept of the "vague torus" to explain their success in finding semiclassical eigenvalues for the Henon-Heiles system in a largely chaotic domain. They found that the Henon-Heiles' trajectories became fully chaotic only by jumping occasionally from one type of pseudo quasiperiodic motion to another. Consider this together with the notion that eigenfunctions are determined by finite-time dynamics via a Fourier transform of the time dependent wavefunction (38). Times only long enough to resolve eigenstates from their neighbors are important, and if within that time an initially localized wavepacket spends much time in a given pseudo quasiperiodic state, then the eigenfunction will likewise be localized to that quasiperiodic region.

Level Spacing Distributions

McDonald & Kaufman (105) have found that the quantum level spacing distribution for the classically chaotic stadium problem (106) can be adequately described by a Wigner surmise distribution. Haller et al (24) analyzed both calculated and experimental molecular spectra for NO_2 and $C_2H_4^+$ in terms of statistical fluctuation measures and compared their results to those predicted by random matrix theory (RMT) (18–21). They found good agreement. Heller & Sundberg (40) computed the quantum level spacing distribution for two energy regions of the "demonic" potential described above. The classical mechanics corresponding to their second energy region is predominantly chaotic. The distribution agreed well with that expected of a Wigner surmise. Finally, in a careful analysis, Bohigas et al (26) made a detailed comparison of the level spacing distribution for the quantum Sinai billiard system with that predicted from RMT. Of course, the classical Sinai billiard is known to be strongly chaotic (7a–c, 8). By computing a large number of eigenvalues (to achieve statistical significance) Bohigas et al (26) indeed once again verified the predictions of RMT.

AN EXPERIMENTAL STUDY OF QUANTUM CHAOS

At high or even intermediate excitation energies, molecular anharmonicities mix normal (and local) modes, making it possible for a host of intramolecular processes to occur. This high (and intermediate) energy region is rather important in the study of unimolecular reactions and to photodissociation (UV and IR). Almost all of the experimental knowledge of the intramolecular processes comes from unimolecular and reaction rate data, as opposed to spectroscopic measurements. Typically, RRKM theory (93–96) has been employed with much success to interpret a large body of data on unimolecular reactions and on chemical activation. Many of these experimental results have been reviewed previously. For excellent reviews see, for example, Refs. (51, 107, 108).

For the purposes here, we discuss one very recent spectroscopic study (109) that is directly relevant to the theoretical literature discussed herein. Abramson et al (109) measured the stimulated emission pumping from a well prepared excited state of acetylene. The spectral resolution was approximately 0.027 cm^{-1}. The observed eigenvalue distribution was found to be well approximated by the Wigner surmise rule; this, as noted above, is expected of a quantum chaotic spectrum. Proper statistical analysis of their intensity distribution requires a careful treatment of the finite (although quite high) spectral resolution. Their data indicates that $P(a, a) = 0.02138$, and (if every line is present) that $P^{STO}(a, a) = 0.01513$, or $P^{STO}(a, a)/P(a, a) = 0.71$. Apparently the prepared state explores 71% or less of its available space. (If there are unresolved lines then P^{STO} will decrease.) Furthermore, if we try to fit this to a normalized chi-squared distribution we find a best fit with four degrees of freedom. Evidently, some smoothing has occurred in the spectral deconvolution. Nonetheless, this appears to be the best direct experimental evidence of the existence of quantum chaotic motion in a molecule to date. We expect and hope that this will stimulate more high resolution spectroscopic experiments and a closer association between the theorists and the experimentalists in this field.

CONCLUDING REMARKS

We believe that the issue of chaos in quantum mechanics is nowhere more relevant than in chemistry. Thus it is not surprising that theoretical chemists have taken a leading role in examining quantum chaos. The modern topic of "quantum chaos" is closely related to the statistical (quantum) theories of nuclear reactions developed more or less continuously since the 1940s. The Wigner level spacing distribution and the Porter-Thomas chi-squared distribution (then for channel widths, now for spectral intensities) are emerging as important concepts in the chemical literature.

Most of all, it is significant that some relevant experiments are now being done. Such experiments will help to resolve the controversy surrounding quantum chaos by forcing attention to measurable aspects and to real systems.

ACKNOWLEDGMENTS

We thank Dr. M. E. Coltrin for his critical reading of this manuscript and his helpful suggestions.

Literature Cited

1. Halmos, P. R. 1956. *Lectures on Ergodic Theory*. New York: Chelsea
2. Halmos, P. R. 1950. *Measure Theory*. Princeton, NJ: Van Nostrand
3a. Siegel, C. L., Moser, J. L. 1971. *Lectures in Celestrial Mechanics*. Berlin: Springer-Verlag
3b. Moser, J. 1973. *Stable and Random Motions in Dynamic Systems*. Ann. Math. Studies, No. 77. Princeton, NJ: Princeton Univ. Press
4a. Arnol'd, V. I., Avez, A. 1968. *Ergodic Problems in Classical Mechanics*. New York: Benjamin
4b. Arnol'd, V. I. 1978. *Mathematical Methods of Classical Mechanics*. New York: Springer-Verlag
5a. Ford, J. 1975. *Fundamental Problems in Statistical Mechanics*, ed. E. G. D. Cohen, 3:215. Amsterdam: North Holland
5b. Jorna, S., ed. 1978. *Topics in Nonlinear Dynamics*. AIP Conf. Proc. No. 46. New York: Am. Inst. Phys.
6. Casati, G., Ford, J., eds. 1979. *Stochastic Behavior in Classical and Quantum Hamiltonian Systems*. Lect. Notes in Physics, Vol. 93. Proc. Volta Mem. Conf. Como, Italy, 1977. New York: Springer
7a. Sinai, Ya. G. 1963. *Sov. Math. Dokl.* 4:1818
7b. Anosov, D. V., Sinai, Ya. G. 1967. *Russ. Math. Surv.* 22:103
7c. Sinai, Ya. G. 1970. *Russ. Math. Surv.* 25:137
8. Sinai, Ya. G. 1976. *Introduction to Ergodic Theory*. Princeton, NJ: Princeton Univ. Press
9a. Haar, D. ter. 1954. *Elements of Statistical Mechanics*. New York: Rinehart
9b. Uhlenbeck, G. E. 1968. *Fundamental Problems in Statistical Mechanics II*, ed. E. G. D. Cohen. Amsterdam: North Holland
9c. Farquhar, I. E. 1964. *Ergodic Theory in Statistical Mechanics*. New York: Interscience
10a. Boltzmann, L. 1871. *Sitzber. Akad. Wiss. Wien.* 63:397, 679
10b. Boltzmann, L. 1887. *J. Reine Angew. Math.* 100:201
11. Gibbs, J. W. 1902. *Elementary Principles of Statistical Mechanics*. New Haven, Conn: Yale Univ. Press
12. Hopf, E. 1934. *J. Math. Phys.* 13:51
13a. Shirts, R. B., Reinhardt, W. P. 1982. *J. Chem. Phys.* 77:5204
13b. Reinhardt, W. P. 1982. *J. Phys. Chem.* 86:2158
14a. Stechel, E. B. 1984. *J. Chem. Phys.* Submitted
14b. Kay, K. G. 1983. *J. Chem. Phys.* 79:3026
15a. Percival, I. C. 1973. *J. Phys. B* 6:L229
15b. Percival, I. C. 1977. *Adv. Chem. Phys.* 36:1
16. Berry, M. V., Tabor, M. 1977. *Proc. R. Soc. London Ser. A* 356:375
17. Maslov, V. P. 1962. *Comp. Math. Math. Phys.* 1:123
18a. Mehta, M. L. 1960. *Nucl. Phys.* 18:395
18b. Mehta, M. L. 1967. *Random Matrices and the Statistical Theory of Energy Levels*. New York: Academic
19. Gaudin, M. 1961. *Nucl. Phys.* 25:447

20a. Dyson, F. J. 1962. *J. Math. Phys.* 3:140
20b. Dyson, F. J. 1962. *J. Math. Phys.* 3:166
21a. Brody, T. A., Flores, J., French, J. B., Mello, P. A., Pandey, A., Wong, S. S. M. 1981. *Rev. Mod. Phys.* 53:385
21b. Porter, C. E., ed. 1965. *Statistical Theories of Spectra: Fluctuations.* New York: Academic
22. Wigner, E. P. 1932. *Phys. Rev.* 40:749
23. Wigner, E. P. 1967. *SIAM Rev.* 9:1
24. Haller, E., Koppel, H., Cederbaum, L. S. 1983. *Chem. Phys. Lett.* 101:215
25. Pechukas, P. 1983. *Phys. Rev. Lett.* 51:943
26. Bohigas, O., Goannoni, M. J., Schmit, C. 1984. *Phys. Rev. Lett.* 52:1
27. Noid, D. W., Marcus, R. A. 1977. *J. Chem. Phys.* 67:559
28. Noid, D. W., Koszykowski, M. L., Marcus, R. A. 1980. *Chem. Phys. Lett.* 73:269
29. Marcus, R. A. 1980. *Horizons in Quantum Chemistry,* ed. K. Fukui, B. Pullman, p. 107. Dordrecht: Reidel
30. Marcus, R. A. 1980. *Ann. NY Acad. Sci.* 357:159
31. Ramaswamy, R., Marcus, R. A. 1981. *J. Chem. Phys.* 74:1379
32. Chirikov, B. V., Keil, E., Sessler, A. M. 1971. *J. Stat. Phys.* 3:307
33. Chirikov, B. V., Izrailev, F. M., Tayursky, V. A. 1973. *Comput. Phys. Commun.* 5:11
34. Zaslavsky, G. M., Chirikov, B. V. 1972. *Sov. Phys. Usp.* 14:549
35. Walker, G. H., Ford, J. 1969. *Phys. Rev.* 188:416
36. Ford, J., Lundsford, G. H. 1970. *Phys. Rev. A* 1:59
37. Berry, M. V. 1977. *J. Phys. A* 10:2083
38. Heller, E. J. 1980. *J. Chem. Phys.* 72:1337
39a. Davis, M. J., Stechel, E. B., Heller, E. J. 1980. *Chem. Phys. Lett.* 76:21
39b. Heller, E. J., Stechel, E. B. 1982. *Chem. Phys. Lett.* 90:484
40. Heller, E. J., Sundberg, R. L. 1984. *Proc. NATO Adv. Res. Workshop on Chaotic Behavior in Quantum Systems, Como, Italy, 1983.* In press
41. Heller, E. J., Davis, M. J. 1982. *J. Phys. Chem.* 86:2118
42. Heller, E. J. 1983. *Faraday Discuss. Chem. Soc.* 75:141
43. Nordholm, K. S. J., Rice, S. A. 1974. *J. Chem. Phys.* 61:203
44. Nordholm, S., Rice, S. A. 1974. *J. Chem. Phys.* 61:768
45. Nordholm, S., Rice, S. A. 1975. *J. Chem. Phys.* 62:157
46. Porter, C. E., Thomas, R. G. 1956. *Phys. Rev.* 104:483
47. Lynn, J. E. 1968. *The Theory of Neutron Resonance Reactions.* Oxford: Clarendon
48a. Wigner, E. 1932. *Phys. Rev.* 40:749
48b. Moyal, J. E. 1949. *Proc. Cambridge Philos. Soc.* 45:99
49a. Hopkins, J. B., Powers, E., Smalley, R. E. 1980. *J. Chem. Phys.* 73:683
49b. Mukamel, S., Smalley, R. E. 1980. *J. Chem. Phys.* 73:4156
50. Dolson, D. A., Parmenter, C. S., Stone, B. M. 1981. *Chem. Phys. Lett.* 81:360
51. Smalley, R. E. 1982. *J. Phys. Chem.* 86:3504, and references therein
52a. Berry, M. V., Tabor, M. 1976. *Proc. R. Soc. London Ser. A* 349:101
52b. Berry, M. V., Tabor, M. 1977. *J. Phys. A* 10:371
53. Tabor, M. 1979. See Ref. 6, p. 293
54. Gutzwiller, M. C. 1967. *J. Math. Phys.* 8:1979
55. Gutzwiller, M. C. 1970. *J. Math. Phys.* 11:1971
56. Gutzwiller, M. C. 1971. *J. Math. Phys.* 12:343
57a. Gutzwiller, M. C. 1980. *Phys. Rev. Lett.* 45:150
57b. Gutzwiller, M. C. 1982. *Physica D* 5:183
58. Berry, M. V. 1981. *Ann. Phys. NY* 131:217, and references therein
59. Delos, J. B., Swimm, R. T. 1977. *Chem. Phys. Lett.* 47:76
60. Swimm, R. T., Delos, J. B. 1979. *J. Chem. Phys.* 71:1706
61. Swimm, R. T., Delos, J. B. 1979. See Ref. 6, p. 306
62. Jaffe, C., Reinhardt, W. P. 1979. *J. Chem. Phys.* 71:1862
63. Birkoff, G. D. 1966. *Dynamical Systems.* New York: Am. Math. Soc.
64. Gustavson, F. G. 1966. *Astron. J.* 71:670
65. Casati, G., Chirikov, B. V., Izrailev, F. M., Ford, J. 1979. See Ref. 6, p. 334
66a. Berry, M. V., Balazs, N. L., Tabor, M., Voros, V. 1979. *Ann. Phys. NY* 122:26
66b. Korsch, J., Berry, M. V. 1981. *Physica D* 3:627
67. Berman, G. P., Zaslavsky, G. M. 1978. *Physica A* 91:450
68a. Izraelev, F. M., Shepelyansky, D. L. 1979. *Sov. Phys. Dokl.* 24:996
68b. Izrailev, F. M., Shepelyansky, D. L. 1980. *Theor. Math. Phys.* 43:553
69. Hogg, T., Huberman, B. A. 1982. *Phys. Rev. Lett.* 48:711
70. Fishman, S., Grempel, D. R., Prange, R. E. 1982. *Phys. Rev. Lett.* 49:509
71. Zaslavsky, G. M. 1981. *Phys. Rep.* 80:157

72. Zaslavsky, G. M. 1977. *Sov. Phys. JETP* 46:1094
73. Kolmogorov, A. 1954. *Dokl. Akad. Nauk. SSSR* 98:527
74. von Neumann, J. 1929. *Z. Phys.* 57:30
75. Kosloff, R., Rice, S. A. 1981. *J. Chem. Phys.* 74:1340
76. Rice, S. A., Kosloff, R. 1982. *J. Phys. Chem.* 86:2153
77. Pechukas, P. 1982. *J. Phys. Chem.* 86:2239
78. Husimi, K. 1940. *Proc. Phys. Math. Soc. Jpn.* 22:264
79. Heller, E. J. 1977. *J. Chem. Phys.* 66:5777
80. Weissman, Y., Jortner, J. 1981. *Chem. Phys. Lett.* 78:224
81. Stechel, E. B., Schwartz, R. N. 1981. *Chem. Phys. Lett.* 83:350
82. Kay, K. G. 1980. *J. Chem. Phys.* 72:5955
83a. Stechel, E. B. 1984. *Chem. Phys. Lett.* Submitted
83b. Pechukas, P. 1982. *Chem. Phys. Lett.* 86:553
84. Hose, G., Taylor, H. S. 1981. *J. Chem. Phys.* 76:5356
85. Hose, G., Taylor, H. S. 1983. *Phys. Rev. Lett.* 51:947
86. Fermi, E., Ulam, J., Pasta, S. 1955. *Studies of Nonlinear Problems*, Sci. Rep. LA-1940, Los Alamos Sci. Lab., Los Alamos, NM
87. Henon, M., Heiles, C. 1964. *Astron. J.* 69:73
88. Contopoulos, G., Moutsoulas, M. 1965. *Astron. J.* 70:817
89. Jeffreys, W. H. 1966. *Astron. J.* 71:306
90. Barbanis, B. 1966. *Astron. J.* 71:415
91. Waite, B. A., Miller, W. H. 1980. *J. Chem. Phys.* 73:3713
92. Waite, B. A., Miller, W. H. 1981. *J. Chem. Phys.* 74:3910
93. Robinson, P. J., Holbrook, K. A. 1972. *Unimolecular Reactions.* New York: Wiley
94. Forst, W. 1973. *Unimolecular Reactions.* New York: Academic
95. Hase, W. L. 1976. *Dynamics of Molecular Collisions*, ed. W. H. Miller, Pt. B, Chap. 3. New York: Plenum
96. Oref, I., Rabinovitch, B. S. 1979. *Acc. Chem. Res.* 12:166
97. Bai, Y. Y., Hose, G., McCurdy, C. W., Taylor, H. S. 1983. *Chem. Phys. Lett.* 99:342
98. Shapiro, M., Child, M. S. 1982. *J. Chem. Phys.* 76:6176
99. Sundberg, R. L., Heller, E. J. 1984. *J. Chem. Phys.* 80:3680
100. Taylor, R. D., Brumer, P. 1983. *Faraday Discuss. Chem. Soc.* 75: In press
101a. Davis, M. J., Heller, E. J. 1984. *J. Chem. Phys.* 80:5036
101b. Moiseyev, N., Certain, P. R. 1982. *J. Phys. Chem.* 86:1149
102. Moiseyev, N., Peres, A. 1983. *J. Chem. Phys.* 79:5945
103. Noid, D. W., Koszykowski, M. L., Tabor, M., Marcus, R. A. 1980. *J. Chem. Phys.* 72:6169
104. Wyatt, R. E., Hose, G., Taylor, H. S. 1984. *Phys. Rev. A* 28:815
105. McDonald, S. W., Kaufman, A. N. 1979. *Phys. Rev. Lett.* 42:1189
106. Bunimovich, L. A. 1974. *Funct. Anal. Appl.* 8:254
107. McDonald, J. D. 1979. *Ann. Rev. Phys. Chem.* 30:29
108. Noid, D. W., Koszykowski, M. L., Marcus, R. A. 1981. *Ann. Rev. Phys. Chem.* 32:267
109. Abramson, E., Field, R. W., Imre, D., Innes, K. K., Kinsey, J. L. 1984. *J. Chem. Phys.* 80:2298

RELAXATION AND VIBRATIONAL ENERGY REDISTRIBUTION PROCESSES IN POLYATOMIC MOLECULES

V. E. Bondybey

AT&T Bell Laboratories, Murray Hill, New Jersey 07974

Introduction

Various types of energy transfer and energy redistribution processes in molecules have been extensively studied for many decades. Most of the earlier studies have emphasized electronic relaxation processes, internal conversion, and intersystem crossing (1, 2). These extensive studies resulted in a steady progress in this field and brought about a good qualitative understanding of electronic relaxation processes. Particularly significant was the realization of the importance of the intramolecular density of states upon the relaxation process (3–6).

In recent years more interest has been directed toward understanding vibrational relaxation and intramolecular vibrational energy redistribution. This interest has particularly intensified with the introduction and application of lasers to studies of molecular chemistry and physics. On the one hand, lasers provided an extremely useful and versatile tool permitting very detailed and unambiguous vibrational relaxation studies, as was discussed in numerous previous reviews. On the other hand, the availability of these intense, monochromatic excitation sources provided an additional motivation and need for understanding the mechanisms and rates of relaxation and energy redistribution.

The possibilities of using laser sources to drive selective chemical reaction were realized quite early after their emergence, and the feasibility of "molecular engineering" or "molecular surgery" was extensively discussed. In an idealized scenario, the laser deposits energy in a particular bond or functional group; this vibrational excitation then causes this bond or group to undergo a selective chemical reaction and give a desired product. Unfortunately, dissipative relaxation processes will tend to redistribute the energy and this will result in the loss of specificity. In the limit of complete energy randomization the excitation by the laser will be equivalent to thermal bulk heating. Similarly, one might imagine depositing enough energy in a particular vibrational mode to cause the corresponding bond to break. Again, relaxation processes will tend toward thermal equilibrium. With thermal, statistical energy redistribution, the weakest bond in the molecule will typically break, and not necessarily the one into which the energy was deposited.

Clearly the key question is whether the excitation energy can remain localized in the desired bond or functional group for a sufficiently long time for the selective chemical reaction to take place. From this point of view, the importance of understanding the rates and pathways of vibrational relaxation and energy redistribution is quite clear. It is therefore not surprising that the theoretical and experimental interest in this field has been increasing steadily in recent years. Because of this lively interest the volume of the published work is very large (7–24), and a comprehensive review would be way beyond the scope of this manuscript. Of necessity I therefore focus selectively upon only a few of the more recent or particularly promising results and techniques.

Infrared Studies of Collisional Ground State Relaxation

Collisional relaxation processes in ground state molecules have been studied rather extensively, and the extensive work has been ably reviewed on several occasions (8–11). Most of the earlier studies have concentrated upon relatively small molecules, such as CO_2, COS, N_2O, and similar triatomics (25–39) as well as a variety of derivatives of methane and several other simple species (40–55). These studies often relied upon coincidences between the wavelengths of various gas lasers, such as CO_2, CO, HF, or He–Ne, and molecular frequencies. In more recent experiments tunable optical parametric oscillators have been used for excitation. In a typical experiment, powerful laser pulses are used to excite the molecular transition, and one follows the time development of the molecular fluorescence. The fluorescence is usually viewed through suitable narrow bandpass filters that isolate individual molecular transitions—or groups of transitions. Vibrational manifolds of the molecules are, in the range studied, usually rather sparse, and unimolecular relaxation can not take place. The relaxation behavior observed is therefore entirely a result of collisional processes.

Owing to the intense laser sources the vibrational excitation is rather efficient, and substantial concentrations of the excited species are often produced. As a result, collisions of two excited molecules are reasonably probable and the relaxation mechanism is often dominated by a "ladder climbing" process of the type

$$A(v = 1) + B(v = n) \to A(v = 0) + B(v = n+1) + \Delta E$$

which places energy in higher vibrational levels. While on the one hand this process is interesting in itself and it permits population of vibrational overtones not directly accessible with the laser source, on the other hand it can complicate the overall relaxation and make unambiguous interpretation of the data much more difficult. Nonetheless, for several molecules, such as methyl fluoride, CH_3F and its isotopic variants, rather detailed energy flow maps are now available. Particularly fruitful and extensive in this area are the studies of Flynn, Weitz and their co-workers (8, 10, 11).

Although we are still rather far from detailed understanding of the relaxation process, in particular with respect to predictive ability, several general observations can be made based on the available data. In general, one finds that $V \to V$ vibrational redistribution processes are much more efficient than $V \to T, R$ (vibration $\to$ translation, rotation) processes, typically by several orders of magnitude. The efficiency of the $V-V$ processes increases with decreasing energy defect, ΔE. For the same ΔE, one finds that processes involving smallest overall changes in vibrational quantum numbers are strongly favored. Also important in the relaxation is intramolecular coupling between different vibrational modes. When, for instance, two nearly resonant levels are known, from spectroscopic observations, to be mixed by Fermi resonance, a greatly enhanced energy transfer is invariably found.

Highly Excited Ground State Molecules

All the experiments described above deal with the low energy, sparse region of vibrational manifold. In this region the molecular motion will be periodic, and the molecule is well described in terms of normal modes. For much higher energies, the density of states becomes progressively larger, and molecular anharmonicity leads to increasing mixing of the individual normal modes. Eventually the molecular motion becomes ergodic and the normal mode description fails. The energy will flow freely among modes; hence the molecule is better described by statistical models (56–59). While this region is not as readily accessible as the low energy levels, numerous ingenious techniques were recently devised permitting probing of highly excited ground state molecules.

Zare and co-workers (60, 61) have developed and applied an elegant, albeit less general, technique for studying molecules at relatively high levels of excitation. Their pump-probe technique involves electronic excitation, under collision free conditions in an effusive beam, of states known to undergo efficient internal conversion into the ground state. The ground state vibrational population is then probed after a suitable time delay by recording the fluorescence excitation spectra using a second dye laser. From their study of glyoxal they concluded that the vibrational energy distribution in the ground electronic state with ~ 22000 cm^{-1} vibrational energy is quite nonstatistical, in spite of the high ($\sim 10^8$/cm^{-1}) level density at this energy.

A conceptually similar technique has been devised by Smalley and co-workers (62) and has been used to study several alkylbenzenes. In their method S_0 levels are populated by spontaneous fluorescence following S_1 excitation. After a suitable delay, the vibrationally excited ground state molecules are two-photon ionized by a second, probe, laser, tuned in the "hot band" region, below the S_1 origin. Resonance enhancement of the ionization process takes place only if the originally excited, optically active S_0 levels are still populated. They indeed observe strong ionization signal with methyl and ethyl substituents. The signal intensity decreases with the length of the side chain and disappears completely for n-pentyl and larger substituents, suggesting that intramolecular vibrational relaxation (IVR) has spread most of the energy among the optically inactive modes.

An indirect, rather general, spectroscopic method of investigating the dynamics of highly excited small polyatomic molecules has been used by Field, Kinsey and co-workers to study acetylene and formaldehyde (63–65). In their experiments one laser pumps a particular S_1 rovibronic level. A second, tunable "dump" laser of longer wavelength stimulates emission into the ground state vibronic levels. The resonances are detected as a decrease in spontaneous fluorescence intensity. In this way the ground state structure can be probed at laser-limited resolution. Their results indicate, even in these four small atomic molecules, a rather strong vibrational mixing due to Fermi resonances and Coriolis interactions. Above ~ 7000–10000 cm^{-1}, these effects destroy the normal mode character of the spectra, and they become more complicated and chaotic. The importance of Coriolis interactions suggests that the onset of IVR and threshold for unimolecular decay may depend on the rotational quantum number.

Another technique useful for studies of molecules with even higher levels of vibrational excitation has been developed by Berry & co-workers (66, 67). They use cw dye lasers and optoacoustic techniques to measure the gas phase absorption spectra of a variety of hydrocarbons in the region of high overtones (up to $\sim v = 7$) of the C-H vibrations. They find that absorption lines in this range are usually characterized by Lorentzian lineshapes and little or no fine structure. Their line widths, in the range of 50–100 cm^{-1} are interpreted in terms of intramolecular vibrational energy redistribution, suggesting a time scale of $\sim 10^{-12}$ sec for this process.

The first overtones of the C–H stretching vibration of several hydrocarbons were examined by Nesbitt & Leone (68); the C–H fundamental region was studied by McDonald and co-workers (69, 70). The latter investigators have studied the infrared fluorescence of a large number of hydrocarbons in molecular beams. The molecules examined ranged in size from methane to norbornene. For some of the molecules they observed relaxed fluorescence and concluded that a threshold density of ~ 10 states $1\ \text{cm}^{-1}$ is needed for IVR to take place.

Bimolecular chemical reaction provides an alternative, nonthermal way of producing molecular species with a controlled, well defined level of excitation. Such "chemical activations" by addition of hydrogen atoms or by similar bimolecular recombination reactions have been used extensively by Rabinovitch and co-workers (71–74) to investigate molecules at high levels of excitation (~ 40–100 kcal/mole). They have concluded that in such highly excited systems nonstatistical behavior is only observed on a time scale shorter than $\sim 10^{-12}$ s.

Multiphoton Infrared Dissociation

Of considerable relevance to the question of intramolecular vibrational energy redistribution are studies of multiphoton infrared dissociation. This phenomenon, in which an isolated molecule absorbs a sufficient number of infrared photons to cause its fragmentation, has received a great deal of attention in the more than ten years since its discovery. Typical experiments in this field employ powerful TEA CO_2 lasers and take advantage of coincidences between the CO_2 laser transitions and molecular vibrations. In view of the high energies and fluences used in these experiments, secondary absorption by a primary product may readily occur and complicate the interpretation of the experiment.

Extensive studies of simple molecules suggest that regardless of which vibrational mode is pumped by the laser source, the primary step involves breakage of the weakest bond in the absorber. Important information about the dissociation process may be gained by probing the internal energy distribution in the fragment, and numerous such experiments have been carried out. Stephenson, King and co-workers (75–78) have probed the energy content and distribution in various dihalo carbenes, such as CF_2 or $CFCl$, produced by fragmentation of small halocarbon molecules. They find that the product fragments and their vibrational state distribution are independent of the CO_2 laser wavelength and of the vibrational mode of the parent pumped. These results appear to be in agreement with statistical theories and with the assumption of a rather complete energy randomization in the parent molecule prior to dissociation.

Electronic Fluorescence and Excited State Studies

In spite of recent rapid progress in this field, infrared studies of ground state vibrational energy redistribution still present considerable experimental difficulties. Widely tunable intense, monochromatic sources are not yet as readily available as visible and ultraviolet lasers, and the sensitivities of infrared detectors are, in general, many orders of magnitude lower than those of photomultipliers. Many organic molecules, particularly aromatic species, possess excited electronic states with bonding and potential surfaces quite similar to those of the ground state. Unlike the ground state levels, vibronic levels of excited electronic states can be readily populated using a variety of pulsed or CW dye lasers, and their population can be monitored by means of their electronic fluorescence. In this way, one can study vibrational relaxation, energy redistribution, and similar phenomena in excited electronic states and obtain information quite similar to that obtainable from ground state studies, while avoiding the difficulties inherent in infrared work.

With the readily available, widely tunable dye laser one can, with a suitable choice of target molecules, readily access levels with a wide range of vibrational energies. With a suitable choice of molecules one can often excite levels with many thousands cm^{-1} of vibrational energy, and span regions where the intramolecular density of vibrational state changes by many orders of magnitude.

As in the ground electronic state, one finds that relaxation behavior is critically dependent upon the density of vibrational states. At the low energy end, where the vibrational manifold is sparse, vibrational relaxation can occur only as a result of collisional processes. At the other end of the scale, in the region where the density of states is high so that the average spacing is small compared with natural linewidth, unimolecular relaxation can take place. Here one observes the intramolecular vibrational redistribution process (IVR). Transition between these two regions is, however, not sharp, and it does not occur at the same vibrational energy in all molecules. The energy at which this transition will take place will depend on a variety of factors, such as the number of atoms in the molecule, its rigidity (i.e. the absence or presence of very low frequency modes), molecular symmetry, and the intrinsic lifetime of the state involved.

Collisional Relaxation in Excited States

Collisional relaxation in the excited electronic state has been studied by a number of investigators. In a typical experiment the laser is used to excite a selected single vibrational level. The energy redistribution is followed by studies of the dispersed fluorescence. Typically, one first records spectra under collision-free conditions to obtain a complete listing of individual vibronic transitions. Subsequently, a selected vibrational level is excited, and spectra are recorded with varying pressures of a suitable buffer gas, usually one of the rare gases. By comparing the collision-free spectra with those obtained in the presence of collisions, one can identify levels populated by relaxation and follow the energy flow within the molecule. If information about the radiative lifetime is available, one can also deduce from the observed yields of relaxed emission quantitative information about relaxation rates and cross sections. Using these techniques, it is usually possible to obtain a fairly detailed map of energy flow in the molecule even without direct time-resolved experiments.

Extensive studies of this nature have been performed by Parmenter and co-workers and several other groups (79–87). Molecules investigated in this way include glyoxal, pyrazine, benzene, aniline, and other species. While detailed understanding of the collisional relaxation processes is still lacking, certain empirical conclusions and patterns can be deduced based on the available data.

One immediate conclusion emerging out of most of these studies is the greatly enhanced cross sections of the relaxation process when compared with the ground state studies. This is probably due to the open shell configuration and lower ionization potential of the excited electronic states, which result in higher polarizability and much stronger interactions. The relaxation rates also usually increase with increasing polarizability of the collision partner. Even relaxation with helium usually requires only a few collisions, as opposed to many thousand hard sphere collisions typically required for deactivation of ground state vibrations.

Another conclusion is that the vibrational relaxation is not random but obeys fairly restrictive propensity rules. Relaxation is also dependent upon the energy gap ΔE and, usually, nearby vibrational levels are favored. As in the ground electronic states, the rates can be greatly enhanced in cases of strong mixing, for instance by Fermi resonance. Also, processes involving only small changes in the individual vibrational quantum numbers are usually favored.

Parmenter suggests (12), based on his observations, a rather specific set of rules. In particular, he proposes a factor of $10^{-\Delta v}$ for each change in vibrational quantum number and an exponential dependence on the energy gap when $\Delta E \geq 50 \text{ cm}^{-1}$. An additional exponential factor $\exp(-\Delta E/RT)$ is included for endoergic processes. Using this set of somewhat arbitrary selection rules, he achieves a good semiquantitative agreement for observed relaxation in benzenes and several of its derivatives.

A more direct way to obtain information about relaxation rates is to use pulsed laser excitation and time-resolved, intensity versus time, measurements to observe the decay and growth of population in the individual vibronic' levels. Rice and co-workers (83) have used such techniques combined with spectroscopic measurements to study the collisional relaxation in aniline. Their conclusions are quite similar to those of Parmenter. They point out that endoergic processes are often important, even when exoergic channels are available. Even in the low symmetry aniline molecule they find that relaxation proceeds through very selective pathways. More specifically, the low vibrational levels are found to fall into two groups, with the relaxation being much more efficient between levels of the same group. The energy equilibration between the individual groups is found, however, to be inefficient.

Free Jet Expansion Studies of Intramolecular Vibrational Energy Redistribution

In a typical laboratory experiment it is often difficult to assure truly collision-free conditions and to differentiate between collisional and intramolecular processes. This is the main reason that early studies concentrated on collisional relaxation and avoided the issue of IVR. In recent years new, powerful, and very convenient techniques for studies of collisionless processes became available with the introduction by Levy and co-workers (88, 89) and other groups of ultra cold supersonic molecular beams into chemical research. The extreme cooling results in great simplification of the spectra and facilitates selective vibrational level excitation. The cold high vacuum environment of the molecular beam then virtually eliminates collisions and permits studies of truly intramolecular processes.

Because the intramolecular density of states is of key importance in IVR, in an ideal experiment one would like to study the fluorescence while varying this density. Classical experiments, very close to this ideal, were performed by Smalley and his co-workers (90–92) in exploring the spectra of several alkyl substituted benzenes. The electronic transition in these species is the property of the aromatic ring. Only a few in-plane ring vibrations are usually active in the spectrum. These ring frequencies are found to be relatively independent of the length of the side chain, and the overall structure of the spectra remains unchanged. The vibrations of the alkyl side chain, many of them low frequency modes, are, on the other hand, "silent." The side chain provides a manifold of background states whose density is determined by the chain length.

Dispersed fluorescence spectra obtained by Smalley et al (91) for a series of alkyl benzenes in supersonic beam are shown in Figure 1. In each case the

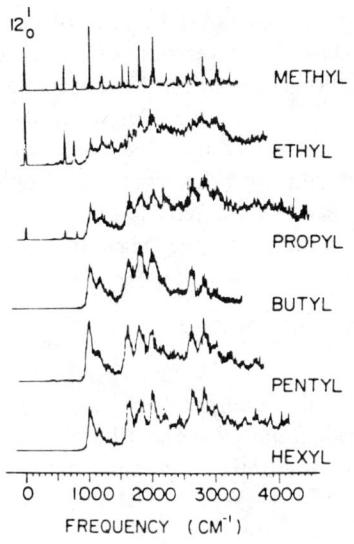

Figure 1 Dispersed $S_1 \to S_0$ fluorescence spectra of several n-alkylbenzenes resulting from 12_0^1 excitation. From Ref. (92).

12^1 level at 932 cm^{-1} involving an in-plane ring mode has been excited. For toluene, the bands in the emission spectrum are quite sharp and can be assigned to transitions originating from the 12^1 level of the fluorescing S_1 state. In all the other cases, however, qualitatively different, broadly structured spectra are present, resulting from the IVR process. IVR spreads the energy among other modes of the molecule, particularly among the low frequency chain modes, and gives rise to a large number of sequence bands. Because, however, the potential surfaces for the various modes are slightly different in the S_1 state than in the ground state, the corresponding sequence transitions will be shifted. Overlapping of a large number of spectrally shifted sequence bands gives rise to the observed broad spectrum.

The broadest spectra are observed for ethyl and n-propyl benzene, with sharper bands for the longer chain species. This has been interpreted as evidence of a nearly complete energy redistribution through the full length of the side chain. In the larger species much of the energy has propagated far from the ring into modes that are less affected by the $\pi-\pi^*$ excitation, resulting in sharper appearance of the fluorescence bands.

When the lower energy $6b^1$ level at 530 cm^{-1} is excited, IVR is again observed, but only when n-butyl or a longer side chain is present. Extensive IVR persists even when the alkyl chain is separated from the ring by an acetylenic group in a variety of phenyl alkynes (93) or by an oxygen atom in phenoxylalkynes (94). Similarly, in p-alkylanilines IVR also takes place even when modes located on the amino-group and isolated from the alkyl chain by the aromatic ring are pumped (95).

The onset of IVR will occur at different energies in different molecules. Amirav et al have observed that the threshold for unrestricted IVR occurs at 1280 cm^{-1} in tetracene (96) and near 1900 cm^{-1} in the 10 ring ovalene (97) system ($C_{32}H_{14}$). In spite of the large size of these molecules, the thresholds are relatively high. These planar π-electron systems are, unlike the alkyl-benzenes, relatively rigid and contain fewer low frequency modes. Numerous other reported examples of IVR in the excited S_1 states include (98–103) anthracene, pentacene, free base phthalocyanine, Mg-tetraphenyl porphin, and, more recently, free base porphin.

The Intermediate Region: Quantum Beats and Restricted IVR

In the low density of states limit the molecule undergoes periodic motion and vibrational relaxation occurs only collisionally. At the other extreme, the behavior is ergodic; complete energy randomization in the molecule takes place. Particularly interesting is the intermediate region between these two extremes. As noted above, only a few vibrational modes are usually active in an electronic spectrum of a large molecule. Vibrational levels associated with these modes appear in the spectrum and are immersed in a much denser manifold of "dark" states, inaccessible in absorption from the S_0 vibrationless level. In the intermediate region, a selective interference between the absorbing level and a small number of the "background states" can take place, and phenomena such as quantum beats and restricted IVR can be observed. These are conceptually similar to reversible intersystem crossing and observations of quantum beats in situations in which an S_1 level interacts with nearly isoenergetic triplet levels (104–107). Whereas in the latter case the coupling proceeds via spin-orbit interaction, in the former Fermi and Coriolis type interactions are more important.

Although in the case of intersystem crossing only the S_1 level carries the oscillator strength, in mixing within the S_1 state none of the levels is truly dark. Each of them will fluoresce, albeit with different Franck-Condon factors and not necessarily into the vibrationless level of the S_0 ground state. Provided both adequate spectral and temporal resolution are available, one can view directly in the time domain the "energy flow" between the interfering levels. A beautiful example was recently reported by Felker & Zewail (108). They excited cold antracene 1380 cm^{-1} above the S_1 origin using a synchronously pumped picosecond dye laser. They monitored the dispersed fluorescence by means of time-correlated single photon counting. Selective observation of the emissions from the two interfering levels produced modulated decays phase-shifted by 180°, as shown in Figure 2.

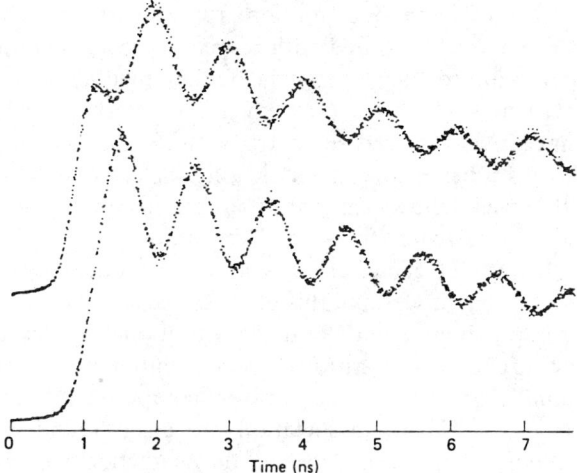

Figure 2 Time-resolved decay profiles resulting from excitation of the S_1 state of anthracene 1420 cm^{-1} above the origin. Vibrational bands near 390 and 1750 cm^{-1}, respectively, were monitored to obtain the upper and lower decay curves, respectively. Reproduced from Ref. (98).

When more than two levels interfere, more complex beat patterns are observed (98, 108, 109). In such cases, a Fourier transform of the decay profile will provide information about the spectral pattern of the interfering levels. Observation of the beats requires an excitation source broad enough to excite coherently the interfering levels. If a sufficiently narrow source exciting just one of the components of the resonant group of levels is used, only an exponential decay will be observed.

At still higher energies at which the density of states and the number of interfering levels become exceedingly high, it may no longer be experimentally possible to monitor quantum beats, and unrestricted IVR between all the available degrees of freedom will take place. Under such circumstances the absorption lines exhibit a simple Lorentzian profile, and one will again observe just simple exponential decay.

Relaxation and Predissociation of Van Der Waals Complexes

In conventional relaxation studies, the available kinetic energies and impact geometry for any particular collision are not known. The observed phenomena represent an average over the entire available range of these parameters. Molecules cooled in supersonic jets, including aromatics such as benzene or tetrazine, are known to form readily Van der Waals type complexes (110–115) with one or more rare gas atoms. These exhibit absorption bands slightly shifted with respect to corresponding bands of the free molecule, due to differences of the binding energies of the complex in the S_0 and S_1 states. As shown by Levy and co-workers (111–113), such complexes have a well-defined geometry, with the rare gas atoms usually centered above or below the aromatic ring. Selective excitation of the complex with a well-defined amount of vibrational energy permits a very controlled way of studying "collisional" relaxation.

The introduction of each rare gas atom introduces three very low frequency vibrational modes and this increases greatly the density of states and hence the possibility of IVR. If the vibrational energy exceeds the complex binding energy, a vibrational predissociation can take place. Both processes actually do occur in rare gas atom complexes with tetrazine and benzene (112, 114). The predissociation process is found to be highly selective, and only a very small subset of the energetically accessible levels are actually populated.

Should one view the dissociative process as a vibrational predissociation, or as a sequential process, in which IVR first redistributes the vibrational energy, followed by dissociation of the "hot" complex? A key experiment studying this fragmentation directly in the time domain has been performed by Rettschnick and co-workers (116). They excited selected vibrational levels of Ar-tetrazine complex in supersonic expansion using picosecond pulses of a synchronously pumped dye laser, and followed the time evolution of the dispersed fluorescence using single photon counting. The measured rise and decay curves of selected fluorescence bands showed that relaxation of the Ar-T complex, when excited into the $6a^1$ level, proceeds via two consecutive steps. The first step, IVR, which populates the $16a^2$ and $16a^1 16b^1$ levels of the complex, is followed by dissociation, yielding finally tetrazine in the $16a^1$ and $16b^1$ levels. Excitation of the $16a^2$ level of the complex, however, leads to direct vibrational predissociation.

Smalley and co-workers (117) have examined complexes of various alkylbenzenes with argon atoms, excited into the $6b^1$ level, 530 cm^{-1} above the S_1 origin. Since the Ar-benzene binding energy is ~ 400 cm^{-1}, one might expect the dissociation to be extremely slow if the energy were completely randomized by IVR among the large number of low frequency side-chain modes. Actually, all complexes, regardless of chain length, dissociate rapidly, suggesting that a vibrational predissociation without preceding IVR occurs in these species.

The vibrational predissociation of the Van der Waals complexes can be viewed as a half-collision with a rare gas atom. It is therefore of interest to compare it with collisional relaxation experiments. While there are distinct similarities, there are instances when channels important in collisional relaxation appear to be absent in complex relaxation. This may be because, unlike the fixed configuration of the complex, the collisional process will sample all possible relative orientations of the aromatic molecule and its collision partner.

Relaxation Processes in Rare Gas Matrices

An experimentally simple alternative approach to cooling molecules in supersonic jets involves studies in low-temperature matrices. Certain important differences exist between relaxation of a molecule in the gas phase and in the low temperature matrix. The sharp distinction between the low-energy region with sparse vibrational manifold and the high energy region with high density of states is not so clear in the solid. The matrix itself provides a continuous density of states and can therefore accept energy and induce vibrational relaxation even in the sparse region. Vibrational relaxation in this region will simulate collisional relaxation in the gas phase. Unlike rare gas atoms in complexes discussed in the preceding section, the host atoms in the matrix sample the guest interaction potential from all directions. In this sense, the matrix process may provide a more realistic model of collisional relaxation than does vibrational predissociation of Van der Waals complexes.

Even diatomic molecules often relax efficiently in low-temperature matrices. This process and related phenomena have been studied rather extensively (21–23). The relaxation rates are dependent upon the "energy gap" or vibrational spacing; the observed values span some 15 orders of magnitude. In heavy diatomics (118, 119) with small vibrational spacing like Ca_2 or in molecules near their dissociation limit, where the spacing approaches the lattice frequencies (120), one observes lifetime broadening of the vibronic bands implying relaxation in the sub-picosecond range. On the other hand, in light molecules with high vibrational frequencies, the true multiphonon relaxation is often very slow and is therefore often masked by alternative, more efficient processes. In many hydrides the relaxation was found to be essentially a $V \to R$ process with high rotational levels accepting the energy (121–123). In excited electronic states the relaxation is often governed by an interelectronic cascade process, involving the vibrational levels of two or more electronic states (124–126).

Vibrational relaxation of diatomics in solid matrix is very interesting in that it represents a very simple, prototypal system of radiationless transition. As such it lends itself to relatively simple theoretical modeling. It is therefore not surprising that diatomic relaxation in rare gas matrices has been very extensively studied theoretically (127–135). For several such systems (e.g. HCl or NH in argon) a good agreement now exists between experiment and quantitative theoretical predictions.

Intermode Coupling and Polyatomic Relaxation

Vibrational relaxation of polyatomics has been much less extensively studied. In the absence of good experimental data it is difficult to make *a priori* predictions of the relaxation pathways, and no theoretical models are available. Figure 3 shows in a greatly simplified way several possible limiting cases of polyatomic relaxation. In the absence of an anharmonic coupling between individual modes one might expect each of the modes to relax independently, as shown in Figure 3a. The earliest theories of matrix relaxation processes have predicted a strong exponential energy gap law for the relaxation rates. Molecular vibrational frequencies (~ 400–3000 cm^{-1}) are typically much higher than the phonon frequencies of the rare gas lattice (< 50 cm^{-1}). In order to dissipate one vibrational quantum, a large number of lattice phonons have to be created. The order of the process will increase with the increasing size of the energy gap, leading to the prediction of exponential rate decrease. In the limit of very strong energy gap law one might expect the relaxation to proceed via a pathway requiring small (in rare gas solids ~ 20–50 cm^{-1}) energy increments without regard for vibrational content and dynamical properties of individual levels, as shown in Figure 3b. A variety of gas phase experiments at moderately high energies of excitation have been interpreted using statistical theories and assuming complete energy randomization. If such redistribution proceeds fast compared with draining of the energy into the lattice, one might expect to see also in the matrix fast establishment of some "vibrational temperature" within the molecule, followed by subsequent slower decay, as shown schematically in Figure 3c.

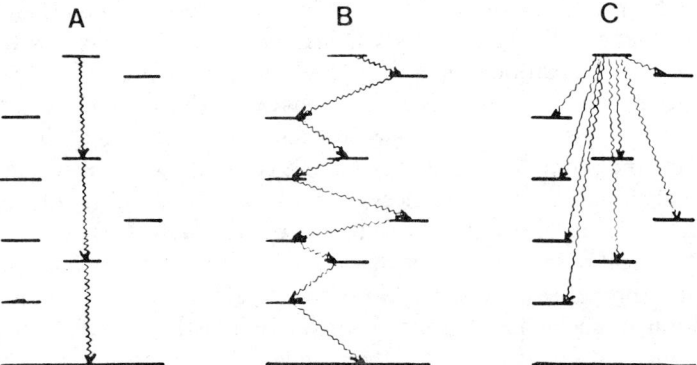

Figure 3 Diagram showing schematically three possible limiting models of relaxation in a polyatomic, matrix isolated molecule (*A*) "harmonic"; (*B*) "energy gap"; (*C*) "statistical"; see text.

Early studies at very low energies of excitation have indeed found cases of inefficient energy exchange between different vibrational modes. Thus, in CClF triatomic (136) the stretching modes and the bending frequency relax independently. On the other hand, in NCO and CNN molecules, clear evidence was obtained (137, 138) for the importance of intermode coupling, with fast relaxation between groups of levels known from the spectroscopy to be in Fermi resonance. Particularly extensively and carefully studied was methyl fluoride, FCH_3, and its deuterated derivative (139–143). A rather efficient mode-to-mode relaxation was again found, with the v_4 and v_1 vibrations, as well as the $2v_2$ and $2v_3$ overtone levels all relaxing into the $2v_3$ level in less than ~ 5 ns, independent of the host. The $2v_3$ and v_3 levels relax on a time scale several orders of magnitude slower, with rates strongly matrix-dependent and increasing from Xe to Ar. Clearly, relaxation of even small polyatomic molecules is much more complex than either of the limiting cases shown schematically in Figure 3.

Large Polyatomic Molecules

Vibrational relaxation in larger polyatomic molecules in condensed phase is usually fast compared with typical radiative rates. Regardless of which vibrational level of an excited electronic state has been populated initially, one typically observes an identical, vibrationally relaxed emission. In several recent studies of species with fully allowed electronic transitions in the visible region we have shown that under favorable circumstances the vibrationally unrelaxed fluorescence is experimentally observable. An advantage of matrix studies is the sharp appearance of the individual vibronic bands and the well-resolved nature of the spectra that makes unambiguous assignments of the individual transitions relatively straightforward. From the spectroscopic information one can gain understanding of the energy flow within the molecule. Typically, vibrational relaxation is much faster than other processes depopulating excited vibrational levels, and there is a direct proportionality between the emission intensity from a given level and its lifetime. If the relaxed fluorescence lifetime is known one can, even in the absence of direct time-resolved measurements, deduce the lifetimes of excited vibrational levels from the quantum yield of the unrelaxed fluorescence.

The vibrational relaxation in the excited electronic states of a variety of halogenated benzene cations in solid neon was studied in this way, using pulsed dye laser induced fluorescence (144, 145). A clear trend apparent in the data was the increasing relaxation rates with decreasing symmetry of the molecule. In the most symmetric of these species, $C_6F_6^+$ (D_{6h} symmetry), the lowest vibrational levels exhibit lifetimes of the order of 100 ps, and many levels up to ~ 2000 cm^{-1} are readily observable in emission. Also in the D_{3h} species extensive unrelaxed emission is observed. A less extensive unrelaxed emission is observed for the lower symmetry cations, such as the isomers of $C_6H_2F_4^+$ (D_{2h} or C_{2v} symmetry) or pentafluorobenzene cation. Fastest relaxation occurs in the lowest symmetry molecule studied, 1,2,4-trifluorobenzene cation (C_s symmetry). In this molecule no unrelaxed emission was observed; a lower limit of 2×10^{11} s^{-1} can be deduced for its relaxation rates, i.e. the vibrationally excited levels relax in less than 5 ps. It is interesting to note that no similar pronounced symmetry effects are apparently present in gas phase studies. The symmetry in the gas phase is effectively lowered during a collision occurring with random relative orientations of the relaxing molecule and its collision partner. In the matrix, however, the guest is isolated in a high symmetry substitutional site, and this may give the intrinsic guest symmetry an increased importance. It would be interesting to compare vibrational predissociation of Van der Waals complexes of rare gas atoms with substituted benzenes of various symmetries.

Vibrational relaxation in even larger molecules can be studied using similar techniques. Thus, napthazarin (146, 147) (1,4-dihydroxy, 5,8-naphthoquinone), hydroxyphenalenone, and several hydroxy anthroquinones were examined. In each case rather extensive vibrationally unrelaxed emission was seen, suggesting that their relaxation proceeds at rates comparable to those observed in the benzenoid cations. Most extensively studied among these molecules was naphthazarin. Typical spectra obtained for excitation of several upper upper state levels are shown in Figure 4. The unrelaxed emission is quite intense; the lifetimes of some levels are in excess of 250 ps. Unrelaxed fluorescence is readily observable even for some levels with over 2000 cm^{-1} of vibrational energy. By examining spectra such as shown in Figure 4 one can deduce a fairly detailed qualitative map of the energy flow in naphthazarin, as shown in Figure 5.

A common feature of the relaxation in all the molecules studied is that it does not follow any of the oversimplified models. In each case an overall trend of increasing relaxation rates with increasing vibrational energy can be seen; the increase, however, is not monotonic, and large fluctuations are superimposed upon the general trend. The relaxation process is certainly not a random, statistical energy redistribution, but proceeds via specific pathways. Of the large number of modes and levels energetically accessible only a small subset is actually populated during the relaxation of any given vibrational level. One can also note generally that overtone and combination bands usually relax faster than fundamental vibrations at the same overall energy. The relaxation in naphthazarin exhibits a weak temperature dependence in solid Xe and only small medium effects. The relative insensitivity of the observed relaxation rates to the temperature and medium suggests that the process is governed by the intramolecular intermode coupling. One might therefore expect collisional gas phase relaxation also to proceed along similar pathways.

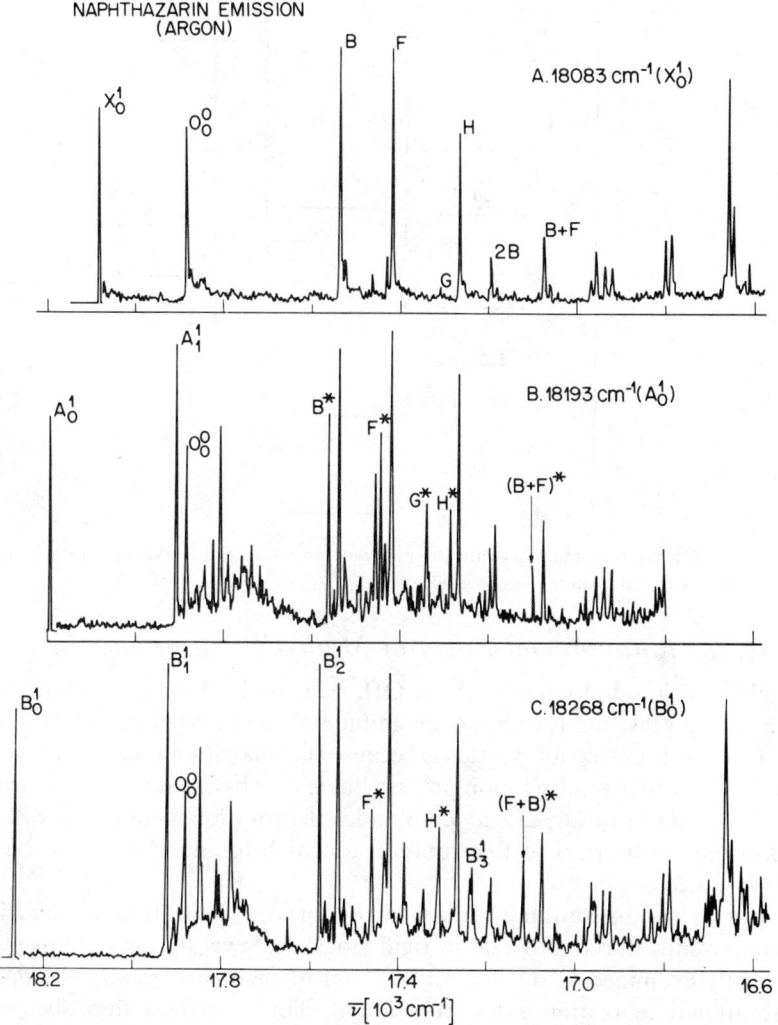

Figure 4 Representative emission spectra of naphthazarin in solid Ar following excitation of several S_1 levels. Spectra show presence of unrelaxed fluorescene. The letters denote vibrational modes of naphthazarin.

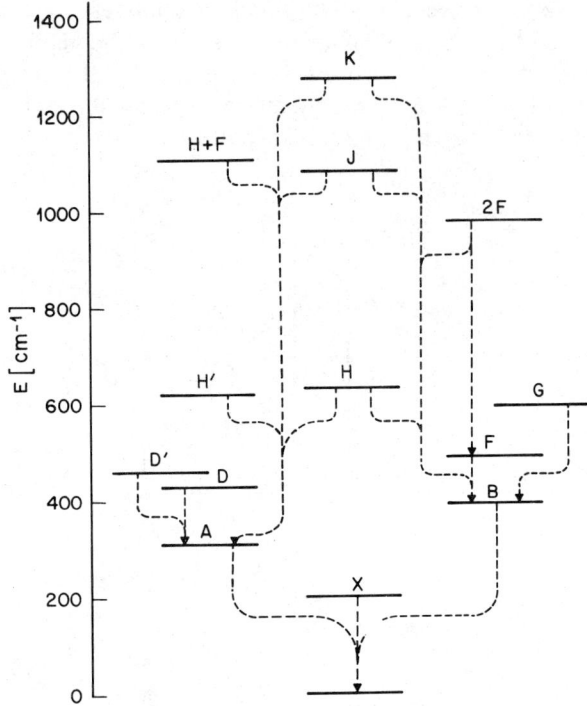

Figure 5 Energy flow map showing the pathways of relaxation in the S_1 state of naphthazarin as deduced from the fluorescence spectra.

Internal Rotation and Effect of Methyl Substitution

Light hydride diatomics, such as OH, NH, or HCl, relax vibrationally surprisingly fast, many orders of magnitude faster than for instance N_2, CO, or CN. As discussed above, this is because the relaxation does not proceed via a high order multiphonon process but rather by the much lower order V–R transfer. For large, heavy molecules the rotational constants will be too small or barriers to free rotation too high for this process to be of importance.

An interesting situation arises when a nearly freely rotating hydride group is introduced into a large, rigid molecule. Several such systems were recently examined in our laboratory, and in each case greatly enhanced vibrational relaxation rates were found. Thus, whereas fluorobenzene cations show extensive unrelaxed emission, no unrelaxed fluorescence was observed for any of the corresponding toluene derivatives. Similarly, no unrelaxed emission was detected for methyl-substituted hydroxy phenalenones, suggesting at least a factor of 50 increase in the relaxation rates when compared with the parent compound. It appears that vibrational excitation of the aromatic framework is more effectively coupled to the methyl rotation than to the delocalized lattice phonons.

It is possible that nearly free internal rotation will be similarly effective in promoting vibrational relaxation and IVR in gas phase molecules. One may note that the threshold for intramolecular relaxation in the S_1 states of molecules such as methyl- or ethyl-benzene appears to be lower than in much larger but rigid molecules like coronene or ovalene.

Conclusions

A result common to collisional gas phase experiments and to low temperature matrix studies is the highly selective nature of the relaxation in the low energy region. Collisional energy redistribution proceeds via specific pathways and is subject to rather restrictive propensity rules.

Another conclusion common to numerous experiments is the existence of a rapid and rather pervasive IVR at higher vibrational energies. The threshold density of states for this process is reported to be surprisingly low, in some cases as low as 10 states per cm^{-1}. Clearly, efficient energy redistribution could be a serious obstacle to mode-specific chemistry. It is perhaps significant that despite extensive research in this field, no clear-cut examples exist today. IVR will, of course, not necessarily adversely affect prospects for species or isotope selective chemistry. Although IVR is rather widespread, it is not quite clear whether it encompasses all levels, or whether longer lived, "vibrationally metastable" levels can exist. Unlike gas phase studies, which seem to indicate a rather monotonic decrease in relaxation rates with increasing energy, some low temperature matrix studies suggest presence of relatively long-lived states in regions where other levels exhibit strong lifetime broadening.

An impressive progress has occurred in our understanding of vibrational energy flow in polyatomic molecules in the last few years. Many questions, however, still remain unanswered, and intensive research will undoubtedly continue in this field for some time to come.

Literature Cited

1. Avouris, P., Gelbart, W. M., El-Sayed, M. A. 1977. *Chem. Rev.* 77:793–833
2. Robinson, G. W. 1974. In *Excited States*, 1:1–34. New York: Academic
3. Robinson, G. W. 1967. *J. Chem. Phys.* 47:1967–79
4. Douglas, A. E. 1966. *J. Chem. Phys.* 45:1007–15
5. Watts, R. J., Strickler, S. J. 1966. *J. Chem. Phys.* 44:2423–26
6. Whitten, G. Z., Rabinovitch, B. S. 1963. *J. Chem. Phys.* 38:2466–73
7. Lemont, S., Flynn, G. W. 1977. *Ann. Rev. Phys. Chem.* 28:261–82
8. Weitz, E., Flynn, G. W. 1974. *Ann. Rev. Phys. Chem.* 25:275–315

9. Yardley, J. T. 1980. Introduction. *Molecular Energy Transfer.* New York: Academic
10. Weitz, E., Flynn, G. W. 1981. *Adv. Chem. Phys.* 47(2): 185–235
11. Flynn, G. W. 1981. *Acc. Chem. Res.* 14: 334–41
12. Parmenter, C. S. 1982. *J. Phys. Chem.* 86: 1735–50
13. Rice, S. A. 1981. *Adv. Chem. Phys.* 47(2): 237–89
14. Jortner, J., Levine, R. D. 1981. *Adv. Chem. Phys.* 47: 1–114
15. McDonald, J. D. 1979. *Ann. Rev. Phys. Chem.* 30: 29–49
16. Noid, D. W., Koszykowski, M. L., Marcus, R. A. 1981. *Ann. Rev. Phys. Chem.* 32: 267–309
17. Brunner, P. 1981. *Adv. Chem. Phys.* 47: 201–6
18. Freed, K. F. 1980. *Adv. Chem. Phys.* 42: 207–69
19. Oref, I., Rabinovitch, B. S. 1979. *Acc. Chem. Res.* 12: 166–75
20. Smalley, R. E. 1982. *J. Phys. Chem.* 86: 3504–12
21. Rice, S. A. 1981. *Adv. Chem. Phys.* 47: 117–200
22. Bondybey, V. E., Brus, L. E. 1980. *Adv. Chem. Phys.* 41: 269–320
23. Legay, F. 1977. In *Chemical and Biological Application of Lasers,* Vol. 2. New York: Academic
24. Bondybey, V. E. 1981. *Adv. Chem. Phys.* 47: 521–33
25. Siebert, D. R., Grabiner, F. R., Flynn, G. W. 1974. *J. Chem. Phys.* 60: 1564–74
26. Siebert, D., Flynn, G. W. 1975. *J. Chem. Phys.* 62: 1212–20
27. Buchwald, M. I., Wolga, G. J. 1975. *J. Chem. Phys.* 62: 2828–32
28. Starr, D. F., Hancock, J. K. 1975. *J. Chem. Phys.* 63: 4730–34
29. Davis, C. C., McFarlane, R. A. 1976. *J. Chem. Phys.* 65: 3708–14
30. Slater, R. C., Flynn, G. W. 1976. *J. Chem. Phys.* 65: 425–37
31. Allen, D. C., Price, T. J., Simpson, C. J. S. M. 1977. *Chem. Phys. Lett.* 45: 182–87
32. West, G. A., Weston, R. E., Flynn, G. W. 1977. *J. Chem. Phys.* 67: 4873–79
33. Starr, D. F., Hancock, J. K. 1975. *J. Chem. Phys.* 62: 3747–53
34. Kung, R. T. V. 1975. *J. Chem. Phys.* 63: 5305–12
35. Siebert, D. R., Flynn, G. W. 1976. *J. Chem. Phys.* 64: 4973–83
36. Casleton, K. H., Flynn, G. W. 1977. *J. Chem. Phys.* 67: 3133–37
37. Gueguen, H., Yzambart, F., Chakroun, A., Margottin-Maclou, M., Doyennette, L., Henry, L. 1975. *Chem. Phys. Lett.* 35: 198–201
38. Doyennette, L., Margottin-Maclou, M., Chakroun, A., Gueguen, H., Henry, L. 1975. *J. Chem. Phys.* 62: 440–47
39. Hastings, P. W., Osborn, M. K., Sadowski, C. M., Smith, I. W. M. 1983. *J. Chem. Phys.* 78: 3893–98
40. Yardley, J. T., Moore, C. B. 1966. *J. Chem. Phys.* 45: 1066–67
41. Apkarian, V. A., Weitz, E. 1979. *J. Chem. Phys.* 71: 4349–68
42. Hess, P., Moore, C. B. 1976. *J. Chem. Phys.* 55: 2339–44
43. Drozdoski, W. S., Bates, R. D., Siebert, D. R. 1978. *J. Chem. Phys.* 69: 863–67
44. Siebert, D. R., Grabiner, F. R., Flynn, G. W. 1974. *J. Chem. Phys.* 60: 1565–74
45. Apkarian, V. A., Weitz, E. 1978. *Chem. Phys. Lett.* 59: 414–19
46. Weitz, E., Flynn, G. W., Ronn, A. M. 1972. *J. Chem. Phys.* 56: 6060–67
47. Sheorey, R. S., Slater, R. C., Flynn, G. W. 1978. *J. Chem. Phys.* 68: 1058–63
48. Preses, J. M., Flynn, G. W., Weitz, E. 1978. *J. Chem. Phys.* 69: 2782–87
49. Kneba, M., Wolfrum, J. 1977. *Ber. Bunsenges. Phys. Chem.* 81: 1275–83
50. Earl, B. L., Ronn, A. M. 1976. *Chem. Phys.* 12: 113–21
51. Langsam, Y., Lee, S. M., Ronn, A. M. 1976. *Chem. Phys.* 14: 375–83
52. Gamss, L. A., Kohn, B. H., Pollack, M. I., Ronn, A. M. 1976. *Chem. Phys.* 18: 85–91
53. Fujimoto, G. T., Weitz, E. 1976. *J. Chem. Phys.* 65: 3795–96
54. Truhlar, D. G., Duff, J. W. 1975. *Chem. Phys. Lett.* 36: 551–54
55. Chen, J. D., Hager, J., Krieger, W. 1982. *Chem. Phys. Lett.* 90: 366–69
56. Nordholm, K. S. J., Rice, S. A. 1975. *J. Chem. Phys.* 62: 157–68
57. Noid, D. W., Marcus, R. A. 1977. *J. Chem. Phys.* 67: 559–67
58. Forst, W. 1973. *Theory of Unimolecular Reactions.* New York: Academic
59. Heller, E. J. 1979. *Chem. Phys. Lett.* 60: 338–40
60. Sander, R. K., Soep, B., Zare, R. N. 1976. *J. Chem. Phys.* 64: 1242–43
61. Naaman, R., Lubman, D. M., Zare, R. N. 1979. *J. Chem. Phys.* 71: 4192–4200
62. Hopkins, J. B., Langridge-Smith, P. R. R., Smalley, R. E. 1983. *J. Chem. Phys.* 78: 3410–14
63. Reisner, D. E., Vaccaro, P. H., Kittrell, C., Field, R. W., Kinsey, J. L., Dai, H. L. 1982. *J. Chem. Phys.* 77: 573–75
64. Vaccaro, P. H., Kinsey, J. L., Field, R. W., Dai, H. L. 1983. *J. Chem. Phys.* 78: 3659–64
65. Abramson, E., Field, R. W., Innes, K. K., Kinsey, J. L. Unpublished

66. Bray, R. G., Berry, M. J. 1979. *J. Chem. Phys.* 71:4909-22
67. Reddy, K. V., Heller, D. F., Berry, M. J. 1982. *J. Chem. Phys.* 76:2814-37
68. Nesbitt, D. J., Leone, S. R. 1982. *Chem. Phys. Lett.* 87:123-27
69. Stewart, G. M., McDonald, J. D. 1981. *J. Chem. Phys.* 75:5949-50
70. Stewart, G. M., McDonald, J. D. 1983. *J. Chem. Phys.* 78:3907-15
71. Wrigley, S. P., Rabinovitch, B. S. 1983. *Chem. Phys. Lett.* 95:363-18
72. Wrigley, S. P., Smith, K. W., Rabinovitch, B. S. 1983. *Chem. Phys.* 75:453-62
73. Trentwith, A. B., Rabinovitch, B. S., Wolters, F. C. 1982. *J. Chem. Phys.* 76:1586-87
74. Wolters, F. C., Rabinovitch, B. S., Ko, A. N. 1980. *Chem. Phys.* 49:65-75
75. Grimley, A. J., Stephenson, J. C. 1981. *J. Chem. Phys.* 74:447-52
76. King, D. S., Stephenson, J. C. 1977. *Chem. Phys. Lett.* 51:48-52
77. Stephenson, J. C., King, D. S. 1978. *J. Chem. Phys.* 69:1485-92
78. Stephenson, J. C., Bialkowski, S. E., King, D. S. 1980. *J. Chem. Phys.* 72:1161-69
79. Parmenter, C. S., Tang, K. Y. 1979. *Am. Chem. Soc. Symp. Ser.* 56:1175-78
80. Parmenter, C. S., Tang, K. Y. 1978. *Chem. Phys.* 27:127-50
81. Atkinson, G. H., Parmenter, C. S., Tang, K. Y. 1978. *J. Chem. Phys.* 71:68-72
82. Chernoff, D. A., Rice, S. A. 1979. *J. Chem. Phys.* 70:2521-41
83. Van der Waal, M., Chernoff, D. A., Rice, S. A. 1981. *J. Chem. Phys.* 74:4888-92
84. Tang, K. Y., Parmenter, C. S. 1983. *J. Chem. Phys.* 78:3922-34
85. McDonald, D. B., Rice, S. A. 1981. *J. Chem. Phys.* 74:4907-17
86. Spears, K. G., Rice, S. A. 1971. *J. Chem. Phys.* 55:5561-81
87. de Leeuw, G., Langeaar, J., Rettschnick, R. P. H. 1980. *J. Mol. Struct.* 61:101-14
88. Levy, D. H. 1981. *Ann. Rev. Phys. Chem.* 31:197-225
89. Levy, D. H., Wharton, L., Smalley, R. E. 1977. *Chemical and Biochemical Applications of Lasers*, 2:1-41. New York: Academic
90. Hopkins, J. B., Powers, D. E., Smalley, R. E. 1980. *J. Chem. Phys.* 73:683-87
91. Hopkins, J. B., Powers, D. E., Smalley, R. E. 1980. *J. Chem. Phys.* 72:5039-48
92. Hopkins, J. B., Powers, D. E., Mukamel, S., Smalley, R. E. 1980. *J. Chem. Phys.* 72:5049-61
93. Powers, D. E., Hopkins, J. B., Smalley, R. E. 1981. *J. Chem. Phys.* 74:5971-76
94. Hopkins, J. B., Powers, D. E., Smalley, R. E. 1981. *J. Chem. Phys.* 74:6986-88
95. Powers, D. E., Hopkins, J., Smalley, R. E. 1980. *J. Chem. Phys.* 72:5721-30
96. Amirav, A., Even, U., Jortner, J. 1980. *Chem. Phys. Lett.* 71:12-16
97. Amirav, A., Even, U., Jortner, J. 1980. *Opt. Commun.* 32:266-68
98. Lambert, W. R., Felker, P. M., Zewail, A. H. 1981. *J. Chem. Phys.* 79:5958-60
99. Amirav, A., Even, U., Jortner, J. 1979. *J. Chem. Phys.* 71:2319-21
100. Fitch, P. S. H., Haynam, C. A., Levy, D. H. 1981. *J. Chem. Phys.* 74:6612-20
101. Even, U., Magen, Y., Jortner, J., Levanon, H. 1981. *J. Am. Chem. Soc.* 103:4583-85
102. Even, U., Jortner, J. 1982. *J. Chem. Phys.* 77:4391-99
103. Even, U., Magen, Y., Jortner, J., Levanon, H. 1982. *J. Chem. Phys.* 76:5684-92
104. Chaiken, J., Gurnick, M., McDonald, J. D. 1981. *J. Chem. Phys.* 74:106-26
105. Jameson, A. K., Okajima, S., Lim, E. C. 1981. *J. Chem. Phys.* 75:480-82
106. Van der Meer, B. J., Jonkman, H. T., ter Horst, G. M., Kommandeur, J. 1982. *J. Chem. Phys.* 76:2099-2100
107. ter Horst, G., Pratt, D. W., Kommandeur, J. 1981. *J. Chem. Phys.* 74:3616-18
108. Felker, P. M., Zewail, A. H. 1983. *Chem. Phys. Lett.* 102:113-18
109. Zewail, A. H., Lambert, W. R., Felker, P. M., Perry, J., Warren, W. 1982. *J. Phys. Chem.* 86:1184-92
110. Kenny, J. E., Johnson, K. E., Sharfin, W., Levy, D. H. 1980. *J. Chem. Phys.* 72:1109-19
111. Brumbaugh, D. V., Kenny, J. E., Levy, D. H. 1983. *J. Chem. Phys.* 78:3415-34
112. Kenny, J. E., Brumbaugh, D. V., Levy, D. H. 1979. *J. Chem. Phys.* 71:4757-58
113. Haynam, C. A., Brumbaugh, D. V., Levy, D. H. 1983. *J. Chem. Phys.* 79:1581-91
114. Stephenson, T. A., Rice, S. A. 1984. *Int. Conf. on Radiationless Transitions*, Newport Beach, Jan. 3-8
115. Bernstein, E. R., Law, K., Schauer, M. 1984. *J. Chem. Phys.* 80:207-20
116. Rettschnick, R. P. H., Ramaekers, J. J. F., Langelaar, J., Lips, H. J. 1984. See Ref. 114
117. Hopkins, J. B., Powers, D. E., Smalley, R. E. 1981. *J. Chem. Phys.* 74:745-47
118. Bondybey, V. E., Albiston, C. 1978. *J. Chem. Phys.* 68:3172-76
119. Bondybey, V. E., English, J. H. 1977. *J. Chem. Phys.* 67:3405-11
120. Bondybey, V. E., Fletcher, C. 1976. *J. Chem. Phys.* 64:3615-20

121. Bondybey, V. E., Brus, L. E. 1975. *J. Chem. Phys.* 63:794–804
122. Bondybey, V. E. 1976. *J. Chem. Phys.* 65:5138–40
123. Wiesenfeld, J., Moore, C. B. 1979. *J. Chem. Phys.* 70:930–46
124. Bondybey, V. E. 1977. *J. Chem. Phys.* 66:995–1001
125. Bondybey, V. E., Nitzan, A. 1977. *Phys. Rev. Lett.* 38:889–92
126. Bondybey, V. E. 1976. *J. Chem. Phys.* 65:2296–2304
127. Berkowitz, M., Gerber, R. B. 1979. *Chem. Phys.* 37:369–88
128. Freed, K. F., Yeager, D. L., Metiu, H. 1977. *Chem. Phys. Lett.* 49:19–23
129. Diestler, D. J., Knapp, E. W., Ladouceur, H. D. 1978. *J. Chem. Phys.* 68:4056–65
130. Knittel, D., Lin, S. H. 1978. *Mol. Phys.* 36:893–906
131. Kono, H., Lin, S. H. 1983. *J. Chem. Phys.* 79:2748–55
132. Weissman, I., Nitzan, A., Jortner, J. 1977. *Chem. Phys.* 26:413–19
133. Fletcher, D., Fujimura, Y., Lin, S. H. 1978. *Chem. Phys. Lett.* 57:400–4
134. Shugard, M., Tully, J. C., Nitzan, A. 1978. *J. Chem. Phys.* 69:336–45
135. Ladouceur, H. D., Diestler, D. J. 1979. *J. Chem. Phys.* 70:2620–30
136. Bondybey, V. E. 1977. *J. Chem. Phys.* 66:4237–39
137. Bondybey, V. E., English, J. H. 1977. *J. Chem. Phys.* 67:664–68
138. Bondybey, V. E., English, J. H. 1977. *J. Chem. Phys.* 67:2868–73
139. Abouaf-Marguin, L., Gauthier-Roy, B. 1980. *Chem. Phys.* 51:213–21
140. Abouaf-Marguin, L., Gauthier-Roy, B., Legay, F. 1977. *Chem. Phys.* 23:443–50
141. Apkarian, V. A., Weitz, E. 1980. *Chem. Phys. Lett.* 76:68–74
142. Janiesch, W., Apkarian, V. A., Weitz, E. 1982. *Chem. Phys. Lett.* 85:505–7
143. Young, L., Moore, C. B. 1982. *J. Chem. Phys.* 76:5869–77
144. Bondybey, V. E., Miller, T. A., English, J. H. 1980. *Phys. Rev. Lett.* 44:1344–47
145. Bondybey, V. E., English, J. H., Miller, T. A. 1983. *J. Phys. Chem.* 87:1300–5
146. Bondybey, V. E., Milton, S. V., English, J. H., Rentzepis, P. M. 1983. *Chem. Phys. Lett.* 97:130–34
147. Rentzepis, P. M., Bondybey, V. E. 1984. *J. Chem. Phys.* 80:4727–37

ELECTRONIC PROCESSES IN ORGANIC SOLIDS

Martin Pope

Radiation and Solid State Laboratory and the Department of Chemistry, New York University, New York, New York 10003

Charles E. Swenberg

Radiation Sciences Department, Armed Forces Radiobiology Research Institute, Bethesda, Maryland 20814

INTRODUCTION

In recent years, the number of studies of electronic processes in organic solids has assumed explosive proportions. In retrospect, it was inevitable. The increasing ability of organic chemists to design and synthesize molecules and structures almost to order has made it possible to provide an experimental testing ground for what were once figments of the imaginations of theorists. Thus excellent approximations of ideal one- and two-dimensional solids can be prepared with electrical properties varying from those of an insulator, to those of a superconductor. Disordered systems can be prepared with a large range of nearest-neighbor interaction energies, with and without additional trapping sites, making it possible to test almost any conceivable theory of energy or charge migration in a random or regular network. Increasingly sophisticated computer simulations are providing unusual insights into the microscopic details of dynamical processes, such as carrier or energy transport. These serve as a check on analytical theories and even a guide to indicate which assumptions are likely to be reasonable. The simulations can probe situations that are as yet not readily accessible to experiment, such as time domains in the femtosecond range. All of this new interest has been superimposed on what has been an orderly growth in a relatively isolated field, the study of the electrical and optical properties of polycyclic aromatic hydrocarbons exemplified in anthracene. In the middle of 1982, our book appeared (1) in which most of the material to be presented herein approximately 40 pages was discussed in 821 pages and, at that, with apologies for significant omissions. It is thus obvious that this review represents an even more drastic condensation and omission of important work.

Books and reviews that have appeared since the beginning of 1982 include that edited by Mort & Pfister (2) dealing with triboelectricity, charge storage, piezoelectricity, pyroelectricity, energy transfer, photoconductivity, and polyacetylene. Organic semiconductors are discussed by Haddon et al (3). A wide ranging conference on the organic solid state was held in 1982 covering such topics as solid state photorearrangement, chemistry, semiconductivity, superconductivity, the polyacetylenes, electronic structure, and many others (4). Several lucid review articles have been written by Duke and his colleagues. These discuss conductivity (5), electronic structure (6, 7), and electronic excitation (8). The preparation of good crystals of pure materials is fundamental to the development of meaningful theory. Two books dealing with this subject are the recent work of Sloan & McGhie (9) and the earlier work of Karl (10).

Mention must be made of the remarkable results that are being observed by the combination of tunable lasers and supersonic molecular beams of isolated molecules in the study of intramolecular dynamics such as collision-free vibrational energy decay. A review of this important field is given by Smalley (11). The study of condensed phase molecular dynamics, including transport and trapping, has been reviewed by Fayer (12). The effect of pressure on molecular luminescence in solution is reviewed by Drickamer (13), and a critique of high pressure studies of molecular crystal spectra has been given by Berry et al (14). A review of luminescence in nucleic acids has been prepared by Callis (15). Additional references to reviews and books are made in the text.

In the course of this review we attempt to provide references to subjects that cannot be adequately covered due to severe space limitations. In addition, we make many references to our recent book (1).

EXCITON PROCESSES

Exciton Transport

Discussion is limited to exciton diffusion, percolation, and exciton annihilation; the reader may refer to recent reviews or papers for discussions of the effect of electric fields on exciton transport (15a,b) polaritons (16–18), exciton-phonon interaction (18, 19), and exciton luminescence (19, 20, 20a,b). In ordered molecular crystals, exciton transport is generally characterized by a diffusion constant, D (1, p. 102 et seq). Sensitized luminescence (21) produced by exciton diffusion to a suitable trap and delayed fluorescence resulting from exciton-exciton annihilation (1, pp. 135–47) have been used to determine D. In the usual interpretation, the exciton diffusion constant (D) is related to γ, the annihilation rate constant, and K, the exciton transfer rate, by the relationships $\gamma = v\gamma' = 8\pi R_d D$ and $K = \rho v K' = 4\pi R_c D$, where v is the volume per molecule of the crystal, ρ the relative guest concentration, and R_d and R_c are the effective radii for annihilation by another exciton and capture by a trap, respectively. Kenkre & Schmid (22) have shown that the use of the above relationships, which are derived from coagulation analysis, would yield the same values for the diffusion constant only under special conditions; values of D inferred from either γ or K could therefore be incorrect. A proper theoretical analysis of the experimental data can be obtained by noting that for trapping at a guest or defect site, the trapping time is the sum of the time to get to a trap, $1/M$, plus the time, $1/c$, needed for the (local) capture process to occur; M is the motion rate defined as (22)

$$M = \left\{ \int_0^\infty \exp(-t/\tau)\Psi_0(t) \, dt \right\}^{-1} \qquad 1.$$

where $\Psi_0(t)$ is the exciton self-propagator, and τ the exciton lifetime. For exciton annihilation a similar sum relationship holds, except that the mutual annihilation time, $1/b$, replaces the local capture time $1/c$. In terms of rates (γ' and K') the following general relationships have been shown to be valid (22):

$$(\gamma')^{-1} = M^{-1} + b^{-1}, \quad (K')^{-1} = M^{-1} + c^{-1}. \qquad 2.$$

It is only in temperature regimes where either γ' or K' is dominant that information about the magnitude of D can be inferred. These formulae were applied by Kenkre & Schmid (22) to naphthalene and anthracene, where both γ' and K' are known over a broad temperature range. For the special case of incoherent three-dimensional motion, where $M = 4F$, F being the nearest neighbor transfer rate, a lower bound on the singlet exciton diffusion constant, $D = Fa^2/6$ can be inferred, where a is the nearest neighbor distance. In the particular case of the naphthalene singlet exciton, at $T = 300$ K, $D \geq 2.1 \times 10^{-5}$ cm^2 s^{-1}, while for anthracene $D \geq 4.3 \times 10^{-3}$ cm^2 s^{-1}. A similar type of analysis is possible for triplet excitons provided there are temperature domains where either $\gamma' > K'$ or $K' > \gamma'$; under these conditions, it may be inferred from Eq. 2 that $D > K'$ or $D > \gamma'$.

An important technique for measuring D, particularly for triplet excitons, involves the use of Ronchi rulings (23; 1, p. 173). It lends itself easily to the study of the anisotropy of D, and may be used to test for coherence in exciton motion. Kenkre et al (24) have adopted the generalized master equation (GME) approach to derive expressions for steady state delayed fluorescence as a function of Ronchi ruling period that hold for any degree of coherence. The GME approach has also been adapted for use in sensitized luminescence studies (25).

Energy transport in mixed molecular crystals has attracted considerable attention, in part because these crystals, with their random distribution of components, provide useful models for testing theoretical concepts of disordered solids. The most extensively studied system has been the naphthalene-(guest)-perdeuteronaphthalene (host) mixed crystal doped with a third component, β-methylnaphthalene (BMN), present in concentrations generally less than 10^{-3} mol fraction (26). The third component has fluorescing levels sufficiently low to serve as a sensor or supertrap for excitons migrating among guest molecules. The experimental parameters are the time dependence and integrated intensities of the luminescence from both guests and supertraps, as a function of guest concentration. Measurements are performed at low temperatures ($\sim 4°K$) so that the host sites act as barriers (antitraps) to energy transport among the guest molecules. For both triplet and singlet exciton transport in $C_{10}H_8/C_{10}D_8$/BMN the sensor emission depends in an almost step-function fashion upon guest concentration, (C_g), increasing rapidly over a narrow (critical) range (27; 1, pp. 125–34).

Several theoretical models have been suggested to account for these results, each of which has some validity. In the percolation model, the sharp transition is attributed to the formation of a minimal macroscopic connecting network of guest sites at the critcal guest concentration, C_c. Percolation analysis requires extensive computer simulations. Analytical solutions have been proposed by Blumen & Silbey (29) and by Loring & Fayer (30). Blumen & Silbey use a continuum kinetic equation formalism, and Loring and Fayer use a Green's function approach to solve the mixed-crystal kinetic master equation.

Several percolation models have been considered (26, 31, 32). *Static* percolation assumes that there is a cutoff on the allowable transfer distance and neglects dynamics within clusters; thus for any cluster containing a supertrap, the probability is unity for capturing a guest exciton that lands anywhere in the cluster. The latter approximation is called the *supertransfer limit* (26). If the supertransfer condition is relaxed so that not all excitons within a cluster containing a sensor are trapped because of limitations imposed by the finite lifetime of the exciton and its finite rate of transfer, then *quasistatic* percolation ensues. For singlet excitons, Gentry & Kopelman (31, 33) have shown that the quasistatic percolation model agrees with experiment at 1.8 K, but fails to account for data acquired at

4.2 K. On the other hand, the two-dimensional continuum models of both Loring & Fayer (30) and Blumen & Silbey (29) are consistent with the experimental results at 4.2 K but not at 1.8 K. In the Loring & Fayer approximation (30) the trapping probability is given by $P = 1 - K_g \tilde{G}(0, \varepsilon = K_g)$, where K_g^{-1} is the guest lifetime, and $\tilde{G}$ is the Laplace transform of the time-dependent part of the system Green's function for the probability of finding excitons somewhere within the guest manifold. This relationship can account for the $T = 4.2$ K data only if the Förster distance (1, p. 100) $R_0 = 8$ Å and the transfer rate, W, has an octupole-octupole distance dependence, i.e. $W \propto R^{14}$. A weakness of the model is that these parameters imply a transfer rate of 5×10^9 s^{-1}, whereas expected nearest neighbor rates are the order of 10^{12} sec^{-1}. The lack of agreement with the 1.8 K data has been discussed by Gentry & Kopelman (33) and is attributed to the implicit assumption in the Loring & Fayer model of the equality of the forward and reverse transfer rates, $W_{ij} = W_{ji}$; this equality is known not to hold for naphthalene singlet excitons at low temperatures (34).

The Blumen & Silbey model (29) calculates the average probability for an exciton to land on a supertrap site and the average trapping time. Their results, in the absence of back transfer from the supertrap, are

$$K_{et}^{-1} \propto \{(C_s/C_g)C_g^{n/d}\} \qquad 3.$$

$$P \propto \{1 + C_{1/2}/C_g\}^{-1} \qquad 4.$$

where the energy transfer rate constant is K_{et}, the sensor concentration is C_s, the guest concentration is C_g and $C_{1/2}$ is that concentration at which $P = 0.5$; n is the multipole transfer exponent ($n = 14$ for octupole-octupole transfer) and d is the dimensionality of the system. Equation 4 provides an excellent fit to the 4.2 K data; however, it does not account for the sharpening of the critical transition upon lowering the temperature. Furthermore, recent time-dependent emission data of Parson & Kopelman (32) performed on the same mixed crystal system at 1.7 K demonstrate that the sensor emission profile is more rapid than expected from rate equation analysis (29) for guest concentrations near the transition regime (guest concentrations between 0.42 to 0.57 mol fraction).

The temperature and guest dependence of luminescence from the mixed crystal system $C_{10}H_8/C_{10}D_8$/BMN has also been studied by Brown et al (35), who found evidence for a transition from coherent triplet exciton motion at very low temperatures to hopping motion for $T > 10$ K. Transitions from band-like to hopping motion of triplet excitons have also been studied by Gentry & Kopelman (36).

In addition to protonated and deuterated isotopically mixed crystals, orientationally disordered solids (37, 38) are excellent model systems for studying the static and dynamic properties of electronic excitations in disordered solids. In these systems local disorder is provided by the random distribution of orientations of the substituent groups, e.g. the chloride and bromide ligands of 1-bromo-4 chloronaphthalene (BCN) crystals. Orientational static disorder was studied by Morgan & El-Sayed (39), who measured the zero phonon $S_1 \to T_1$ transition at ≈ 4.2 K in BCN single crystals. The absorption line shape was Gaussian, which is characteristic of inhomogeneous broadening; the line width (HMHW) was ~ 32 cm^{-1}, which is approximately two orders of magnitude larger than the linewidths of the corresponding transition in single crystals of 1,4-dibromonaphthalene and 1,4-dichloronaphthalene. In addition, Morgan & El-Sayed (39) have shown how the temporal behavior of the phosphorescence emission at different laser excitation energies in the long wavelength region of the absorption band provides a measure of both the energy transfer rate and its mechanism.

Exciton Interactions

Excitons can interact with other excitons, and with carriers, defects, photons, phonons, and essentially any other entity that can be reached by a mobile electronically excited state (1, pp. 89–180). Exciton-exciton annihilation in neat molecular solids is characterized by an annihilation rate constant γ. The limitation of the assumption that the depletion of excitons (n) due to bimolecular annihilation is given simply by $-\gamma n^2$ has been reviewed by Kenkre (25). The role of boundaries and domain (or cluster) size on exciton kinetic rates is of considerable current interest [see Hatlee & Kozak (40) and references quoted]; confinement to small domains is expected to enhance the number of collisions with other excitons, defects, or traps. In the particular case of tetracene triplet excitons created in pairs in small tetracene domains by singlet exciton fission, confinement results in an increase in the fluorescence yield and singlet exciton lifetime as the domain size decreases (41).

Mixed crystals are good model systems for the study of size effects on exciton reactions. Recently Klymko & Kopelman (42) have shown that when guest cluster formation occurs, the delayed fluorescence (D) resulting from guest-guest exciton annihilation is no longer proportional to the square of the guest phosphorescence (P), even at low triplet exciton concentrations. Using mixed $C_{10}H_8/C_{10}D_8$ crystals with dopant levels of 0.1% to 20% $C_{10}H_8$ they found $D \propto P^x$, where x is a function of excitation intensity, time after excitation, and guest concentration. Values of x ranged from 2 to ~ 30 at $T = 1.7$ K and increased monotonically with decreasing naphthalene concentration. As noted by the authors, these results are inexplicable in terms of standard diffusion kinetics (43). Klymko &

Kopelman (44) interpret their data in terms of fractal kinetics, which is appropriate for bimolecular processes in small domains. A discussion of fractal properties is beyond the scope of this review; see Alexander & Orbach (45) and DeGennes (46) for more information.

Excitons interact with carriers (1, pp. 164 et seq), and in the process of interacting, it may happen that a transient intermediate state is produced between the exciton and its collision partner. More recently the bound exciton + hole (excitonic ion) case was treated by Schilling & Mattis (47, 48), who call the ion a trion as initially suggested by Thomas & Rice (49). Here, the exciton could be either the Frenkel exciton or the Wannier exciton. The case of N excitons and one hole was solved by the same authors (48). Excitonic ions with Frenkel exciton parentage were discussed by Singh (50, 51) and ions based on excitons intermediate between Frenkel and Wannier excitons were treated by Gumbs & Mavroyannis (52). Experimental evidence for the existence of excitonic ions in benzene solid films is given by Sanche et al (53). Excitons interacting with trapped electrons was discussed by Agranovich & Zakhidov (54), who deduced an attractive interaction, which is the order of $e^2\Delta\alpha/2\varepsilon r$, where $\Delta\alpha$ is the change in the molecular polarizability of the excited molecule, $\sim 10^{-23}$ cm^3, r is the separation of the charge e from the exciton, and ε is the local dielectric constant.

One intriguing possibility that has not yet been ruled out theoretically is that of a Bose-Einstein condensation of Frenkel excitons in organic crystals, particularly triplet excitons (55); attempts are being made to detect this phenomenon, or at least the formation of exciton aggregates. A review of these experiments, particularly in the Soviet Union, is given by Brikenshtein et al (56).

Exciton Trapping

Trapping of excitons at either impurity or guest sites has been a subject of considerable study because of the insight gained on the mechanism of exciton diffusion (21, 25). Studies by Powell and co-workers (21) had suggested that the exciton transfer rate K was time-dependent, in significant departure from earlier results obtained in sensitized luminescence experiments (57). A careful study of the time-dependence of K has recently been made by in tetracene-doped anthracene by Braun et al (58). They demonstrated that for temperatures between 1.6 and 300 K and over a dopant concentration range of 10^{-6} to 2.3×10^{-3} mol fraction, that K was time-independent at least for times greater than a few picoseconds. This is in agreement with previously reported studies of Campillo et al (59) and Al-Obardi et al (60).

A recent study of singlet exciton trapping in the system p-terphenyl (host)/tetracene (guest) (61) also showed that the transfer rate from host to host was time-independent but, more interestingly, that it was thermally activated. It was concluded that the energy of activation was to be associated with singlet exciton motion on the host lattice sites, and that the transfer from host to host was facilitated by a phonon-assisted increase in co-planarity of their donor and acceptor molecules (62). This phenomenon is reminiscent of that observed by Meyer et al (63) described herein. The concentration of sites at which an increase in co-planarity might be facilitated (predimer or incipent dimer) can reach 10^{-3} mol fraction (64).

The theory of trapping of excitons at guest sites in the presence of low trap concentrations has been considered by several authors (64–67). An analytic theory that accounts for the guest and host fluorescence yield for all guest concentrations has recently been given by Kenkre (68) and Kenkre & Parris (69), who use the general master equation (GME) governing excitation transfer (1, p. 108). They gave explicit expressions for both guest and host luminescence yield as a function of trap concentration (69).

In addition to exciton trapping at impurities and guests that have lower excited state energies than host molecules, it is also possible for self-trapping to occur in the host lattice (69a). Self-trapping of Frenkel excitons in homomolecular aromatic organic crystals was unlikely, according to Toyozawa & Shinozuka (70). However, under certain circumstances, the formation of an excimer that lies energetically beneath the exciton band may be viewed as a self-trapped exciton (1, pp. 85–89, 257–67).

The association of broad excimer-like luminescence with the existence of a self-trapped state necessitates (a) simultaneous observation of both free exciton and excimer luminescence, (b) an activated temperature dependence of the emission bands, i.e. the existence of a small energy barrier between the free state and the self-trapped state, and (c) the association of the emission of the trapped state with the intrinsic property of the crystal and not with defects. Previously reported excimer emission in β-9-10 dichloroanthracene (71), initially identified with a self-trapped exciton luminescence, is now known (72) not to satisfy the above criteria, since the monomer emission arises from defects. This conclusion was reached after it was found that crystal annealing greatly reduced the monomer emission. Thus, in this case, excimer formation was not a thermally activated process.

Evidence of exciton self-trapping in α-perylene was initially provided by von Freydorf et al (73), who observed a weakly structured emission (called the Y-emission) at low temperature (at slightly higher energy than the broad excimer fluorescence band) that had a long first-order decay time (≈ 38 ns). In β-perylene crystals the temperature dependence of the luminescence spectrum provides evidence for at least two types of self-trapped states; at least one of them is in thermal equilibrium with free excitons (74). Additional evidence for self-trapping in both crystal phases comes from the work of Matsui et al (75).

SPECTROSCOPY

Lineshape Analysis

A theoretical analysis of EPR, NMR, and optical lineshape data can provide a wealth of information about structural and dynamical properties of solids. The inhomogeneous linewidth of optical lines is determined primarily by the degree of static disorder, and for this reason it is sometimes used as an indicator of the structural quality of crystalline materials. The homogeneous linewidth reflects intrinsic properties such as lifetime effects (T_1) (76) and the coupling strength of a particular excited electronic state to the local dynamical (phonon) modes (T_2). The homogeneous linewidth of optical transitions is not an easily accessible parameter; however, studies of fluorescence line-narrowing (77), hole-burning (78), fluorescence saturation (79), and optical coherence such as the photon-echo (80) and stimulated photon-echo (81) are methods that can yield the homogeneous linewidth of a particular vibronic state [see supersonic studies of (82, 82a)]. Hole-burning spectroscopy is a static method for investigating the dynamics of molecular vibronic states. In non-photochemical hole burning (NPHB) a particular set of molecules tuned to the laser frequency are excited but not chemically altered. The excited chromophore produces a change in its microenvironment, converting it rapidly to a new structural arrangement before deactivation; this decreases the number of chromophores having excitation energies at the laser frequency and increases the number of chromophores having excitation energies at other frequencies. The net result is that a hole is burned in the absorption band; the profile can be probed by measuring the fluorescence excitation spectra. When spectral diffusion processes are negligible, the hole shape is Lorentzian and its shape will be a measure of the true homogeneous lineshape [FWHM = $(\pi T_2)^{-1}$].

NPHB has been observed in a wide variety of amorphous matrices at low temperatures (83). Jankowiak & Bässler (84, 85) studied NPHB in tetracene ($\sim 10^{-5}$ mol fraction) incorporated into amorphous anthracene at 4 K. Holes burned at $T = 2.5$ K at $\lambda = 491.5$ nm show a pronounced zero-phonon hole and two almost symmetric one-phonon sideband holes; in some materials multiple phonon sidebands can be detected (86, 87). The ratio of the zero- to the one-phonon sideband intensities gave an electron-phonon coupling constant of ~ 0.9 and a phonon energy of ≈ 40 cm^{-1} for those phonons which interact most strongly with the electronic state. The holes disappeared after annealing at 40 K for one hour. The homogeneous linewidth Γ (at FWMH) of the zero-phonon hole varied with temperature as $\Gamma(T) = \Gamma_0 + aT^2$ where $\Gamma_0 \approx 4.0$ cm^{-1}; this corresponds to a dephasing time on the order of 2.5 ps, which is 3×10^3 times faster than the radiative decay of tetracene in solution [for a study of vibrational dephasing in naphthalene, see (87a)]. The temperature dependence was rationalized in terms of the Reineker & Morawitz theory (88), which considers a distribution of tunneling states (TLS) coupled to the matrix acoustic phonons.

In photochemical hole burning, a photochemical process, such as proton-transfer (detachment), occurs; the photoproducts in general have absorption bands that lie outside those of the inhomogeneous band of the reactant; and if the photochemistry is irreversible, there will be no loss in the integrated hole intensity with time.

Fluorescence saturation methods have recently been applied by Treshchalov & Rozman (79) to determine the homogeneous linewidths of the first two inhomogeneously broadened vibronic states (400 cm^{-1} and 1400 cm^{-1} phonons) of molecular anthracene, at 10^{-5} mol fraction in naphthalene. The technique involves a measurement of the fluorescence intensity (F) as a function of the excitation intensity (I_n) and is applicable to guest molecules that are photochemically stable. For both steady-state and pulse excitation conditions, $F \propto I_n$ in the low intensity regime. At high excitation intensities $F \propto \sqrt{I_n}$. Treshchalov & Rozman (79) reported an energy relaxation time T_1 equal to 40 and 25 ps for the 400 and 1400 cm^{-1} vibrational modes of anthracene; for comparison we note that in a fluorene host, hot luminescence spectral measurement gave 28 and 22 ps, respectively, for these states (90).

The introduction of femtosecond time-resolution capabilities into the field of spectroscopy will probably provide the ultimate measure of relaxation dynamics. So far, the shortest-lived molecular process that has been directly time-resolved is the decay of the Rydberg $3R_g$ state in gas-phase benzene; this is 70 ± 20 fs, as measured by Wiesenfeld & Greene (91). The decay mechanism was not established, but may involve transitions directly into the high vibrational levels of the ground state. This time scale is so short that intermolecular collisions could be neglected. The ionization rate is comparable to the relaxation rate.

Photoelectron Spectroscopy

With the advent of synchrotron light sources, the energy region between far UV and X-ray has been made accessible to experimenters. A discussion of the principles of this light source and a description of recent photoemission experiments on phthalocyanines (Pc) is given by Koch & Gürtler (92). An examination of photoelectron energy distribution curves in Zn-Pc shows that the excitation of $3d$ to $4s$ states in the metal Pc is enhanced relative to that in the free metal atom because 4s states in the metal Pc are transferred from the metal site to the ligand, increasing the availability of empty 4s states for occupation from 3d levels. In a similar way it was found that there is a smaller number of empty 4s states in Cu-Pc than in a Zn-Pc, because the $3d \to 4s$ transitions are stronger in the Zn-Pc. This hybridization of metal 3d electrons with the π-electron structure of phthalocyanine ligand is particularly important in understanding the high conductivity of the porphyrinic molecular metals (93). In many studies of photoemission, the three-stage model of Berglund & Spicer is used (1, p. 533) because of its simplicity. A good discussion of this model has appeared, using anthracene data, showing that structural defects must be taken into consideration (94).

A singularly clear and powerful demonstration of the utility of photoelectron spectroscopy coupled with the application of a simple theoretical model to determine the electronic structure of solids, particularly polymers, continues to emerge from the laboratory of C. B. Duke. A CNDO/S3 (complete neglect of differential overlap, self-consistent field method) model, originally developed to rationalize the photoemission and UV absorption spectra from polyacenes, was extended (95) to include infinite (periodic) macromolecules, as for example polyacetylene. In a companion paper (96) this method was used to rationalize the ultraviolet absorption and emission spectra of pyrrole and polypyrrole. In the calculation, it was found by varying the nymber of pyrrole units in the polymer that some of the essential features of the polypyrrole photoemission spectrum appeared when 4 to 6 pyrrole units were connected, implying that the photoemission hole state extends over at least 4 to 6 pyrrole units, or ~ 14 Å or more. In a similar way, it was found that a bonding $\pi \to \pi^*$ transition extending over at least 4 pyrrole units is responsible for the 3.0 eV peak in the absorption spectrum. By simulating the twisting of the pyrrole units, it was found that there would be no major changes in the photoemission or absorption spectra relative to that of a planar molecule, so no information about twisting may be deduced from the spectra. In another paper (97) the electronic structure of poly(p-phenylene) (PPP), poly(p-phenylene vinylene) (PPV), and poly(p-xylylene) (PPX), as well as the previously studied polyacetylene and polypyrrole was calculated. [For ab initio calculations, see (97a).]

CARRIER GENERATION MECHANISMS

Intrinsic

Among the homomolecular polynuclear aromatic hydrocarbons (PAH) compounds, anthracene has long been the hydrogen-atom for theoretical calculations. As one proceeds from naphthalene to pentacene, for example, the optical absorption spectrum shows a significant increase in the contribution of states that have ionic character (98). This increase in ionic character is paralleled by an increase in the quantum efficiency of photogeneration. However, there is not yet unanimity on the association of this increased ionic character with a specific mode of ionization.

One hypothesis has it that carrier generation requires the excitation of a precursor Frenkel exciton that dissociates when its energy is degenerate with that of a pair of free carriers. This process is referred to as *autoionization* (AI). It does not follow that all AI transitions lead to completely uncorrelated carrier pairs. The electron thermalization distance is ~ 60 Å (1, p. 489). Thus, most electrons that are created inside the solid will thermalize within the Coulomb capture radius of the geminate positive ion, forming a transient charge-transfer (CT) state that can either decay to the ground state or dissociate by the absorption of ambient energy. For excitation energies exceeding that of crystal ionization, electrons are photoemitted with the maximum kinetic energy (100, 101) as calculated by the Einstein photoelectric equation, or with kinetic energies diminished by amounts equal to the energy of excitation of the remaining positive ion. This is proof that no intermediate CT state is necessary for free carrier generation.

However, there is strong experimental and theoretical evidence that direct optical excitation to a CT state takes place in homomolecular solids (102, 103), particularly in the more highly colored members of the polyacene family and particularly for excitation energies less than that of the band gap. The proposition that direct CT generation was the primary mechanism for photoconductivity in most, if not all, PAH crystals was put forth in a series of theoretical papers by Bounds & Siebrand (102, 103) and more recently by Bounds, Petelenz & Siebrand (BPS) (104, 105).

The fundamental premise of the theoretical work of (BPS) (based on anthracene) is that the direct excitation of the CT exciton is made possible by the coupling of this low oscillator strength state, with that of the high oscillator strength $(f \approx 1)S_3(^1B_u)$ Frenkel state located at 4.63 eV. In anthracene the S_3 state is about 0.5 eV higher in energy than the anthracene band gap $E_g \approx 4.1$ eV, and all optical excitations in the energy region up to 4.63 eV would produce CT states (albeit, vibrationally excited); in other words, there would be an energy of activation for photoconductivity due to the dissociation of the intermediate CT state even when the optical energy $hv > E_g$; this is observed (106). The energy E_{CT} of CT states of varying separation distance between the charges was calculated quantum mechanically by BPS; the CT energies were capable of being described qualitatively by a simple Rydberg-like expression of the form

$$E_{CT}(n) = E_g - \mu e^4/8(h\varepsilon\varepsilon_0 n)^2 \qquad 5.$$

where μ is the effective mass of the electron-hole pair, $\varepsilon\varepsilon_0$ the dielectric constant parameters, and n is a principal quantum number, related to the distance r between electron and hole. The $E_{CT}(n)$ values can be related to the different activation energies $E_a(hv)$ found for carrier generation by light of energy hv by the equation $E_a = E_g - E_{CT}(n)$. The explicit assumption is made here that while it is possible to excite vibrational states of CT excitons for each value of E_n, only the 0–0 vibrational state is available for thermal dissociation, the excess energy being dissipated in a time shorter than that for dissociation or recombination. The agreement between theory and experiment (107) for the E_a values is good, but not so good for the quantum efficiencies of carrier generation $\phi_0(hv)$. It was possible for BPS to estimate the contribution of CT states to the overall absorption spectrum; in anthracene this contribution was small but significant, while in tetracene and pentacene, the CT contribution was dominant.

Considerable experimental support for the CT exciton mechanisms for carrier generation has been provided by work on electric field modulated absorption spectroscopy in pentacene (108, 1, p. 574). Similar studies were made on anthracene by Sebastian et al (109), although it proved to be more difficult to detect the CT states. These authors correlated the energies of the peaks that were attributed to the CT exciton electroabsorption spectra with the size of the CT exciton using a Coulombic equation. The extrapolated value of E_g proved to be 4.4 eV, in contrast with the accepted value of about 4.1 eV; they justified this discrepancy by identifying their CT values with the vertical (Franck-Condon) transitions from a Frenkel ground state to vibrationally excited CT states. Furthermore, in opposition to a basic premise of the BPS treatment, they postulated that the excess vibrational energy for any particular CT state could be used to separate further the members of the ion-pair state. This postulate, in a sense, combines the processes of direct CT exciton generation with the ballistic carrier separation mechanism that is implied in the AI mechanism. The identification of the structures found by Sebastian et al (109) in the electroabsorption spectrum with CT states of different separations r_{CT} has been questioned by Siebrand & Zgierski (109a). These authors calculated the spectrum of CT states in anthracene and attributed the structure observed by Sebastian et al (109) to the vibrational overtones of the nearest-neighbor CT state (110). A resolution of the discrepancies between theory and experiment is not yet at hand.

The AI mechanisms of ionization coupled with the ballistic model of electron-hole separation has been discussed in detail by Silinsh et al (111), who studied tetracene and pentacene. Here, AI steps are succeeded by the escape of the hot electrons into the lattice, where they thermalize by acoustic phonon scattering, producing the series of bound CT states. Using a rather simple scattering theory, Silinsh et al (111) calculated the optical energy dependence of the thermalization distances, $r_0(h\nu)$, which in turn correspond to specific separations of CT states. They found good agreement between their calculated and experimental values. Calculations were also made of the photoconductivity quantum efficiency, and agreement was found between calculated (using ballistic theory and a modified Onsager theory) and measured values in the energy region $E(h\nu) > E_g$. In the region $E(h\nu) \sim E_g$, they found a discrepancy that they attributed to a contribution from direct CT state absorption. In the higher regions of photon energy, they found another discrepancy that they attributed to excitation to a different AI state.

In addition to the cases considered above there is another interesting experimental result in which it appears that the efficiency of photoconduction is independent of photon energy (112). This work was carried out with X-metal-free phthalocyanine crystals embedded in a polymeric matrix. It was found that all the ionization proceeded from the first excited singlet state, S_1.

Summarizing the results presented herein regarding the relative roles of AI and direct CT exciton formation, it appears that there are materials and energy ranges in which one or the other, or both, mechanisms prevail; this point has been made by Silinsh et al (111). The actual determination of the relative roles of AI and direct CT exciton formation might be resolvable if photocurrent rise time measurements could be made. This suggestion was made by Bounds et al (105).

Extrinsic

Carrier injection by means of a suitably chosen electrode is a convenient and widely used method for producing a mobile carrier density in organic compounds (1, pp. 273 et seq). The time-resolved study of the injection process, which makes it possible to expose its microscopic details, can be facilitated by the fact that the rate constant for the recombination of the injected carrier with the electrode can be made arbitrarily small by choosing an electrolyte as the injection electrode. Using this fact, Eichhorn et al (113) have carried out the first time-resolved measurement of the escape of charge carriers by diffusional motion out of a Coulombic well near the injecting electrode. This well is created by the attraction of the injected carrier to its image in the electrode. The complete model consisted of a hole injected into the top layer of molecules in the crystal, which could recombine with the electrode with a rate constant k_{rec}, or decay to a trap level with a rate constant k_{tr}. The trapped hole could either recombine with the electrode with a rate constant $k_{t(rec)}$ or be activated back into the free state with a rate constant k_d. The rate of escape of the electron out of its Coulombic well was k_{es}. A fairly good agreement with experiment was found using 4 Å as the effective boundary distance from which carrier injection proceeded, and the values $k_d = 5 \times 10^7 \text{ s}^{-1}$, $k_{t(rec)} = 4 \times 10^7 \text{ s}^{-1}$, $k_{tr} = 3 \times 10^8 \text{ s}^{-1}$, $k_{rec} \approx 0$. Other conclusions that were made were that the hole mobility in the c' direction was independent of electric field up to at least $5 \times 10^5 \text{ V cm}^{-1}$ and also at $1 \times 10^6 \text{ V cm}^{-1}$ (114). The recombination was assumed to be nongeminate, and to involve an OH^- ion. There is a problem with this assumption because the energy required to discharge an OH^- ion in a neutral aqueous solution is ≈ 6.8 eV, referenced to the vacuum zero (115), whereas only 5.8 eV are available as a result of the discharge of the anthracene hole. Chemical attack by a water molecule is a more likely route for this process.

CARRIER RECOMBINATION

Time Independent (Theory)

The Onsager theory of geminate recombination (116, 117) was designed as a steady state solution for a pair of oppositely charged geminate particles moving in a condensed phase continuum in the presence of an external electric field. Nevertheless, it has been widely and rather successfully used to interpret the electric field dependence of the quantum yield of photo-generated carriers in crystals, where the carriers undoubtedly are moving on a discrete lattice (118) (see 1, pp. 481 et seq). Recently, the Onsager continuum theory was extended to include the transient case (119), and the original Onsager assumption of a point-sink at $r = 0$ was refined into a recombination sphere of finite radius and recombination velocity (120). A seminal theory has recently been introduced by Rackovsky & Scher (121a) in which the problem of geminate recombination in a face-centered cubic (fcc) lattice is treated as a random walk on lattice sites in the presence of both an external field and a Coulomb field. The solution is given in terms of lattice Green's functions, modified by the presence of an applied electric field. In this model, one carrier is fixed at the origin and the other, oppositely charged, carrier moves on the lattice in accordance with a function $\psi(1, t)$ that gives the probability per unit time that it will leave a site 1 in a given direction. This function is determined by relative transition rates to the neighbors, and the transition rates in turn are determined by variables such as wavefunction overlap, energetic differences, and electron-vibration interactions. An important factor is the role of competing processes, such as the decay rate of the electronically excited precursors to the ion-pair state that dissociates. What emerges from this treatment will have profound effects on the interpretation of experiments using the Onsager formalism. The Onsager theory has two parameters that are specific to the system under study. These are the electron thermalization distance r_0, and the initial quantum yield of geminate pair generation ϕ_0. These parameters can be evaluated if an assumption is made regarding the initial distribution of geminate pair distances (1, pp. 489 et seq). Scher & Rackovsky show that depending on the relative magnitudes of some molecular parameters described in Figure 1 and using a *fixed* initial distribution of charges (hence, a fixed r_0), the values of ϕ_0 that would have been calculated using the Onsager formalism on the yield versus field plots can be considerably different from each other. Thus, consider the calculation shown in Figure 1 illustrating the photogeneration efficiency η dependence on reduced electric field strength. Here $\varepsilon_{ex} = E_{ex}/kT$ where E_{ex} is the difference in energy between the first electronically excited state at the origin and the state consisting of a hole at the origin and an excess electron at a nearest neighbor. This could be a few tenths of an eV. The term χ is the strength of the Coulomb field between the carriers in the geminate pair, R gives the ratio of the relaxation rate of the initially excited state to the rate of intermolecular charge transfer, and the reduced field strength $\gamma = eFa/2kT$ where F is the electric field and the lattice spacing is a. Notice that in the top two curves, only ε_{ex} has changed; this could be produced for example by local energy fluctuations, which would be significant in the case of

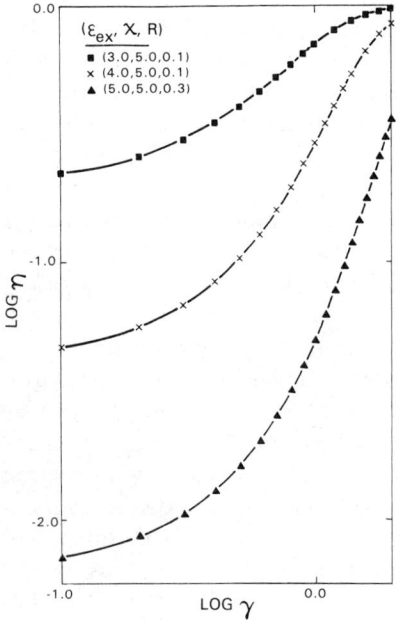

Figure 1 Log-log plot of the photogeneration efficiency η vs the reduced field strength γ illustrating the effect of varying ε_{ex} with χ and R essentially fixed. From Scher & Rackowsky (121).

molecularly doped polymers. In this calculation, the initial distribution is fixed (i.e. the average value of r_0 is *constant*). At low fields, it may be seen that for the lower value of ε_{ex}, η has decreased by an order of magnitude. If the Onsager formalism were used then for the same value of ϕ_0 the apparent value of r_0 would have to be altered in order to compensate for the differences in η. For a greater ε_{ex}, and a slightly larger value of R, η changes by another factor of 10, r_0 would have to change even more, and no saturation is evident at high field, in contrast with what is seen for the top curve.

Another important experiment is the measurement of the temperature dependence of the yield. According to the Onsager theory, the temperature dependence of η at zero field should follow an Arrhenius relationship in which the activation energy is $E_a = e^2/4\pi\varepsilon\varepsilon_0 r_0$. In the notation of Scher & Rackovsky, this would read $E_a = a\chi kT/2r_0$, or $r_0/a = \chi kT/2E_a$. This prediction is not realized. When R is large, a linear Arrhenius plot is obtained but with E_a a function of applied field. When R is small there is no linear plot.

In addition to the analytical treatment of Scher & Rackovsky, a Monte Carlo simulation was carried out by Ries et al (122) of field and temperature assisted dissociation on a perfectly isotropic 3D lattice, and perfectly linear (1D) systems. It was found that for a perfectly 1D system, the zero-field intercept of the dissociation probability would be zero, and the extent to which this intercept is not zero is a measure of the degree of anisotropy of the carrier motion. The power of the simulation approach is that any degree of anisotropy between 1D and 3D can easily be inserted.

An important, and not self-evident, assumption made in the simulations is that of field-independent hopping rates in 1D and 3D systems, i.e. the field does not affect either the initial pair distribution, or the prefactor to the expression containing the barrier height (which is modified by the field). For intersite hopping energies $<kT$, the Onsager result emerges.

An important test of the field dependence of the hopping rates was made by Seiferheld et al (123) using crystalline polydiacetylenes (see 1, pp. 673–99 for a description of these compounds) as model compounds. These compounds (PTS and DCH) are formed by the solid-state polymerization of crystalline acetylenic monomers, as a result of which a highly crystalline solid is formed composed of long conjugated polymeric chains held together by van der Waals forces. The strong covalent forces along the chain and the weak van der Waals forces between the chains lead to a markedly anisotropic electrical behavior. This makes these crystals good examples of quasi-1D systems. The degree of anisotropy may be quantified by means of the ratio $\mu_{\|}/\mu_{\perp}$, which are respectively, the mobilities parallel and transverse to the molecular chain. The deduced ratio for the microscopic anisotropy was $>10^5$ in opposition to the value of $\sim 10^3$ indicated by current flow measurements (125, 126). The conclusion that the anisotropy is so high follows from an application of the Onsager theory of geminate recombination as modified for true 1D behavior. In a true 1D system, in the absence of an external electric field, the degree of dissociation of a hole-electron pair must be zero (127) because a particle executing a random walk on a 1D lattice must return to the origin with unit probability. In the present case, the experimental results are in accordance with the Onsager theory for a 1D system for applied fields down to 30 V/cm. Furthermore, according to a computer simulation carried out by Ries et al (122), the low-field intercept of the dissociation yield is a direct measure of the mobility anisotropy, i.e. $\lim_{F \to 0} \phi^{\text{esc}}(F) \propto \mu_\perp/\mu_\parallel$, where ϕ^{esc} is the probability of an electron escaping geminate recombination at the applied field F; no deviation from the Onsager 1D behavior appeared even when $\phi^{\text{esc}}(F) < 10^{-4}$.

Another important conclusion of this paper follows from the low-field agreement with the predictions of the Onsager 1D theory. According to the computer simulation results (122) the Onsager equations will be followed as long as the mean free path of the carrier between scattering events is comparable to the distance between lattice sites (≤ 10Å). The low-field agreement with the Onsager predictions then imply that diffusive transport in PTS and DCH is accompanied by strong scattering. However, for DCH, a mean free path of ≈ 100 Å for carrier scattering follows from an $m^* = 0.05\, m_e$ (128) and a scattering time of 8×10^{-14} s. This apparent contradiction may be removed by the application of theoretical findings of Movaghar & Cade (129, 130), who have shown that carrier transport in PTS occurs via localized polaron states whose mean free path for scattering could be about the distance between polymer repeat units. During the initial hole-electron formation, during which the Onsager formalism is not operative, the electron could travel in the wide conduction band for a distance of about 100 Å before falling into a polaron state. This distance could be associated with the thermalization distance, which is calculated by ~ 65 Å from other results.

The proposal that a polaron state is created in $\sim 10^{-14}$ s has other important ramifications. According to Donovan & Wilson (131), electron mean free paths are ~ 1 μm in length, and the electron mobility is $>2 \times 10^5$ cm^2 V^{-1} s^{-1}. If the mean free paths are so long, then geminate recombination should not be observed, whereas it is. The high mobilities (132) also become puzzling. This paradox was resolved by Movaghar et al (133), who showed that standard linear response theory as applied to transport was not applicable in these 1D systems.

ELECTRONIC PROCESSES 631

A most significant result is that at high-field ($F > 10^4$ V cm^{-1}) the anticipated saturation does not occur. This failure to saturate had already been noticed in anthracene by Geacintov & Pope (134). Seiferheld et al (123) concluded that their results can only be explained by a primary ionization yield ϕ_0 that is field dependent.

Time Independent (In Polymers)

Evidence for exciton fusion as a mechanism for the intrinsic generation of free carriers in molecularly doped polymers has been reported by Orlowski & Scher (135, 136). The system studied consisted of solid solutions of the amine TTA [tri-(p-tolyl)amine] in the polymer Lexan (bisphenol-A-polycarbonate) in concentration between 30 and 40% by weight. Bulk (as distinguished from surface) photoconductivity was induced in 10 μm thick films. The singlet exciton decays by intersystem crossing ($\Phi_{isc} \approx 0.95$) to the lowest triplet state (T) of TTA. The phosphorescence of TTA peaks at 2.8 eV, so two triplet excitons can supply 5–6 eV; this is sufficient to induce photoionization. It was found that the photogeneration yield, q/A, varied smoothly as the light intensity I^2 at low light intensity, as I at high light intensity, and at some intermediate value in between. With an increase in field strength F from 5 V/μm to 20 V/μm, q/A increased by a factor of about 250 at $I = 10^{13}$ photons cm^{-2}, and in general the I dependence itself was F dependent. The scheme used to explain the results was as follows:

$$T + T \underset{\xi}{\overset{\gamma}{\rightleftarrows}} (TT)^* \xrightarrow{k_1} P \xrightarrow{k_{cp}} C$$

with k_2 above $(TT)^*$, and $\downarrow k$ to S_0 from T, and $\downarrow k_R$ to S_0 from P.

$(TT)^* = T^*$ is an associated triplet-pair state, S_0 is the ground electronic state, and C is the hole on the TTA molecule generated at the rate K_{CP} from P which is a charge-separated state in which the polymer contributes an acceptor state. The constant γ refers to the diffusion-limited process of triplet exciton fusion, and ξ applies to the diffusion-limited separation of T^*; K_1 refers to the creation of P, K_2 describes the annihilation of T^* to produce T and S_0, and K_R represents the recombination of hole and electrons to produce a ground state. The quantity η_{eff} is to be compared with the quantum yield obtained from data representing the linear response of q/A to I, and γ_{eff} corresponds to what would be measured as the diffusion-limited rate constant for triplet-triplet fusion to produce T^*. It was found that γ_{eff} increased from 0.16×10^{-12} cm^3 s^{-1} at $F = 0$, to

160×10^{-12} cm^3 s^{-1} at $F = 20 \times 10^4$ V cm^{-1}. At $F = 5 \times 10^4$ V cm^{-1}, η_{eff} was 1×10^{-4}, while at $F = 38 \times 10^4$ V cm^{-1}, η_{eff} was 24×10^{-4}.

The Onsager field effect found in many systems (1, pp. 484–95) applies to the relative magnitudes of K_{CP} and K_R, with P given by some initial efficiency ϕ_0, and says nothing about any of the other quantities in the schematic diagram. However, the Onsager theory cannot explain the magnitude of the observed results even if ϕ_0 is made field dependent (111). The picture that emerges is that the T^* state is probably a configuration consisting of a mixture of a triplet exciton pair, an excited S^* state, an excited charge-transfer complex and some other configurations as well. The electric field would have the effect of increasing the tunneling rate of the CT state to the P state, thus enhancing the value of γ_{eff}. The assignment of CT character to the intermediate triplet-pair state has been made in the past (137, 138). In systems in which singlet excitons play a dominant role in photoionizations, an extension of this mechanism would propose the existence of an intermediate S^* state that has CT character; the electric field should have a similar effect in this system and, indeed, had already been recognized by others (139, 140), who found this field effect on the ratio of carrier generation yield to fluorescence yield. Popovic (140) studied X-metal free phthalocyanene powders dispersed in Lexan. He concluded that the electric field increased the rate of dissociation of the lowest-lying excited singlet state S_1 into a CT state, which subsequently dissociated thermally, assisted by the external field. His work also provided an example of a case in which the same S_1 state acted as a precursor to CT state formation even though the initially excited state was higher in energy.

It has now become clear from the work of Popovic (140), Scher & Orlowski (135, 136), Silinsh and co-workers (141), and Rackovsky & Scher (121a) that the field dependence of photocarrier generation in organic molecular solids is not capable of being interpreted entirely by the continuum Onsager model (116, 117). Depending on the energy of photoexcitation, a state is produced in these materials that has a complex character including, for example, singlet-singlet (142), triplet-triplet (143), and a considerable charge-transfer component. The effect of the high external field is to shift the character of the intermediate state to one that is more ionic, and easier to ionize. Following the dissociation of the intermediate state, the usual Onsager formalism applies.

Time Dependent

Monte Carlo computer simulations of physical processes have also played a major role in the elucidation of the time-dependent geminate recombination. This is particularly important because it is difficult to carry out experiments in the picosecond time scale. The time dependence of geminate recombination was simulated by Ries, Schönherr, Bässler, and Silver (144) and compared with the analytical solution given for the isotropic continuum model by Hong & Noolandi (HN) (119, 120) and Hong, Noolandi & Street (145). The simulation was not restricted to the isotropic continuum case, but considered 1, 2, and 3D motion on a discrete lattice of variable size and graininess. One of the conclusions of HN was that at long times, the survival probability of a geminate pair in 3-D follows a $t^{-3/2}$ decay law. That is, according to HN,

$$R^{3D}(t \gg \tau_0) = r_c \exp(-r_c/r_0)/(4\pi D t^3)^{1/2}$$

where

$$\tau_0 = r_c^2/4D, \qquad r_0 = e^2/4\pi\varepsilon\varepsilon_0 kT$$

and r_c is the Coulomb capture radius, D is the diffusion coefficient, R^{3D} is the geminate recombination rate in 3D. The simulation indicates that the $t^{-3/2}$ behavior, if realizable at all in an actual experiment, would be observed only when the signal amplitude (as for example, that for recombination luminescence) has decayed to 10^{-4} of its peak value. For practical situations, the exponent x of the decay function t^{-x} is $>3/2$; the value of x will depend also on the dielectric constant of the medium (through r_c) and the dimensionality of the motion. In anthracene, $r_c/r_0 \approx 8$ (118), whereas for α-Si, where $\varepsilon = 11.5$ (compared to $\bar{\varepsilon} = 3.2$ for anthracene), r_c/r_0 is small enough so that the HN theory works well in explaining the α-Si luminescence decay rate (121).

Another startling result of the simulation is that in the first decade of the decay process ($t > \tau_0$) the simulated decay rate exceeds the HN results if extrapolated to short time limits. Thus, by using the HN theory to rationalize an observed recombination luminescence life time, one would have to use an inordinately large diffusion coefficient. The simulated and analytic results converge when the number of surviving particles has decayed to 10^{-6} of its peak value. Again, the HN theory predicts that for $t \gg \tau_0$

$$N^{3D} = \exp(-r_c/r_0)|1 + r_c/(4\pi D t)^{1/2}|$$

where N^{3D} is the fraction of surviving pairs, whereas the simulation results show a more complex function of time.

Another parameter that was varied in the simulation was the frequency of the final jump to the recombination center. This is particularly relevant because of the accumulating experimental evidence for the existence of a long-lived electron-hole pair state in organic crystals (142, 146–152). So far no attempts have been made to rationalize such behavior theoretically. Ordinarily one would expect the final recombination step to be rapid.

Although no picosecond time-resolved geminate recombination studies have been made on anthracene or its homologs, Braun & Scott (152) did carry out such a study on a hexane solution of anthracene at a concentration of 4.5×10^{15} cm^{-3}. At room temperature, the decay rate of photogenerated geminate cation-electron pairs is characterized by a first half-life of no more than 9 ps. The recombination process is strongly nonexponential, as would be predicted by the theory of Hong & Noolandi (120) and the recent work of Ries et al (144). Thus, the second half-life is 28 ps and the third half-life is 125 ps. The work in liquids indicates that geminate recombination can be rapid. However, there are conditions under which the final jump can be slow. To explore the consequences of such a situation, Ries et al (144) generated a "waiting factor" f_W for the final jump such that the inverse probability of the final step is $t_{rec} = f_W \tau_0$, where τ_0 is the hop time for an isoenergetic or exoenergetic jump. Thus, if the final recombination took place at a defect site, there might be an energetic barrier U_B to overcome, to which could be related an $f_W = \exp(U_B/kT)$. Carriers trapped in defect sites may exhibit abnormally long dwell times (see this review under CARRIER TRAPPING).

CARRIER TRANSPORT

In Molecular Crystals

Carrier transport in van der Waals molecular crystals, such as anthracene and naphthalene, has been the subject of intensive study during the last several years. The adequacy of band theory as a proper framework for interpreting experiments had been questioned as a result of its failure to explain electron mobility in the c' direction in anthracene and naphthalene. For a review of pertinent experimental data and theoretical models prior to 1982 see (1, pp. 337–78). A collection of data on mobilities in organic molecular crystals was published by Schein & Brown (153). In this review, we present the most recent experimental data and theoretical model. Requiring explanation are (a) the band-to-hopping transition observed (154) in naphthalene at $T \approx 100$ K for electron transport (μ^-) along the c' crystallographic direction, and (b) the almost complete lack of temperature dependence in μ^- in anthracene from 78–479 K (155) and in naphthalene from 100–325 K (156).

In contrast to electrons, hole transport (μ^+) characteristics are in accordance with band theory predictions (1, p. 349); μ^+ increases with decreasing temperature in the temperature region, in which shallow trapping is not dominant. Because of the weak intermolecular interactions in van der Waals crystals, the nearest-neighbor transfer energies, which are on the order of 10^{-2} eV (157, 158), are comparable to the conduction band broadening due to dynamic disorder; strong dynamic disorder tends to localize carriers. A major theoretical problem has been to isolate the dominant terms in the complete transport Hamiltonian, which so far has been intractable. Sumi (159–162), in a series of papers, assumed that μ^- was dominated by electron-libron scattering. A recent analysis of mobility data suggests that a more conventional approach is fully capable of accounting for the temperature dependence of the electron mobility for $T < 100$ K. Specifically Andersen et al (163) have demonstrated that a Boltzmann equation analysis with longitudinal acoustic phonon-electron scattering can account quantitatively for the electron mobility in all crystallographic directions, in which case $\mu \propto T^{-3/2}$. Figure 2 illustrates how well $\mu \propto T^{-3/2}$ fits the experimental data. The lack of a temperature dependence in μ for $T > 100$ K is, however, still unexplained. Furthermore, as noted by Andersen et al (163), there is still the difficulty of reconciling the conventional band picture with the short mean free path (λ) of the carriers in the c' direction at temperatures $40 \leq T \leq 100$ K ($\lambda \approx a$).

Although saturation of the electron velocity along the c' direction has not been observed (164), field-dependent hole mobilities for F parallel to the a-axis at low temperatures have been reported by Warta & Karl (165).

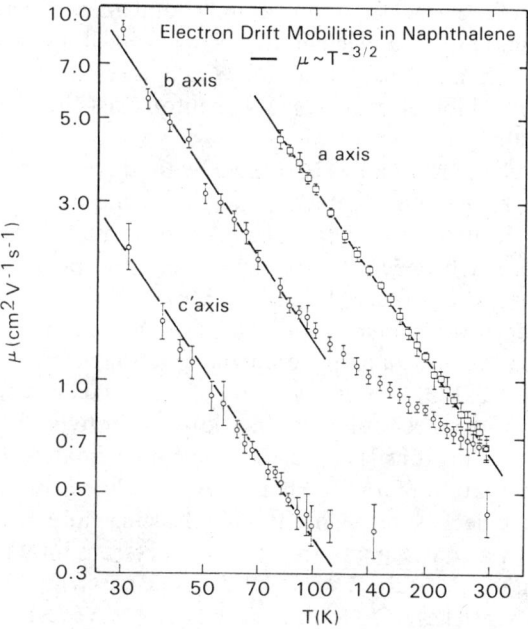

Figure 2 Measured electron drift mobilities along the **a**, **b**, and **c'** directions in naphthalene as a function of temperature. From Anderson, Duke & Kenkre (163).

Presumably the freezing out of phonon modes at low temperatures enhances the carrier mean free path sufficiently so that momentum states near the edge of the Brillouin zone are significantly occupied.

In Poly-π-electron Polymers

Interest in the electrical conductivity of polymers is high and will certainly become greater with time. Several conferences have been devoted almost entirely to the subject of these poly-π-electron polymers (166, 167). A review of polyacetylene has been given by Etemad et al (168). Doped polyacetylenes have conductivities that are metallic ($\sim 1000 \ \Omega^{-1} \ cm^{-1}$) and they have been fashioned into p-n junctions (169). In this review, we discuss only two polymeric systems that consist of long chains of conjugated molecules; these are the polyacetylenes and the polydiacetylenes, which have been the model systems used for the study of carrier transport.

POLYACETYLENES The polymer polyacetylene has received the greatest attention both theoretically and experimentally due, partly because it is the simplest conjugated polymer and also because the *trans* form can support mobile solitons. A *soliton* is an uncharged defect or a kink in the *trans*-polyacetylene (*t*-PA) chain consisting of two single bonds adjacent to each other, instead of the usual single and double bond alternation. This kink, or soliton, has one unpaired electron associated with it and is free to move along the *t*-PA chain. A charged soliton would have no unpaired electrons, and would not exhibit electron spin. No soliton exists in *cis*-polyacetylene (1, pp. 701, 780).

Structural changes usually associated with the *cis*- to *trans*-thermal isomerization have been observed in polyacetylene by Robin et al (170), who used synchrotron X-radiation (1.34 Å); the isomerization process was found to proceed homogeneously throughout the polymer and not in isolated amorphous or selected crystallite regions. That doping also induces a *cis*-to-*trans* isomerization has been demonstrated by Hoffman et al (171). The soliton has an EPR Lorentzian line-shape with a linewidth less than 1 Gauss (172, 173); that is suggestive of motional narrowing at least at room temperature since estimated linewidths from hyperfine interactions with neighboring protons is expected to be approximately 23 Gauss (174, 175). The observation of an Overhauser effect (176) is consistent with the paramagnetic defect being highly mobile. Furthermore ENDOR studies (177, 178) show a coalescing of the two-line spectrum into a single line on increasing the temperature. The ENDOR 2-line spectrum has been shown by Heeger & Schrieffer (179) to express the qualitative features of a mobile soliton. Evidence for solitons, independent of optical studies (180), has been provided by elastic tunneling conductance measurements (181).

A direct measurement of the time evolution of a photogenerated electron-hole state into charged kink-antikink states in cis- and trans-PA has been made by Shank et al (182). They showed that the electron-hole state decays in less than 1.5×10^{-13} s. This result supports the theoretical calculation of Su & Schrieffer (182a). From the decay curve of the charged kink state in cis-PA, it appears that the photogenerated state converts into a metastable state with a lifetime greater than 10 μsec. The long lifetime for recombination in cis-PA is reminiscent of that found in the polyacenes (146–151). In t-PA, the recombination is complete in ~ 200 ps, in which nearest neighbor jump times of $\sim 10^{-13}$ s at 300 K and 10^{-12} s at 20 K were calculated.

Soliton theory has been quite successful in explaining a wide variety of physical properties in cis- and trans-PA. Some of the subtle difficulties with the soliton theory in explaining EPR and NMR measurements and photo-induced changes in optical absorption have recently been summarized by Baeriswyl (183). We note here only the problem in explaining the transport characteristics. The Kivelson theory (184) of phonon-assisted hopping between charged and neutral solitons, although quite successful in making specific predictions regarding conductivity as a function of temperature, frequency, and concentration, is not without difficulties. The theory requires the presence of both neutral and charged solitons and is valid only at low concentrations, in the regime where $Y < 0.005$. This assumption is hard to reconcile with the observation of spinless conductivity at high dopant levels (at 2% dopant, all neutral solitons have been converted to charged solitons). Furthermore, spinless conductivity is not a unique property of trans-polyacetylene; it has been observed in poly(p-phenylene) (PPP) and for this polymer it is known that solitons cannot exist since the benzenoid form is significantly more stable than the quinoid form, i.e. PPP does not possess a degenerate ground state. In fact, PPP doped with AsF_5 has been reported to have a conductivity comparable to that of doped-PA (~ 500 Ω^{-1} cm^{-1} at dopant levels of 0.42 mole of AsF_5 per mole of monomer (185)). Thus, a theory of spinless conductivity that is based solely on charged solitons, even for trans-$(CH)_x$ is considered highly artificial.

A more promising model for spinless conductivity is the polaron-bipolaron theory of Bredas, Chance & Silbey (186, 187). This model does not exclude solitons; in factor neutral solitons are the main extrinsic defects in undoped trans-$(CH)_x$. In $(CH)_x$ the polaron is viewed as a pair consisting of a free radical (neutral soliton) and an ion (charged soliton) with an estimated binding energy of 0.05 eV, a value quite comparable to the pair binding energy of 0.03 eV calculated for PPP. An antisoliton is the mating kink to a soliton. An antisoliton will annihilate a soliton, producing a defect-free chain. An antipolaron is a combination of a charged soliton and a neutral antisoliton. The evolution of states within the gap as a function of dopant concentration can be summarized as follows: Low doping introduces both bonding and antibonding localized polaron states within the gap. At higher doping levels the localized polaron states coalesce to form two polaron bands. In PA the interaction of a polaron with an antipolaron of the same charge results in two free charged solitons. This process occurs only for PA since the neutral soliton-antisoliton pair annihilates and the degenerate ground state allows the charged solitons to move apart. In PPP and other conducting polymers that lack a degenerate ground state, the two charged defects cannot separate and bipolaron formation ensues. Bipolarons are correlated charged soliton-charged antisoliton pairs. At even higher dopant concentrations bipolaron bands are formed. Evidence to support the existence of bipolarons and a polaron band in doped-PPP has been provided by electron energy loss spectroscopy (188).

Within the bipolaron model of transport, the neutral solitons that are required on neighboring chains are produced by the formation on a neighboring polymer chain of a virtual neutral soliton-antisoliton pair. After the charged portion of the bipolaron jumps to the virtual soliton-antisoliton pair on a neighboring chain, the residual soliton-antisoliton pair annihilates. This mechanism requires only a small activation energy, is applicable not only to *trans*-$(CH)_x$ but also to other conducting polymers, and does not eliminate the phonon-assisted interchain hopping mechanism of Kivelson in *trans*-$(CH)_x$ in dopant regimes where sufficient neutral solitons are available.

First principle calculations have shown that as in PPP, polaron states form within the band gap in polypyrrole at low doping levels (189). Optical absorption due to polarons in polyacetylene and in other polymers (polypyrroles, poly-paraphenylenes) has been considered theoretically by Bredas et al (186) and Fesser, Bishop & Campbell (190). Experimental support for bonding and antibonding polaron states in the low dopant limit in *t*-PA is reported by Etemad et al (191). It appears that in *t*-PA both polarons and solitons coexist at low doping concentrations and are the current carriers, whereas in polymers lacking a degenerate ground state and thereby precluding soliton formation, polarons are the exclusive carriers.

POLYDIACETYLENE   The unusual features of carrier transport in PDA-TS are the apparent saturation of the drift velocity v_d at $\sim 2 \times 10^5$ cm s^{-1} down to fields as low as 1 V cm^{-1} [giving a lower bound for $\mu \approx 2 \times 10^5$ cm^2 V^{-1} s^{-1} (131)] and the photocurrent decay rate following a short light flash; this decaying photocurrent $I(t)$ obeys a relationship of the form $t^{-\alpha}$ where $\alpha \sim 1$ over 6 decades in time (10^{-6} s to 1 s) (144). This decay law is often found in amorphous materials, whereas in PDA-TS one would expect very few scattering or recombination centers. Additional experiments by the authors in the time domain for which $I(t) \propto t^{-\alpha}$ shows that v_d is still field independent and $\alpha \sim 0.85$.

According to Seiferheld, Ries & Bässler (123), the anomalously high carrier mobility ($\mu \approx 10^5$ cm^2 V^{-1} s^{-1}) deduced for PDA-TS (131) is inconsistent with the strong scattering required to explain their successful use of the Onsager 1D carrier recombination formalism. The paper by Movaghar et al (133) resolves these difficulties and shows that there need be no qualitative conflict between the results of Donovan & Wilson (131) and those of Seiferheld et al (123).

The explanation for all of the observed results lies in the breakdown of linear response theory as applied to transport problems in a broad class of 1D systems (192). A model is used that is suitable for materials like PDA-TS, with its low concentration of crystal defects. Using the 1D diffusion theory of Alexander et al (193) one may deduce a time (t) dependent current $j(t)$ relationship of the form

$$j(t) \sim ea\eta t^{-\alpha/(2-\alpha)}, \qquad t \to \infty. \qquad 6.$$

Here η is the applied reduced electric field and need not be defined here other than to state that for small fields, $2\eta = eFa/kT$, where a is the lattice spacing. However, as explained by Movaghar et al (192), linear-response theory breaks down in the long time domain and therefore Eq. 6 is wrong. The correct form for $j(t)$ at long times is

$$j(t) \sim 2ea\eta W_0^{1-\alpha} C(\alpha)/|2\eta t^\alpha| \qquad 7.$$

where $C(\alpha)$ is constant and W_0 is the hopping rate in a defect free 1D chain. Since $j(t)$ is proportional to the carrier drift velocity v_d, then according to Eq. 7 and the definition of η, v_d will vary sublinearly with the electric field. Furthermore, since $\alpha \sim 1$, it follows that the drift velocity will be almost independent at F.

The authors make the important point that behavior of the type described by Eq. 7 should appear in a wide class of 1D transport problems. Thus, for PDA-TS, the authors derived, for $t > t_c$

$$v_d(t, \eta) \sim (2\alpha\eta C(\alpha, x)/x) W_0^{1-\alpha}/|2\eta t|^\alpha \qquad 8.$$

where x is the fraction of jumps that are difficult and t_c is the time required for the carrier to drift to a defect; $t_c \sim (2\eta x W_0)^{-1}$. Equation 8 applies to the PDA-TS experiments where $\alpha = 0.85$, and $v_d \propto F^{0.15}$, which looks like a saturated drift velocity. In addition, the photocurrent will decay as $t^{-\alpha}$ until carrier recombination (or collection at the electrode) takes place. It can be also shown that a lower bound for the mobility along the regular portion of the chain will be $\mu_f > 10^8 a/2xF$. If v_d remains saturated at 2×10^5 cm s^{-1} down to $F = 1$ V cm^{-1}, as stated by Donovan & Wilson (131), then $\mu_f > 2 \times 10^5$ cm^2 V^{-1} s^{-1}. Thus, with $a = 1.2$ Å x could be $\sim 10^{-5}$ or fewer than 1 defect per 10^5 Å; this was also stated by Donovan & Wilson (131). Using the data of Seiferheld et al (123), who only went down to $F = 50$ V cm^{-1}, and taking $v_d = 2 \times 10^5$ cm s^{-1}, one gets $\mu_f \sim 10^3$ cm^2 V^{-1} s^{-1} and $x \sim 10^{-5}$, which is still a high mobility. Another interesting calculation is the time required to cross half the sample length, or $L_0/2$. This turns out to be

$$t_0 = (2\eta W_0)^{-1}(L_0 x/2a)^{1/(1-\alpha)}. \qquad 9.$$

With $v_d = 2 \times 10^5$ cm s^{-1}, $F = 1$ V cm^{-1}, $L_0 = 0.3$ cm, $\alpha = 0.85$, $a \sim 5$ Å, and $x = 10^{-5}$, then $t_0 \sim 10^{-2}$ s. For $x = 10^{-4}$, one obtains $t_0 \sim 10^5$ s. Thus, t_0 is extraordinarily sensitive to x and α. It is possible to increase x by creating radiation-induced defects; this has been done and very long relaxation times have indeed been observed (194).

In Amorphous Systems

As mentioned above, the introduction of computer modeling as a quasi-experimental technique is having a profound effect on the course of both experimental and theoretical studies. Consider the study of carrier transport in amorphous systems, of which polymeric conductors are a prime example. In amorphous systems, carrier transport clearly is by hopping. It has been shown by Bässler (195) and Schönherr et al (196) that by assuming a Gaussian distribution of hopping site energies, a wave function overlap factor $\exp(-2\alpha r)$, where r is the distance between neighboring sites, and a Boltzmann factor to accommodate hops to energetically unfavorable sites, one obtains the following: a temperature dependent trap-free mobility of the form $\mu(T) = \mu_\infty \exp\{-(T_0/T)^2\}$ where μ_∞ is the ideal mobility in the absence of disorder, and a field dependent mobility of the form $\mu(F) = \mu(0) \exp(F/F_0)$. The field dependence is of interest because it shows that no charged traps need be invoked to explain a field dependent mobility (1, p. 78); such a field dependence had already been observed, and charged traps had been conjectured (197). In a series of experiments (198) carried out with hole-transporting solid solution of a phenylmethane derivative in a plastic matrix, complete agreement was found between experiment and simulation, without invoking charged traps. As for the temperature dependence of the mobility in amorphous systems, this was studied experimentally by Lange & Bässler (199), who used amorphous tetracene films. The use of amorphous films of an otherwise crystalline material permits a study of the effects of disorder without changing chemical composition. Tetracene films deposited on substrate held at 120–180 K are amorphous. A trap-free mobility obeys a relationship of the form $\mu(T) = \mu_\infty \exp\{-(T_0/T)^2\}$, where $\mu_\infty \approx 0.4$ to $0.8 \text{ cm}^2 \text{V}^{-1}\text{s}^{-1}$, validating the Monte Carlo computer simulation. The Gaussian width given by Γ_0 is ≈ 0.1 eV, and is a function of sample preparation temperature T_F and hence of the degree of disorder. Important results of this paper are that the Gaussian width is a consequence of disorder-induced splitting of the transport band into a distribution of localized site energies; this effect is larger than that produced by the dielectric relaxation of the environment around a carrier. The success of the computer simulation strengthens support in its assumptions, and provides an alternate explanation for the results reported by Mort & Pfister (200).

Another interesting result was the observation of an almost temperature independent trapping time, although μ varied by more than an order of magnitude. Their explanation is based on the notion of trapping at an incipient dimer site. The density-of-states' profile in tetracene as deduced in these experiments exhibit a 0.1 eV width; this is in contrast with widths of 0.5 to 1 eV found in photoelectron spectra of molecular glasses (6). An explanation may be found in the differences between the processes of photoemission (which depends not only on the inhomogeneous broadening of the valence states, but lifetime broadening of the final state) and transport. In addition, transport probes the volume density of state distribution, whereas photoemission is essentially a surface phenomenon and the polarization energy changes markedly in this region (201, 202; see also 1, p. 63).

Another example of the power of a computer simulation is that carried out by Silver et al (203), who showed that while a Gaussian distribution of hopping site energies gives little dispersion, an exponential energy distribution gives rise to dispersive transport indistinguishable from multiple trapping. This introduces another mechanism for the interpretation of experimental results.

CARRIER TRAPPING

Experimental Techniques

A general technique for the determination of the concentration, energetic distribution, and composition of carrier trapping sites in organic solids does not yet exist. It remains as a challenge, and the rewards for solving the problem are great, since relatively small concentrations of traps radically modify the transport properties of the carriers. Present techniques include space charge limited currents (SCLC), thermally stimulated currents (TSC), optical detrapping, and electric field detrapping (1, pp. 267 et seq).

An important comparison of the results obtained on trapping states by

SCLC and TSC was made by Taure et al (204). They studied thin films of tetracene (Tc) and pentacene (Pc) in the form of oriented crystallites. The SCLC results were consistent with the existence of at least two sets of shallow traps of Gaussian distribution: In Tc the binding energy of one was $E_t = 0.1$ eV, with a distribution parameter ranging from $\sigma = 0.08$ to 0.14 eV in different samples, and a total trap density $N_c = \approx 4 \times 10^{15}$ cm^{-3}; for the other, $E_t \approx 0.3$ eV, $\sigma = 0.05$ to 0.1 eV and $N_t \approx 10^{14}$ cm^{-3}. The TSC results were in fair agreement with those of SCLC as shown in Figure 3, but gave higher resolution and probed deeper traps. Trapping cross-sections for the hole traps in Pc and Tc range from 10^{-16} to 10^{-14} cm^2 for the energy range 0.2 to 0.5 eV, and are about 10^{14} cm^2 for deep electron traps in Pc.

Some recent work involving the use of thermally stimulated currents (TSC) is that of Samoc et al (205), which also contains many references to important work in this field. This paper tests a recent theory of Plans et al (206) that deals with the field and sample thickness dependence of the peak at T_m, the temperature of the maximum that appears in the TSC glow curve. Samoc et al (205) added phenothiazine in a concentration of 5×10^{-5} mol fraction to anthracene. Using their technique, they found that phenothiazine created a hole trap centered at 0.62 eV above the valence band and that there was sufficient distribution of energy around this value to induce dispersive transport in the doped material. This trap depth may be

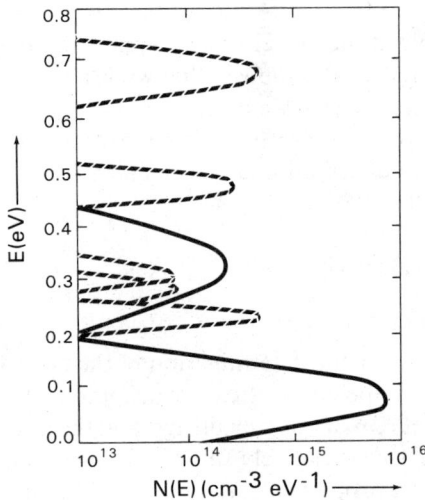

Figure 3 Density of electron trapping states *N(E)* in pentacene layers as a function of trap depth *E* as determined by SCLS (*solid curves*) and TSC (*dashed curves*) techniques. From Taure et al (204).

compared with the value of 0.57 eV deduced by Karl et al (207) from photoemission studies.

An experimental technique for studying multiple deep trapping levels using only a single thermal scan was developed by Yoshie & Kamihara (208), and several intuitively appealing theoretical treatments of multiple trapping have appeared (209, 210). In these papers, an exponential distribution of traps with relatively fixed parameters was assumed. In a recent paper, Monroe & Kastner (211) relax these restrictions by allowing variation in capture cross-section, release times, trap distribution, and energy. The final result is a classification scheme in which the current transients may fall into any of five basic types; this scheme may be used as a diagnostic tool to identify the relevant trapping parameters operative in the experimental situation. The analytical results of Monroe & Kastner can also serve as a check on the results of increasing popular computer simulations. Mention should also be made of another comprehensive treatment of multiple trapping in amorphous systems by Arkhipov et al (212).

An interesting aspect of electron trapping was exposed by Meyer et al (63). They found that the electron trapping lifetime for transport in the c' direction over a range of temperatures (81–374 K) and applied electric fields (10^3 to 4.7×10^4 V cm^{-1}) was activated, and that this activation energy is independent of the field. The activation energies, E_a, ranged from 30 to 84 meV on crystals all prepared from the same sample, indicating that the behavior is extrinsic. The electron transport in the c' direction was not shallow-trap controlled; it was nearly independent of temperature. The authors conclude that the most likely source of the activation energy was the cross-section of capture. A parallel experiment was carried out by Arnold & Hassan (213), who measured the effect of pressure on the triplet exciton lifetime in anthracene. They found that the cross-section of capture was increased markedly (by a factor of four, in going from 1 atm to 6.4 kbar), without changing the trap concentration. Arnold & Hassan concluded that specific mechanical defects, referred to as preexcimer sites, were the source of these traps, and that triplet excimers were formed at these sites. They attributed the pressure effect to the creation of a more favorable intermolecular orientation. The role of predimer or incipient dimer sites as trapping centers was discussed by Pope & Kallmann (214; 1, pp. 283 et seq) and the existence of such sites was supported by the work of Thomas et al (215) and was also discussed by Lange and Bässler (199) and, at great length, by Williams & Thomas (216; 1, pp. 50–53). More recently, Zboinski (217) has examined the possibility that a deep trap can be created by the localization of a carrier at a pair of approximately parallel anthracene molecules.

A novel explanation of what can account for apparently deep (~ 0.7 eV) traps in pure molecular crystals has been presented by Petelenz (218). In pure crystals, one can account for shallow (~ 0.3 eV) traps by the increase in polarization energy surrounding a defect in the crystal lattice. However, one cannot use lattice distortions to explain a trap of depth ~ 0.7 eV. Petelenz points out that the detrapping of a trapped electron involves not only supplying the binding energy E_t of the trap, but the electron must transfer itself away from the trap in a time short compared to the vibrational relaxation time of the excess electron. This requirement is the same as that for autoionization of a neutral state. In a sample calculation, Petelenz showed that detrapping could be accomplished by exciting vibrational degrees of freedom, rather than the carrier itself. He also showed that for a trap only 10^{-3} eV deep by virtue of lattice distortion, the apparent detrapping energy could be ≥ 0.75 eV due to the requirement of exciting effective vibrational modes.

Field Effects

One of the many differences between 3D systems and those of lesser dimensionality (1, p. 617) is the electric field-dependent reduction in the charge carrier trapping time in the 1 and 2D systems. In a 3D system, trapping is generally a first-order process with a time-independent rate constant. This is a consequence of the trapping probability increasing linearly with the number of new sites visited. The number of new sites visited by a randomly diffusing particle varies as n in 3D, $n/\ln(n)$ in 2D, and $n^{1/2}$ in 1D, where n is the number of steps taken (1, p. 122). If the hopping rate is constant, then the trapping rate will become time dependent in 2D and 1D motion. Movaghar et al (219, 220) showed analytically that in a 1D system, at long times, the relaxation of an excitation follows an $\exp[-(t/\tau)^{1/3}]$ law. This law was shown to apply to 1D trapping kinetics by Hunt et al (221) in a disordered polydiacetylene polymer, which shall be referred to as PDA-10H; this polymer contains a pendant CH_2OH group in every repeat unit of four carbon lengths. This paper reports the first measurement of a 1D relaxation law. The specific phenomenon is the decay of a transient photocurrent pulse in the PDA-10H in the time range 1 to 2 $\times 10^4$ s. Although the more widely studied PTS has an exceedingly low dislocation density (222), the concentration of deep traps is ~ 1 per mm of polymer chain (131) in PDA-10H; due to the presence of extensive hydrogen bonding between the CH_2OH groups, there is a closer packing and greater internal strain in the final crystal, which is not as perfect as the PTS single crystals. A sample was excited by a square wave pulse of light and it was found that the decay follows an $\exp(-bt^{1/3})$ law.

In the presence of an electric field, the motion of the particle will be anisotropic and thus, in a 1D system, more new sites are likely to be visited than would otherwise be encountered. This implies that the trapping rate will increase with field strength. This effect was first observed by Haarer & Möhwald (223; 1, p. 616), in the charge-transfer complex phenanthrene-pyromellitic acid dianhydride. An exact solution of the time dependence of the trapping probability in the presence of an electric field F was derived by B. Movaghar, B. Pohlmann, and D. Würtz (MPW) (unpublished). They showed that as $t \to \infty$, a simple exponential law should be approached, the free carrier concentration decay time going as F^{-2} below a critical field F_c, and as F^{-1} above F_c. This theory was tested by Seiferheld et al (194), who used PTS; this polymer is an excellent model for a 1D crystal because its mobility anisotropy ratio $\mu_\parallel/\mu_\perp$ is $\sim 10^5$, where $\mu_\parallel$ is the mobility along the chain and $\mu_\perp$ is the mobility transverse to the chain direction. The PTS polymer is also unusual because it contains essentially no recombination centers for carriers (131), so the decay of a transient carrier population is governed by the kinetics of hopping and detrapping from traps of ~ 0.7 eV and discharge at the electrodes. In this experiment, traps and recombination centers were produced in a PTS crystal by bombardment with 100 KeV He$^+$ ions. It was found that at long times, and for fields of $\leq 1.7 \times 10^3$ V cm^{-1}, the current relaxation went as $\exp[-(t/\tau_1)^{1/3}]$. As the field increased, the long time relaxation curve became steeper, approaching a simple exponential, $\exp[-(t/\tau_2)]$. Moreover, τ_2 varied as F^{-2} below $F_c \approx 10^4$ V cm^{-1} and as F^{-1} above F_c. The effective carrier jump rate W can be calculated from τ_1, giving ~ 1 s^{-1}; this implies that the trap depth is ≈ 0.8 eV.

SUPERCONDUCTIVITY

The dramatic discovery of superconductivity in organic crystals was made by Bechgaard et al (225) and Jerome et al (226) in a family of isostructural compounds of the general formula (TMTSF)$_2$X (also referred to as Bechgaard salts) where X is an anion (ClO$_4^-$, TaF$_6^-$, AsF$_6^-$, SbF$_6^-$, and ReO$_4^-$) and the cation is tetramethyltetraseleno-fulvalene. Only the ClO$_4^-$ derivative exhibits superconductivity at atmospheric pressure; this occurs at a critical temperature $T_c \approx 1$ K. The other compounds all require the application of external pressure (8–12 kbar). Reviews of this field have been written by Jerome & Schulz (227) and by Friedel & Jerome (228). In addition, the proceedings of two conferences are particularly useful (229).

In the search for a general mechanism for superconductivity in organic materials, attention has been focused on the unit cell volume V_c and on the interstack Se–Se distances (230). As has been found to be the case with the highly conducting ion-radical salts of the TTF-TCNQ class (1, pp. 581–91), the nearly planar, almost parallel TMTSF molecules are arranged in stacks along the a crystal axis (1, p. 638). In addition, the TMTSF molecules in neighboring stacks are sufficiently close to each other along the b direction so that infinite sheets are formed in the ab plane, separated by columns of anions. The short interstack, and intrastack Se–Se distances ($d < 4.0$ Å; this is less than the van der Waals radius sum for Se–Se) lead to strong interactions that result in the high conductivity in these materials. As the temperature is lowered from 298 to 125 K, anisotropic structural changes take place in which the interstack distances can decrease almost twice as much as the intrastack distances (231). A linear correlation was found between the unit cell volume, V_c, and the average interstack Se–Se distance (d_{avg}); this is shown in Figure 4. The correlation shows ClO_4^-, FSO_3^-, and BF_4^- salts clustering around the minimum values for V_c and d_{avg}, with almost identical values for d_{avg}. This suggests that these compounds will have a similar Se atom network geometry and similar low temperature electrical properties barring the onset of anion ordering (232); such ordering introduces a new crystal symmetry, and is considered to be a prerequisite for the attainment of superconductivity in $(TMTSF)_2$ ClO_4 (233), in direct opposition to previous beliefs. In addition, those compounds requiring external pressure to achieve superconductivity have d_{avg} values greater than $d_{avg}(ClO_4^-)$. One may therefore venture to predict the anion size most likely to produce the desired V_{cp} (V_{cp} = predicted cell volume V_c). These results suggest that good results should be obtainable with those anions whose sizes are comparable to ClO_4^-; these include not only BF_4^-

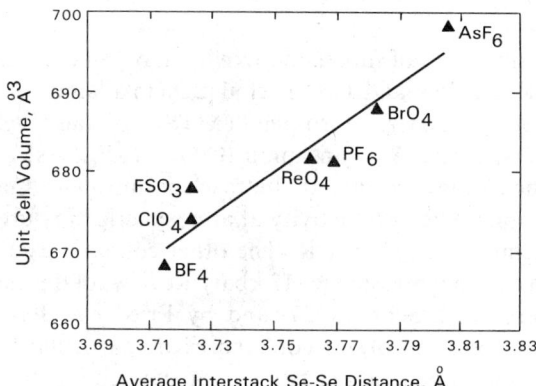

Figure 4 Plot of observed unit cell volume (V_c) vs the average interstack Se–Se distance for various $(TMTSF)_2X$ metals at 125 K. From Williams et al (231).

and FSO_3^-, but anion alloys embodying all possible combinations of these two anions.

Most organic molecules in crystals that exhibit metallic conductivity are planar, of D_{2h} symmetry, and pack in parallel layers in a stack, along which the conductivity is a maximum (1, pp. 588 et seq). In this quasi-1D crystal, it would therefore be expected that there would be intrastack bonding in this direction; this expectation is supported by the observation that in the organic metal TTF-TCNQ, the TCNQ-TCNQ distances are ≈ 3.17 Å as compared with 3.45 Å in the pure TCNQ crystal. In the case of the superconductors in the $(TMTSF)_2X$ family, it has been concluded that Se–Se bonding within a stack and between stacks creates a quasi-2D structure, which is chiefly responsible for the high conductivity. This conclusion was tested by making X-ray diffraction studies on perfect crystals of $(TMTSF)_2AsF_6$ prepared by Wudl (234). An accumulation of electron density was found above and below the TMTSF molecular plane in the region between the Se atoms, implying that there is bonding between the TMTSF molecules within the stack, probably as a result of the Se atom interactions. In addition, there was considerable electron density between Se atoms at the edge of adjacent stacks in the stacking direction. This could be evidence of a one-dimensional conduction band. In view of the substantial conductivity observed in the direction (b axis) perpendicular to the stacking axis (a) of the TMTSF molecules, electron density was looked for between Se atoms in neighboring stacks. An electron density was found between the Se atoms at the edge of adjacent stacks, but surprisingly, only between Se atoms with the longest interstack distance, ~ 4.15 Å. This was predicted by Grant (235). The continuum of electron density from Se to Se between the stacks may represent a conduction band. Thus, there exists a bonding originating from the Se atoms, between molecules within and between the stacks of $(TMTSF)_2AsF_6$ and $(TMTSF)_2PF_6$.

In the search for other superconducting compounds a new, and only the second, family of organic conductors was discovered (235a) based on the substitution of S for Se in the cation. The newly discovered compound is called $(BEDT-TTF)_4 (ReO_4)_2$ where the cation is bis(ethylenedithiolo)-tetrathiafulvalene, and the anion is the perrhenate ion. This compound becomes superconductive above 4 kbar for $T_c \approx 2$ K. A metal-insulator transition takes place at a pressure <7 kbar, and it may be associated with an anion rearrangement. These BEDT-salts have a variety of crystal structures and stoichiometries (236), in distinction to the Bechgaard salts, so it may be easier to locate the origin of superconductivity in these materials. These discoveries are particularly valuable because the mechanism for organic superconductivity is still poorly understood. One arduously sought for goal is to find a compound that has a high T_c.

MISCELLANEOUS

In this section, we would like to mention work that cannot be adequately discussed due to space limitations.

An example of the power of picosecond spectroscopy to unravel mechanisms of chemical and biological reactions is given in a review article by Rentzepis (237). In the wings stands the even more potent tool of femtosecond spectroscopy, as outlined by Shank & Greene (238). Optical pulses as short as 30 fs have been attained, which is shorter than some vibrational periods. One can anticipate the use of such pulses to follow the evolution of energy transfer in a coherent fashion between two degenerate states. Another impressive experiment showing the fission of a singly excited state into multiple excitonic states from which quantum yields and autoionization efficiencies are obtained has been carried out by Klein (239).

A discussion of electroluminescence in organic crystals is given by Kalinowski (240); the effect of pressure and temperature on the luminescence of tetracene single crystals was also studied by Kalinowski et al (241). In this latter study, a surprising feature of earlier work on tetracene was reexamined; this was the observation of an unusually large Stokes red shift (~ 500 cm^{-1}) between the 0–0 fluorescence and absorption transitions, whereas in anthracene it is 100–200 cm^{-1}. It now appears that this shift is an artifact caused by the compensation of the red shift resulting from the temperature modification of the exciton state and a blue shift caused by decreased overlap of the fluorescence and absorption spectrum. The Stokes shift of the reabsorption free 0–0 transition at 528 nm is 260 cm^{-1}, in good agreement with the value 280 cm^{-1} found at 4.2 K (242).

A recent review of the important subject of energy transfer has been given by Klöpffer (243). This review discusses basic concepts, measuring techniques, and results, mainly in polymeric systems. In the same book is an excellent discussion of triboelectricity in organic materials (244). This subject has not yet been put on a sound theoretical footing because of the enormous experimental difficulties in creating reproducible contacting surfaces. Important insights have been provided by Duke (245) and Duke & Fabish (246). Triboelectricity and triboluminescence may one day be studied in a more straightforward manner in outer space. Triboluminescence is thought to arise from the creation and annihilation of mobile cracks (247, 248), which in piezoelectric crystals can result in the creation of oppositely charged neighboring surfaces. The intense electric field at the tips of the cracks may facilitate charge recombination. In Mort & Pfister's book, a review of piezoelectricity and pyroelectricity by Wada (249) brings the field up to date from the last reviews prepared by Kepler (250) and Davies (251).

In the first of a series of photoconductivity experiments on PDA-TS crystals, Donovan & Wilson (131) found that the low field mobility was high, $\mu > 2 \times 10^5$ cm^2 V^{-1} s^{-1} and that the drift velocity saturated at a low value $v_s = 2 \times 10^5$ cm s^{-1} even for fields down to 1 V cm^{-1}. On the other hand, Spannring & Bässler (252) measured SCLC in PDA-DCH using ohmic electrodes; they found that $J \propto V^2$ for $F < 26$ V cm^{-1}, $v_d \propto F$, and $\mu = 6 \times 10^3$ cm^2 V^{-1} s^{-1}. This large discrepancy has been removed by the development of a SCLC theory for 1D materials. The conclusions are as follows:

1. A trap-limited SCLC shows a $J \propto V^2$ dependence even if v_s is saturated.
2. If F increases, then for some critical value F_c if $F > F_c$ the SCLC will become trap-free instead of trap-limited if v_d is saturated.
3. A trap-free SCLC in a 1D material shows a $J \propto V$ dependence if v_d is saturated.

With these findings, the discrepancies between the Donovan & Wilson results and those of Spannring & Bässler can be reconciled. The properly interpreted data of Spannring & Bässler yield a calculated μ for PDA-DCH of 1.6×10^5 cm^2 V^{-1} s^{-1}, very similar to that in PDA-TS.

CONCLUSIONS

This has been an active period for this field and promises to become even more so. The enormous skill of the organic chemist is being harnessed for the creation of a cornucopia of compounds with novel electronic properties. The discovery of superconductivity in more than one type of ion-radical salt greatly increases the prospects of determining the mechanism(s) of superconductivity, and hence of synthesizing compounds of higher transition temperature. The synthesis of compounds that behave as quasi 1- and 2D materials has provided a field day for theorists who can now find exact solutions to transport problems. The use of computer simulations has assumed major proportions,[1] and is already dominating fields such as amorphous solids, in which transport takes place by hopping. The continued development of ultra-short laser pulses of precisely defined wavelength has made possible the excitation of specific vibrational modes, and the study of their intrinsic relaxation rates; homogeneous linewidths are being measured and the mechanisms of line broadening elaborated.

Carrier generation in the homomolecular polyacenes has become much

---

[1] Journals devoted to a discussion of simulation techniques include *Mathematics and Computers in Simulation*, published by North Holland, and *Simulation*, published by Society of Computer Simulation.

better understood with the recognition of the importance of direct optical excitation of charge-separated (CT) states, and of precursors to CT states. It was certainly satisfying to view the evidence that at least at low temperatures, all carrier transport processes in anthracene and undoubtedly in essentially all of the polyacenes, is understandable in terms of a band theory of mobility. There is still the problem of the almost zero temperature dependence of electron mobility in the c' direction in anthracene, but this should give way before the next review of this field. The development of a generalized master equation (GME) approach to the study of exciton transport has revealed instances in which significant errors have been made in interpreting experimental data. The study of carrier recombination was enlivened by the discovery of novel high field effects that point to the existence of a field sensitive process for producing CT states, and by indications that the evaluation of the thermalization distance and the initial ionization yield from Onsager theory is a more delicate operation than previously thought. This latter conclusion followed upon the development of analytical and computer simulation techniques for following the recombination of geminate carriers on a discrete lattice.

The surface has not been scratched in the study of electronic processes in organic solids.

ACKNOWLEDGMENTS

One of us (M. P.) wishes to acknowledge support of the Department of Energy. We have benefitted from correspondence with H. Bässler, C. B. Duke, A. J. Epstein, and W. Siebrand. We express our appreciation to Ms. M. Menzel and Ms. A. Lunsford for their cheerful and indefatigable typing effort.

Literature Cited

1. Pope, M., Swenberg, C. E. 1982. *Electronic Processes in Organic Crystals.* New York: Clarendon Press, Oxford. 821 pp.
2. Mort, J., Pfister, G., eds. 1982. *Electronic Properties of Polymers.* New York: Wiley
3. Haddon, R. C., Kaplan, M. L., Wudl, F. 1982. In *Kirk-Othmer Encyclopedia of Chemical Technology,* 20:674–98. New York: Wiley. 3rd ed.
4. Adler, G., ed. 1983. *Proc. 6th Int. Conf. on Chem. of Organic Solid State. Mol. Cryst. Liquid Cryst.,* Vols. 93, 96
5. Duke, C. B. 1982. *Festkorperprobleme* 22:21–34
6. Duke, C. B. 1982. In *Extended Linear Chain Compounds,* ed. J. S. Miller, 2:59–125. New York: Plenum
7. Ford, W. K., Duke, C. B. 1983. See Ref. 4, 93:327–54
8. Duke, C. B. 1983. In *Electronic Excitations and Interaction. Processes in Organic Molecular Aggregates,* ed. P. Reineker, H. Haken, H. C. Wolf, 49:14–19. Berlin/Heidelberg/New York: Springer-Verlag
9. Sloan, G. J., McGhie, A. R. 1984. *Melt*

Crystallization Techniques. New York: Wiley
10. Karl, N. 1980. *Crystals: Growth Properties and Applications*, pp. 1–100. Berlin: Springer-Verlag
11. Smalley, R. E. 1983. *Ann. Rev. Phys. Chem.* 34:129–53
12. Fayer, M. D. 1982. *Ann. Rev. Phys. Chem.* 33:63–87
13. Drickamer, H. G. 1982. *Ann. Rev. Phys. Chem.* 33:25–47
14. Berry, D. E., Tompkins, R. C., Williams, F. 1982. *J. Chem. Phys.* 76:3362–70
15. Callis, P. R. 1983. *Ann. Rev. Phys. Chem.* 34:329–57
15a. Hanson, D. M. 1983. In *Molecular Electronic Devices*, ed. F. L. Carter, pp. 89–111. New York: Dekker
15b. Hanson, D. M., Patel, J. S., Winkler, I. C., Morrobel-Sosa, N. 1983. In *Spectroscopy and Excitation Dynamics of Condensed Molecular Systems*, ed. V. M. Agranovich, R. M. Hochstrasser, pp. 621–79. Amsterdam: North Holland
16. Johnson, C. K., Small, G. J. 1982. In *Excited States*, ed. E. C. Lim, 6:97–216. New York: Academic
17. Turlet, J. M., Kottis, P., Philpott, M. R. 1983. *Adv. Chem. Phys.* 54:303–468
18. Agranovich, V. M., Hochstrasser, R. M. 1983. *Spectroscopy and Excitation Dynamics of Condensed Molecular Systems*. Amsterdam: North Holland
19. Agranovich, V. M. 1982. *Electronic Excitation and Energy Transfer*. Amsterdam: North Holland
20. Knox, R. S. 1983. In *Collective Excitations in Solids*, ed. B. DiBartolo, pp. 183–245. New York: Plenum Press
20a. Aaviskoo, J., Liidja, G., Saari, P. 1982. *Phys. Status Solidi B* 110:69–73
20b. Saari, P., Rebane, K. 1981. *J. Phys. Colloq.* 42(C6):502–4
21. Powell, R. C., Soos, Z. G. 1975. *J. Lumin.* 11:1–45
22. Kenkre, V. M., Schmid, D. 1983. *Chem. Phys. Lett.* 94:603–8
23. Ern, V., Avakian, P., Merrifield, R. E. 1966. *Phys. Rev.* 148:862–67
24. Kenkre, V. M., Fort, A., Ern, V. 1983. *Chem. Phys. Lett.* 96:658–66
25. Kenkre, V. M. 1982. In *Exciton Dynamics in Molecular Crystals and Aggregates*, ed. G. Hohler. Berlin: Springer-Verlag
26. Kopelman, R. 1976. In *Topics in Applied Physics*, ed. F. K. Fong, 15:297–346. Berlin: Springer-Verlag
27. Kopelman, R., Monberg, E. M., Ochs, F. W., Prasad, P. 1975. *Phys. Rev. Lett.* 34:1506–9
28. Deleted in proof
29. Blumen, A., Silbey, R. 1979. *J. Chem. Phys.* 70:3707–14
30. Loring, R. F., Fayer, M. D. 1982. *Chem. Phys.* 70:139–47
31. Gentry, S. T., Kopelman, R. 1982. *Chem. Phys. Lett.* 93:264–66
32. Parson, R. P., Kopelman, R. 1982. *Chem. Phys. Lett.* 87:528–32
33. Gentry, S. T., Kopelman, R. 1983. *J. Chem. Phys.* 78:373–82
34. Monberg, E. M., Kopelman, R. 1980. *Mol. Cryst. Liquid Cryst.* 57:271–312
35. Brown, R., Lemaistre, J. P., Megel, J., Pee, P., Dupuy, F., Kottis, P. 1982. *J. Chem. Phys.* 76:5719–26
36. Gentry, S. T., Kopelman, R. 1983. *Phys. Rev. B* 27:2579–82
37. Prasad, P. M., Morgan, J. R., El-Sayed, M. A. 1981. *J. Phys. Chem.* 85:3569–71
38. Morgan, J. R., El-Sayed, M. A. 1983. *J. Phys. Chem.* 87:383–85
39. Morgan, J. R., El-Sayed, M. A. 1983. *J. Phys. Chem.* 87:2178–85
40. Hatlee, M. D., Kozak, J. J. 1981. *Phys. Rev. B* 23:1713–18
41. Arnold, S., Alfano, R. R., Pope, M., Yo, H. P., Selsby, R., Tharrats, J., Swenberg, C. E. 1976. *J. Chem. Phys.* 64:5104–14
42. Klymko, P. W., Kopelman, R. 1982. *J. Phys. Chem.* 86:3686–88
43. Blumen, A., Klafter, J., Silbey, R. 1980. *J. Chem. Phys.* 72:5320–32
44. Klymko, P. W., Kopelman, R. 1983. *J. Phys. Chem.* 87:4565–67
45. Alexander, S., Orbach, R. 1982. *J. Phys. Lett.* 43:L625–L631
46. DeGennes, P. G. 1983. *C.R. Acad. Sci. Paris* 296(2):881–85
47. Schilling, R., Mattis, D. C. 1982. *Phys. Rev. Lett.* 49:808–11
48. Schilling, R., Mattis, D. C. 1983. *Phys. Rev. B* 27:3318–23
49. Thomas, G. A., Rice, T. M. 1977. *Solid State Commun.* 23:359–63
50. Singh, J. 1981. *Phys. Status Solidi B* 1036:423–28
51. Singh, J. 1981. *J. Chem. Phys.* 75:4603–11
52. Gumbs, G., Mavroyannis, C. 1982. *Solid State Commun.* 41:237–40
53. Sanche, L., Bader, G. B., Caron, L. 1982. *J. Chem. Phys.* 76:4017–27; 77:3166–70
54. Agranovich, V. M., Zakhidov, A. A. 1979. *Chem. Phys. Lett.* 68:86–89
55. Sugakov, V. I. 1979. *Sov. Phys. Solid State* 21:332–35
56. Brikenshtein, V. Kh., Benderskii, V. A., Filippov, P. G. 1983. *Phys. Status Solidi B* 117:9–39

57. Wolf, H. C. 1967. In *Advances in Atomic and Molecular Physics*, Vol. 3. New York: Academic
58. Braun, A., Mayer, U., Auweter, H., Wolf, H. C., Schmid, D. 1982. *Z. Naturforsch. Teil A* 37:1013–1023
59. Campillo, A. J., Shapiro, S. L., Swenberg, C. E. 1977. *Chem. Phys. Lett.* 52:11–15
60. Al-Obaidi, S. J., Birks, J. B., Birch, D. J. S., Hallam, A., Imhoff, R. E. 1978. *J. Phys. B* 11:2301–11
61. Davies, M. J., Jones, A. C., Williams, J. O., Munn, R. W. 1983. *J. Phys. Chem.* 87:541–43
62. Shinohara, H., Kotani, M. 1980. *Bull Chem. Soc. Jpn.* 53:3171–75
63. Meyer, K. E., Schein, L. B., Anderson, R. W., Narang, R. S., McGhie, A. R. 1983. *Mol. Cryst. Liquid Cryst.* 101:199–218
64. Crisp, G. M., Walmsley, S. H. 1982. *Chem. Phys.* 68:213–22
65. Huber, D. L. 1980. *Phys. Rev. B* 22:1714–21; 24:1083–86
66. Kenkre, V. M., Wong, Y. M. 1981. *Phys. Rev. B* 23:3748–55
67. Lakatos-Lindenberg, K., Hemenger, R. P., Pearlstein, R. M. 1972. *J. Chem. Phys.* 56:4852–67
68. Kenkre, V. M. 1982. *Chem. Phys. Lett.* 93:260–63
69. Kenkre, V. M., Parris, P. E. 1983. *Phys. Rev. B* 27:3221–34
69a. Khizhnyakov, V. V., Sherman, A. V. 1980. *Sov. Phys. Sol. State* 32:1904–9
70. Toyozawa, Y., Shinozuka, Y. 1980. *J. Phys. Soc. Jpn.* 48:472–78
71. Mayer, U., Auweter, H., Braun, A., Wolf, H. C., Schmid, D. 1981. *Chem. Phys.* 59:449–65
72. Ludmer, Z., Berkovic, G. E., Zeiri, L., Muhle, W., Wolf, H. C. 1982. *Chem. Phys. Lett.* 90:245–46
73. Von Freydorf, E., Kinder, J., Michel-Beyerle, M. E. 1978. *Chem. Phys.* 27:199–209
74. Matsui, A., Nishimura, H. 1982. *J. Phys. Soc. Jpn.* 51:1711–12
75. Matsui, A., Mizuno, K., Iemura, M. 1982. *J. Phys. Soc. Jpn.* 51:1871–77
76. Berk, N. F., Rosenthal, J., Yarmus, L. 1983. *Phys. Rev. B* 28:4963–69
77. Szabo, A. 1971. *Phys. Rev. Lett.* 27:323–26
78. DeVries, H., Wiersma, D. A. 1976. *Phys. Rev. Lett.* 36:91–94
79. Treshchalov, A. B., Rozman, M. G. 1983. *Opt. Commun.* 47:262–67
80. Olson, R. W., Patterson, F. G., Lee, H. W. H., Fayer, M. D. 1981. *Chem. Phys. Lett.* 77:403–7
81. Hessenlink, W. H., Wiersma, D. A. 1978. *Chem. Phys. Lett.* 50:51–56
82. Amirav, A., Even, O., Jortner, J. 1982. *J. Phys. Chem.* 86:3345–58
82a. Godzik, K., Hays, T. R., Henkre, W. E., Selzle, H. L., Schlag, E. W. 1982. In *Laser Chemistry* 1:59–75. London: Harwood Headem Press
83. Small, G. J. 1982. In *Molecular Spectroscopy; volume of Modern Problems in Solid State Physics*, ed. V. M. Agranovich, A. A. Maradudin. Amsterdam: North-Holland
84. Jankowiak, R., Bässler, H. 1983. *Chem. Phys. Lett.* 95:124–28
85. Jankowiak, R., Bässler, H. 1983. *Chem. Phys. Lett.* 95:310–14
86. Jankowiak, R., Bässler, H. 1983. *Chem. Phys. Lett.* 101:274–78
87. Olson, R. W., Lee, H. W. H., Patterson, F. G., Fayer, M. D., Shelby, R. M., Burum, D. P., Macfarlane, R. M. 1982. *J. Chem. Phys.* 77:2283–89
87a. Hess, L. A., Prasad, P. N. 1983. *J. Chem. Phys.* 78:626–31
88. Reineker, R., Morawitz, H. 1982. *Chem. Phys. Lett.* 86:359–64
89. Deleted in proof
90. Freiberg, A., Saari, P. 1983. *IEEE J. Quantum Electron.* QE19:622–30
91. Wiesenfeld, J. M., Greene, B. I. 1983. *Phys. Rev. Lett.* 51:1754–58
92. Koch, E. E., Gürtler, P. 1983. In *Photophysics and Photochemistry in the Vacuum UV*, ed. S. P. McGlynn, G. L. Findley, R. H. Huebner. Dordrecht, Holland: Reidel
93. Hoffman, B. M., Ibero, J. A. 1983. *Acc. Chem. Res.* 16:15–21
94. Zagrubskii, A. A., Petrov, V. V., Lovcjus, V. A. 1981. *Izv. Akad. Nauk. Lat. SSR Fiz. Tver. Tela.* 5:29–33
95. Ford, W. K., Duke, C. B., Patson, A. 1982. *J. Chem. Phys.* 77:4564–72
96. Ford, W. K., Duke, C. B., Salaneck, W. R. 1982. *J. Chem. Phys.* 77:5030–39
97. Duke, C. B., Ford, W. K. 1983. *Int. J. Quant. Chem.* 17:597–608
97a. Ladik, J. 1981. In *Theoretical Chemistry*, ed. C. Thomson, 4:49-ff. London: R. Soc. Chem. London/Butterworths
98. Rice, S. A., Jortner, J. 1967. In *Physics and Chemistry of the Organic State*, ed. D. Fox, M. M. Labes, A. Weissberger, 3:201–497. New York: Wiley
99. Deleted in proof
100. Pope, M., Kallmann, H., Giachino, J. 1965. *J. Chem. Phys.* 42:2540–43
101. Vilesov, F. I., Zagrubskii, A. A., Garbuzov, D. Z. 1964. *Sov. Phys. Sol. State* 5:1460–64

102. Bounds, P. J., Siebrand, W. 1980. *Chem. Phys. Lett.* 75:414–18
103. Bounds, P. J., Siebrand, W. 1982. *Chem. Phys. Lett.* 85:496–98
104. Bounds, P. J., Petelenz, P., Siebrand, W. 1981. *Chem. Phys.* 63:303–20
105. Bounds, P. J., Petelenz, P., Siebrand, W. 1982. *Chem. Phys. Lett.* 89:1–3
106. Chance, R. R., Braun, C. L. 1976. *J. Chem. Phys.* 64:3573–81
107. Kato, K., Braun, C. L. 1976. *J. Chem. Phys.* 72:172–76
108. Sebastian, L., Weiser, G., Bässler, H. 1981. *Chem. Phys.* 61:125–35
109. Sebastian, L., Weiser, G., Peter, G., Bässler, H. 1983. *Chem. Phys.* 75:103–14
109a. Siebrand, W., Zgierski, M. Z. 1983. In *Electronic Excitations and Interaction Processes in Organic Molecular Aggregates*, ed. P. Reineker, H. Haken, H. C. Wolf, pp. 136–44. Berlin: Springer-Verlag
110. Hinchliffe, A., Munn, R. W., Siebrand, W. 1983. *J. Phys. Chem.* 87:3837–39
111. Silinsh, E. A., Kolesnikov, V. A., Muzikante, I. J., Balode, D. R. 1982. *Phys. Status Solidi* 113b:379–93
112. Popovic, Z. D., Sharp, J. H. 1977. *J. Chem. Phys.* 66:5076–82
113. Eichhorn, M., Willig, F., Charle, K. P., Bitterling, K. 1982. *J. Chem. Phys.* 76:4648–56
114. Eichhorn, M. 1982. PhD thesis. Technical Univ., Berlin
115. Slotnick, K. 1983. PhD thesis. New York: Univ., New York
116. Onsager, L. 1934. *J. Chem. Phys.* 2:599–615
117. Onsager, L. 1938. *Phys. Rev.* 54:554–57
118. Chance, R. R., Braun, C. L. 1976. *J. Chem. Phys.* 64:3573–81
119. Hong, K. M., Noolandi, J. 1978. *J. Chem. Phys.* 69:5026–39
120. Noolandi, J., Hong, K. M. 1979. *J. Chem. Phys.* 70:3230–36
121. Scher, H., Rackovsky, S. 1984. *J. Chem. Phys.* In press
121a. Rackovsky, S., Scher, H. 1984. *Phys. Rev. Lett.* 52:453–56
122. Ries, B., Schönherr, G., Bässler, H., Silver, M. 1983. *Philos. Mag.* 48:87–106
123. Seiferheld, U., Ries, B., Bässler, H. 1983. *J. Phys. C Solid State Phys.* 16:5189–5201
124. Deleted in proof
125. Lochner, K., Reimer, B., Bässler, H. 1976. *Chem. Phys. Lett.* 41:388–90
126. Siddiqui, A. S., Wilson, E. G. 1979. *J. Phys. C Solid State Phys.* 12:4237–43
127. Haberkorn, R., Michel-Beyerle, M. E. 1973. *Chem. Phys. Lett.* 23:128–30
128. Sebastian, L., Weiser, G. 1981. *Phys. Rev. Lett.* 46:1156–59
129. Movaghar, B., Cade, N. A. 1982. *J. Phys. C Solid State Phys.* 15:L807–L813
130. Cade, N. A., Movaghar, B. 1983. *J. Phys. C Solid State Phys.* 16:539–50
131. Donovan, K., Wilson, E. G. 1981. *Philos. Mag. B* 44:9–29
132. Spannring, W., Bässler, H. 1979. *Ber. Bunsenges Phys. Chem.* 83:433–36
133. Movaghar, B., Murray, D., Donovan, K. J., Wilson, E. G. 1984. *J. Phys. C* 17:1247–55
134. Geacintov, N. E., Pope, M. 1971. In *Proc. 3rd Int. Conf. Photoconductivity*, ed. E. M. Pell, pp. 289–95. New York: Pergamon
135. Orlowski, T. E., Scher, H. 1983. *Phys. Rev. B* 27:7691–7702
136. Scher, H., Orlowski, T. E. 1983. *Phys. Rev. Lett.* 50:775–78
137. Jortner, J., Choi, S. I., Katz, J. L., Rice, S. A. 1963. *Phys. Rev. Lett.* 11:323–26
138. Petelenz, P. 1980. *Chem. Phys. Lett.* 76:186–89
139. Yokoyama, M., Matsubara, A., Shimiokihara, S., Mikawa, H. 1982. *Polymer J.* 14:73–75
140. Popovic, Z. D. 1983. *J. Chem. Phys.* 78:1552–58
141. Silinsh, E. A., Kolesnikov, V. A., Muzikante, I. J., Balode, D. R., Gailis, A. K. 1981. *Izv. Akad. Nauk. Latv. SSR Ser. Fiz. Tekhn. Nauk.* 5:14–28
142. Klein, J., Martin, P., Voltz, R. 1981. *J. Lumin.* 24/25:99–102
143. Moller, W., Pope, M. 1973. *J. Chem. Phys.* 59:2760–61
144. Ries, B., Schönherr, G., Bässler, H., Silver, M. 1983. *Philos. Mag. B* 48:87–106
145. Hong, K. M., Noolandi, J., Street, R. A. 1981. *Phys. Rev. B* 23:2967–76
146. Pope, M., Burgos, J. 1966. *Mol. Cryst.* 1:395–415
147. Altwegg, L., Pope, M., Fowlkes, W. 1984. *Chem. Phys.* 86:471–82
148. Fünfschilling, J., Samoc, M., Williams, D. F. 1983. *Chem. Phys. Lett.* 96:157–60
149. Klein, G., Carvalho, M. J. 1977. *Chem. Phys. Lett.* 51:409–12
150. Popovic, Z. D. 1983. *Chem. Phys. Lett.* 100:227–29
151. Altwegg, L., Davidovich, M. A., Fünfschilling, J., Zschokke-Gränacher, I. 1978. *Phys. Lett. Rev. B* 18:4444–53
152. Braun, C. L., Scott, T. W. 1983. *J. Phys. Chem.* 87:4776–78
153. Schein, L. B., Brown, D. W. 1982. *Mol. Cryst. Liquid Cryst.* 87:1–12

154. Schein, L. B., Duke, C. B., McGhie, A. R. 1978. *Phys. Rev. Lett.* 40:197–200
155. Schein, L. B. 1977. *Chem. Phys. Lett.* 48:571–75
156. Schein, L. B., McGhie, A. R. 1979. *Phys. Rev. B* 20:1631–39
157. Katz, J. L., Rice, S. A., Choi, S., Jortner, J. 1963. *J. Chem. Phys.* 39:1683–97
158. Silbey, R., Jortner, J., Rice, S. A., Vala, M. T. Jr. 1965. *J. Chem. Phys.* 42:733–37
159. Sumi, H. 1978. *Solid State Commun.* 28:309–12; 29:495–99
160. Sumi, H. 1979. *J. Chem. Phys.* 70:3775–85
161. Sumi, H. 1979. *J. Chem. Phys.* 71:3403–11
162. Sumi, H. 1979. *J. Chem. Phys.* 75:2987–93
163. Andersen, J. D., Duke, C. B., Kenkre, V. M. 1983. *Phys. Rev. Lett.* 51:2202–5
164. Schein, L. B., Narang, R. S., Anderson, R. W., Meyer, K. E., McGhie, A. R. 1983. *Chem. Phys. Lett.* 100:37–40
165. Warta, W., Karl, N. 1982. *10th Mol. Cryst. Symp.*, St. Jovette, Canada
166. Epstein, A. J., Conwell, E. M., eds. 1981/82. *Proc. Int. Conf. on Low Dimensional Conductors, Mol. Cryst. Liquid Cryst.* Vols. 77, 79, 81, 83, 85, 86
167. *Int. Conf. sur la Phys. Chem. des Polymeres et Conductors,* 1983. *J. Physique* 44:Coll. C-3
168. Etemad, S., Heeger, A. J., MacDiarmid, A. G. 1982. *Ann. Rev. Phys. Chem.* 33:443–69
169. Nigrey, P. J., MacDiarmid, A. G., Heeger, A. J. 1982. *Mol. Cryst. Liquid Cryst.* 81:309–17
170. Robin, P., Pouget, J. P., Comes, R., Gibson, H. W., Epstein, A. J. 1983. *Phys. Rev. B* 27:3938–94
171. Hoffman, D. M., Gibson, H. W., Epstein, A. J., Tanner, D. B. 1983. *Phys. Rev. B* 27:1454–57
172. Francois, B., Bernard, M., Andre, J. J. 1981. *J. Chem. Phys.* 75:4142–52
173. Weinberger, B. R., Ehrenfreund, E., Pron, A., Heeger, A. J., MacDiarmid, A. G. 1983. *J. Chem. Phys.* 72:4749–55
174. Weinberger, B. R., Kaufer, J., Heeger, A. J., Pron, J., MacDiarmid, A. G. 1979. *Phys. Rev. B* 20:223–30
175. Roth, S., Eichinger, K., Menke, K. 1984. In *Quantum Chemistry of Polymers*, ed. J. Ladik, M. André, pp. 165–90. Dordrecht: Reidel
176. Nechtschein, M., Devreux, F., Greene, R. L., Clarke, T. C., Street, G. B. 1980. *Phys. Rev. Lett.* 44:356–59
177. Thomann, H., Dalton, L. R., Tomkiewicz, Y., Shiren, N. S., Clarke, T. C. 1983. *Phys. Rev. Lett.* 50:533–36
178. Kuroda, S., Schrieffer, J. R. 1983. *Solid State Commun.* 43:591–94
179. Heeger, A. J., Schrieffer, J. R. 1983. *Solid State Commun.* 48:207–10
180. Suzuki, N., Ozaki, M., Etemad, S., Heeger, A. J., MacDiarmid, A. G. 1980. *Phys. Rev. Lett.* 45:1209–13
181. Leo, V., Gusman, G., Deltour, R. 1982. *Phys. Rev. R B* 26:3285–88
182. Shank, C. V., Yen, R., Fork, R. L., Orenstein, J., Baker, G. L. 1982. *Phys. Rev. Lett.* 49:1660–63
182a. Su, W. P., Schrieffer, J. R. 1980. *Proc. Natl. Acad. Sci. USA* 77:5626–29
183. Baeriswyl, D. 1983. *Helv. Phys. Acta* 56:639–53
184. Kivelson, S. 1981. *Phys. Rev. Lett.* 47:1549–53; 1982. *Phys. Rev. B* 25:3798–3821
185. Shacklette, L. W., Chance, R. R., Ivory, D. M., Miller, G. G., Baughman, R. H. 1980. *Synth. Metals* 1:307–20
186. Brédas, J. L., Chance, R. R., Silbey, R. 1982. *Phys. Rev. B* 26:5843–54
187. Brédas, J. L., Chance, R. R., Silbey, R. 1982. *Mol. Cryst. Liquid Cryst.* 77:319–32
188. Crecelius, G., Stamm, M., Fink, J., Ritsko, J. J. 1983. *Phys. Rev. Lett.* 50:498–500
189. Brédas, J. L., Themans, B., Andre, J. M. 1983. *Phys. Rev. B* 27:7827–30
190. Fesser, K., Bishop, A. R., Campbell, D. L. 1983. *Phys. Rev. B* 27:4804–25
191. Etemad, S., Feldblum, A., MacDiarmid, A. G., Chung, T. C., Heeger, A. J. 1983. *J. Phys. Coll.* 44(C3):413–22
192. Movaghar, B., Murray, D., Pohlmann, B. Würtz, D. 1984. *J. Phys. C* 17:1677–83
193. Alexander, S., Bernesconi, J., Schneider, W. R. 1981. *Rev. Mod. Phys.* 53(2):175–98
194. Seiferheld, U., Bässler, H., Movaghar, B. 1983. *Phys. Rev. Lett.* 51:813–16
195. Bässler, H. 1981. *Phys. Status Solidi B* 107:9–53
196. Schönherr, G., Bässler, H., Silver, M. 1981. *Philos. Mag. B* 44:369–81
197. Hirsch, J. 1974. *Phys. Status Solidi A* 25:575–80
198. Bässler, H., Schönherr, G., Abkowitz, M., Pai, P. M. 1982. *Phys. Rev. B* 26:3105–13
199. Lange, J., Bässler, H. 1982. *Phys. Status Solidi B* 114:561–69
200. Mort, J., Pfister, G. 1982. In *Electronic Properties of Polymers*, ed. J. Mort, G. Pfister, pp. 215–65. New York: Wiley
201. Salaneck, W. R. 1978. *Phys. Rev. Lett.* 40:60–63
202. Duke, C. B., Fabish, T. J., Paton, A. 1977. *Chem. Phys. Lett.* 49:133–36
203. Silver, M., Schönherr, G., Bässler, H. 1982. *Phys. Rev. Lett.* 48:352–55
204. Taure, L. F., Silinsh, E. A., Muzikante, I. J., Rampens, A. J. 1983. In *Tagungsband*

205. Samoc, M., Samoc, A., Sworakowski, J., Karl, N. 1983. *J. Phys. C Solid State Phys.* 16:171-80
206. Plans, J., Zielinski, M., Kryszewski, M. 1981. *Phys. Rev. B* 23:6557-69
207. Karl, N., Sato, N., Seki, K., Inokuchi, H. 1983. *J. Chem. Phys.* 77:4870-78
208. Yoshie, O., Kamihara, M. 1983. *J. Appl. Phys. Jpn.* 22:629-35
209. Tiedje, T., Rose, A. 1981. *Solid State Commun.* 37:49-52
210. Orenstein, J., Kastner, M. A. 1981. *Solid State Commun.* 40:85-89
211. Monroe, D., Kastner, M. A. 1983. *Philos. Mag. B* 47:605-20
212. Arkhipov, V. I., Popova, J. A., Rudenko, A. I. 1983. *Philos. Mag. B* 48:401-10 and references therein
213. Arnold, S., Hassan, N. 1983. *J. Chem. Phys.* 78:5606-11
214. Pope, M., Kallmann, H. 1972. *Isr. J. Chem.* 10:269-86
215. Thomas, J. M., Evans, E. L., Williams, J. O. 1972. *Proc. R. Soc. Ser. A* 331:417-27
216. Williams, J. O., Thomas, J. M. 1973. In *Surface and Defect Properties of Solids*, pp. 229-49. Vol. 2, Specialist Periodical Report. London: Chemical Society
217. Zboinski, Z. 1983. *Chem. Phys.* 75:297-304
218. Petelenz, P. 1981. *Mat. Sci.* 7:285-90
219. Movaghar, B., Sauer, G. W., Wurtz, D., Huber, D. L. 1981. *Solid State Commun.* 39:1179-82
220. Movaghar, B., Sauer, G. W., Wurtz, D. 1982. *J. Stat. Phys.* 27:473-85
221. Hunt, I., Bloor, D., Movaghar, B. 1983. *J. Phys. C Solid State* 16:2623-28
222. Dudley, M., Sherwood, J. N., Bloor, D., Ando, D. J. 1982. *J. Mat. Sci. Lett.* 1:479-81
223. Haarer, D., Möhwald, H. 1975. *Phys. Rev. Lett.* 34:1447-50
224. Deleted in proof
225. Bechgaard, K., Jacobsen, C. S., Mortensen, K., Pedersen, H. J., Thorup, N. 1980. *Solid State Commun.* 33:1119-25
226. Jerome, D., Mazaud, A., Ribault, M., Bechgaard, K. 1980. *J. Phys. Paris Lett.* 41:L95-L98
227. Jerome, D., Schulz, H. J. 1982. *Adv. Phys.* 31:229-490
228. Friedel, J., Jerome, D. 1982. *Contemp. Phys.* 23:583-624
229. Colloq. Iht. CNR sur la Phys. et la Chem. des Metaux Synthetique et Organiques. 1983. *J. Phys. Colloq.* 44(C.3)
230. Williams, J. M., Beno, M. A., Sullivan, J. C., Banovetz, L. M., Braam, J. M., Blackman, G. S., Carlson, C. D., Greer, D. L., Loesing, D. M. 1983. *J. Am. Chem. Soc.* 105:643-45
231. Williams, J. M., Beno, M. A., Appelman, E. H., Wickel, F., Aharon-Shalom, E., Nalewajek, D. 1982. *Mol. Cryst. Liquid Cryst.* 79:319-26
232. Jacobsen, C. S., Pederson, H. J., Mortensen, K., Rindorf, G., Thorup, N., Torrance, J., Bechgaard, K. 1982. *J. Phys. C* 15:2651-63
233. Pouget, J. P., Shirane, G., Bechgaard, K., Fabre, J. M. 1983. *Phys. Rev. B* 27:5203-6
234. Wudl, F. 1981. *J. Am. Chem. Soc.* 103:7064-69
235. Grant, P. M. 1982. *Phys. Rev. B* 26:6888-95
235a. Parkin, S. S. P., Engler, E. M., Schumacher, R. R., Lagier, R., Lee, V. Y., Scott, J. C., Greene, R. L. 1983. *Phys. Rev. Lett.* 50:270-73
236. Parkin, S. S. P., Engler, E. M., Schumacher, R. R., Lagier, R., Lee, V. Y., Voiron, J., Carneiro, K., Scott, J. C., Greene, R. L. 1983. *J. Phys. Colloq.* 44(C3):791-97
237. Rentzepis, P. M. 1982. *Science* 218:1183-90
238. Shank, C. V., Greene, B. I. 1983. *J. Am. Chem. Soc.* 87:732-34
239. Klein, G. 1983. *Chem. Phys. Lett.* 95:305-9; 97:114-18
240. Kalinowski, J. 1981. *Mat. Sci.* 7:44-50
241. Kalinowski, J., Jankowiak, R., Bässler, H. 1981. *J. Lumin.* 22:397-18
242. Kolendritskii, D. D., Kurik, M. V., Piryatinskii, Yu. P. 1979. *Phys. Status Solidi B* 91:741-51
243. Klöpffer, W. 1982. In *Electronic Properties of Polymers*, ed. J. Mort, G. Pfister, pp. 161-213. New York: Wiley
244. Ritsko, J. J. 1982. In *Electronic Properties of Polymers*, ed. J. Mort, G. Pfister, pp. 13-58. New York: Wiley
245. Duke, C. B. 1983. In *Physicochemical Aspects of Polymer Surfaces*, ed. K. L. Mittle, 1:463-75. New York: Plenum
246. Duke, C. B., Fabish, T. J. 1978. *J. Appl. Phys.* 49:315-21
247. Alzetta, G., Chudacek, I., Scarmozzino, R. 1970. *Phys. Status Solidi* 1:775-85
248. Nowak, R., Krajewska, A., Samoc, M. 1983. *Chem. Phys. Lett.* 94:270-71
249. Wada, Y. 1982. In *Electronic Properties of Polymers*, ed. J. Mort, G. Pfister, pp. 109-60. New York: Wiley
250. Kepler, R. G. 1978. *Ann. Rev. Phys. Chem.* 29:497-18
251. Davies, G. R. 1980. In *Physics of Dielectric Solids*, ed. C. H. L. Goodman, pp. 50-63. Conf. Ser. No. 58. Bristol: Inst. Physics Techno House
252. Spannring, W., Bässler, H. 1981. *Chem. Phys. Lett.* 84:54-58

SELECTIVE EXCITATION STUDIES OF UNIMOLECULAR REACTION DYNAMICS

F. F. Crim

Department of Chemistry, University of Wisconsin, Madison, Wisconsin 53706

INTRODUCTION

Reaction of an isolated, energized molecule is an elementary chemical process whose fundamental and practical importance has motivated extensive experimental and theoretical study for many decades (1–8). The conceptual simplicity of unimolecular reactions makes them particularly attractive for detailed investigation since one can reasonably hope to establish a theoretical framework that is amenable to critical experimental tests. The apparent simplicity of unimolecular reactions belies a subtle process whose complete description is likely to incorporate fundamental notions about the dynamics of intramolecular energy transfer and the potential energy surface for the reacting molecule. Unimolecular processes often play a crucial role in practical chemistry as well. Complex reactions can contain several elementary steps in which an energized molecule, perhaps created in previous steps, decomposes or rearranges to form new reactive species or the final product. These processes are important in a variety of thermal and nonthermal reactions. Two prominent examples are combustion and atmospheric chemistry. Another example of the crucial role unimolecular processes can play arises in infrared multiphoton dissociation (9–14). In this laser-driven process, unimolecular reaction of the absorbing molecule controls the level of excitation by competing with photon absorption.

The possibility of altering the pathways of chemical reactions by selectively exciting a portion of a molecule is an intriguing possibility with obvious chemical consequences. Suitable unimolecular reaction dynamics are essential to realizing such selectivity, since the rate of reaction must favorably compete with the destruction of the initial excitation specificity by intramolecular energy redistribution. Consequently, discussions of unimolecular reaction dynamics, particularly those referring to laser excitation, often focus on the possibility of selective chemistry. However, the fundamental motivation for increasingly detailed studies of unimolecular reactions is the need to understand the chemical dynamics of individual molecules thoroughly. Pursuing this goal is likely to uncover selective unimolecular chemistry or, at least, establish the criteria for such a process, and the successful implementation of selective chemical reactions will certainly signal an improved understanding of unimolecular reaction dynamics.

Statistical theories (15, 16) of unimolecular reactions, which are predicated on the idea that the only important quantity is the total energy content of the molecule and, in some cases, the total angular momentum, enjoy tremendous success in predicting reaction rates in a variety of systems. These theories can also be extended to predict energy disposal in reaction products (17). The time scale for intramolecular energy transfer determines the applicability of statistical theories. Statistical models should be quite applicable to reaction rate calculations, no matter how specific the initial energy deposition, if the excitation energy is indiscriminately spread among the internal modes of a molecule during a time that is short compared to the characteristic reaction time. However, slower energy redistribution invalidates any assumption of intrinsically statistical behavior. Thus, studies of unimolecular reaction dynamics inescapably become involved with the means and consequences of intramolecular energy transfer.

An ideal experimental technique for studying unimolecular reactions would prepare the reactant molecule in a well-characterized and highly selective manner while detecting the rate of product formation in individual quantum states. The excitation scheme at least should be energy selective in order to create the reactants with a narrow distribution of energy. (Because the unimolecular reaction rate is a very strong function of energy content, an experiment that averages over a large range of energies is difficult to use in a definitive comparison with theory.) A monoenergetic site selective excitation scheme that deposits energy in one portion of the molecule is potentially even more informative, while true state selective preparation in which only one or a very few quantum states are initially populated is ideal. The product detection technique should yield the reaction rate of the selectively excited molecule as well as measure the product energy content. Information about energy disposal provides additional insight into the reaction even if statistical behavior completely controls the rate. The distribution of energy among the degrees-of-freedom of the products partly reflects the topology of the potential energy surface on which the reaction occurs. This is most commonly seen as "exit channel" effects, which lead to selective excitation of particular motions in the products. The ideal measurement presents formidable experimental difficulties, and a variety of approaches that incorporate varying degrees of specificity in excitation and detection have provided great insights into unimolecular reaction dynamics. Newly developed techniques promise even more detailed measurements in the future.

This review first provides a survey of a variety of techniques and then describes one of the newer schemes, direct excitation of overtone vibrations, in more detail. A section dealing with overtone vibrations describes their spectroscopy and the experimental considerations for their excitation, and another section discusses recent studies of unimolecular reactions using laser excitation of overtone vibrations in both steady-state and time-resolved experiments.

EXPERIMENTAL APPROACHES

The essential aspect of studies of unimolecular reactions is the technique that deposits the energy required to overcome the barrier to chemical reaction in the molecule. The goal is to produce molecules with a narrow distribution of energy above the barrier, since such preparation leads to the most incisive comparison with theory. Not surprisingly, various techniques offer different degrees of excitation specificity, with highly selective schemes usually being correspondingly inefficient.

Chemical Activation

The overwhelming majority of data on unimolecular reactions comes from chemical activation studies in which an exothermic reaction such as free radical addition creates a highly energized molecule. The enormous body of data generated by this technique has largely shaped current understanding of unimolecular reactions and collisional energy transfer in highly excited polyatomic molecules (15, 16, 18). Chemical activation is potentially a site-selective technique, as elegantly demonstrated by Rabinovitch and co-workers (19–25) in a series of experiments in which highly excited reactants are created by free radical addition to unsaturated hydrocarbons. For example, addition of photolytically generated CH_2 to the double bond in $CF_3CF{=}CF_2$ produces an excited fluoro-alkyl cyclopropane molecule with about 110 kcal mol^{-1} of energy initially localized in the newly formed cyclopropyl ring (21). Rabinovitch has exploited this site selectivity to demonstrate that intramolecular energy redistribution from the initial excitation site occurs in about a picosecond in at least nine different molecules (25, 26).

Statistical models such as Rice-Ramsperger-Kassel-Marcus (RRKM) theory are largely successful in describing chemical activation experiments (15), as one would expect from the demonstrated energy redistribution times in these systems. Rowland and co-workers (27–29) seem to have discovered a notable exception in the decomposition of chemically activated radicals formed by the addition of fluorine atoms to tetraallyl tin and germanium [$M(CH_2CH=CH_2)_4$ where M = Sn or Ge]. The fluorine atom adds at the double bond to create a fluorinated radical with the reaction exothermicity initially localized on a single chain. The decomposition rates inferred from the pressure dependence of the product yield are much faster than an RRKM calculation predicts for such a large molecule and are actually about the same as those observed for radicals that are the size of a single side chain. Rowland and co-workers concluded that the heavy central metal atom blocks the transfer of energy to other allyl groups for as much as a nanosecond and allows the initially excited portion of the molecule to react alone.

A classical trajectory calculation by Lopez & Marcus (30) on a pair of linear three carbon atom chains connected by a heavy central atom indeed shows isolation of energy in one chain for a sufficiently massive central atom and high level of excitation. This "local group mode" behavior is consistent with the experiments on tetraallyl tin and germanium. However, the factors that lead to this restricted behavior are apparently rather subtle. A subsequent measurement by Wrigley & Rabinovitch (31) on radicals formed by H-atom addition to 4-(trimethyl tin) butene-1 [$(CH_3)_3Sn(CH_2)_2CH=CH_2$] showed no localization of the initial energy in the butene moiety on the timescale of the radical decomposition. The excitation levels do not differ drastically between this species and the tetraallyl tin (about 46 and 40 kcal mol^{-1}, respectively), and their barrier heights are similar (about 32 and 29 kcal mol^{-1}) (31). The origin of this behavior may well lie in the detailed dynamics of the excited molecule, and its elucidation merits continued experimental and theoretical work.

Bulk chemical activation experiments are subject to averaging over the initial energy content of the reactants and are unable to provide any information on energy disposal into the products. An extensive series of chemical activation experiments in molecular beams by Lee and co-workers (32, 33) have largely overcome these limitations by using supersonic expansions to prepare precursors with narrow distributions of initial translational and internal energy such that the total energy of the chemically activated molecules is uncertain by less than 5%. Because the decomposition fragments fly unimpeded into a movable mass spectrometer, angularly resolved time-of-flight detection provides the final translational energy and angular distributions of the products. The translational energy release can be compared to predictions of statistical theories, and the angular distribution often reflects the lifetime of the excited reactant. The consequences of potential energy surface topology appear in the translational energy distributions where, for example, an exit channel barrier leads to more energy in translation than a statistical theory would predict in the absence of a barrier (32). Finer dynamical effects that may mirror restricted coupling among vibrational modes and more extensive exit channel interactions can be inferred from the translational energy distributions as well.

Measurements by McDonald and co-workers (34–37) are a different type of molecular beam chemical activation experiment. They have used effusive precursor beams and detected infrared emission from the polyatomic decomposition fragments. An important observation in this work is that the density of vibrational states in the fragments can control the extent of energy redistribution on the millisecond time scale of the observation of infrared emission. Small fragments retain nonrandom distributions of energy, which seem to arise from interactions occurring as the fragments separate. A particularly important comparison (33–35) is possible between the experiments of Lee and co-workers and those of McDonald and co-workers, since they have studied several systems in common using techniques that are sensitive to different aspects of the unimolecular reaction dynamics.

Infrared Multiphoton Excitation

Infrared multiphoton excitation is a process in which a molecule successively absorbs a large number of infrared photons from an intense laser pulse (Figure 1a) (9–14). Sufficiently intense pulses excite the molecule above its dissociation limit, where the competition between photon absorption and unimolecular reaction determines the energy content of the dissociating molecule. As the molecule absorbs more photons, it decays more rapidly because its internal energy increases until eventually it reaches a level at which the decomposition rate is equal to the excitation rate. Most studies of infrared multiphoton excitation are directed toward understanding the excitation process itself, but some do offer insight into unimolecular decomposition dynamics as well.

The crucial aspect of using infrared multiphoton excitation as a preparation technique in selective studies of unimolecular reactions is knowledge of the excitation level. Although the average excitation energy is controlled by the laser fluence, the resulting distribution of energy in the excited molecules precludes infrared multiphoton excitation being a very precisely energy selective technique. A number of studies employing state-resolved and, in some cases, time-resolved product detection reveal important features of infrared multiphoton dissociation (38–53). For example, King & Stephenson (46–48) and Reyner & Hackett (53) have measured the vibrational and rotational temperatures of CF_2 fragments from the decomposition of several chlorofluorocarbons and made estimates of the mean translational energy using laser-induced fluorescence. Wittig and co-workers (49, 50) have studied the unimolecular decomposition of CF_3CN using laser-induced fluorescence to detect the CN and have interpreted the product energy partitioning in terms of the shape of the potential energy surface in the region of the transition state. Hancock and co-workers have investigated this same system using tailored infrared laser pulses with constant intensity and a 200 ns duration (41). The two groups together (52) have shown that the statistical adiabatic channel model (54, 55) nicely explains the different vibrational ($T_v \sim 2000$ K) and rotational ($T_R \sim 1200$ K) temperatures that they observe. In their calculations, the infrared multiphoton-generated distribution of energy in the CF_3CN molecules is taken as being of Gaussian form with a width of about 40 to 50% of the average energy. Thus, despite the lack of specificity in the excitation, these infrared multiphoton decomposition measurements de-

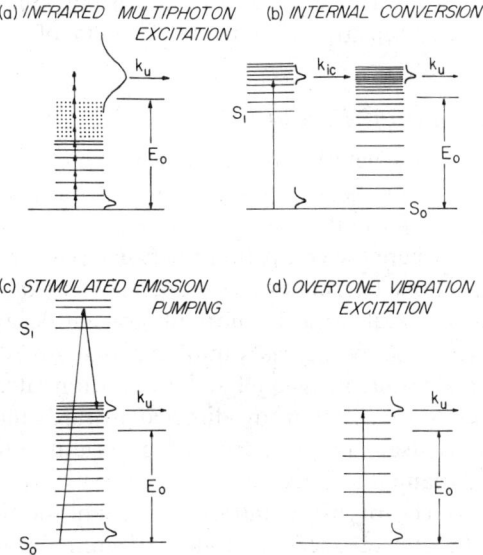

Figure 1 Schematic representation of four techniques for preparing highly vibrationally excited molecules. E_0 is the barrier to unimolecular reaction, and k_u is the unimolecular reaction rate constant. In (*b*), k_{ic} is the internal conversion rate constant.

monstrate the ability of a modified statistical model to explain the inequivalent temperatures of two product degrees-of-freedom.

Molecular beam techniques again provide more detailed information because the reactants are excited and the products are detected in isolation. Lee and co-workers (56–62) have exploited the angular and translational energy resolution in their molecular beam infrared multiphoton dissociation experiments to establish the statistical nature of the unimolecular decomposition of the highly vibrationally excited molecules created by infrared multiphoton absorption. Translational energy distributions for some systems clearly reflect the presence of exit channel energy barriers (58, 60), and mass-resolved detection allows the study of competing channels in unimolecular decomposition (60). For example, Lee and co-workers studied the atomic and molecular chlorine elimination from CF_2Cl_2 (60), which has also been investigated using laser induced fluorescence of CF_2 (46, 47, 53) and molecular beam sampling mass spectrometry (63). The molecular beam study shows that the atomic chlorine elimination accounts for only about 10% of the decomposition and that the translational energy distribution of Cl_2 from the dominant molecular elimination channel has a sharp maximum that arises from an 8 kcal mol^{-1} exit barrier. These results as well as those from similar competitive elimination studies of ethers (61, 62) are consistent with statistical unimolecular decay of the excited molecule, as comparisons with RRKM calculations show.

Another example showing the potential scope of infrared multiphoton excitation studies of unimolecular decomposition is the work of Rockney & Grant (64), who detected both the CH_3 and NO_2 fragments from CH_3NO_2 decomposition using multiphoton ionization mass spectrometry. This spectroscopic detection technique provides product state information as well as spatially resolved data from which angular and translational energy distributions can be inferred. The rotational cooling in the supersonic expansion of CH_3NO_2 simplifies their comparison with statistical calculations by reducing the distribution of total angular momentum states from which the excited molecule decomposes. Rockney & Grant assumed the initial total angular momentum is zero and used phase space theory to calculate product state distributions for comparison with the observation. The energy disposal into rotation seems statistical, whereas the translational energy release may indicate an unexpected barrier in the exit channel.

The great efficiency of infrared multiphoton excitation is its primary advantage for preparation of highly vibrationally excited molecules, but the accompanying distribution of excitation energies limits the energy selectivity. A recent experiment by Haas and co-workers (65, 66) has provided a first experimental assessment of the energy distribution in a molecule excited beyond its dissociation limit by infrared multiphoton absorption. The key to the analysis is comparison of the time-evolution of the products of an infrared multiphoton induced decomposition with that of the products of an overtone vibration induced decomposition (discussed below). Because the latter excitation technique does not broaden the internal energy distribution beyond the initial thermal energy spread, it serves as a basis for comparison to the infrared excitation results. At sufficiently low fluences, the time-evolution of visible chemiluminescence of electronically excited acetone from the infrared induced decomposition of tetramethyldioxetane is comparable to that from overtone vibration induced decomposition. Using data from the three different overtone vibration excitations, Haas and co-workers concluded that the distributions produced by infrared multiphoton excitation are surprisingly narrow, about 3 to 6 kcal mol^{-1}, for excess energies of about 11 kcal mol^{-1}. These data are for relatively low fluences that produce decomposition rates that are slower than usually observed. In more typical experiments, the energy spread is probably greater. Measurements that help describe the excitation level and distribution in an ensemble prepared by infrared multiphoton excitation are extremely valuable in its application to selective studies of unimolecular reactions.

Internal Conversion

Internal conversion creates highly vibrationally excited molecules by the isoenergetic crossing of electronically excited molecules into high vibrational levels of the ground electronic state (Figure 1b). In systems with well-characterized and favorable photophysics, laser excitation followed by rapid internal conversion is a useful means of preparing energized molecules for collisional energy transfer (67–73) and unimolecular decay studies (74–78). This is an energy selective preparation technique with only the distribution of initial thermal energy in the molecule introducing uncertainty in the total energy content. In some cases, the internal conversion might be partly mode selective with only certain vibrations being excited initially (79), but in general the photophysics is not known at that level of detail.

Troe and co-workers have pioneered the application of internal conversion to time-resolved unimolecular decay studies with work on cycloheptatrienes (74–76), cyclooctatetrene (77), and toluene (78). The experiments on the cycloheptatrienes are the most extensive direct measurement of the energy dependent unimolecular reaction rate constant $k(E)$ thus far. In this work, continuous ultraviolet absorption monitors the time evolution of the substituted benzene products of the unimolecular isomerization of highly vibrationally excited cycloheptatriene molecules, which are prepared by internal conversion following excitation with a 10 to 20 ns laser pulse. Shock tube data (80) and laser excitation results (81) provide the requisite spectroscopic information on the energized cycloheptatrienes and the substituted benzenes. Analysis of the time dependence of the product appearance yields a direct measure of the unimolecular reaction rate constant at the total energy determined by the excitation laser wavelength and the initial thermal energy.

Troe and co-workers have studied methyl, ethyl, and isopropyl cycloheptatriene isomerization using internal conversion in combination with time-resolved ultraviolet absorption (76). Different fixed excitation wavelengths provide $k(E)$ for two energies in these experiments. Figure 2 compares the direct measurements with a statistical (RRKM) calculation and with steady-state photoisomerization results (82). The RRKM calculation, which is a fit to thermal isomerization data, recovers the energy dependence of the rate constant well. A common scaling factor of about 2.5, which is comparable to the uncertainty in the analysis of the thermal data, is applied to all the calculations. The steady-state photoisomerization results agree well with the direct measurements when the efficiency of competing energy transfer processes is included in the analysis. This illustrates the care required to analyze systems in which collisional relaxation is important. An earlier analysis (82) that used a larger value of energy transferred on each collision led to an overestimate of the rate constant by a factor of two. Subsequent energy transfer measurements (67, 68) gave a smaller relaxation efficiency, which brought the steady-state and time-resolved data into

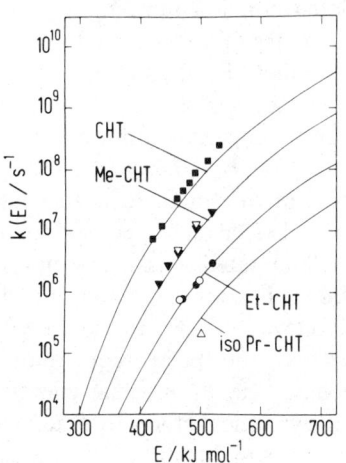

Figure 2 Rate constants for the unimolecular isomerization of cycloheptatrienes. The *open symbols* are the results of time-resolved studies using internal conversion, and the *closed symbols* come from steady-state photoisomerization work. The *solid curves* are RRKM calculations. [Reproduced with permission from Ref. (76).]

agreement. Troe and co-workers concluded that the thermal, steady-state, and time-resolved isomerization studies agree and are consistent with a statistical model of the unimolecular reaction rate. They emphasize that both statistical and nonstatistical models of $k(E)$ could explain their data.

The most extensive application of internal conversion preparation coupled with time-resolved absorption has been to isomerization reactions of the cycloheptatrienes, but Troe and co-workers have also extended this technique to simple bond fission reactions by monitoring the evolution of the energized substituted benzene formed by the isomerization (78). Thus, for example, they have studied methyl elimination from highly vibrationally excited toluene formed by the isomerization of cycloheptatriene. The crucial innovation in the work of Troe is the coupling of an energy selective technique with a direct measure of the unimolecular decay rate. These time-domain experiments provide access to $k(E)$ without the necessity of deconvoluting the effects of collisional energy transfer that competes with reaction. Such approaches hold great promise for detailed unimolecular reaction rate measurements on systems with suitable photophysics.

Stimulated Emission Pumping

Stimulated emission pumping (Figure 1c) produces highly vibrationally excited molecules by a two-photon process in which the first photon excites a molecule to an upper electronic state and the second transfers it to a high vibrational level of the ground electronic state by stimulated emission. Field, Kinsey, and co-workers (83) first demonstrated stimulated emission pumping on molecular iodine and have extended it to formaldehyde (84, 85) and acetylene (86). Lawrance & Knight have selectively excited p-difluorobenzene, as well (87–90). Stimulated emission pumping is a highly selective process that can excite individual states of polyatomic molecules with adequate efficiency for dynamical studies. Field, Kinsey, and co-workers (85) estimate that they transfer about 0.02% of all the formaldehyde molecules in a 3 mm diameter laser beam into a single rovibrational level in the electronic ground state.

The use of an electronically excited state as an intermediate in stimulated emission pumping requires well characterized electronic spectroscopy for the molecule of interest. By the same token, stimulated emission pumping is a powerful spectroscopic tool, and its initial applications have been directed toward the spectroscopy of high vibrational levels (83–86). In the acetylene study (86), Field, Kinsey, and co-workers used the spectroscopic data to assess the extent of quantum chaotic behavior at an internal energy of almost 80 kcal mol^{-1}. The most detailed dynamical study to date is the collisional energy transfer work of Lawrance & Knight (90). They have used stimulated emission pumping to prepare p-difluorobenzene with a distribution of rotational states in a single vibrational state $5_2 30_2$ that has 2036 cm^{-1} (5.8 kcal mol^{-1}) of vibrational energy. Using a third, time-delayed probe laser pulse, they interrogated the prepared state with time-resolved laser induced single vibronic level fluorescence. The time evolution of the fluorescence intensity provided the collisional energy transfer rate with each of 23 different partners. These are beautifully detailed energy and vibrational state selected measurements for which only the rotational state remains uncertain. No examples of applying this technique to selective unimolecular reactions are available yet, although such information may well come from the work on formaldehyde spectroscopy (85, 91). Stimulated emission pumping is quite an important and potentially versatile preparation scheme for highly vibrationally excited molecules with suitable, well-characterized electronic spectroscopy. In particular, it can provide access to all of the vibrational levels that have adequate Franck-Condon factors and is a means of preparing molecules with many quanta of a low frequency vibration.

Overtone Vibration Excitation

Overtone vibration excitation generates vibrationally energized molecules by single photon preparation of high vibrational states (Figure 1d). This is a mode selective technique that retains the initial distribution of energy in the degrees-of-freedom that do not interact with the excitation photon. For molecules with sufficiently sparse rotational structure, excitation of an individual angular momentum state is possible. The combination of either continuous or pulsed laser excitation with a variety of product detection techniques makes excitation of overtone vibrations a very informative approach for studying selectively initiated unimolecular reactions.

Continuous excitation experiments use gas chromatographic (92–98) or spectroscopic (99, 100) techniques to analyze the products of an extended irradiation and use the pressure dependence of the product yield in a Stern-Volmer analysis to extract the unimolecular reaction rate. This approach relies on the "collisional clock" provided by quenching of the energized species prior to reaction to establish the time-scale of the reaction. Pulsed measurements produce the excited molecules with a short (∼ 10 ns) pulse of laser light and directly monitor the products with time-resolved spectroscopy. These experiments have used both product chemiluminescence (101) and laser-induced fluorescence (102, 103a,b). The latter technique has the important feature for dynamical studies of providing internal state as well as time information. Overtone vibration excitation is the technique that is singled out for more extensive discussion in the remainder of this review. A subsequent section describes the excitation technique in detail, and the last section describes its application in both continuous and pulsed irradiation experiments.

Comparisons

Table 1 compares the energy selectivity and total energy content for a few examples of the five excitation techniques discussed in this section. These are typical values but a large range is possible in most cases. Chemical activation, infrared multiphoton absorption, and internal conversion generally produce large (>100 kcal mol^{-1}) energy contents although the first two techniques have been used for substantially smaller excitations (25, 65, 66). Stimulated emission pumping can yield a large range of excitation energies as illustrated by the two examples, and the flexibility to reach quite high excitation energies selectively promises to be a major advantage of this technique. Intermediate excitation levels (~50 kcal mol^{-1}) are typical in unimolecular decay studies using overtone vibration excitation since the strong decrease in overtone transition strength with increasing excitation energy makes excitation above 60 kcal mol^{-1} difficult.

The absolute uncertainty in the energy content (δE), along with the fractional uncertainty in the total energy ($\delta E/E$) and in the energy above the decomposition barrier [$\delta E/(E-E_0)$], appear in Table 1. Because chemical activation usually produces very high levels of excitation, its fairly large absolute energy uncertainty yields relatively good energy selectivity. This is especially true for molecular beams, where favorable kinematics can make the uncertainty as low as 2% (33). In bulk chemical activation experiments,

Table 1 Energy selectivity for preparing highly vibrationally excited molecules

Technique	Total energy, E (kcal mol^{-1})	Uncertainty, δE (kcal mol^{-1})	$\delta E/E$	$\delta E/(E-E_0)$[a]	Ref.
Chemical activation					
Bulk	110	10	0.10	0.16	15
Beam	35	2	0.05	0.09	33
Infrared multiphoton	150	19	0.13	0.50	52
excitation	37	4	0.11	0.35	65
Internal conversion	120	3	0.02	0.04	75
Stimulated emission	6	0.6	0.10	—[b]	90
pumping	80	0.01	<10^{-4}	—[b]	85
Overtone vibration	40	3	0.08	0.12	101
excitation	50	1	0.02	0.20	103
	40	3	0.08	0.10	98

[a] E_0 is the barrier to unimolecular reaction. This column gives the fractional uncertainty in the excess energy.
[b] This technique has not been applied to a unimolecular reaction.

collisions as well as the initial thermal distribution of internal energies control the distribution of energy in the reacting molecules, and the corresponding estimate of δE is less accurate. Infrared multiphoton excitation has the worst selectivity because the distribution of energy is controlled by the dynamics of the excitation process rather than the initial thermal energy content of the molecules. Because the competition between infrared excitation and unimolecular decay does not allow the molecules to reach levels far above threshold, the fractional uncertainties in the energy above the decomposition barrier are particularly large.

Internal conversion and overtone vibration excitation project the initial thermal internal energy distribution to a higher energy determined by the photon wavelength when applied to systems with dense or continuous spectroscopic structure. In molecules having less dense spectra, the excitation is potentially more selective, and one or only a few rotational and vibrational states can be excited. The energy uncertainties for internal conversion and overtone vibration excitation in Table 1 are estimates of the thermal distribution widths, but these are not the fundamental limits of the energy selectivity achievable with these methods. The key to greater selectivity in the preparation is either to study molecules with resolvable ground state structure or with very narrow thermal energy distributions such as are obtainable in the cold environment of a supersonic expansion. This latter approach also improves the possible state selectivity by reducing the population to fewer quantum states. Stimulated emission pumping has largely dealt with molecules having isolated or nearly isolated rotational transitions. The energy transfer study of Lawrance & Knight (90) did prepare a distribution of rotational states, however, and the considerations for improving the selectivity of internal conversion and overtone vibration excitation apply to stimulated emission pumping in such molecules as well. An important complementary feature of stimulated emission pumping to overtone vibration excitation is that the former can excite low frequency vibrations that are inaccessible in the latter.

The laser excitation techniques that permit time-resolved detection can directly measure unimolecular decomposition rates in the range of 10^5 to 10^8 s^{-1}, although one technique, infrared multiphoton excitation, is not particularly well suited to determining the energy dependent rate constant, as discussed above. The laser pulse duration determines the upper limit on the measurable rate constant, and, although laser pulses of about 10 ns or longer have been used to date, there is no fundamental reason that overtone vibration excitation and stimulated emission pumping could not be carried into the picosecond regime. The rate of crossing into the high vibrational levels of the ground state might limit internal conversion excitation to studies of slightly slower processes. Certainly extension of these techniques to shorter times will be quite informative. The lower limit on the rate constant of $k(E) \sim 10^5 \text{ s}^{-1}$ comes from typical diffusion or flight times in bulb or molecular beam experiments. Chemical activation in bulbs and continuous overtone vibration excitation cover a much larger range by using the collisional quenching of vibrationally excited molecules as a clock for determining unimolecular decomposition rates. Thus, these techniques potentially probe rate constants in the range of 10^{11} s^{-1} to 10^4 s^{-1} or slower. Chemical activation in molecular beams is a collisionless process, but the rotation time of the complex created by the chemical activation provides a measure of the timescale for decompositions that occur in a rotation period ($\sim 10^{-12}$ s). The angular distribution of the fragments reflects the complex lifetime if it is comparable to the rotational period but carries no rate information on slower reactions.

OVERTONE VIBRATIONS

Single photon preparation of highly vibrationally excited molecules requires a direct transition from a low, usually ground, vibrational state to a higher state containing several quanta of vibrational energy. These overtone transitions do not exist in a truly harmonic system, but the electrical and mechanical anharmonicities of real molecules give them a finite, albeit rather small, probability (104). The utility of the usual normal mode description of molecular vibrations is limited for high vibrational levels of polyatomic molecules where it requires many parameters to predict overtone transition frequencies (105–107). Stretching motions involving light atoms are most prominent in vibrational overtone excitation spectra, and they seem to be described efficiently by a local mode picture in which each bond to a light atom is treated as an isolated anharmonic oscillator. Several reviews (105–108) discuss the formal basis of the local mode model of overtone vibrations and its applications, and this section discusses only a few aspects that bear on selective studies of unimolecular reactions.

Spectroscopy

The small transition probability for overtone vibrations necessitates using sensitive techniques to examine their spectroscopy. Direct absorption measurements are possible for up to three or four quanta of excitation in liquid samples (106), but studying higher levels or more dilute samples, such as gases, requires other techniques. Approaches using laser excitation are suitable for the more demanding measurements, and two, thermal lensing for liquids (109) and photoacoustic detection for gases (110), have provided a great deal of the overtone vibration spectroscopic data now available. In the former approach, a probe laser detects the change in index of refraction caused by the degradation of energy deposited in the overtone vibration into heat, and, in the latter, a microphone detects the acoustical signal generated by the relaxation of an overtone vibration that has been excited by a pulsed laser or a modulated continuous laser. For cases in which overtone vibration excitation initiates a unimolecular decomposition, product detection provides spectroscopic data as well. The wavelength dependence of the product yield is the absorption spectrum of molecules that eventually react. This approach can be quite sensitive when the products are detected by an efficient technique such as laser-induced fluorescence (102, 103) or visible chemiluminescence (101). The result is a convolution of the absorption spectrum and the wavelength dependence of the reaction yield into the interrogated state and, thus, potentially carries information on the influence of selective excitation on product state distributions. Product detection is sufficiently sensitive in favorable systems that West et al (111) were able to use the product chemiluminescence from overtone vibration initiated decomposition of tetramethyldioxetane (101) to obtain its spectrum in a molecular beam.

The frequencies of the transitions in overtone vibration spectroscopy support the local mode picture. The most prominent transitions in a large variety of polyatomic molecules occur at frequencies corresponding to those of diatomic anharmonic oscillators. The usual Birge-Sponer relationship, $v = Av + Bv^2$, gives the transition frequencies where $\omega_e = A - B$ is the mechanical frequency, $\omega_e x_e = -B$ is the anharmonicity, and v is the quantum number of the excited local mode vibration. Figure 3 shows photoacoustic spectra in the region of six quanta of CH stretching vibration ($6v_{CH}$) and in the region of five quanta of OH stretching vibration ($5v_{OH}$) for t-butylalcohol (t-BuOH). The Birge-Sponer plots for the CH and OH stretches (Figure 4) reveal several characteristics of these overtone vibrations. The mechanical frequency for the OH oscillator ($A = 3725$ cm^{-1}) is much larger than that for the CH oscillator ($A = 2975$ cm^{-1}) in accord with the greater strength of the OH bond. (The anharmonic oscillator has a dissociation energy of $D = -A^2/4B$.) The values of B show that the OH bond is substantially more anharmonic than the CH bond. Intensity data from overtone vibration spectra add further support to the local mode picture. The transition intensity increases linearly with the number of equivalent light atom containing oscillators in a molecule.

Comparing the data on t-butylhydroperoxide (t-BuOOH) shown in Figure 4 to those for t-BuOH illustrates the sensitivity of high overtone vibration transitions to the local environment of the anharmonic oscillator. The CH oscillators in the hydroperoxide and alcohol have quite similar local environments since in both cases they are part of a tertiary butyl

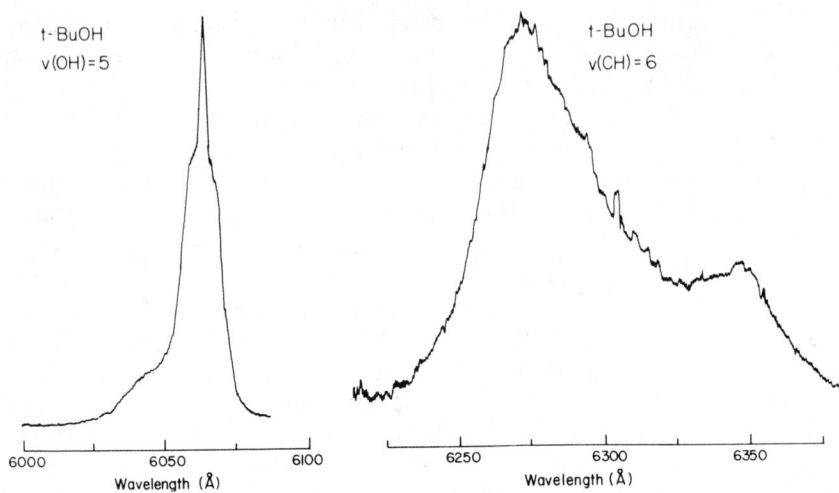

Figure 3 Photoacoustic overtone vibration absorption spectra for *t*-butylhydroperoxide in the region of $5\nu_{OH}$ and $6\nu_{CH}$. The intensities for the two transitions are arbitrary in this figure. The $6\nu_{CH}$ transition is much weaker than that for $5\nu_{OH}$.

moiety linked to an oxygen atom. Consequently, the local mode parameters in the two cases are essentially identical, as the open (*t*-BuOH) and closed (*t*-BuOOH) symbols in Figure 4a illustrate. The situation is quite different for the OH oscillator (Figure 4b). In the alcohol, the hydrogen is connected to a C–O linkage, but, in the hydroperoxide, it is adjacent to an O–O bond. Different electron density distributions in the two cases cause the two local oscillators to have dissimilar potentials and, as shown by the Birge-Sponer plots in Figure 4b, the OH oscillator in the hydroperoxide has a lower mechanical frequency but greater anharmonicity than the alcohol. Even more subtle environmental influences can appear in overtone vibration spectra. Orientational effects in which CH oscillators have different transition frequencies depending on their location relative to a concentration of electron density certainly occur (101, 112, 113a,b), and overtone vibration spectra clearly reflect the different environments for methyl, methylenic, olefinic, and other CH bonds. Moore and co-workers (113a,b) extensively discuss such behavior in terms of the bonding and internal interactions in hydrocarbon molecules. They show, for example, distinctly different transitions for methyl hydrogens lying in the plane of a nearby carbon skeleton or out of it. Such features appear in the high overtone vibration spectra because rotation of the methyl group does not average out the environmental differences on the time scale of the photon interaction with the molecule. The sensitivity of the overtone vibration to the local environment of the oscillator is important for selective studies of unimolecular reactions, since it potentially provides a means of localizing energy at individual sites in a polyatomic molecule.

Simple local mode models describe many aspects of overtone vibration spectroscopy. Child & Halonen (108) review an approach in which the light atom containing anharmonic oscillators are taken as a basis set for

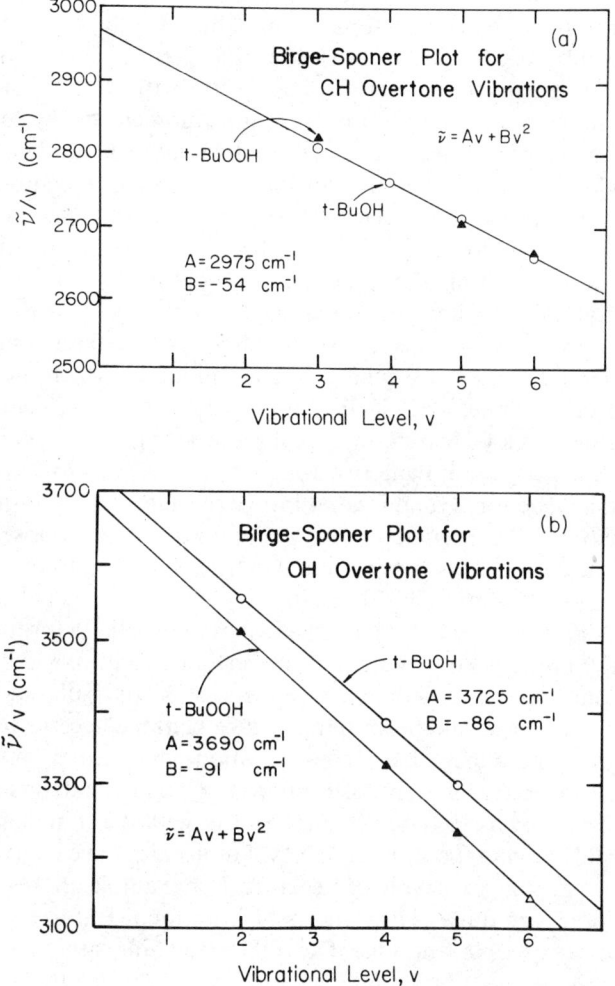

Figure 4 Birge-Sponer plots for CH and OH overtone vibrations in *t*-butylalcohol and *t*-butylhydroperoxide. The data come from photoacoustic measurements, except for the *open triangle* in (*b*), which is the result of a vibrational overtone initiated decomposition.

describing the overtone vibrations in a symmetric molecule. By constructing appropriate combinations of equivalent oscillators that reflect the symmetry of the molecule and considering potential and kinetic energy coupling among the oscillators, they fit the frequencies of overtone vibrations in H_2O, CH_4, and several other molecules. The fitting parameters are the frequency, anharmonicity, and anharmonic coupling matrix elements. These are all calculable from the molecular potential, if available, and simple expressions relating the parameters to normal mode frequencies and anharmonicities apply in some cases. Child & Lawton (114) fit more than a dozen transitions in H_2O to better than 7 cm^{-1} deviation with this model and compare the numerical calculation with model potentials. A simple bond dipole model for the transition moment in the local mode picture predicts the relative amplitudes of different local mode transitions. In particular, it shows that a transition producing a combination of excitation in two local mode oscillators (three quanta in one and one quantum in another, for example) is significantly weaker than a single local mode transition with all the quanta in one mode. This agrees with the experimental observation that local-local combination transitions are weak (106). Experiments most commonly find the strongest combinations to be those of a local mode with a normal mode to which it is strongly coupled. For example, a CCH bending normal mode is often found in combination with a CH stretching local mode (115).

A crucial aspect of applying overtone vibration excitation to studies of unimolecular reaction dynamics and intramolecular energy transfer is the exact nature of the prepared state. An important and closely related concern is the type of dynamical information that might come from the spectroscopy of overtone vibrations. Photoacoustic spectra of benzene and its deuterated derivatives (116) are the experimental basis for models that describe the initially prepared state in overtone vibration excitation and that predict the spectroscopic consequences of the subsequent intramolecular dynamics. The room temperature benzene spectra have broad and smooth lineshapes that grow to about 100 cm^{-1} at $5v_{CH}$ and become narrower at high levels of excitation. Partially deuterated benzenes exhibit different overtone vibration linewidths, with the completely deuterated molecule having the smallest. Not all molecules have overtone vibration transitions that are devoid of structure. For example, in acetylene (117) the transitions are quite sharp and have rotational features suitable for detailed spectroscopic analysis. The OH stretching region of t-BuOH shown in Figure 3 is not smooth but shows a sharp central feature that suggests a Q-branch transition.

The observed lineshapes are the key to obtaining dynamical information from overtone vibration spectra. Stannard & Gelbart (118) have discussed overtone vibration transitions using a picture in which the stretching vibration carries the oscillator strength for the absorption, but coupling of this zero order local mode state to background states spreads the transition probability over the collection of strongly coupled states to generate a set whose members contain fractional local mode character. The distribution of the resulting states is determined by the coupling strengths and controls the observed overtone vibration linewidth. Stannard & Gelbart also have considered how the weaker interaction of the strongly coupled states with the remaining background states influences the lineshape and establishes the rate of energy redistribution throughout the molecule. They applied this model to benzene and water as examples of large and small molecule limiting behavior. This viewpoint, which is strongly reminiscent of radiationless transition theory (119), is central to several models of vibrational overtone excitation. Identifying the strongly coupled background states and the nature of the coupling is the challenging aspect of the theory. One approach (107, 116) takes the collection of local mode states (combinations of CH stretches, for example) as being strongly coupled to the zero order state, and another approach (120, 121) takes the states that are in Fermi resonance with the initially excited local mode state as interacting strongly with it. Sibert et al (120, 121) concluded that kinetic energy coupling between a CH stretch and the Fermi resonant CCH bend is dominant in benzene, and they made quantitative predictions of lineshapes that agree with many features of the experimental lineshapes. The details of these coupling schemes, which may vary from molecule to molecule, are important in assessing the exact nature of the prepared state. In all these models, however, the notion that the initially excited state is mixed with background states implies that excitation with a nanosecond pulsed laser or a continuous laser cannot prepare a pure local mode state, but rather that a coherent excitation of the entire linewidth by a transform limited laser pulse is required (95, 121). As Jasinski et al (95, 98) have argued, one can expect nuclear motions that are close to the local mode oscillator in space or frequency to contribute most strongly to the transition. Vibrational overtone excitation certainly prepares an initial state having strongly nonstatistical stretching excitation in a large molecule, as the transition frequencies and intensities attest. However, knowing the exact nature of this state requires detailed information about the molecular Hamiltonian. Studies such as those of Sibert et al (121), which attempt to identify the crucial elements and compare predictions with experimental data, are essential in this regard. In general, molecules with sparse structure arising from a few strongly coupled background states (the "small molecule" limit) offer the best possibility for detailed spectroscopic analysis of the excitation of their overtone vibrations.

Excitation

Only stretching vibrations involving light atom motions have sufficient electrical and mechanical anharmonicity to possess significant overtone transition probabilities. Several considerations are important in using these transitions to prepare highly vibrationally excited molecules for dynamical studies. Child & Halonen give a formal discussion of the role of mechanical and electrical anharmonicity in determining the magnitude of the transition probability using a bond dipole approximation (108). The transition probability for a lower overtone vibration of a particular motion (e.g. a CH stretch) is larger than that for a higher one, and a rough guide is that the integrated absorption cross section decreases by about an order-of-magnitude for each higher vibrational level. In benzene, for example, the decrease is about a factor of 12 from $3v_{CH}$ to $4v_{CH}$ and a factor of seven from $8v_{CH}$ to $9v_{CH}$ (116). A lower overtone transition is stronger, but the number of molecules excited also depends on the intensity of laser sources available in various spectral regions. A more fundamental constraint than the number of excitation photons available is the energy requirement of the dynamical process of interest. In the case of a unimolecular reaction, the excitation of the overtone vibration must deposit enough energy to overcome the barrier to reaction. The lowest barriers to simple bond fission reactions are typically ~ 45 kcal mol^{-1} (16,000 cm^{-1}), while rearrangements or complex bond fissions can be ~ 25 kcal mol^{-1} (11,500 cm^{-1}) or lower. The data in Figure 4 show that the larger barrier corresponds to about five quanta of OH stretching excitation ($5v_{OH}$) or six quanta of CH stretching excitation ($6v_{CH}$). Given the efficiency of visible dye lasers, excitation in the region of 13,000 cm^{-1} to 20,000 cm^{-1} is most common, particularly in continuous irradiation experiments (92–100). One pulsed experiment has used excitation of only $3v_{CH}$ to initiate a complex bond fission (101). Also from a practical viewpoint, the excitation must prepare the molecule with enough excess energy to make the reaction rate fast compared to competing processes such as collisional relaxation and flight out of the observation region.

The number of photons absorbed by an overtone vibration in a sample of number density n and length l is

$$N_{abs} = nlN \int g(v)\sigma(v) \, dv \qquad 1.$$

where $g(v)$ is the normalized laser lineshape, N is the total number of photons entering the sample, and $\sigma(v)$ is the absorption cross section. For a homogeneously broadened transition that is much wider than the laser line, the number of photons absorbed at the center of the transition (v_0) is

$$N_{abs} = nlN\sigma(v_0) \sim nlN \frac{S}{\Delta v} \qquad 2.$$

where $S = \int \sigma(v) \, dv$ is the integrated cross section and Δv is the linewidth of the transition. (For the less common limiting case of a transition that is narrower than the laser linewidth, the bandwidth simply becomes that for the laser Δv_L.) This result applies to inhomogeneously broadened lines as well, provided the appropriate peak absorption cross section is used. The fraction of the molecules in a laser beam of cross-sectional area a excited by a pulse of energy E is

$$f = \frac{N_{abs}}{nal} = \frac{E}{ahv}\sigma(v_0) \sim \frac{E}{ahv}\frac{S}{\Delta v} \qquad 3.$$

where we have used $N = E/hv$. In a continuous experiment, the fraction excited depends on the rate constant k_l for loss processes such as reaction, collisional quenching, and flight out of the beam. In the limit of low excitation, the fraction excited in a continuous experiment is just the ratio of the excitation rate, $R_{ex} = I\sigma(v_0)$, to the loss rate. The fraction excited is $f = I\sigma(v_0)/k_l$ where I is the total intensity (photons cm^{-2} s^{-1}). Table 2 collects calculations of fractional excitations for several different CH and OH stretching overtone vibrations in pulsed experiments. The cross sections for CH are for a single oscillator in benzene (116), and those for OH are estimates based on the relative strength of the $6v_{CH}$ and $5v_{OH}$ transitions in t-BuOOH (122).

The small cross sections and resulting weak excitations for overtone vibration transitions demand sensitive product detection in unimolecular reaction studies. In continuous experiments, product accumulation during a long irradiation provides the signal averaging required to achieve adequate sensitivity, and, in pulsed experiments, efficient detection schemes such as chemiluminescence and laser-induced fluorescence are able to

Table 2 Excited fractions for homogeneously broadened CH and OH vibrational overtone transitions[a]

Transition	$\tilde{v}_0$ (cm^{-1})	$\sigma_0 \times 10^{24}$ (cm^2)	$E(mJ)$[b]	f[c]
$4v_{CH}$	11,500	3.8	20	1×10^{-5}
$5v_{CH}$	14,000	0.35	25	1×10^{-6}
$6v_{CH}$	16,500	0.07	40	3×10^{-7}
$4v_{OH}$	13,500	~ 10[d]	25	3×10^{-5}
$5v_{OH}$	16,000	1.0	40	4×10^{-6}
$6v_{OH}$	19,000	~ 0.1[d]	15	1×10^{-7}

[a] Typical linewidths for CH stretches are $\Delta\tilde{v}_{CH} \sim 100$ cm^{-1} and for OH stretches are $\Delta\tilde{v}_{OH} \sim 50$ cm^{-1}. The integrated absorption cross section is $S \sim \sigma_0 \cdot \Delta v$.
[b] Typical energies for a Nd:YAG/dye laser.
[c] Beam diameter of 2 mm.
[d] Calculated by scaling the estimate for $5v_{OH}$ by an order of magnitude.

measure the small product yields. Because so few molecules are excited, overtone vibration initiated unimolecular reaction experiments are susceptible to competition from other inefficient processes. Very slow dark and wall reactions can provide a product yield that is comparable to that from overtone vibration excitation in continuous experiments. Dark reactions are generally not a problem in pulsed experiments, but two photon excitation directly to a dissociative electronic state can have an effective cross section that approaches that for an overtone vibration. In both types of experiments, very weak one-photon excitations to excited electronic states are potentially comparable to excitation of the overtone vibration. Even though an electronic transition may be many orders-of-magnitude weaker than its maximum value in the region of a vibrational overtone, its cross section can substantially exceed that for an overtone vibration. Thus, reactions initiated by excitation in the long wavelength "tail" of an electronic transition can compete with overtone vibration initiated reactions. The cross sections in Table 2 correspond to molar extinction coefficients as small as $\varepsilon \sim 10^{-5}$ l cm^{-1} mol^{-1}, and any comparable process that leads to products is a potential interference. Fortunately, a variety of experimental diagnostics can uncover these competitive artifacts.

OVERTONE VIBRATION INITIATED UNIMOLECULAR REACTIONS

Exciting an overtone vibration prepares molecules with a well defined energy increment whose deposition is site selective and, in favorable cases, state selective. Table 3 lists the nine reported unimolecular reaction studies using continuous and pulsed laser excitation along with their local mode transitions. These few experiments are among the most detailed investigations of unimolecular reaction dynamics yet achieved.

Continuous Excitation

Continuous irradiation studies (92–100) use the product yield from an extended irradiation as a function of pressure to obtain the ratio of the unimolecular reaction rate constant to the collisional quenching rate constant, $k(E)/k_d$, from a Stern-Volmer analysis. This is based on a kinetic scheme,

$$A \underset{k_d[M]}{\overset{k_a[h\nu]}{\rightleftarrows}} A^\dagger \xrightarrow{k(E)} \text{Product}$$

in which collisional deactivation competes with unimolecular reaction. The linear plot of the reciprocal of the apparent rate constant,

$$k^{-1} = k_a^{-1} + k_d[M]/k_a k(E),$$

versus pressure yields the desired ratio. The value of k_d comes from the strong collision assumption, which postulates that every collision deactivates the vibrationally excited molecule or, at least, removes sufficient energy to make the decay rate of the relaxed molecule negligible compared to that of the excited molecule. Long irradiation times and detection of the stable products by gas chromatography (92–98) or photoacoustic spectroscopy (99, 100) provide good sensitivity. The absolute value of the unimolecular reaction rate constant depends on the exact value of k_d, but

Table 3 Reactions studied using overtone vibration excitation

Reaction	Excitation region[a]	Barrier E_0 (kcal mol^{-1})	Rate constant $k(E) \times 10^7$ (s^{-1})	Ref.
		Continuous excitation		
$CH_3NC \rightarrow CH_3CN$	$4\nu_{CH}$	38	3.3	94
	$5\nu_{CH}$		20	
	$6\nu_{CH}$		~800	
⌇NC → ⌇CN	$5\nu_{CH}$	~36	0.27	93
	$6\nu_{CH}$		6.0	
	$7\nu_{CH}$		46	
□ → ⌇	$5\nu_{CH}$	~33	3.5	95
	$6\nu_{CH}$		49	
◁▱ → ⌇⫟	$6\nu_{CH}$	~34	0.10	96
⌂ → ⌂⌇	$5\nu_{CH}$	~27	4.2	97
t-BuOOH → t-BuO + OH	$5\nu_{OH}$	44	0.04[b]	99
	$6\nu_{OH}$		0.4[b]	100
		Pulsed excitation		
⫟⫟ → 2 ⩘	$3\nu_{CH}$	27	—[c]	101
	$4\nu_{CH}$		0.012	
	$5\nu_{CH}$		0.35	
t-BuOOH → t-BuO + OH	$5\nu_{OH}$	44	—[d]	122
	$6\nu_{OH}$		0.40	102
HOOH(D) → OH + OH(D)	$5\nu_{OH}$	50	>10[e]	122
	$6\nu_{OH}$		>10[e]	103a,b

[a] The entries in this column indicate the general region of excitation. In many cases, several transitions or combination bands lie in this region and are excited separately.
[b] Best estimate used in fitting a nonlinear Stern-Volmer plot.
[c] Collisional relaxation of the excited molecule dominates the time evolution.
[d] Rotational energy transfer in the product alters the observed time evolution.
[e] The decomposition time is less than the 10 ns experimental time resolution.

discerning trends in the rates arising from excitation of different local modes or the presence of an extremely rapid component often requires only a knowledge of the relative rate constants. Because the products are detected after many collisions and without state resolution, continuous measurements provide no information about partitioning of energy among the product degrees-of-freedom.

The work of Reddy & Berry (92, 94) on the isomerization of methylisocyanide (CH_3NC) is the first study of an overtone vibration initiated unimolecular decomposition. They irradiated the samples inside a dye laser cavity, where the photon density is more than two orders-of-magnitude larger than outside the cavity, for several hours and determined the product yield gas chromatographically. The resulting Stern-Volmer plots are quite linear, except for the highest level of excitation ($6v_{CH}$) where Reddy & Berry invoked a relaxation model, and yield rate constants that are consistent with, but in some cases five times larger than, previous RRKM calculations based on thermal data. Methylisocyanide is a particularly good molecule to study because a substantial body of thermal isomerization data is available. As illustrated in the internal conversion work discussed above (76), obtaining a consistent statistical analysis of data from thermal reactions and from more selective excitation techniques requires careful consideration of the collisional energy transfer efficiencies in the thermal reaction and, perhaps in this case, in the selective excitation measurement. Agreement within a factor of five seems consistent with a statistical model given the possible uncertainties in the two measurements, but Reddy & Berry have suggested alternative nonstatistical interpretations, as well.

Applying the same methodology to a larger molecule, allylisocyanide ($H_2C=CHCH_2NC$), which has three distinct types of hydrogens, they found that the rate did not increase monotonically with increasing energy content. Exciting the nonterminal olefinic CH stretch produces a larger isomerization rate constant than exciting a terminal olefinic stretch, even though the excitation of the latter adds more energy to the molecule. The difference in rate constants is greater than the quoted uncertainties for excitation in the $6v_{CH}$ and $7v_{CH}$ overtone vibrations. This result stands alone as an example of vibrational overtone excitation of two different molecular motions yielding unimolecular reaction rate constants that are inconsistent with statistical use of the deposited energy. The apparent selective energy use is particularly surprising since both of the vibrational overtone excitations are associated with sites that are well removed from the NC group, whose motion is the reaction coordinate (123). Preferential collisional relaxation of one of the initially prepared states is a possible explanation of the observed differences. As Reddy & Berry mentioned, this in itself would reflect rather special (nonergodic) intramolecular energy transfer dynamics in which the initial energy is incompletely redistributed in the molecule on the time scale of collisional relaxation.

The most extensive studies of overtone vibration initiated unimolecular isomerization are those of Jasinski, Frisoli & Moore (95–98) on cyclobutene, 1-cyclopropylcyclobutene, and 2-methylcyclopentadiene (see Table 3). Because these molecules have distinguishable CH oscillators in different locations relative to the reaction coordinate, they are excellent vehicles for exploring the generality and interpretation of the intriguing allylisocyanide results. Jasinski et al excited the fourth ($5v_{CH}$) and fifth ($6v_{CH}$) overtone vibrations of both the methylenic and olefinic CH stretches in cyclobutene and found a monotonic increase of the rate with energy within their experimental uncertainties. The energy dependence agrees well with a RRKM calculation, although the rate constants are a factor of two to three larger than a calculation by Elliott & Frey (124) that is designed to reproduce the thermal rate data. Using the same parameters but with a slightly looser transition state having five frequencies reduced by ~ 200 cm^{-1}, Jasinski et al calculated rate constants that quantitatively agree with their vibrational overtone measurements.

A similar measurement on 1-cyclopropylcyclobutene monitors reactions induced by excitation of the fifth overtone ($6v_{CH}$) of the methylenic CH stretch, both in the cyclobutene ring and in the cyclopropyl ring (see Table 3) (96, 98). This is similar in spirit to chemical activation experiments that excite different structures within a molecule. Exciting the higher energy methylenic overtone transition in the cyclopropyl ring (and possibly a coincident cyclobutenyl olefinic transition) produces a larger rate constant than exciting the lower energy methylenic overtone vibration in the cyclobutenyl ring. The relative rates are in accord with indiscriminant use of the energy deposited in the molecule and agree quantitatively with an RRKM calculation. Because breaking the bond between the methylenic carbons is an essential step in the isomerization, excitation of the methylenic stretch in the cyclobutenyl ring deposits energy in a portion of the molecule where it could selectively accelerate the reaction. To search for such behavior, Jasinski et al monitored reactions initiated by excitation of the fifth overtone of the cyclobutenyl methylenic CH stretch up to pressures of 700 torr in cyclobutene and 10 torr in cyclopropylcyclobutene. A prompt reaction due to molecules rapidly isomerizing prior to intramolecular energy redistribution would produce a characteristic curvature in the Stern-Volmer plot. In such a case, the reciprocal of the apparent rate constant k^{-1} does not increase linearly with pressure, as in the simple kinetic scheme discussed above, but rather approaches a value determined by the relative rates of prompt reaction and energy redistribution (99, 100). Jasinski et al found linear behavior over the entire pressure range they studied. Thus, excitation of the stretches closest to the reaction coordinate does not produce a detectable nonrandom component, and the 100 ps collision interval at the highest pressure establishes an upper limit to the intramolecular energy redistribution time.

An isomerization involving significant hydrogen atom motion should be most easily enhanced by vibrational overtone excitation, and Jasinski et al (97, 98) have studied the [1, 5] sigmatropic hydrogen shift in methylcyclopentadiene with this in mind (Table 3). Excitation of $5\nu_{CH}$ for the methylenic hydrogen as well as of distinct methyl hydrogen and the olefinic hydrogen yields rate constants that are consistent with statistical behavior. The Stern-Volmer plot for excitation of the methylenic transition is linear up to 100 torr of added n-pentane quencher. The measurements by Jansinski et al on carefully chosen molecules are consistent with energy randomization in all three cases and yield rates that agree with RRKM calculations. Comparisons with thermal data are adequate as well. These continuous overtone vibration excitation measurements show the utility of statistical unimolecular reaction rate theories even in describing experiments that incorporate rather selective excitation. The failure of the very special behavior found in the allylisocyanide system to appear in molecules that seem more likely to react selectively points to the subtle nature of the dynamics involved. These experiments particularly emphasize the exceptional nature of nonstatistical behavior.

Zare and co-workers (99, 100) have applied continuous vibrational overtone excitation to the simple bond fission of t-butylhydroperoxide (t-BuOOH $\rightarrow$ t-BuO + OH). They detected the t-butylalcohol formed by subsequent reaction of the t-butyl radical from the initial decomposition using in situ photoacoustic detection. This hydroperoxide is particularly suitable for overtone vibration studies since it has a relatively weak oxygen-oxygen bond of ~ 44 kcal mol^{-1} and contains both OH and CH bonds that are possible excitation sites. These advantages, along with the OH product being detectable by laser induced fluorescence, make t-BuOOH suitable for time and state resolved studies as well (102). This system is unique in having been investigated using both continuous and pulsed vibrational overtone excitation.

The Stern-Volmer plots for both $5\nu_{OH}$ and $6\nu_{OH}$ excitation are curved in keeping with the presence of a prompt (nonstatistical) reaction component of about 1% in the former case and about 30% in the latter case. This result is consistent with some of the excited molecules decomposing prior to energy redistribution, and t-BuOOH does seem to be a reasonable molecule in which to expect that energy might be used to break the O–O bond before flowing into the t-butyl group. Zare and co-workers emphasize the possibility of other origins of the curvature in the Stern-Volmer plots and suggest, for example, that collisional production of an electronically excited state could be responsible for the behavior that they observe. Comparison with the time-resolved measurements on t-BuOOH (125), which are discussed subsequently, provides another possible explanation. The rate constant from the direct measurement on $6\nu_{OH}$ excitation agrees well with the observation and RRKM calculation by Zare et al. The time evolution in the direct measurement indeed shows a prompt component, but it persists when the excitation laser is tuned away from the overtone vibration transition. The linear power dependence of this nonresonant component seems to indicate that it is a one-photon excitation of the weak underlying electronic transition discussed above. However, Zare and coworkers (126) find no reaction above background when their laser is not coincident with an overtone transition. It is certainly the case that the nonresonant prompt component prevents any measurement of the resonant prompt component in the direct excitation experiment, if there is one. Two photon excitation is a possible artifact in the pulsed experiment, but the power dependence (125) and product state distributions (122) argue against it. Further experimental work is likely to resolve this point, but it highlights the complementary nature of continuous experiments (which can be made sensitive to small rapid components by exploiting collisional competition) and time and state resolved measurements (which directly determine rates and product energy partitioning largely unperturbed by collisions).

Pulsed Excitation

Pulsed excitation studies (101–103a,b) extract the unimolecular reaction rate from the time evolution of the products of an overtone vibration initiated reaction. In these experiments, a pulsed laser prepares the highly vibrationally excited molecule, and a time-resolved spectroscopic technique such as visible chemiluminescence (101) or laser-induced fluorescence (102, 103a,b) monitors the products. As discussed above, the wavelength dependence of the product yield provides data on the vibrational overtone absorption spectrum with sufficient sensitivity to study very dilute samples. The product energy partitioning data available in pulsed excitation measurements using state resolved detection are particularly important in that they provide a view of the unimolecular reaction dynamics that is unavailable in the rate data alone. Even in experiments in which the unimolecular reaction is too fast for the appearance time of the products to be obtained, the time resolution of the measurement is important. Observing products at times substantially less than the collision interval yields state distributions that are uncontaminated by collisional energy transfer. In these cases, the measured time evolutions serve as an excellent diagnostic of the unrelaxed character of the observed product state distributions (103a,b).

The first reported time-resolved overtone vibration initiated unimolecular reaction is the decomposition of tetramethyldioxetane, a cyclic peroxide that in solution decomposes over a barrier of about 27 kcal mol^{-1} to form acetone (Table 3) (101). The reaction barrier and exothermicity provide sufficient energy to excite one of the product molecules electronically, and about 30% of the decompositions produce an excited, chemiluminescent product. Efficient time-resolved detection of the product emission permits the study of gas phase decompositions for excitation of $3v_{CH}$, $4v_{CH}$, and $5v_{CH}$ at pressures of only a few millitorr. The longer time behavior of the product emission reflects the chemiluminescence kinetics and collisional quenching of the products (127), but the initial growth in the emission carries the desired information on the unimolecular decay rate of the selectively excited molecule. Treating the time evolution as the sum of exponentials arising from the unimolecular reaction and subsequent emission and quenching of the products gives rate constants for decomposition at the $4v_{CH}$ and $5v_{CH}$ levels of excitation that agree with a conventional RRKM calculation. However, because the entire time evolution, rather than just a single rate constant, is measured, it contains still more detailed information. The products do not appear with a single rate constant, as the nonexponential growth of the chemiluminescence shows. Because the unimolecular reaction rate constant $k(E)$ depends strongly on the total energy content, the time evolution for molecules with different energies, which come from the distribution of initial vibrational energy in the reactant molecules, must be considered. Indeed, an average over the thermal vibrational energy distribution using $k(E)$ calculated from RRKM theory reproduces the time evolution for $4v_{CH}$ and $5v_{CH}$ (101). The effects of the initial energy distribution are present in any selective unimolecular reaction experiment using molecules that have significant thermal populations in vibrationally excited states, but studies that directly observe the time evolution of the products demonstrate them most clearly. Collapsing the distribution by studying the reaction of molecules cooled in a supersonic expansion will remove this thermal averaging and provide an even sharper comparison with theory.

Another time-resolved measurement on a relatively large molecule, and the first to use state-resolved product detection, is that on t-butylhydroperoxide by Rizzo & Crim (102). Using laser-induced fluorescence to monitor individual rotational states of the OH product at varying intervals between the visible laser, which excites the overtone vibration, and the ultraviolet laser, which probes the products, they directly determined the unimolecular rate constant at the level of $6\nu_{OH}$. The observed rate constant agrees with the continuous measurement and RRKM calculation of Zare et al (100), but the prompt off-resonant component, described above, prevents an unambiguous determination of the state distribution of products arising from the unimolecular reaction (122). Preliminary data on $5\nu_{OH}$ illustrate a complication that can arise in determining rates from observing a single state in an experiment in which collisions occur. The unimolecular decay following $5\nu_{OH}$ excitation is sufficiently slow (~ 2.5 μs) that some rotational relaxation of the OH product occurs on the time-scale of the reaction. At pressures of only 20 mTorr, rotational energy transfer, which can occur at more than the gas kinetic collision rate, moves population in or out of the observed state during the reaction. This process contributes significantly to the observed time evolution of the product and must be considered in extracting the rate constant (122).

Small molecules offer several advantages in studies of unimolecular dynamics. Because their density of states is less than that in larger molecules, the uncertainty in initial energy content following vibrational overtone excitation is reduced. The accompanying reduction in spectral congestion increases the possible selectivity as well, and theoretical studies using realistic potential energy surfaces are certainly more likely to be possible on a molecule containing only a few atoms. The disadvantage of studying small molecules is that directly measuring their shorter unimolecular lifetimes requires greater time resolution. Rizzo et al (103a,b) have studied the vibrational overtone initiated unimolecular decomposition of HOOH and HOOD to exploit the advantages of experimentally investigating a four-atom system. As in the t-BuOOH study, they detected individual states of the OH (or OD) product, but the decomposition is too fast to be measured with their ~ 20 ns time resolution. Thus, the information available from their experiment is the overtone vibration excitation spectrum, obtained by monitoring the relative product yield as a function of excitation laser wavelength, and the distribution of OH(OD) products among their quantum states, obtained by varying the probe laser wavelength. Figure 5 shows a portion of the vibrational overtone excitation spectrum in the region of the $6\nu_{OH}$ excitation. An additional feature, which is not shown in Figure 5, occurs at 385 cm^{-1} higher energy than the pure overtone vibration and seems to arise from a local mode-normal mode combination transition involving six quanta of the OH stretching vibration and one quantum of a torsional motion. The main transition shows definite rotational structure in which the sharp features rising out of the congested absorption are likely to be members of ^PQ_K and ^RQ_K progressions (in symmetric top notation). Simulations recover these features faithfully (122, 128) and explain changes in the rotational contour in going from HOOH to HOOD. This spectrum illustrates two important points: Vibrational overtone excitation will probably allow the unimolecular decay dynamics of molecules prepared in one or a small subset of initial angular momentum states to be studied, particularly in higher resolution molecular beam experiments; and product detection is a sensitive means of measuring the spectra of high overtone vibrations.

Excitation in the lower energy region near the $5v_{OH}$ transition of H_2O_2 produces a highly structured spectrum (122). The photon energy is less than the O–O bond energy, but molecules with a few thousand wavenumbers of initial thermal energy can dissociate. In this case, the excitation spectrum arises from a relatively small number of energetic initial states and is, consequently, less congested than that for $6v_{OH}$. Signal calculations indicate that measurements in the region of $7v_{OH}$ are feasible; however, the nonresonant one photon dissociation dominates the signal and obscures the overtone vibration.

The population of the rotational states of the OH products resulting from $6v_{OH}$ excitation increases from $N = 1$ to a maximum near $N = 6$ and sharply decreases to essentially zero at $N = 10$ (103a,b). This is the highest level that can be populated by the photon energy plus the average initial thermal energy of the molecule. The populations are not in a Boltzmann distribution, but such a distribution is not necessarily the outcome of even a statistical decomposition. A phase space calculation provides a useful point of comparison (122). The initial energies and angular momenta of H_2O_2 for

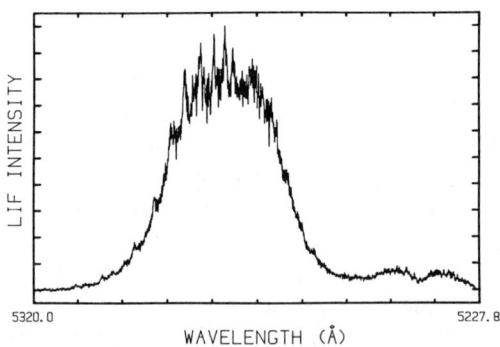

Figure 5 Overtone vibration excitation spectrum for H_2O_2 in the region of $6v_{OH}$. The probe laser is coincident with the $Q_1(4)$ transition of the OH product of the unimolecular decomposition of H_2O_2.

the calculation come from a thermal distribution to which the photon energy is added, and phase space theory determines the probability of all sets of final product states that conserve energy and angular momentum. This statistical calculation predicts the shape of the product rotational state distribution quite well. Thus, the results of $6v_{OH}$ excitation are consistent with statistical energy disposal in the decomposition of H_2O_2 (103a,b). A further test of the consequences of excitation specificity is the detection of OD following excitation of $6v_{OH}$ in HOOD, since the excited portion and detected fragment of the molecule are distinguishable. The rotational energy distribution for OD from HOOD is the same as that for OH from HOOH and also agrees with the phase space calculation, in keeping with statistical energy redistribution and the absence of site-selective behavior. Exciting the combination feature at 385 cm^{-1} higher energy that $6v_{OH}$ does yield an unexpected product state distribution, however. The maximum in the rotational state distribution moves to lower energy, in contradiction to the outcome predicted by statistical theories. This may reflect restricted or special coupling of the torsion, which is the normal mode component of the combination, to rotation of the OH fragments (129).

New data for $5v_{OH}$ excitation of H_2O_2 illustrate the influence that preparing a restricted range of initial angular momentum states can have on the product state distribution (122). Excitation of features that are separated by only 10 cm^{-1} in the structured overtone vibration spectrum produces significantly different rotational state distributions, all having their greatest populations in the lowest energy state. The differences that arise from changing the excitation slightly demonstrate the potential of sufficiently selective measurements to uncover correlations between the initially selected state in the reactant and particular states in unimolecular decomposition products.

CONCLUSION

A variety of new techniques and clever applications of established ones are revealing many features of unimolecular reaction dynamics in previously unattainable detail. These data serve as critical tests of theoretical descriptions of the behavior of energized, isolated molecules. Unimolecular reaction rates and product energy distributions in keeping with statistical descriptions are most common, but chemical intuition certainly suggests that preparation schemes in which energy is not spread indiscriminately about the molecule prior to reaction are possible. The nature and strength of coupling of the internal degrees-of-freedom, which determines the energy redistribution times, are the crucial features in such selective reactions.

Observing apparently nonstatistical behavior in a few cases while finding statistical reactions in similar systems emphasizes the subtle nature of unimolecular processes. These results call for more detailed experimental investigation and continued theoretical study of unimolecular reactions with the goal of an improved understanding of the underlying dynamics. Measuring product state distributions provides a slightly different perspective on unimolecular reaction dynamics since they are often more sensitive to the shape of the potential energy surface in the region of product formation than are the unimolecular reaction rates. Thus, product state distributions, particularly from reactants that are initially in a single or very few quantum states, potentially offer a view of both the unimolecular reaction dynamics and potential energy surface that control the reaction. Still more revealing studies of unimolecular reactions promise to enhance our understanding in the near future.

ACKNOWLEDGMENTS

It is a pleasure to acknowledge the contributions of several people to our continuing studies of unimolecular decay dynamics. B. D. Cannon, T. R. Rizzo, C. C. Hayden, E. S. McGinley, and S. M. Penn are past and present co-workers who have carried our time-resolved unimolecular decomposition measurements forward. I particularly thank T. R. Rizzo for critically reading parts of this manuscript. Different aspects of this work are supported by the Office of Basic Energy Sciences of the Department of Energy and by the Army Research Office.

Literature Cited

1. Spicer, L. D., Rabinovitch, B. S. 1970. *Ann. Rev. Phys. Chem.* 21:349–86
2. Setser, D. W. 1972. *MTP Int. Rev. Sci. Phys. Chem.* 9:1–43
3. Robinson, P. J. 1975. *Reaction Kinetics*, ed. P. G. Ashmore, 1:93–160. London: Specialist Period. Rep., Chem. Soc.
4. Quack, M., Troe, J. 1977. *Gas Kinetics and Energy Transfer*, ed. P. G. Ashmore, R. J. Donovan, 2:175–238. London: Specialist Period. Rep., Chem. Soc.
5. Frey, H. M., Walsh, R. 1978. *Gas Kinetics and Energy Transfer*, ed. P. G. Ashmore, R. J. Donovan, 3:1–41. London: Specialist Period. Rep., Chem. Soc.
6. Hase, W. L. 1981. *Potential Energy Surfaces and Dynamics Calculations*, ed. D. G. Truhlar, pp. 1–35. New York: Plenum
7. Quack, M., Troe, J. 1981. *Int. Rev. Phys. Chem.* 1:97–147
8. Holbrook, K. A. 1983. *Chem. Soc. Rev.* 12:163–211
9. Ambartzumian, R. V., Letokhov, V. S. 1977. *Acc. Chem. Res.* 10:61–67
10. Schulz, P. A., Sudbø, A. S., Krajnovich, D. J., Kwok, H. S., Shen, Y. R., Lee, Y. T. 1979. *Ann. Rev. Phys. Chem.* 30:379–409
11. Golden, D. M., Rossi, M. J., Baldwin, A. C., Barker, J. R. 1980. *Acc. Chem. Res.* 14:56–62
12. Ashford, M. N. R., Hancock, G. 1981. *Gas Kinetics and Energy Transfer*, ed. P. G. Ashmore, R. J. Donovan, 4:73–136. London: Specialist Period. Rep., Chem. Soc.
13. King, D. S. 1982. *Adv. Chem. Phys.* 50:105–89
14. Quack, M. 1982. *Adv. Chem. Phys.* 50:395–474

15. Robinson, P. J., Holbrook, K. A. 1972. *Unimolecular Reactions.* New York: Wiley-Interscience
16. Smith, I. W. M. 1980. *Kinetics and Dynamics of Elementary Gas Reactions.* London/Boston: Butterworths
17. Safron, S. A., Weinstein, N. D., Herschbach, D. R., Tully, J. C. 1972. *Chem. Phys. Lett.* 12:564–68
18. Tardy, D. C., Rabinovitch, B. S. 1977. *Chem. Rev.* 77:369–408
19. Rynbrandt, J. C., Rabinovitch, B. S. 1971. *J. Chem. Phys.* 54:2275–78
20. Rynbrandt, J. C., Rabinovitch, B. S. 1971. *J. Chem. Phys.* 75:2164–76
21. Meagher, J. F., Chao, K.-J., Barker, J. R., Rabinovitch, B. S. 1974. *J. Phys. Chem.* 78:2535–43
22. Wang, F. M., Rabinovitch, B. S. 1976. *Can. J. Chem.* 54:943–48
23. Ko, A. N., Rabinovitch, B. S., Chao, K.-J. 1977. *J. Chem. Phys.* 66:1374–76
24. Ko, A. N., Rabinovitch, B. S. 1978. *Chem. Phys.* 30:361–74
25. Trenwith, A. B., Rabinovitch, B. S. 1982. *J. Phys. Chem.* 86:3447–53
26. Oref, I., Rabinovitch, B. S. 1979. *Acc. Chem. Res.* 12:166–75
27. Rodgers, P., Montague, D. C., Frank, J. P., Tyler, S. C., Rowland, F. S. 1982. *Chem. Phys. Lett.* 89:9–12
28. Rodgers, P., Selco, J. I., Rowland, F. S. 1983. *Chem. Phys. Lett.* 97:313–16
29. Rowland, F. S. 1983. *Faraday Discuss. Chem. Soc.* 75:158–61
30. Lopez, V., Marcus, R. A. 1982. *Chem. Phys. Lett.* 93:232–34
31. Wrigley, S. P., Rabinovitch, B. S. 1983. *Chem. Phys. Lett.* 98:386–92
32. Farrar, J. M., Lee, Y. T. 1976. *J. Chem. Phys.* 65:1414–26
33. Buss, R. J., Coggiola, M. J., Lee, Y. T. 1979. *Faraday Discuss. Chem. Soc.* 67:162–72
34. Moehlman, J. G., McDonald, J. D. 1975. *J. Chem. Phys.* 63:3052–60
35. Durana, J. F., McDonald, J. D. 1976. *J. Chem. Phys.* 64:2518–32
36. Moss, M. G., Hudgens, J. W., McDonald, J. D. 1980. *J. Chem. Phys.* 72:3486–89
37. Moss, M. G., Ensminger, M. D., Stewart, G. M., Mordaunt, D., McDonald, J. D. 1980. *J. Chem. Phys.* 73:1256–64
38. Lesiecki, M. L., Guillory, W. A. 1978. *J. Chem. Phys.* 69:4572–79
39. Hicks, K. W., Lesiecki, M. L., Guillory, W. A. 1979. *J. Phys. Chem.* 82:1936–39
40. Schmiedl, R., Boettner, R., Zacharias, H., Meier, U., Welge, K. H. 1980. *Opt. Commun.* 31:329–33
41. Ashford, M. N. R., Hancock, G., Ketley, G. 1979. *Faraday Discuss. Chem. Soc.* 67:204–11
42. Ashford, M. N. R., Hancock, G., Hardaker, M. L. 1980. *J. Photochem.* 14:85–88
43. Miller, C. M., Zare, R. N. 1980. *Chem. Phys. Lett.* 71:376–80
44. Miller, C. M., McKillop, J. S., Zare, R. N. 1982. *J. Chem. Phys.* 76:2390–98
45. Campbell, J. D., Yu, M. H., Mangir, M., Wittig, C. 1978. *J. Chem. Phys.* 69:3854–57
46. King, D. S., Stephenson, J. C. 1977. *Chem. Phys. Lett.* 51:48–52
47. Stephenson, J. C., King, D. S. 1978. *J. Chem. Phys.* 69:1485–92
48. Stephenson, J. C., King, D. S. 1980. *J. Chem. Phys.* 72:1161–69
49. Reisler, H., Kong, F., Renlund, A. M., Wittig, C. 1982. *J. Chem. Phys.* 76:997–1006
50. Reisler, H., Kong, F., Wittig, C., Stone, J., Thiele, E., Goodman, M. F. 1982. *J. Chem. Phys.* 77:328–36
51. Reisler, H., Pessini, F. B. T., Wittig, C. 1983. *Chem. Phys. Lett.* 99:388–93
52. Beresford, J. R., Hancock, G., MacRobert, A. J., Catanzarite, J., Radhakrishan, G., et al. 1983. *Faraday Discuss. Chem. Soc.* 75:211–22
53. Rayner, D. M., Hackett, P. A. 1983. *J. Chem. Phys.* 79:5414–22
54. Quack, M., Troe, J. 1977. *Ber. Bunsenges. Phys. Chem.* 78:240–53
55. Quack, M., Troe, J. 1975. *Ber. Bunsenges. Phys. Chem.* 79:170–83
56. Grant, E. R., Schulz, P. A., Sudbø, A. S., Shen, Y. R., Lee, Y. T. 1978. *Phys. Rev. Lett.* 40:115–18
57. Sudbø, A. S., Schulz, P. A., Grant, E. R., Shen, Y. R., Lee, Y. T. 1978. *J. Chem. Phys.* 68:1306–7
58. Sudbø, A. S., Schulz, P. A., Grant, E. R., Shen, Y. R., Lee, Y. T. 1979. *J. Chem. Phys.* 70:912–22
59. Schulz, P. A., Sudbø, A. S., Grant, E. R., Shen, Y. R., Lee, Y. T. 1980. *J. Chem. Phys.* 72:4985–95
60. Krajnovich, D., Huisken, F., Zhang, Z., Shen, Y. R., Lee, Y. T. 1982. *J. Chem. Phys.* 77:5977–89
61. Huisken, F., Krajnovich, D., Zhang, Z., Shen, Y. R., Lee, Y. T. 1983. *J. Chem. Phys.* 78:3806–15
62. Butler, L. J., Buss, R. J., Brudzynski, R. J., Lee, Y. T. 1983. *J. Phys. Chem.* 87:5106–12
63. Hudgens, J. W. 1978. *J. Chem. Phys.* 68:777–78
64. Rockney, B. H., Grant, E. R. 1983. *J. Chem. Phys.* 79:708–19

65. Ruhman, S., Anner, O., Gerjhumi, S., Haas, Y. 1983. *Chem. Phys. Lett.* 99: 281–86
66. Ruhman, S., Anner, O., Haas, Y. 1983. *Faraday Discuss. Chem. Soc.* 75: 239–50
67. Hippler, H., Troe, J., Wendelken, H. J. 1981. *Chem. Phys. Lett.* 84: 257–59
68. Hippler, H., Troe, J., Wendelken, H. J. 1983. *J. Chem. Phys.* 78: 6709–17
69. Hippler, H., Troe, J., Wendelken, H. J. 1983. *J. Chem. Phys.* 78: 6718–24
70. Smith, G. P., Barker, J. R. 1981. *Chem. Phys. Lett.* 78: 253–58
71. Rossi, M. J., Barker, J. R. 1982. *Chem. Phys. Lett.* 85: 21–26
72. Rossi, M. J., Pladziewicz, J. R., Barker, J. R. 1983. *J. Chem. Phys.* 78: 6695–6708
73. Nakashima, N., Yoshihara, K. 1982. *J. Chem. Phys.* 77: 6040–50
74. Hippler, H., Luther, K., Troe, J., Walsh, R. 1978. *J. Chem. Phys.* 68: 323–25
75. Hippler, H., Luther, K., Troe, J. 1979. *Faraday Discuss. Chem. Soc.* 67: 173–79
76. Hippler, H., Luther, K., Troe, J., Wendelken, J. 1983. *J. Chem. Phys.* 79: 239–46
77. Dudek, D., Glänzer, K., Troe, J. 1979. *Ber. Bunsenges. Phys. Chem.* 83: 788–98
78. Hippler, H., Schubert, V., Troe, J., Wendelken, H. J. 1981. *Chem. Phys. Lett.* 84: 253–56
79. Naaman, R., Lubman, D. M., Zare, R. N. 1979. *J. Chem. Phys.* 71: 4192–4200
80. Astholz, D. C., Brouwer, L., Troe, J. 1981. *Ber. Bunsenges. Phys. Chem.* 85: 559–64
81. Hippler, H., Troe, J., Wendelken, H. J. 1983. *J. Chem. Phys.* 78: 5351–57
82. Troe, J., Wieters, W. 1979. *J. Chem. Phys.* 71: 3931–41
83. Kittrell, C., Abramson, E., Kinsey, J. L., McDonald, S. A., Reisner, D. E., et al. 1981. *J. Chem. Phys.* 75: 2056–59
84. Reisner, D. E., Vaccaro, P. H., Kittrell, C., Field, R. W., Kinsey, J. L., et al. 1982. *J. Chem. Phys.* 77: 573–75
85. Vaccaro, P. H., Kinsey, J. L., Field, R. W., Dai, H.-L. 1983. *J. Chem. Phys.* 78: 3659–64
86. Abramson, E., Field, R. W., Imre, D., Innes, K. K., Kinsey, J. L. 1984. *J. Chem. Phys.* 80: 2298–2300
87. Lawrance, W. D., Knight A. E. W. 1982. *J. Chem. Phys.* 76: 5637–39
88. Lawrance, W. D., Knight, A. E. W. 1982. *J. Chem. Phys.* 77: 570–71
89. Lawrance, W. D., Knight, A. E. W. 1983. *J. Phys. Chem.* 87: 389–91
90. Lawrance, W. D., Knight, A. E. W. 1983. *J. Chem. Phys.* 79: 6030–42
91. Kinsey, J. L. 1983. *Bull. Am. Phys. Soc.* 28: 478
92. Reddy, K. V., Berry, M. J. 1977. *Chem. Phys. Lett.* 52: 111–16
93. Reddy, K. V., Berry, M. J. 1979. *Chem. Phys. Lett.* 56: 223–29
94. Reddy, K. V., Berry, M. J. 1979. *Faraday Discuss. Chem. Soc.* 67: 188–203
95. Jasinski, J. M., Frisoli, J. K., Moore, C. B. 1983. *J. Chem. Phys.* 79: 1313–18
96. Jasinski, J. M., Frisoli, J. K., Moore, C. B. 1983. *J. Phys. Chem.* 87: 3826–29
97. Jasinski, J. M., Frisoli, J. K., Moore, C. B. 1983. *J. Phys. Chem.* 87: 2209–13
98. Jasinski, J. M., Frisoli, J. K., Moore, C. B. 1983. *Faraday Discuss. Chem. Soc.* 75: 289–300
99. Chandler, D. W., Farneth, W. E., Zare, R. N. 1982. *J. Chem. Phys.* 77: 4447–58
100. Chuang, M.-L., Baggott, J. E., Chandler, D. W., Farneth, W. E., Zare, R. N. 1983. *Faraday Discuss. Chem. Soc.* 75: 301–14
101. Cannon, B. D., Crim, F. F. 1981. *J. Chem. Phys.* 75: 1752–61
102. Rizzo, T. R., Crim, F. F. 1982. *J. Chem. Phys.* 76: 2754–56
103a. Rizzo, T. R., Hayden, C. C., Crim, F. F. 1983. *Faraday Discuss. Chem. Soc.* 75: 223–37
103b. Rizzo, T. R., Hayden, C. C., Crim, F. F. 1984. *J. Chem. Phys.* 81: In press
104. Herzberg, G. 1945. *Infrared and Raman Spectra of Polyatomic Molecules.* New York: Van Nostrand Reinhold
105. Henry, B. R. 1977. *Acc. Chem. Res.* 10: 207–13
106. Henry, B. R. 1981. *Vibrational Spectra and Structure,* ed. J. R. Durig, 10: 269–20. New York: Elsevier
107. Sage, M. L., Jortner, J. 1981. *Adv. Chem. Phys.* 47: 293–322
108. Child, M. S., Halonen, L. 1984. *Adv. Chem. Phys.* In press
109. Long, M. E., Swofford, R. L., Albrecht, A. C. 1976. *Science* 191: 183–85
110. West, G. A., Barrett, J. J., Siebert, D. R., Reddy, K. V. 1983. *Rev. Sci. Instrum.* 54: 797–817
111. West, G. A., Mariella, R. P. Jr., Pete, J. A., Hammond, W. B., Heller, D. F. 1981. *J. Chem. Phys.* 75: 2006–7
112. Henry, B. R., Greenlay, W. R. A. 1980. *J. Chem. Phys.* 72: 5516–23
113a. Wong, J. S., Moore, C. B. 1982. *J. Chem. Phys.* 77: 603–15
113b. Wong, J. S., MacPhail, R. A., Moore, C. B., Strauss, H. L. 1982. *J. Phys. Chem.* 86: 1478–84
114. Child, M. S., Lawton, R. T. 1981. *Faraday Discuss. Chem. Soc.* 71: 273–85
115. Fang, H. L., Swofford, R. L. 1980. *J. Phys. Chem.* 73: 2607–17
116. Reddy, K. V., Heller, D. F., Berry, M. J. 1982. *J. Chem. Phys.* 76: 2814–37
117. Scherer, G. J., Lehmann, K. K., Klem-

perer, W. 1983. *J. Chem. Phys.* 78:2817–32
118. Stannard, P. R., Gelbart, W. M. 1981. *J. Chem. Phys.* 85:3592–99
119. Avouris, P., Gelbart, W. M., El-Sayed, M. A. 1977. *Chem. Rev.* 77:793–833
120. Sibert, E. L., Reinhardt, W. P., Hynes, J. T. 1982. *Chem. Phys. Lett.* 92:455–58
121. Sibert, E. L., Reinhardt, W. P., Hynes, J. T. 1984. *J. Chem. Phys.* 81: In press
122. Rizzo, T. R. 1983. *State specific studies of unimolecular reactions of small polyatomic molecules.* PhD thesis. Univ. Wis., Madison
123. Liskow, D. H., Bender, C. F., Schaefer, H. F. 1972. *J. Chem. Phys.* 57:4509–11
124. Elliott, C. S., Frey, H. M. 1966. *Trans. Faraday Soc.* 62:895–909
125. Rizzo, T. R., Hayden, C. C., Crim, F. F. 1983. *Faraday Discuss. Chem. Soc.* 75:350–53
126. Chuang, M.-L., Baggott, J. E., Chandler, D. W., Farneth, W. E., Zare, R. N. 1983. *Faraday Discuss. Chem. Soc.* 75:353–55
127. Cannon, B. D., Crim, F. F. 1981. *J. Am. Chem. Soc.* 103:6722–26
128. Dübal, H. R., Quack, M. 1983. *Faraday Discuss. Chem. Soc.* 75:272–76
129. Rizzo, T. R., Hayden, C. C., Crim, F. F. 1983. *Faraday Discuss. Chem. Soc.* 75:276–77

AUTHOR INDEX

A

Aaviskoo, J., 614
Abarenkov, I. V., 361, 362
Abe, H., 304, 312, 320, 524, 527
Abkowitz, M., 640
Abragam, A., 266
Abramson, E., 586, 666
Abrash, S., 467
Ackerman, M., 300, 304, 305, 309, 312, 313, 317
Ackerman, T. P., 483, 502, 503
Adam, G., 150, 244
Adam, M., 8
Adams, A., 516-18, 521, 523, 527
Adams, J. E., 171, 179, 180, 182
Adler, G., 614
Adrian, F. J., 512, 524
Agarwal, G. S., 518, 519
Agmon, N., 176, 182
Agranovich, V. M., 614, 619
Agrawalla, B. S., 112, 115, 118
Aharon-Shalom, E., 646
Ahmed, F., 299, 309
Aikawa, M., 138, 148
Akimoto, K., 67
Akins, D. E., 524, 526
Akiyoshi, K., 151
Al-Obaidi, S. J., 619
Albery, W. J., 438
Albrecht, A. C., 670
Albrecht, M. G., 507, 524, 526
Albritton, D. L., 397
Aleixo, R. M. V., 138, 144-46
Alexander, M. H., 130
Alexander, S., 619, 639
Alexandrov, I. V., 455
Alfano, R. R., 618
Alfrey, T., 16
Aliaga Guerra, D., 247
Alkemade, C. Th. J., 267, 274, 275, 278
Allan, G., 542
Allara, D. L., 514
Allen, F. S., 341-43
Allen, L., 267, 269
Allgen, L. G., 80-82, 86
Allison, A. C., 137
Allison, J., 268, 277, 278
Allison, S. A., 43
Almgren, M., 138, 146
ALTKORN, R., 265-89; 122, 123
Altman, R. S., 403, 407
Altona, C., 199, 201-4
Altwegg, L., 633, 637
Alverez, M. S., 524, 526, 529
Alzetta, G., 648

Amamou, A., 247
Amano, T., 387, 390, 393, 394, 402, 404, 405, 410
Ambartumian, R. V., 657, 661
Ames, L. L., 326
Amirav, A., 621
Andersen, J. D., 635
Andersen, O. K., 549
Andersen, P., 118, 121, 129
Anderson, A. B., 296, 298, 299, 301, 306, 319, 321, 472
Anderson, D., 116
Anderson, J. B., 165
Anderson, P. W., 467
Anderson, R. W., 620, 635, 643
Anderson, S. L., 119, 120
Anderson, T. G., 393, 395
Andersson, S., 53, 55, 59-62
Ando, D. J., 644
Ando, R., 151, 154
Andre, J. J., 636
Andre, J. M., 381, 638
Andresen, P., 130
Andrews, L., 388
Angell, J. K., 487, 488, 501
Animalu, A. O. E., 362
Anner, O., 663, 668
Anosov, D. V., 564, 572, 586
Anton, A. B., 61, 65
Antoniewicz, P., 223, 232
Aoiz, F. J., 123, 127
Aoiz, F. J. L., 123
Aoyama, Y., 151
Apell, P., 526
Appelbaum, J. A., 539, 549
Appelman, E. H., 646
Applequist, J., 339
Arakawa, E. T., 524
Aravind, P. K., 513, 516, 518, 519, 521, 523, 526, 529, 530
Argon, A. S., 260
Arias, J., 517, 523
Arkhipov, V. I., 643
Arlinghaus, F. J., 549
Armstrong, R. C., 40, 41
Armstrong, S., 304, 313, 320
Arnold, G. S., 129, 284
Arnold, S., 618, 643
Arnol'd, V. I., 564, 566, 579
Arnott, S., 343-45, 347, 349-51
Arpin, M., 35, 38
Arrowsmith, P., 130, 131
Arya, K., 518, 519
Asai, K. W., 257
Asakawa, E. T., 518
Ashford, M. N. R., 657, 661, 662
Aslund, N., 300, 309
Aspnes, D. E., 531

Astholz, D. C., 665
Astumian, R. D., 150
Asubiojo, O. I., 175
Atha, P., 297, 303, 312
Atha, P. M., 303
Atherton, ♂. J., 468
Atik, S. S., 138
Atkins, R. M., 113, 302
Attwood, D., 139
Auer, P. L., 42
Ausloos, M., 531
Austin, B. J., 361, 364
Auweter, H., 619, 620
Avakian, P., 615
Avan, P., 267-69
Avery, N. R., 61, 63, 65
Avez, A., 564, 566, 579
AVOURIS, P., 49-73; 55, 56, 61, 62, 65-70, 520, 675

B

Baase, W. A., 343, 346
Babamov, V. K., 180
Baberschke, K., 229-31
Bacastow, R. B., 483
Bachackashvilli, A., 513
Bachelet, G. B., 369
Bachmann, C., 301, 321
Backes, U., 61
Backx, C., 61, 65
Bacon, F., 261
Bacskay, G. B., 206
Baden, A. D., 58
Bader, G. B., 619
Bader, S. D., 67
Baer, M., 183
Baer, T., 119
Baerends, E. J., 301, 310, 321, 373
Baeriswyl, D., 637
Baetzold, R. C., 292, 299, 301, 303-5, 313, 321
Baggott, J. E., 667, 676, 678, 679, 681-83, 685
Bagley, B. G., 253
Bagus, P. S., 206, 321
Bai, Y. Y., 581
Bailey, S. P., 487, 501
Baillie, M. B., 300, 309
Baker, A. D., 461, 462, 465
Baker, G. L., 637
Baker, J. C., 255
Baker, J. M., 68
Balabaev, N. K., 420
Balamuta, J., 127
Balasubramanian, K., 376, 378
Balazs, N. L., 331, 333, 571
Balducci, G., 378
Baldwin, A. C., 657, 661
Baldwin, D., 345

694 AUTHOR INDEX

Baldwin, R. L., 35
Balint-Kurti, G. G., 362
Ball, R. C., 433
Ballard, D. G. H., 34
Balode, D. R., 625, 626, 632
Balzani, V., 469, 475
Banavar, J. R., 425
Bangham, A. D., 140
Bánhegyi, Gy., 203, 204
Banovetz, L. M., 646
Bansal, R., 424
Barbanis, B., 579, 583
Barber, P. W., 524, 526
Bard, A. J., 475
Bardeen, J., 538
Bardsley, J. N., 360, 362, 380
Bare, S. R., 62
Barker, J. R., 657, 659, 661, 664
Barkley, M. D., 36, 43
Barnes, M. R., 61
Baro, A. M., 60, 62
Baronavski, A. P., 267, 283
Barraclough, C. G., 524, 526
Barrera, R. G., 517, 527, 531
Barrett, E. W., 486
Barrett, J. J., 670
Barrett, P. H., 298, 307, 308, 316
Barron, L. D., 332
Barrow, R. F., 300, 309, 326
Barsky, A., 544
Barthelat, J. C., 301, 310, 321, 368-72
Bartlett, R. J., 206, 207
Bartoszek, F. E., 122, 123, 131
Bartovics, A., 16
Bartsch, H., 151
Basch, H., 300, 303-5, 309, 312, 313, 317, 319, 321, 371, 374, 378, 379
Baseman, R. J., 124
Basilevsky, M. V., 184
Bass, L. M., 176
Bässler, H., 621, 625, 629-31, 633, 634, 638-41, 645, 648, 649
Batchelor, G. K., 93
Bates, D. R., 375
Bates, J. K., 302, 311
Bauer, E., 484
Baughman, R. H., 637
Baumann, C. A., 297, 307-9, 316, 317, 319, 321
Baumgarten, E., 138, 144-46
BAUMGÄRTNER, A., 419-35; 419, 423-28, 430, 433
Bauschlicher, C. W. Jr., 293, 297, 301, 306, 310-12, 378, 380
Baybutt, P., 365, 366, 368
Bayles, B. J., 253

Bayley, P. M., 349, 350
Beals, E., 36
Beasley, J. K., 18
Beattie, J. K., 472, 473
Bechgaard, K., 645, 646
Beck, H., 242
Becker, H. U., 304, 312, 313, 320
Beckmann, C. O., 11
Beckmann, H.-O., 301
Bedeaux, D., 101
Begemann, M. H., 398-403, 405-7
Behe, M., 347
Beitz, J. V., 462, 463, 466, 467, 475, 476
Bell, R. P., 179
Bender, C. F., 680
Bender, M. L., 137
Benderskii, V. A., 619
Benedek, G. B., 6
Benedek, R., 372
Benham, C. J., 36
Benndorf, C., 235-38
Benner, R. E., 522, 524, 526
Bennett, W. R. Jr., 268
Bennetti, P., 267
Beno, M. A., 646
Benoit, H., 28, 34, 35
Beratan, D. N., 461-63, 465, 467
Berberian, J. G., 79
Beresford, J. R., 662, 668
Berezin, I. V., 138, 141
Berg, J. O., 267
Berg, O. G., 104
Bergman, D. J., 531
Bergman, J. G., 524, 526
Bergmann, K., 284
Berk, N. F., 621
Berkovic, G. E., 620
Berman, G. P., 571
Bernal, J. D., 241, 247
Bernard, M., 636
Bernardi, F., 204
Bernasek, S. L., 61, 65, 116
Bernath, P., 393, 402, 404
Bernath, P. F., 394
Bernengo, J. C., 79
Bernesconi, J., 639
Bernhard, P., 472, 473
Bernholc, J., 297, 303, 306, 311
Bernstein, R. B., 109, 123, 127, 128, 266
Berquist, B. M., 115
Berreman, D. W., 524
Berry, D. E., 614
Berry, G. C., 33, 35, 38
Berry, M. B., 571
Berry, M. J., 667, 674-80
Berry, M. V., 567-71, 577

Bertel, E., 69, 216, 217, 224, 226-28, 231
Best, S. P., 472, 473
Bethe, H. A., 373, 380
Bhat, S. V., 297, 307, 321
Bianchi, E., 38
Bicerano, J., 182
Bidani, M., 121
Bienenstock, A., 247
Bierbaum, V. M., 113-15, 123, 176
Biermann, L., 380
Biewenga, K. J., 274, 278
Billmeyer, F. W., 18
Biloen, P., 61, 65
Bilotta, R. M., 119, 120
Binder, K., 423, 425-28, 430, 433
Binkley, J. S., 196-99, 201, 206
Binnig, G., 546
Birac, C., 257
Birch, D. J. S., 619
Bird, R. B., 40, 41
Birdsall, D. L., 347, 349
Birke, R. L., 524, 526
Birkoff, G. D., 571
Birks, J. B., 619
Birks, J. W., 481, 482, 487, 502
Bishop, A. R., 638
Bishop, M., 424, 426, 427, 430, 431
Bittenson, S., 115
Bitterling, K., 626
Bixon, M., 40, 420, 452, 455-58, 460, 461, 464, 474, 475
Bizzigotti, G. O., 151
Blackman, G. S., 646
Blais, N. C., 169, 172
Blake, D. R., 483
Blakely, J. M., 67
Blanchet, G. B., 55
Blaser, S., 244, 246
Blatchford, C. G., 513, 524, 526
Bleicher, W., 201
Blint, R. J., 175
Blom, C. E., 199, 201-4, 206, 209
Blomberg, C., 104
Bloomfield, P., 487, 501
Bloomfield, V. A., 33, 38
Bloor, D., 644
Blough, N. V., 468
Blume, M., 376
Blumen, A., 616-18
Bly, S. H. P., 130, 131
Blyholder, G., 62, 68, 298, 319
Bobrowicz, F. W., 299, 300, 309, 377, 378
Bock, C. W., 204, 205
Bockris, J. O'M., 465

AUTHOR INDEX 695

Boettinger, W. J., 254, 255
Boettner, R., 662
Boeyens, J. C. A., 301, 304, 305, 313
Bogan, D. J., 111, 112
Boggs, J. E., 196, 199-210
Bohigas, O., 568, 585, 586
Boletta, F., 475
Bolotina, I. A., 337
Boltzmann, L., 565
Bonczyk, P. A., 267
Bondi, D. K., 168, 169, 174, 181
BONDYBEY, V. E., 591-612; 296, 300, 306, 310, 320, 387, 390, 393
Boni, H., 471
Bonifacic, V., 362, 363, 370, 377
Bonilha, J. B. S., 138, 145, 146
Bonner, N. A., 472
Bonzel, H. P., 62
Boots, H., 422, 431
Bordignon, E., 468
Born, M., 5, 14, 16, 17, 509
Botschwina, P., 196, 201, 204, 206, 415
Bottcher, C., 380
Böttcher, W., 472-76
Boudreaux, D. S., 248
Boughner, R. E., 482, 496
Bounds, P. J., 624, 626
Bourdon, J., 292
Bourguignon, B., 126
Bowers, M. T., 119, 160, 165, 174, 176, 181, 414
Bowman, D. R., 231, 232
Bowman, J. M., 177, 178, 183
Bowman, R. L., 340
Bowman, W. C., 396
Boyce, J. B., 247
Boyer, K., 268
Bozso, F., 517
Bozzelli, J. W., 115
Braam, J. M., 646
Bradbury, J. H., 38
Bradley, J. A., 514, 517-20
Bradshaw, A., 68
Bradshaw, A. M., 235
Brahms, J., 337, 339, 346, 349
Brahms, J. G., 345
Brahms, S., 337, 339, 345, 349
Brandon, J. R., 472
Brandt, D., 115
Brant, D. A., 25
Bratož, S., 197
Brault, J. W., 394, 405
Brauman, J. I., 175, 438
Braun, A. M., 138, 619, 620
Braun, C. L., 624, 625, 627, 633, 634
Brechthold, P. S., 524
Breckenridge, W. H., 122, 126

Brédas, J. L., 381, 637, 638
Brender, C., 422
Brennan, S., 218, 219, 229
Brenner, H., 36, 102
Brewer, L., 302, 303, 310-14, 318, 320
Brewer, R. G., 268
Brikenshtein, V. Kh., 619
Brillante, A., 517, 521, 522, 527
Briscoe, B. J., 425, 429
Brochard, F., 77, 105, 425
Brochard-Wyart, F., 77
Brocki, T. R., 331, 333
Brodsky, M. B., 67
Brody, T. A., 567, 568, 585
Broersma, S., 45
Broida, H. P., 517, 527
Brooks, B. R., 196, 206
Brooks, P. R., 122, 130
Brophy, J. H., 129, 284
Brosteaux, J., 19
Brouwer, L., 665
Browett, R. J., 124
Brown, C. M., 304, 312
Brown, D. W., 634
Brown, G. M., 472, 473
Brown, J. K., 517
Brown, R. C., 181, 617
Brown, W. L., 257
Browning, R., 119
Broyer, M., 267, 272
Brubaker, C. H. Jr., 472
Brubaker, G. R., 461, 462, 473
Bruchmann, D., 59, 61, 62
Bruchmann, H. D., 60
Brudno, S., 346
Brudzynski, R. J., 663
Brumer, P., 582
Bruno, J. B., 128
Bruns, W., 424
Brunschwig, B. S., 442, 452-54, 456, 458-61, 463, 468, 472-76
Brus, L. E., 517, 527-29
Bucher, H., 139, 522
Bucher, Ph., 345
Buckel, W., 253
Buckingham, A. D., 332
Bueche, A. M., 13
Buenker, R. J., 206
Bugl, P., 26
Buhks, E., 452, 456-58, 461, 474
Bulmer, J. T., 470
Bumm, A., 524, 526
Bunimovich, L. A., 585
Bunker, D. L., 162
Bunker, P. R., 415
Bunton, C. A., 144-46
Burak, I., 266
Burchard, W., 7
Burgers, J. M., 36

Bürgi, H. B., 472, 473
Burgos, J., 633, 637
Burke, L. A., 381
Burnham, R. E., 130
Burshtein, A. I., 455
Burstein, E., 521, 522, 524, 526
Bursten, B. E., 296, 303, 306, 312
Burum, D. P., 621
Busby, P. E., 258, 259
Busby, R., 292-95
Busch, G. E., 233
Bush, C. A., 349
Buss, R. J., 117, 121, 124, 660, 661, 663, 668
Bustamante, C., 330, 333-35, 349
Butler, J. E., 117, 118, 121
Butler, L. J., 663
Butscher, W., 376
Buttenshaw, J., 390, 404-6
Butz, R., 61
Bywater, S., 9

C

Cabannes, J., 17
Cade, N. A., 630
Cahn, J. W., 255
Cahn, R. W., 242, 256, 258
Calcaterra, L. T., 467, 468
Calder, V., 297, 298, 307
Caldwell, C. D., 279
Calef, D. F., 455, 456, 461, 471
Callaghan, R., 524, 526
Callaway, J., 380
Callcott, T. A., 524
Callis, L. B., 482, 496
Callis, P. R., 614
Calvin, M., 137
Cammarata, R. C., 257
Campbell, D. L., 638
Campbell, F. M., 119, 120
Campbell, J. D., 662
Campbell, J. R., 513
Campillo, A. J., 619
Campion, A., 517, 527
Cannon, B. D., 667, 668, 671, 672, 676, 679, 683, 684
Cannon, R. D., 438, 453, 454, 475
Cantor, B., 256, 258
Cantor, C. R., 342
Cargill, G. S. III , 242, 246-48
Carney, G. D., 402, 415
Caron, L., 619
Carothers, W. H., 4
Carr, C. I. Jr., 18
Carrington, A., 390, 392-94, 404-6
Carrington, T., 131
Carroll, H. F., 121

AUTHOR INDEX

Čarsky, P., 205
Carter, W. L., 247
Carvalho, M. J., 633, 637
Casassa, E. F., 5
Casati, G., 564, 566, 571, 579
Casavecchia, P., 117, 121, 124
Case, D. A., 266, 273, 279
Caspar, J. V., 475
Cassel, J., 473
Castner, D. G., 63
Castro, M., 303
Catanzarite, J., 662, 668
Cates, R. D., 174
Cavin, O. B., 256
Cech, C. L., 349-51
Cech, R. E., 241
Cederbaum, L. S., 567, 585
Celli, F., 178
Celli, V., 518, 519
Ceperley, D., 420, 424, 426, 427, 430, 431
Cerjan, C. J., 179
Certain, P. R., 583
Chabay, I., 522, 524, 526
Chadi, D. J., 542, 544, 546, 547
Chaimovich, H., 138, 144-46
Chakravorty, K. K., 128
Chalmers, B., 253
Chan, A., 346
Chan, K. H., 337
Chan, M.-S., 472
Chance, B., 438, 439, 443, 447, 456
Chance, R. R., 381, 509, 514, 516, 517, 624, 627, 633, 637, 638
Chandesris, D., 554
Chandler, D. W., 667, 676, 678, 679, 681-83, 685
Chandrasekaran, R., 343-45, 347, 349-51
Chandraskhar, S., 420
Chang, C. T., 337
Chang, J. S., 491, 492, 496
Chang, R. K., 522, 524, 526
Chang, T. C., 362
Chao, K.-J., 659
Chapman, S., 164
Charle, K. P., 626
Charney, E., 104
Charters, P. E., 130, 131
Chaturvedi, B. K., 184
Chaudhari, P., 242, 252, 255
Chebotayev, V. P., 267, 268
Chekhov, V. O., 337
Chelikowsky, J. R., 540-43, 547-49
Chemla, D. S., 524, 526
Chen, C. K., 522
Chen, C. Y., 524
Chen, H. S., 242, 244, 245, 257, 261

Chen, W. P., 521, 522
Chen, Y. C., 517
Chen, Y. H., 337
Chen, Y. J., 521, 522
Cheng, K. T., 374
Chesnavich, W. J., 160, 164, 165, 176, 177
Chesters, M. A., 67, 68
Cheung, A. S.-C., 379
Chew, H., 512, 524
Chew, W. C., 90, 92, 99-101, 103
Chiang, C., 543
Child, M. S., 184, 451, 581, 584, 670, 673, 674, 676
Chipman, D. M., 206
Chirikov, B. V., 568, 571
Choi, S. I., 632, 634
Chou, C.-C., 483, 517
Chou, M., 474-76
Chou, P., 337
Christe, K. O., 201
Christiansen, P. A., 365, 367, 370, 371, 375, 376, 378, 380
Christman, K., 514, 518, 520
Christov, S. G., 178
Chuang, M.-L., 667, 676, 678, 679, 681-83, 685
Chuang, T. J., 524
Chudacek, I., 648
Chung, J., 362
Chung, T. C., 638
Chung, T. J., 524
Chung, Y. W., 58
Ciano, M., 475
Cicerone, R. J., 482, 494, 498
Ciferri, A., 38
Clark, D. T., 205
Clarke, G. A., 293-99, 301, 306, 321
Clarke, T. C., 636
Clary, D. C., 168-70, 185
Clemo, A. R., 124
Clinton, W. L., 233
Closs, G. L., 467, 468
Clyne, M. A. A., 266
Cocke, D. L., 313
Coggiola, M. J., 660, 661, 668
Cohen, D., 303, 304, 312, 379
Cohen, K. H., 350, 351
Cohen, M. H., 244, 245, 252, 361
COHEN, M. L., 537-62; 538, 540-44, 546-49, 551, 555
Cohen, N., 171, 175
Cohen, R. J., 6
Cohen-Tannoudji, C., 267-69
Cole, K. S., 81, 82, 119
Cole, R. H., 78, 79, 81, 82, 84, 86
Colligan, G. A., 253

Collins, R. J., 268
Collins, W. E., 217, 220, 226, 231
Coltrin, M. E., 179, 180
Colvin, M. E., 415
Comes, F. J., 117, 121
Comes, R., 636
Coniglio, A., 8
Connor, J. N. L., 168, 169, 174, 181, 185
Conrad, H., 553, 555
Contopoulos, G., 579
Conway, T. J., 126
Conwell, E. M., 636
Cook, H. E., 257
Cooke, D. D., 524
Cooper, W. F., 293-99, 301, 306, 321
Córdova, J. F., 118, 121, 268, 274, 282, 286
Coriell, S. C., 253, 255
Corset, J., 524, 526
Cosse, C., 293, 295
Costes-Puech, E., 371
Costley, J., 167
Cotton, F. A., 291, 296, 303, 306, 312
Cotton, J. P., 34
Cowan, R. D., 374
Cox, D. M., 298, 307
Cox, J. W., 126
Cox, P. A., 58, 464
Craig, H., 483, 484
Cram, D. J., 137
Crawford, J. L., 343
Crecelius, G., 638
Creighton, J. A., 507, 513, 524, 526
Creutz, C., 461, 462, 467-69, 471-76
Crick, F. H. C., 35
CRIM, F. F., 657-91; 667, 668, 671, 672, 676, 679, 682-87
Crisp, G. M., 620
Crofton, M. W., 400, 402, 403, 407
Crosley, D. R., 266, 267
Cross, J. B., 128
Crosswhite, H. M., 402
Crowell, J., 522
Cruse, H. W., 284
Crutzen, P. J., 481, 482, 490, 496, 499, 502
Csaszar, P., 198
Cuccovia, I. M., 138, 144-46
Cummerow, R., 6
Cunningham, J. A., 517
Curl, R. F., 130
Curtis, E. C., 201
Curtis, J. C., 467
Curtiss, C. F., 40, 41

AUTHOR INDEX 697

D

Dagdigian, P. J., 126, 130, 266, 276, 284
Dahl, J. P., 300, 309
Dahl, K. S., 350
Dai, H.-L., 666-68
Daily, J. W., 274
Dakhnovskii, Yu. I., 453-55
Dalgarno, A., 380
Dalton, L. R., 636
Damgaard, A., 375
Daniels, H. E., 24, 32
Danon, J., 119
Darinskii, A. A., 420
Das, G., 293, 294, 374
DAS, P., 507-36
Das, P. C., 524
Dasgupta, B. B., 526
Da Silva, E., 149
Datta, S. N., 374
Daudey, J. P., 380
Davenport, J. W., 55, 57
Davidovich, M. A., 633, 637
Davidovits, P., 126
Davidson, B., 337
Davidson, E. R., 198, 206, 207
Davidson, F. E., 124
Davidson, N., 11
Davidson, R. E., 124
Davies, B. M., 55
Davies, D. R., 345, 350, 351
Davies, H. A., 252
Davies, L. B., 247
Davies, M. J., 620
Davies, P. B., 393, 405
Davis, G. R., 648
Davis, J. P., 177
Davis, L. A., 261
Davis, M. J., 569, 576, 583, 584
Davis, S. C., 292
Davis, S. P., 394, 405
Davoli, I., 524
Davydov, A. S., 27
Davydov, B., 538
Dayan, S., 38
Debye, P., 13, 19, 90, 95
Debye, P. J. W., 13, 14, 16, 18
de Castro, A. R. B., 522
Decker, F., 34
Decker-Freyss, D., 9
Decomps, B., 267, 272
DeFrees, D. J., 197, 199, 201
De Gennes, P. G., 11, 77, 105, 419, 420, 422, 425-31, 433, 619
de Goede, J., 78
deGroot, C. P. M., 61, 265
de Groot, S. R., 90, 94
de Haën, C., 38
Dehalle, J., 381
Dehaven, J., 122, 123, 126

de Keizer, A., 90
de Lacey, E. H. B., 90, 93-95, 98, 100, 101, 103
Delahay, P., 452
Delanaye, F., 51
Delbrüch, M., 150
del Condë, G., 321
Delley, B., 297, 301, 303, 306, 311
Delos, J. B., 571
Delpech, J. F., 130
Del Sole, R., 542
Deltour, R., 636
DeLucia, F. C., 396
Delwiche, C. C., 483, 484
Demtröder, W., 268
DEMUTH, J. E., 49-73; 55, 56, 61-63, 65, 67-70, 514, 518, 520, 553, 557, 558
Demuynck, J., 301, 321, 322
DePaola, R. A., 63
DePristo, A. E., 465
dePuy, C. H., 176
Derjaguin, B. V., 90
Desclaux, J. P., 359, 373-75, 377
Detrich, J., 374
Dettenmaier, M., 34
Deutch, J. M., 38, 75, 422, 431
Deutsch, J. M., 430, 431
DeVault, D. C., 438, 439, 443, 447, 456, 466
DeVoe, H., 349-51
DeVore, T. C., 297, 298, 307
Devreux, F., 636
DeVries, H., 621
deVries, M. S., 129
Di Capua, E., 345
Dickerson, R. E., 343
Diercksen, G. H., 206
Diercksen, G. H. F., 415
Dietz, T. G., 300, 310
DiGuiseppe, T. G., 126
DiLella, D. P., 292-99, 306-9, 316, 319, 320, 513, 524, 526, 527
Dill, B., 111, 112
DiMaria, D. J., 523
DiMarzio, E. A., 244, 245
Dimpfl, W. L., 129, 284
DiNardo, N. J., 55, 61, 62, 65, 67-70
Dinglinger, A., 2
Dintzis, H. M., 80, 81, 87
Dirscherl, R., 126
Dispert, H. H., 122
Ditchfield, R., 199
Dixmier, J., 247
Dixon, D. A., 320, 321, 380
Dixon, R. N., 301, 362
Dixon, T. A., 393, 395, 415
Dodiuk, H., 471

Dodson, R. W., 472, 473
Doering, D. L., 216, 226, 235, 237
Dogonadze, R. R., 454, 456, 457, 459
Doherty, P. M., 267
Doi, M., 40, 429, 433
Dolg, M., 301, 310
Dolin, S. P., 465
Dolson, D. A., 571
Donaldson, D. J., 111, 112, 117, 121
Dongonadze, R. R., 465
Donovan, K. J., 630, 631, 638-40, 644, 645, 649
Donovan, R. J., 118, 127
Doran, M., 294, 296, 302, 306
Dorfman, L. M., 472
Dornhaus, R., 522, 524, 526
Dose, V., 68
Doty, P. M., 16, 18, 28, 38
Douglas, D. J., 111
Drehman, A. J., 252
Dreiling, T. D., 126
Drew, H., 343
Drickamer, H. G., 614
Drolshagen, G., 170
Drowart, J., 292, 294, 300, 304, 305, 309, 312, 313, 317
Drude, P., 509
Druger, S. D., 524
Dübal, H. R., 686
Dubois, L. H., 61-63
Duchovic, R. J., 175
Ducloy, M., 266, 267, 272
Dudek, D., 664
Dudley, M., 644
Dudluk, H., 471
Duewer, W. H., 491, 492
Duff, J. W., 183
Duke, C. B., 542, 614, 623, 634, 635, 641, 648
Dukhin, S. S., 89, 90, 93-95, 99
Dulz, G., 474
Dumond, J., 257
Dumont, M., 267, 272
Duncan, G. L., 124
Duncan, J. L., 202, 203
Duncan, M. A., 118
Dung, M. H., 138, 146
Dunlap, B. I., 297
Dunn, K. M., 203
Dunning, T. H. Jr., 171, 172, 178, 180, 181, 377, 379
Dupuis, M., 197, 198, 205, 415
Dupuy, F., 617
Durana, J. F., 661
Durand, J., 247
Durand, Ph., 368-71, 381
Durham, B., 473
Durkin, A., 124

Durup, J., 119, 120
Dušek, K., 7
Duwez, P., 242
Dwarakanath, K., 473
Dwyer, F. P., 472, 473
Dykstra, C. E., 206
Dyson, F. J., 567, 585
Dyson, W., 298, 316, 319
Dzelzkalns, L. S., 113, 115

E

Eagen, C. F., 517, 522, 524, 527
Eastman, D. E., 68, 517, 544
Eberhardt, W., 62, 554, 556, 558, 560
Eberly, J. H., 267, 269
Eckbreth, A. C., 267
Edmonds, A. R., 27
Edwards, S. F., 420, 426-29, 433
Eesley, G. L., 513, 523
Efremov, Y. M., 293, 296, 302, 310, 311, 318
Efrima, S., 455-57, 460, 464, 474, 475, 508, 509, 511, 513, 514, 516, 517, 531
Egdell, R. G., 58
Ehhalt, D. H., 483
Ehinger, K., 636
Ehrenfreund, E., 636
Ehrlich, D. J., 127
Ehrlich, P. R., 483
Eichhorn, M., 626, 627
Eicke, H. F., 139
Eigen, M., 150
Einstein, A., 13, 18, 269
Eisenberg, H., 33, 38, 39, 81, 82
Eizner, Yu. E., 36
Elgersma, H., 171
El Gomati, M. M., 67, 69
Eliason, M. A., 162
Elliott, C. S., 681
Ellis, D. E., 297, 301, 303, 306, 311
Ellison, G. B., 113, 114, 118, 123
El-Sayed, M. A., 617, 618, 675
Elson, J. M., 522
Emin, D., 224, 229-31
Enderby, J. E., 464
Endicott, J. F., 461, 462, 473
Endo, Y., 410
Endriz, J. G., 522
Engel, N., 313
Engelhardt, R., 284
Engler, E. M., 647
English, J. H., 296, 300, 306, 310, 320
Ensminger, M. D., 116, 661

Epstein, A. J., 636
Eri, T., 351
Erickson, W. R., 5
Erley, W., 61, 62
Ermler, W. C., 206, 305, 313, 370, 371, 373-76, 378
Ern, V., 615, 616
Ernst, M. H., 7
Ernst, R. R., 524
Erpenbeck, J. J., 40, 41
Erskine, J. L., 55, 61
Ertl, G., 553, 555
Escabi-Perez, J. R., 148, 150
Esser, M., 376
Estler, R. C., 122, 129
Etemad, S., 636, 638
Evans, E., 50, 58, 59
Evans, E. L., 643
Evans, G. T., 42
Evans, K. E., 426-28, 433
Evans, M. G., 161
Even, O., 621
Everett, W. W., 33
Ewart, R. H., 13
Ewen, B., 430
Ewig, C. S., 207, 362, 374
Ewing, A., 297, 298, 307
Eyers, A., 329
Eyring, H., 161, 162

F

Fabish, T. J., 641, 648
Fabre, J. M., 646
Falkenhagen, H., 90, 95
Fang, H. L., 674
Fano, U., 280
Farina, R., 472, 473
Farley, J. W., 390, 406
Farneth, W. E., 175, 667, 676, 678, 679, 681-83, 685
Farnoux, B., 34
Farnworth, H. E., 546
Farquhar, I. E., 565, 572, 574, 575
Farrar, J. M., 119, 120, 660, 661
Farrell, H. H., 547
Fasman, G. D., 337
Fassler, D., 524, 526
Fauster, Th., 68
Fayer, M. D., 614, 616, 617, 621
Fearing, V. L., 258, 259
Feher, G., 467
Feibelman, P. J., 215, 216, 222, 223, 226, 517, 527, 531, 555
Feld, M. S., 267
Feldblum, A., 638
Feldhaus, J., 218, 219, 229
Feldman, B. J., 267
Fellner, W. H., 487, 501

Felsenfeld, G., 347
Felter, T. E., 63
Fendler, E. J., 137, 138, 143
FENDLER, J. H., 137-57; 137-41, 143, 144, 146, 148, 150-54
Fenn, J. B., 116, 524, 526
Fermi, E., 579
Fernie, D. P., 118, 124
Ferrell, T. L., 524
Ferry, J. D., 78, 429, 433
Feshbach, H., 524
Fesser, K., 638
Feulner, P., 239
Feynman, R. P., 25
Field, R. W., 380, 586, 666
Filippov, P. G., 619
Fink, J., 638
Finney, J. L., 241, 247
Fisch, H. L., 426, 427, 430, 431
Fischer, C. F., 366
Fischer, E. W., 34
Fischer, S. F., 457, 460
Fisher, G. B., 61
Fishman, S., 571
Fitzgerald, G., 198, 206
Fixman, M., 9, 17, 25, 32, 40-42, 76, 90, 92, 94-104, 420
Fleischmann, M., 507
Flesch, J., 380
Fletcher, D., 429
Flodstrom, S. A., 229, 231, 544
Florence, A. T., 139
Flores, J., 567, 568, 585
Flory, P. J., 11, 13, 23, 25, 29, 32, 34, 35, 40
Fluendy, M. A. D., 118
Flynn, C. P., 517
Fock, V., 358
FOGARASI, G., 191-213; 193, 196, 198-206
Foosnaes, T., 302, 311
Ford, G. W., 517, 527, 531
Ford, J., 564, 566, 568, 571, 572, 579
Ford, T. A., 295
Ford, W. K., 542, 614, 623
Ford-Smith, M. H., 474
Fork, R. L., 637
Forman, P., 1
Forst, W., 179, 580, 586
Fort, A., 616
Foster, J. F., 33
Fotakis, C., 127
Fouassier, M., 293, 295
Fournier, P. G., 390, 404
Fowler, R. H., 15
Fowlkes, W., 633, 637
Fox, D. J., 198, 206
Fox, T. G. Jr., 13
Franchy, R., 231
Franci, M. M., 199

AUTHOR INDEX 699

Francois, B., 636
Frank, A. J., 148, 150
Frank, F. C., 241
Frank, J. P., 660
Franzen, H. F., 297, 298, 307
Fraser, P. J., 483
Frauenfelder, H., 438, 439, 443, 447, 456
Fredkin, D. R., 467
Freed, K. F., 25
Freeman, A. J., 297, 301, 303, 306, 311, 549
Freeman, M. P., 333
Freeouf, J. L., 67, 68
Freiberg, A., 622
Freire, J. J., 32
French, J. B., 567, 568, 585
Frenkel, D., 433
Frese, K. W. Jr., 475
Frey, H. M., 657, 681
Fridberg, J., 258
Friddle, R. J., 67
Fried, F., 38
Friedel, J., 645
Friedman, H. L., 438-42, 454, 455, 459, 461, 463, 464, 468, 470, 471
Friedrich, H. B., 206
Frisch, H. L., 424
Frisch, M. J., 175
Frisoli, J. K., 667, 668, 675, 676, 678, 679, 681, 682
Fritsch, P., 151
Froben, F. W., 524
Fröhlich, H., 119, 469, 470
Froitzheim, H., 61, 62, 66
Frost, H. J., 247, 248
Froyen, S., 543
Fuchs, R., 57, 517, 526, 527, 531
Fuentealba, P., 301, 310, 380
Fuhrhop, J. H., 151
Fujii, M., 26-29, 31-36, 38, 93, 102
Fujii, M., 33, 38
Fujita, H., 32, 33, 37, 38
Fujita, S., 26
Fujiyoshi, Y., 524, 526
Fulcher, G., 244
Fuller, F. B., 35
Funasaki, N., 146
Fünfschilling, J., 633, 637
Fuoss, P. H., 247
Fuoss, R. M., 80, 81, 87

G

Gadzuk, J. W., 56
Gailis, A. K., 632
Galla, H. J., 150
Gallo, A. R., 517, 527
Gans, R., 18
Garbuzov, D. Z., 624

Garcia de la Torre, J., 38
Garcia-Prieto, J., 300, 320, 321
Garland, D. A., 318, 320
Garmire, E. M., 268
Garnett, R. W., 517
Garoff, S., 524, 526, 528, 529
Garrell, R. L., 524, 526
GARRETT, B. C., 159-89; 68, 160, 162-64, 167-72, 174, 176-85
Garscadden, A., 413
Garvin, D., 488
Gaskell, P. H., 247, 248
Gaswick, D. C., 461, 462
Gates, J. A., 58, 63, 65
Gaudin, M., 567, 585
Gauyacq, J. P., 123
Gaw, J. F., 198, 206
Gay, J. G., 549
Geacintov, N. E., 631
Geddes, J., 109
Geiger, L. C., 126
Geis, M. W., 122
Gelb, A., 267
Gelbart, W. M., 674, 675
Gelich, D., 119, 120
Genack, A. Z., 524
Gentry, S. T., 616, 617
Gentry, W. R., 392
Gény, F., 42, 423
George, P. M., 63, 204, 205, 461, 462
Geraedts, J., 268
Gerardy, J. M., 531
Gerber, Ch., 546
Gerber, W. H., 318
Gerbstein, Y. M., 522
Gericke, K.-H., 117, 121
Gerischner, H., 524, 526
Gerjhumi, S., 663, 668
German, E. D., 453, 454, 465, 472
Gerratt, J., 196, 197
Gerrity, D. P., 125
Gershfeld, N. C., 139
Gersten, J. I., 61, 512, 514, 524, 526, 528, 529
Geselowitz, D., 472, 473
Geusic, M. E., 296, 299, 300, 306, 309, 310
Giachino, J., 624
Gibbs, J. H., 79, 81, 82, 84, 86, 103, 104, 244, 245
Gibbs, J. W., 565
Gibson, H. W., 636
Gidel, L. T., 499
Giese, C. F., 392
Giessen, B. C., 242
Gingerich, K. A., 113, 293-95, 298-305, 307-10, 312-15, 317, 318, 320
Ginter, M. L., 304, 312
Girerd, J.-J., 464

Girlando, A., 518-21
Giuliano, C. R., 266, 268
Gland, J. L., 61, 62, 65
Glänzer, K., 664
Glass, A. M., 524, 526
Glass, G. P., 184
Glass, N. E., 518, 519
Glen, R. M., 118
Glockner, J., 337, 338
Glynn, C. P., 517
Gö, N., 40, 42
Goannoni, M. J., 568, 585, 586
Gobush, W., 32
Goddard, J. D., 196, 206
Goddard, W. A. III, 297, 300, 303, 306, 309, 311, 312, 363, 364, 366, 368-70
Godfrey, J. E., 33, 38, 39
Godzik, K., 621
Gohin, A., 513
Golde, M. F., 127
Golden, D. M., 657, 661
Goldstein, M., 244
Goldstein, N., 118
Goldstein, Y., 61
Gole, J. L., 126, 292, 300, 310, 320, 321, 380
Gombas, P., 360
Gomer, R., 215, 216, 222-25, 231, 232
Goncher, G. M., 529
Gonser, U., 247
Gonzalez Urena, A., 123
Goodenough, J. B., 467
Goodgame, M. M., 297, 303, 306, 311, 312
Goodman, M. F., 662
Goodwin, E. T., 538
Gordon, J. G. II, 518
Gordon, J. P., 524
Gordon, M., 7, 9
Goreau, T. J., 483
Gorry, P. A., 124
Gotlib, Yu. Ya., 40, 420
Gottscho, R. A., 266
Gouedard, G., 267, 272
Govers, T. R., 119
Gower, M. C., 127
Grabowski, J. J., 176
Graessley, W. W., 419, 425, 429
Gramila, T. J., 524, 526, 529
Grant, E. R., 663
Grant, P. M., 647
Grätzel, M., 137, 138, 146-48, 150, 151
Gray, D. M., 340-43, 346, 347, 349
Gray, H. B., 302, 311, 468
Gray, S. K., 179
Green, A. K., 217, 226
Green, G. J., 126, 346

Greene, B. I., 622, 648
Greene, C. H., 266, 279, 280
Greene, R. L., 636
Greenfield, N., 337
Greenlaw, D. K., 517
Greenlay, W. R. A., 672
Greenler, R. G., 524
Greer, A. L., 252, 256, 257, 260
Greer, J., 337
Gregor, J. M., 248
Gregoriadis, G., 137
Gregory, A. R., 205
Grempel, D. R., 571
Grest, G. S., 245
Grev, R. S., 160, 163, 164, 167, 170, 172, 177, 182, 184, 185
Greve, J., 350
Gribkovskii, V. P., 270, 286
Grice, R., 124
Grieneisen, H. P., 130
Grieser, F., 138, 524, 526
Griffin, K. P., 344, 345, 347
Griffith, J. S., 299, 461, 462
Grimmelmann, E. K., 164
Grinter, R., 304, 313, 320
Grizzle, V. M., 517
Gropen, O., 363, 370, 372
Gross, L., 140, 151
Gross, P. M., 78
Grote, R. F., 438, 439, 471
Grover, J. R., 123
Groves, K. S., 496
Gruen, D. M., 296, 302, 306, 311
Grundy, P. J., 247
Guarr, T., 468, 474
GUDEMAN, C. S., 387-418; 396, 398-403, 405-9, 413
Guelachvili, G., 394
Guenzburger, D., 298, 301, 308, 316
Guermonprez, R., 79
Guest, M. F., 294, 296, 302, 303, 306
Guichar, G., 544
Guillory, W. A., 117, 121, 662
Gumbs, G., 619
Gunthard, H. H., 206, 209
Guntherodt, H. J., 242
Gupta, A., 120, 123
Gupta, D., 257
Gupta, S. K., 113, 302, 303, 310, 312, 315, 318
Gürtler, P., 622
Gurvich, L. V., 293, 296, 302, 310, 311, 318
Guschlbauer, W., 341, 342, 347, 349
Gusman, G., 636
Gustafsson, T., 551
Gustavson, F. G., 571

Guterman, L., 152-54
Guth, E., 2
Gutzwiller, M. C., 571
Guyer, D. R., 118, 119, 268
Guyon, P.-M., 119

H

Ha, T. K., 206, 209
Haarer, D., 645
Haas, Y., 663, 668
Haasen, P., 261
Haberkorn, R., 467, 630
Habib, H. S., 472, 473
Habitz, P., 362, 376
Hackett, P. A., 662, 663
Haddon, R. C., 614
Haenel-Immendorfer, I., 19
Haese, N. N., 397, 403-5, 407, 408, 413
Hafner, P., 374, 376, 381
Hagerman, P. J., 38
Hagstrum, H. D., 233, 517
Hahn, J., 484
Haim, A., 472, 475
Hall, D. G., 78
Hall, K. B., 347, 348, 352
Hall, M. B., 297, 303
Hallam, A., 619
Haller, E., 567, 585
Haller, G. L., 116
Halmos, P. R., 564, 575
Halonen, L., 670, 673, 676
Halperin, B. I., 253
Halpern, B. L., 116
Halpern, J., 461, 462
Hamann, D. R., 369, 539, 543, 549, 555
Hamilton, C. E., 114, 118
Hamilton, E., 202, 203
Hamilton, F. D., 342
Hamilton, J. F., 292
Hamilton, P. A., 393, 405
Hammond, W. B., 671
Hampson, R. F., 488
Hanabusa, M., 267, 271
Hanai, T., 78
Hanamura, E., 522
Hancock, G., 122, 123, 657, 661, 662, 668
Handy, N. C., 171, 179, 180, 182, 196
Haneman, D., 66
Hanlon, S., 346
Hanrahan, C. P., 129, 517
Hänsch, T. W., 268
Hansen, G. P., 293, 295, 299, 309
Hansen, S. G., 296, 300, 306, 310
Hansma, P. K., 516-18, 521, 523, 527
Hanson, C. D., 524, 526, 529

Hanson, D. M., 216, 218, 219, 226, 231-33, 235, 614
Hanson, J., 482, 500
Hanss, M., 79
Hansson, G. V., 544
Happel, J., 36, 102
Harada, S., 150
Hardaker, M. L., 662
Hare, C. R., 293-99, 301, 306, 321
Härri, H. P., 318
Harris, C. B., 69, 517, 527, 529
Harris, F. E., 298
Harris, J., 61, 295-301, 311, 312
Harris, R. A., 25, 36, 40, 331, 332
Harrison, S. F., 2, 5
Harrison, W. A., 360
Hart, L. P., 267
Hart, R. M., 524
Harte, M., 483
Hartman, K. W., 467
Härtner, H., 201, 204
Harwell, M. A., 483
Hase, W. L., 160, 164, 165, 168, 175, 177, 451, 580, 586, 657
Hassager, O., 40, 41
Hassan, N., 643
Hatlee, M. D., 138, 147, 148, 150, 151, 618
Hauser, J., 472, 473
Hawryluk, A. M., 524, 526
Hay, P. J., 299, 300, 309, 366, 374, 377-79
Hayakawa, R., 78, 79, 81-84, 86, 87
Hayden, C. C., 125, 667, 668, 671, 679, 683, 687
Hayes, E. F., 178
Hayes, T. L., 333-35
Hayes, T. M., 247
Hays, T. R., 621
Hayter, J., 77, 430
Hazi, A., 360-62
He, G., 124
Head, J., 298
Heard, S. M., 524, 526
Hearst, J. E., 25, 33, 36, 40, 346
Heckenkamp, C., 329
Heeg, M. J., 461, 462
Heeger, A. J., 636, 638
Hefter, U., 284
Hegedüs, A., 201
Hehre, W. J., 197, 199, 206
Heidt, L. E., 496
Heiles, C., 579
Heimann, P., 544
Heine, S., 34
Heine, V., 360-62, 364, 540
Heinzmann, U., 329

AUTHOR INDEX 701

Heismann, F., 123
Heitmann, D., 518-21
Helfand, E., 21, 42, 91, 420
Heller, D. F., 671, 674-77
HELLER, E. J., 563-89; 181, 569, 571, 573, 574, 576, 581, 583-85
Hellmann, H., 360
Helman, A. B., 455
Helminger, P., 396
Helminiak, T. E., 38
Hemenger, R. P., 620
Hemminger, J. C., 517
Henderson, D., 109
Hendra, P. J., 507
Hendriks, E. M., 7
Henglein, A., 148, 150
Henkre, W. E., 621
Hennessey, J. P. Jr., 337, 338
Hennessy, R. J., 121, 127, 129
Hennig, P., 206, 415
Henon, M., 579
Henry, B. R., 670, 672, 674
Henry, J. M., 165
Hepburn, J. W., 126, 127
Herbst, E., 396
Hering, P., 130, 284
Heritage, J. P., 524, 526
Herman, I. P., 121
Hermans, J. J., 24, 32
Herrero, V. J., 123
Herschbach, D. R., 123, 266, 273, 279, 658
Hertel, I. V., 268
Hertz, H. G., 464
Herzberg, G. H., 388, 670
Heskett, D., 62
Hess, B. D., 207
Hess, L. A., 622
Hess, L. D., 266, 268
Hesse, R., 67
Hessenlink, W. H., 621
Hester, R. E., 518, 524, 526
Heuer, W., 2
Heuts, M. J. G., 275
Heydtmann, H., 111, 112, 115, 122
Hibbs, A. R., 25
Hicks, K. W., 662
Higgins, J., 34
Hilborn, R. C., 271
Hildebrandt, B., 111, 112, 115, 122
Hilhorst, H. J., 422, 431
Hillert, M., 258
Hilliard, J. E., 257
Hillier, I. H., 294, 296, 297, 302, 303, 306, 312
Hilpert, K., 320
Hilsch, R., 253
Hilton, P. R., 517, 531
Himpsel, F. J., 68, 229, 517, 544

Hinch, E. J., 90, 92, 100, 101, 103
Hinchliffe, A., 625
Hinze, J., 195
Hinze, W. L., 146
Hinzmann, G., 524, 526
Hippler, H., 664-66, 668, 680
Hirakawa, S., 151, 154
Hirata, F., 464
Hirooka, T., 117, 121, 124
Hirota, E., 410
Hirsch, J., 640
Hirschfelder, J. O., 160, 162
Hiskes, J. R., 380
Ho, K. M., 547, 549, 551
Ho, W., 57, 61
Ho, Y.-S., 127
Hobson, J. H., 124
Hochstrasser, R. M., 614
Hoffman, B. M., 468, 623
Hoffman, D. M., 636
Hoffman, J. D., 433
Hoffman, R., 465
Hoffmann, F. M., 63, 235
Hoffmann, M. R., 198, 206
Hoffmann, P., 62
Hoffmann, S. M. A., 124
Hoffmeister, M., 122
Hofmann, M., 318
Hogg, T., 571
Hohenberg, P., 544
Höjer, G., 362
Holbrook, K. A., 580, 585, 657-60, 668
Hollander, Tj., 274, 275, 278
Hollis, J. M., 395
Hollstein, U., 343
Holmes, B. E., 109
Holstein, T., 456-60
Holt, H. K., 267
Holtzer, A. M., 38
Holz, M., 464
Holzwarth, N. A. W., 297, 303, 306, 311
Honda, K., 33, 38
Hong, K. M., 627, 633, 634
Hong, L. S., 145
Honijk, D. D., 78
Hood, E., 516, 518, 519, 521
Hooker, T. M. Jr., 339
Hoover, W. G., 245
Hopf, E., 565
Hopfield, J. J., 461-63, 465-67
Hopkins, B. J., 67, 68
Hopkins, J. B., 300, 302, 310, 311, 320, 321, 571
Hopper, J. F., 483
Hopster, H., 60, 62
Horiuti, J., 161
Horn, P., 34, 35
Hornstein, S. M., 164
Horowitz, C. J., 169
Hose, G., 579, 581, 584

Hoshikawa, H., 25
Hoshino, Y., 26
Houle, F. A., 119, 120
Houston, J. E., 216, 218, 226, 232, 233, 235
Hout, R. F. Jr., 197, 206
Houzay, F., 544
Howard, J. A., 320-22
Høye, J. S., 75
Hrbek, J., 63
Hsu, Y. C., 126
Huang, K. N., 374
Hubbard, J. B., 101, 469, 470
Huber, D. L., 620, 644
Huber, H., 295, 301, 304, 305, 313
Huber, K. P., 388
Huberman, B. A., 571
Huddleston, R. K., 467, 475, 476
Hudgens, J. W., 115, 661, 663
Hug, W., 349-51
Huggins, M. L., 2
Hugo, J. M. V., 362
Huisken, F., 121, 663
Hukins, D. W. L., 345
Hulse, J. E., 299, 309, 319-21
Hunt, I., 644
Hunt, J. P., 472, 473
Huntress, W. T. Jr., 174, 407
Hupp, J. T., 438, 474-76
Hurley, A. C., 206
Husemann, E., 2
Hush, N. S., 206, 440, 451, 465, 466, 475
Husimi, K., 573
Husimi, Y., 79
Hutchinson, J. W., 260
Hüwel, L., 118, 119, 268
Huzinaga, S., 301, 362, 363, 370, 371, 377
Hynes, J. T., 39, 160, 168, 177, 438, 439, 451, 455, 461, 471, 675

I

Ibach, H., 49, 50, 57-66
Ibel, K., 34
Ibero, J. A., 623
Ibers, J. A., 472, 473
Ichikawa, T., 247
Iemura, M., 621
Ihara, H., 151, 152, 154
Ihara, Y., 151
I'Haya, Y. J., 351
Ihm, J., 542-44, 546, 547
Ikeda, S., 38
Ikeda, Y., 40, 41
Illies, A. J., 181
Imai, N., 81, 82, 86
Imhoff, R. E., 619
Imre, D., 586, 666

AUTHOR INDEX

Inagaki, T., 518
Indelli, M. T., 469
Infelta, P. P., 138, 146-48, 151
Inman, R. B., 38
Innes, K. K., 586, 666
Inokuchi, H., 643
Inoue, G., 130
Inoue, M., 524
Irish, D. E., 470
Irvin, J. A., 126
Isaacson, A. D., 160, 167-71, 177, 178, 180-82, 184
Isaacson, R. A., 467
Ise, N., 146
Ishikawa, A., 78
Ishikawa, Y., 374, 376
Ishiwatari, T., 151
Isied, S. S., 468
Itakura, K., 343
Ito, H., 351
Itou, S., 33, 38
Ivanov, V., 347, 352
Ivanov, V. I., 350, 352
Ivory, D. M., 637
Iwata, K., 40, 422
Izrailev, F. M., 568, 571

J

Jackson, J. D., 524, 525
Jacobsen, C. S., 645, 646
Jacobson, H., 10, 11, 35
Jacquet, R., 206
Jaeger, R., 218, 219, 229-31, 239
Jaffe, C., 571
Jaffe, R. L., 165, 205
Jagannathan, S., 100, 102
James, R. B., 268
Janik-Czachor, M., 517
Jankowiak, R., 621, 648
Jannick, G., 77
Jansson, K., 304
Jarrold, M. F., 176, 181
Jasinski, J. M., 175, 667, 668, 675, 676, 678, 679, 681, 682
Jaszunski, M., 197
Javan, A., 268
Jayasoorinya, U. A., 304, 313, 320
Jeanmaire, D., 507, 511
Jeffrey, G. B., 37
Jeffreys, W. H., 579
Jenard, A., 78, 81-83, 86
Jennison, D. R., 215, 216, 224, 225, 229-31
Jepsen, O., 549
Jernigan, R. L., 32
Jerome, D., 645
Jeung, G. H., 301, 310, 321, 372, 380
Jezequel, G., 544

Jha, S. S., 518, 519, 521, 523
Joannopoulos, J. D., 543, 547
Johansen, H., 300, 309
Johns, J. W. C., 394, 405
Johnson, B. B., 350
Johnson, C. K., 614
Johnson, J. F., 78
Johnson, M. A., 268, 277, 278
Johnson, P. D., 554, 556
Johnson, W. C. Jr., 335, 337, 338, 340, 343, 345, 346
Johnson, W. L., 247, 257
JOHNSTON, H. S., 481-505; 482, 487, 490, 493
Jonathan, N. B. H., 112, 115
Jones, A. C., 620
Jones, D. S., 511
Jones, M. B., 343
Jones, R. O., 295-301, 311, 312
Jones, V. O., 219
Jorna, S., 564, 566, 572, 579
Jortner, J., 452, 453, 456-58, 460, 461, 465, 474, 573, 621, 624, 632, 634, 670, 675
Josse, J., 341
Jouvet, C., 131
Jovin, T. M., 343, 345, 347, 348, 471
Joyes, P., 301
Ju, G.-Z., 177, 178
Judson, R. S., 130
Jugner, G., 80
Jugner, I., 80
Julienne, P. S., 378, 379
Jung, P., 78
Junge, C., 484
Junge, C. E., 496

K

Kahler, C. C., 121
Kahn, L. R., 364, 365, 366, 368, 372, 374, 377
Kaiser, A. D., 341
Kaldor, A., 298, 307
Kalinowski, J., 648
Kallmann, H., 624, 643
Kalos, M. H., 420, 424, 426, 427, 430, 431
Kamihara, M., 643
Kanda, H., 79, 81, 82, 84-86
Kanety, H., 471
Kant, A., 293-99, 306-9, 317, 320
Kaplan, M. L., 614
Kapral, R., 101
Karl, N., 614, 635, 642, 643
Karny, Z., 120, 129
Karo, A. M., 380
Karplus, M., 420
Karplus, R., 266

Kashiwagi, Y., 33, 38
Kassatotschkin, W., 360
Kastner, M. A., 643
Kasuya, T., 118
Katchalsky, A., 81, 82
Kato, K., 625
Kato, S., 196
Kato, T., 119, 120
Katz, B., 513
Katz, J. L., 632, 634
Kaufer, J., 636
Kaufman, A. N., 585
Kaufman, F., 113, 115
Kaufman, J. J., 370
Kauzmann, W., 244
Kay, K. G., 567, 570, 571, 576, 577
Kayser, R. F., 101
Keck, J. C., 162
Keck, P. C., 344, 347
Keeling, C. D., 483
Keene, F. R., 472, 473
Keil, E., 568
Kelber, J. A., 225
Keller, D., 330, 332
Keller, J., 303, 318
Keller, M., 483
Kellerman, M., 340
Kemper, P. R., 176
Kenkre, V. M., 615, 616, 618-20, 635
Kennedy, R. A., 390, 392, 404-6
Kennedy, S. J., 11
Kepler, R. G., 648
Kerker, G., 543
Kerker, G. P., 551, 555
Kerker, M., 512, 524, 526
Kesmodel, L. L., 58, 63, 65
Kestner, N. R., 452, 453, 456, 457, 460
Ketley, G., 662
Kettler, U., 524
Khairutdinov, R. F., 467
Khalil, M. A. K., 483
Khan, S. U. M., 465
Khizhnyakov, V. V., 620
Kidd, K. G., 205
Kiefer, J., 36
Kikuchi, J., 151
Kilkuskie, R., 346
Killinger, D. K., 267, 271
Kilpatrick, N. J., 166
Kim, K.-J., 329
Kim, Y. K., 359, 374
Kimelberg, H. K., 137
Kimerling, L. C., 257
Kimura, H., 261
Kinder, J., 620
King, A. P., 524
King, B., 320
King, D. A., 62
King, D. S., 657, 661-63

AUTHOR INDEX 703

King, H. F., 198
King, M. C., 160
Kinsey, J. L., 118, 121, 129, 266, 268, 271, 274, 276, 282, 284, 286, 586, 666-68
Kirkwood, J. G., 13, 36, 40-42
Kirste, R. G., 34
Kirtley, J. R., 517-19, 521, 523
Kitahara, A., 139
Kitaura, K., 377, 381
Kittrell, C., 666
Kivelson, S., 637
Klabunde, K. J., 292
Klafter, J., 618
Klein, F. S., 173
Klein, G., 633, 637, 648
Klein, J., 425, 429, 433, 632, 633
Kleinermanns, K., 118, 121, 125, 129
Kleinman, L., 360, 374
Kleinwachter, V., 349
Kleman, B., 300, 304, 305, 309, 312
Kleman, M., 248
Klement, W., 242
Klemperer, W., 674
Kley, D., 490, 496
Kliewer, K. L., 57
Klimek, D., 126
Klobukowski, M., 305, 313, 377
Klopffer, W., 648
Kloss, U., 524
Klotzbücher, W., 292-96, 302, 303, 305, 306, 310, 311, 313, 317, 318
Klotzbücher, W. E., 297
Klymko, P. W., 618, 619
Klysik, J., 347, 348
Kneba, M., 109, 173
Kneipp, K., 524, 526
Knight, A. E. W., 666, 668, 669
Knight, E. A. S., 524, 526
Knight, L. B. Jr., 292, 294, 318, 319, 322
Knights, J. C., 394
Knoll, W., 518
Knotek, M. L., 216, 219-21, 223, 225, 226, 229
Knox, D. G., 138, 146
Knox, R. S., 614
Ko, A. N., 659
Kober, E. M., 475
Koch, C. C., 256
Koch, E. E., 622
Koel, B. E., 65, 67
Koene, R. S., 77
Koeppel, G. W., 164
Kohn, W., 544, 548
Kok, R. A., 297

Kolari, H. J., 296, 302, 303, 306, 311, 318
Kolb, D. M., 517, 522
Kolendritskii, D. D., 648
Kolesnikov, V. A., 625, 626, 632
Koller, Th., 345
Kolmogorov, A., 572
Kolpak, F. J., 343
Kolts, J. H., 127
Komhyr, W. D., 486
Komiya, M., 137
Komornicki, A., 205
Kompa, K. L., 130
Kong, F., 662
Konowalow, D. D., 379, 380
Kopelman, R., 616-19
Koppel, H., 567, 585
Korbitzer, B., 115
Kordis, J., 305, 313
Kori, M., 116
Kornberg, A., 341
Korsch, J., 571
Korshover, J., 487
Korzeniewski, G., 517, 527, 531
Kosa, K., 201
Kosloff, R., 572
Kosower, E. M., 141, 471
Koszykowski, M. L., 568, 584, 586
Kotani, M., 620
Kotlar, A. J., 267
Kotler, Z., 531
Kottis, P., 614, 617
Kotz, R., 522
Kovac, J., 25, 40, 41
Kovacic, R. T., 38
Kovalenko, L. J., 115
Kowakski, A., 126
Koyama, R., 34
Koyano, I., 119, 120
Kozak, J. J., 138, 146-48, 150, 151, 618
Kozhukhovsky, V. B., 293, 296, 302, 310, 311
Kraemer, W. P., 206, 207, 415
Krahl-Urban, B., 218
Krajewska, A., 648
Krajnovich, D., 121, 663
Krajnovich, D. J., 657, 661
Krakauer, H., 549
Kramers, H. A., 6, 40
Kranbuehl, D. E., 421, 422
Krasnec, J. P., 496
Krasser, W., 524
Kratky, O., 23, 24, 28, 34
Kraus, J. S., 217, 220, 226, 231
Krause, J. T., 242
KRAUSS, M., 357-85; 371, 375, 376, 378-80
Krauss, W. A., 524
Kreevoy, M. M., 160, 438

Kremer, K., 423, 425-28, 430, 431, 433
Krenos, J., 128
Kretschmann, E., 521, 522
Krigbaum, W. R., 38
Krimm, S., 524, 526
Krishnan, R., 196, 201, 206
Kroger, P., 471
Kron, A. K., 424
Kruse, W. A., 34
Kryszewski, M., 642
Ku, J. K., 130
Kuhlmann, K. F., 42
Kuhn, H., 139, 522
Kuhn, W., 2, 10
Kumar, K., 473
Kündig, E. P., 295, 296, 306
Kunitake, T., 151, 152, 154
Kuntz, P. J., 109
Kunugi, S., 146
Kunz, R. E., 522
Kuppermann, A., 167, 170, 178
Küppers, J., 67, 553, 555
Kurata, M., 422
Kurihara, K., 151
Kurik, M. V., 648
Kuroda, S., 636
Kurtz, R. L., 224, 226-28, 231
Kutzelnigg, W., 197, 206
Kuznetsov, A. M., 453, 454, 456-59, 461, 462, 472
Kwei, G. H., 128
Kwok, H. S., 657, 661
Kypr, J., 347, 349
Kyrala, G. A., 390, 404

L

LaRoe, P. R., 217, 226
Laaksonen, L., 375
Lacis, A., 482, 500
Ladik, J., 623
Lagier, R., 647
Laidig, W. D., 196, 206
Laidler, K. J., 160, 162
Lain, L., 127
Lakatos-Lindenberg, K., 620
Laks, B., 518, 519, 523
Lamb, W. E. Jr., 267-69, 389, 390, 404, 406
Lamontagne, R. A., 483
Landau, L. D., 26, 90-93, 101
Lang, G. L., 337
Lang, N. D., 517, 527, 531, 548
Lange, J., 640
Langevin, P., 411
Langhoff, S. R., 206, 207
Langowski, J., 35
Langridge-Smith, P. R. R., 293, 295, 299, 302, 309, 311, 320, 321
Lannoo, M., 542

AUTHOR INDEX

Laor, U., 531
Larsen, P. K., 547
Larson, J. E., 347, 348
Larsson, S., 451, 459, 462, 463, 465
Laskowski, B. C., 380
Latimer, C. J., 119, 120
Latta, E. E., 553, 555
Laube, B. L., 524, 526
Laughlin, W. T., 244, 246
Laurendeau, N. M., 267, 278
Lavollee, M., 119
Lawrance, W. D., 666, 668, 669
Lawton, R. T., 674
Lax, M., 422
Leak, G. M., 258, 259
Lebowitz, J. L., 420, 424
Le Bret, M., 36
LeCante, J., 554
Lee, C. S., 342
Lee, C.-Y., 465
Lee, D. H., 547
Lee, E. H., 524, 526
Lee, H. U., 126
Lee, H. W. H., 621
Lee, J.-G., 199, 206-10
Lee, K. C., 524, 526
Lee, K. T., 177, 178, 183
Lee, N., 379
Lee, P., 138, 148, 150, 151, 482, 500
Lee, P. C., 524, 526
Lee, T. J., 415
Lee, V. Y., 647
Lee, Y. S., 305, 313, 365, 367, 370, 371, 373-78
Lee, Y. T., 117, 119-21, 124, 125, 657, 660, 661, 663, 668
Lehmann, H., 472, 473
Lehmann, J. C., 267, 272
Lehmann, K. K., 674
Lehn, J. M., 137
Lehwald, S., 59-65
Leleyter, M., 301
Lemaistre, J. P., 617
Lemmer, R. H., 301, 304, 305, 313
Lenac, Z., 51, 52
Leo, V., 636
LEONE, S. R., 109-35; 113-16, 118, 119, 123, 125, 128, 268
Lepp, A., 524, 526
LeRoy, R. J., 181, 392
Lesiecki, M. L., 662
Leslie, A. G. W., 343, 344, 347, 349
Lesueur, D., 257
Letokhov, V. S., 267, 268, 657, 661
Levashov, A., 138, 141

Levenson, M. D., 268
Levi, B. A., 197, 206
Levich, V. G., 453, 454, 456
Levin, A. I., 350
Levine, R. D., 109, 117, 121, 177
Levitt, M., 337
Levy, M. R., 109, 266
Lewald, S., 61
Lewis, D. J., 346
Lewis, N. A., 461, 462, 465
Lewis, R., 488, 501
Lewis-Bevan, W., 393, 405
Li, C. H., 54, 55
Li, T. T.-T., 472
Liao, P. F., 524, 526
Liao, T.-P., 42
Lichtman, D., 217
Liddy, J. P., 115
Lie, G. C., 206
Liebsch, A., 531
Lieser, G., 34
Lifshitz, A., 121
Lifshitz, E. M., 26, 90-93, 101
Light, J. C., 160, 177
Liidja, G., 614
Lillie, H. R., 259
Lim, Y. Y., 146
Limm, W., 292, 294, 296, 297, 306, 307, 319
Lin, C. J., 252, 253, 255, 257
Lin, C.-T., 474-76
Lin, G. H., 119
Lin, J., 177
Lin, M. C., 117, 121
Lin, S.-S., 293-95, 297, 298, 307, 320
Lindemann, F. A., 5
Linderberg, J., 465
Lindinger, W., 397
Lindkvist, S., 304, 305, 312
Lindman, B., 138, 141
Lindqvist, S., 300, 309
Lindsay, D. M., 318, 320
Lindsey, J. S., 468
Linschitz, H., 468
Lippitsch, M. E., 524, 526
Lipson, R. H., 296, 306, 319, 513
Lischka, H., 373
Liskow, D. H., 680
Little, J. W., 524
Liu, B., 169, 206, 297, 303, 306, 307, 311, 312
Liu, J.-J., 342
Liu, K., 126, 127, 268, 274, 282
Liu, S. C., 490, 494, 496, 498
Liu, W.-K., 392
Lloyd, D. A., 342
Lochet, J., 300, 309
Lochet, R., 18
Lochner, K., 630

Loeb, A. L., 90
Loesch, H. J., 122, 123
Logan, J., 452, 453, 456-58, 460, 463-65, 472
Lohr, L. L., 164
Lomax, T. D., 139
Lombardi, J. R., 524, 526
London, J., 486
Long, M. B., 524, 526
Long, M. E., 670
Lopez, V., 180, 660
López-Sancho, J. M., 68
López-Sancho, M. P., 68
Lord, R. C., 343
Loring, R. F., 616, 617
Loubriel, G., 229
Loudon, R., 269
Loughran, T., 150
LOUIE, S. G., 537-62; 540-43, 548, 549, 551, 554-58, 560
Lovcjus, V. A., 623
Love, A. E. H., 26
Lovelock, J. E., 483
Lowe, R. S., 130
Lowry, H. H., 5
Lubman, D. M., 664
Luborsky, F. E., 242, 261
Lucas, A. A., 50, 51, 53
Lucht, R. P., 267, 278
Ludi, A., 472, 473
Ludmer, Z., 620
Lugauskas, V. Y., 337
Luisi, P. P., 139
Lukosz, W., 522
Lumry, R. W., 442, 445
Lundqvist, S., 524
Lundsford, G. H., 568
Luntz, A. C., 117, 118, 121, 129
Lüth, H., 58, 60, 66, 67
Luther, F. M., 492, 496, 497
Luther, K., 664, 665, 668, 680
Lutz, H. P., 524
Lynn, J. E., 570
Lyon, S. A., 524
Lysov, Yu. P., 352
Lyyra, A. M., 379

M

Maass, O., 5
Mabuchi, M., 524, 526
Macartney, D. H., 468, 472-75
MacCarthy, J. E., 138
MacDiarmid, A. G., 636, 638
MacDonald, M., 127
MacDonald, R. G., 110, 111, 117, 121, 126, 127
Macedo, P. B., 244, 246
Macek, J. H., 280
Mack, J. V., 69
Mack, R. E., 303, 321
Mackenzie, P., 304

AUTHOR INDEX 705

Macomber, J. D., 268
MacPhail, R. A., 672
MacRobert, A. J., 662, 668
Maczuk, J., 101
MADEY, T. E., 215-40; 215-20, 224, 226-29, 231-37
Madhava, M., 242
Madison, V., 349, 350
Madix, R. J., 61, 65
Madsen, J., 549
Maeda, H., 29
Maestre, M. F., 330, 332-35, 343, 344, 346-49, 352
Maestri, M., 146, 148, 475
Magee, W. S., 32
Maggs, R. J., 483
Magnuson, A. W., 163, 167, 170, 171, 182
Maguire, J. F., 433
Maguire, T. C., 130
Mahajan, D., 472, 473
Mahan, B. H., 119
Mahan, G. D., 51, 53
Mahler, H. R., 346
Mailänder, M., 267, 274
Malins, R. J., 115
Malinson, P. D., 202, 203
Malli, G., 373, 374
Malloy, D. E., 123
Malrieu, J. P., 371, 380
Man, C.-K., 122
Mandel, F., 424
MANDEL, M., 75-108; 75, 77, 78, 81-88, 103-5
Mandy, F., 300, 304, 312, 313
Mangir, M., 662
Maniv, T., 517, 527, 531
Manning, G. S., 103, 146
Manocha, A. S., 111, 112, 115, 121
Mansfield, M. L., 29
Mantella, D., 116
Many, A., 61
Manz, J., 117, 121, 185
Manzel, K., 304, 313, 320, 524, 527
Maradudin, A. A., 518, 519, 522
Marck, C., 341, 342, 347, 349
Marcus, M., 252
Marcus, R. A., 162, 179, 180, 182, 438-40, 442, 443, 446, 447, 451-57, 463, 467, 468, 474, 475, 568, 584, 586
Mareca, P., 303
Maricq, M. M., 113, 114, 123
Mariella, R. P. Jr., 671
Marinero, E. E., 125
Mark, F., 373
Mark, H., 16, 18
Markovitz, H., 5
Marsagishvili, T. A., 459

Martilla, C. M., 350, 351
Martin, D. L., 164, 165
Martin, J. C., 347, 348
Martin, P., 632, 633
Martin, R. L., 299-301, 309, 310, 377, 378
Martin, R. M., 129, 517
Martinek, K., 138, 141
Martinez, H. M., 337
Martins-Faranchetti, S., 138, 144-46
Martner, C., 119
Martner, C. C., 408, 409, 413
Marvin, A., 518, 519
Marx, R., 119
Maslov, V. P., 567
Mason, E. A., 395, 411, 414
Massoudi, H., 524
Masumoto, T., 261
Mathieu, J., 151
Matsubara, A., 632
Matsubara, T., 471
Matsui, A., 621
Matsumi, Y., 118
Matsuo, K., 42
Matsuoka, H., 267
Mattar, S. M., 300, 320
Matthew, J. A. D., 67, 69, 524, 526
Mattingly, S. R., 496
Mattis, D. C., 619
Matz, R., 58, 60, 66
Mauclaire, G., 119
Mauzerall, D. C., 468
Mavroyannis, C., 619
Mayer, E. W., 483
Mayer, J. E., 2, 5, 6, 10, 14, 15
Mayer, M. G., 2, 6, 10, 14
Mayer, U., 619, 620
Mayhew, E. G., 137
Mazaud, A., 645
Mazur, P., 90, 94
Mazzioti, A., 362
McBreen, P., 513
McCall, S. L., 512, 524
McCammon, J. A., 38, 420
McCarthy, S. L., 517, 522, 527
McClain, W. M., 331, 332
McClellan, M. R., 61, 62
McClelland, G. M., 266, 273, 279
McCombie, J., 304, 313, 320
McConnell, H. M., 461, 462
McCurdy, C. W., 581
McDaniel, E. W., 395, 411, 414
McDermid, I. S., 266
McDonald, J. D., 115, 116, 586, 661
McDonald, J. R., 118, 121, 267, 283
McDonald, S. A., 666

McDonald, S. W., 585
McElhiney, G., 67
McElroy, M. B., 483
McFarland, M., 490, 496
McFarlane, R. A., 268
McFeeley, F. R., 61, 62
McGhie, A. R., 614, 620, 634, 635, 643
McGill, T. C., 366, 369, 370
McGinn, G., 362, 380
McGourty, J. L., 468
McGuire, M., 468
McIntosh, D. F., 300, 301, 305, 313, 320
McKamey, C. G., 256
McKean, D. C., 202, 203
McKellar, A. R. W., 394
McKillop, J. S., 662
McLafferty, F. J., 164
McLean, A. D., 297, 303-7, 311-13, 370, 373, 374, 376-78
McLendon, G., 468, 474
McMahan, M. A., 123, 127
McMillan, W. G. Jr., 15
McNab, T. K., 298, 307, 308
McNulty, P. J., 524
McNutt, J. F., 125
McQuillan, A. J., 507, 524, 526
McTague, J. P., 103, 104
Meagher, J. F., 659
Measures, R. M., 274
Megel, J., 617
Mehta, M. L., 567, 585
Meier, F., 292
Meier, M., 522
Meier, U., 662
Meijer, H. A. J., 275
Meisel, D., 524, 526
Mejean, T., 293, 295
Melander, L., 173
Melius, C. F., 175, 300, 309, 364, 369
Mello, P. A., 567, 568, 585
Melngailis, J., 524, 526
Menke, K., 636
Menzel, D., 67, 68, 216, 218, 219, 222, 224-26, 229, 231, 239
Menzinger, M., 126
Merer, A. J., 293, 295, 379
Merkulov, I. A., 522
Merrifield, R. E., 615
Messinger, B. J., 524
Messmer, R. P., 62
Metcalfe, K., 518
METIU, H., 507-36; 508, 509, 511, 513, 514, 516-19, 521-23, 526, 527, 529-31
Meyer, K. E., 620, 635, 643
Meyer, P. I., 104
Meyer, R., 185

AUTHOR INDEX

Meyer, R. B., 252, 253
Meyer, T. J., 467, 471-73, 475
Meyer, W., 196, 200-2, 206, 209, 380
Meyerhoff, G., 38
Michael-Beyerle, M. E., 467
Michalopoulos, D. L., 296, 299, 300, 306, 309, 310
Michel-Beyerle, M. E., 620, 630
Micklitz, H., 298, 307, 308, 319
Miedema, A. R., 303, 312, 314-17
Mikawa, H., 632
Mile, B., 320-22
Miles, H. T., 347, 348
Miles, S. L., 61, 65
Millaud, B., 38
Miller, C. M., 662
Miller, G. G., 637
Miller, J. A., 164, 175
Miller, J. R., 462, 463, 466-68, 475, 476
Miller, J. W., 259
Miller, T. A., 266, 387, 390, 393
Miller, W. G., 25
Miller, W. H., 164, 171, 176-80, 182, 580, 581
Mills, D. L., 49, 50, 54, 55, 58, 59, 61, 65, 66, 518, 519, 522, 523
Mills, I. M., 196, 197
Minakata, A., 81-83, 86, 88, 103, 104
Minkwitz, R., 304, 313, 320
Minyat, E., 347, 352
Mirlin, D. A., 522
Miskovic, Z., 233, 234
Mislow, K., 204
Mita, K., 146
Mitchell, D. J., 176
Mitchell, G. E., 65
Mitchell, J. B. A., 123
Mitchell, J. W., 305, 313, 321
Mitchell, S. A., 292, 300, 301, 305, 313, 319, 320
Mittal, K. L., 138
Miyake, A., 26
Miyoshi, E., 301, 321
Mizuno, K., 621
Mobius, D., 139, 517, 522, 527
Mocker, H. W., 268
Moehlman, J. G., 661
Möhwald, H., 645
Moiseyev, N., 583, 584
Mok, C. Y., 462, 469
Moleres, F. J. A., 123
Molina, M. J., 482, 494
Molinari, R. J., 79, 81, 82, 84, 86
Moller, W., 632

Molt, K., 201, 202
Mommaerts, W. F. H. M., 346
Monberg, E. M., 616, 617
Monnerie, L., 423
Monroe, D., 643
Montague, D. C., 660
Montano, P. A., 298, 307, 308, 316, 319
Montroll, E. W., 150
Moore, C. B., 403, 405, 407, 667, 668, 672, 675, 676, 678, 679, 681, 682
Moore, D. S., 125, 330, 333, 350-52
Moore, L. D. Jr., 12
Morawetz, H., 42
Morawitz, H., 622
Mordaunt, D., 116, 661
Moreland, J., 516, 518, 521
Morgan, A. R., 346
Morgan, J. R., 617, 618
Morokuma, K., 196, 377, 381
Morrison, T., 298, 308
Morrobel-Sosa, N., 614
Morse, M. D., 293, 295, 299, 302, 309, 311, 320, 321
Morse, P. M., 524
Mort, J., 614, 641
Mortensen, K., 645, 646
Mortola, A. P., 300, 309
Moscovits, M., 513
Moser, J. L., 564, 566, 579
Moskovits, M., 292-99, 306-9, 316, 319-21, 512, 524, 526, 527
Moskowitz, J. W., 300, 309, 319, 369
Moss, M. G., 115, 116, 661
Moss, R. A., 151
Motosuga, M., 518
Motowoka, M., 33, 37, 38
Mott, N. F., 538
Moutsoulas, M., 579
Movaghar, B., 630, 631, 639, 640, 644, 645
Moyal, J. E., 570, 573
Mrosan, E., 302, 310
Muhle, W., 620
Mukamel, S., 571
Muller, C. H. III, 267
Muller, D. F., 268
Muller, G., 81-85, 88
Müller, H., 302, 310
Müller, J., 205
Müller, M., 7
Müller, W., 380
Mulliken, R. S., 466
Mullins, D. R., 517
Munakata, T., 118
Munn, R. W., 620, 625
Murakami, H., 32, 33, 38
Murakami, Y., 151
Murday, J. S., 224, 228

Murphy, E. J., 129, 284
Murray, C. A., 514
Murray, D., 631, 639
Musho, M. K., 138, 148, 150
Muthukumar, M., 425
Muzikante, I. J., 625, 626, 632, 642

N

Naaman, R., 664
Nagai, K., 32, 34, 35, 410
Nagarathna, H. M., 298, 307, 308, 316
Nagasaka, K., 28, 29, 31, 33, 35
Nagayama, K., 25
Nagel, S. R., 242
Naik, V. M., 298, 307, 308, 316
Nakajima, H., 38
Nakamura, H., 79
Nakamura, T., 301, 321
Nakano, A., 151
Nakashima, N., 664
Nalewajek, D., 646
Nappi, B. M., 303, 312, 315, 318
Narang, R. S., 620, 635, 643
Narayanaswamy, O. S., 246
Natarajan, M., 482, 496
Natason, G. A., 167
Navon, G., 452, 456, 458, 461
Nayfeh, A., 99
Nazar, M. A., 111, 121
Nechtschein, M., 636
Needs, R., 428
Needs, R. J., 433
Neelov, I. M., 420
Neftel, A., 483
Neilson, G. W., 464
Nelin, C. J., 293, 297, 306, 311, 312
Nelson, D. R., 248, 249, 253
Nelson, R. G., 346
Nesbet, R. K., 297, 307
Nesbitt, D. J., 128, 403, 405, 407
Nettles, S. M., 474
Netzel, T. L., 471
Netzer, F. P., 67, 69, 216, 218, 226, 232, 233, 235-37
Neumann, H. M., 472, 473
Neumark, D. M., 125
Neviere, M., 518, 519
Newns, D. M., 50, 557
NEWTON, M. D., 437-80; 299, 300, 309, 319, 378, 438-42, 448, 452-59, 461, 463-65, 470, 472, 474
Nicholls, J. M., 544
Nicolai, T., 77
Nicolas, G., 381

AUTHOR INDEX 707

Nicovich, J. M., 171, 172
Niehus, H., 216, 218
Nielsen, E. B., 349, 350
Nigrey, P. J., 636
Nikitin, E. E., 177, 448-50, 459
Nishida, K., 151
Nishijima, M., 67
Nishimura, H., 621
Nishioka, N., 33, 38
Nitzan, A., 512, 514, 524, 526-29, 531
Nixon, E. R., 299, 309
Nobl, C., 237, 238
Nocera, D. G., 468
Noda, I., 25
Noell, J. O., 299, 300, 309, 377, 378, 472
Noid, D. W., 568, 584, 586
Noodleman, L., 301, 305, 313
Noolandi, J., 627, 633, 634
Nordholm, K. S. J., 569, 576
Norisuye, T., 32, 33, 37, 38
Norman, J. G. Jr., 296, 301-3, 305, 306, 311, 313, 318
Norris, D., 304, 313, 320
North, C. S., 206
Northrup, F. J., 126, 127
Northrup, J. E., 542, 544
Northrup, S. H., 438, 439, 455, 461, 471
Novaro, O., 321
Nowak, R., 648
Nowikow, C. V., 124
Numata, H., 518
Nyberg, C., 61

O

Obara, S., 377
O'Brien, E. V., 63
O'Brien, R. W., 90, 93, 94, 100, 101, 103
Ochs, F. W., 616
O'Connor, C. J., 139
O'Connor, M. T., 126
Ödberg, L., 470
ODIJK, T., 75-108; 76, 77, 81, 83, 86, 88, 105
Oeschger, H., 483
O'Grady, B. V., 127
Ogura, H., 127
Ohnuma, H., 35, 38
Ohta, K., 381
Ohtaka, K., 522, 524
Oka, T., 393, 397, 400, 402, 403, 405-8, 413
Okahata, Y., 151, 152, 154
Okamura, M., 393, 405
Okamura, M. Y., 467
Okano, K., 40
O'Keefe, A., 119
Okita, K., 38
Okubo, T., 146

Olafson, B. D., 369
Oliveros, E., 371
Olmstead, W. N., 175
Olson, D. H., 524, 526
Olson, R. W., 621
Oltmans, S. J., 486
O'Malley, T. F., 445, 448
Omenetto, N., 267
Onchi, M., 67
Oncley, J. L., 78, 80, 81, 87
Ondrechen, M. J., 461, 462
Ondrey, G. S., 130
Ono, Y., 127, 129
Onsager, L., 90, 469, 627, 632
Oosawa, F., 103, 104
Oprysko, M. M., 123, 127
Optiz, Ch., 302, 310
Orbach, R., 619
Orchard, A. F., 58
Oref, I., 580, 586, 660
Oreg, J., 373
Orenstein, J., 637, 643
Orgel, L. E., 461, 462
Orlandi, G., 469
Orlowski, T. E., 631, 632
Ortega-Blake, I., 371
Orwoll, R. A., 422
Osamura, Y., 197, 198, 206
Osgood, R. M. Jr., 127
Oskam, A., 203
Oster, G., 34, 35
Otto, A., 68, 508, 514, 521, 522
Otto, L. P., 203
Outer, P., 18
Ovchinnikov, A. A., 453-55
Overbeek, J. T. G., 90, 91, 103
Overberger, C. G., 5
Owen, J. F., 524, 526
Oxtoby, D. W., 517, 531
Ozaki, M., 636
Ozin, G. A., 292-97, 299-306, 309-13, 317-20
Ozturk, B., 258, 259

P

Pacansky, J., 205, 206
Pace, S. A., 300, 310
Pai, P. M., 640
Palay, P., 198
Palecek, E., 349
Palke, W., 517, 531
Pan, F.-S., 397, 405, 408, 413
Pandey, A., 567, 568, 585
Pandey, K. C., 66, 544
Pang, F., 196, 199, 200
Panissod, P., 247
Panosh, R. L., 67
Pantano, C. G., 216
Paoletti, S., 81-83, 88
Papp, H., 67
Parent, D., 119

Parker, D. H., 128
Parkin, S. S. P., 647
Parks, C. C., 217, 226, 229
Parks, E. K., 296, 306
Parmenter, C. S., 571
Parris, P. E., 620
Parson, J. M., 126, 268, 274, 282
Parson, R. P., 616, 617
Parsons, J. M., 110, 111, 112
Partridge, H., 380
Pasta, S., 579
Pasternack, L., 267
Pasternak, M., 319
Patel, J. S., 614
Patey, G. N., 75
Paton, A., 641
Patson, A., 623
Pattengill, M., 162
Patterson, F. G., 621
Patterson, L. K., 138, 150
Paul, R. L., 507
Pauling, L., 301
Pauly, H., 101
Pear, M. R., 42
Pearce, E. M., 5
Pearlstein, R. M., 620
Pechukas, P., 160, 163, 164, 167, 170, 177, 184, 568, 572, 573, 578, 582
Pedersen, H. J., 645
Pederson, H. J., 646
Pee, P., 617
Peebles, D. E., 67
Peeples, J. A., 467
Pelissier, M., 301, 310, 321, 371, 372, 374, 376, 378, 381
Pellerite, M. J., 175, 438
Pellin, M. J., 296, 302, 306, 311
Penn, D. R., 526
Penner, J. E., 492, 496, 497
Pennington, D. E., 438, 475
Perchak, D., 42
Percival, I. C., 567-70
Perenboom, J. A. A. J., 292
Perepezko, J. H., 252
Peres, A., 584
Perrin, F., 37
Perrin, J., 394
Perry, D. S., 120, 122, 123
Persky, A., 173
Persson, B. N. J., 51-53, 60, 66, 67, 517, 527, 531
Persson, M., 59
Pesić, D. S., 304, 312
Pessini, F. B. T., 662
Pete, J. A., 671
Petek, H., 403, 405, 407
Petelenz, P., 624, 626, 632, 644
Peter, G., 625

Peterlin, A., 36
Petit, R., 519, 521
Petraschen, M., 358
Petroff, Y., 554
Petrov, V. V., 623
Pettersson, L., 363, 370
Pettersson, L. G. M., 372
Pettinger, B., 517, 522
Peyerimhoff, S. D., 206
Pfaff, J., 398-402, 405, 406
Pfister, G., 614, 641
Phillips, D. H., 377
Phillips, J. C., 360
Philpott, M. R., 514, 518-21, 524, 614
Piacente, V., 378
Pian, T. R., 217, 220, 226, 231
Piepmeier, E. H., 267
Pierotti, D., 484
Pietro, W. J., 199
Piller, J., 261
Pinchaux, R., 544, 554
Pincus, P., 77, 105
Piper, L. G., 115
Piper, T. C., 49
Pirug, G., 62
Piryatinskii, Yu. P., 648
Pitaevskii, L. P., 90
Pittel, B., 362
Pitz, R. W.
Pitzer, K. S., 305, 313, 365, 367, 370, 371, 373-79
Pladziewicz, J. R., 664
Plans, J., 642
Platzman, P. M., 512, 524
Plummer, E. W., 55, 57, 61, 62, 551, 554, 556, 558, 560
Poate, J. M., 257
Pobo, L. G., 296, 306
Pockrand, I., 514, 517, 521, 522, 527
Pohl, F. M., 343, 345, 347, 352
Pohlmann, B., 639
Polak-Dingels, P., 130
Poland, D., 11
Polanyi, J. C., 109, 111, 115, 121, 126, 127, 130, 131
Politi, M. J., 138, 144-46, 151
Polk, D. E., 242, 247
Pollak, E., 160, 163, 164, 167, 168, 177, 182-85
Pongor, G., 199, 202-5
Pope, C. G., 524, 526
POPE, M., 613-55; 614, 618-20, 623-29, 631-34, 636, 637, 640, 641, 643-47
Popkie, H. E., 370
Pople, J. A., 196, 197, 199, 201, 206
Popova, J.A., 643

Popovic, Z. D., 626, 632, 633, 637
Porod, G., 23, 24, 28, 34
Porter, C. E., 567, 568, 570, 585
Porter, G., 137
Porter, R. N., 402, 415
Post, D., 301, 321
Posternak, M., 549
Potthast, L., 122
Pouget, J. P., 636, 646
Powell, C. J., 50
Powell, R. C., 615, 619
Powers, D. E., 296, 300, 306, 310
Powers, E., 571
Prager, S., 37
Prange, R. E., 571
Prasad, P., 616
Prasad, P. M., 617
Prasad, P. N., 622
Preston, J., 38
Preston, K. F., 320, 321
Preston, R., 247
Preuninger, F. N., 119
Preuss, D. R., 300, 310
Preuss, H., 301, 310, 380
Priel, Z., 513
Prince, K., 68
Prinn, R. G., 482, 483
Pritchard, J., 67
Prock, A., 509, 514, 516, 517
Pron, A., 636
Pron, J., 636
Propst, F. M., 49, 61
Proske, Th., 148, 150
Provencher, S. W., 337, 338
Pruett, J. C., 126
Pruett, J. G., 120, 122
Ptitsyn, O. B., 36, 337, 424
Puiu, A. C., 300, 310
PULAY, P., 191-213; 193, 195, 196, 198-210
Purdum, H., 298, 308
Purvis, G. D. III, 206, 207
Putzeys, P., 19
Pyper, N. C., 373, 374
Pyykko, P., 373, 375, 377

Q

Quack, M., 162, 175, 657, 661, 662, 686
Quick, C. R. Jr., 125
Quigley, G. J., 343
Quina, F. H., 138, 144-46
Quina, F.H., 138, 144-46

R

Rabinovitch, B. S., 580, 586, 657, 659, 660, 668

Rackovsky, S., 627, 628, 632, 633
Radhakrishan, G., 662, 668
Rae, A. G. A., 118
Raff, L. M., 164, 165
Raffenetti, R. C., 377
Raghavachar, K., 197
Raghavachari, K., 175
Rahman, T. S., 59, 61
Rai, S. N., 175, 176
Raine, G. P., 415
Ramage, R. E., 139
RAMAKER, D. E., 215-40; 215, 216, 222-28, 230, 231
Raman, C. V., 18
Ramanathan, R., 18
Ramanathan, V., 482
Ramasami, T., 461, 462, 473
Ramaswamy, R., 568
Ramker, D. E., 224, 231
Rampens, A. J., 642
Ranade, A., 343
Rappé, A. K., 368, 370
Rasmussen, R. A., 482-84
Ratcliff, L. B., 380
Ratliff, R. L., 342-44, 346, 347, 349
Ratner, M. A., 300, 309, 461, 462, 465
Rau, R. C., 104
Ravishankara, A. R., 171, 172
Rayleigh, J. W. S., 518
Rayner, D. M., 662, 663
Raziel, A., 81, 82
Rebane, K., 614
Record, M. T. Jr., 38
Reddy, K. V., 667, 670, 674-80
Redhead, P. A., 222
Redi, M., 467
Redmann, S. M., 349-51
Redmon, M. J., 125
Redondo, A., 366, 369, 370
Ree, F. H., 245
Reed, W., 152-54
Regis, A., 524, 526
Rehn, V., 217, 219, 226
Reid, E. S., 507
Reihl, B., 577, 544
Reimer, B., 630
Reineker, R., 622
Reinhardt, W. P., 566, 571, 585, 675
Reinisch, R., 518, 519
Reinsel, G. C., 487, 488, 501
Reisler, H., 662
Reisner, D. E., 666
Rempp, P., 9
Rendell, R. W., 517, 526, 527, 529
Renlund, A. M., 662
Rentzepis, P. M., 648

AUTHOR INDEX 709

Rettner, C. T., 118, 121, 125, 127, 129, 268, 274, 282, 286
Reuss, J., 268
Reynolds, W. L., 442, 445
Rheinwine, M., 514
Rhodes, C. K., 268
Rhodes, W., 349-51
Rhodin, T. N., 300
Ribault, M., 645
Rice, S. A., 360-62, 465, 569, 572, 576, 624, 632, 634
Rice, T. M., 619
Rich, A., 343
Richard, E. G., 390, 404
Richards, W. G., 300, 309
Richardson, D. E., 465
Richter, D., 430
Richter, P. H., 150
Richtsmeier, S. C., 320, 321
Ridge, D. P., 178
Riedl, W., 239
Ries, B., 629-31, 633, 634, 638-40
Riley, S. J., 296, 306
Rindorf, G., 646
Ringsdorf, H., 140, 151
Riseman, J., 13, 36
Ritchie, G., 521, 522, 524, 526
Ritchie, R. H., 522
Ritsko, J. J., 638, 648
Ritz, A., 66
Rivoal, J.-C., 297, 307
Rizzo, T. R., 667, 668, 671, 677, 679, 682-87
Rizzo, V., 352
Roberts, G. J., 259
Roberts, G. P., 341, 342
Roberts, J. P., 259
Robertson, I. L., 301
Robiette, A. G., 401, 407
Robin, P., 636
Robinson, G. N., 125
Robinson, P. J., 580, 585, 657-60, 668
Robota, H. J., 69, 517, 527
Rockney, B. H., 663
Rodgers, M. A. J., 149
Rodgers, P., 660
Rodrigo, A. B., 274
Rodrigo, M. M., 32
Rogozik, J., 68
Rohlfing, E. A., 298, 307
Rohmer, M., 321, 322
Rohrer, H., 546
Römelt, J., 184, 185
Romero, A., 137
Romsted, L. S., 138, 141, 144-46
Ron, S., 183
Ronchetti, M., 249
Roos, B. O., 198, 206, 207, 293, 297, 306, 311, 312

Rosano, W. J., 126
Rösch, N., 300
Rose, A., 643
Rosen, P., 8
Rosenberg, B. J., 206
Rosenberg, R. A., 217, 226
Rosenblum, M. P., 257
Rosenfeld, L., 331
Rosenheck, K., 31
Rosenkrantz, M. E., 378, 379
Rosenkranz, H., 339
Rosenthal, A., 61
Rosenthal, J., 621
Rosetti, R., 517, 527
Rosicky, F., 373
Rosseinsky, D., 442
Rossi, M. J., 657, 661, 664
Rostas, J., 126, 268
Roswall, S., 81, 82, 86
Roth, S., 636
Rothenberger, G., 138, 146-48, 151
Rothschild, M., 268
Rotne, J., 37
Rouse, P. E., 420, 422
Rousset, A., 18
Rowe, J. E., 514
Rowland, F. S., 482, 494, 660
Roy, K. B., 347, 348
Rozman, M. G., 621, 622
Ruamps, J., 304, 312
Rubingh, D. N., 33, 38
Rubinstein, M., 249
Rubio, J., 68
Rubloff, G. W., 67, 68
Rudenko, A. I., 643
Ruette, F., 298
Ruff, G. A., 389, 404
Ruff, I., 472, 473
Ruhman, S., 663, 668
Runnels, J. H., 42
Ruppin, R., 524
Russel, W. B., 90
Ruth, J. A. Jr., 258, 259
Ryaboy, V. M., 184
Ryali, S., 116
Rydholm, R., 146
Rynbrandt, J. C., 659
Rys, J., 198

S

Saari, P., 614, 622
Sabbatini, N., 469
Sabelli, N. H., 372
Sachs, S. B., 81, 82
Sackmann, E., 150
Sadanobu, J., 38
Sadlej, A. J., 197
Sadoc, J. F., 247, 248
Saez Rabanos, V., 123
Saffman, P. G., 260
Safron, S. A., 658

Sage, M. L., 670, 675
Sagiv, J., 140
Sahyun, M. R. V., 305, 313, 321
Saito, N., 25, 28, 29, 37, 40
Saitovich, E. M. B., 298, 308, 316
Sakai, Y., 362, 370, 371
Sakamoto, M., 79, 81, 82, 84-86
Sakamoto, S., 83, 84, 87
Sakamoto, T., 151, 154
Sakimoto, K., 178
Sakisaka, Y., 67
Sakoda, K., 522
Salahub, D. R., 322
Salama, M., 260
Salaneck, W. R., 623, 641
Salpeter, E. E., 373
Samoc, A., 642
Samoc, M., 633, 637, 642, 648
Samoilova, A. N., 293, 296, 302, 310, 311, 318
Sanche, L., 619
Sanda, P. N., 514, 518-20
Sanders, N. D., 118, 121
Sando, K. M., 206
Santo, R., 518-21
Sargent, M. III, 267, 269
Sargeson, A. M., 472, 473
Sasaki, S., 78
Sasaki, Y., 522
Sastry, K. V. L. N., 396
Sato, N., 643
Sattler, K., 292
Sauer, G. W., 644
Sauer, H., 15
Saunders, W. H. Jr., 173
Saville, D. A., 90, 94
Sawodny, W., 201, 202, 204
Saxe, P., 196-98, 205, 206
Saxena, V. P., 337
SAYKALLY, R. J., 387-418; 387, 390, 393, 395, 398-403, 405-9, 413, 415
Scandola, F., 469
Scantlebury, G. R., 7
Scarbrough, J. O., 256
Scarmozzino, R., 648
Scarsdale, J. N., 199
Schaad, L. J., 207
Schaefer, H. F., 680
Schaefer, H. F. III, 179, 182, 196-98, 205, 206, 415
Schäfer, E., 401, 406, 407
Schäfer, L., 199
Schatz, G. C., 167, 170, 171, 177, 178, 183-85, 524, 531
Schatz, P. N., 451
Schaurte, J. A., 332
Schawlow, A. L., 412
Scheider, W., 78

AUTHOR INDEX

Scheidt, H., 68
Schein, L. B., 620, 634, 635, 643
Schellman, J. A., 93, 349, 350, 352
Schelly, Z. A., 150
Schelten, J., 34
Scher, H., 457, 459, 627, 628, 631-33
Scheraga, H. A., 11, 31, 40, 42
Scherer, G. J., 674
Schilling, R., 619
Schinke, R., 125, 129
Schissel, P., 300, 304, 305, 309, 312, 313, 317
Schlag, E. W., 621
Schlegel, H. B., 175, 176, 196, 197, 201, 204, 206
Schlesinger, L., 518
Schlier, R. E., 546
Schlüter, M., 369, 540-43, 547-49
Schmeisser, D., 55, 56, 61, 67, 69, 520
Schmid, D., 615, 619, 620
Schmidt, P. P., 460
Schmidtke, H.-H., 294-301, 306, 317
Schmiedl, R., 662
Schmit, C., 568, 585, 586
Schmitt, J. P. M., 394
Schmitt, M. J., 467
Schmitz, P. J., 34
Schnatterly, S. E., 49
Schneider, W. R., 639
Schofers, F., 329
Schofield, K., 267
Scholl, E., 318
Schonhammer, K., 557
Schonhense, G., 329
Schönherr, G., 629, 630, 633, 634, 638, 640, 641
Schottky, W., 538
Schreck, R. P., 151
Schreiber, J. L., 109
Schrieffer, J. R., 57, 438, 439, 443, 447, 456, 636, 637
Schubert, V., 664, 666
Schultz, A., 123
Schultz, S. G., 517
Schulz, G. J., 49, 55
Schulz, G. V., 2, 19
Schulz, H. J., 645
Schulz, P. A., 657, 661, 663
Schulze, W., 304, 312, 313, 320, 524, 527
Schumacher, E., 318
Schumacher, R. R., 647
Schupp, H., 140, 151
Schurr, J. M., 43, 103
Schwan, H. P., 78, 79, 101
Schwartz, G. P., 61
Schwartz, M. E., 362

Schwartz, R. N., 573
Schwartz, S., 178
Schwartz, W. H. E., 381
Schwarz, G., 101, 103
Schwarz, H. A., 388, 406
Schwarz, R. B., 257
Schwarz, W. H. E., 362, 374, 376
Schwenke, D. W., 184, 185
Schwenz, R. W., 126
Schwinger, J., 266
Schworder, J., 483
Scott, J. C., 647
Scott, T. W., 633, 634
Sculley, M., 524
Scullman, R., 304
Scully, M. O., 267, 269
Searby, G. M., 38
Sebastian, L., 625, 630
Secrest, D., 206
Seeger, R., 197
Seiferheld, U., 629, 631, 639, 640, 645
Seifert, G., 302, 310
Seitz, G., 461, 462, 465
Sekerka, R. F., 255
Seki, H., 514, 518, 524
Seki, K., 643
Selco, J. I., 660
Selin, L. E., 305
Sellers, P. V., 112, 115
Selsby, R., 618
Selzle, H. L., 621
Semancik, S., 237
Semkov, A. M., 201
Semlyen, J. A., 11, 25
Sen, P. N., 90, 92, 99, 100, 101, 103
Serafini, A., 368-71
Sessler, E. L., 568
Setser, D. W., 109, 111, 112, 115, 118, 121, 126, 127, 130, 657
Seurin-Vellutini, M. J., 38
Sexton, B., 60
Sexton, B. A., 61, 65
Seyse, R. J., 305, 313
Shackleford, W. L., 267
Shacklette, L. W., 637
ShaksEmampour, J., 297, 307
Sham, L. J., 361, 364, 544
Sham, T.-K., 468, 472-74
Shamai, S., 319
Shanfield, Z., 298, 307, 308, 316
Shank, C. V., 514, 637, 648
Shapiro, M., 184, 581, 584
Shapiro, S. L., 619
Sharp, J. H., 626
Sharp, P., 33
Shavitt, I., 206, 207
Shaw, K. D., 524, 526
Shaw, T. H., 78

Shelby, R. M., 621
Shen, Y. R., 121, 522, 657, 661, 663
Sheng, P., 518, 519
Shenoy, G. K., 298, 307, 308
Shepelyansky, D. L., 571
Sheppard, J. C., 483
Sherman, A. V., 620
Sherwood, J. D., 90, 92, 100, 101, 103
Sherwood, J. N., 644
Shi, S.-H., 179
Shilov, V. N., 89, 90, 93-95, 99-103
Shim, I., 298-301, 303, 304, 307-9, 311-14, 316, 318
Shimada, J., 26-29, 31-36, 42, 43
Shimazu, M., 267
Shimoda, K., 268
Shimokihara, S., 632
Shimomura, M., 154
Shin, J. J., 151
Shinkai, S., 151, 154
Shinohara, H., 620
Shinozuka, Y., 620
Shirane, G., 646
Shiren, N. S., 636
Shirley, D. A., 229
Shirley, J. A., 267
Shirts, R. B., 566, 571, 585
Shmueli, U., 31
Shobatake, K., 125
Shockley, W., 538
Shoemaker, R. L., 268
Shore, D., 35
Shushin, A. I., 455
Shy, J.-T., 390, 406
Sibener, S. J., 117, 121, 124
Sibert, E. L., 675
Siddiqui, A. S., 630
Siders, P., 443, 452, 456, 457, 475
Siebert, D. R., 670
Siebrand, W., 624-26
Siegbahn, P., 169
Siegbahn, P. E. M., 198, 206, 207, 301, 310
Siegel, A., 123
Siegel, C. L., 564, 566, 579
Siegel, J. B., 337
Siegel, S., 6
Sigety, E. A., 242
Siiman, O., 524, 526
Silbey, R., 381, 465, 509, 514, 516, 517, 616-18, 634, 637, 638
Silinsh, E. A., 625, 626, 632, 642
Silver, M., 629, 630, 633, 634, 638, 640, 641
Silverman, J., 472, 473
Simha, R., 37

AUTHOR INDEX 711

Simkovich, G., 258, 259
Simmonds, P. G., 482, 483
Simons, G., 362
Simons, J. P., 121, 127, 129
Simpson, C. J. S. M., 113, 114, 123
Sinai, Ya. G., 564, 572, 586
Singer, L. A., 138
Singh, J. P., 118, 619
Sinha, M. P., 279
Sittel, K., 78
Sixou, P., 38
Skelton, B. W., 472, 473
Skinner, D. M., 342
Skodje, R. T., 160, 168, 167, 170, 171, 178-80, 184, 185
Skolnick, J., 21, 32, 42, 76, 420
Sleight, T. P., 301, 321
Slepian, J., 6
Slichter, C. P., 266
Slingerland, P. J., 199
Sloan, G. J., 614
Sloan, J. J., 110-12, 114, 117, 121
Sloane, C. F., 331, 332
Slocum, G., 486
Slotnick, K., 627
Smagellotti, A., 205
Small, G. J., 614, 621
Smalley, R. E., 293, 295, 296, 300, 302, 306, 310, 311, 320, 321, 571, 586, 614
Smedley, T. A., 368, 370
Smith, D. J., 124
Smith, D. L., 268
Smith, G. K., 117, 121
Smith, G. P., 268, 664
Smith, I. W. M., 109, 658, 659
Smith, J. R., 549
Smith, M. A., 113-15, 123
Smith, N. V., 547, 554, 556
Smith, P. J. C., 345
Smith, R. A., 267
Smoes, S., 292, 294, 300, 304, 312, 313
Snider, D. R., 524
Snijders, J. G., 301, 310, 373
Snyder, L. E., 395
Snyder, R. W., 339
Soda, K., 40
Soep, B., 131
Softley, T. P., 390, 393, 394, 404, 405
Sokcevic, D., 51, 52
Solal, F., 544
Solc, K., 33
Somorjai, G. A., 62, 63, 65
Sondergaard, N. C., 452
Soos, Z. G., 615, 619
Sorbello, R. S., 524
Sorriso, S., 75

SPAEPEN, F., 241-63; 244, 248-53, 255, 257, 259, 260
Spannring, W., 630, 649
Sparks, R. K., 125
Spezeski, J. J., 389, 404
Spicer, L. D., 657
Spicer, W. E., 522
Spiegelmann, F., 374, 376, 381
Spirko, V., 415
Spitzer, A., 67
Sponar, J., 347, 349
Sprecher, C. A., 343, 345, 346
Springall, J. P., 304, 313, 320
Squires, R. R., 176
Srdanov, V. I., 129, 304, 312
Stace, A. J., 112, 115
Stacy, A. M., 511, 514
Staemmler, V., 206
Stafford, F. E., 300, 304, 305, 309, 312, 313, 317
Staib, P., 67
St. Amand, R., 242
St. John, D. S., 487, 501
Stamm, M., 638
Stannard, P. R., 674
Starkloff, T., 543
Staudinger, H., 2, 19
Stauffer, D., 8
STECHEL, E. B., 563-89; 566, 569, 573, 575, 576, 584
Steif, P. S., 260, 261
Stein, C. A., 461, 462, 465
Stein, G. P., 164
Stein, L., 118
Steinart, W., 121
Steinberg, M., 267
Steinert, W., 184
Steinfeld, J. I., 266
Steinhardt, P. J., 249
Steininger, H., 60, 61, 65
Steinmetz, W. E., 337
Stell, G., 7, 75
Stenholm, S., 267
Stepanka, S., 347, 349
Stepanov, B. I., 270, 286
Stephens, R. B., 246
Stephenson, J. C., 662, 663
Stepleman, R. S., 518, 519
STEVENS, W. J., 357-85; 371, 375, 376, 378-80
Stewart, G. M., 116, 661
Stigter, D., 90, 93
Stirdivant, S. M., 347, 348
STOCKBAUER, R., 215-40; 216-20, 224, 226-29, 231-33, 235
Stockbauer, R. L., 231
Stockburger, M., 343
STOCKMAYER, W. H., 1-21; 6, 7, 9-11, 17, 18, 29, 32, 33, 35, 36, 40, 42, 419, 420, 422, 428, 431

Stoffel, N. G., 217, 220, 226, 231
Stohr, J., 218, 219, 229-31, 239
Stolarski, R. S., 482
Stoll, H., 301, 310, 380
Stoll, W., 268
Stolow, A., 111, 112
Stolte, S., 128, 268
Stone, A. J., 322
Stone, B. M., 571
Stone, J., 662
Stratton, J. A., 525
Straugh, S., 468
Strauss, B., 296, 297, 299, 306-9, 320
Strauss, H. L., 672
Strazielle, C., 35, 38
Street, G. B., 636
Street, R. A., 633
Strey, G., 206
Strich, A., 321, 322
Striker, G., 471
Strong, R. L., 61
Struik, D. J., 26
Stuhlen, R. H., 229
Stwalley, W. C., 292
Stynes, H. C., 472, 473
Su, C. F., 414
Su, T., 165, 176, 414
Su, W. P., 637
Subirana, J. A., 333-35
Sudbø, A. S., 657, 661, 663
Sugakov, V. I., 619
Suh, J. S., 524, 526
Sullivan, B. P., 475
Sullivan, J. C., 646
Sumi, H., 635
Summerville, R. H., 465
Sundberg, R. L., 569, 574, 576, 581, 583, 585
Sung, J. P., 111, 112
Sunjic, M., 50-52
Surowiec, A., 75
Sutcliffe, R., 320-22
Sutherland, J. C., 344, 345, 347
SUTIN, N., 437-80; 438, 439, 442, 443, 446, 447, 451-56, 458, 461-63, 468, 469, 472-76
Sutton, D. G., 266
Suzuki, N., 636
Sverdlik, D. I., 164
Svetlov, Yu. Ye., 40
Swalen, J. D., 518-21
Swamy, K. N., 165
Swanson, E., 38
Swarc, M., 162
Sweeney, D. W., 267, 278
SWENBERG, C. E., 613-55; 614, 618-20, 623-29, 632, 634, 636, 640, 641, 643-47
Swimm, R. T., 571

AUTHOR INDEX

Switalski, J. D., 362
Swofford, R. L., 670, 674
Sworakowski, J., 642
Szabo, A., 621
Szalda, D. J., 472, 473
Szanto, P. G., 393, 395, 415
Szasz, L., 362, 380
Szeftel, J. M., 59, 61
Szentpály, L. V., 301, 310
Szöke, A., 268

T

Tabor, M., 567, 571, 584
Tachiya, M., 149
Tadjeddine, A., 522
Tagami, Y., 25, 34
Taieb, G., 126
Takacs, P. Z., 344, 347
Takahashi, K., 25, 28, 40
Takano, T., 343
Takashima, S., 81-83, 85, 86
Takenada, T., 524, 526
Takubo, Y., 267
Talarico, M. A., 316
Talley, L. D., 117, 121
Tamagake, K., 112, 127
Tamilarasan, R., 461, 462
Tamm, I., 538
Tamura, K., 150
Tanaka, G., 38, 40
Tanaka, K., 119, 120
Tanaka, S., 343
Tanford, C., 139
Tanner, D. B., 636
Tanner, D. W., 38
Tanner, J. E., 425, 429
Tarantelli, F., 205
Tardy, D. C., 115, 659
Tasker, P. W., 362
Tatewaki, H., 301, 321
Taub, A. I., 257, 260
Taube, H., 465, 472, 473
Taure, L. F., 642
Taylor, A. W., 379
Taylor, G. I., 260
Taylor, H. S., 579, 581, 584
Taylor, K. V., 296, 306, 319, 320
Taylor, P. R., 206
Taylor, R. D., 582
Tayursky, V. A., 568
Tebbe, F. N., 63
Teichteil, C., 371, 374, 376, 381
Teller, D. C., 38
Tembe, B. L., 438-42, 454, 459, 461, 464
Tengstal, C. G., 61
Teramoto, A., 33, 38
ter Haar, D., 565, 572, 574, 575
Terhune, R. W., 517, 522, 527
Tesche, B., 524

Teubner, M., 102
Thamann, T. J., 343
Tharrats, J., 618
Theis, T. N., 518, 523
Themans, B., 638
Thiel, P. A., 63
Thiele, E., 662
Thiry, P. A., 60, 554
Thomann, H., 636
Thomas, G. A., 619
Thomas, G. E., 53, 61, 62
Thomas, J. K., 138
Thomas, J. M., 643
Thomas, R. G., 570
Thomas, W. R., 258, 259
Thompson, C. V., 252
Thompson, G. A., 320
Thompson, J. O., 13
Thompson, M. L., 487, 501
Thompson, R. L., 171, 172
Thornman, R. P., 116
Thorup, N., 645, 646
Tiao, G. C., 488, 501
Tiedje, T., 643
Tiee, J. J., 125
Tien, T. H., 137, 140
TINOCO, I. JR., 329-55; 330, 332-35, 340-43, 349-51
Tirado, M. M., 38
Tischler, F., 320
Titze, B., 130
Tobias, I., 331, 333
Toby, B. H., 65
Toigo, F., 518, 519
Tolk, N. H., 215-17, 220, 226, 231
Tolliver, D. E., 390, 404
Tom, H. W. K., 522
Tomkiewicz, Y., 636
Tompkins, R. C., 614
Tong, S. Y., 54, 55
Toon, O. B., 483, 502, 503
Topiol, S., 303, 304, 312, 369, 374, 379
Torndahl, L. E., 258
Török, F., 201, 204
Torrance, J., 646
Torres-Filho, A., 120, 122
Townes, C. H., 266, 412
Toyozawa, Y., 620
Trachtman, M., 204, 205
Trajmar, S., 49
Tranquille, M., 293, 295
Traub, W., 31
Traum, M. M., 215-17, 220, 226, 231
Trautwein, A., 298
Travis, D. N., 300, 309
Treichler, R., 229-31, 239
Trenwith, A. B., 659, 660, 668
Treshchalov, A. B., 621, 622
Trickl, T., 118
Troe, J., 162, 175, 657, 662, 664-66, 668, 680

Trogler, W. C., 302, 311
Troxell, T. C., 31
TRUHLAR, D. G., 159-89; 160, 162-64, 167-72, 174-85
Tsang, J. C., 514, 517-21, 523
Tsao, S. S., 244, 259
Tse, J. S., 322
Tse, R., 124
Tsuji, T., 38
Tu, K. N., 257
Tuck, A. F., 496
Tully, F. P., 171, 172
Tully, J. C., 215, 216, 221, 222, 226, 233, 234, 658
Tundo, P., 140, 151-54
Tung, M. S., 79, 82-86
Tunis, M.-J., 346
Tunis-Schneider, M. J. B., 343, 344, 346, 347
Turco, R. P., 483, 502, 503
Turlet, J. M., 614
TURNBULL, D., 241-63; 241-45, 249-53, 255-57, 260
Turner, T., 119, 120
Turro, N. J., 138, 148, 149
Tweedale, A., 162
Tyler, S. C., 483, 660
Tymen, S., 347, 349

U

Udagawa, M., 517
Uhlenbeck, G. E., 565, 572, 574, 575
Uhlmann, D. R., 244, 246, 249
Uhrberg, R. I. G., 544
Ulam, J., 579
Ullman, R., 24, 32, 36
Ulstrup, J., 438, 443, 445, 447, 448, 452, 454, 456-58, 461, 462
Umemoto, H., 122, 126
Umemura, S., 78, 79
Upton, T. H., 300, 309
Urushadze, Z. D., 454
Ushioda, S., 517, 522
Utiyama, H., 34
Uyeda, N., 524, 526

V

Vaccaro, P. H., 666-68
Vala, M., 297, 307
Vala, M. T., 465
Vala, M. T. Jr., 634
Valentini, J. J., 125, 128
Van Alsenoy, Ch., 199, 203
van Beek, W. M., 78, 81, 86
van Boom, J. H., 343
Van Calcar, R. A., 274, 275, 278
van de Hulst, H. C., 331
van den Ende, D., 128

AUTHOR INDEX 713

Vander Auwera-Mahieu, A., 300, 304, 312, 313
van der Drift, W. P. J. T., 90
van der Marel, G., 343
VanderSande, J., 242
van de Sande, J. H., 347, 348
van der Touw, F., 75, 78, 81-88, 104, 105
Van der Veen, J. F., 229
Van der Zwan, G., 455, 471
Van de Ven, M. J. M., 274, 278
van Dijk, W., 104
Van Duyne, R. P., 457, 460, 507, 511, 514, 517
van Dyke, M., 99
Van Holde, K. E., 38
van Koppen, A. M., 176
Vanni, H., 111, 112, 115, 122
Van Uitert, B. K., 274, 275, 278
van Wazer, J. R., 362, 374
VAN ZEE, R. J., 291-327; 292, 294, 297, 307-9, 314-19, 321, 322
Vargha, A., 199, 202-4
Vasile, M. J., 233
Vasmel, H., 350
Vaughan, W. E., 104
Vaughn, M. R., 342
Veillard, A., 301, 321, 322
Velasco, R. M., 77, 105
Verdieck, J. F., 267
Verdier, P. H., 421, 422, 428, 431
Vergne, J., 345
Verhaegen, G., 292, 294
Verhoof, H. G. F., 78
Vigué, J., 267, 272
Vilesov, F. I., 624
Vincent, P., 518, 519
Vogel, H., 244
Voigt-Martin, I., 34
Voiron, J., 647
Voltz, R., 632, 633
von Carlowitz, S., 203
von Engel, A., 395, 411
Von Freydorf, E., 620
von Neumann, J., 572, 574, 575
Von Raben, K. U., 524, 526
von Smoluchowski, M., 13
Vorlickova, M., 347, 349
Voros, V., 571
Vreugdenhil, T., 81-85, 88
Vukanic, J., 233, 234

W

Waayer, M., 268
Wada, A., 79
Wada, Y., 38, 78, 79, 81-84, 86, 87, 648
Waddill, G. D., 63
Wade, R. J., 483
Wadt, W. R., 366, 371, 376, 377, 379
Wagner, A. F., 172, 177, 178, 180, 181, 183
Wagner, C. N. J., 242, 246
Wagner, H., 61, 62
Wagner, I., 61
Wagner, R. H., 12
Wahl, A. C., 374, 472
Wahlgren, U., 206, 363, 370, 372
Waite, B. A., 580, 581
Walch, S. P., 171, 293, 297, 301, 306, 310-12, 380
Wales, M., 13
Walker, G. H., 568
Walker, J. L., 253
Walker, R. B., 160, 167, 178, 184, 185
Wall, F. T., 424
Wallace, W. L., 475
Walmsley, S. H., 620
Walsh, R., 657, 664
Walsh, S. P., 378
Walter, J. L., 261
Walters, S., 494, 498
Walther, H., 118
Walton, R. A., 291
Wang, A. H.-J., 343
Wang, C. C., 267, 271
Wang, C. S., 549
Wang, D. S., 512, 524, 526
Wang, F. M., 659
Wang, J. C., 11
Wang, S. L., 551
Wang, S. W., 379
Wanna, J., 115
Wanner, J., 118
Wannier, G. H., 6
Warburton, W. K., 247, 256
Warga, M. E., 258, 259
Warlaumont, J. M., 514, 518, 520
Warta, W., 635
Waseda, Y., 246, 247
Wassel, P. J., 111
Wasserman, Z. R., 42, 420
Watson, D. G., 110-12, 114
Watson, G. S., 487, 501
Watson, R. E., 376
Watts, L., 345
Weaver, M. J., 438, 455, 474-76
Webb, D. A., 177
Weber, M., 518, 519, 523
Weber, T. A., 42, 420
Weber, W. H., 517, 522, 527, 531
Webman, I., 452, 456, 457
Webster, C. R., 268
Wechsel-Trakowski, E., 376
Weddle, G., 178
Weeks, J. D., 360-62
Weibel, E., 546

Weil, L. L., 8
Weiler, R., 244, 246
Weill, G., 35
Weinberg, W. H., 53, 61-63, 65
Weinberger, B. R., 636
Weiner, J. H., 42, 130
Weinert, C. M., 69
Weinstein, N. D., 658
Weiser, G., 625, 630
Weiss, A. W., 374
Weiss, G. H., 36
Weiss, R. F., 483, 484
Weisshaar, J. C., 113, 114, 118, 123
Weissman, Y., 573
Weitz, D. A., 524, 526, 528, 529
Welge, K. H., 662
Wells, C., 258, 259
Wells, R. D., 347, 348
WELTNER, W. JR., 291-327; 292, 294, 297, 307-9, 314-19, 321, 322
Wendelken, H. J., 664, 666
Wendelken, J. F., 61, 664, 665, 680
Wendoloski, J. J., 196
Wendorff, J. H., 34
Wennerström, H., 141
Wenning, U., 517
Wesenberg, G. E., 104
Wesslov, W., 358
West, G. A., 670, 671
West, W. P., 517, 527
Westberg, H., 483
Westberg, K. R., 171, 175
Westheimer, F. H., 173
Weston, R. E. Jr., 173
Wetlaufer, D. B., 337
Wetzel, H., 524, 526
Wexler, S., 296, 306
Wheeler, J. C., 11
Wheeler, J. R., 118
Wheeler, M. E., 149
White, A. H., 472, 473
White, C. T., 224, 228
White, C. W., 544
White, J. M., 67
White, L. R., 90, 93-95, 98, 100, 101, 103
Whitehead, J. C., 268
Whitmore, P. M., 69, 517, 527
Whitney, R. S., 7
Whitten, G., 487
Whorf, T. P., 483
Wicke, B. G., 126
Wickel, F., 646
Wickramaaratchi, M. A., 111, 112, 115, 121
Wiersema, P. H., 90
Wiersma, D. A., 621
Wiesenfeld, J. M., 622
Wiesenfeld, J. R., 118
Wieters, W., 665

AUTHOR INDEX

Wight, C. A., 125
Wignall, G. D., 34
Wigner, E., 161, 179
Wigner, E. P., 567, 570, 573
Wilcomb, B. E., 130
Wild, U. P., 524
Wilemski, G., 40
Wilkins, R. G., 472, 473
Wilkinson, J. P. T., 127
Willens, R. H., 242
WILLIAMS, A. L. Jr., 329-55; 350-52
Williams, A. R., 517
Williams, D. F., 633, 637
Williams, F., 614
Williams, J. M., 646
Williams, J. O., 620, 643
Williams, J. W., 13
Williamson, J., 111
Williamson, J. M., 112, 114
Willie, R. A., 67
Willig, F., 626
Willis, R. F., 57, 60, 61
Willmer, R. J. D., 58
Wilson, E. G., 630, 631, 638-40, 644, 645, 649
Wilson, K. R., 233
Winefordner, J. D., 267
Wing, W. H., 389, 390, 404, 406
Winkler, I. C., 614
Winkler, J. R., 468
Winn, J. S., 302, 303, 310-12, 314
Winter, N. W., 374
Winton, R. I., 67, 68
Witko, M., 301
Witt, D., 524
Witt, J., 284
Wittig, C., 662
Wodtke, A. M., 125
Wofsy, S. C., 483
Wokaun, A., 524, 526
Wolf, A., 294-301, 306, 317
Wolf, B., 346
Wolf, E., 16, 17, 509
Wolf, H. C., 619, 620
Wolf, R. J., 164
Wolfe, S., 176, 204
Wolff, P. A., 512, 524
Wolfrum, J., 109, 125, 129, 173
Wolynes, P. G., 75, 91, 455, 456, 461, 471
Wong, C.-P., 35, 38
Wong, J. S., 672
Wong, K. Y., 451
Wong, M., 393, 402, 404
Wong, S. S. M., 567, 568, 585
Wong, W. H., 162
Wong, Y. M., 620
Wood, C., 294, 296, 302, 306
Wood, C. P., 373, 374

Wood, T. H., 514
Woodbury, C. P., 38
Woods, R. C., 387, 390, 393, 395, 415
Woodward, R. W., 318, 319, 322
Woody, R. W., 339, 340, 349
Worlock, J. M., 524
Worosila, G., 468
Woste, L., 127
Wright, J. S., 112, 114
Wright, P., 465
Wrigley, S. P., 660
Wu, C. S. C., 337
Wu, T. T., 346
Wudl, F., 614, 647
Wuebbles, D. J., 486, 487, 491, 492, 494-98, 501
Würtz, D., 639, 644
Wyatt, R. E., 125, 184, 584
Wyder, P., 292
Wyss, J. C., 523

X

Xue-Jing, H., 130

Y

Yakobson, B. I., 455
Yalkovski, S. H., 146
Yamada, C., 410
Yamaguchi, S., 524
Yamaguchi, Y., 179, 196-98, 205, 206, 415
YAMAKAWA, H., 23-47; 23-29, 31-39, 41, 42
Yamamori, K., 518
Yamashita, T., 38
Yanagawa, S., 524
Yanaki, T., 38
Yang, E. S., 472
Yang, H. W.-H., 42
Yang, J. T., 337
Yano, H., 267
Yariv, A., 268
Yarmus, L., 621
Yates, J. H., 370, 371
Yates, J. T. Jr., 216, 218, 226, 232, 233, 235, 517
Yatsimiskii, A. K., 138, 141
Yee, E. L., 455, 475
Yekta, A., 138, 148, 149
Yen, R., 637
Yin, M. T., 543, 544, 555
Yo, H. P., 618
Yokom, K. M., 468
Yokoyama, M., 632
Yoon, D. Y., 25, 34
Yoshie, O., 643
Yoshihara, K., 664
Yoshioka, Y., 381

Yoshizaki, T., 27, 29, 31, 33, 34, 36, 38, 40, 42, 43
Young, R. C., 472, 473
Youtz, J. P., 257
Yu, H., 33, 38
Yu, M. H., 662
Yung, Y. L., 483
Yunoki, Y., 25, 28, 40

Z

Zacharias, H., 662
Zacharias, W., 347, 348
Zagrubskii, A. A., 623, 624
Zahradnik, R., 205
Zakaraya, M. G., 456, 457
Zakhidov, A. A., 619
Zamaraev, K. I., 467
Zander, R. J., 483
Zanella, A. W., 462, 469
Zanette, D., 138, 144-46
ZARE, R. N., 265-89; 120, 122, 123, 125, 127-29, 266, 268, 276-80, 284, 662, 664, 667, 676, 678, 679, 681-83, 685
Zaslavsky, G. M., 568, 571, 572
Zboinski, Z., 643
Zehner, C., 544
Zeil, W., 203
Zeiri, L., 620
Zeiri, Y., 184
Zellner, R., 121, 184
Zeringue, K. J., 297, 307
Zeyher, R., 518, 519
Zgierski, M. Z., 625
Zhang, Q. J., 231, 232
Zhang, Z., 121, 663
Zheng, L.-S., 299, 309
Zhurkin, V. B., 350, 352
Ziegler, T., 301, 310, 373
Zielinski, M., 642
Ziesche, P., 302, 310
Ziff, R. M., 7
Ziman, J. M., 360
ZIMM, B. H., 1-21; 9, 16-18, 36, 38, 40, 43, 420
Zimmer, C., 347, 349
Zimmerman, S. B., 350, 351
Zimonyi, M., 472, 473
Zinsli, P. E., 275
Zografi, G., 146
Zschokke-Gränacher, I., 633, 637
Zunger, A., 543
Zusman, L. D., 455
Zwanzig, R., 17, 36, 39, 40, 420
Zwemer, D. A., 514
Zwier, T. S., 113, 114, 118, 123
Zwolle, S., 81-85, 88

SUBJECT INDEX

A

Acetic acid
  adsorbed surface products on metals, 65
Acetone
  adsorbed surface products on metals, 65
Acetylene
  adsorption on metals, 63
  electron energy loss spectroscopy and, 62
  excited state of
    stimulated emission pumping from, 586
  force constant calculations and, 206
  highly excited ground state molecules in, 594
  stimulated emission pumping and, 666
  vibrational modes on nickel, 55
Actinides
  core-valence correlation effects in, 380-81
Actinide series
  start at element 93, 2
Adiabatic barriers
  variational transition state theory and, 182-84
Adiabatic collision theory
  variational transition state theory and, 166-67
Adipic acid
  mixture with glycerol
    tree formation and, 4
Adsorbate-adsorbate interaction
  ionic desorption and, 226
Adsorbates
  electron scattering by, 51-57
  electronic excitations of, 66-69
  optical response of
    chemisorption and, 507-8
  spectroscopic signals of
    electromagnetic resonances and, 515
  vibrational spectroscopy of, 60-65
Adsorbate-substrate effects
  ionic desorption and, 226

Aldehydes
  $O(^3P)$ reactions with
    laser-induced fluorescence and, 118
Alkali/alkaline earth elements
  valence and core electron correlation in, 380
Alkali halides
  ionic desorption from, 226
Alkenes
  electron energy loss spectroscopy and, 62
Alkylanilines
  intramolecular vibrational energy redistribution in, 599
Alkylbenzenes
  dispersed fluorescence spectra of, 598-99
  highly excited ground state molecules in, 594
  relaxation and predissociation of, 602
Alkyl halides
  supersonic O atom reactions with, 124
Alkynes
  electron energy loss spectroscopy and, 62
Allylisocyanide
  isomerization of, 680-81
Aluminum
  subsurface oxygen vibrations in, 61
  xenon adsorption on, 69
Amines
  $O(^3P)$ reactions with
    laser-induced fluorescence and, 118
Amino acids
  cyclic dimers of
    circular dichroism of, 340
Ammonia
  protonated
    infrared laser spectroscopy and, 400-1
  valence angle of
    correlation effect on, 209
Ammonium ions
  solvated
    infrared laser spectroscopy and, 389

Amylose
  parameters for code 26, 29-31
  transport properties in dimethyl sulfoxide, 38
Anharmonicity
  Morse approximation for, 168-69
Aniline
  collisional relaxation processes in, 597
Anthracene
  carrier generation in, 623-25
  carrier transport in, 634-36
  charge-transfer states in, 625
  electrical and optical properties of, 613-14
  geminate recombination in, 634
  intramolecular vibrational energy redistribution in, 600
Anthroquinones
  vibrational relaxation in, 606
Aqueous micelles
  formation of, 138-39
Aromatic chromophores
  circular dichroism structural analysis and, 339
Aromatic hydrocarbons
  adsorption on metals, 65
  electron energy loss spectroscopy and, 63
  homonuclear polynuclear
    carrier generation in, 623
  polycyclic
    electrical and optical properties of, 613-14
Associative detachment reactions
  infrared chemiluminescence and, 114-15
Atmosphere, 481-503
  catalytic cycles and, 489-90
  methane-smog reactions and, 490
  nuclear war and, 502-3
  ozone production in, 490-91
  photochemical reactions in, 484-85
  trace species in, 483-86
Atom-diatom reactions
  collinear

715

SUBJECT INDEX

microcanonical variational
transition states for,
163-64
variational transition state
theory and, 172
Atomic core electrons
see Effective core potentials
Atomic-polyatomic reactions
infrared chemiluminescence
and, 110-11
Atomic transport
in metallic glasses, 256-59
Attenuated total reflection
surface-enhanced spectroscopy
and, 521-22
Auger decay
ionic desorption and, 225-26
Auger process
ion desorption and, 223-24
Autoionization, 624

B

Benzene
adsorption on metals, 65
collisional relaxation processes
in, 597
electron energy loss spectroscopy and, 63
electron-stimulated ion desorption from, 225
excitations on silver, 68
Hartree-Fock force constant
calculations and, 201
photoacoustic spectra of, 674-75
relaxation and predissociation
of, 601-2
Benzene solid films
excitonic ions in, 619
Bessel functions, 457
Bilayer lipid membranes
formation of, 140
Bilayers
diffusion-controlled reaction
rates and, 150
Bimodal rotational distribution,
121-22
Bimolecular electron transfer
steady-state, 438-40
Bimolecular reactions
micellar
kinetics of, 144
Birge-Sponer relationship, 671
Blanc's law, 414
Blyholder model
of CO chemisorption bond, 68
Bond polarity
ionic desorption and, 226
Born-Oppenheimer approximation, 191
Born-Oppenheimer potential
energy surface

electron transfer processes
and, 444
Bose-Einstein condensation, 5
double-helix theory a, 11
Bose-Einstein gas
pressure and number density
of, 10-11
Bragg diffraction, 54
Branched polyesters
viscosities of, 9
Branched polymers
mean square radii of gyration
for, 9
Brody distribution, 568
Bromine
electronic excitations on palladium, 69
Brownian motion
single polymer chains and,
419-29
Bugl-Fujita chain, 26
Butylhydroperoxide
bond fission of
vibrational overtone excitation and, 682
t-Butylhydroperoxide
overtone vibration excitation
and, 684-85
overtone vibrations and, 671-72

C

Candidate laser molecules
assignments of transitions and
states in, 378
Canonical variational theory,
163
Carbon dioxide
atmospheric
increase in, 483
mass of, 481
force constant calculations
and, 206
Carbon monoxide
metastable states of
electron and photon desorption and, 217
Cartesian degrees of freedom
short-time polymer motion
and, 420-21
Cascade theory, 7
Catalytic reaction dynamics
surface
infrared chemiluminescence
and, 116
Cellulose
random hydrolytic degradation
of, 2
Cellulose acetate
transport properties in trifluoroethanol, 38
Cellulose derivatives
transport properties of, 38

CEPA-2, 197
Charge transfer reactions
photoelectron/product ion
coincidence and, 119
Chemical activation
unimolecular reactions and,
659-61
Chemical statistics, 4-11
Chemiluminescence
overtone vibration excitation
and, 667
see also Infrared chemiluminescence
Chemisorption
optical response of adsorbed
molecules and, 507-8
Chiral amides
circular dichroism of, 340
Chiral object
definition of, 329
Chlorine
catalytic destruction of stratospheric ozone and, 482
Chlorine nitrate
atmospheric formation of, 494
Chlorine oxide
catalytic destruction of stratospheric ozone and, 482
Chlorofluorocarbons
catalytic destruction of stratospheric ozone and, 482
Chromium
triatomic molecule of, 319
Chromium diatomics
bonding in, 316-17
Chromium molecule
bonding in, 306-7
ground state of, 306
Chromium-molybdenum
bonding in, 318
Cini-Sawatzky theory
ionic desorption and, 224
Circular dichroism, 329-53
classical polarizability theory
and, 350
double-stranded polynucleotides and, 343-45
Kronig-Kramers transform of,
333
large molecule
orientation average for,
331-32
linear response theory and,
350
matrix method and, 350
Mueller matrix and, 331-32
nearest-neighbor analysis by,
340-43
perturbation theory and, 349
polymer
coupled oscillator theory
and, 349
polynucleotide, 340-53
polypeptide, 336-40

SUBJECT INDEX 717

protein, 336-40
purine-pyrimidine copolymers
  and, 348-49
vacuum ultraviolet
  DNA and, 344-45
Circular dichroism spectra
  analysis of, 337-38
  polynucleotide, 341
  unordered polypeptide
    red-shift of, 340
Circular intensity differential
    scattering
  Mueller matrix and, 331-32
  orientation average for, 332
Circularly polarized light
  differential absorption of, 330
  differential scattering of, 331
Classical polarizability theory
  circular dichroism and, 350
Climate
  nuclear war and, 502-3
Closed bilayer aggregates
  formation of, 140
Cobalt molecule
  dissociation energy of, 308
Cole-Cole curve, 81-82
Colloidal solutions
  Raman scattering in, 526-27
  surface-enhanced spectroscopy
    and, 523-24
Conductivity
  dielectric response of polyions
    and, 90
Configuration interaction theory
  ionic desorption and, 224
Configurational entropy
  metallic glasses and, 244-45
Copper
  chemisorbed CO on
    electron energy loss spec-
      troscopy and, 62
    pentatomic molecule of, 322
    Raman spectra enhancement
      and, 512
  surface of
    electronic properties of,
      554
  triatomic molecule of, 320-21
Copper molecule
  bonding in, 309-10
Copper-silver
  bonding in, 317-18
Coriolis interactions
  intramolecular vibrational re-
    laxation and, 594, 600
Coronene
  intramolecular relaxation in,
    609
Counterion condensation theory
  reactivities in polyelectrolytes
    and, 146
Coupled oscillator theory
  polymer circular dichroism
    and, 349

Coupled-perturbed Hartree-Fock
  equations, 197
Covalent interaction
  ionic desorption and, 226
Critical micelle concentration,
    138-39
  micellar pseudophase and, 141
Cryogenic matrices
  molecular ions in
    infrared laser spectroscopy
      and, 388
Crystalline solids
  melting points of
  shear modulus and, 6
Cyclobutane
  rocking frequency of, 204
Cyclobutene
  overtone vibration initiated
    unimolecular isomeriza-
      tion in, 681
Cycloheptatrienes
  internal conversion and, 664-
    66
Cyclohexane
  chemisorbed on metals
    electron energy loss spec-
      troscopy and, 63
  electron-stimulated ion desorp-
    tion from, 225
Cyclooctatetrene
  internal conversion and, 664
1-Cyclopropylcyclobutene
  overtone vibration initiated
    unimolecular isomeriza-
      tion in, 681

D

Daniels approximation, 35
DC discharges
  Doppler shifts and, 396
  tunable infrared laser spectros-
    copy and, 393-94
Debye equation, 81
  dieletric relaxation of polar
    solvents and, 469
Debye-Falkenhagen dielectric increment, 95
de Gennes-Doi-Edwards theory
  polymer chain dynamics and,
    430
Dense random packing
  metallic glasses and, 241, 247
Density force, 195
Detergent dialysis
  small unilamellar vesicle
    preparation and, 140
Diatomic molecules
  effective core potentials and,
    371-72
  heteronuclear, 313-18
  homonuclear, 292-13
  light hydride
    internal rotation and, 608-9

resonance scattering and, 55
Dielectric continuum theory
  electron transfer reactions and,
    453-54
Differential absorption
  of circularly polarized light,
    330
Differential scattering
  of circularly polarized light,
    331
Diffusivities
  in metallic glasses, 256-59
p-Difluorobenzene
  stimulated emission pumping
    and, 666-67
DIIS, 198
Dimethyl sulfoxide
  amylose in
    transport properties of, 38
Dipole scattering
  adsorbates and, 51-54
  characteristics of, 52-53
Dirac delta functions, 457, 458
Dirac-Fock method
  all-electron, 373
  effective spin-orbit potentials
    and, 375-76
Dissociation energy
  chromium molecule, 306
  cobalt molecule, 308
  iron molecule, 307
  manganese molecule, 307
  nickel molecule, 309
  palladium molecule, 312
DNA
  nearest-neighbor frequencies
    for, 342
  vacuum ultraviolet circular
    dichroism and, 344-45
Doi-Edwards theory
  Monte Carlo simulations and,
    433
Doppler shift
  velocity modulation infrared
    laser spectroscopy and,
      395-97
Double-helix theory
  Bose-Einstein condensation
    and, 11

E

Effective core potentials, 363-69
  pseudo-orbitals and, 370-71
Effective potentials, 357-82
  all-electron methods and, 369-
    72
  averaged relativistic, 359-60
  applications of, 376-80
  development of, 360-69
  model, 362-63
  relativistic, 359-60, 373-80
  spin-orbit contributions to,
    374

Effective spin-orbit potentials, 375-76
Efrima-Metiu scheme
 for electromagnetic resonances in spectroscopy, 511-12
Eigenfunctions
 semiclassical, 568-70
Eigenvalue equation
 for valence electrons, 360-61
Einstein coefficients
 directional
  spectroscopic transitions and, 269-71
Elasticity
 see specific type
Electrode effect
 polyelectrolyte permittivity and, 78
Electroluminescence
 organic crystal, 648
Electromagnetic resonances
 spectroscopic signals of adsorbed molecules and, 515
Electromagnetism
 surface-enhanced spectroscopy and, 508-15
Electron energy loss spectroscopy, 49-70
 adsorbates and, 60-65
 chemisorbed CO on metals and, 62-63
 Fuchs-Kliewer phonons and, 58
 metal surface phonons and, 59
Electronic absorption bands
 iron molecule, 307
Electronic excitations
 adsorbate, 66-69
 clean surfaces 66
Electronic relaxation
 collisional in ground state molecules, 592-93
 intramolecular density of states and, 591
 see also Vibrational relaxation
Electronic states
 molecular reaction dynamics and, 126-28
Electron jump mechanism, 126-28
Electron scattering
 adsorbates and, 51-57
Electron-stimulated desorption, 215-39
 angle-resolved, 235-37
 experimental characteristics of, 216-18
 experimental methods in, 218-20
 mechanisms of, 220-35
Electron-stimulated desorption
 ion angular distributions, 235-37
Electron transfer reactions, 437-76
 adiabatic, 447-51
 cross-relations and, 474-76
 electronic factors in, 461-69
 exchange, 471-74
 free energy and, 474-76
 inner- and outer-shell reorganization and, 440-42
 nonadiabatic, 447-51
 nonadiabatic rate constants and, 460-61
 precursor complex and, 442-43
 preexponential frequencies and, 469-71
 quantum mechanical, 456-60
 redox centers and, 438
 relaxation times and, 469-71
 semiclassical, 456-60
 steady-state, 438-43
 thermal, 446-47
 time-dependent rate constants and, 443
 transition state theory and, 451-56
Electrophoresis
 dielectric response of polyions and, 90
Electroviscous effect
 dielectric response of polyions and, 90
Endoergic reactions
 variational transition state theory and, 175
Energy derivatives
 calculation of, 193-98
Energy level spacings
 semiclassical ergodic theory and, 567-68
Equilibrium
 stiff-chain macromolecules and, 31-36
Ergodicity
 Boltzmann's intuitive notion of, 565
Ergodic theory
 classical, 564-66
 quantum, 572-79
 semiclassical, 566-72
Ethylene
 electron energy loss spectroscopy and, 62
 force constant calculations and, 206
 force field of, 202-3
Excimer laser molecule spectra
 analysis of, 379
Excitons
 interactions of, 618-19
 transport of, 614-18
trapping of, 619-21
Exoergic reactions
 variational transition state theory and, 174-75

F

Femtosecond spectroscopy, 648
Fermi interactions
 intramolecular vibrational relaxation and, 600
Flames
 laser-induced fluorescence measurements in, 267
Flexible chains
 conformational statistics of, 2
Flow-tube devices
 thermal energy ion-molecule reactions in, 113-15
 vibrational state distributions in, 112
Fluorescence
 surface-enhanced, 527-28
 see also Laser-induced fluorescence
Fluorinated analogs
 electron-stimulated ion desorption from, 225
Fluorobenzene cations
 unrelaxed emission of, 608
Fluoronium ion
 infrared laser spectroscopy and, 401
Flux
 laser-induced fluorescence and, 284
Force constants
 calculation from wave functions, 205-10
 diagonal deformation correlation contributions to, 209
 dynamic correlation effects and, 207-10
 empirical scaling of, 200
 experimental determination of, 191
 Hartree-Fock, 200-5
 quadratic stretching correlation contributions to, 209
Force fields
 scaled quantum mechanical, 202-4
 vibrational, 191-210
Formaldehyde
 adsorbed surface products on metals, 65
 force constant calculations and, 206
 Hartree-Fock force constant calculations and, 201

SUBJECT INDEX 719

highly excited ground state molecules in, 594
stimulated emission pumping and, 666
unimolecular decomposition of second-order tunneling method and, 179
Formic acid
  adsorbed surface products on metals, 65
Forster distance, 617
Fourier transform infrared spectroscopy
  stable gas phase molecules and, 394
Franck-Condon excitation
  ionic desorption and, 222-23
Franck-Condon factors, 582-85
  intramolecular vibrational relaxation and, 600
Free energy
  electron transfer reactions and, 474-76
Free volume theory
  metallic glasses and, 244-45
Friedel oscillations, 548
Fröhlich equation, 469
Fuchs-Kliewer phonons, 57-58
Fulcher-Vogel law, 244

G

Gas chromatography
  overtone vibration excitation and, 667
Gaseous molecules
  dissociation and ionization of electron- and photon-induced, 216-17
Gas-phase spectroscopy
  scandium molecules and, 292-93
Gaussian molecules
  radii of gyration of calculation of, 7
Gaussian orthogonal ensemble, 568
Gel filtration
  small unilammelar vesicle preparation and, 140
Gel points
  dilution and, 8
Geminate recombination
  Monte Carlo simulations and, 632-33
  Onsager theory of, 627-31
Generalized master equation, 616
Generalized transition state theory, 162
  recrossing corrections and, 177-78
Glycerol
  mixture with adipic acid tree formation and, 4
Glyoxal
  collisional relaxation processes in, 597
Gold
  Raman spectra enhancement and, 512
Gold molecule
  bonding in, 313
Gratings
  surface-enhanced spectroscopy on, 517-21
Greenhouse effect, 482, 499-502
Ground state molecules
  collisional relaxation processes in, 592-93

H

Halogens
  ground and excited state metal atom reactions with, 126
  supersonic O atom reactions with, 124
Hamiltonian operators
  electron transfer processes and, 443
Harmonic oscillator model
  electron transfer reactions and, 453
Hartree-Fock core
  orthogonality constraint and, 358
Hartree-Fock equations
  coupled-perturbed, 197
Hartree-Fock force constants, 200-5
Hartree-Fock orbitals
  G1 orbitals and, 364
Hartree-Fock wavefunctions, 192, 195
  effective core potentials and, 366-67
Hazard analysis
  short-time polymer motion and, 421
Hearst-Stockmayer equation
  Kratky-Porod chain and, 33
Hellmann-Feynman force, 195
Hellmann-Feynman forces
  semiconductor surfaces and, 544
Henon-Heiles oscillator
  intramolecular and dissociation rates in, 581
Henon-Heiles potential energy surface, 580
Heterolytic reactions
  ion-exchange models for, 141-46
Hilbert space, 570
Hückel model, 459
Hybrid rate constant
  microcanonical variational theory and, 166-67
Hydrocarbons
  C-H stretching vibration of, 595
  effective core potentials and, 371-72
  electron energy loss spectroscopy and, 62-63
  gas phase absorption spectra of, 594
  Hartree-Fock force constant calculations and, 201
  $O(^3P)$ reactions with laser-induced fluorescence and, 118
  see also Aromatic hydrocarbons
Hydrogen
  infrared predissociation spectrum of, 390-92
Hydrogen bonding
  ionic desorption and, 226
Hydrogen cyanide
  force constant calculations and, 206
Hydrogen-transfer reactions
  intramolecular
  transition state theory with tunneling and, 181-82
Hydroxyl radicals
  catalytic ozone destruction and, 489
Hydroxyphenalenone
  vibrational relaxation in, 606

I

Image fields, 509
Impact scattering
  adsorbates and, 54-55
Improved canonical variational theory, 163
Inert gases
  desorption from metals, 225
Infrared chemiluminescence, 110-16
  associative detachment reactions and, 114-15
  surface catalytic reaction dynamics and, 116
  thermal energy ion-molecule reactions and, 113-14
  translational kinetic energies and, 115-16
  vibration-rotation states and, 110-11
  see also Chemiluminescence
Infrared laser spectroscopy, 387-416
  collisional relaxation processes and, 592-93

SUBJECT INDEX

DC discharges and, 393-94
fast ion beams and, 389-93
Fourier transform, 394
pulsed radiolysis and, 388-89
velocity modulation, 395-410
theory of, 410-15
Infrared multiphoton excitation
  unimolecular reactions and, 661-64
Infrared spectroscopy
  chemisorbed CO on metals and, 62
Integral force, 195
Internal conversion
  unimolecular reactions and, 664-66
Intramicellar reactions
  excited state and radical, 146-51
Invariant tori theory
  ergodicity and, 566
Iodine
  stimulated emission pumping and, 666
Ion beams
  infrared laser spectroscopy and, 389-93
Ion cyclotron resonance traps
  translational energies of products in
    laser-induced fluorescence and, 119
Ion desorption
  Auger process and, 223-24
  Knotek-Feibelman model for, 223
  measurement of, 216
  mechanisms of, 220-35
  Menzel-Gomer-Redhead model of, 222-23
  neutral, 231-32
  sequence of, 221
Ion-exchange models
  for heterolytic reactions, 141-46
Ionic compounds
  surface phonons of vibrations of, 57-58
Ionization potential
  iron molecule, 307
Ions
  collisions of polar bodies with capture rate constants in, 165
Iron
  chemisorbed CO on
    electron energy loss spectroscopy and, 62
  diatomic molecule of
    bonding in, 316
  triatomic molecule of, 319
Iron molecule
  dissociation energy of, 307-8

Isomerization
  methylisocyanide, 680
  overtone vibration initiated unimolecular, 680-83
  polyacetylene, 363
Isotropic scattering function
  stiff-chain macromolecules and, 33

K

KAM theorem
  quantum analog to, 579
Kirkwood procedure, 36
Kirkwood-Riseman approximations, 36-37
Knotek-Feibelman model
  ionic desorption and, 223
Kolmogorov entropy, 572
  quantum analog of, 572-73
Kramers chain, 40
Kratky-Porod chain
  definition of, 24-25
  Hearst-Stockmayer equation for, 33
  steady-state transport coefficients and, 36
Kronecker delta functions, 458
Kronig-Kramers transform
  circular dichroism and, 333
Kron model
  polymer chains and, 424-25

L

Landau-Zener model
  electron transfer reactions and, 448
Langevin equations
  short-time polymer motion and, 420-21
Langmuir-Blodgett layers
  surface-enchanced spectroscopy and, 522
Lanthanides
  core-valence correlation effects in, 380-81
Lanthanum
  triatomic molecule of, 318-19
Laser-induced fluorescence
  intensity of, 271-75
  molybdenum molecule and, 311
  overtone vibration excitation and, 667
  vibrational and rotational states and, 116-19
Laser-induced fluorescence saturation, 265-87
  angular factors and, 282-83
  definition of, 266
  directional Einstein coefficients and, 269-71
  flux measurement and, 284

polarization measurements and, 279-82
population measurements and, 275-79
transition dynamics and, 268-69
Lasers
  reactive dynamics and, 125-26
  vibrational relaxation studies and, 591-92
Lattice statistical theory
  intramicellar reactions and, 148
Lennard-Jones potential, 424
Ligand binding
  circular dichroism and, 343
Light scattering
  angular dependence of, 19
  molecular weights and, 11-19
Linear condensation polymers
  ring-chain equilibrium in, 10
Linear response theory
  circular dichroism and, 350
Lineweaver-Burke equation, 143
Liouville equation, 40
Liquids
  molecular ions in
    infrared laser spectroscopy and, 388
Lithium
  metastable states of
    electron and photon desorption and, 217
Low energy electron diffraction, 54
  electron-stimulated desorption and, 219

M

Macromolecules
  stiff-chain, 23-43
    dynamic properties of, 40-43
    equilibrium properties of, 31-36
    models of, 24-31
    transport properties of, 36-40
Malonaldehyde
  hydrogen atom transfer between equivalent sites in, 182
Manganese
  pentatomic molecule of, 321
Manganese diatomics
  bonding in, 317
Manganese molecule
  dissociation energy of, 307
Marcus-Coltrin path, 179
Marcus expression, 454
Maslov index, 567
Mass spectrometry
  iron molecule and, 307-8

SUBJECT INDEX 721

nickel molecule and, 309
Mass spectroscopy
  copper molecule and, 309
  silver vapor and, 312-13
Matrix Raman spectroscopy
  silver molecule and, 313
Maxwell-Wagner effects, 87
Measure algebra
  quantum, 575-777
Melt-glass transitions, 243-46
Membrane filtration
  small unilamellar vesicle preparation and, 140
Membrane mimetic systems, 137-55
  see also specific component
Menzel-Gomer-Redhead model
  ionic desorption and, 222-23
Mercury halides
  spectroscopic properties of
    relativistic effective potentials and, 377
Metal alloy glasses
  formation of
    kinetic resistance to partitioning in, 254-55
Metal clusters
  assignments of transitions and states in, 378
  substrate interactions with
    relativistic effective potentials and, 377
Metal-halogen reactions, 126-27
Metallic binding
  effective potential method and, 360
Metallic glasses, 241-61
  atomic transport in, 256-59
  binary
    models for, 247-48
  deformation of, 259-61
  diffraction pattern of, 246-47
  diffusivities in, 256-59
  ductile fracture of, 260-61
  formation and kinetic stability of, 249-56
  homogeneous flow in, 259-60
  inhomogeneous flow in, 260
  melt-glass transition and, 243-46
  properties of, 256-61
  structure of, 246-49
Metallic liquids
  nucleation of crystals in
    kinetic resistance to, 241
Metalloids
  diffusivities in, 259
Metallo-organic complexes
  relativistic effective potentials and, 380
Metal melts
  nucleation onset in
    minimum scaled undercooling at, 251

Metal-metal bonded systems
  electronic structure of, 378-79
Metal-metalloid alloys
  melt-quenched to glasses, 242
Metal-metalloid glasses
  short-range order in, 248
Metal-oxidant systems
  electron jump mechanism in, 126
  electronically excited states in, 126
Metal oxide spectra
  assignments of transitions and states in, 378
  relativistic effective potentials and, 379
Metal particles
  isolated
    surface-enhanced spectroscopy and, 523-29
Metals
  aromatic hydrocarbon adsorption on, 65
  inert gas desorption from, 225
  molecularly absorbed oxygen on
    O-O stretching frequency of, 61
  noble gases physisorbed on, 67
  surface phonons of
    vibrations of, 59-60
  surfaces of, 547-60
Methane
  collisional ground state relaxation in, 592
  force constant calculations and, 206
  greenhouse effect and, 482
  primary source of, 483
  stratospheric ozone and, 496
Methane-smog reactions, 490
Methanol
  adsorbed surface products on metals, 65
  circular dichroism of polynucleotides and, 347
Methylamine
  frequencies of, 204
  Hartree-Fock force constant calculations and, 201
2-Methylcyclopentadiene
  overtone vibration initiated unimolecular isomerization in, 681
Methyl fluoride
  energy flow maps and, 593
  force constant calculation and, 210
Methylisocyanide
  isomerization of, 680
Micellar compartments
  excited state and radical reactions in, 146-51

Micellar pseudophase
  critical micelle concentration and, 141
Micellar surface potentials
  reactive counterion distribution and, 146
Micelle-catalyzed reactions
  electrolyte effects on rates of, 144
Micelles
  diffusion-controlled reaction rates and, 150
  hydrodynamic diameters of, 139
  kinetic stability of, 140
Micellization
  surfactants and, 138-39
Michaelis-Menten equation, 141-43
Microcanonical variational theory, 162-63
Microcanonical variational transition states
  for collinear atom-diatom reactions, 163-64
Microwave rotational spectroscopy, 192
Microwave spectroscopy
  Doppler shift measurements and, 395-96
Model potentials, 362-63
Molecular crystals
  carrier transport in, 634-36
Molecular ions
  infrared laser spectroscopy and, 388-89
  infrared spectroscopy and, 397-410
  velocity modulation spectroscopy and, 398-99
Molecular reaction dynamics, 109-31
  angular distributions and, 123-26
  electronic states and, 126-28
  infrared chemiluminescence and, 110-16
  laser-induced fluorescence and, 116-19
  orientation effects and, 128-30
  rotational effects and, 121-23
  transition state and, 130-31
  translational excitation and, 123-26
  vibrational enhancement and, 119-21
  vibrational excitation and, 110-21
Molecular reagents
  alignment and orientation of, 128-30
Molecular transitions
  laser-induced fluorescence saturation of, 278-79

SUBJECT INDEX

Molecular weights
 light scattering and, 11-19
Møller-Plesset perturbation
 theory, 197
 second-order, 206
MOLPRO, 196
Molybdenum
 surface of
  electronic properties of, 549-51
Molybdenum molecule
 bonding in, 310-12
Monatomic liquids
 atomic short-range order in preferred topology of, 241
Monatomic materials
 existence in metastable amorphous solid form, 242
Monolayers
 diffusion-controlled reaction rates and, 150
Monte Carlo simulations
 Doi-Edwards theory and, 433
 excluded volume and isolated polymer chains and, 421-22
 geminate recombination and, 632-33
 Kratky-Porod chains and, 38
 polymer chains and, 432
 reptation polymer models and, 426-27
 short-time polymer motion and, 420
Morse approximation, 183
 for anharmonicity, 168-69
Mössbauer spectroscopy
 iron diatomics and, 316
 metallic glasses and, 247
 iron molecule, 308
Mueller matrix, 331
Multilamellar vesicles
 formation of, 140
Multiphoton infrared dissociation
 polyatomic molecules and, 595-96

N

Naphthalene
 carrier transport in, 634-36
 Hartree-Fock force constant calculations and, 201
Naphthalene-perdeuteronaphthalene mixed crystal
 exciton transport in, 616
Naphthalenesulfonate (phenylamino) derivatives
 intramolecular electron transfer rates in, 471
Naphthazarin
 in large polyatomic molecules, 606

Navier-Stokes equation, 37
 quasistatic version of, 90
Neopentane
 electron-stimulated ion desorption from, 225
Nickel
 acetylene adsorption on, 63
 benzene adsorption on, 65
 binding configuration of pyridine on, 69
 chemisorbed CO on
  electron energy loss spectroscopy and, 62
 cyclohexane chemisorbed on
  electron energy loss spectroscopy and, 63
 surface of
  electronic properties of, 560
 triatomic molecule of, 319-20
 vibrational modes of acetylene on, 55
Nickel-copper
 bonding in, 316
Nickel molecule
 dissociation energy of, 309
Niobium
 surface of
  electronic properties of, 549-51
Niobium-iridium
 bonding in, 318
Niobium molecule
 ground state and bonding in, 310
Nitrogen oxides
 stratospheric
  mass of, 481
  stratospheric ozone and, 482
Nitrous oxide
 atmospheric
  increase in, 483-84
  greenhouse effect and, 482
NMR spectroscopy
 metallic glasses and, 247
Noble gases
 physisorbed on metals, 67
Normalization force, 195
Nuclear war
 climatic effects of, 502-3

O

Onsager theory, 627-31
Optical anisotropy
 stiff-chain macromolecules and, 34
Optical pumping experiments
 nonlinear effects in, 266-67
Optical spectroscopy
 copper molecule and, 309
Organic molecules

Hartree-Fock force constant calculations and, 202
Organic solids
 carrier generation mechanisms in, 623-27
 carrier recombination in, 627-34
 carrier transport in, 634-41
 carrier trapping in, 641-45
 exciton interactions in, 618-19
 exciton transport in, 614-18
 exciton trapping in, 619-21
 spectroscopy and, 621-23
 superconductivity in, 645-47
Oseen-Burgers procedure, 36
Oseen tensor, 37-38, 424
Osmometry
 molecular weights of polymers and, 12
Ovalene
 intramolecular relaxation in, 609
 intramolecular vibrational energy redistribution in, 600
Overtone vibrations
 unimolecular reactions and, 667-68, 670-87
Oxetane
 rocking frequency of, 204
Oxides
 ionic desorption from, 226
Oxonium hydrates
 infrared laser spectroscopy and, 388-89
Ozone
 atmospheric
  mass of, 481
 stratospheric, 492-99
  greenhouse effect and, 500
  nitrogen oxides and, 482
  trends in, 486-88

P

Pair correlation functions
 metallic glasses and, 246-47
Palladium
 acetylene adsorption on, 63
 bromine adsorption on, 69
 surface of
  electronic properties of, 549-58
Palladium molecule
 dissociation energy of, 312
Particulate matter
 atmospheric, 496
Pentacene
 charge-transfer states in, 625
 intramolecular vibrational energy redistribution in, 600

SUBJECT INDEX 723

Pentatomic molecules
  transition metal, 321-22
Percolation theory, 245
Perturbation theory
  circular dichroism and, 349
Phenalenones
  methyl-substituted hydroxy
    unrelaxed emission and, 608-9
Phenoxylalkynes
  intramolecular vibrational energy redistribution in, 599
Phenylalanine side chains
  circular dichroism and, 339
Phenylalkynes
  intramolecular vibrational energy redistribution in, 599
Phillips-Kleinman method
  one-electron, 361-62
Phonons
  Fuchs-Kliewer, 57-58
  see also Surface phonons
Photoacoustic spectroscopy
  benzene, 674-75
Photodissociation
  high energy region and, 586
Photoelectron spectroscopy
  organic solids and, 622-23
Photometry
  polystyrene solutions and, 17-18
Photon-stimulated desorption
  angle-resolved, 235-37
  experimental characteristics of, 216-18
  experimental methods of, 218-20
  mechanisms of, 220-35
Photopolymerization, 152-54
Phthalocyanine
  intramolecular vibrational energy redistribution in, 600
Picosecond spectroscopy, 648
Platinum
  acetylene adsorption on, 63
  benzene adsorption on, 65
  chemisorbed CO on
    electron energy loss spectroscopy and, 62
  cyclohexane chemisorbed on
    electron energy loss spectroscopy and, 63
  hydrogen chemisorbed on
    photoemission properties of, 379
  molecularly absorbed oxygen on
    O-O stretching frequency of, 61
  surface of
    electronic properties of, 560
Platinum complexes
  conformational and energy surface behavior of
    relativistic effective potentials and, 377
Platinum molecule
  bonding in, 312
Poisson-Boltzmann equation, 99
  equilibrium, 91
Poisson distribution, 567-68
Polar molecules
  collisions of ions with
    capture rate constants in, 165
Polar solvents
  dieletric relaxation of, 469
Polaritons, 614
Polarizability tensor
  linear spectroscopic processes and, 509-10
Polarization
  laser-induced fluorescence saturation and, 279-82
Polarized light
  chiral object interactions with, 329
  see also Circularly polarized light
Polaron-bipolaron theory
  spinless conductivity and, 637-38
Polyacetylenes
  carrier transport in, 636-38
Polyamides
  transport properties of, 38
Polyatomic atom abstraction reactions
  secondary reactions with products of, 112
Polyatomic molecules, 591-609
  collisional ground state relaxation in, 592-93
  collisional relaxation in excited states in, 596-98
  electronic fluorescence and, 596
  excited state studies and, 596
  Hartree-Fock force constant calculations and, 200
  highly excited ground state, 593-95
  intermode coupling and, 604-5
  internal rotation and, 608-9
  intramolecular vibrational energy redistribution in, 598-600
  multiphoton infrared dissociation and, 595-96
  quantum beats in, 600-1
  rare gas
    relaxation processes in, 603
  van der Waals relaxation and predissociation in, 601-3
Polyatomic reactions
  variational transition state theory and, 171-72
Polyatomic systems
  averaged relativistic effective potentials and, 376
Polycarbonate
  transport properties of, 38
Polycondensation, 2
Polydiacetylene
  carrier transport in, 638-40
Polydimethylsiloxane
  characteristic ratio of, 25
Poly-DL-alanine
  characteristic ratio of, 25
Polyelectrolyte permittivity, 75-106
  counterion polarization and, 102-5
  double-layer polarization and, 90-95
Polyelectrolytes
  counterion condensation theory and, 146
  definition of, 76
Polyethylene
  band structure calculations in, 381
Polyions
  rod-like
    polarization of, 101-2
Polymer chains
  entangled
    dynamics of, 429-33
  flexible
    local modes in, 420-21
  Kron model for, 424-25
  long-time relaxation effects and, 421-29
  reptation model for, 425-28
  Rouse model for, 423-24
  star, 428-29
  Verdier-Stockmayer model for, 421-23
Polymerization
  free-radical chain, 2
  vesicle stability and, 140
Polymer melts
  dynamics of
    simulation of, 430-31
Polymer motion
  simulation of, 419-33
Polymers
  carrier generation in, 631-32
  carrier transport in, 636-40
  lattice
    equilibrium configurations of, 425
  molecular weights of, 11-19
  see also specific type
Polymer solutions

viscosities of, 2
Polymethylene
  characterization of, 23
Poly(n-hexyl isocyanate)
  transport properties of, 38
Polynomial approximations
  least-squares
    stiff-chain macromolecules and, 32
Polynucleotides
  circular dichroism of, 340-53
  double-stranded conformations of
    circular dichroism of, 343-45
Polypeptides
  alpha-helical
    transport properties of, 38
  circular dichroism of, 336-40
Polyproline I and II
  circular dichroism spectra of, 339-40
Polystyrene
  characterization of, 23
Polystyrene solutions
  photometry and, 17-18
Porphin
  intramolecular vibrational energy redistribution in, 600
Porter-Thomas distribution, 570
Proteins
  circular dichroism of, 336-40
Pseudo-orbitals, 361
  definition of, 367
  effective core potentials and, 370-71
Pseudophase partitioning
  heterolytic reactions and, 141-46
Pseudopotentials, 361
  metal surfaces and, 547-49
  semiconductor surfaces and, 540, 547
Pulsed radiolysis
  infrared laser spectroscopy and, 388-89
Purine-pyrimidine copolymers
  circular dichroism and, 348-49
Pyrazine
  collisional relaxation processes in, 597
  excitations on silver, 68
Pyridine
  binding configuration on nickel, 69
  excitations on silver, 68

Q

Quadrupole mass spectrometer
  ESD ion identification and, 219

Quantum chaos
  experimental study of, 586
Quantum ergodicity, 563-87
Quantum mechanics
  electron transfer reactions and, 456-60
  variational transition state theory and, 166-76
Quenching
  intramicellar reactions and, 148-49

R

Raman spectra
  carbon monoxide molecule, 527
  enhancement of, 512
  roughened surface, 507
  see also Resonance Raman spectra
Random matrices
  energy level spacings in, 567-68
Random matrix theory, 585-86
Rare gas atoms
  electronically excited metastable
    reactive behavior of, 127
Rare gas excimers
  spectroscopic properties of
    relativistic effective potentials and, 377
Rare gas matrices
  relaxation processes in, 603
Rate constants
  electron transfer reactions and, 460-61
Reaction rates
  laser-induced fluorescence and, 284
Redox centers
  electron transfer reactions and, 438
Relaxation times
  electron transfer reactions and, 469-71
  Rouse-Zimm, 42
Reptation model
  polymer chains and, 425-28
Resonance Raman scattering
  colloidal solution, 526-27
  Raman cross-section enhancement through
    chemisorption and, 507
Resonance Raman spectra
  iron molecule, 307
  see also Raman spectra
Resonances
  variational transition state theory and, 184
Resonance scattering
  adsorbates and, 55-57

Resonant decay
  ionic desorption and, 225-26
Rhodium
  acetylene adsorption on, 63
  benzene adsorption on, 65
  chemisorbed CO on
    electron energy loss spectroscopy and, 62
Ribonuclease S
  circular dichroism of, 339
Rice-Ramsperger-Kassel-Marcus theory, 159, 164
  chemical activation experiments and, 660
  unimolecular reactions and, 586
Rigid-body ensemble approximation, 40
Ring chain equilibrium
  in linear condensation polymers, 10
Ring closure probabilities
  stiff-chain macromolecular, 35-36
Ronchi rulings, 615
Rotational state distributions
  chemical reaction mechanisms and, 121-23
Rotational states
  laser-induced fluorescence and, 116-19
Rouse model
  polymer chains and, 423-24
RRKM theory
  see Rice-Ramsperger-Kassel-Marcus theory
Rubber elasticity
  molecular theory of, 2
Runge-Katta methods
  short-time polymer motion and, 421
Ruthenium
  chemisorbed CO on
    electron energy loss spectroscopy and, 62
  chemisorbed cyclohexane on
    electron energy loss spectroscopy and, 63
Ruthenium molecule
  bonding in, 312

S

Salts
  molecular ions in
    infrared laser spectroscopy and, 388
  polyelectrolyte permittivity and, 86
Scandium
  triatomic molecule of, 318-19

SUBJECT INDEX 725

Scandium molecules
 spectroscopic detection of, 292-93
Scandium-nickel
 bonding in, 314-15
Schizophyllan
 transport properties of, 38
Schrödinger equation, 192
 electron transfer processes and, 444
 valence electrons and, 357-58
Sedimentation
 dielectric response of polyions and, 90
Semiconductors
 organic, 614
 surface phonons of vibrations of, 58-59
Semiconductor surfaces, 538-47
 ionic desorption from, 226
Shear viscosity
 crystallization of liquid metals and, 256
Silicon surfaces
 ball-and-stick model of, 539
 electronic properties of, 538-47
Silination, 140
Silver
 benzene adsorption on, 65
 diatomic adsorption on
  resonance scattering and, 55
 excitations of aromatic adsorbates on, 68
 molecularly absorbed oxygen on
  O-O stretching frequency of, 61
 pentatomic molecule of, 322
 pyridine and pyrazine adsorption on
  excited state formation and, 520
  Raman spectra enhancement and, 512
 triatomic molecule of, 320-21
 xenon adsorption on, 69
Silver-gold
 bonding in, 317-18
Silver molecule
 bonding in, 312-13
Silver surfaces
 Raman cross sections and, 507
Sinai billiard system, 585-86
Single valence electron systems
 core-valence correlation and, 380
Slater determinant, 193
Small unilammellar vesicles
 formation of, 140
Smoluchowski equation, 424

Smoluchowski level, 40
Sodium
 metastable states of
  electron and photon desorption and, 217
Sodium atoms
 LIF saturation in flames, 278
Soliton theory
 polyacetylene and, 637
Solvents
 circular dichroism of polynucleotides and, 346-47
 surfactant self-association in, 139
Spanning tree approximation, 7
Spectroscopic transitions
 directional Einstein coefficients and, 269-71
 laser-induced fluorescence and, 268-69
Spectroscopy
 light emission process in electric field due to, 508-9
 organic solids and, 621-23
 polarizability tensor and, 509-10
 see also specific type
Spin-orbit interactions
 intramolecular vibrational relaxation and, 600
Spin-orbit potentials
 effective, 375-76
Stiff-chain macromolecules
 see Macromolecules
Stimulated emission pumping
 unimolecular reactions and, 666-67
Stokes equation, 90
Stratosphere
 catalytic destruction of ozone in, 482
Superconductivity
 in organic crystals, 645-47
Surface dipole selection rule, 53
Surface emission fields, 508-9
Surface-enhanced Raman effect, 68
Surface-enhanced spectroscopy, 507-31
 attenuated total reflection and, 521-22
 electromagnetic effects and, 511-15
 electromagnetic model for, 508-11
 electromagnetic resonances and, 529-31
 on flat surfaces, 515-17
 on gratings, 517-21
 image fields in, 509
 on isolated particles, 523-29
 small roughness and, 522-23

 surface emission fields in, 508-9
Surface phonons
 of elemental semiconductors, 58-59
 of ionic compounds, 57-58
 of metals, 59-60
Surface plasmons
 emitting dipole interactions with, 516
 excitation of, 515-17
 optical, 517-18
Surface polarization
 local laser field enhancement and, 507
Surfaces, 537-61
 effects of, 507
 electronic structures of
  calculation of, 540-41
 of metals, 547-60
 of semiconductors, 538-47
Surfactant assemblies
 organization of, 138-41
Surfactant molecules
 behavior in water, 138-39
Surfactants
 self-association in apolar solvents, 139
Surfactant vesicles
 reactivity control in, 151-54
Syndiotactic poly(methyl methacrylate)
 characteristic ratio of, 25

T

Temperature
 polyelectrolyte permittivity and, 86
Tetracene
 charge-transfer states in, 625
 intramolecular vibrational energy redistribution in, 600
Tetramethyldioxetane
 decomposition of, 684
 overtone vibration initiated decomposition of, 671
Tetrazine
 relaxation and predissociation of, 601-2
TEXAS, 196
Thin double layer approximation, 96-102
Threshold photoelectron/product ion coincidence
 charge transfer reactions and, 119
Time-of-flight analyzer
 electron- and photon-stimulated desorption and, 219-20

SUBJECT INDEX

Titanium-cobalt
  bonding in, 315
Titanium molecule
  ground state of, 293
Titanium-vanadium
  bonding in, 315
Toluene
  internal conversion and, 664-66
Topological linking number
  stiff-chain macromolecular, 35-36
Transition metal molecules, 291-323
  heteronuclear diatomics, 313-18
  homonuclear diatomics, 292-313
  pentatomics, 321-22
  triatomics, 318-21
Transition metal oxides
  ionic desorption from, 226-29
Transition metals
  core-valence correlation effects in, 380-81
  early-late
    melt-quenched to glasses, 242
  effective potentials and, 372
Transition state theory, 162
  electron transfer reactions and, 451-56
Translational energy
  excess
    infrared chemiluminescence and, 115-16
Translational excitation
  reaction dynamics and, 123-26
Transport coefficients
  stiff-chain macromolecular, 36
Trees
  cascade theory and, 7
  formation of, 4-5
Triatomic molecules
  collisional ground state relaxation in, 592
  transition metal, 318-21
Triboelectricity, 614
Trifluoroethanol
  cellulose acetate in
    transport properties of, 38
  circular dichroism of polynucleotides and, 347
Troposphere
  hydroxyl radicals and, 489
Tunneling
  reaction-coordinate motion and, 167
  variational transition state theory and, 178-82
Tyrosine side chains
  circular dichroism and, 339

U

Ultracentrifugation
  molecular weights of polymers and, 12-13
  small unilammelar vesicle preparation and, 140
Ultracentrifuge
  development of, 2
Unified statistical theory, 176-77
Unimolecular reactions, 657-88
  chemical activation and, 659-61
  excitation and, 676-78
  infrared multiphoton excitation and, 661-64
  internal conversion and, 664-66
  micellar effects on, 143
  overtone vibration excitation and, 667-68
  overtone vibration initiated, 678-87
  RRKM theory and, 586
  spectroscopy and, 670-75
  stimulated emission pumping and, 666-67
Uranium fluorides
  conformations and orbital energies of
    relativistic effective potentials and, 377

V

Vacuum ultraviolet
  circular dichroism of proteins and, 337
Valence electrons
  eigenvalue equation for, 360-61
  Schrödinger equation and, 357-58
Valence wavefunctions
  determination of, 358
Vanadium molecule
  dissociation energy of, 293
Vanadium-nickel
  bonding in, 315-16
Van der Waals complexes
  relaxation and predissociation of, 601-3
Van der Waals crystals
  carrier transport in, 634-36
Vapor-liquid condensation
  Mayer theory of, 6
Vapor-liquid equilibrium
  determination of, 5
Variational transition state theory, 159-85
  classical, 163-66
  derivation of, 161
  dynamical approximations and, 176-78
  quantum mechanics and, 166-76
  resonances and, 184
  state-selected reactions and, 184
  tunneling and, 178-82
  vibrational bonding and, 185
  vibrationally adiabatic barriers and, 182-84
Verdier-Stockmayer model
  polymer chains and, 421-23
Vesicles
  kinetic stability of, 140
  see also specific type
Vibrational bonding
  variational transition state theory and, 185
Vibrational excitation
  reactivity enhancement and, 120-21
Vibrational relaxation
  intramolecular, 594-95
    free jet expansion and, 598-600
    quantum beats and, 600-1
    large polyatomic molecule, 605-7
  lasers and, 591-92
  see also Electronic relaxation
Vibrational spectroscopy
  adsorbates and, 60-65
Vibrational states
  laser-induced fluorescence and, 116-19
Vibration-rotation states
  infrared chemiluminescence and, 110-11
Viscosity
  of metallic glasses, 244-45
Viscosity-molecular weight rule, 2

W

Water
  force constant calculations and, 206
  surfactant molecules in, 138-39
Water-insoluble compounds
  micelle solubilization of, 139
Wave functions
  force constant calculations and, 205-10
Weighting function method
  stiff-chain macromolecules and, 32-33

Wigner surmise, 567, 568, 572, 586
WKB approximation, 169, 183
Wolfram
　chemisorbed CO on
　　electron energy loss spectroscopy and, 62

X

Xenon
　electronic excitations on metals, 69
X-ray absorption
　near-edge structure
　　of iron molecule, 307

Y

Yttrium
　triatomic molecule of, 318-19
Yttrium-palladium
　bonding in, 318

CUMULATIVE INDEXES

CONTRIBUTING AUTHORS, VOLUMES 31–35

A

Albery, W. J., 31:227–63
Albrecht, A. C., 33:353–76
Alder, B. J., 32:311–29
Altkorn, R., 35:265–89
Anderson, C. F., 33:191–222
Angell, C. A., 34:593–630
Avouris, P., 35:49–73

B

Ballard, S. G., 33:377–407
Bartlett, R. J., 32:359–401
Baumgärtner, A., 35:419–35
Bersohn, R., 33:409–42
Bondybey, V. E., 35:591–612
Borejdo, J. 33:319–51
Botts, J., 33:319–51
Boxer, S. G., 34:389–417
Brauman, J. I., 34:187–215
Brochard, F., 32:433–51
Bunker, P. R., 34:59–75
Burch, R. R., 33:89–118

C

Calef, D. F., 34:493–524
Callis, P. R., 34:329–57
Cardillo, M. J., 32:331–57
Case, D. A., 33:151–71
Champion, P. M., 33:353–76
Chidsey, C. E. D., 34:389–417
Clouthier, D. J., 34:31–58
Cohen, M. L.,35:537–62
Cooke, R., 33:319–51
Crim, F. F., 35:657–91

D

Dacol, D., 34:419–61
Das, P., 35:507–36
Debrunner, P. G., 33:283–99
de Gennes, P. G., 33:49–61
Demuth, J., 35:49–73
Deslattes, R. D., 31:435–61
Deutch, J. M., 34:493–524

Djeu, N., 34:557–91
Drickamer, H. G., 33:25–47
Duncan, J. L., 34:245–72
Dunning, F. B., 33:173–89
Durup, J., 32:53–76
Dykstra, C. E., 32:25–52

E

Ehrlich, G., 31:503–37
Eichinger, B. E., 34:359–87
Enderby, J. E., 34:155–85
Etemad, S., 33:443–69

F

Fanconi, B., 31:265–91
Fayer, M. D., 33:63–87
Fendler, J. H., 35:137–57
Fogarasi, G., 35:191–213
Frauenfelder, H., 33:283–99
Freeman, G. R., 34:463–92
Frenkel, D., 31:491–521
Friedman, H. L., 32:179–204
Friedman, J. M., 33:471–91
Friedrich, D. M., 31:559–77
Frisch, H. L., 32:433–51

G

Gardiner, W. C. Jr., 31:377–99
Garrett, B. C., 35:159–89
Golden, D. M., 33:493–532
Green, S., 32:103–38
Greene, C. H., 33:119–50
Greer, S. C., 32:233–65
Gudeman, C. S., 35:387–418
Gutowsky, H. S., 31:1–27

H

Heeger, A. J., 33:443–69
Heller, E. J., 35:563–89
Hildebrand, J. H., 32:1–23
Hirschfelder, J. O., 34:xi–xvi;
  1–29
Hoover, W. G., 34:103–27
Hyde, J. S., 31:293–317

J

Jaynes, E. T., 31:579–601
Johnson, P. M., 32:139–57
Johnston, H. S., 35:481–505
Jonas, J., 31:1–27

K

Karplus, M., 31:29–45
Kivelson, D., 31:523–58
Kneba, M., 31:47–79
Koszykowski, M. L., 32:267–309
Kramer, M., 34:419–61
Krauss, M., 35:357–85

L

Lèger, L., 33:49–61
Legon, A. C., 34:275–300
Leone, S. R., 35:109–35
Levy,. D. H., 31:197–225
Light, J. C., 31:401–33
Lin, M. C., 34:557–91
Louie, S. G., 35:537–62

M

MacDiarmid, A. G., 33:443–69
Madden, P. A., 31:523–58
Madey, T. E., 35:215–40
Mandel, M., 35:75–108
Marcus, R. A., 32:267–309
Mauzerall, D., 33:377–407
Mayer, J. E., 33:1–23
McCammon, J. A., 31:29–45
McClain, W. M., 31:559–77
McMillen, D. F., 33:493–532
McTague, J. P., 31:491–521
Mendelson, R. A., 33:319–51
Metiu, H., 35:507–36
Miller, T. A., 33:257–82
Moldover, M. R., 32:233–65
Moore, C. B., 34:525–55
Morales, M. F., 33:319–51
Moseley, J., 32:53–76
Moylan, C. R., 34:187–215
Muetterties, E. L., 33:89–118

728

N

Nachtrieb, N. H., 31:131–56
Nagle, J. F., 31:157–95
Newton, M. D., 35:437–80
Noid, D. W., 32:267–309

O

Odijk, T., 35:75–108
Olson, D. B., 31:377–99
Ondrias, M. R., 33:471–91
Opella, S. J., 33:533–62
Osgood, R. M. Jr., 34:77–101
Otis, C. E., 32:139–57
Oxtoby, D. W., 32:77–101

P

Parr, R. G., 34:631–56
Pechukas, P., 32:159–77
Philpott, M. R., 31:97–129
Pollock, E. L., 32:311–29
Pope, M., 35:613–55
Pulay, P., 35:191–213

R

Rabitz, H., 34:419–61
Ramaker, D. E., 35:215–40
Ramsay, D. A., 34:31–58
Record, M. T. Jr., 33:191–222

Reinhardt, W. P., 33:223–55
Rousseau, D. L., 33:471–91
Robiette, A. G., 34:245–72
Roelofs, M. G., 34:389–417

S

Saykally, R. J., 32:403–31; 35:387–418
Scoles, G., 31:81–96
Shapiro, M., 33:409–42
Smalley, R. E., 34:129–53
Spaepen, F., 35:241–63
Stebbings, R. F., 33:173–89
Stechel, E. B., 35:563–89
Stevens, W. J., 35:357–85
Stockbauer, R., 35:215–40
Stockmayer, W. H., 35:1–21
Stolt, K., 31:603–37
Stolzenberg, A. M., 33:89–118
Strauss, H. L., 34:301–28
Sutin, N., 35:437–80
Swenberg, C. E., 35:613–55

T

Takashi, R., 33:319–51
Thomas, D. D., 31:293–317
Tinoco, I. Jr., 35:329–55
Truhlar, D. G., 35:150–89
Tully, J. C., 31:319–43
Turnbull, D., 35:241–63

U

Umstead, M. E., 34:557–91

V

Van Zee, R. J., 35:291–327
Vilches, O. E., 31:463–90

W

Walker, R. B., 31:401–33
Weinberg, W. H., 34:217–43
Weisshaar, J. C., 34:525–55
Weissman, M. B., 32:205–32
Weissman, S. I., 33:302–18
Weltner, W. Jr., 35:291–327
Williams, A. L., 35:329–55
Williams, C., 32:433–51
Wolfrum, J., 31:47–79
Wolynes, P. G., 31:345–76
Woods, R. C., 32:403–31

Y

Yamakawa, H., 35:23–47

Z

Zare, R. N., 33:119–50; 35:265–89
Zimm, B. H., 35:1–21

CHAPTER TITLES, VOLUMES 31–35

BIOPHYSICAL CHEMISTRY

Simulation of Protein Dynamics	J. A. McCammon, M. Karplus	31:29–45
Theory of the Main Lipid Bilayer Phase Transition	J. F. Nagle	31:157–95
Saturation-Transfer Spectroscopy	J. S. Hyde, D. D. Thomas	31:293–317
Polyelectrolyte Theories and Their Applications to DNA	C. F. Anderson, M. T. Record, Jr.	33:191–222
Dynamics of Proteins	P. G. Debrunner, H. Frauenfelder	33:283–99
Some Physical Studies of the Contractile Mechanism in Muscle	M. F. Morales, J. Borejdo, J. Botts, R. Cooke, R. A. Mendelson, R. Takashi	33:319–51
Time-Resolved Resonance Raman Studies of Hemoglobin	J. M. Friedman, D. L. Rousseau, M. R. Ondrias	33:471–91
Solid State NMR of Biological Systems	S. J. Opella	33:533–62
Electronic States and Luminescence of Nucleic Acid Systems	P. R. Callis	34:329–57
Magnetic Field Effects on Reaction Yields in the Solid State: An Example From Photosynthetic Reaction Centers	S. G. Boxer, C. E. D. Chidsey, M. G. Roelofs	34:389–417
Interactions and Kinetics in Membrane Mimetic Systems	J. H. Fendler	35:137–57
Differential Absorption and Differential Scattering of Circularly Polarized Light: Applications to Biological Macromolecules	I. Tinoco, Jr., A. L. Williams, Jr.	35:329–55

CHEMICAL KINETICS—GAS PHASE

Chemical Kinetics of High Temperature Combustion	W. C. Gardiner Jr., D. B. Olson	31:377–99
Transition State Theory	P. Pechukas	32:159–77
Collisions of Rydberg Atoms with Molecules	F. B. Dunning, R. F. Stebbings	33:173–89
Sensitivity Analysis in Chemical Kinetics	H. Rabitz, M. Kramer, D. Dacol	34:419–61
Variational Transition State Theory	D. G. Truhlar, B. C. Garrett	35:159–89
Selective Excitation Studies of Unimolecular Reaction Dynamics	F. F. Crim	35:657–91

CHEMICAL KINETICS—PHOTOCHEMISTRY AND RADIATION CHEMISTRY

I. Electrons in Fluids II. Nonhomogeneous Kinetics	G. R. Freeman	34:463–92
Formaldehyde Photochemistry	C. B. Moore, J. C. Weisshaar	34:525–55

CHEMICAL KINETICS—REACTION DYNAMICS

Reactive Molecular Collisions	R. B. Walker, J. C. Light	31:401–33
Fast Ion Beam Photofragment Spectroscopy	J. Moseley, J. Durup	32:53–76
Quasiperiodic and Stochastic Behavior in Molecules	D. W. Noid, M. L. Koszykowski, R. A. Marcus	32:267–309
Photofragment Alignment and Orientation	C. H. Greene, R. N. Zare	33:119–50
Theories of the Dynamics of Photodissociation	M. Shapiro, R. Bersohn	33:409–42
Nonequilibrium Molecular Dynamics	W. G. Hoover	34:103–27
Dynamics of Electronically Excited States	R. E. Smalley	34:129–53
State-Resolved Molecular Reaction Dynamics	S. R. Leone	35:109–35

CHAPTER TITLES

CHEMICAL KINETICS—SOLUTIONS (CONDENSED PHASE)
The Application of the Marcus Relation to Reactions in Solution	W. J. Albery	31:227–63
Diffusion-Controlled Reactions	D. F. Calef, J. M. Deutch	34:493–524
Electron Transfer Reactions in Condensed Phases	M.D. Newton, N. Sutin	35:437–80

ELECTROCHEMISTRY
Dynamics of Electrolyte Solutions	P. G. Wolynes	31:345–76
Ionization in Solution by Photoactivated Electron Transfer	D. Mauzerall, S. G. Ballard	33:377–407

GEOCHEMISTRY AND COSMOCHEMISTRY
Interstellar Chemistry: Exotic Molecules in Space	S. Green	32:103–38
Human Effects on the Global Atmosphere	H. S. Johnston	35:481–505

LASER CHEMISTRY, ENERGY TRANSFER AND RELAXATION
Bimolecular Reactions of Vibrationally Excited Molecules	M. Kneba, J. Wolfrum	31:47–79
Molecular Multiphoton Spectroscopy with Ionization Detection	P. M. Johnson, C. E. Otis	32:139–57
Dynamics of Molecules in Condensed Phases: Picosecond Holographic Grating Experiments	M. D. Fayer	33:63–87
Chemical Lasers	M. C. Lin, M. E. Umstead, N. Djeu	34:557–91
Relaxation and Vibrational Energy Redistribution Processes in Polyatomic Molecules	V. E. Bondybey	35:591–612

LIQUID STATE—SIMPLE FLUIDS
Renormalized Kinetic Theory of Dense Fluids	S. Yip	30:547–77

LIQUID STATE—SOLUTIONS OF ELECTROLYTES; FUSED SALTS
Conduction in Fused Salts and Salt-Metal Solutions	N. H. Nachtrieb	31:131–56
Electrolyte Solutions at Equilibrium	H. L. Friedman	32:179–204
Neutron Scattering from Ionic Solutions	J. E. Enderby	34:155–85
Dielectric Properties of Polyelectrolyte Solutions	M. Mandel, T. Odijk	35:75–108

LIQUID STATE—STRUCTURE
Computer Simulations of Freezing and Supercooled Liquids	D. Frenkel, J. P. McTague	31:491–521
Simulation of Polar and Polarizable Fluids	B. J. Alder, E. L. Pollock	32:311–29
Supercooled Water	C. A. Angell	34:593–630

MAGNETIC RESONANCE (ELECTRON SPIN, NUCLEAR, QUADRUPOLE)
Recent Developments in Electron Paramagnetic Resonance: Transient Methods	S. I. Weissman	33:301–18

MISCELLANEOUS
The Avogadro Constant	R. D. Deslattes	31:435–61
Laser Microchemistry and Its Application to Electron-Device Fabrication	R. M. Osgood, Jr.	34:77–101
Metallic Glasses	F. Spaepen, D. Turnbull	35:241–63

MOLECULAR STRUCTURE
High Resolution Spectroscopy of Molecular Ions	R. J. Saykally, R. Claude Woods	32:403–31
Light and Radical Ions	T. A. Miller	33:257–82
Transition Metal Molecules	W. Weltner, Jr., R. J. Van Zee	35:291–327

PHYSICAL ORGANIC
Gas Phase Acid-Base Chemistry	C. R. Moylan, J. I. Brauman	34:187–215

PHYSICAL PHENOMENA—MISCELLANEOUS
Fluctuation Spectroscopy	M. B. Weissman	32:205–32
Thermodynamic Anomalies at Critical Points of Fluids	S. C. Greer, M. R. Moldover	32:233–65

POLYMERS AND MACROMOLECULES
Molecular Vibrations of Polymers	B. Fanconi	31:265–91
Dynamics of Entangled Polymer Chains	P. G. de Gennes, L. Lèger	33:49–61
Polyacetylene, $(CH)_x$: The Prototype Conducting Polymer	S. Etemad, A. J. Heeger, A. G. MacDiarmid	33:443–69
The Theory of High Elasticity	B. E. Eichinger	34:359–87
Stiff-Chain Macromolecules	H. Yamakawa	35:23–47
Simulation of Polymer Motion	A. Baumgärtner	35:419–35

PREFATORY CHAPTERS
NMR in Chemistry—An Evergreen	J. Jonas, H. S. Gutowsky	31:1–27
A History of Solution Theory	J. H. Hildebrand	32:1–23
The Way It Was	J. E. Mayer	33:1–23
My Adventures in Theoretical Chemistry	J. O. Hirschfelder	34:1–29
When Polymer Science Looked Easy	W. H. Stockmayer, B. H. Zimm	35:1–21

QUANTUM CHEMISTRY
Potential Energy Barriers in Unimolecular Rearrangements	C. E. Dykstra	32:25–52
Many-Body Perturbation Theory and Coupled Cluster Theory for Electron Correlation in Molecules	R. J. Bartlett	32:359–401
Electronic Structure Calculations Using the $X\alpha$ Method	D. A. Case	33:151–71
Complex Coordinates in the Theory of Atomic and Molecular Structure and Dynamics	W. P. Reinhardt	33:223–55
Quasilinear and Quasiplanar Molecules	P. R. Bunker	34:59–75
Density Functional Theory	R. G. Parr	34:631–56
Ab Initio Vibrational Force Fields	G. Fogarasi, P. Pulay	35:191–213
Effective Potentials in Molecular Quantum Chemistry	M. Krauss, W. J. Stevens	35:357–85

QUANTUM MECHANICS
Quantum Ergodicity and Spectral Chaos	E. B. Stechel, E. J. Heller	35:563–89

SCATTERING PHENOMENA—DYNAMICAL
Light Scattering Studies of Molecular Liquids	D. Kivelson, P. A. Madden	31:523–58

SCATTERING PHENOMENA—STRUCTURAL
Developments in Extended X-Ray Absorption Fine Structure Applied to Chemical Systems	D. R. Sandstrom, F. W. Lytle	30:215–38

SOLIDS AND ORDERED ARRAYS—STRUCTURE AND DYNAMICS
Electronic Processes in Organic Solids	M. Pope, C. E. Swenberg	35:613–55

SPECTROSCOPY—ELECTRONIC AND PHOTOELECTRONIC
Optical Reflection Spectroscopy of Organic Solids	M. R. Philpott	31:97–129
Two-Photon Molecular Electronic Spectroscopy	D. M. Friedrich, W. M. McClain	31:599–677
High Pressure Studies of Molecular Luminescence	H. G. Drickamer	33:25–47
The Spectroscopy of Formaldehyde and Thioformaldehyde	D. J. Clouthier, D. A. Ramsay	34:31–58

Effects of Saturation on Laser-Induced Fluorescence Measurements of Population and Polarization	R. Altkorn, R. N. Zare	35:265–89

SPECTROSCOPY—INFRARED AND RAMAN

Laser Spectroscopy of Cold Gas-Phase Molecules	D. H. Levy	31:197–225
Resonance Raman Scattering: The Multimode Problem and Transform Methods	P. M. Champion, A. C. Albrecht	33:353–76
High Resolution Vibration-Rotation Spectroscopy	A. G. Robiette, J. L. Duncan	34:245–72
Pseudorotation: A Large Amplitude Molecular Motion	H. L. Strauss	34:301–28
Velocity Modulation Infrared Laser Spectroscopy of Molecular Ions	C. S. Gudeman, R. J. Saykally	35:387–418
The Electromagnetic Theory of Surface Enhanced Spectroscopy	H. Metiu, P. Das	35:507–36

SPECTROSCOPY—MICROWAVE

Pulsed-Nozzle, Fourier-Transform Microwave Spectroscopy of Weakly Bound Dimers	A. C. Legon	34:275–300

STATISTICAL MECHANICS

Two-Body, Spherical, Atom-Atom, and Atom-Molecule Interaction Energies	G. Scoles	31:81–96
The Minimum Entropy Production Principle	E. T. Jaynes	31:579–601
Vibrational Relaxation in Liquids	D. W. Oxtoby	32:77–101
Simulation of Polar and Polarizable Fluids	B. J. Alder, E. L. Pollock	32:311–29
Polymer Collapse	C. Williams, F. Brochard, H. L. Frisch	32:433–51

SURFACES—ADSORPTION AND CATALYSIS

Kinetic Processes on Metal Single-Crystal Surfaces	R. J. Madix, J. Benziger	29:285–306

SURFACES—STRUCTURE AND DYNAMICS

Theories of the Dynamics of Inelastic and Reactive Processes at Surfaces	J. C. Tully	31:319–43
Phase Transitions in Monomolecular Layer Films Physisorbed on Crystalline Surfaces	O. E. Vilches	31:463–90
Surface Diffusion	G. Ehrlich, K. Stolt	31:603–37
Gas-Surface Interactions Studied with Molecular Beam Techniques	M. J. Cardillo	32:331–57
Molecular Features of Metal Cluster Reactions	E. L. Muetterties, R. R. Burch, A. M. Stolzenberg	33:89–118
Order-Disorder Phase Transitions in Chemisorbed Overlayers	W. H. Weinberg	34:217–43
Electron Energy Loss Spectroscopy in the Study of Surfaces	P. Avouris, J. Demuth	35:265–89
Characterization of Surfaces Through Electron and Photon Stimulated Desorption	T. E. Madey, D. E. Ramaker, R. Stockbauer	35:215–40
Electronic Properties of Surfaces	M. L. Cohen, S. G. Louie	35:537–62

THERMOCHEMISTRY AND THERMODYNAMICS

Hydrocarbon Bond Dissociation Energies	D. F. McMillen, D. M. Golden	33:493–532

Please list the volumes you wish to order by volume number. If you wish a standing order (the latest volume sent to you automatically each year), indicate volume number to begin order. Volumes not yet published will be shipped in month and year indicated. All prices subject to change without notice. Prepayment required from individuals. Telephone orders charged to VISA, MasterCard, American Express, welcomed.

ANNUAL REVIEW SERIES

		Prices Postpaid per volume USA/elsewhere	Regular Order Please send: Vol. number	Standing Order Begin with: Vol. number
Annual Review of ANTHROPOLOGY				
Vols. 1-10	(1972-1981)	$20.00/$21.00		
Vol. 11	(1982)	$22.00/$25.00		
Vol. 12	(1983)	$27.00/$30.00		
Vol. 13	(avail. Oct. 1984)	$27.00/$30.00	Vol(s). _____	Vol. _____
Annual Review of ASTRONOMY AND ASTROPHYSICS				
Vols. 1-19	(1963-1981)	$20.00/$21.00		
Vol. 20	(1982)	$22.00/$25.00		
Vol. 21	(1983)	$44.00/$47.00		
Vol. 22	(avail. Sept. 1984)	$44.00/$47.00	Vol(s). _____	Vol. _____
Annual Review of BIOCHEMISTRY				
Vols. 29-50	(1960-1981)	$21.00/$22.00		
Vol. 51	(1982)	$23.00/$26.00		
Vol. 52	(1983)	$29.00/$32.00		
Vol. 53	(avail. July 1984)	$29.00/$32.00	Vol(s). _____	Vol. _____
Annual Review of BIOPHYSICS AND BIOENGINEERING				
Vols. 1-10	(1972-1981)	$20.00/$21.00		
Vol. 11	(1982)	$22.00/$25.00		
Vol. 12	(1983)	$47.00/$50.00		
Vol. 13	(avail. June 1984)	$47.00/$50.00	Vol(s). _____	Vol. _____
Annual Review of EARTH AND PLANETARY SCIENCES				
Vols. 1-9	(1973-1981)	$20.00/$21.00		
Vol. 10	(1982)	$22.00/$25.00		
Vol. 11	(1983)	$44.00/$47.00		
Vol. 12	(avail. May 1984)	$44.00/$47.00	Vol(s). _____	Vol. _____
Annual Review of ECOLOGY AND SYSTEMATICS				
Vols. 1-12	(1970-1981)	$20.00/$21.00		
Vol. 13	(1982)	$22.00/$25.00		
Vol. 14	(1983)	$27.00/$30.00		
Vol. 15	(avail. Nov. 1984)	$27.00/$30.00	Vol(s). _____	Vol. _____

SEE ORDERING INFORMATION ON PAGE 4.

		Prices Postpaid per volume USA/elsewhere	Regular Order Please send: Vol. number	Standing Order Begin with: Vol. number
Annual Review of ENERGY				
Vols. 1-6	(1976-1981)	$20.00/$21.00		
Vol. 7	(1982)	$22.00/$25.00		
Vol. 8	(1983)	$56.00/$59.00		
Vol. 9	(avail. Oct. 1984)	$56.00/$59.00	Vol(s). _____	Vol. _____
Annual Review of ENTOMOLOGY				
Vols. 7-16, 18-26	(1962-1971; 1973-1981)	$20.00/$21.00		
Vol. 27	(1982)	$22.00/$25.00		
Vol. 28	(1983)	$27.00/$30.00		
Vol. 29	(avail. Jan. 1984)	$27.00/$30.00	Vol(s). _____	Vol. _____
Annual Review of FLUID MECHANICS				
Vols. 1-13	(1969-1981)	$20.00/$21.00		
Vol. 14	(1982)	$22.00/$25.00		
Vol. 15	(1983)	$28.00/$31.00		
Vol. 16	(avail. Jan. 1984)	$28.00/$31.00	Vol(s). _____	Vol. _____
Annual Review of GENETICS				
Vols. 1-15	(1967-1981)	$20.00/$21.00		
Vol. 16	(1982)	$22.00/$25.00		
Vol. 17	(1983)	$27.00/$30.00		
Vol. 18	(avail. Dec. 1984)	$27.00/$30.00	Vol(s). _____	Vol. _____
Annual Review of IMMUNOLOGY				
Vol. 1	(1983)	$27.00/$30.00		
Vol. 2	(avail. April 1984)	$27.00/$30.00	Vol(s). _____	Vol. _____
Annual Review of MATERIALS SCIENCE				
Vols. 1-11	(1971-1981)	$20.00/$21.00		
Vol. 12	(1982)	$22.00/$25.00		
Vol. 13	(1983)	$64.00/$67.00		
Vol. 14	(avail. Aug. 1984)	$64.00/$67.00	Vol(s). _____	Vol. _____
Annual Review of MEDICINE: Selected Topics in the Clinical Sciences				
Vols. 1-3, 5-15	(1950-1952; 1954-1964)	$20.00/$21.00		
Vols. 17-32	(1966-1981)	$20.00/$21.00		
Vol. 33	(1982)	$22.00/$25.00		
Vol. 34	(1983)	$27.00/$30.00		
Vol. 35	(avail. April 1984)	$27.00/$30.00	Vol(s). _____	Vol. _____
Annual Review of MICROBIOLOGY				
Vols. 17-35	(1963-1981)	$20.00/$21.00		
Vol. 36	(1982)	$22.00/$25.00		
Vol. 37	(1983)	$27.00/$30.00		
Vol. 38	(avail. Oct. 1984)	$27.00/$30.00	Vol(s). _____	Vol. _____
Annual Review of NEUROSCIENCE				
Vols. 1-4	(1978-1981)	$20.00/$21.00		
Vol. 5	(1982)	$22.00/$25.00		
Vol. 6	(1983)	$27.00/$30.00		
Vol. 7	(avail. March 1984)	$27.00/$30.00	Vol(s). _____	Vol. _____
Annual Review of NUCLEAR AND PARTICLE SCIENCE				
Vols. 12-31	(1962-1981)	$22.50/$23.50		
Vol. 32	(1982)	$25.00/$28.00		
Vol. 33	(1983)	$30.00/$33.00		
Vol. 34	(avail. Dec. 1984)	$30.00/$33.00	Vol(s). _____	Vol. _____

SEE ORDERING INFORMATION ON PAGE 4.

DATE DUE

NOV 3 0 1984